# The Fig

## Botany, Production and Uses

# The Fig

## Botany, Production and Uses

---

*Edited by*

**Ali Sarkhosh**

*Horticultural Sciences Department, University of Florida, Gainesville, FL 32611, USA*

**Alimohammad Yavari**

*Pomologist, Private Consultant, Tehran, Iran*

**Louise Ferguson**

*Department of Plant Sciences, University of California, Davis, CA 95616, USA*

CABI

**CABI is a trading name of CAB International**

CABI
Nosworthy Way
Wallingford
Oxfordshire OX10 8DE
UK

Tel: +44 (0)1491 832111
Fax: +44 (0)1491 833508
E-mail: info@cabi.org
Website: www.cabi.org

CABI
WeWork
One Lincoln St
24th Floor
Boston, MA 02111
USA

Tel: +1 (617)682-9015
E-mail: cabi-nao@cabi.org



A catalogue record for this book is available from the British Library, London, UK.



ISBN-13: 978 1 78924 288 1 (hardback)
978 1 78924 289 8 (ePDF)
978 1 78924 290 4 (ePub)

DOI: 10.1079/9781789242881.0000

Commissioning Editor: Rebecca Stubbs
Editorial Assistant: Emma McCann
Production Editor: Marta Patiño

Typeset by SPi, Pondicherry, India
Printed and bound in the UK by CPI Group (UK) Ltd, Croydon, CR0 4YY

# Contents

# Contributors


**Abdolahipour, Mohammad,** *Department of Water Engineering, College of Aburaihan, University of Tehran, Tehran, Iran*

**Aljane, Fateh,** *Laboratoire d'Aridoculture et Cultures Oasiennes, Institut des Régions Arides (IRA), 4119 Medenine, Université de Gabès, Tunisia.*

**Amiri, Azam,** *Department of Horticultural Science, Faculty of Agriculture, Shahid Chamran University of Ahvaz, Ahvaz, Iran.*

**Amodio, Maria L.,** *Department of Agriculture, Food, Natural Resources and Engineering (DAFNE), University of Foggia, Via Napoli 25, 71122, Foggia, Italy.*

**Balas, Francisco C.,** *Junta de Extremadura, Finca La Orden-Valdesequera Research Centre (CICYTEX), Horticulture, Autovía Madrid-Lisboa, s/n, 06187 Badajoz, Spain.*

**Ben Mimoun, Mehdi,** *University of Carthage, Institut National Agronomique de Tunisie (INAT), Laboratory GREEN-TEAM (LR17AGR01), 43 avenue Charles Nicolle, 1082 Tunis, Tunisia.*

**Brown, Patrick,** *Department of Plant Sciences, University of California, Davis, CA 95616, USA.*

**Bolboli, Zeinab,** *Department of Plant Protection, Shiraz University, Shiraz, Iran.*

**Chawla, Anshul,** *College of Pharmacy, Shahpur, Jalandhar-144020 (Punjab), India.*

**Colasuonno, Pasqualina,** *Department of Environmental and Territorial Sciences, University of Bari "Aldo Moro", Via G. Amendola 165/A, 70126 - Bari, Italy.*

**Colelli, Giancarlo,** *Department of Agriculture, Food, Natural Resources and Engineering (DAFNE), University of Foggia, Via Napoli 25, 71122, Foggia, Italy.*

**Crisosto, Carlos H.,** *Department of Plant Sciences, University of California, Davis, CA 95616, USA.*

**Crivelli, Megan,** *The Produce Nerd, Cloverdale, CA, USA.*

**Dalir, Neda,** *Department of Soil Science, Tarbiat Modares University, Tehran, Iran.*

**Davatgar, Naser,** *Soil and Water Research Institute, Agricultural Research, Education and Extension Organization (AREEO), Karaj, Iran.*

**Ertan, Birgül,** *Fig Research Institute, Aydın, Turkey.*

**Ertan, Engin,** *Department of Horticulture, Faculty of Agriculture, University of Aydin Adnan Menderes, Aydın, Turkey.*

**Ferguson, Louise,** *Department of Plant Sciences, University of California, Davis, CA 95616, USA.*

**Ferrara, Giuseppe,** *Department of Soil, Plant and Food Sciences, University of Bari "Aldo Moro", Via G. Amendola 165/A, 70126, Bari, Italy.*

**Gaaliche, Badii,** *Laboratory of Horticulture, National Agricultural Research Institute of Tunisia (INRAT), IRESA-University of Carthage, Hédi Karray Street, 1004 El Menzah, Tunis, Tunisia.*

**Galán A.J.**, *Junta de Extremadura, Finca La Orden-Valdesequera Research Centre (CICYTEX), Fruticulture, Autovía Madrid-Lisboa, s/n, 06187 Badajoz, Spain.*

**Galván, Ana I.**, *Junta de Extremadura, Finca La Orden-Valdesequera Research Centre (CICYTEX), Fruticulture, Autovía Madrid-Lisboa, s/n, 06187 Badajoz, Spain.*

**Garza-Alonso, Carlos A.**, *Facultad de Agronomía – Universidad Autónoma de Nuevo León, Francisco Villa s/n, Ex Hacienda El Canadá, Gral, Escobedo, N.L., 66050, México.*

**Giliomee, Jan**, *Department of Botany and Zoology, University of Stellenbosch, Matieland, South Africa.*

**Gordon, Phoebe**, *Cooperative Extension, Madera and Merced Counties, Division of Agriculture and Natural Resources, University of California, Madera, CA, USA.*

**Grieshop, Matthew J.**, *Department of Entomology, Michigan State University, East Lansing, USA.*

**Günver Dalkılıç, G.**, *Department of Horticulture, Faculty of Agriculture, University of Aydın Adnan Menderes, Aydın, Turkey.*

**Hormaza, José I.**, *IHSM La Mayora – CSIC – UMA, Avda. Dr. Wienberg s/n, 29750 Algarrobo-Costa, Malaga, Spain.*

**Hosomi, Akihiro**, *Toyo Institute of Food Technology, Kawanishi 666-0026, Japan.*

**Ikegami, Hidetoshi**, *Fukuoka Agriculture and Forestry Research Center, Buzen Branch. 2-4-1, Nishiizumi, Yukuhashi, 824-0038, Japan.*

**Kamgar-Haghighi, Ali Akbar**, *Water Engineering Department, Shiraz University, Shiraz, Iran.*

**Kjellberg, Finn**, *CEFE, CNRS, University of Montpellier, EPHE, IRD, 1919 route de Mende, F-34293 Montpellier Cédex 5, France.*

**Jafari, Moslem**, *Fig Research Station, Fars Agricultural and Natural Resources Research and Education Center, Agricultural Research, Education and Extension Organization (AREEO), Estahban, Fars, Iran.*

**Kokaj, Tatjana**, *Institute of Plant Genetic Resource, Agriculture University of Tirana, Albania.*

**Li, Zhihong**, *Department of Entomology, China Agricultural University, Beijing, China.*

**López-Corrales, Margarita**, *Junta de Extremadura, Finca La Orden-Valdesequera Research Centre (CICYTEX), Fruticulture. Autovía Madrid-Lisboa, s/n, 06187, Badajoz, Spain.*

**Maghoumi, Mahshad**, *Department of Agriculture, Food, Natural Resources and Engineering (DAFNE), University of Foggia, Via Napoli 25, 71122, Foggia, Italy.*

**Mahmoudi, Souhila**, *Department of Agronomic Sciences, University "Mohamed BOUDIAF", PB-166 Msila 28000, Algeria.*

**Marcotuli, Ilaria**, *Department of Environmental and Territorial Sciences, University of Bari "Aldo Moro", Via G. Amendola 165/A, 70126 - Bari, Italy.*

**Massimino Cocuzza, Giuseppe E.**, *Dipartimento di Agricoltura, Alimentazione e Ambiente, Università di Catania, Italy.*

**Mazri, Chafiaa**, *Département des Sciences Agronomiques, Faculté des Sciences de la Nature et de la Vie, Université Le Colonel M'hand Akli Oulhade, Bouira, Algéria.*

**Mazzeo, Andrea**, *Department of Soil, Plant and Food Sciences, University of Bari "Aldo Moro", Via G. Amendola 165/A, 70126, Bari, Italy.*

**Michailides, Themis J.**, *Department of Plant Pathology, University of California-Davis, Kearney Agricultural Research and Extension Center, Parlier, CA 93648, USA.*

**Mostowfizadeh-Ghalamfarsa, Reza**, *Department of Plant Protection, Shiraz University, Shiraz, Iran.*

**Narzary, Diganta**, *Department of Botany, Gauhati University, Guwahati 781 014, Assam, India.*

**Neily, Mohamed H.**, *Laboratoire d'Aridoculture et Cultures Oasiennes, Institut des Régions Arides (IRA), 4119 Medenine, Université de Gabès, Tunisia.*

**Nogata, Hitoshi**, *Fukuoka Agriculture and Forestry Research Center, Buzen Branch, 2-4-1, Nishiizumi, Yukuhashi, 824-0038, Japan.*

**Polat, A. Aytekin**, *Department of Horticulture, Faculty of Agriculture, Mustafa Kemal University, 31060 Alahan-Antakya, Hatay, Turkey.*

**Ramezanian, Asghar**, *Department of Horticultural Science, School of Agriculture, Shiraz University, Shiraz, Iran.*

**Rasplus, Jean Yves**, *INRAE, IRD, CIRAD, SupAgro, University Montpellier, CBGP, 755 Avenue du Campus Agropolis - CS 30 016, 34988 Montferrier-sur-Lez, France.*

**Rodrigues, Maria G.F.,** *Fruticulture and Floriculture Applied Laboratory, Faculty of Agricultural and Technological Sciences (FCAT - Dracena), São Paulo State University (UNESP), São Paulo, Brazil.*

**Sabanadzovic, Sead,** *Department of Biochemistry, Molecular Biology, Entomology and Plant Pathology, Mississippi State University, Mississippi, USA.*

**Salmaninezhad, Fatemeh,** *Department of Plant Protection, Shiraz University, Shiraz, Iran.*

**Sarkhosh, Ali,** *Horticultural Sciences Department, University of Florida, Gainesville, FL 32611, USA.*

**Sandhu, Rajwinder K.,** *Department of Fruit Science, Punjab Agricultural University, Ludhiana, India.*

**Sepaskhah, Ali Reza,** *Water Engineering Department and Drought Research Institute, Shiraz University, Shiraz, Iran.*

**Serradilla, Manuel J.,** *Junta de Extremadura, Agri-Food Technological Institute of Extremadura (INTAEX-CICYTEX), Plant Food. Avda. Adolfo Suárez, s/n, 06007, Badajoz, Spain.*

**Shokoohi, Zeinab,** *Department of Agricultural Economics, School of Agriculture, Shiraz University, Fars, Iran.*

**Singh, Sandeep,** *Department of Fruit Science, Punjab Agricultural University, Ludhiana, India.*

**Spagnoli, Federica,** *Department Italian Institute of Oriental Studies - ISO, Sapienza University of Rome - Archaeological Expedition to Motya, Italy.*

**Stournaras, Vasileios,** *Department of Olive and Horticultural Crops, Hellenic Agricultural Organization (DEMETER) Institute of Olive Tree, Subtropical Crops and Viticulture, Kalamata, Greece.*

**Tarazkar, Mohammad H.,** *Department of Agricultural Economics, School of Agriculture, Shiraz University, Fars, Iran.*

**Tzanetakis, Ioannis,** *Department of Entomology & Plant Pathology, Division of Agriculture, University of Arkansas System, Fayetteville, AR, USA.*

**van Noort, Simon,** *Research and Exhibitions Department, South African Museum, Iziko Museums of South Africa, P.O. Box 61, Cape Town, 8000, South Africa; Department of Biological Sciences, University of Cape Town, Private Bag, Rondebosch, 7701, South Africa.*

**Yavari, Alimohammad,** *Pomologist, Private Consultant, Tehran, Iran.*

**Yakushiji, Hiroshi,** *Division of Fruit Tree Production Research, Institute of Fruit Tree and Tea Science, NARO, Higashihiroshima, Hiroshima 739-2494, Japan.*

**Zhang, Yue,** *Department of Entomology, China Agricultural University, Beijing, China.*

# About the Editors

Ali Sarkhosh is an Assistant Professor and Extension Specialist (fruit crops) in the Horticultural Sciences Department at the University of Florida, where he has developed research and extension programmes in fruit crop production, physiology, and diversification. His research focus is on developing and extending sustainable fruit crops production in Florida. Dr Sarkhosh has over 15 years of experience with research and production of different fruit crops in the US, Australia, New Zealand and Iran. His research with fruit crops has focused on assessment of cultural practices to optimize yield and production efficiency, rootstocks and cultivars evaluations, and breeding. Toward improving fruit crops production, he has published many sources of both technical and referred articles in production, physiology, breeding and genetic diversity.

Alimohammad Yavari is a private pomologist and consultant. Mr Yavari received his MS in horticulture (Pomology) from the University of Urmia, Iran. Mr Yavari has expertise in all aspects of fruit crops production as well as interacting worldwide with commercial growers and researchers in the field of fruit production, mainly tropical and subtropical fruits. His experiences in fig production are mostly harvesting and exporting issues, and training courses for orchardists. He provides public and private sectors with project design and writing proposals related to fig production and other tropical and subtropical fruits.

Louise Ferguson is an Extension Specialist in the Department of Plant Sciences, University of California in Davis (UC-Davis). Dr Ferguson has developed research and extension programmes covering pistachios, olives, figs, citrus, persimmons and pomegranates. Her research focus is on developing and extending long-term applications for sustainable tree production. Specific applied research areas are alternate bearing of tree crops, rootstock effects on multiple scion characteristics, salinity tolerance of pistachios, fig breeding, and mechanical harvesting and pruning of olives and pistachios. She has served as Director of the Fruit and Nut Research and Information Centre at University of California Davis, President of the Plant Growth Regulation Society of America and American Society of Horticultural Science and Core Faculty for the California Agricultural Leadership Foundation's Ag Leadership programme.

# Preface

*Ficus carica* L. is a member of the Moraceae family. The number of species in the genus *Ficus* varies from 600 to more than 1900; making *Ficus* among the plant genera with most individual species; however only a few species produce edible fruits. Most species are located in the tropics, subtropics, Middle East and Mediterranean region.

It is one of the oldest fruits domesticated by humans. The fig is the third classical fruit crop, along with pomegranate and olive, associated with the beginning of horticulture in the Mediterranean basin 6000 years ago. The fig, marketed as processed, dried or fresh, is now cultivated in most warm and temperate climates. Growth is more limited by winter low temperatures than by summer heat. Climate markedly affects the size, shape, skin and pulp color of figs. Rain, hail and wind can reduce fruit quality and production.

Figs now constitute an important part of the Mediterranean diet, fresh and dried. Fig fruit, particularly dried figs, are especially rich in fibre, trace minerals, polyphenols, proteins, vitamin K, protein, carbohydrates, magnesium, potassium, calcium and sugars. Reports in the literature document the efficacy of fresh and dried figs in treating cardiovascular diseases, cancer and constipation. In traditional Mediterranean medicine, the fruits are reported to have antipyretic, purgative, aphrodisiac, anti-inflammatory and anti-paralytic properties. Other parts of the tree, leaves, latex, bark and roots, are also used for medicinal purposes. *F. carica* leaves, shoots, bark and latex have been evaluated for their bioactivity as antimicrobial, anticancer, anti-inflammatory, antipyretic, antidiabetic, anti-obesogenic, antihyperlipidemic, hepatoprotective, neurological and antidiarrheal activity.

Given that the fig is so nutrient dense and versatile, it could play a more important role in global markets. Over 1,315,588 metric tons were produced on 289,818 ha in 2020 (FAOSTAT, 2021, *World Fig Production*. Available at: www.fao.org/faostat/en/#data/QC, accessed 2021). Approximately 90% of this production was in the Mediterranean basin and Middle East. The world's largest fig producer is Turkey with ~26% of the world's fig production. Turkey, Egypt, Iran, Greece, Algeria and Morocco are the top six global producers, accounting for ~70% of the world's annual production.

In comparison to the major fruit crops like apple, banana and citrus, global fig production is small. However, the fig is approaching a third to half the apricot and sweet cherry global production and therefore is no longer a minor crop. Most current commercial fig production is dried. The greatest growth potential is for top-quality fresh figs. Currently, due to short transport and shelf life, fresh figs are a high-value fruit with limited demand. Improving the postharvest life of high-quality fresh figs is necessary for increasing fresh fig markets.

*The Fig: Botany, Production and Uses*, was authored by over 60 international scientists, who developed much of the information summarized. This comprehensive summary of the scientific literatures covers fig botany, production, processing, nutrition and markets. This book is for anybody who wants a better understanding of the fig tree and its fruits; educators, researchers, students, agriculture extension farmers, industry personnel, health professionals and consumers.

The editors gratefully acknowledge all the authors who generously gave their time and shared expertise. We also acknowledge the scientists whose reviews greatly improved this book. And we thank all who supported us in this effort.

Ali Sarkhosh
Alimohammad Yavari
Louise Ferguson

*"I cannot do all the good that the world needs. But the world needs all the good that I can do."*
*– Jana Stanfield*

# 1 History, Archaeology and Culture

**F. Spagnoli[1]* and A. Yavari[2]**
[1]*Department Italian Institute of Oriental Studies - ISO, Sapienza University of Rome, Italy;* [2]*Pomologist and Consultant, Tehran, Iran*

## 1.1 Introduction

It is said that a basket of figs was enough for Cato the Censor to convince the reticent Senators of Rome of the need to destroy Carthage. Libyan figs were sold fresh in the city market. 'The land that produced them is only three days away from Rome', Cato said, coming back from a diplomatic mission to the African metropolis in 150 BCE (Pliny the Elder, *Naturalis Historia* 20:1–3). With this demonstration, Cato did not just want to show the proximity of Carthage to Rome. He also wanted to represent the vast wealth due to the flourishing agricultural activity in the fertile North African hinterland of the Punic city symbolized by the sumptuousness of these fruits. The fig (*Ficus carica* L.) is a fruit that evokes, in the collective imagination, the generosity and fruitfulness of nature that nourishes man with its spontaneous fruits even before man domesticated and cultivated plants and fruit trees.

## 1.2 Etymology

The names given through the centuries to the fig have a direct relationship to their origin and distribution. The Latin *Ficus* comes from an earlier Indian stem word such as fag or the Hebrew feg, the Italian fico, the Portuguese figo, the Spanish higo, the French figue, the German Feigen, the Dutch vijg and the early English figge or fegge, later simplified to 'fig'. In Greece, the wild fig was called erineos, and the edible fig was known as sykon from which is derived 'syconium', the botanical name of the fruit. The edible fig was called Teena in Hebrew and in Aramaic, tena, Arabic tin, Phoenician paggim, North Syrian pagga, and Persian anjir (Condit, 1947).

## 1.3 Domestication, Dispersal and Archaeological Evidence of the Fig

The fig, *Ficus carica* L. (Moraceae), is the third classical fruit crop associated with the beginning of horticulture in the Mediterranean basin and south-west Asia (Zohary and Spiegel-Roy, 1975). This common fig (based on the evidence available) has been a part of food production in this region since the Early Bronze Age period, providing fresh summer fruit and storable, sugar-rich, dry figs throughout the year. The domesticated fig tree exhibits a close morphological resemblance, striking similarities in climate requirements and close genetic interconnections with an

*Corresponding author: federica_spagnoli@yahoo.it

 *The Fig: Botany, Production and Uses* (eds. Sarkhosh, A., Yavari, A. and Ferguson, L.)
DOI: 10.1079/9781789242881.0001

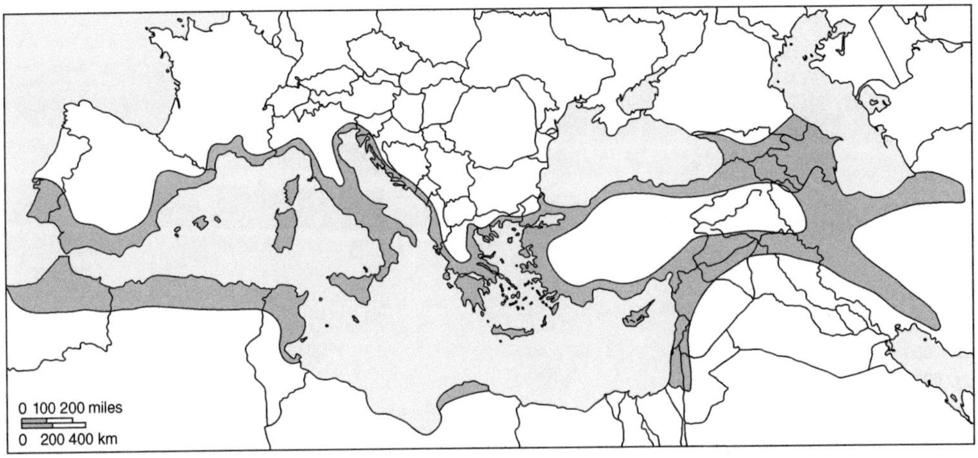


**Fig. 1.1.** Geographical distribution of the wild fig, *Ficus carica* L. (figure used with permission from Zohary *et al.,* 2012).

aggregate of wild and weedy fig types distributed throughout the Mediterranean basin (Fig. 1.1) (Zohary *et al.*, 2012).

Under domestication, the plants grew vegetatively through the rooting of winter dormant twigs of female individuals selected for their desirable characteristics, and occasionally through grafting. The main changes in this fruit crop under domestication were the shift to the vegetative propagation of female clones for: (i) an increase in syconia size, flesh and sugar content; (ii) the introduction of artificial pollination and caprification; and (iii) the successful selection for parthenocarpy, which eliminates the pollination required for fruit set (Zohary *et al.*, 2012).

By the fourth millennium BCE *F. carica* was cultivated in the Nile Delta, although it remains unclear where in the Nile Valley native sycamore fig (*Ficus sycomorus*) was originally distributed, or how early it was cultivated (most likely through vegetative propagation) (Fig. 1.2) (Fuller and Stevens, 2019).

## 1.4 Fig in Ancient Egypt

Cultivation of the sycamore fig, *Ficus sycomorus* L. (Moraceae), was almost exclusively an Egyptian specialty. Compared with the common fig, *F. carica*, it is a larger and taller tree, but produces smaller, inferior syconia. Since the Early Dynastic period, at end of the 4th millennium BCE, the sycamore fig was (and still is) a common fruit crop and a valued timber source in the lower Nile Valley (Zohary *et al.*, 2012). In Egypt, the earliest archaeobotanical record of figs is from Tell el-Fara'in (ancient Buto) in the Nile Delta in the Predynastic period (ca. 3650–3450 BCE: Thanheiser, 1991).

One of the most ancient paleobotanic attestations of fig fruits comes from the oasis of Kharga in Egypt, where early cultivation of figs was documented at the end of the Palaeolithic Era (Caton Thompson, 1952, p. 117).

After its complete domestication, at the end of the 4th millennium BCE, the fig acquired a strong symbolic value alluding to fertility during the First Egyptian Dynasty (3100–2900 BCE). With the palm tree, it was considered the tree of life and it was linked to Osiris, one of the most important deities of the Egyptian pantheon (De Rachewiltz, 1982, p. 147). For this reason, fig trees (*F. sycomorus*) adorned the courtyards of royal palaces and temples of the Pharaoh and temple and tomb decorations.

In Memphis there is a sycamore fig tree near the temple of Ptah in Karnak (Luxor), where, according to tradition, the goddess Hathor feeds figs to the dead as she rises from the trunk (Gillam, 1995, pp. 221–222). In Egypt, the sycamore was called *nehet*, a term whose root, *neh*, indicates protection while the female *nehetet* designated the goddess Hathor. In the Book of the Dead (ch. CLXXXIV), we read: 'Let me feed on the Sycamore

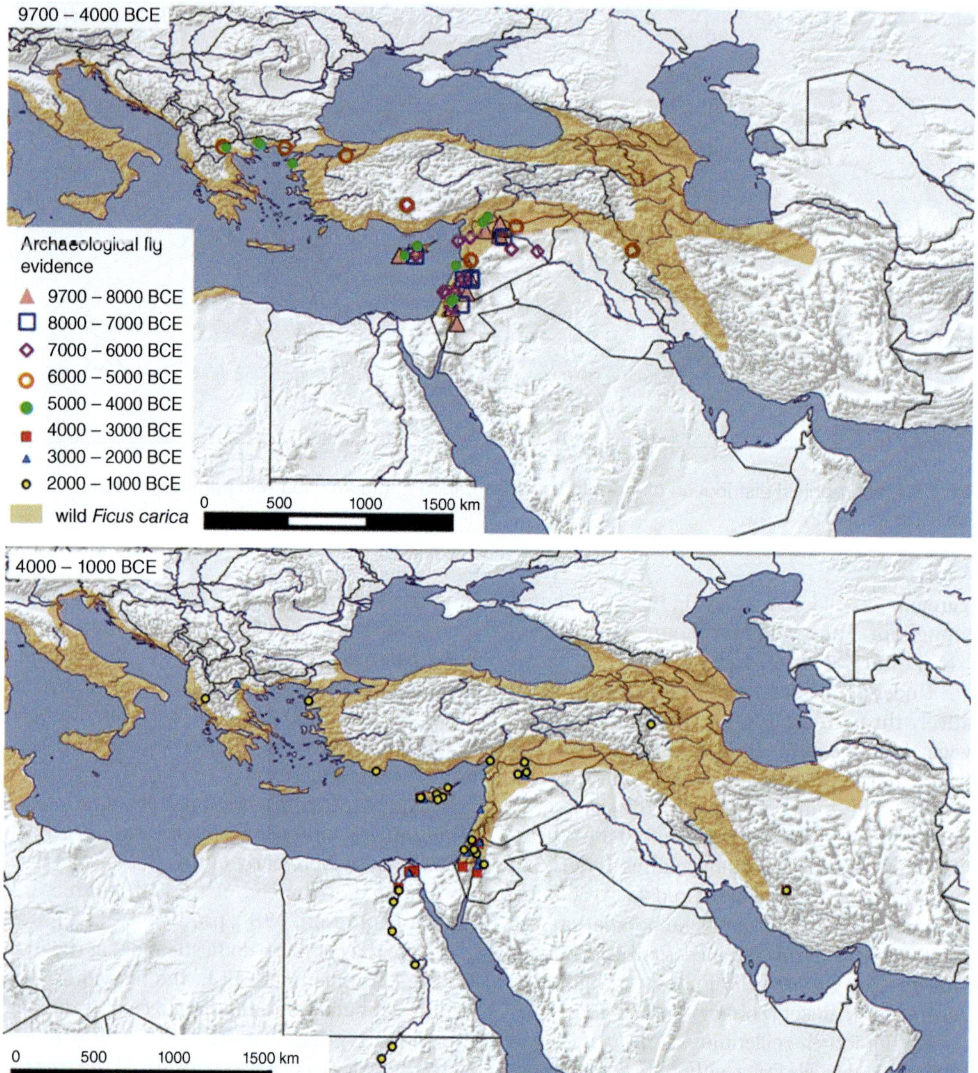


**Fig. 1.2.** Archaeobotanical finds of the fig in West Asia, North Africa and adjacent regions (mainly *Ficus carica* but potentially also *Ficus sycomorus* especially in North Africa) compared to the common fig's (*Ficus carica*) wild distribution. Sites are shown in millennium blocks from 9700 to 4000 BCE above, and from 4000 to 1000 BCE below. (Figure used with permission from Fuller and Stevens, 2019.)

of the goddess Hathor'. This scene is often reproduced with the holy buckets of milk and water used in fertility cult ceremonies.

A beautifully preserved drawing of a fig harvest has been found in the Tomb of Khnumhotep (Early 12th Dynasty, ca. 1950 BCE) at Beni Hasan depicting farmers training monkeys to climb trees and harvest them (Darby *et al.*, 1977).

The *F. sycomorus* tree is depicted in a decorative fragment from the tomb of Nebamon, an official who lived at Thebes under the reign of Thutmose IV (1401–1391 BCE), and Amenhotep III (1391–1353 BCE). The fragment, now in Londons's British Museum (TT146), depicts the garden to cheer the souls of the deceased. The garden, in which trees, flowers, birds and fish are meticulously represented, develops around a swimming pool. Among the trees, arranged in an orderly sequence, one can recognize the palm and the sycamore fig. This representation is perhaps inspired by the bas-reliefs of the earlier 'Syrian Garden' of Thutmose III (1479–1424 BCE) at Karnak. The fig is also one of the preferred foods for the journey to the afterlife; often consumed at banquets honoring the deceased. Such a banquet is shown in the almost contemporaneous wall painting of the Tomb of Nakht, a scribe who lived during the reign of Thutmose IV (TT152, Necropolis of Sheikh Abdel-Qurna: Seidel and Ghaffar Shedid, 1991).

In the Tell el-Amarna period (second half of 14th century BCE), representations of figs can be seen in numerous pharaonic reliefs describing scenes of everyday life of the Pharaoh and family. Here figs are served with pomegranates as both indicate the high status of their consumers and wealth and abundance of the land that produced them. Egypt was known for its figs.

A basket of figs held the fate of Cleopatra, the last Pharaoh of Egypt. Plutarch wrote the Queen had hidden the snakes with which she intended to kill herself in a basket of big and juicy figs: 'It is said that the asp was brought to Cleopatra in the basket together with the figs and that she had given orders to hide it among the leaves so that the reptile would bite her without her noticing. But when she removed the figs, she saw it, and said: "There it is, you see". She, therefore, bared her arm and offered it to the bite of the animal' (Plutarch, *Life of Antonius*, 86).

## 1.5 At the Beginnings of Civilization: The Fig in Neolithic Levant and East Mediterranean

In the Neolithic period, wild fig fruits were an essential part of the human diet. Therefore, man rapidly successfully cultivated them. One of the earliest proofs of domestication is attested to the Levant. In the Neolithic village of Gilgal, in the Lower Jordan Valley, numerous remains of stored carbonized figs dating to the 8th millennium BCE have been found. Gilgal figs belong to a mutant parthenocarpic variety, lack embryonic seeds and were probably vegetatively propagated through planting cuttings. Figs are, therefore, the first domesticated and cultivated plant, preceeding even cereals. Gilgal figs were grown with other plant foods; wild barley (*Hordeum spontaneum*), wild oats (*Avena sterilis*), and acorns (*Quercus ithaburensis*). The early Gilgal farmers, therefore, had a double subsistence strategy, the collection and conservation of wild plants, and early domestication of the fig tree. This type of economy was widely practiced during the Neolithic period in Mesopotamia and in the Levant (Kislev *et al.*, 2006).

However, along the coasts of the eastern Mediterranean, domestication of the fig occurred centuries later. These regions had the closest wild ancestor of the cultivated fig, *Ficus palmata* L., which is the wild *F. carica* variety of the Levant, South Turkey and Aegean area. It is thought to be the most likely ancestral species of the early domesticated fig. The temperate climate with abundant autumn and spring rains was a favorable element for fig, vine and olive cultivation. The association with grapes is meaningful, as figs were needed to trigger wine fermentation. For this reason, the fig played a cultural role in this area, being considered an essential element of the daily diet, nutritious, and easily preserved. In the later Minoan, Mycenaean and Phoenician maritime societies, dried figs were food on long Mediterranean voyages.

## 1.6 The Role of the Phoenicians: Figs in Greece and West Mediterranean in the Iron Age

From the Near East and the Levant, Phoenicians carried *F. carica* to the western world via two distinct routes. Following the Phoenician route via the larger islands of the Mediterranean; Cyprus, Rhodes, Crete, Malta, Sicily, Sardinia, and Corsica; fig culture reached the central and western regions of the Mediterranean and beyond through the Pillar of Hercules on the Portugal coast. From the coasts, the material and immaterial knowledge, customs and traditions of the Phoenician culture seeped inland. Multiple fig tree species had been carried since the beginnings of the Iron Age. In the early stages of diffusion, the tree and its fruit were initially a luxury for the élites. Later, as figs are rich in sugar and carbohydrates, and able to be consumed fresh, dried, in jams or with honey, they were an important component of the daily diet (Mata Parreño *et al.*, 2010a). In Phoenician and Punic sites, whole figs and partial figs have been found in funeral offerings as symbolic nourishment in the afterlife, and in domestic contexts. Fig seeds have also found in the same domestic contexts and in pits. The presence of seeds may have been accidental. The fleshy, juicy fig fruit decomposes easily. Perhaps the seeds remained in archaeological deposits or in silos after ritual funeral or domestic fires.

Generally, in archaeological layers, it is more frequent to find the fruits rather than the wood, suggesting the great importance of fig production and less use of the wood itself as firewood. Fig is low quality, brittle, coarse-grained, pithy wood, and smokes with burning. Its bark secretes a white latex that irritates the skin but which, if mixed with milk, produces vegetable rennet. In Phoenician and Punic art, iconographic representations of the fig tree or its fruits are quite rare. Nevertheless, they played an important role in the Phoenician imagination, as they were appreciated in everyday life and also as nourishment and gifts for the afterlife, as evidenced by the offering of whole figs in graves. Figs were widely spread throughout the Mediterranean by the Phoenicians. For them figs were not only a luxury when fresh and ripe but also as an important dried diet staple during the winter months (Bartoloni and Guirguis, 2017).

The first mention of figs in Greece is in the Odyssey; in Book 7 figs are described in the land of Phaeacians. However, these verses citing the figs were probably interpolated much later. The earliest genuine mention of figs is by the 7th century BCE poet Archilochus. He tells of figs cultivated, and greatly enjoyed, in Paros. The fig was introduced to Greece from the Levant in the 8th century BCE. It arrived via the same island route of the eastern and central Cyclades, as other fruits and plants of oriental origin, including the pomegranate and the palm tree (Mata Parreño *et al.*, 2010b, p. 43). Fresh figs are found in tombs as food for the deceased, often with fresh grapes, pomegranates, olives and hazelnut.

The fig was an essential component in the ancient Greek diet, consumed fresh, dry or as a condiment in cooked dishes. Athens had laws regulating the export of the best quality figs. Exporting figs meant stealing the main food of poor people, and therefore was prohibited by law. The ancient Greek language had a specific word for those who reported the fig exporters of Attica and for those who stole sacred figs: *sycophant*, from σῦκον, fig, and φαίνειν [to show, to denounce]. This reference to fig thieves, highlights another important facet of the fig tree in Ancient Greece. The fig tree and its fruit was consecrated to Dionysus and Demeter, deities linked to the mystery cults, the *Bacchanalian* and the Eleusinian mysteries.

Athenaeus of Naucratis, a Greek rhetorician and grammarian, lived between the second and third century CE. His major work *Deipnosophistae* (3:78a) refers to the association of the fig and Dionysos: 'Sosibos the Lakedaimonian, by way of proving that the fig-tree is a discovery of Dionysos, says that for that reason the Lakedaimonians even worship Dionysos *Sykites* (of the fig). And the Naxians, according to Andriskos and again Aglaosthenes, write that Dionysos is called *Meilikhios* (gentle) because he bestowed the fruit of the fig. For this reason, among the Naxians the face of the god called Dionysos *Bakkheos* is made of grapevine, whereas that of Dionysos *Meilikhios* is of fig-wood. For, they say, figs are called *meilikha*, mild fruit' (Gulick, 1941). The fig was one of Greece's sacred plants in addition to the grapevine and wild ivy. In Greek mythology Dionysus was first to discover the fig, and for this reason he is often addressed with the epithets Sykites and Meilikhios.

The fig and the fig tree are also linked to Demeter, the goddess of agriculture, nature and the seasons. According to Pausania (*Description of Greece*, 1:37, 2), the hero Phytalus of Laciadae in Attica was said to have been taught the culture of figs by Demeter: 'A little way past the grave of Themistocles is a precinct sacred to Lacius, a hero, a parish called after him Laciadae, and the tomb of Nicocles of Tarentum, who won a unique reputation as a harpist. There is also an altar of Zephyrus and a sanctuary of Demeter and her daughter. Within them Athena and Poseidon are worshipped. There is a legend that in this place Phytalus welcomed Demeter in his home, for which act the goddess gave him the fig tree. This story is borne out by the inscription on the grave of Phytalus: "Hero and king, Phytalus here welcome gave to Demeter, August goddess, when first she created fruit of the harvest; Sacred

Fig is the name which mortal men have assigned it".' To thank Pythalus for his hospitality, Demeter made the fig tree grow out of the ground the same way Athena sprouted the olive tree from the ground. Thus the fig tree became a sacred tree for the Greeks, second only in importance to the olive.

## 1.7 The Fig in Roman Culture

The fig tree plays an important role in Roman mythology. The legend of Romes's founding tells of the wolf that suckled Romulus and Remus resting under a fig tree, that has since been called *ficus ruminalis*. Plutarch (*Life of Romulus*) examines three hypotheses about the meaning of this definition. The adjective *ruminalis* descends from Romulus; ruminalis designates the 'ruminants', cows that stopped there to rest in his shadow; or it refers to the breastfeeding of the two babies, as the ancients called the breast *ruma* and *Rumina* was the goddess who Romans believe cares for children (Carandini, 2007).

The sycamore fig has been an element in the Roman religion since ancient times. As Pliny the Elder tells in his *Naturalis Historia* (20:4–11), the fig tree was undoubtedly the most important sacred tree of Rome because it was a spectator and guardian of its origins. In the heart of the Roman Forum, near the *Comitium*, the fig tree had a place of honor with the olive tree and grapevine as the sacred plants in the Roman religion. Sumptuous baskets of figs and luxuriant fig trees were represented on the frescoes of the patrician *villae* of Pompeii, as symbols of health and prosperity.

The fig had a prominent place in the eating habits of Romans. Literary sources, in particular Columella and Cato, devote ample space to the preservation of figs and other precious fruits: walnuts, hazelnuts, almonds, pine nuts, grapes, apples, pears, quinces, prunus, rowanberries, but above all drying and storing figs. The fig, due to its high nutritional value, was also integrated the diet of the plebeian classes and slaves (Cato, *De Re Rustica*, 56). Columella said that in *Hispania* people kept figs in boxes and used them to make a fig bread (Schulten, 1959, p. 422). As figs are very sugary, they could be also used as a condiment in more elaborate foods. According to Pliny, eating figs 'increases the strength of young people, improves the health of old people and reduces wrinkles', and thanks to its caloric intake and easy digestibility, they were loved by athletes and convalescents. According to Pliny (*Nat. Hist.* 15. 19), the Romans knew twenty-nine fig varieties, coming from all over the Mediterranean. Many of these varieties were lost with the end of the Roman Age. In Medieval times memory of the fig and knowledge of it and its cultivation were lost.

## 1.8 The Fig Tree in the Holy Books

The figs are mentioned in the scriptural texts of the three monotheistic religions (Judaism, Christianity and Islam), the Hebrew Bible, the Christian Bible and the Quran.

In the Bible the first mention is in the book of Genesis (*Gen.* 3, 5), when Adam and Eve, after eating the fruits of the tree of the knowledge of good and evil, discover themselves naked, and cover themselves with fig leaves: 'And the eyes of them both were opened, and they knew that they were naked; and they sewed fig leaves together, and made themselves aprons' (*Gen.* 3:7). But generally, in the Old Testament the tree and its fruits represent the richness and fruitfulness of the Holy Land ('a land with wheat and barley, vines and fig trees, pomegranates, olive oil and honey', *Deut.* 8:8), as they are symbols of fertility, and the emblem of the joyful life in the messianic kingdom. Many fig citations document the nutritional importance of figs in the daily diet as healthful foods, consumed both fresh and dried[1], as testified in many passages of the Bible[2].

Biblical references to the fig also indicate the extent of its culture in ancient times: 'And the trees said unto the fig-tree, come and reign over us. But the fig tree said to them, should I forsake my sweetness, and my good fruit, and go to be promoted over the trees?' (*Judges* 9:10–11). 'I was not a prophet or a prophet's son but I was an herdsman, and a piercer [mistranslated as "gatherer" in the King James Bible] of sycomore fruit' (*Amos* 7:14). 'And seeing a fig tree afar off having leaves, he came, if haply he find anything thereon: And when he came to it, he found nothing but leaves; for the time of figs was not yet, and in the morning as they passed by, they saw the fig tree dried up from the roots' (*Mark* 11:13, 20).

'A certain man had a fig tree planted in his vineyard; and he came and sought fruit thereon, and found none. Then said he unto the dresser [cultivator] of his vineyard, Behold, these three years I come seeking fruit on this fig tree, and find none; cut it down; why cumbereth it the ground? And he answering said unto him, Lord, let it alone this year also, till I shall dig about, and dung it: And if it bear fruit, well: and if not, then after that thou shalt cut it down' (*Luke* 13:69).

### Quran (Koran)

A number of plants are mentioned in various verses of the Quran (Koran) as well as horticultural and agricultural products. The fig is one of the fruits mentioned in the Quran, along with olives, grapes, pomegranate and dates. The medicinal advantages of the fig is discussed in the first Quranic verses in sura Al-Tin, the Allah says: 'I swear by the Fig and the Olive' (Quran sura Al-Tin 95:1). For both figs and olives, Allah swore that they were foods with high nutritional value for humans. He also noted the sanctity of their native lands as well.

## Notes

[1] Samuel 25:18: 'Abigail acted quickly. She took two hundred loaves of bread, two skins of wine, five dressed sheep, five seahs of roasted grain, a hundred cakes of raisins and two hundred cakes of pressed figs, and loaded them on donkeys'.

[2] Samuel 30:12: 'And they gave him a piece of a cake of figs, and two clusters of raisins: and when he had eaten, his spirit came again to him: for he had eaten no bread, nor drunk any water, three days and three nights'. 'Moreover those who were near to them, even as far as Issachar and Zebulun and Naphtali, brought food on donkeys, camels, mules and on oxen, great quantities of flour cakes, fig cakes and bunches of raisins, wine, oil, oxen and sheep. There was joy indeed in Israel'. 2 Kings 20:7: 'Then Isaiah said, "Prepare a poultice of figs." They did so and applied it to the boil, and he recovered'.

## References

Bartoloni, P. and Guirguis, M. (2017) *I Fenici del mare e le vie dei tonni. Un'inchiesta storico-archeologca dal Mediterraneo orientale all'Atlantico*. Quaderni stintinesi, 7. Sassari, Editrice Democratica Sarda, Sassari, Italy.

Carandini, A. (2007) *Roma. Il primo giorno*. Laterza, Rome-Bari.

Caton-Thompson, G. (1952) *Kharga Oasis in Prehistory*. Athlone Press, University of London, London.

Condit, I.J. (1947) *The fig*. Chronica Botanica, Waltham, MA.

Darby, W.J., Ghalioungui, P., and Grivetti, L. (1977) *Food: the gift of Osiris*. Academic Press, London; New York.

de Rachewiltz, B. (1982) *Il libro egizio degli inferi*. Atanor, Rome.

Fuller, D.Q. and Stevens, C.J. (2019) Between domestication and civilization: the role of agriculture and arboriculture in the emergence of the first urban societies. *Vegetation History and Archaeobotany* 28, 263–282.

Gillam, R.A. (1995) Priestesses of Hathor: Their Function, Decline and Disappearance. *Journal of the American Research Center in Egypt* 32, 211–237.

Gulick, C.B. (ed.) (1941) *Athaenaeus, The Deipnosophists*. Harvard University Press, Cambridge, MA.

Kislev, M.E., Hartmann, A. and Bar-Yosef, O. (2006) Early domesticated fig in the Jordan Valley. *Science* 312(5778), 1372–1374. DOI: 10.1126/science.1125910.

Mata Parreño, C., Badal García, E., Bonet Rosado, H., Collado Mataix, E., Fabado Alós, F.J., Fuentes Albero, M., Izquierdo Peraile, I., Moreno Martín, A., Ntinou, M., Quixal Santos, D., Ripollès Alegre, P.P. and Soria Combadiera, L. (2010a) Comida para la eternidad. Saguntum: *Papeles del Laboratorio de Arqueología de Valencia-Extra*, 9.

Mata Parreño, C., Badal García, E., Collado Mataix, P.P. and Ripollès, A. (eds) (2010b) *Flora Iberica. De lo real a l'imaginario*. Servicio de Investigación Prehistórica. Serie de Trabajos Varios, Núm. 111. Diputación de Valencia, Valencia, Spain.

Schulten, A. (1959) *Geografía y Etnografía Antigua de la Peninsula Ibérica*, 2 vols. Consejo Superior de Investigaciones Científicas, Instituto "Rodrigo Caro" de Arqueología, Madrid.
Seidel, M. and Ghaffar Shedid, A. (1991) *Das Grab des Nacht. Kunst und Geschichte eines Beamtengrabes der 18. Dynastie in Theben-West*. von Zabern, Mainz, Germany.
Thanheiser, U. (1991) Untersuchungen zur Landwirtschaft der vor- und Fruhdynastischen Zeit in Tell-el-Fara'in Buto. *Agypten und Levante* 2, 39–45.
Zohary, D. and Spiegel-Roy, P. (1975) Beginnings of fruit growing in the Old World. *Science* 187(3),19327.
Zohary, D., Hopf, M. and Weiss, E. (2012) *Domestication of Plants in the Old World: The Origin and Spread of Domesticated Plants in Southwest Asia, Europe, and the Mediterranean Basin*. 4th edn. Oxford University Press, Oxford, UK.

# 2 Taxonomy, Botany and Physiology

**B. Gaaliche[1]*, D. Narzary[2], M. Ben Mimoun[3] and A. Sarkhosh[4]**
[1]*Laboratory of Horticulture, National Agricultural Research Institute of Tunisia (INRAT), Tunis, Tunisia;* [2]*Department of Botany, Gauhati University, Assam, India;* [3]*University of Carthage, National Agronomic Institute of Tunisia (INAT), Laboratory GREEN-TEAM (LR17AGR01), Tunis, Tunisia;* [4]*Horticultural Sciences Department, University of Florida, Gainesville, USA*

## 2.1 Introduction

The species, *Ficus carica* L., which is generally known as the 'fig', 'common fig' or 'edible fig', belongs to the genus *Ficus* (family Moraceae). Botanically the fruit is a 'syconium'. The term 'figs' can denote single or multiple species of *Ficus* (Janzen, 1979). Reported to have been cultivated since the Neolithic era (~5000 BCE), the fig's multiple uses are frequently mentioned in ancient historical records of the arts, commerce and agriculture (Morton, 1987). The fig appears to have been domesticated from a group of diverse spontaneous figs occurring in the southeastern Mediterranean (Zohary and Hopf, 2000). The fig is interesting to botanists, entomologists and evolutionary biologists because of its specialized inflorescence 'syconium' and distinctive pollination biology. The fig tree is unique among the *Ficus* species spread throughout tropical and subtropical countries in that the edible fruits are commercially valuable (Ferguson *et al.*, 1990).

This chapter will focus on the botany and physiology of the horticulturally important fig species. The global demand for figs is increasing due to their nutritional and medicinal value. Understanding the botany, biology and physiology of figs will enhance crop management and provide germplasm improvement to meet the demand with increased production.

## 2.2 Origin, Distribution and Habitat

Figs are thought to have originated in southern Arabia, where wild fig and caprifig trees were found (Condit, 1947; Storey, 1975). Jona and Gribaudo (1991) suggested that *F. carica* might have originated in the jungles of the Caspian foreshores, northwest Turkey and other surrounding areas. According to Sinha (2003), the fig is native to the southern Arabian Peninsula, Italy, the Balkan Peninsula and Russia. The fig is considered the first domesticated plant of the early Neolithic Revolution, because it could be easily propagated through cuttings (Kislev *et al.*, 2006). Through migration, the fig has dispersed widely outside of its natural range. The fig is among the ancient fruit trees that began horticulture in the Mediterranean basin (Zohary and Spiegel-Roy, 1975). Human interaction in Mediterranean regions dispersed figs to Persia, Asia Minor and Syria (Datwyler and Weiblen, 2004). It was cultivated by Phoenicians, Syrians, Egyptians, Greeks and other surrounding Mediterranean groups (Storey, 1975). The fig

*Corresponding author: gaalichebadii@gmail.com

 *The Fig: Botany, Production and Uses* (eds. Sarkhosh, A., Yavari, A. and Ferguson, L.)
DOI: 10.1079/9781789242881.0002

was introduced into England between 1525 and 1548, after which European types were exported to China, Japan, India, South Africa and Australia. Dehydration facilitated fig distribution, resulting in many distinctive cultivars dispersed to Europe (Aksoy, 1998). The fig was naturalized in many places and was established extensively in Western Asia, Crimea and Transcaucasia (Khatib and Vaya, 2010). Nevertheless, its movement towards the East was rather slow due to the rainy and humid climate of Southern and Southeastern Asia (Aksoy, 1998).

*F. carica* is found growing in diverse habitats, including infertile rocky land, woodland, scrubland and even places with hot, dry soil (Lansky and Paavilainen, 2010). Fig trees grow naturally in open places, mixed forests, fissures of rocks and stony slopes in river valleys (Browicz, 1982). Fig is especially well-adapted to Mediterranean climates, with moderate winters and hot dry summers (Morton, 1987). Both cultivated and wild or semi-wild figs (Fig. 2.1) grow throughout the Middle East and Mediterranean region (De Candolle, 1967).

## 2.3 Taxonomy and Systematics

The genus *Ficus* L. comprises about 735 species in the world. The taxonomy of the genus *Ficus* is highly complex, with 6 subgenera, 19 sections, and 27 subsections (Berg and Corner, 2005). Co-speciation between specialized pollinators and their hosts (Weiblen and Bush, 2002) might account for the extreme species diversity of *Ficus* compared to other genera of the family. Pollination in *Ficus* is species-specific; changes in the chemical composition of the attractants may have played a role in speciation. The presence of latex (milky sap) and the syconium uniquely distinguish the genus *Ficus* from other families. The genus can be recognized by the presence of waxy glandular spots on the leaf lamina beneath and on the base of the midrib. Taxonomically *F. carica* has been placed in the subsection *Ficus* (section *Ficus*) with *F. palmata* Forssk and *F. iidaiana* Wilson (Berg and Corner, 2005). The term '*carica*' comes from Caria in Asia Minor where excellent figs were grown in ancient times (Condit, 1947).

*F. carica* has some primitive features. It is in the dioecious subgenus *Ficus*, with some cultivated varieties which appear to be throwbacks, producing gall-figs and seed-figs on the same plant. Fig is a small tree with a distinctly pachycaulous habit, with a disproportionately thick trunk for its height, and few branches. The word is derived from the Greek word 'pachy'; meaning thick or stout. It has variously palmatifid leaves, ranging from deeply dissected in some varieties to practically entire in others. In some varieties, the mid lobe itself can be lobed, the central leaf blade can be pinnately lobed with a palmately lobed base (Corner, 1985). The pedunculate syconium of *F. carica* is advanced in its absence of lateral bracts and sclerotic cells in the fig wall, probably as a result of selection for edibility. Internal bristles are short but abundant in syconia. The male flowers are advanced in their ostiolar position with several rows, retaining the primitive plurality of 2–4 stamens, often with a sterile, rudimentary pistil, a pistillode. The female flowers retain the primitive bifid stigma. Pollination by the pollinating wasp (*Blastophaga psenes*) takes the more primitive means of copious pollen (Galil and Meiri, 1981).

### Phylogeny

An antecedent for *F. carica* is unknown. Fig is typically not found in forest and tropical environments like the strangling figs which have exploited tropical forest canopies (Corner, 1985). *F. carica* appears to have originated outside the great forest. Although truly wild specimens have not been discovered, it is thought to have originated in the Middle East. Cultivated figs are thought to have been derived naturally from wild figs through a chance mutation. Or, the subtropical species *F. carica* may have evolved from *F. palmata*, distributed throughout the tropical and subtropical regions from Asia to Africa.

According to Trabut (1922), *F. carica* is a synthetic species that may have been derived from several wild species through natural hybridization. It is found near other closely related species or subspecies, such as *F. persika*, *F. virgata* and *F. johannis*; grown in southwestern Persia, Mesopotamia and Arabia. Natural hybridization occurs between all these species, the resulting

**Fig. 2.1.** Tree habits of the common fig (*F. carica*). (a) wild fig; (b, c) cultivated fig with multi-trunk system; (c) cultivated fig with weeping tree habit; (d) deciduous or dormant stage. (Figures used with permission of Ali Sarkhosh and Alimohammad Yavari.)

seedlings being exceedingly variable. *F. palmata*, though generally regarded as the Indian form of *F. carica*, is widely distributed from northern India to Afghanistan, Arabia, Egypt and Abyssinia. The small bitter fruits of *F. palmata* have little commercial value.

### *Distinguishing features of F. carica*

In general, the comparatively large and palatable fruits are what distinguish *F. carica* from the other members of the genus *Ficus*. Sometimes the form with lobed leaves is described as *F. carica*, whereas the form with entire leaves is described as *F. palmata*, while all variations can be seen on the same plant from different regions (Tiwari *et al.*, 2014). Since these variations are continuous with numerous intermediate forms, delimiting the taxa based on leaf morphology is not always easy. Misidentification between *F. palmata* and *F. carica* is common in herbaria.

*F. carica* differs from its closely allied species, *F. palmata*, in the degree of pubescence and thickness of young stems, color and pubescence of tepals and fruit size. In *F. carica*, the twigs are glabrous or glabrescent and stouter, the figs more than 2 cm in diameter and tepals often glabrous or rarely glabrescent and white to pink.

*F. palmata* twigs are tomentose and slender, the figs up to 2 cm in diameter and tepals puberulous and white (Corner, 1965). *F. palmata* figs are hard and unpalatable compared are sweet, tasty, and edible ripe *F. carica* figs. Although the palmatifid leaf is found in other members of subgenus when non-bearing Ficus, *F. carica* is the only species to perpetuate the palmately lobed leaf as an adult (Corner, 1985).

## 2.4 Cytology

*F. carica* is diploid, and the somatic chromosome number ($2n = 2x$) is 26 (Storey, 1975; Jona and Gribaudo, 1991). Falistocco (2009) reported triploid ($2n = 3x = 39$) cytotypes in a cultivated common fig. The two tree forms, i.e., male and female of *F. carica* are thought to be controlled by genetic factors. Males are the heterogametic (GA/ga) and females are the homogametic (ga/ga) (Storey *et al.*, 1977). Short-style female flowers and male flowers are determined by a pair of dominant genes on the same chromosome (GA), while recessive alleles result in long-style female flowers and suppression of male flowers (ga) (Singh *et al.*, 2015). Mori *et al.* (2017) identified *RAN1*, an orthologue gene associated with sex determination in figs.

## 2.5 Horticultural Classification

The fig tree is a dioecious species, present in two sexual forms, i.e., the domesticated fig (female tree producing edible fruit) and the caprifig (male tree, source of pollen for commercial fig plantings) (Stover *et al.*, 2007).

### Domesticated fig or female fig

The domesticated fig can produce two crop types within the same year. The first crop (also called breba crop) is formed laterally from buds produced in the lateral leaf axils of the previous season's shoot growth (Fig. 2.2a). These buds

**Fig. 2.2.** Different seasonal crops of *F. carica*. (a) Breba (first crop) and main (second) crop of an edible fig; (b) matured mamme crop and developing profichi crop of a caprifig. (Figures used with permission of Badii Gaaliche and Alimohammad Yavari.)

develop in the following spring and the fruits mature between June and July. The second crop (also called the main crop) is produced laterally in the axils of leaves on the current season's shoot growth (Fig. 2.2b) (Gaaliche *et al.*, 2011a). Fruit maturation starts from July and can last until temperatures drop between October and December (Crane and Brown, 1950; Petrucci and Crane, 1950).

Female fig cultivars are classified into three horticultural types, i.e., Smyrna, San Pedro and Common-types (Fig. 2.3) based upon their cropping and pollination characteristics (Condit, 1955; Storey, 1975). The Smyrna-type usually does not produce a breba crop (first crop), and the syconia of the main crop (second crop) will form and ripen only if successfully pollinated. Typical examples of Smyrna-type include the cultivars, 'Sarilop', 'Marabout', and 'Zidi'. The San Pedro-type will initiate a breba crop that persists to maturity (parthenocarpic), while the main crop requires pollination to develop to maturity. Typical examples of San Pedro-type include the cultivars, 'San Pedro', 'Dauphine', 'King', and 'Bither Abiadh'. The Common-type may or may not initiate a breba crop while the main crop is abundant and develops to maturity; neither requires pollination. Typical examples of Common-type include the cultivars, 'Brown Turkey', 'Mission' and 'Adriatic'.

**Fig. 2.3.** Fruits of four fig types: (a) Caprifig, (b) Smyrna ('Calimyrna'), (c) San Pedro ('Beyaz orak'), and (d) Common ('Brown Turkey'). (Figures used with permission of Alimohammad Yavari, Megan Crivelli and Birgül Ertan.)

### Caprifig or male fig

The caprifig produces syconia with flowers, which contain pollen that is destined to pollinate the Smyrna and San Pedro-types (Stover *et al.*, 2007). The cultivars 'Poozdombali', 'Amellal' and 'Jrani' are typical varieties of this group. Caprifig trees produce three crops of syconia a year. The first crop is identified as the mamme crop (Fig. 2.2b). This crop originates from the current season's growth in the fall, after which it overwinters on the tree and comes to maturity in the spring. The second or profichi crop originates in the spring from latent buds on shoots developed the previous year. The third, mammoni crop of caprifigs develop on the current season's growth and come to maturity in the fall (Gaaliche *et al.*, 2017a).

## 2.6 Germplasm Collection and Conservation

Due to habitat destruction and over-exploitation of resources, germplasm collection and conservation of crop plants are important for bioresource safety and security. A list of fig germplasm collection centers compiled from the secondary sources is presented in Table 2.1. The Nikita Botanical Gardens (NBG), Yalta, Crimea, Ukraine has the most fig accessions (334), followed by the Regional Center for Research in Oasis Agriculture (CRRAO) of Degache in Tunisia (300 accessions), and the Erbeyli Fig Research Institute, Turkey with 285 accessions. In the National Clonal Germplasm Repository, Davis, CA, USA, there are presently 194 accessions of fig germplasm (Aradhya *et al.*, 2010; Hummer *et al.*, 2012) a decrease from the earlier 275 accessions (Flaishman *et al.*, 2008). If researchers cease collecting from unexplored regions, accessions will decrease over time. DNA fingerprinting, using standard molecular markers, has proved useful in assessing current fig collections.

Because fig it is so easily propagated it quickly spread from its center of origin (somewhere in the Middle East) to the neighboring countries throughout the Mediterranean region and western Asia, extending as far as North Africa and India. Local landraces and wild relatives of the fig are still found in Egypt and Libya (Abdelsalam *et al.*, 2019), Turkey (Gozlekci, 2010; Caliskan and Polat, 2012; Caliskan *et al.*, 2018), Italy (Ciarmiello *et al.*, 2015), Croatia (Marcotuli *et al.*, 2019), Egypt (Mustafa and Abou-Ellail, 2013; Abou-Ellail *et al.*, 2014), Algeria (Boudchicha *et al.*, 2018), Slovenia (Knap *et al.*, 2017; Knap *et al.*, 2018), Iraq (Ali *et al.*, 2013; Ali, 2019), Iran (Khadivi-Khub and Anjam, 2014), Jordan (Almajali *et al.*, 2012), Tunisia (Saddoud *et al.*, 2007; Gaaliche *et al.*, 2012; Ben Abdelkrim *et al.*, 2015; Haffar *et al.*, 2017), Spain (Perez-Jimenez *et al.*, 2012) and Morocco (Achtak *et al.*, 2009a). The diversity among fig cultivars may be a result of plants acclimatizing and adapting to new environments, becoming new ecotypes. Fig germplasm collection from diversified agro-climatic zones could be a source of material for adapting to climate change.

Germplasm collection and maintenance is a challenging task in agriculture. The majority of germplasm maintenance is tedious and demands sustained land, trained manpower, and support. Therefore, germplasm screening to select the unique and genetically distinct accessions is important to ensure non-redundant germplasm collection and effective bioresource management in collection centers or gene banks. DNA fingerprinting is incredibly useful for screening a high volume of samples. It can provide identification of genetically distinct accessions within a brief period. The co-dominant markers such as simple sequence repeats (SSRs) and amplified fragment length polymorphism (AFLP) are more reproducible and informative than the dominant markers such as RAPD, ISSR and DAMD. However, designing SSR primers needs genome sequence information from closely allied species if not from the same species. Giraldo *et al.* (2004) developed microsatellite, or SSRs, in *F. carica* from a CT-enriched genomic library of a Spanish fig cultivar to evaluate the cross-species transferability in 14 *Ficus* and 2 *Morus* species where 5 of 8 primer pairs showed successful amplification in all the *Ficus* and *Morus* species. Giraldo *et al.* (2005) also successfully assessed the cross-species transferability of microsatellites from the African fig (*F. sycomorus*) to the common fig (*F. carica*). Khadari *et al.* (2003) and Ikegami *et al.* (2009) compared the ISSR, RAPD and SSR marker analysis in fig cultivars and concluded SSR markers were more effective at distinguishing cultivars than the other

**Table 2.1.** Fig germplasm collection centers with 20 or more accessions

| Germplasm center | Country | No. of accessions | Reference |
|---|---|---|---|
| Nikita Botanical Gardens (NBG), Yalta, Crimea, Ukraine | Ukraine | 334 | Yezhov *et al.*, 2005 |
| Erbeyli Fig Research Institute, Turkey | Turkey | 285 | Tanriver, 2019 |
| Conservatoire Botanique National Mediterraneen de Porquerolles, Castel Sainte Claire, France | France | 277 | Roger and Khadari, 2003 |
| Servicio de Investigación y Desarrollo Tecnológico (SIDT), Finca La Orden, Guadajira, Badajoz, Spain | Spain | 220 | Giraldo *et al.*, 2008 |
| Instituto Valenciano de Investigaciones Agrarias (IVIA), Moncada (Valencia), Spain | Spain | 164 | www.ueresgen29.unifi.it/fig.htm (accessed 26 July 2020) |
| USDA National Clonal Germplasm Repository (NCGR), Davis, California | USA | 194 | Aradhya *et al.*, 2010; Hummer *et al.*, 2012 |
| Caserta field Station, Istituto Sperimentale per la Frutticoltura, Rome | Italy | 169 | Monastra *et al.*, 1995 |
| Plant Production Department (DPV-PZ), University of Basilicata, Potenza, Italy | Italy | 25 | www.ueresgen29.unifi.it/fig.htm (accessed 26 July 2020) |
| Experimental Institute for Fruit Growing - Sec. Caserta (ISF-CE), Caserta, Italy | Italy | 51 | www.ueresgen29.unifi.it/fig.htm (accessed 26 July 2020) |
| Institute of Arboreal Cultivations (ICA-SS), University of Sassari, Sassari, Italy | Italy | 31 | www.ueresgen29.unifi.it/fig.htm (accessed 26 July 2020) |
| Arab Center for the Studies of Arid Zones and Dry Lands (ACSA), Jillin (Southern Syria) | Syria | 100 | El Rayes, 1995 |
| Mahtumkuli Scientific Production Center of Plant Genetic Resources (MSPCPGR), Turkmenistan | Turkmenistan | 92 | Turkmenistan-CBD Report 2009 |
| Ain Taoujdate Experimental Station (INRA, Meknes), Morocco | Morocco | 75 | Achtak *et al.*, 2009a |
| Institute of Olive and Horticultural crops of Kalamata (ELGO 'Demeter') | Greece | 90 | Ganopoulos *et al.*, 2015 |
| Direcção Regional de Agricultura do Algarve, Tavira, Algarve | Portugal | 55 | Elisiário *et al.*, 1998 |
| Faculty of Agrarian and Technological Sciences FCAT/UNESP, Campus of Dracena, São Paulo, Brazil | Brazil | 45 | Rodrigues *et al.*, 2019 |
| BERC-Til Botanic Gardens, Til, Nablus, Palestine | Palestine | 24 | Ali-Shtayeh *et al.*, 2014 |
| Regional Center for Research in Oasis Agriculture (CRRAO), Degache | Tunisia | 300 | Saddoud *et al.*, 2007 |
| Arid Regions Institute (IRA), Médenine | Tunisia | 139 | Aljane *et al.*, 2018 |
| Higher Agronomic Institute of Chott-Mariem (ISA-CM), Sousse | Tunisia | 75 | Mars *et al.*, 2008 |
| Regional Commissariat for Agricultural Development (CRDA) of Gafsa | Tunisia | 50 | Saddoud *et al.*, 2007 |
| Fruit tree Field Genebank, National Genebank of Tunisia (BNG) | Tunisia | 45 | Saddoud *et al.*, 2007 |
| National Agricultural Research Institute of Tunisia (INRAT) | Tunisia | 34 | Gaaliche *et al.*, 2018 |

methods. Later, other researchers used SSR markers to evaluate the genetic diversity among different fig germplasm collections (Achtak *et al.*, 2009b; Almajali *et al.*, 2012; Perez-Jimenez *et al.*, 2012; Abou-Ellail *et al.*, 2014; Essid *et al.*, 2015; Boudchicha *et al.*, 2018; Caliskan *et al.*, 2018; Marcotuli *et al.*, 2019). For example, molecular data in combination with the morphological data, recommended by Gil *et al.* (2011), made germplasm evaluation for the cultivar 'Albacor' possible. 'Albacor', which has a narrow genetic diversity, but distinct horticultural traits, is grown worldwide, and is known by different names in different localities. The development of trait-specific proteins or DNA markers is also useful for germplasm tagging, variety identification and marker-aided selection in crop breeding. Recently, Ikten and Yilmaz (2019) used gender-specific CAPS markers for the identification of male or female figs in the Turkish fig collection and F1 progenies. Collection of caprifig (male fig) germplasm is equally important for the parthenocarpic fruit-bearing female figs. Caprifigs have no economic value (Condit, 1932), except for pollen production and as a host for life-cycle completion of the pollinating agent (*B. psenes*). Fertile seeds produced from cross-pollination produce the variability among progenies necessary for natural selection and adaptation in changing environments.

Germplasm conservation of figs can be performed either *in situ* or *ex situ*, where they are grown, or in germplasm collections. The traditional cultivars maintained by farmers in the Mediterranean basin and Western Asian countries are quite diverse. Although *ex situ* conservation in germplasm collection centers or experimental gardens have a certain advantage over *in situ* or on-farm conservation, maintaining an immense stock of live plants is challenging and resource-dependent. Therefore, on-farm conservation is an effective alternative for research institutes, local non-governmental organizations, development agencies and farmers who want to co-operate in germplasm conservation. In the last decade, Tunisia has implemented an on-farm conservation policy to conserve the traditional fig cultivars identified by local farmers (Aljane *et al.*, 2018).

*In vitro* conservation, used for several other crop plants, is also used for fig germplasm conservation. There are two methods of *in vitro* conservation, cryopreservation and the slow-growth method (Engelmann, 2000). In cryopreservation, the plant tissues are treated with a cryoprotectant, such as dimethyl sulfoxide, ethylene glycol or polyethylene glycol. The samples are preserved long term in liquid nitrogen at −196°C. Plant materials such as shoot tips, buds, meristems, cell cultures and somatic embryos can be maintained in cool but not in freezing conditions using the slow-growth technique which significantly slows growth while keeping the plant material alive. Engelmann, (2000) suggested the slow-growth technique is best for *in vitro* conservation of shoot tips; he reported a 50% plant survival and regeneration rate by treating shoot tips with cryoprotectant for 20 min then cooling to −40°C. However, genetic instability and low revival rates are associated with *in vitro* techniques.

## 2.7 Botany

### Tree habit

The fig is deciduous in most regions but is evergreen in tropical regions (Condit, 1947). It is a shrub or small tree, 10–15 m tall, with numerous spreading branches (Fig. 2.1) and a trunk rarely more than 18 cm in diameter (Yaman, 2014). Fig trees can range from open to compact, or pendulous to upright in growth habit (Morton, 1987). Complete descriptions are given in the IPGRI and CIHEAM (2003) publication. Tree growth habits and branching patterns differ by variety (Condit, 1947). The 'Brown Turkey' has a weeping willow habit with compact, down-spreading branches, while 'Sierra' and 'Autumn Honey' have vigorous upright branches. Individual trees grown in favorable environments are often large, while orchard trees are usually more compact (Flaishman *et al.*, 2008). Mirheidari *et al.* (2020) described five tree growth habits Iranian fig accessions; weeping, spreading, open, semi-erect, and erect. Tree vigor, height, branching, branch density, branch flexibility and trunk diameter were similar and high in most accessions. Khadivi-Khub and Anjam (2014) also observed considerable variability in the growth habit of Iranian caprifig accessions. Essid *et al.* (2017) reported erect, semi-erect, weeping and spreading tree growth habits in 15 Tunisian caprifig ecotypes. Podgornik *et al.* (2010) reported that the majority of fig accessions from Slovenia had an open and spreading, not weeping habit. Fig tree develops a bushy habit when basal reiteration and natural layering is allowed (Lauri and Caraglio, 1993). Gaaliche *et al.* (2016) distinguished three principal tree shapes among nine Tunisian fig cultivars based on their growth habit and branch ramification. The first one, Type I, bears orthotropic branches found at the basal and median parts of the bearer axis with dense and continuous ramification. The second group, Type II, has diffuse branching differed in lateral shoot distribution. The third group, Type III, has an open and diffuse branching located on the basal zone of the carrier axis.

### Root system

Fig trees lack taproots but have a fibrous, widespread root system that can grow to 11–15 m laterally and 7 m vertically (Flaishman *et al.*, 2008). This extensive and wide-ranging root system enables optimal exploitation of available soil water and facilitates drought and salinity tolerance. The ease of rooting fig cuttings has made it the preferred method of propagation for establishing new orchards for centuries (Soliman and Abd Alhady, 2017). Atzmon and Henkin (1998) demonstrated that the fig root system responds to soil rhizosphere modification. Fig tree roots can easily spread two to three times wider than canopy diameter (Himelrick, 1999). Despite the typically shallow nature of fig tree roots, they can extend to surprisingly great depths in select soils (Abdolahipour *et al.*, 2020). Because rainfed fig trees have extensive and wide-ranging root systems that facilitate water and mineral absorption (Leonel and Damatto Junior, 2007) they can tolerate dry regions (Gholami *et al.*, 2012; Karimi *et al.*, 2012).

### Stem wood and bark

Fig wood is soft, has a homogeneous pith, and is light and uniform in color with poorly distinguishable annual rings (Condit, 1941). The bark is smooth with few fissures. The outer bark is ash-colored, exfoliated through asymmetrical, rounded flakes. The sections of middle bark look light reddish brown. The layers of light yellowish or orange-brown-colored granular tissue are present in the inner bark (Badgujar *et al.*, 2014). Some fig varieties have burrknots, tubers and nodal swellings on the trunk bark and larger branches. Several varieties have prominent lateral enlargements at the branch nodes that develop the second season in the ground and continue to enlarge indefinitely (Condit, 1947). Ali *et al.* (2013) reported differences in the number of stem epidermal layers (2–4), and xylem clusters (3–9), among seven Egyptian fig cultivars. Yaman (2014) studied the stem and wood anatomy of *F. carica* grown in Turkey. He found no qualitative difference, except for prismatic crystals in the ray cells, but considerable quantitative differences in fiber length, vessel element length, vessel frequency and ray width. The fiber length and vessel element length in branch wood were about 16% and 3% shorter, respectively. The vessel frequency was about 52% higher, whilst the ray width was about 18% narrower in branch wood than in the stem wood of *F. carica*. Such detailed anatomical information could be useful in developing selection criteria for abiotic stress tolerance.

### Buds

Both vegetative and fruiting buds of the fig form in axils of leaves during the late summer and fall and remain on the tree during the winter. Vegetative buds are smaller and conical with an attenuated tip and are apical. The fruiting buds are large, plump and located laterally (Condit, 1947). Caprifig trees, which have a light summer and fall crop, commonly produce an enormous number of fruit buds that swell in the early spring and mature a crop in June. Terminal bud color is similar to the fruit bud and fruit color. The apical meristem of the terminal bud elongates in the spring producing scale meristems, leaves, inflorescences and lateral vegetative buds (Flaishman *et al.*, 2008). The terminal bud holds four to five primordial leaves lined on either side by a scale. Near the base of the bud are one vegetative and two inflorescence primordia.

A vegetative primordium has three scales covering the bud axis. As the bud elongates, the cover scales abscise, and the apical meristem develops into a shoot that produces leaves and new inflorescences (Condit, 1947). The number and size of fruit buds are closely correlated with the vegetative growth and determine the size of the summer and fall crops (Flaishman *et al.*, 2008). Gaaliche *et al.* (2011a) reported the number of reproductive buds and fruits; both breba and main crop were closely related to shoot length, and more specifically the portion of shoots that corresponded to growth cycles within each year.

## Leaves

### *Leaf morphology*

Fig has bright green, single, alternate and large leaves. The leaves are petiolate and palmate. They can have 3 to 5 or sometimes 7 lobes, a cordiform shape with regular toothing, or nearly entire margins (IPGRI and CIHEAM, 2003; Abdelsalam *et al.*, 2019). Condit (1947) described eight different leaf shapes (Fig. 2.4). The blade is up to 25 cm in length and width, thick and rough on the upper surface, softly hairy on the underside

**Fig. 2.4.** Leaf shapes in *F. carica*. (A) base calcarate, lobes linear; (B) base cordate, five-lobed, lobes spatulate; (C) base calcarate, lobes lyrate; (D) base calcarate, lobes latate; (E) base cordate, three-lobed; (F) base truncate; (G) base decurrent; (H) leaf not lobed. (Figures used with permission of University of California, Division of Agriculture and Natural Resources, and Zeinab Bolboli.)

(Morton, 1987). Leaf characteristics are variable with 3, 5, and 7 lobed leaves often occurring simultaneously on the same tree. The leaf commonly has five main veins, each originating at the petiole and supplying a corresponding lobe. Between these lobes are the five sinuses, i.e., two upper, two lower, and one petiolar sinus. The upper and lower sinuses vary considerably and are categorized according to depth and form (Condit, 1947).

The petioles can be thick, thin, long, short, slender, stout or flattened. They are typically green in color, or sometimes tinted with rose-carmine. Petiole length widely varies, with shaded leaves having longer petioles than those exposed to direct sun. Petiole color is closely correlated with fruit and terminal bud color. For example, green is the predominant color of fig petioles on trees with green or yellow fruits (Condit, 1947).

Fig leaf color ranges from light green to deep green. Due to the numerous epidermal hairs present on the underside of the leaf, the lower leaf surface appears to be a lighter green than the upper surface. The five main veins of a typical fig leaf are light green, or almost white, contrasting sharply with the deep green of the leaf surface. Leaves of 'Fraga', 'Adriatic', and 'Constantine' have a glossy or shiny surface compared to the dull surface of 'Dottato' and 'Lob Injir'. The epidermis on both the upper and the lower side of the leaf is studded with minute hairs or spicules. On the upper surface, the hairs are stiff and widely scattered, feeling like sandpaper to the touch. The lower surface has abundant soft hairs of various lengths, some capitated with 3–4 celled hairs and others with unicellular hairs. Capitates hairs are scarce and almost identical in shape and size to those on the fruit surface (Condit, 1947).

Leaves of the common fig are generally thick and stiff, with considerable variation in thickness and texture. 'Adriatic' fig leaves are stiff and rough; harvesting the fresh fruit is more difficult than harvesting 'Dottato' and 'Lob Injir' figs, both of which have more pliable leaves. 'Panache', 'Fraga', and 'Pastiliere' commonly have thick and brittle leaves (Condit, 1947).

Leaf characteristics are quite stable and serve as an important parameter in cultivar identification (Ferguson *et al.*, 1990) as there is a great deal of variation in leaf traits among species. Leaf shape, margin, dentation, dimensions, petiole color, and depth of basal and lateral sinus are used to distinguish cultivars (Gaaliche *et al.*, 2012; Baziar *et al.*, 2018). Fig begins to grow in early spring, continuing to form new leaves until temperatures drop in the fall. Toward the end of the growing season, the low temperatures, shorter photoperiod, wind and rain cause leaf drop.

### *Leaf anatomy*

In cross-section, the fig leaf petiole consists of epidermis, a differentiated cortex, and the two rings of conductive tissue, phloem and xylem elements (Fig. 2.5a) (Bercu and Popoviciu, 2014). The bifacial blade has a heterogeneous, hypostomatic mesophyll with a fascicular vascular system (Fig. 2.5b). The upper epidermis sits atop a one-layered hypodermis often bearing lithocysts with cystolith inside in mature leaves (Fig. 2.5c). The lower epidermal cells are smaller than the upper epidermal cells (Mamoucha *et al.*, 2016). The large cystolith within lithocysts differs in structure compared to the crystal-bearing idioblasts of the palisade tissue visible under polarized light. The cystolith arises from the epidermal portion of the cell wall as a stalked protrusion on which successive layers of calcium carbonate are gradually deposited (Fig. 2.5f). The stalk is strongly silicified and ordinarily extends beyond the surface of the cell into a sharp nipple-like protuberance. Cystoliths are excretory bodies which serve as reservoirs for the surplus of calcium carbonate from a metabolic process (Condit, 1947).

There are two types of trichomes, or hairs, found on fig leaves: (a) pointed protective trichomes (Figs 2.5b and 2.5c) found on both leaf surfaces, and (b) capitate secreting trichomes found only on the abaxial surface (Fig. 2.5e). Leaf hairs, laticifers, trichome secretions, solid inclusions and silicification of the fig leaf play a significant role in leaf protection from high-intensity radiation, herbivores and pathogens (Giordano *et al.*, 2019). Stomata appear only on the lower (abaxial) surface (Fig. 2.5g) (Mamoucha *et al.*, 2016).

## Latex cells

Latex cells, called laticifers, are the single tubular cells or series of connected cells which produce the milky exudates, characteristic of all fig cultivars (Evert, 2006; Lazreg-Aref *et al.*, 2018). Latex

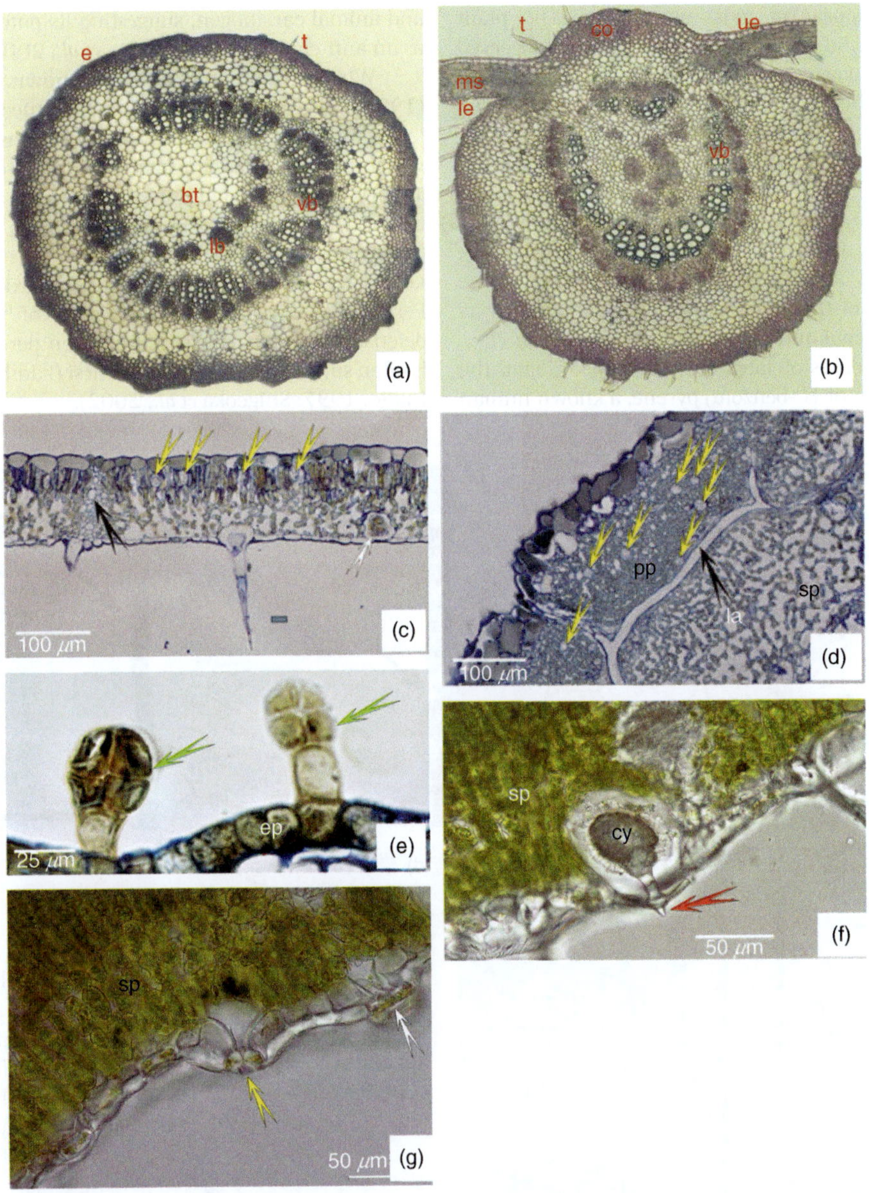


**Fig. 2.5.** Leaf anatomy of *F. carica*. (a) Cross-section of a petiole; (b) cross-section of the blade in the midrib area-ensemble (x100); (c) cross-section of a mature leaf showing the idioblasts (yellow arrows) on the thick upper epidermis, the laticifer (black arrow) within the leaf vein, the lithocyst with a large sized cystolith (white arrow), and the long protective trichome on lower epidermis; (d) leaf paradermal section showing idioblasts (yellow arrow), and a laticifer (black arrow); (e) capitate secretive trichomes of the abaxial epidermis; (f) cross-section of a lithocyst with a characteristic spiked end protruding from the outer part of the cell wall; (g) cross-section of leaf showing the adaxial epidermis with large (yellow arrow) and medium (white arrow) stomata. bt: basic tissue, c: cortex, co: collenchyma, cy: cystolith, ep: epidermis, la: laticifer, lb: leptocentric bundle, le: lower epidermis, ms: mesophyll, pp: palisade parenchyma, sp: spongy parenchyma, t: trichome, ue: upper epidermis, vb: vascular bundle. (Figures used from Bercu and Popoviciu, 2014 with permission from *Annals of the Romanian Society for Cell Biology* and from Mamoucha *et al.,* 2016 with Permission from RightsLink.)

cells appear in most parts of the fig plant (Figs 2.5d and 2.6). Laticifers can be observed in the cortex, secondary phloem, and pith of both the current-year branches and the trunk (Fig. 2.6). They are also found in the secondary xylem and cambium of the trunk (Kajii *et al.*, 2014). Fig latex contains a proteolytic enzyme known as ficin, used for tenderizing meat, rendering fat and clarifying beverages (Aksoy, 1998). Aref *et al.* (2010) have screened the effects of the organic extracts of fig latex for its antifungal and antibacterial properties *in vitro*. Small doses of latex significantly inhibited the binding of $H^3$-benzo(α) pyrene, a known human and animal carcinogen, suggesting its potential as an anti-cancer agent (Rubnov *et al.*, 2001).

Why plants produce latex is unknown. John (1993) suggests it is a defense against mechanical wounding and/or herbivores such as insects, vertebrates, microorganisms and fungi (John, 1993). The ficins in latex are thought to be associated with the plant defense system. Kim *et al.* (2003) reported three latex genes from *F. carica*, i.e., a trypsin inhibitor, chitinase and peroxidase. Peroxidase plays a role in a variety of plant defense mechanisms, where hydrogen peroxide is often supplied by an oxidative burst (Lamb and Dixon, 1997; Shigeoka *et al.*, 2002).

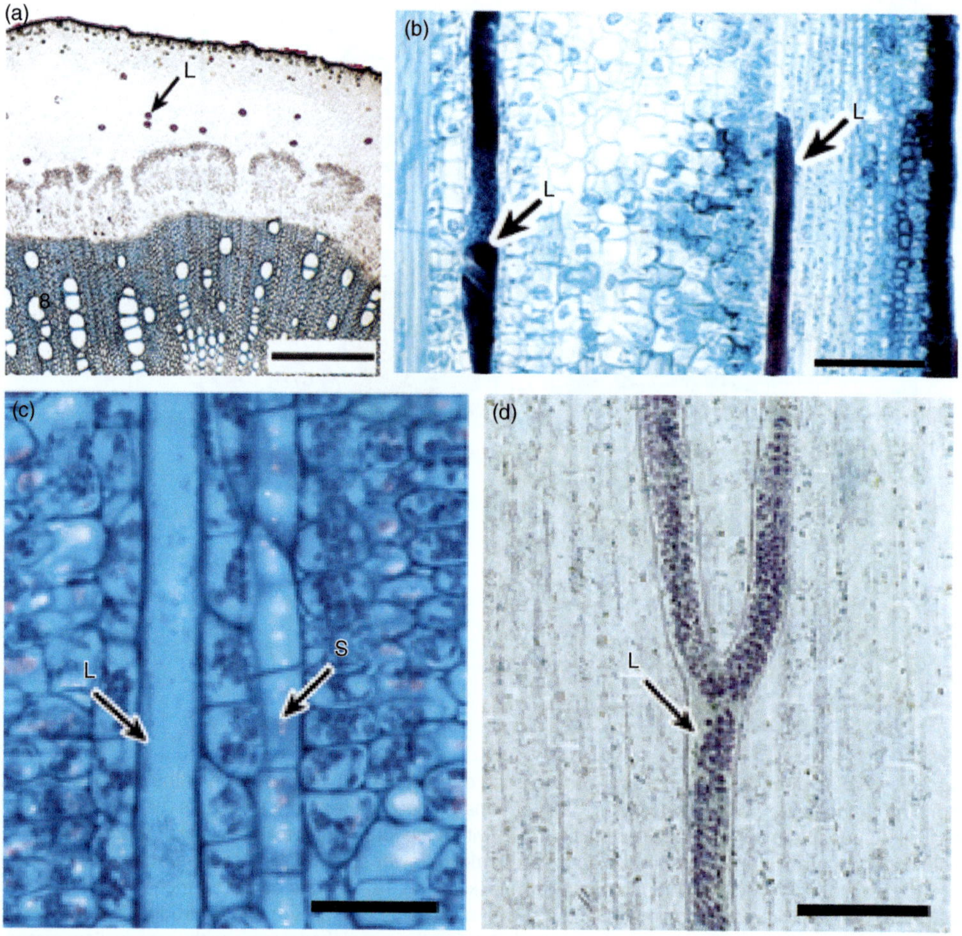


**Fig. 2.6.** Latex cells (laticifers) in fig stem. (a) Cross-section of stem showing laticifers; (b) axial laticifers in the cortex; (c) axial laticifers in secondary phloem; (d) branched laticifers in secondary phloem. L: laticifers, S: sieve tube. Scale: 500 μm in (a); 200 μm in (b); 50 μm in (c) and (d). (Figures used from Kajii *et al.* 2014 with permission from RightsLink.)

## Inflorescence

The botanical term for the fig inflorescence is 'syconium'; a fleshy, hollow receptacle with a small opening at the apex partly closed by small scales (Fig. 2.7). Syconia of *F. carica* are produced in the leaf axils. The true tiny flowers are massed on the inside wall of the receptacle. The syconium may be obovoid, turbinate or pear-shaped, 1–4 inches (2.5–10 cm) long and vary from yellowish-green to coppery, bronze or dark purple. *F. carica* has two types of syconia grown on different trees. One is a gynodioecious or functionally dioecious plant. The other has staminate flowers and short-styled pistillate flowers. These are the caprifigs, which are hermaphrodite but functionally a male fig. In the fruit-producing female trees the syconia are lined with tiny long-styled, pistillate flowers (Condit, 1947). Syconia of the fig can be either persistent or caducous (deciduous) which is genetically regulated (Storey, 1975).

## Flowers

The flowers are typically greenish, pentamerous, fleshy and imperfect. Fundamentally, the flowers are of two kinds, staminate or male and pistillate or female (Fig. 2.8). Pistillate flowers can be either short-styled, characteristic of caprifigs, or long-styled, characteristic of edible figs, but never occur together within a syconium. Figs having short-styled flowers are considered more primitive than figs having long-styled flowers. In caprifig varieties, flower numbers can range from 1350 pistillate and 220 staminate to 169 pistillate and 39 staminate flowers. The normal proportion is 7 or 8 pistillate to 1 staminate flower (Condit, 1947). Heterostyly is genetically fixed (Valdeyron and Lloyd, 1979). Female wasps can oviposit only in the ovaries of the short-styled flowers. Therefore, the oviposited short-styled flowers do not produce seeds, although the ovaries appear to be in perfect state at anthesis. Seed production only occurs in the syconia with long-styled flowers, of which the ovaries are not oviposited and thus are not involved in the breeding of larvae.

Mature pistillate flowers within a syconium typically have the following parts, i.e., a pedicel of varying length, a gynoecium, 5–8 perianth lobes partly enclosing the spherical ovary, a simple style, and a bifid stigma usually cleft into either equal or unequal limbs. In long-styled flowers, the stigma is typically bifid, having 1–3 lobes. It can be papillate and characterized as wet due to the presence of copious secretions at receptivity. The style length is about 1.64 mm in long-styled flowers, while it is about 0.86 mm in short-styled

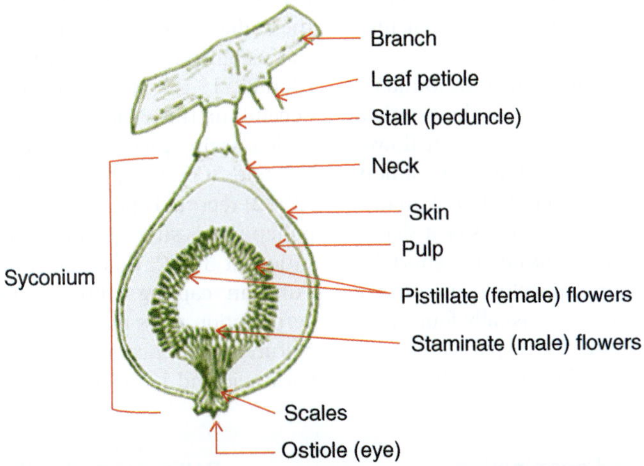


**Fig. 2.7.** A typical syconium of common fig (*F. carica*). (Diagram: University of California, Division of Agricultural Science, Leaflet 21051. 1978, used with permission from University of California, Division of Agricultural Science.)

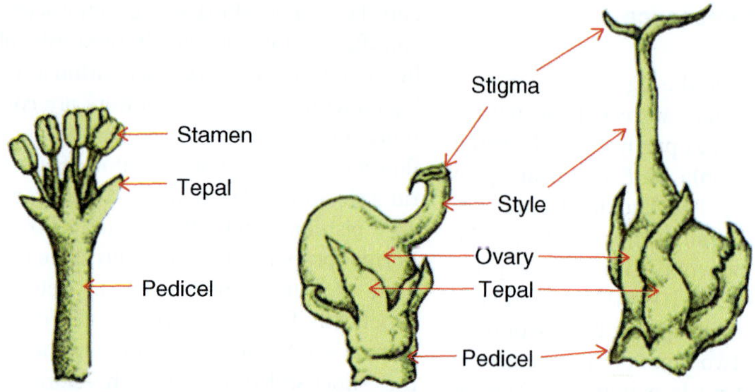


(a) Staminate flower (b) Short-styled pistillate flower (c) Long-styled pistillate flower

**Fig. 2.8.** Enlarged view of male and female flowers of *F. carica*. (a) Male flowers and (b) short-styled female flowers are found together in the syconium of a male fig tree, whereas (c) long-styled female flowers are found only in the syconia of female fig trees. (Diagram used with permission from Wayne P. Armstrong with Elaine M. Collins.)

flowers. Short-styled pistillate flowers, which are typical of caprifig are adapted to oviposition by the Blastophaga. The papillate stigmatic cells of both long and short-styled flowers are receptive to pollen and both types of flowers can develop fertile seeds. Long-styled pistillate flowers are typical of edible figs. A single ovule develops in each ovary. The pollination and fertilization in the ovule lead to a normal endosperm development (Condit, 1947). Parthenogenetic endosperm development in short-styled flowers is often stimulated by the presence of a Blastophaga larva in the ovule. Similar parthenogenetic endosperm development in long-styled flowers sometimes occurs without any apparent stimulus.

Staminate flowers are commonly found in a mass just below the scales lining the ostiole of the caprifig. In some caprifigs, staminate flowers are found scattered indiscriminately among the pistillate flowers (Condit, 1947). A mature staminate flower generally consists of a slender pedicel bearing five perianths and five short filaments; each filament supporting an introrse, two-lobed anther. There are usually four or five stamens in a male flower. The tepals can be pale yellow or pinkish and glabrous or ciliolate.

## Pollen and germination

Fig pollen is dimorphic, without surface markings, and is classified as either triporate and spherical or diporate and ellipsoidal (Beck and Lord, 1988). The pollen is tectate and is often covered with pollenkitt (Stanley and Linskens, 1974). The average pollen grain diameter is 12 μm. Both pollen types are produced within a single anther and can germinate on a receptive stigma (Beck and Lord, 1988). The function of the pollen dimorphism is unknown, but pollen type may be correlated with segregation of genes which determines whether seed from a single syconium will be either a caprifig or an edible fig (Storey, 1975). Fig pollen stored at room temperature can remain viable for 3–20 days, but the radiation exposure at 470 R can prolong the viability of the pollen for up to 105 days (Condit, 1947). Successful pollen germination depends on the media (Zeybekoğlu *et al.*, 1998). Pollen grains germinate readily in a 5% sugar solution (Condit, 1947). Vego and Miljković (2012) reported optimal germination of wild fig pollen in 3% sucrose and 0.01% boric acid solution at 30 °C. Gaaliche *et al.* (2013), using Tunisian caprifig cultivars, reported optimal germination rates and pollen tube growth in a culture medium containing 5% sucrose, 5 ppm boric acid and 1% agar at 25°C.

## Pollinator and pollination

The female fig is pollinated by an obligate symbiotic wasp, *B. psenes* (synonym *B. grossorum*).

*B. psenes* belonging to the family *Agaonidae*, the order Hymenoptera. It is a free-living wasp, without a colony or nest, and a short life cycle; up to two weeks. The wasps breed within the annual cycle of the three caprifig crops (Fig. 2.9). They locate the female figs through their olfactory sense. The Blastophaga completes its life cycle within the three caprifig cycle of the syconia. The insect can only oviposit on the short-styled female flowers of caprifigs. Male wasps inseminate the female while she is still inside the ovary of short-style flowers (Fig. 2.9e) (Armstrong, 2012). The female wasps enter the fig through the ostiole to oviposit (Figs 2.9a and 2.9c). A pistillate fig flower ovary containing the larva of a fig wasp is called a gall flower (Fig. 2.9d). The female wasp varies in size according to the caprifig from which she emerges, but the average size is 2.5 mm long. The body is glossy and black; the wings have very few veins (Figs 2.9g–i). The males are wingless and have an attenuated abdomen much longer than the head and thorax combined (Fig. 2.9f). The female lays 300–400 eggs, each deposited separately in a short-styled flower (Condit, 1947). The female fig wasp lives its larval life cycle in a short-styled flower in a syconium of a caprifig; when mature, it exits the flower through a minimally sclerified area of the fruit wall and ovule. The female wasp leaves the caprifig syconium through the ostiole and enters an adjacent receptive syconium. The wasp loses its wings and antennae when entering the bracts of the syconium (Al-Khalaf *et al.*, 2018).

No pistillate flowers holding both an embryo and a Blastophaga larva together have been reported in the fig. Pollen tube growth leading to a fertilization fails in oviposited female flowers, because during oviposition the female Blastophaga's insertion and withdrawal of the ovipositor injure the cells of the stigmatic surface and the stylar canal. Embryo development is successful if it happens parthenogenetically, or if the pollen tube growth and fertilization of the ovule take place before Blastophaga oviposition.

Beck and Lord (1988) studied the details of pollination in Smyrna and second crop of San Pedro-type figs. The female fig wasp carries pollen from staminate flowers of caprifigs to the pistillate flowers of another plant. When the fig wasp emerges from the gall flower and leaves the syconium, it may be dusted with pollen if the staminate flowers near the ostiole are mature at that time (Condit, 1932). Pollens are carried by the fig wasp only in intersegmental and pleural invaginations of the thorax (Beck and Lord, 1988). Only the caprifigs bear functional staminate flowers and thus are the only source of pollen in *F. carica*. Caprifig flowers display a distinct dichogamy where staminate flowers mature pollen several weeks after the pistillate flowers in the same fig are receptive to pollen. For this reason, self-pollination is impossible and thus, *F. carica* is a cross-pollinated plant. Biologically, pollination is necessary for fertile seeds, or drupelet, production in figs (Gaaliche *et al.*, 2011b). Without pollination, the immature Smyrna figs usually shrivel and drop when about an inch in diameter (Condit, 1932). Pollination is not always necessary for fig production in horticulture due to parthenocarpy in cultivated figs, except for Smyrna figs which rarely produce parthenocarpic fruits.

Pollination of the stigmatic surface is followed almost immediately by germination of the microspores. The pollen tube grows through the style and into the ovule rapidly. The stigmatic surface of short-styled caprifig flowers is as favorable for pollen germination as that of long-styled flowers found in other fig types. However, Beck and Lord (1988) reported that the copious secretions of stigma and stylar cells in long-styled female flowers facilitate pollen tube growth and subsequent penetration of the micropyle. The lack of sufficient secretions in short-styled flowers makes it difficult for the pollen tube to reach the micropyle. Although seed set can occur in the short-styled flowers, its stigma is not a good receptive site for pollination as pollen retention is poor.

## Caprification and impacts on fruit set

Fig pollination, known as caprification, is customary practice in all fig-growing regions where 'Smyrna' type figs are produced (Gaaliche *et al.*, 2011b; Trad *et al.*, 2012). The caprifigs (with wasps inside) are collected and placed in small bags or wire baskets hung in the fruiting 'Smyrna' trees. The emerging female wasps covered with pollen, enter the syconia of 'Smyrna' or 'San Pedro' figs and pollinate the long-styled pistillate flowers. To obtain profitable

**Fig. 2.9.** *Blastophaga psenes* at different stages of life. (a) A mature female wasp approaching a fig ostiole, (b) female wasps lose their wings when entering a syconium, (c) a female wasp inside a syconium looking for female flowers in which to oviposit, (d) wasp larva (arrow marked) inside a gall flower, (e) a male wasp looking for a female wasp inside a gall flower for mating, (f) a matured male wasp, (g) female wasp emerging from a gall flower, (h) a matured female wasp inside a syconium, (i) a female wasp leaving a syconium. (Figures recreated by authors from www.figweb.org; Wayne P. Armstrong and www.alchetron.com/Blastophaga-psenes).

yields, caprification must be repeated two or three times because the female fig flowers become receptive gradually (Gaaliche *et al.*, 2011b). It is essential to have two or three cultivars of male figs (caprifigs) in fig orchards to extend the caprification process (Zare, 2008). But the caprifig trees should be grown separately from the female figs such as 'Smyrna' or 'San Pedro' types to control caprification and avoid diseases in commercial fig production. A current fig production problem is the availability of viable caprifigs, synchronization between female receptivity and pollen availability, and the high cost and efficiency of caprification (Gaaliche *et al.*, 2017a; Mars *et al.*, 2017).

Gaaliche *et al.* (2011b) reported that pollination frequency and intensity had significant effects on fruit set for fig cultivars 'Bouhouli' and 'Zidi' where four times caprification with an interval of 4 days between two consecutive caprifications improved the fruit set and fruit quality. It corroborated the earlier report of Ateyyeh and Sadder (2006) on Jordanian cultivars. Caprification on four dates, 4 days apart, doubled the fruit number per shoot, consistent with the results of Ateyyeh and Sadder (2006) for Jordanian cultivars. Poor caprification, possibly weather-related, may lead to fruit drop (Oukabli *et al.*, 2003). Pollen source (caprifig cultivar) has also been reported as having a significant effect on the fruit size of 'Sarilop' (Aksoy *et al.*, 2003). Although there were no significant differences among the pollen sources in north Tunisia, results showed the superiority of some caprifigs over others. The 'Djebba 2' caprifig cultivar produced a high fruit set (Gaaliche *et al.*, 2011b). In Iran, caprification done every three days significantly increased the ratio of early ripening fruits (Zare, 2008). Other Iranian cultivars showed differences in fruit ripening period, but the effects of the pollen source and the interaction 'pollen source × female cultivars' were not significant (Pourghayoumi *et al.*, 2012). Flaishman *et al.* (2008) reported that caprified fruits of the cultivar 'Autumn Honey' were bigger, with darker purple skin color and red pulp, and had longer storage life compared to non-caprified fruits. In Tunisian cultivars, fruit weight, fruit length, fruit diameter, ostiole diameter, flesh thickness and the number of fertile seeds increased significantly with pollination rate (Gaaliche *et al.*, 2011b).

Caprification has also been reported to increase the fruit dry matter content (Trad *et al.*, 2014), an important quality parameter in dried fig production (Piga *et al.*, 2003). Trad *et al.* (2013) found that caprification significantly affected polyphenol composition in Tunisian fig cultivars. Pollen source affects total fig fruits soluble solids by influencing the concentration of flavonoids and anthocyanins (Pourghayoumi *et al.*, 2012). This suggests 'metaxenia' occurs in caprified figs, affecting color, texture, chemical composition, hormonal and enzymatic activity (Pourghayoumi *et al.*, 2012; Trad *et al.*, 2014).

## Seed morphology and germination

Following pollination and fertilization, the ovaries develop into minute one-seeded drupelets, with an endocarp surrounding the seed (Condit, 1955; Galil, 1977). If pollinated, both long-styled and short-styled flowers can produce seeds. However, the seed set is low in short-styled flower presumably due to inefficient oviposition or damage by the wasp. Fig seeds vary in size, small, medium, large, conspicuous or indistinct, and number, from 30 to 1600, by variety (Morton, 1987). Seeds of 'Marseilles' fig are unusually conspicuous, as they stand out sharply against the background of the white pulp. Normal seeds are slightly flattened, pointed or protruding at the hilum, minutely pitted over the surface, and light chestnut in color (Condit, 1947). According to Lisci and Pacini (1994), fig seeds consist of integument and embryo only which contradicts the earlier reports of Condit (1932) and Storey (1985) who claimed that the seeds of edible figs were endospermic and the endosperm development was either normal or parthenogenic. The mature endosperm makes up about half of the bulk of the drupelets and consists of thick-walled, polygonal cells (Condit, 1932). The curved shape of the mature embryo allows the hypocotyl and the cotyledons to lie side-by-side during germination. The embryo and the endosperm of mature achenes account for the excellent quality of Smyrna figs.

Gaaliche *et al.* (2011b) reported that the pollination rate significantly affected the seed number of cultivars 'Bouhouli' and 'Zidi'. The average number of fertile seeds ranged from 229 (non-caprified) to 804 (caprified) per fig in the case of 'Bouhouli', and from 155 (non-caprified) to 617 (caprified) per fig in the case of 'Zidi'.

A study of 18 varieties of Kabylian figs reported the average number of fertile seeds ranged from 716 to 1831 per fig, and that sterile seeds ranged from 15 to 218. Studies of eleven caprified 'Adriatic' figs in California reported 986 sterile seeds and 1597 fertile seeds (Condit, 1947). Too many fertilized seeds due to caprification may cause ostiole cracking (Ferguson *et al.*, 1990; Crisosto *et al.*, 2010). Pollens received from the caprifig cultivars can affect the quality and ripening of fig fruits too. Two caprifig cultivars, 'Daneh-Sefid' and 'Shah Anjiri' were reported to contribute in the early ripening of edible figs (Rahemi and Jafari, 2008). Pollen grains of different caprifig cultivars can have different amounts of tryptophan and methionine. These amino acids are the substrates for auxin and ethylene biosynthesis, respectively. The fruit size and the ostiole opening increase in syconia of 'Sabz' due to higher auxin and gibberelic acid ($GA_3$) production, respectively, if it is caprified with 'Daneh-Sefid' (Zare, 2008). Lisci and Pacini (1994) studied the germination ecology of fig drupelets. The seeds germinate best between 10 and 30°C at constant humidity of the growing substrate with or without light. Germination is enhanced in constant humidity with a cycle of 30°C in 8 hours of light and 20°C in 16 hours darkness. Soaking dried seeds for 10 min in hot water (40–60°C) enhances seed germination. Temperatures below 10°C inhibit germination. Fig seed dormancy can be broken with $GA_3$ (Çalişkan *et al.*, 2012).

The high sugar content in the fig fruit receptacle, the syconium, has an inhibitory effect on seed germination due to its high osmotic pressure (Lisci and Pacini, 1994). Seeds need to be removed from the syconium to germinate. In nature fruit-eating birds and animals do this. The fig seed integument is rich in lignin, resistant to enzymatic damage and protects the seed while passing through the birds' or animal digestive tracts. Rainwater promotes syconia drop, degradation of syconium wall and dilution of high sugar content in syconia, so some drupelets may germinate without being eaten.

## Fruit

Botanically the true fig fruits are the minute one-seeded drupelets composed of a sclerified endocarp surrounding the seed that line the inside of the syconium (Armstrong, 2012). Condit (1932) called them achenes. The fig resembles multiple fruits but differs in that the individual fruits are not adherent. Caprifig and Smyrna figs produce drupelets with fertile seeds. In mature caprifig syconia, fertile drupelets are often found among the gall flowers. An empty drupe that develops parthenocarpically is a cenocarp, while a wasp-bearing drupe in the caprifig is a psenocarp (Condit, 1932). Horticulturally, the syconium is the fig fruit (Fig. 2.10). Ontogenetically, the syconium is a shortened, fleshy branch composed of a series of fused internodes, the scales, or reduced leaves of which remain distinct. Both unicellular attenuate hairs and multicellular capitate hairs occur on the surface of the fig fruit (Condit, 1941). The color, shape and size of fig fruits may vary from variety to variety (Fig. 2.11).

Caprifigs bear three crops annually, the *profichi* spring crop, the *mammoni* summer crop, and the *mamme* winter crop. The Smyrna, San Pedro, and Common-type figs produce two

**Fig. 2.10.** A typical fig fruit anatomy. (Figure used with permission from Ali Sarkhosh.)

**Fig. 2.11.** Fig fruits with different shapes. (a) Oval (var. Pingo-de-Mel), (b) elongated (Irradiated Plant 440), (c) oblong pyriform (Roxo-de-Valinhos), and (d) flattened (White Genova). (Figures used from Rodrigues *et al.* 2019 with permission from SciELO Brazil and Cenap Yilmaz.)

crops annually, the breba first crop, corresponding to the caprifig *profichi* and the second main crop corresponding to the caprifig *mammoni* (Condit, 1932; Morales and Gil, 2014). Caprifig fruits are non-edible due to impregnation of the female flowers with larvae of the pollinating wasp, so they are salvaged for animal feed. The other fig types, Smyrna, San Pedro and Common, produce edible fruits. Smyrna fig fruits have a nutty flavor due to the seeds produced by caprification. The first crop of San Pedro is parthenocarpic, develops without pollination; the main crop, like Smyrna figs, requires caprification. The Common-type fig is the world's primary commercial *F. carica* cultivar. Cultivars of this type do not require caprification, the drupelets inside develop parthenocarpically and the endocarps are hollow and without fertile seeds (Condit, 1955; Galil, 1977).

#### Parthenocarpy

Parthenocarpic fruit development in figs is either stimulative or vegetative. In most caprifigs, the syconia do not set unless the pistillate flowers are stimulated by oviposition and by the larval development of the Blastophaga in the ovary. Both partly parthenocarpic and completely parthenocarpic caprifigs exist. Smyrna-type figs are completely non-parthenocarpic, the syconia maturing only after the pollination of their flowers and resulting development of the ovules. Occasionally, brebas of 'Lob Injir' variety develop parthenocarpically. Trees of San Pedro-type show a very peculiar and interesting behavior; the brebas set are parthenocarpically but the second crop are non-parthenocarpic like Smyrna figs. In this case parthenocarpy and non-parthenocarpy are on the same branch in the same growing season (Condit, 1947).

Common-type figs are completely parthenocarpic. Select Common-type figs are incompletely parthenocarpic and drop fruits if not caprified. For example, 'Bourjassotte Grise' drops at least 90% of its crop unless the figs are caprified. The factor for parthenocarpy, therefore, appears to be unstable or only incompletely fixed in some common figs. Chandler (1934) reported a significant percentage of fruits naturally abscised on trees of the 'Brunswick' fig, with fruit buds forming and dropping all summer.

## 2.8 Phenology and Reproductive Biology

#### Phenology

Understanding the reproductive and vegetative growth cycle of a crop is important for managing cultural practices like pruning, training, irrigation, fertilization and application of growth regulators (Ateyyeh and Sadder, 2006). Additionally, understanding the bearing habit of a fruit species, specifically where the species produces flowers and fruits, aids in understanding how a tree responds to pruning cuts. In figs, fruiting buds occur laterally in the leaf axils on the current or previous season's shoots. Removing the terminal bud stimulates lateral fruiting and branch buds into activity, suppressing leaf development and accelerating fruit development (Condit, 1926).

## Phenology of *F. carica* in naturally growing conditions

Valdeyron and Lloyd (1979) studied the phenology of naturally growing populations of *F. carica* found in the Montpellier region of southern France and described five distinctive phases:

Phase A: Pre-receptive phase with two substages, an initial 'bud' phase and a subsequent 'enlarging' phase which follows a variable period of developmental arrest.
Phase B: Receptive phase when adult female wasps (loaded with pollen) are received into a syconium.
Phase C: Interim phase when a new generation of wasps is developing in female flowers.
Phase D: Donor phase when female wasps leave a syconium.
Phase E: Post-donor phase.

Syconia of male trees (caprifigs) usually drop off soon after the donor phase and the post-donor phase is brief. Syconia of female fig trees lack a donor stage and have only one extended 'post-receptive' phase after receptivity, during which the seeds and syconia mature.

As the terminal bud unfolds and growth occurs in the spring, fig fruits are borne in the axils of leaves (Flaishman *et al.*, 2008). Two syconia and one vegetative bud are present at the same lateral position in the leaf axils. A sequence of syconia is initiated in nodes formed in succession after the first leaf unfolds around April 1 at the top of the previous year's shoot (Valdeyron and Lloyd, 1979). The subsequent leaves unfold sequentially every 10–12 days until around July 1. The growth of the deciduous shoots and initiation of syconia occur within approximately three months. During this period, a typical fruiting shoot has approximately nine leaves, each with both a syconium bud and a shoot bud in its axil (Fig. 2.12). To maintain a suitably spaced sequence of developing syconia, some caprifig syconia must be delayed in their development between initiation and receptivity. One crop, the *mamme*, must remain dormant at the interim stage throughout the winter. The variation between syconia in the duration of the delay in enlargement results in coordinated sequences of crops in the two sexes.

### *Syconium development in caprifig trees*

The full sequence of crops in caprifig trees forms an annual cycle of three generations of crops and wasps. The *mamme*, *profichi*, and *mammoni* crops reach the donor stage in the spring, summer, and fall, respectively at the same time as the next crop is receptive. Thus, wasps emerging from one caprifig crop enter and lay eggs in the next. Of the three caprifig crops, only the *mammoni* syconia enlarge soon after they are initiated and reach the donor stage in the same season. The *mamme* crop is receptive in the season in which it is initiated, but it overwinters in the interim stage and reaches the donor stage in the subsequent spring. The buds of *profichi* syconia occupy most nodes of an annual shoot increment, i.e., these remain dormant during the season in which they are initiated and develop only in the following spring (Valdeyron and Lloyd, 1979).

The production of *mammoni* syconia is irregular among the wild populations of *F. carica* found in France and neighboring areas. The receptive stage of *mammoni* coincides with the donor stage of *profichi*. Therefore, reduction or complete elimination of the *mammoni* crop endangers the reproductive cycle of the wasp, since the *mamme* crop is normally receptive one or two months after the *profichi* syconia are at the donor stage (Valdeyron and Lloyd, 1979). In the cold climate of the Montpellier region, caprifig trees mature their *profichi* as late as the beginning of September. As the insects can fly long distances (more than 10 km), those released from late *profichi* may enter the earlier receptive syconia of the *mamme* crop anywhere in the region. The frequent omission of the *mammoni* crop from the annual cycle of wasp reproduction around Montpellier results in a mixture of two- and three-generation cycles of these annual events (Valdeyron and Lloyd, 1979).

### *Syconium development in female fig trees*

Most syconia of female figs enlarge quickly and are receptive during summer, in July and August, when they are pollinated by wasps that emerge from the *profichi* crop. The main crop of edible figs produced after the first weeks of September contains numerous seeds, but no emerging wasps (Valdeyron and Lloyd, 1979). Female fig trees can set fruit with or without pollination.

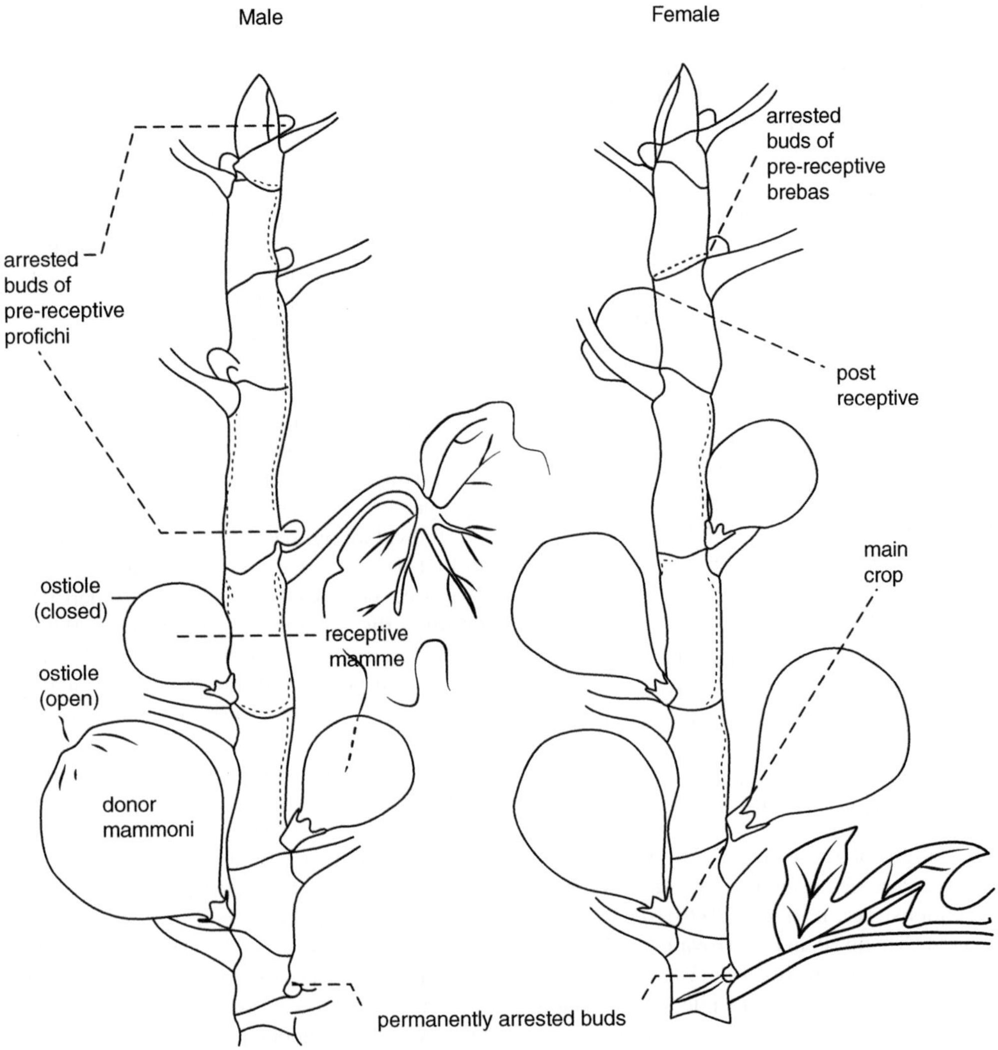


**Fig. 2.12.** Production of syconia on typical male and female shoots of naturally growing caprifigs in France. The nodes and buds are initiated in September. The male shoot bears three crops of syconia, whereas the female shoot bears only two crops. (Diagram used with permission from Valdeyron and Lloyd, 1979.)

Breba and main crop figs have different patterns of growth and development (Flaishman *et al.*, 2008). Shortly after initiation, the breba syconium enters dormancy. In spring, the breba syconium resumes growth for 7–8 weeks. The syconia of Smyrna and the main crop of San Pedro-type are persistent only if they are pollinated at the time of anthesis, while the persistence of syconia in the case of the Common-types and the breba crop of San Pedro-type are dependent on the prevailing weather conditions at the time of anthesis.

Delayed syconia are less abundant in female trees, where they usually occur at only two or three nodes. A majority of delayed female syconia fall off three to four weeks before they would become receptive in the spring. Syconia that do reach receptivity cannot be fertilized, since there is no pollen available to fertilize their flowers, causing them to fall in May and June. Parthenocarpic female fig trees do not drop their first crop of syconia, producing breba crop of edible figs (Valdeyron and Lloyd, 1979). This 'parthenocarpic' growth of syconia is the result of artificial selection and is the basis of edible fig production in many regions where the pollinator species is absent.

### Phenology of Common-type figs

Common-type figs do not require pollination for fig production. Therefore, the phenological phases described by Valdeyron and Lloyd (1979) for caprifigs cannot be applied as the phases B to E are irrelevant to the Common-types. Ferraz *et al.* (2020) described another phenological cycle appropriate for the Common-type figs that include three stages beginning from pruning:

Budding stage (BS): the interval between pruning and the beginning of sprouting.
Growth stage (GS): the interval between sprouting and the beginning of harvest.
Harvest stage (HS): the interval between the beginning and the end of the harvest.

The length of these stages may vary among the fig varieties grown in the same climatic conditions. Ferraz *et al.* (2020) reported the longest GH and HS in the 'Troyano' variety, and the shortest BS in 'Roxo de Valinhos' (13.75 days), while studying the phenology of four Common-type fig varieties in Brazil (Fig. 2.13).

### Reproductive biology

*F. carica* can be propagated either through sexual or vegetative means. In general, unpollinated parthenocarpic cultivars with the persistent gene must be propagated by vegetative means because the drupelets in their syconia are hollow cenocarps without seeds (Armstrong, 2012). Vegetative propagation of fig can be performed in several ways; using burrs, plunging, hardwood or softwood cuttings, grafting and budding onto the rootstock of another type of fig, tissue culture, air-layering, and ground-layering (Boliani *et al.*, 2019). However, taking hardwood cuttings is the most common method, as it can be done easily.

Fertile seeds produced through sexual means are necessary for breeding and selection purposes in horticulture. The sexual reproduction of fig is complex, consisting of two tree morphs, three functional floral forms and an ovipositing/pollinating wasp (Beck and Lord, 1988). Synchronized growth and maturation of the male flower, female flower and the pollinating wasp are necessary for successful pollination and fertile seed production (Fig. 2.14). Caprifig trees participate to some extent in all three activities, namely pollen production, seed production and the nurturing of wasps; whereas, the female figs never participate in pollen production and the nurturing of the wasp, although certain types participate in seed production.

Individual fig trees may continue to thrive up to 100 or more years of age, but most orchard trees decline after 50 to 75 years, depending upon the soil and the cultural conditions. A chilling requirement is necessary for fig growth and production, but the length of the dormant period is limited by local climatic conditions (Flaishman *et al.*, 2008). Fig trees growing under hot climatic conditions can grow continuously, remaining evergreen. Trees growing in colder weather, being deciduous, enter dormancy after defoliation and develop typical terminal buds. Kawamata *et al.* (2002) classified the endodormancy of the fig bud in three phases, *viz.* (1) introductory (early October to early November), (2) deepest (late November), and (3) awaking phases (middle December to late January). The fig shoot can be heated or headed back at any season to get fruit-bearing shoots for double cropping or year-round production (Ateyyeh and Sadder, 2006).

Two tree morphs (male and female) are generally found in the fig and both the morphs occur in wild as well as cultivated states (Valdeyron and Lloyd, 1979). One tree morph, the caprifig, produces syconia containing male and short-styled female flowers. The other tree morph, the edible fig, produces syconia containing long-styled female flowers. Seeds from pollinated long-styled female flowers produce both tree morphs (Beck and Lord, 1988). Although the short-styled flower primarily acts as an oviposition site for the pollinator wasp and the long-styled flowers as a seed producer, the functions

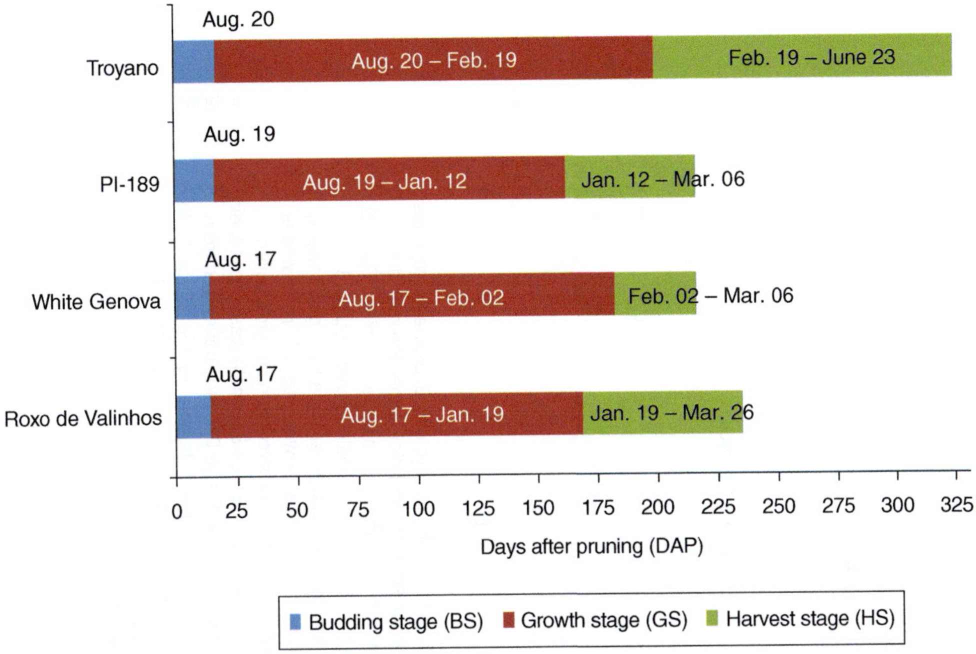


**Fig. 2.13.** Phenological stages of Common-type fig varieties grown in Brazil. Three stages as recorded in Brazilian climatic conditions are shown. (Figure used from Ferraz *et al.* 2020 with permission from SciELO Brazil.)

are not completely separated. In some species, long-styled flowers occasionally receive wasp eggs, and a few short-styled flowers produce seeds (Valdeyron and Lloyd, 1979).

The caprifig tree typically produces three crops of fruit annually, each harboring the larvae, pupae, and temporarily the adult *Blastophaga* wasps (Flaishman *et al.*, 2008). The spring crop, *profichi*, produced in large numbers on wood from the previous season, is the pollen source for the edible figs. The summer crop, *mammoni*, produces single or double fruits in the axils of leaves on branches of the current season. They mature during October when the *Blastophaga* wasps leave them and enter young *mamme* that develop on current growth. Cool temperatures in October and November retard the development of *mamme* fruit and their attendant wasp larvae, which overwinter and develop into pupae in March.

In early April, the adult male wasp continues its life cycle by emerging through the ovary wall (Flaishman *et al.*, 2008). When free in the fig cavity, the male searches for female wasps and copulates with them. The females emerge from the *mamme* fruit to search for developing *profichi* fruits. The females then lay eggs in the ovaries of the short-styled pistillate flowers of the *profichi* spring crop. An important botanical component of this coevolution is the protogynous nature of the caprifig, allowing pistillate flowers to become receptive six to eight weeks before anthers mature in the same syconium (Condit, 1932). This allows wasps to enter, pollinate and oviposit within a syconium. The syconium will later contain matured pollen, as the next wasp generation emerges.

If pistillate flowers (either short-styled or long-styled) are not oviposited by the female wasp, they can be pollinated and produce viable seeds. Most seeds originate from long-styled pistillate flowers. Where *B. psenes* wasps are present, both caprified and non-caprified fruits may develop on the same branch of female figs (Flaishman *et al.*, 2008). Thus, caprification can be applied in all cultivated fig types to facilitate fertile seed production.

Climatic constraints may limit the seed production ability through phenological mismatches that could lead to local extinction of

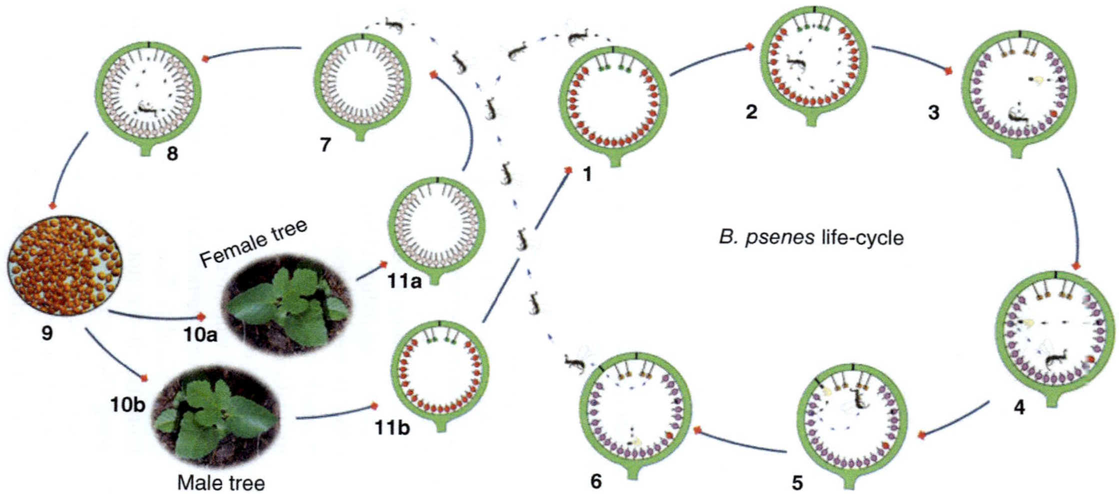


**Fig. 2.14.** Life cycle and symbiotic relationship between *F. carica* and *B. psenes*. The life cycle of *B. psenes* completes and continues from Stage 1 to 6. The life cycle of *F. carica* which starts from Stage 9 can follow either the male tree cycle or female tree cycle depending on the sex of the seedling. Stage 1, female wasps visit a male syconium; Stage 2, female wasps lose their wings, enter the male syconium and oviposit in the short-styled female flowers; Stage 3, female wasps die after oviposition, wasp larvae develop inside the gall flowers, male wasps emerge from gall flowers; Stage 4, male wasps inseminate female wasps while they are still inside the ovary of gall flower; Stage 5, male wasps cut open the syconium wall to make escape holes; Stage 6, female wasps leave the male syconium carrying pollen from male flowers; Stage 7, female wasps carrying pollen visit syconia of female trees; Stage 8, female wasps lose their wings, enter the syconium and pollinate the long-styled female flowers and die inside; Stage 9, seeds developed from fertilized female flowers; Stage 10 (a and b), seeds germinate to give either male fig tree or female fig tree; Stage 11 (a and b), male tree produces syconia containing functional male flowers and short-styled non-functional female flowers, female tree produces syconia containing long-styled functional female flowers. (Figure recreated by authors, from sources: David Johson (syconia), Armstrong 2012 (seeds), and Joseph M. DiTomaso, University of California (seedling)).

the short-lived pollinators (Chen *et al.*, 2018) as well as non-productivity of fig cultivars that require pollination. The receptive stage of female fig syconia resembles those of caprifig syconia. This 'female mimicry' deceives the pollinating wasps (Gilbert, 1975) as they cannot oviposit into the long-styled flowers of the female figs (Valdeyron and Lloyd, 1979). The occurrence of short-styled female flowers is essential for the completion of the pollinating wasp's life cycle.

## 2.9 Growth and Reproductive Physiology

Figs are found in diverse habitats. However mild-temperate climates favor both the plant and its pollinator. Fig also grows well in tropical and subtropical regions (Morton, 1987). It fares well in humid regions, including Southern and Eastern Europe. Its high adaptability to diverse climatic conditions and its easy propagation through seeds and vegetative cuttings can threaten other native plant species in newly introduced areas (Holmes *et al.*, 2014). Once established, the fig is drought tolerant, requiring less water than other fruit species. However, it requires irrigation to achieve higher yields (Singh *et al.*, 2015). The frequency of irrigation depends on tree size, vigor, soil type and rainfall. Fig trees may become stressed during dry periods because of their shallow root systems (Flaishman *et al.*, 2008). Foliar application of the anti-transpirant, brassinolide at a concentration of 200 ml/L, can ameliorate the effects of water stress and enhance drought resistance by minimizing water loss through stomatal control (Zulkarnaini *et al.*, 2020).

The vegetative growth cycle of fig trees varies depending on the weather conditions and localities. In warm climatic regions where the winter temperature is above 10°C, fig trees do not defoliate and or enter winter dormancy. Fig trees growing in cooler regions defoliate and enter dormancy. The length of the dormant period depends on local climatic conditions (Ersoy *et al.*, 2008). Fully dormant fig trees can withstand temperatures as low as −6°C (Sarkhosh and Andersen, 2020), but young trees are much more susceptible to injury from low temperatures (Condit, 1933). Frostbitten aerial parts can later regenerate new shoots from underground protected buds when the temperatures rise (Kawamata *et al.*, 2002). Fig leaf development begins in early spring, and the production of young leaves continues until midsummer. The lifespan of fig leaves is from three to six months depending on when they first flushed. Simultaneously, young fig leaves are developing at the tip of the shoots through summer. Therefore, both leaf types are directly exposed to high irradiation and increased temperature (Mlinaric *et al.*, 2017).

The annual vegetative growth cycle of the fig tree has three distinct growth stages, i.e., Stage I, the first growth in spring; Stage II, no growth in summer and Stage III, second growth in fall (Ateyyeh and Sadder, 2006; Gaaliche *et al.*, 2011a); a double sigmoid growth curve separated by Stage II. Dormancy is weaker on the basal nodes than on the upper nodes. Starch accumulates during the deepest phase of endodormancy and converts to sugar during the awakening phase to alleviate frost damage to buds (Kawamata *et al.*, 2002). In figs, it is possible to break endodormancy at any stage by increasing the temperature in a greenhouse environment, inducing shoot growth and fruit production (Kawamata *et al.*, 2002).

Andel-Razik and El-Darier (1991) reported that growth initiation of fig is dependent on carbohydrate reserves, estimated at approximately 62% reserves, during the early growth in spring, leaving storage of 20% of net assimilate produced for later use. The major soluble sugars in fig leaves and fruits are sucrose, glucose and fructose. Sucrose is always higher than other sugars in the leaves. However, soluble sugar and starch concentrations decrease in leaves after attaining their maximum size. Sucrose is low in the fruits throughout development and maturation (Vemmos *et al.*, 2013).

The fig tree is moderately resistant to salinity (Golombek and Lüdders, 1993). Salinity stress tolerance/sensitivity levels may vary from zero to as high as 171.1 mM NaCl, depending on variety (Pakyurek and Dumanoglu, 2019). NaCl concentrations of ≥200 mM result in extensive leaf necrosis, leaf abscission and a dramatic decrease in leaf photosynthesis, as the leaf chlorophyll content is impeded by high salt concentrations (Caruso *et al.*, 2017).

Vangelisti *et al.* (2019) reported relatively low transcriptome changes after 24 days of 100 mM NaCl treatment and drastic transcriptome changes after 48 days. They identified several salt-regulated genes involved in plant salinity stress response.

Management of various cultural practices, pruning, training, irrigation, soil fertilization and growth regulators can control reproductive and vegetative growth in fig trees (Ateyyeh and Sadder, 2006). The dormancy breaking agent hydrogen cyanamide ($H_2CN_2$), Dormex®, can synchronize the development of breba fig syconia (Yablowitz *et al.*, 1998). In warm areas, the application of $H_2CN_2$ is an effective cultural practice to stimulate the vegetative growth of fig trees cv. Zidi, resulting in a higher main crop figs load and early fruit ripening. Applications of 1.5% $H_2CN_2$ significantly increased vegetative growth and fig yield (Gaaliche *et al.*, 2017b).

An *et al.* (2016) reported 5-aminolevulinic acid (ALA), a natural plant hormone, can remediate waterlogging stress in fig cultivation. An ALA pretreatment reduces oxidative stress in waterlogged conditions by increasing the leaf photosynthetic rate, root respiration and antioxidant production in fig leaves. Fig trees exhibit efficient water uptake and the ability to draw nutrients from leaves before abscission. Significant resources are harvested with the fruits.

## Floral induction and fruit growth

While other plants may undergo floral induction annually given amenable environmental conditions, figs experience floral induction from spring through fall (Ikegami *et al.*, 2013). Reproductive bud development begins with meristem broadening and elongation, leading to axis elongation and initiation of the scales that surround the ostiole of the syconium (Flaishman *et al.*, 2008). The apex of the developing syconium enlarges around the periphery forming a cup-shaped structure lined with floral primordia. The cup-shaped structure grows toward the center, forming an ostiole partly enclosed in numerous scales (Crane and Baker, 1953). Ikegami *et al.* (2013) identified a single copy of *FcFT1*, a flower-inducing gene, that functions under long and short-day conditions, but never under continuous darkness.

As already described, fig fruit is a syconium composed of multiple drupelets which develop from the ovaries enclosed within a fleshy receptacle (Storey *et al.*, 1977). Chessa *et al.* (1992) studied the details of flower development in Common-type fig syconia. Flowers develop from flower primordia within the syconium that undergoes sequential development giving rise to four or five tepals, one pedicel, one long style and one bifid stigma in each flower. After anthesis, further changes occur in the flower's internal structures.

Fig fruit development, like vegetative bud development, shows three distinct stages with a double-sigmoid growth curve (Marei and Crane, 1971). Phase I is the initial rapid growth phase, and Phase III is the ripening phase with fruit growth and color development, softening of the pulp, and palatability. Phase II shows no visible changes; the fruit remains the same size, color and firmness as in Phase I. During the first stage, fruit growth is a function of intensive cell division. During the second stage, mitotic activity ceases. The final stage is a stage of cell expansion, and changes in color, texture and aroma. The cell wall of the fig fruit is also modified as tissues mature.

Measurements of whole fruits showed that Phase II was characterized by increased firmness, titratable acidity and total pectin concentration (Tsantili, 1990). Acidity and total pectin reached their maxima at the end of Phase II. Concentrations of soluble pectin, glucose, fructose and sucrose remained low up to Phase III. During the rapid growth in Phase III, firmness, titratable acidity and total pectin decreased, while soluble pectin, glucose, fructose and sucrose concentration increased considerably. However, sucrose was always a minor component. At all stages, starch concentrations were extremely low (Tsantili, 1990).

## Fruit maturation and postharvest physiology

Temperature and humidity play major roles in fruit quality and yield. Temperatures higher than 37.8°C (100°F) lead to premature fruit ripening, tough skin and decreased pulp (Magness and Traub, 1941). Sudden change in humidity may cause splitting of immature fruit in certain fig varieties. A slight increase in heat units and greater relative humidity can also cause splitting of the fruit and spoilage-related injuries in figs (Condit, 1933). During ripening, fruit quality can be diminished due to scarring from intense winds. The stimulation of fruit maturation is of great interest to researchers and has been the topic of several studies. As early as the 3rd century BCE, growers applied a drop of olive oil on the ostiole to stimulate fruit growth, termed oleification (Chessa, 1997). Refined olive oil was observed to be more effective than unrefined olive oil or other vegetable oils. However, to date, there is no clear connection between oleification and the stimulation of maturation and ethylene production in figs. Applying ethylene to fig fruits during the late second stage of development has the potential to stimulate ripening and growth. Although, ethylene application to fig fruit may only be effective in inducing ripening when physiological conditions are acceptable for autocatalytic ethylene production (Marei and Crane, 1971). Treating fig fruits with linolenic acid promoted rapid ethylene production (Hirai *et al.*, 1966). Fig fruit growth can also be stimulated by auxin applications which potentially induce ethylene production in fruits and leaves (Marei and Crane, 1971).

Fig fruits are considered climacteric, as ethylene production increases when respiration rates increase at the onset of fruit ripening. Commercial fig crops must be harvested within the cultivar's specific commercial maturity range as they do not continue to ripen postharvest like other climacteric fruit such as apple, pear and banana (Flaishman *et al.*, 2008). The fig is a highly perishable fruit. The primary metabolites in fresh fruits are subject to further changes due to respiration. Such metabolic changes in figs can be controlled by applying edible coatings, such as one extracted from the cladodes of *Opuntia ficus-indica*, to the harvested fruits (Allegra *et al.*, 2018). Edible coatings can be used to maintain fruit quality, fresh weight and firmness. Such edible coating on fresh fruits also helps in maintaining the carbohydrate amounts and other key qualitative metabolites, such as beta-sitosterol and glycerol in figs (Allegra *et al.*, 2018).

Fruit storage at low temperatures can extend the postharvest life of the fruits for 7–10 days. If held at 0–5°C in atmospheres enriched with 15% or 20% $CO_2$ figs can be stored for up to 4 weeks (Colelli *et al.*, 1991). The $CO_2$ reduces ethylene production and maintains the fruits' bright external appearance. Preharvest 1-methyl cyclopropene (1-MCP) applications inhibit ripening and improve fig storage quality (Freiman *et al.*, 2012). Owino *et al.* (2006) isolated and studied the expression patterns of fig ethylene-synthesis genes (three *ACC-synthase* and one *ACC oxidase*) in postharvest figs treated with 1-MCP. The treated figs displayed downregulation of *FcACS1*, *FcACS3* and *FcACO1*, as opposed to upregulation of *FcACS2* when ethylene was present in tissues.

Freiman *et al.* (2015) compared fruit ripening-related gene expression between the receptacle and the female flower tissue during on-tree fig ripening. The expression pattern of the ripening-gene in female flowers was similar to those in tomato and apple but differed in the receptacle of fig syconium. During on-tree fruit ripening, all fruit ripening-related genes (*MADS-box*, *ACS*, *EOL*, *ACO*, *ETR*, *EIN4*, *ERS1*, *CTR1/EDR1*, *EIN2*, *EBF1*, *EIN3/EIL* and *ERF*) were upregulated in the female flowers, in the receptacle some of the genes (*MADS-box*, *EOL*, *ETR*, *EIN4*, *ERS1* and *CTR1/EDR1*) were downregulated. For most fig fruit-ripening genes, the expression pattern in the fig receptacle is dissimilar to that in the female flowers. The female flowers developed inside the syconium regulate the ripening processes for the whole accessory fruit, the syconium.

Anthocyanins are responsible for the color change in ripening fig fruits. Light deprivation significantly represses anthocyanin biosynthesis in the peel, leading to yellow-green figs, while the inside of the figs remains strawberry-colored. This suggests varying regulation of fig peel and female flower tissue pigmentation. The female flower tissue and the peel differ in anthocyanin content. Flower tissues contain pelargonidin-3-glucoside and cyanidin-3-rutinoside, whereas peel tissues contain cyanidin 3-O-glucoside and cyanidin-3-rutinoside. Before ripening, anthocyanin content is higher in the female flower than in the peel, but when ripening is initiated anthocyanin content in the peel is about five times greater than that of the female flower (Wang *et al.*, 2019).

In a variety of naturally ripening fig fruit cultivars, transcriptome analysis has shown high expression levels of anthocyanin-biosynthesis pathway genes (Ikegami *et al.*, 2013; Wang *et al.*, 2017). Applying exogenous ABA to 'Brown Turkey' fig accelerates ripening, whereas application of the ABA inhibitors NDGA and fluoridone delays ripening (Lama *et al.*, 2019). In addition to enhancing ripening, exogenous ABA and ethephon treatments also improved the fig fruit's color via early accumulation of cyanidin 3-O-glucoside and cyanidin 3-O-rutinoside. These two cyanidins have been reported to be the major anthocyanins in ripened fig fruit, responsible for the skin color intensity (Dueñas *et al.*, 2008). Similar to the ripening process in the receptacle where ABA-dependent non-climacteric ripening occurs, ABA has been shown to initiate color development by enhancing anthocyanin accumulation in the fig fruit (Lama *et al.*, 2020).

During anthesis of female fig syconium, or fruit, when the fruit's transverse diameter reaches 18–20 mm, there are no apparent differences between breba and main crop flowers in either Common-type fig (Beck and Lord, 1988) or San Pedro-type figs (Chai *et al.*, 2017). However, during the post-anthesis stage when the fruit's transverse diameter is approximately 25 mm, the young main crop fruit yellowed, shriveled slightly, and abscised at the stem connection to the branch (Chai *et al.*, 2017). San Pedro-type fig produces parthenocarpic breba crops and non-parthenocarpic main crop on the same tree. Pollination is not required for breba crop production, whereas pollination is required for caprifig in main crop production in San Pedro-types. The absence of pollination in main crops can cause syconia abscission shortly after flower anthesis. Based on the transcriptomic analysis and hormone assays, Chai *et al.* (2017) attempted to understand the regulatory mechanisms of parthenocarpic breba and non-parthenocarpic main crop production in the San Pedro-type fig. Differential expressions of plant hormone signal transduction, zeatin biosynthesis and diterpenoid biosynthesis pathways between breba crop and main crop were reported. Simultaneously, gibberellin was also significantly lower in main crop flowers, while zeatin and ABA were higher than in breba. They suggested that in absence of pollination, auxin might play a key role in parthenocarpy of breba and main crops in the San Pedro-type fig.

Cui *et al.* (2019) elucidated the fig ripening process based on quantitative proteomic and transcriptomic analyses at the two harvest stages, commercial and tree-ripe. They reported a remarkable increase in fig size, weight, sugar content and flavor from commercial-ripe to tree-ripe stages. However, they also reported a decrease in fruit firmness. They identified differentially regulated proteins and genes related to ficin, sugar accumulation, cell-wall modification, ethylene pathway and other biological processes involved in fruit quality and ripening.

## Conclusion

The fig is a promising fruit crop due to its high adaptability to diverse climatic conditions, reported drought tolerance and ease of vegetative propagation. Interesting features such as gynodioecious plants with functionally dioecious behavior, a short climacteric period with a non-climacteric postharvest fruit behavior, highly variable leaf morphology, heterostyly, dichogamy and obligate association of pollinating wasps have resulted in multiple investigations by researchers and students of biology. While earlier research provided basic information regarding the developmental biology of the fig, modern tools and techniques can aid in verifying the controversial statements in the literature, for example, conflicting reports of fig seeds as endospermic versus non-endospermic or fig fruits reported as climacteric versus non-climacteric. Decoding the fig's genetic information using high throughput next-generation sequencing and advanced bioinformatical tools to address some of its physiological and developmental processes is the next stage in fig investigations. Through understanding of its reproductive complexity and developmental biology, complete genome sequencing could place the fig as a model plant in horticulture and evolutionary biology. And, in Mediterranean regions and Western Asia, where both cultivated and semi-wild figs are found, a systematic germplasm collection of the traditional fig cultivars and wild relatives is imperative for proper germplasm management and conservation.

## References


Abdelsalam, N.R., Awad, R.M., Ali, H.M., Salem, M.Z.M., Abdellatif, K.F. and Elshikh, M.S. (2019) Morphological, pomological and specific molecular marker resources for genetic diversity analyses in fig (*Ficus carica* L.). *HortScience* 54, 1299–1309. DOI: 10.21273/HORTSCI14091-19

Abdolahipour, M., Kamgar-Haghighi, A.A., Sepaskhah, A.R., Zand-Parsa, S. and Honar T. (2020) Root length density of rainfed fig trees under different times, amounts, and positions of supplemental irrigation. *Journal of Agricultural Science and Technology* 22, 1137–1150. www.jast.modares.ac.ir/article-23-27902-en.html (Accessed 3 August 2021).

Abou-Ellail, M., Mahfouze, S.A., El-Enany, M.A.M. and Mustafa, N.S.A. (2014) Using biochemical and simple sequence repeats (SSR) markers to characterize (*Ficus carica* L.) cultivars. *World Applied Sciences Journal* 29, 313–321. DOI:10.5829/idosi.wasj.2014.29.03.13835

Achtak, H., Ater, M., Oukabli, A., Santoni, S., Kjellberg, F. and Khadari, B. (2009a) Moroccan fig genetic structure suggesting a local selection. *Acta Horticulturae* 814, 81–86. DOI: 10.17660/ActaHortic.2009.814.6

Achtak, H., Oukabli, A., Ater, M., Santoni, S., Kjellberg, F. and Khadari, B. (2009b) Microsatellite markers as reliable tools for fig cultivar identification. *Journal of the American Society for Horticultural Science* 134, 624–631. DOI: 10.21273/JASHS.134.6.624

Aksoy, U. (1998) Why figs? An old taste and a new perspective. *Acta Horticulturae* 480, 25–26. DOI: 10.17660/ActaHortic.1998.480.1

Aksoy, U., Balci, B., Can, H.Z. and Hepaksoy, B. (2003) Some significant results of the research-work in Turkey on fig. *Acta Horticulturae* 605, 173–180. DOI: 10.17660/ActaHortic.2003.605.26

Ali, S.H., Jubrael, J.M.S. (2013) Aflp assessment of *Ficus* cultivars for identification and conservation. *Science Journal of University of Zakho* 1(2), 642–650. www.sjuoz.uoz.edu.krd/index.php/sjuoz/article/view/322

Ali, S.H. (2019) Characterization of microsatellite loci in different fig (*Ficus carica* L.) landraces in Duhok and Erbil provinces in Kurdistan region-Iraq. *Zanco Journal of Pure and Applied Sciences* 31, 48–56. DOI: 10.21271/ZJPAS.31.2.7

Ali, Z.A., Mustafa, N.S., Abdel-Raouf, H.S., El-Shazly, S.M. and El-Berry, I.M. (2013) Characterization of some fig cultivars by anatomical traits on both leaves and stems. *World Applied Sciences Journal* 24, 1065–1071. DOI: 10.5829/idosi.wasj.2013.24.08.13259

Ali-Shtayeh, M.S., Jamous, R.M., Abu Zaitoun, S.Y., Mallah, O.B. and Mubaslat, A.K. (2014) Genetic diversity of the Palestinian fig (*Ficus carica* L.) collection by pomological traits and RAPD markers. *American Journal of Plant Sciences* 5, 1139–1155. DOI: 10.4236/ajps.2014.59127

Aljane, F., Essid, A. and Nahdi, S. (2018) Improvement of fig (*Ficus carica* L.) by conventional breeding and biotechnology. In: Al-Khayri J., Jain S. and Johnson D. (eds) *Advances in Plant Breeding Strategies: Fruits.* Springer, Cham, Switzerland. pp. 343–375. DOI: 10.1007/978-3-319-91944-7_9

Al-Khalaf, A.A., van Noort, S. and Compton, S.G. (2018) *Fig wasps and fig trees in Saudi Arabia*. King Saud University Press, Riyadh, pp. 75.

Allegra, A., Gallotta, A., Carimi, F., Mercati, F., Inglese, P. and Martinelli, F. (2018) Metabolic profiling and post-harvest behavior of “Dottato” fig (*Ficus carica* L.) fruit covered with an edible coating from *O. ficus-indica*. *Frontiers in Plant Science* 9, 1321. DOI: 10.3389/fpls.2018.01321

Almajali, D., Abdel-Ghani, A.H. and Migdadi, H. (2012) Evaluation of genetic diversity among Jordanian fig germplasm accessions by morphological traits and ISSR markers. *Scientia Horticulturae* 147, 8–19. DOI: 10.1016/j.scientia.2012.08.029

An, Y., Qi, L. and Wang, L. (2016) ALA Pretreatment improves waterlogging tolerance of fig plants. *PLoS ONE* 11, e0147202. DOI: 10.1371/journal.pone.0147202

Andel-Razik, M.S. and El-Darier, S. (1991) Functional adaptations of fig trees (*Ficus carica* L.) in agroecosystems of the western Mediterranean desert of Egypt. *Qatar University Science Journal* 11, 183–199.

Aradhya, M.K., Stover, E., Velasco, D. and Koehmstedt, A. (2010) Genetic structure and differentiation in cultivated fig (*Ficus carica* L.). *Genetica* 138, 681–694. DOI: 10.1007/s10709-010-9442-3

Aref, H.L., Salah, K.B., Chaumont, J.P., Fekih, A., Aouni, M. and Said, K. (2010) In vitro antimicrobial activity of four *Ficus carica* latex fractions against resistant human pathogens (antimicrobial activity of *Ficus carica* latex). *Pakistan Journal of Pharmaceutical Sciences* 23, 53–58.

Armstrong, W.P. (2012) Sex Life of Figs: Coevolution of a Tree & Minute Wasp (Part 1). www2.palomar.edu/users/warmstrong/arbimg10.htm (accessed 13 April 2020).

Ateyyeh, A.F. and Sadder, M.T. (2006) Preliminary study on the vegetative and reproductive growth of six Common Fig (*Ficus carica* L.) cultivars in Jordan. *Jordan Journal of Agricultural Sciences* 2, 1–7. arch.emarefa.net/detail/BIM-285252

Atzmon, N. and Henkin, Z. (1998) Establishing forest tree species on peatland in reflooded area of the Huleh Valley, Israel. *Forestry* 71, 141–146. DOI: 10.1093/forestry/71.2.141

Badgujar, S.B., Patel, V.V., Bandivdekar, A.H. and Mahajan, R.T. (2014) Traditional uses, phytochemistry and pharmacology of *Ficus carica*: A review. *Pharmaceutical Biology* 52, 1487–1503. DOI: 10.3109/13880209.2014.892515

Baziar, G., Jafari, M., Noori, M.S.S. and Samarfard, S. (2018) Evaluation of genotic diversity among Persian fig cultivars by morphological traits and RAPD markers. *HortScience* 53, 613–619. DOI: 10.21273/HORTSCI11306-16

Beck, N.G. and Lord, E.M. (1988) Breeding system in *Ficus carica*, the common fig. II. Pollination events. *American Journal of Botany* 75, 1913–1922. DOI: 10.1002/j.1537-2197.1988.tb11272.x

Ben Abdelkrim, A., Baraket, G., Essalouh, L., Achtak, H., Khadari, B. and Salhi-Hannachi, A. (2015) Use of morphological traits and microsatellite markers to characterize the Tunisian cultivated and wild figs (*Ficus carica* L.). *Biochemical Systematics and Ecology* 59, 209–219. DOI: 10.1016/j.bse.2015.01.026

Bercu, R. and Popoviciu, D. (2014) Anatomical study of *Ficus carica* L. leaf. *Annals of the Romanian Society for Cell Biology* 1, 33–36.

Berg, C.C. and Corner, E.J.F. (2005) Moraceae: Ficeae. *Flora Malesiana*, Series I, Vol. 17 (Part 2), 1–70.

Boliani, A.C., Ferreira, A.F.A., Monteiro, L.N.H., Cordeiro da Silva, M.S. and Rombola, A.D. (2019) Advances in propagation of *Ficus carica* L. *Revista Brasileira de Fruticultura* 41, e-026. DOI: 10.1590/0100-29452019026

Boudchicha, R.H., Hormaza, J.I. and Benbouza, H. (2018) Diversity analysis and genetic relationship among local Algerian fig cultivars (*Ficus carica* L.) using SSR markers. *South African Journal of Botany* 116, 207–215. DOI: 10.1016/j.sajb.2018.03.015

Browicz, K. (1982) Moraceae, pp. 1–8. In: Rechinger, K.H. (ed.), *Flora Iranica.* 153, Akademische Druck-U. Verlagsanstalt Graz-Austria. pp. 15.

Caliskan, O. and Polat, A. (2012) Morphological diversity among fig (*Ficus carica* L.) accessions sampled from the Eastern Mediterranean region of Turkey. *Turkish Journal of Agriculture and Forestry* 36, 179–193. DOI: 10.3906/tar-1102-33

Çalişkan, O., Mavi, K. and Polat, A. (2012) Influences of presowing treatments on the germination and emergence of fig seeds (*Ficus carica* L.). *Acta Scientiarum* 34, 293–297. DOI: 10.1590/S1807-86212012000300009

Caliskan, O., Bayazit, S., Ilgin, M., Karatas, N. and Ergul, A. (2018) Genetic diversity and population structure in caprifigs (*Ficus carica* var. *caprificus*) using SSR markers. *Spanish Journal of Agricultural Research* 16, e0703. DOI: 10.5424/sjar/2018163-11662

Caruso, G., Gennai, C., Ugolini, F., Marchini, F., Quartacci, M.F. and Gucci, R. (2017) Tolerance and physiological response of young *Ficus carica* L. plants irrigated with saline water. *Acta Horticulturae* 1173, 137–142. DOI: 10.17660/ActaHortic.2017.1173.23

Chai, L., Wang, Z., Chai, P., Chen, S. and Ma, H. (2017) Transcriptome analysis of San Pedro-type fig (*Ficus carica* L.) parthenocarpic breba and non-parthenocarpic main crop reveals divergent phytohormone-related gene expression. *Trees Genetics & Genomes* 13, 83. DOI: 10.1007/s11295-017-1166-4

Chandler, W.H. (1934) Dry matter residue of trees and their products in proportion to leaf area. *Proceedings of the American Society for Horticultural Science* 31, 39–56.

Chen, H., Zhang, Y., Peng, Y. and Corlett, R.T. (2018) Latitudinal effects on phenology near the northern limit of figs in China. *Scientific Reports* 8, 4320. DOI: 10.1038/s41598-018-22548-7

Chessa, I. (1997) Fig. In: Mitra, S. (ed.), *Postharvest Physiology and Storage of Tropical and Subtropical Fruits*. CAB International, Wallingford, Oxon, UK, pp. 245–268.

Chessa, I., Nieddu, G. and Schirra, M. (1992) Growth and ripening of main-crop fig. *Advances in Horticultural Science* 6, 112–115.

Ciarmiello, L.F., Piccirillo, P., Carillo, P., De Luca, A. and Woodrow, P. (2015) Determination of the genetic relatedness of fig (*Ficus carica* L.) accessions using RAPD fingerprint and their agro-morphological characterization. *South African Journal of Botany* 97, 40–47. DOI: 10.1016/j.sajb.2014.11.012

Colelli, G., Mitchell, F.G. and Kader, A.A. (1991) Extension of postharvest life of 'Mission' figs by CO2-enriched atmospheres. *HortScience* 26, 1193–1195.

Condit, I.J. (1926) Fruit-bud and flower development in *Ficus carica*. *Proceedings of the American Society for Horticultural Science* 23, 259–263.

Condit, I.J. (1932) The structure and development of flowers in *Ficus carica* L. *Hilgardia* 6, 443–481.

Condit, I.J. (1933) Fig Culture in California. California Agricultural Extension Service, Circular 77, pp. 18–20.

Condit, I.J. (1941) Fig characteristics useful in the identification of varieties. *Hilgardia* 14, 1–69. DOI: 10.3733/hilg.v14n01p001

Condit, I.J. (1947) *The Fig*. Chronica Botanica Co., Walthams, MA. pp. 1–222.

Condit, I.J. (1955) Fig varieties: A monograph. *Hilgardia* 23, 323–538.

Corner, E.J.H. (1965) Check-list of *Ficus* in Asia and Australasia, with keys to identification. *Garden's Bulletin Singapore* 21, 1–186.

Corner, E.J.H. (1985) Essays on *Ficus*. *Allertonia* 4, 125–168.

Crane, J.C. and Baker, R.E. (1953). Growth comparisons of the fruits and fruitlets of figs and strawberries. *Journal of the American Society for Horticultural Science* 62, 257–260.

Crane, J.C. and Brown, J.G. (1950) Growth of the fig fruit *Ficus carica* var Mission. *Proceedings of the American Society for Horticultural Science* 56, 93–97.

Crisosto, C.H., Bremer, V., Ferguson, L. and Crisosto, G.M. (2010) Evaluating quality attributes of four fresh fig (*Ficus carica* L.) cultivars harvested at two maturity stages. *HortScience* 45, 707–710. DOI: 10.21273/HORTSCI.45.4.707

Cui, Y., Wang, Z., Chen, S., Vainstein, A. and Ma, H. (2019) Proteome and transcriptome analyses reveal key molecular differences between quality parameters of commercial-ripe and tree-ripe fig (*Ficus carica* L.). *BMC Plant Biology* 19, 146. DOI: 10.1186/s12870-019-1742-x

Datwyler, S.L. and Weiblen, G.D. (2004) On the origin of the fig: phylogenetic relationships of Moraceae from *ndhF* sequences. *American Journal of Botany* 91, 767–777. DOI: 10.3732/ajb.91.5.767

De Candolle, A. (1967) Origin of cultivated plants. Hafner Publishing, New York (Reprint of the 2nd Edition of 1886).

Dueñas, M., Pérez-Alonso, J.J., Santos-Buelga, C. and Escribano-Bailón, T. (2008) Anthocyanin composition in fig (*Ficus carica*, L.). *Journal of Food Composition and Analysis* 21, 107–115. DOI: 10.1016/j.jfca.2007.09.002

Elisiário, P.J., Neto, M.C., Cabrita, L.F. and Leitão, J.M. (1998) Isozyme and RAPDs characterization of a collection of fig (*Ficus carica* L.) traditional varieties. *Acta Horticulturae* 480, 149–154. DOI: 10.17660/ActaHortic.1998.480.25

El Rayes, R. (1995) The fig tree in the Mediterranean region and in Syria. In: Llácer, G., Aksoy, U. and Mars, M. (ed.) *Underutilized Fruit Crops in the Mediterranean Region*. CIHEAM (Cahiers Options Méditerranéennes; n. 13), Zaragosa, Spain. pp. 79–83.

Engelmann, F. (2000) Importance of cryopreservation for the conservation of plant genetic resources. In: Engelmann, F. and Takagi, H. (eds) *Cryopreservation of Tropical Plant Germplasm: Current Research Progress and Application*. IPGRI, Rome. pp. 8–20.

Ersoy, N., Gözlekçi, S., Kaynak, L., Dal, B. and Yazıcı, K. (2008) Relationships among the shoot, leaf and fruit growth of some Fig (*Ficus carica*) cultivars grown in Antalya conditions. *Acta Horticulturae* 768, 143–148. DOI: 10.17660/ActaHortic.2008.798.19

Essid, A., Aljane, F., Ferchichi, A. and Hormaza, J.I. (2015) Analysis of genetic diversity of Tunisian caprifig (*Ficus carica* L.) accessions using simple sequence repeat (SSR) markers. *Hereditas* 152, 1. DOI: 10.1186/s41065-015-0002-9

Essid, A., Aljane, F. and Ferchichi, A. (2017) Morphological characterization and pollen evaluation of some Tunisian ex situ planted caprifig (*Ficus carica* L.) ecotypes. *South African Journal of Botany* 111, 134–143. DOI: 10.1016/j.sajb.2017.03.001

Evert, R.F. (2006) *Esau's Plant Anatomy: Meristems, Cells, and Tissues of the Plant Body: Their Structure, Function, and Development*, 3rd edn. John Wiley & Sons, New York.

Falistocco, E. (2009) Presence of triploid cytotypes in the common fig (*Ficus carica* L.). *Genome* 52, 919–925. DOI: 10.1139/G09-068

Ferguson, L., Michailides, T.J. and Shorey, H.H. (1990) The California fig industry. *Horticultural Reviews* 12, 409–490. DOI: 10.1002/9781118060858.ch9

Ferraz, R.A., Leonel, S., Souza, J.M.A., Ferreira, R.B., Modesto, J.H. and Arruda, L.L. (2020) Phenology, vegetative growth, and yield performance of fig in Southeastern Brazil. *Pesquisa Agropecuária Brasileira* 55, e01192. DOI: 10.1590/S1678-3921.pab2020.v55.01192

Flaishman, M.A., Rodov, V. and Stover, E. (2008) The fig: botany, horticulture, and breeding. *Horticultural Reviews* 34, 113–197. DOI: 10.1002/9780470380147.ch2

Freiman, Z.E., Rodov, V., Yablovitz, Z., Horev, B. and Flaishman, M.A. (2012) Preharvest application of 1-methylcyclopropene inhibits ripening and improves keeping quality of 'Brown Turkey' figs (*Ficus carica* L.). *Scientia Horticulturae* 138, 266–272. DOI: 10.1016/j.scienta.2012.01.007

Freiman, Z.E., Rosianskey, Y., Dasmohapatra, R., Kamara, I. and Flaishman, M.A. (2015) The ambiguous ripening nature of the fig (*Ficus carica* L.) fruit: a gene expression study of potential ripening regulators and ethylene-related genes. *Journal of Experimental Botany* 66, 3309–3324. DOI: 10.1093/jxb/erv140

Gaaliche, B., Lauri, P.E., Trad, M., Costes, E. and Mars, M. (2011a) Interactions between vegetative and generative growth and between crop generations in fig tree (*Ficus carica* L.). *Scientia Horticulturae* 131, 22–28. DOI: 10.1016/j.scienta.2011.09.022

Gaaliche, B., Trad, M. and Mars, M. (2011b) Effect of pollination intensity, frequency and pollen source on fig (*Ficus carica* L.) productivity and fruit quality. *Scientia Horticulturae* 130, 737–742. DOI: 10.1016/j.scienta.2011.08.032

Gaaliche, B., Saddoud, O. and Mars, M. (2012) Morphological and pomological diversity of fig (*Ficus carica* L.) cultivars in northwest of Tunisia. *ISRN Agronomy*, 1–9.

Gaaliche, B., Majdoub, A., Trad, M. and Mars, M. (2013) Assessment of pollen viability, germination and tube growth in eight Tunisian caprifig (*Ficus carica* L.) cultivars. *ISRN Agronomy* 2013, 1–4. DOI: 10.1155/2013/207434

Gaaliche, B., Aïachi-Mezghani, M., Trad, M., Costes, E., Lauri, P.E. and Mars, M. (2016) Shoot architecture and morphology of different branch orders in fig tree (*Ficus carica* L.). *International Journal of Fruit Science* 16, 378–394. DOI: 10.1080/15538362.2015.1126699

Gaaliche, B., Zarrouk, I. and Mars, M. (2017a) Agro-phenological behaviour of several caprifigs grown in two different ecological areas in Tunisia. *Acta Horticulturae* 1173, 149–156. DOI: 10.17660/ActaHortic.2017.1173.25

Gaaliche, B., Ghrab, M. and Ben Mimoun, M. (2017b) Effect of hydrogen cyanamide on vegetative growth, yield and fruit quality of fig cv. Zidi in a warm production area. *International Journal of Fruit Science* 17, 63–71. DOI: 10.1080/15538362.2016.1202169

Gaaliche, B., Ladhari, A., Mkadmi, M., Ben Hamda, H., Ben Mimoun, M. and Harbi, M. (2018) Propagation and conservation of fig (*Ficus carica* L.) in the Northeast of Tunisia. *Annales de l'INRAT* 91, 200–210.

Galil, J. (1977) Fig biology. *Endeavor* 1, 52–56.

Galil, J. and Meiri, L. (1981) Number and structure of anthers in fig-syconia in relation to behaviour of pollen vectors. *New Phytologist* 88, 83–87. DOI: 10.1111/j.1469-8137.1981.tb04571.x

Ganopoulos, I., Xanthopoulou, A., Molassiotis, A., Karagiannis, E., Moysiadis, T., Katsaris, P., Aravanopoulos, F., Tsaftaris, A., Kalivas, A. and Madesis, P. (2015) Mediterranean basin *Ficus carica* L.: from genetic diversity and structure to authentication of a Protected Designation of Origin cultivar using microsatellite markers. *Trees* 29, 1959–1971. DOI: 10.1007/s00468-015-1276-2

Gholami, M., Rahemi, M. and Rastegar, S. (2012) Use of rapid screening methods for detecting drought tolerant cultivars of fig (*Ficus carica* L.). *Scientia Horticulturae* 143, 7–14. DOI: 10.1016/j.scienta.2012.05.012

Gil, M., Pérez, F., Giraldo, E., Cortés, J. and López-Corrales, M. (2011) Classification and synonymy of 'Albacor' fig in the genebank at Finca la Orden, Spain. *Acta Horticulturae* 918, 961–963. DOI: 10.17660/ActaHortic.2011.918.126

Gilbert, L.E. (1975) Ecological consequences of a coevolved mutualism between butterflies and plants. In: Gilbert, L.E. and Raven, L.E. (eds) *Coevolution of Animals and Plants*. Univ. of Texas Press, Austin, Texas. pp. 211–239.

Giordano, G., Maleci, L., Agati, G. and Petruccelli, R. (2019) *Ficus carica* L. leaf anatomy: Trichomes and solid inclusions. *Annals of Applied Biology* 176, 47–54. DOI: 10.1111/aab.12557

Giraldo, E., López-Corrales, M., Viruel, M.A. and Hormaza, J.I. (2004) Development of microsatellite markers in fig (*Ficus carica* L.). *Acta Horticulturae* 663, 635–638. DOI: 10.17660/ActaHortic.2004.663.113

Giraldo, E., Viruel, M.A., López-Corrales, M. and Hormaza, J.I. (2005) Characterisation and cross-species transferability of microsatellites in the common fig (*Ficus carica* L.). *The Journal of Horticultural Science and Biotechnology* 80, 217–224. DOI: 10.1080/14620316.2005.11511920

Giraldo, E., Lopez-Corrales, M. and Hormaza, J.I. (2008) The use of SSR markers to screen new accessions before their incorporation into fig germplasm collections. *Acta Horticulturae* 798, 165–168. DOI: 10.17660/ActaHortic.2008.798.22

Golombek, S.D. and Lüdders, P. (1993) Effects of short-term salinity on leaf gas exchange of the fig (*Ficus carica* L.). *Plant Soil* 148, 21–27. DOI: 10.1007/BF02185381

Gozlekci, S. (2010) Selection studies on fig (*Ficus carica* L.) in Antalya Province of Turkey. *African Journal of Biotechnology* 9, 7857–7862. DOI: 10.5897/AJB10.1382

Haffar, S., Ben Abdelkrim, A., Salhi-Hannachi, A. and Baraket, G. (2017) Genetic diversity assessment of Tunisian fig (*Ficus carica* L.) using conserved DNA-derived polymorphism. *Acta Horticulturae* 1173, 51–56. DOI: 10.17660/ActaHortic.2017.1173.9

Himelrick, D.G. (1999) *Fig Production Guide*. ANR-1145, Alabama Cooperative Extension System, Alabama A&M and Auburn Universities, USA. (Available at: https://figvarieties.com/wp-content/uploads/2021/03/Alabama-Cooperative-Extension-System-Fig-Production-Guide-ANR-1145.pdf (accessed 2 February 2022).

Hirai, J., Hirata, N. and Horiuchi, H. (1966) Effect of oleification on hastening the maturity of the fig fruits. III. Promoting effect of fatty acids and glycerine on fruit maturation. *Journal of the Japanese Society for Horticultural Science* 36, 147–150.

Holmes, K.A., Greco, S.E. and Berry, A.M. (2014) Pattern and process of fig (*Ficus carica*) invasion in a California Riparian Forest. *Invasive Plant Science and Management* 7, 46–58. DOI: 10.1614/IPSM-D-13-00045.1

Hummer, K.E., Pomper, K.W., Postman, J., Graham, C.J., Stover, E., Mercure, E.W., Aradhya, M., Crisosto, C.H., Ferguson, L., Thompson, M.M., Byers, P. and Zee, F. (2012) Emerging fruit crops. In: Badenes M.L. and Byrnes, D.H. (eds). *Fruit Breeding*. Springer Science+Business Media LLC, New York, NY, pp. 97–147.

Ikegami, H., Nogata, H., Hirashima, K., Awamura, M. and Nakahara, T. (2009) Analysis of genetic diversity among European and Asian fig varieties (*Ficus carica* L.) using ISSR, RAPD, and SSR markers. *Genetic Resources and Crop Evolution* 56, 201–209. DOI: 10.1007/s10722-008-9355-5

Ikegami, H., Nagota, H., Inoue, Y., Himeno, S., Yakushiji, H., Hirata, C., Hirashima, K., Mori, M., Awamura, M. and Nakahara, T. (2013) Expression of *FcFT1*, a *FLOWERING LOCUS T*-like gene, is regulated by light and associated with inflorescence differentiation in fig (*Ficus carica* L.). *BMC Plant Biology* 13, 216. DOI: 10.1186/1471-2229-13-216

Ikten, H. and Yilmaz, Y. (2019) Validation of gender specific CAPS marker in Turkish fig (*Ficus carica* L.) collection and F1 progenies. *Notulae Botanicae Horti Agrobotanici Cluj-Napoca* 47, 867–871. DOI: 10.15835/nbha47311499

IPGRI and CIHEAM (2003) *Descriptors for Fig*. International Plant Genetic Resources Institute, Rome, Italy, and International Centre for Advanced Mediterranean Agronomic Studies, Paris, France.

Janzen, D.H. (1979) How to be a fig. *Annual Review of Ecology and Systematics* 10, 13–51. https://www.jstor.org/stable/2096784

John, P. (1993) *Biosynthesis of the Major Crop Products*. Wiley/New York, pp. 114–126.

Jona, B. and Gribaudo, I. (1991) *Ficus* spp. In: Baja, Y.P.S. (ed.) *Biotechnology in Agriculture and Forestry*. Vol. 16, Trees III. SpringerVerlag, Berlin. pp. 76–93.

Kajii, C., Morita, T. and Kuroda, K. (2014) Laticifers of *Ficus carica* and their potential role in plant defense. *IAWA Journal* 35, 109–115. DOI: 10.1163/22941932-00000052

Karimi, S., Hojati, S., Eshghi, S., Nazary Moghaddam, R. and Jandoust, S. (2012) Magnetic exposure improves tolerance of fig 'Sabz' explants to drought stress induced *in vitro*. *Scientia Horticulturae* 137, 95–99. DOI: 10.1016/j.scienta.2012.01.018

Kawamata, M., Nishida, E., Ohara, H., Ohkawa, K. and Matsui, H. (2002) Changes in the intensity of bud dormancy and internal compositions of current shoot in fig. *Journal of the Japanese Society for Horticultural Science* 71, 177–182. DOI: 10.2503/jjshs.71.177

Khadari, B., Hochu, I., Bouzid, L., Santoni, S., Roger, J.P. and Kjellberg, F. (2003) The use of microsatellite markers for identification and genetic diversity evaluation of the fig collection in CBNMP. *Acta Horticulturae* 605, 77–86. DOI: 10.17660/ActaHortic.2003.605.10

Khadivi-Khub, A. and Anjam, K. (2014) Characterization and evaluation of male fig (caprifig) accessions in Iran. *Plant Systematics and Evolution* 300, 2177–2189. DOI: 10.1007/s00606-014-1038-4

Khatib, S. and Vaya, J. (2010) Fig, carob, pistachio, and health. In: Watson, R.R. and Preedy, V.R. (eds) *Bioactive Foods in Promoting Health Fruits and Vegetables*. Academic Press, pp. 245–263. DOI: 10.1016/B978-0-12-374628-3.00017-7

Kim, J.S., Kim, Y.O., Ryu, H.J., Kwak, Y.S., Lee, J.Y. and Kang, H. (2003) Isolation of stress related genes of rubber particles and latex in fig tree (*Ficus carica*) and their expressions by abiotic stress or plant hormone treatments. *Plant Cell Physiology* 44, 412–414. DOI: 10.1093/pcp/pcg058

Kislev, M.E., Hartmann, A. and Bar-Yosef, O. (2006) Early domesticated fig in the Jordan valley. *Science* 312, 1372–1374.

Knap, T., Aradhya, M., Arbeiter, A.B, Hladnik, M. and Bandelj, D. (2018) DNA profiling of figs (*Ficus carica* L.) from Slovenia and Californian USDA collection revealed the uniqueness of some North Adriatic varieties. *Genetic Resources and Crop Evolution* 65, 1503–1516. DOI: 10.1007/s10722-018-0634-5

Knap, T., Arbeiter, B., Jakse, J., Cizmovic, M., Adakalic, M., Popovic, R., Lazovic, B., Strikic, F., Podgornik, M. and Bandelj, D. (2017) Diversity of figs (*Ficus carica* L.) from the east Adriatic coast. *Acta Horticulturae* 1173, 11–16. DOI: 10.17660/ActaHortic.2017.1173.2

Lama, K., Harlev, G., Shafran, H., Peer, R. and Flaishman, M.A. (2020) Anthocyanin accumulation is initiated by abscisic acid to enhance fruit color during fig (*Ficus carica* L.) ripening. *Journal of Plant Physiology* 251, 153–192. DOI: 10.1016/j.jplph.2020.153192

Lama, K., Modi, A., Peer, R., Izhaki, Y. and Flaishman, M.A. (2019) On-tree ABA application synchronizes fruit ripening and maintains keeping quality of figs (*Ficus carica* L.). *Scientia Horticulturae* 253, 405–411. DOI: 10.1016/j.scienta.2019.04.063

Lamb, C. and Dixon, R.A. (1997) The oxidative burst in plant disease resistance. *Annual Review of Plant Physiology and Plant Molecular Biology* 48, 251–275. DOI: 10.1146/annurev.arplant.48.1.251

Lansky, E.P. and Paavilainen, H.M. (2010) Figs: the genus *Ficus*. In: *Figs: the genus Ficus*. CRC Press Inc. pp. 415.

Lauri, P.E. and Caraglio, Y. (1993) Architectural analysis of fig trees as a tool to study genetic diversity in cultivars. *Acta Horticulturae* 349, 265–267. DOI: 10.13140/2.1.4323.4560

Lazreg-Aref, H., Gaaliche, B., Ladhari, A., Hammami, M. and Hammami, S.O. (2018) Co-evolution of enzyme activities and latex in fig (*Ficus carica* L.) during fruit maturity process. *South African Journal of Botany* 115, 143–152. DOI: 10.1016/j.sajb.2018.01.022

Leonel, S. and Damatto Junior, E.R. (2007) Root distribution of fig trees cultivated with organic manure. *Revista Brasileira de Fruticultura* 29, 191–194. DOI: 10.1590/S0100-29452007000100041

Lisci, M. and Pacini, E. (1994) Germination ecology of drupelets of the fig (*Ficus carica* L.). *Botanical Journal of the Linnean Society* 114, 133–146. DOI: 10.1111/j.1095-8339.1994.tb01927.x

Magness, J.R. and Traub, H.P. (1941) Climatic adaptation of fruit and nut crops. In: *Climate and Man*, Yearbook of Agriculture, US Dept of Agriculture, pp. 400–420. Available at: https://naldc.nal.usda.gov/download/IND43893780/PDF (accessed 2 February 2022).

Mamoucha, S., Fokialakis, N. and Christodoulakis, N.S. (2016) Leaf structure and histochemistry of *Ficus carica* (Moraceae), the fig tree. *Flora* 218, 1–11. DOI: 10.1016/j.flora.2015.11.003

Marcotuli, I., Mazzeo, A., Nigro, D., Giove, S., Giancaspro, A., Colasuonno, P., Prgomet, Ž., Prgomet, I., Tarantino, A., Ferrara, G. and Gadaleta, A. (2019) Analysis of genetic diversity of *Ficus carica* L. (Moraceae) collection using simple sequence repeat (SSR) markers. *Acta Scientiarum Polonorum Hortorum Cultus* 18, 93–109. DOI: 10.24326/asphc.2019.4.9

Marei, N. and Crane, J.C. (1971) Growth and respiratory response of fig (*Ficus carica* L. cv. Mission) fruits to ethylene. *Plant Physiology* 48, 249–254. DOI: 10.1104/pp.48.3.249

Mars, M., Chatti, K., Saddoud, O., Salhi-Hannachi, A., Trifi, M. and Marrakchi, M. (2008) Fig cultivation and genetic resources in Tunisia, an overview. *Acta Horticulturae* 798, 27–32. DOI: 10.17660/ActaHortic.2008.798.2

Mars, M., Trad, M. and Gaaliche, B. (2017) The unique fig caprification system and its effects on productivity and fruit characteristics. *Acta Horticulturae* 1173, 127–136. DOI: 10.17660/ActaHortic.2017.1173.22

Mirheidari, F., Khadivi, A., Moradi, Y. and Paryan, S. (2020) Phenotypic variability of naturally grown edible fig (*Ficus carica* L.) and caprifig (*Ficus carica* var. *caprificus* Risso) accessions. *Scientia Horticulturae* 267. DOI: 10.1016/j.scienta.2020.109320

Mlinaric, S., Dunic, J.A., Babojelic, M.S., Cesar, V. and Lepedus, H. (2017) Differential accumulation of photosynthetic proteins regulates diurnal photochemical adjustments of PSII in common fig (*Ficus carica* L.) leaves. *Journal of Plant Physiology* 209, 1–10. DOI: 10.1016/j.jplph.2016.12.002

Monastra, F., Insero, O. and Tamponi, G. (1995) The present situation of some underutilized fruit crops in Italy. In: Llacer, D., Aksoy, U. and Mars, M. (eds) *Underutilized Fruit Crops in the Mediterranean Region*. CIHEAM (Cahiers Options Mediterraneennes; n. 13), Zaragoza, Spain. pp. 39–46. om.ciheam.org/article.php?IDPDF=96605639 (accessed 3 September 2021).

Morales, J. and Gil, J. (2014) Fruit as staple food: The role of fig (*Ficus carica* L.) during the pre-hispanic period of the Canary Islands, Spain (from the 3rd–2nd Centuries BCE to the 15th Century CE). In: Chevalier, A., Marinova, E. and Pena-Chocarro, L. (eds) *Plants and People - Choices and Diversity through Time* (Vol. 1). Oxbow Books, UK. pp. 182–210.

Mori, K., Shirasawa, K., Nogata, H., Hirata, C., Tashiro, K., Habu, T., Kim, S., Himeno, S., Kuhara, S. and Ikegami, H. (2017) Identification of *RAN1* orthologue associated with sex determination through

whole genome sequencing analysis in fig (*Ficus carica* L.). *Scientific Reports* 7, 41124. DOI: 10.1038/srep41124
Morton, J. (1987) Fig. In: *Fruits of Warm Climates*. Julia F. Morton, Miami, FL. pp. 47–50.
Mustafa, N.S.A. and Abou-Ellail, M. (2013) Genetic discrimination of Egyptian fig cultivars revealed by DNA fingerprints. *International Journal of Agricultural Research* 8, 17–25. scialert.net/abstract/?doi=ijar.2013.17.25 (accessed 3 September 2021).
Oukabli, A., Mamouni, A., Laghezali, M., Ater, M., Roger, J.P. and Khadari, B. (2003) Local caprifig tree characterization and analysis of interest for pollination. *Acta Horticulturae* 605, 61–64. DOI: 10.17660/ActaHortic.2003.605.7
Owino, W.O., Manabe, Y., Mathooko, F.M., Kubo, Y. and Inaba, A. (2006) Regulatory mechanisms of ethylene biosynthesis in response to various stimuli during maturation and ripening in fig fruit (*Ficus carica* L.). *Plant Physiology and Biochemistry* 44, 335–342. DOI: 10.1016/j.plaphy.2006.03.009
Pakyurek, M. and Dumanoglu, H. (2019) Effects of saline stress on *in vitro* seed germination and seedling growth of some Turkish fig cultivars (*Ficus carica* L.). *Applied Ecology and Environmental Research* 17, 13485–13492. DOI: 10.15666/aeer/1706_1348513492
Perez-Jiménez, M., López, B., Dorado, G., Pujadas-Salvá, A., Guzmán, G. and Hernandez, P. (2012) Analysis of genetic diversity of southern Spain fig tree (*Ficus carica* L.) and reference materials as a tool for breeding and conservation. *Hereditas* 149, 108–113. DOI: 10.1111/j.1601-5223.2012.02154.x
Petrucci, V.E. and Crane, J.C. (1950) Fruit bud initiation and differentiation in the fig. *Proceedings of the American Society for Horticultural Science* 56, 86–92.
Piga, A., Agabbio, M. and Farris, G.A. (2003) Dehydration performance of local fig cultivars. *Acta Horticulturae* 605, 241–245. DOI: 10.17660/ActaHortic.2003.605.37
Podgornik, M., Vuk, I., Vrhovnik, I. and Mavsar, D.B. (2010) A survey and morphological evaluation of fig (*Ficus carica* L.) genetic resources from Slovenia. *Scientia Horticulturae* 125, 380–389. DOI: 10.1016/j.scienta.2010.04.030
Pourghayoumi, M., Bakhshi, D., Rahemi, M. and Jafari, M. (2012) Effect of pollen source on quantitative and qualitative characteristics of dried figs (*Ficus carica* L.) cvs 'Payves' and 'Sabz' in Kazerun – Iran. *Scientia Horticulturae* 147, 98–104. DOI: 10.1016/j.scienta.2012.08.026
Rahemi, M. and Jafari, M. (2008) Effect of caprifig type on quantity and quality of Estahban dried fig *Ficus carica* cv. Sabz. *Acta Horticulturae* 798, 249–252. DOI: 10.17660/ActaHortic.2008.798.35
Rodrigues, M.G.F., dos Santos, T.P., Ferreira, A.F.A., Monteiro, L.N.H., Nakanishi, E.S. and Boliani, A.C. (2019) Morphological characterization of active germplasm bank fig tree accessions. *Revista Brasileira de Fruticultura* 41, e-074. DOI: 10.1590/0100-29452019074
Roger, J.P. and Khadari, B. (2003) Fig germplasm of CBNM Porquerolles displays an important genetic variability: evidence from pomological descriptors. *Acta Horticulturae* 605, 51–59. DOI: 10.17660/ActaHortic.2003.605.6
Rubnov, S., Kashman, Y., Rabinowitz, R., Schlesinger, M. and Mechoulam, R. (2001) Suppressors of cancer cell proliferation from fig (*Ficus carica*) resin: isolation and structure elucidation. *Journal of Natural Products* 64, 993–996. DOI: 10.1021/np000592z
Saddoud, O., Chatti, K., Salhi-Hannachi, A., Mars, M., Rhouma, A., Marrakchi, M. and Trifi, M. (2007) Genetic diversity of Tunisian figs (*Ficus carica* L.) as revealed by nuclear microsatellites. *Hereditas* 144, 149–157. DOI: 10.1111/j.2007.0018-0661.01967.x
Sarkhosh, A. and Andersen, P.C. (2020) *The Fig*. University of Florida, IFAS Extension, HS27, Florida.
Shigeoka, S., Ishikawa, T., Tamoi, M., Miyagawa, Y., Takeda, T., Yabuta, Y. and Yoshimura, K. (2002) Regulation and function of ascorbate peroxidase isoenzymes. *Journal of Experimental Botany* 53, 1305–1319.
Singh, A., Prakash, J., Meghwal, P.R. and Ranpise, S.A. (2015) Fig (*Ficus carica*). In: Ghosh, S.N. (ed.) *Breeding of Underutilized Fruit Crops (Part-I)*. Narendra Publishing House, Delhi. pp. 149–179.
Sinha, K.K. (2003) Figs. In: *Encyclopedia of Food Sciences and Nutrition*, Caballero, B. (ed) Academic Press, pp. 2394–2399.
Soliman, H.I.A. and Abd Alhady, M.R.A. (2017) Evaluation of salt tolerance ability in some fig (*Ficus carica* L.) cultivars using tissue culture technique. *Journal of Applied Biology and Biotechnology* 5, 29–39.
Stanley, R.G. and Linskens, H.F. (1974) *Pollen*. Springer-Verlag, New York.
Storey, W.B. (1975) Figs. In: Janick, J. and Moore, J. N. (ed.) *Advances in Fruit Breeding*. Purdue University Press, West Lafayette, IN. pp. 568–589.

Storey, W.B. (1985) *Ficus carica* L. In: Halevy, A.H. (ed.) *Handbook of Flowering (Vol. 3).* CRC Press, Boca Raton, Florida, pp. 9–15.

Storey, W.B., Enderud, J.E., Saleeb, W.F. and Nauer, E.M. (1977) *The Fig: Its Biology, History, Culture and Utilization.* Jurupa Mountains Cultural Center, Riverside, California, pp. 94.

Stover, E.W., Aradhya, M.K., Crisosto, C. and Ferguson, L. (2007) The fig: overview of an ancient fruit. *HortScience* 42, 1083–1087.

Tanriver, E. (2019) Fig production and germplasm in Turkey. In: Kahramanoglu, I. (ed.) *Modern Food Industry*. DOI: 10.5772/intechopen.86997

Tiwari, R., Sudhakar, J.V., Srivastava, A.K., Chaudhary, L.B., Murthy, G.V.S. and Durgapal, A. (2014) Taxonomy, distribution and diversity of *Ficus palmata* Forssk. subsp. *virgata* (Roxb.) Browicz (Moraceae) in India. *Journal of Threatened Taxa* 6, 6172–6185. DOI: 10.11609/JoTT.o3739.6172-85

Trabut, L. (1922) Sur les origines du figuier. *Revue de Botanique Appliquée* 2, 393–396. www.persee.fr/doc/jatba_0370-3681_1922_num_2_12_1412

Trad, M., Ginies, C., Gaaliche, B., Renard, C.M.G.C. and Mars, M. (2012) Does pollination affect aroma development in ripened fig [*Ficus carica* L.] fruit? *Scientia Horticulturae* 134, 93–99. DOI: 10.1016/j.scienta.2011.11.004

Trad, M., Le Bourvellec, C., Gaaliche, B., Ginies, C., Renard, C.M.G.C. and Mars, M. (2013) Caprification modifies polyphenols but not cell wall concentrations in ripe figs. *Scientia Horticulturae* 160, 115–122. DOI: 10.1016/j.scienta.2013.05.037

Trad, M., Ginies, C., Gaaliche, B., Renard, C.M.G.C. and Mars, M. (2014) Relationship between pollination and cell wall properties in common fig fruit. *Phytochemistry* 98, 78–84. DOI: 10.1016/j.phytochem.2013.12.011

Tsantili, E. (1990) Changes during development of 'Tsapela' fig fruits. *Scientia Horticulturae* 44, 227–234. DOI: 10.1016/0304-4238(90)90122-U

Turkmenistan-CBD Report (2009) Turkmenistan Fourth National Report on Implementation of the UN Convention on Biological Diversity at National Level. Ashgabat, Turkmenistan. p. 71.

Valdeyron, G. and Lloyd, D.G. (1979) Sex differences and flowering phenology in the common fig, *Ficus carica* L. *Evolution* 33, 673–685.

Vangelisti, A., Zambrano, L.S., Caruso, G., Macheda, D., Bernardi, R., Usai, G., Mascagni, F., Giordani, T., Gucci, R., Cavallini, A. and Natali, L. (2019) How an ancient, salt-tolerant fruit crop, *Ficus carica* L., copes with salinity: a transcriptome analysis. *Scientific Reports* 9, 2561.

Vego, D. and Miljković, I. (2012) *In vitro* germination of pollen grains of wild fig (*Ficus carica* L. *var caprificus*). *Pomologia Croatica* 18, 19–32. www.hrcak.srce.hr/102046 (accessed 3 September 2021).

Vemmos, S.N., Petri, E. and Stournaras, V. (2013) Seasonal changes in photosynthetic activity and carbohydrate content in leaves and fruit of three fig cultivars (*Ficus carica* L.). *Scientia Horticulturae* 160, 198–207. DOI: 10.1016/j.scienta.2013.05.036

Wang, Z., Cui, Y., Vainstein, A., Chen, S. and Ma, H. (2017) Regulation of fig (*Ficus carica* L.) fruit color: metabolomic and transcriptomic analyses of the flavonoid biosynthetic pathway. *Frontiers in Plant Science* 8, 1990. DOI: 10.3389/fpls.2017.01990

Wang, Z., Song, M., Li, Y., Chen, S. and Ma, H. (2019) Differential color development and response to light deprivation of fig (*Ficus carica* L.) syconia peel and female flower tissues: transcriptome elucidation. *BMC Plant Biology* 19, 217.

Weiblen, G.D. and Bush, G.L. (2002) Speciation in fig pollinators and parasites. *Molecular Ecology* 11, 1573–1578. DOI: 10.1046/j.1365-294X.2002.01529.x

Yablowitz, Z., Nir, G. and Erez, A. (1998) Breba fig production in Israel: regular and pesticide-free systems. *Acta Horticulturae* 480, 137–142. DOI: 10.17660/ActaHortic.1998.480.23

Yaman, B. (2014) Anatomical differences between stem and branch wood of *Ficus carica* L. subsp. *carica*. *Modern Phytomorphology* 6, 79–83.

Yezhov, V.N., Smykov, A.V., Smykov, V.K., Khokhlov, S.Y., Zaurov, D.E., Mehlenbacher, S.A., Molnar, T.J., Goffreda, J.C. and Funk, C.R. (2005). Genetic resources of temperate and subtropical fruit and nut species at the Nikita Botanical Gardens. *HortScience* 40, 5–9. DOI: 10.21273/HORTSCI.40.1.5

Zare, H. (2008) Comparison of fig caprification vessels, period and caprifig cultivar usable in Iran. *Acta Horticulturae* 798, 259–261. DOI: 10.17660/ActaHortic.2008.798.33

Zeybekoğlu, N., Misirli, A. and Gülcan, R. (1998) Researches on pollen germination ability of some caprifig varieties. *Acta Horticulturae* 480, 125–128. DOI: 10.17660/ActaHortic.1998.480.20

Zohary, D. and Hopf, M. (2000) *Domestication of Plants in the Old World*, 3rd edn. Oxford University Press, Oxford, UK.
Zohary, D. and Spiegel-Roy, P. (1975) Beginnings of fruit growing in the old world. *Science* 187, 319–327. DOI: 10.1126/science.187.4174.319
Zulkarnaini, Z.M., Sakimin, S.Z., Mohamed, M.T.M., Jaafar, H.B., Amnah, S. and Mellisa (2020) Anti-transpirant application improves the drought tolerance of fig (*Ficus carica* L.) under optimization of brassinolide. *Asian Journal of Crop Science* 12, 1–11. DOI: 10.3923/ajcs.2020.1.11

# 3 Production and Growing Regions

**G. Ferrara[1*], A. Mazzeo[1], P. Colasuonno[2] and I. Marcotuli[2]**
*[1]Department of Soil, Plant and Food Sciences, University of Bari "Aldo Moro", Bari, Italy; [2]Department of Environmental and Territorial Sciences, University of Bari "Aldo Moro", Bari, Italy*

## 3.1 Introduction

The economic importance of fig production is likely to continue into the future since the growing interest of consumers in this 'minor' fruit species. In the world market, there is an increasing demand for fresh figs and a stable demand for dried figs. The most critical trade concern for fresh fruit is the short shelf-life which limits the long-distance trade, while for dried figs the competition among producers of different countries is very strong (Aksoy and Beghin, 2005). The countries with the largest harvested areas and production in 2019 are shown in Figs. 3.1a and 3.2a, respectively. World fig harvested area and production statistics for the last 50 years are shown in Figs. 3.1b and 3.2b, respectively.

The 2019 world fig production was 1,315,588 metric tonnes on 289,818 ha (FAOSTAT, 2021) and average production 2015–2019 was 1,185,768 on 285,513 ha (Tables 3.1 and 3.2). Approximately 90% of this production was in the countries of Mediterranean basin and Middle East. The most important fig producers are Turkey, Egypt, Morocco, Algeria, Iran, Spain and Syria. The United States (California), Tunisia, Afghanistan, Albania, Brazil, Greece, China, India and Japan are recent (or less recent) and important production areas. General harvested area statistics for the last 50 years are shown in Table 3.1, while the distribution of the production by world regions is reported in Table 3.2. As shown in the tables, the harvested area has dramatically reduced in the last 50 years, by more than 100,000 ha, in particular in countries such as Portugal, Greece, Italy and Spain, while a significant increase in the harvested area is reported in countries such as Morocco, Turkey, Egypt and Iran. World fig production is more stable (~1.0–1.2 million tonnes) with lower fluctuations in the last 50 years with respect to harvested area and with a slight increase in the very last few years. Among the total world production of more than 1.3 million tonnes in 2019, Turkey is the leading country producing 310,000 t (23.6% of the world fig production in 2019), followed by Egypt (225,295 t) and Morocco (153,472 t) (FAOSTAT, 2021). The area of fig cultivation is concentrated in North Africa (Table 3.1), with more than 50% of the entire world's production in five countries (Morocco, Algeria, Egypt, Tunisia and Libya). In the last decade, both the area and fig production were stable, but some trends emerged. There was a reduction in production (and area) in traditional fig producing countries in the Mediterranean basin, Algeria, Tunisia, Italy, Greece, Syria and Lebanon, and an increase in 'new' countries

*Corresponding author: giuseppe.ferrara@uniba.it

 *The Fig: Botany, Production and Uses* (eds. Sarkhosh, A., Yavari, A. and Ferguson, L.)
DOI: 10.1079/9781789242881.0003

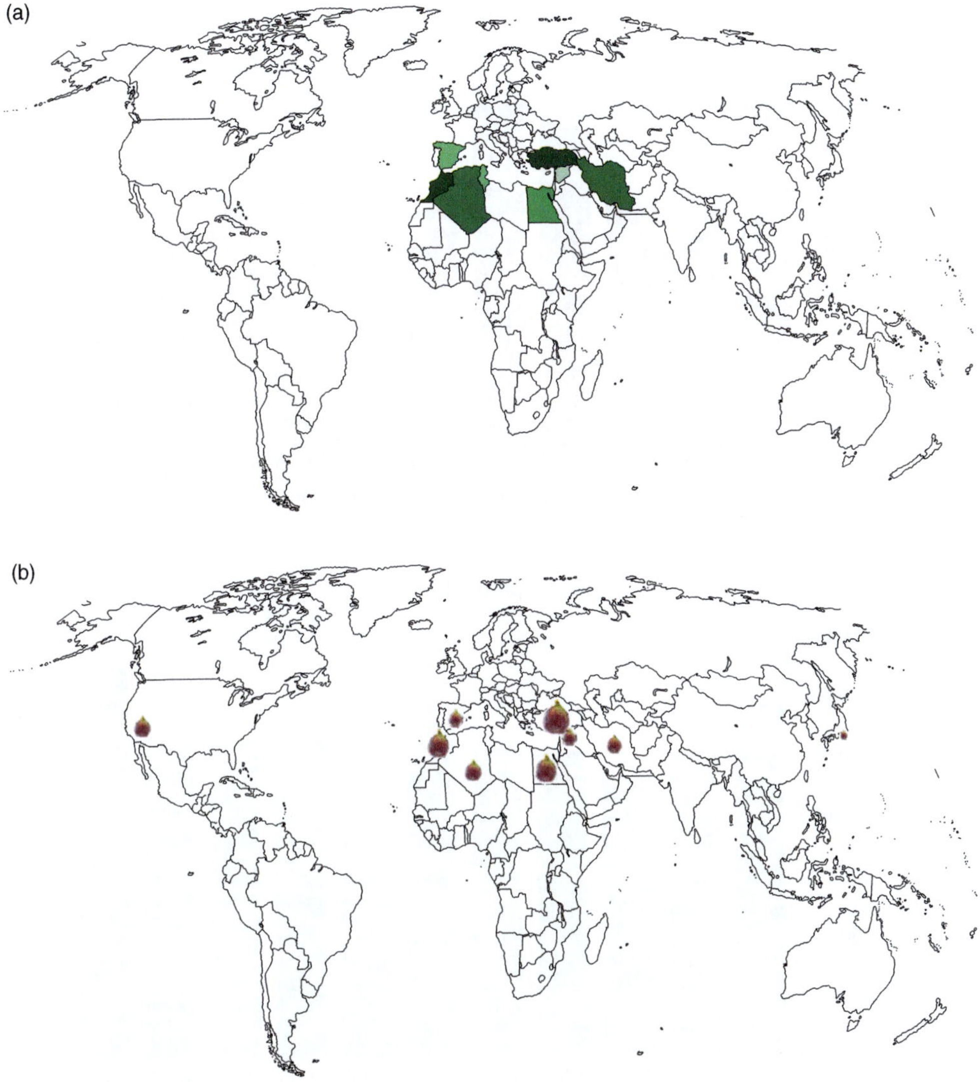


**Fig. 3.1.** (a) Countries with the largest fig harvested area in 2019, (b) countries with the largest fig production in 2019. (Figure created by authors from FAOSTAT data, 2021.)

such as Mexico, China, Colombia, South Africa, Azerbaijan, Uzbekistan, etc. but also in traditional countries such as Turkey, Egypt, Morocco and Iran (Tables 3.1 and 3.2). Data from 2019 indicate a world yield of 4.5 t/ha, a significant increase with respect to 3.1 and 2.7 in 1980 and 1970, respectively.

The world import of fresh figs as an average for 2015–2019 is 63,294 t worth almost 190 million dollars (Table 3.3). The most important countries for imports are in Europe (Germany, France, Austria, Switzerland, United Kingdom) together with the USA, Russia and United Arab Emirates, both for quantity and value in dollars; important importers are also the Netherlands, China, Belgium and Canada (Table 3.3). Only three countries, Germany, France and Austria, import one third of the world's fresh figs both as quantity and value (Table 3.3). The world export of fresh figs as an average for 2015–2019 is 36,904 t worth almost 98 million dollars (Table 3.4). The leading country for world export (quantity)

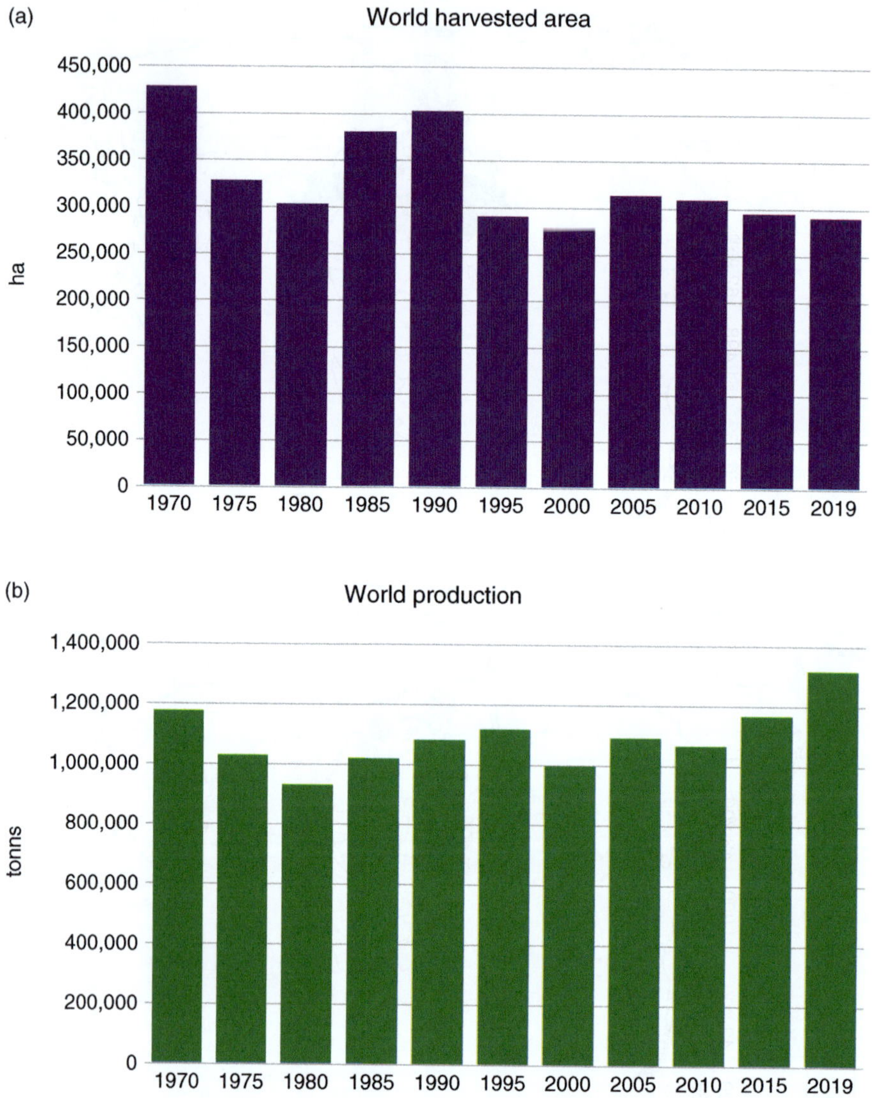


**Fig. 3.2.** (a) World fig harvested area (ha) in the years from 1970 to 2019, (b) world fig production (t) in the years from 1970 to 2019. (Figure created by authors from FAOSTAT data, 2021.)

in 2019 was Saudi Arabia (6699 t) followed by Austria and United Arab Emirates; important exporters are traditional European countries, such as Italy and Spain, but surprisingly also the Netherlands and Germany, where packaging processes probably happen giving a rise in important value of the export (Table 3.4). Countries that export fresh figs (both quantity and value) are Iran, Brazil, Mexico, Afghanistan, Armenia, Bulgaria, Peru and France (Table 3.4) which export fresh figs to different countries in the world (Africa, North America, Middle East). The quantities and values of import/export of dried figs are larger than fresh figs (Tables 3.5 and 3.6). In particular, import of dried figs in 2019 was 87,632 t with a value of almost 400 million dollars (Table 3.5). The most important countries for importing dried figs, both as quantity and value are India, France, Germany, USA, Italy and United Kingdom (Table 3.5). All these countries, except Italy and United Kingdom, show a growing trend in the years 2015–2019,

**Table 3.1.** Countries with the largest areas (ha) of fig production in both the Northern and Southern Hemispheres (FAOSTAT, 2021).

| Country | Area (ha) | | | | | | | | | | | | Mean | Percentage |
|---|---|---|---|---|---|---|---|---|---|---|---|---|---|---|
| | 1970 | 1980 | 1990 | 1995 | 2000 | 2005 | 2010 | 2015 | 2016 | 2017 | 2018 | 2019 | (2015–2019) | |
| Northern Hemisphere | | | | | | | | | | | | | | |
| Morocco | – | – | 39,000 | 42,000 | 43,900 | 43,800 | 49,682 | 55,260 | 58,306 | 60,533 | 61,498 | 62,969 | *59,713* | *20.91* |
| Turkey | 38,206 | 36,875 | 60,338 | 48,993 | 48,300 | 49,000 | 47,857 | 49,718 | 49,987 | 50,330 | 51,389 | 52,116 | *50,708* | *17.76* |
| Algeria | 33,000 | 46,000 | 44,700 | 40,110 | 36,000 | 46,592 | 46,921 | 43,130 | 42,764 | 40,932 | 39,356 | 39,438 | *41,124* | *14.40* |
| Egypt | 2,941 | 3,572 | 13,361 | 18,187 | 22,200 | 28,580 | 31,773 | 27,881 | 27,779 | 27,820 | 29,029 | 31,674 | *28,837* | *10.10* |
| Iran (Islamic Republic of) | 7,000 | 8,000 | 27,549 | 33,447 | 19,786 | 44,804 | 46,581 | 33,921 | 21,167 | 15,209 | 37,749 | 18,655 | *25,340* | *8.88* |
| Tunisia | 13,000 | 17,000 | 14,000 | 13,000 | 19,350 | 18,600 | 17,600 | 19,546 | 15,643 | 17,639 | 17,192 | 16,863 | *17,377* | *6.09* |
| Spain | 30,800 | 23,000 | 26,500 | 19,400 | 19,625 | 19,314 | 11,488 | 12,751 | 12,613 | 13,564 | 13,980 | 14,600 | *13,502* | *4.73* |
| Syrian Arab Republic | 21,629 | 18,793 | 14,600 | 10,700 | 10,650 | 10,000 | 9,714 | 9,400 | 9,400 | 9,429 | 9,422 | 9,435 | *9,417* | *3.30* |
| India | 900 | 870 | 2,100 | 3,349 | 4,543 | 5,366 | 5,407 | 5,649 | 5,632 | 5,665 | 5,700 | 5,734 | *5,676* | *1.99* |
| Portugal | 200,000 | 100,000 | 85,900 | 9,163 | 7,571 | 5,765 | 4,245 | 4,329 | 4,103 | 4,130 | 4,130 | 4,130 | *4,164* | *1.46* |
| Greece | 20,000 | 10,000 | 20,000 | 7,335 | 7,102 | 6,478 | 6,231 | 4,721 | 4,245 | 3,720 | 3,770 | 3,990 | *4,089* | *1.43* |
| Afghanistan | 4,400 | 4,530 | 3,396 | 3,340 | 2,000 | 2,541 | 1,924 | 1,525 | 811 | 1,623 | 1,644 | 3,529 | *1,826* | *0.64* |
| China | – | – | 550 | 600 | 1,200 | 1,300 | 2,225 | 2,583 | 2,687 | 2,773 | 2,856 | 2,937 | *2,767* | *0.97* |
| Libya | – | – | 11,841 | 10,342 | 3,000 | 3,286 | 3,070 | 2,677 | 2,700 | 2,641 | 2,597 | 2,554 | *2,634* | *0.92* |
| United States of America | 6,720 | 5,790 | 6,760 | 5,950 | 6,232 | 5,180 | 3,480 | 2,590 | 2,630 | 2,710 | 2,616 | 2,488 | *2,607* | *0.91* |
| Italy | 39,000 | 13,700 | 12,269 | 6,258 | 8,043 | 3,478 | 2,553 | 2,423 | 2,260 | 2,336 | 2,230 | 2,150 | *2,280* | *0.80* |
| Azerbaijan | – | – | – | 1,372 | 1,164 | 1,336 | 1,539 | 1,738 | 1,820 | 1,808 | 1,826 | 1,880 | *1,814* | *0.64* |
| Albania | – | – | 1,000 | 900 | 1,000 | 1,300 | 1,437 | 1,543 | 1,489 | 1,483 | 1,482 | 1,489 | *1,497* | *0.52* |
| Mexico | 1,597 | 3,988 | 707 | 888 | 825 | 977 | 826 | 1,199 | 1,340 | 1,440 | 1,194 | 1,322 | *1,299* | *0.45* |
| Montenegro | – | – | – | – | – | – | 1,184 | 1,275 | 1,271 | 1,297 | 1,301 | 1,306 | *1,290* | *0.45* |
| Iraq | | | 4,895 | 3,649 | 1,500 | 1,250 | 1,268 | 412 | 447 | 451 | 411 | 1,210 | *586* | *0.21* |
| Japan | – | – | – | – | – | 1,181 | 1,070 | 1,030 | 1,035 | 1,029 | 1,024 | 1,019 | *1,027* | *0.36* |
| Lebanon | 1,376 | 1,800 | 1,200 | 600 | 2,970 | 1,400 | 1,714 | 828 | 456 | 663 | 617 | 757 | *664* | *0.23* |
| Yemen | | 114 | 192 | 374 | 413 | 405 | 499 | 505 | 491 | 485 | 517 | 547 | *509* | *0.18* |
| Bosnia and Herzegovina | – | – | – | 100 | 70 | 150 | 300 | 414 | 415 | 334 | 448 | 500 | *422* | *0.15* |

| | | | | | | | | | | | | | | |
|---|---|---|---|---|---|---|---|---|---|---|---|---|---|---|
| Southern Hemisphere | | | | | | | | | | | | | | |
| Brazil | 3,044 | 5,430 | 3,295 | 2,249 | 2,805 | 2,911 | 2,933 | 2,856 | 2,804 | 2,592 | 2,410 | 2,208 | *2,574* | *0.90* |
| South Africa | 600 | 600 | 560 | 497 | 690 | 700 | 611 | 637 | 682 | 676 | 686 | 610 | *658* | *0.23* |
| Peru | 713 | 646 | 580 | 582 | 598 | 566 | 592 | 541 | 524 | 540 | 563 | 490 | *532* | *0.19* |
| Argentina | – | – | 251 | 274 | 268 | 262 | 268 | 271 | 273 | 274 | 275 | 275 | *274* | *0.10* |
| Bolivia (Plurinational State of) | 70 | 495 | 510 | 535 | 570 | 268 | 248 | 126 | 127 | 131 | 130 | 132 | *129* | *0.05* |
| Colombia | – | – | – | – | 303 | 74 | 108 | 95 | 99 | 112 | 102 | 102 | *102* | *0.04* |
| Australia | 55 | 30 | 30 | 30 | 32 | 43 | 63 | 61 | 53 | 51 | 58 | 67 | *58* | *0.02* |
| Chile | – | – | 0 | 14 | 30 | 56 | 102 | 107 | 104 | 113 | 67 | 65 | *91* | *0.03* |
| **World (Total)** | **427,521** | **302,348** | **402,583** | **290,233** | **277,951** | **312,925** | **308,422** | **294,383** | **278,802** | **277,013** | **287,551** | **289,818** | ***285,513*** | ***100*** |

**Table 3.2.** Countries with the highest fig production (t) in both Northern and Southern Hemisphere (FAOSTAT, 2021).

| | Yield (t) | | | | | | | | | | | | Mean | Percentage |
|---|---|---|---|---|---|---|---|---|---|---|---|---|---|---|
| Country | 1970 | 1980 | 1990 | 1995 | 2000 | 2005 | 2010 | 2015 | 2016 | 2017 | 2018 | 2019 | (2015–2019) | |
| Northern Hemisphere | | | | | | | | | | | | | | |
| Turkey | 214,000 | 205,000 | 300,000 | 300,000 | 240,000 | 285,000 | 254,838 | 300,600 | 305,450 | 305,689 | 306,499 | 310,000 | *305,648* | *25.78* |
| Egypt | 14,000 | 11,000 | 81,485 | 236,915 | 187,698 | 192,250 | 184,972 | 172,474 | 183,794 | 194,202 | 221,369 | 225,295 | *199,427* | *16.82* |
| Morocco | 67,000 | 65,000 | 58,200 | 67,000 | 68,400 | 82,600 | 109,735 | 150,111 | 59,881 | 137,934 | 128,380 | 153,472 | *125,956* | *10.62* |
| Iran (Islamic Republic of) | 13,000 | 15,000 | 77,734 | 69,486 | 57,201 | 87,522 | 52,366 | 72,647 | 53,732 | 50,316 | 111,111 | 130,328 | *83,627* | *7.05* |
| Algeria | 50,411 | 69,410 | 58,390 | 60,008 | 54,326 | 69,799 | 123,763 | 139,137 | 120,348 | 128,684 | 109,214 | 114,092 | *122,295* | *10.31* |
| Spain | 101,900 | 57,300 | 56,961 | 51,884 | 56,014 | 35,295 | 30,351 | 26,479 | 45,718 | 36,380 | 47,750 | 51,600 | *41,585* | *3.51* |
| Syrian Arab Republic | 43,658 | 46,648 | 36,900 | 48,106 | 44,071 | 49,881 | 40,966 | 35,300 | 35,300 | 42,011 | 38,035 | 43,015 | *38,732* | *3.27* |
| United States of America | 44,996 | 41,280 | 45,359 | 47,540 | 50,712 | 44,633 | 37,113 | 28,760 | 29,660 | 28,300 | 28,440 | 28,174 | *28,667* | *2.42* |
| Tunisia | 13,800 | 27,500 | 27,000 | 26,000 | 30,000 | 23,000 | 26,000 | 30,000 | 22,500 | 25,800 | 25,123 | 24,619 | *25,608* | *2.16* |
| Afghanistan | 14,400 | 11,800 | 9,000 | 9,808 | 15,000 | 20,000 | 8,658 | 4,996 | 3,143 | 7,887 | 9,416 | 24,319 | *9,952* | *0.84* |
| Albania | 13,000 | 15,500 | 12,600 | 11,500 | 13,100 | 16,000 | 18,387 | 20,929 | 20,902 | 20,077 | 24,448 | 22,024 | *21,676* | *1.83* |
| Greece | 144,023 | 112,173 | 120,000 | 34,882 | 29,117 | 23,493 | 19,479 | 20,936 | 22,280 | 23,591 | 16,010 | 19,730 | *20,509* | *1.73* |
| China | – | – | 1,000 | 2,000 | 3,000 | 6,000 | 11,051 | 14,389 | 14,957 | 15,713 | 16,469 | 17,225 | *15,751* | *1.33* |
| India | 2,000 | 2,000 | 4,800 | 7,700 | 10,500 | 12,633 | 13,283 | 14,295 | 14,387 | 14,561 | 14,737 | 14,914 | *14,579* | *1.23* |
| Japan | – | – | – | – | – | 15,942 | 14,763 | 14,294 | 14,345 | 14,289 | 14,232 | 14,175 | *14,267* | *1.20* |
| Azerbaijan | – | – | – | 5,791 | 8,283 | 9,503 | 8,534 | 9,701 | 8,820 | 9,263 | 11,215 | 12,074 | *10,215* | *0.86* |
| Italy | 198,000 | 72,700 | 31,510 | 21,439 | 25,000 | 20,091 | 12,022 | 13,112 | 11,297 | 11,363 | 10,650 | 11,830 | *11,650* | *0.98* |
| Libya | 2,363 | 5,301 | 34,665 | 26,920 | 10,000 | 10,350 | 10,291 | 9,674 | 9,855 | 9,760 | 9,717 | 9,673 | *9,736* | *0.82* |
| Mexico | 11,437 | 24,975 | 3,059 | 3,398 | 2,425 | 4,645 | 3,640 | 5,381 | 7,089 | 8,166 | 7,705 | 9,466 | *7,561* | *0.64* |
| Iraq | 6,000 | 12,000 | 18,500 | 16,000 | 7,000 | 7,000 | 8,466 | 3,018 | 3,271 | 3,349 | 3,104 | 9,265 | *4,401* | *0.37* |
| Yemen | – | 620 | 1,435 | 2,874 | 3,482 | 4,014 | 5,338 | 5,127 | 5,028 | 4,875 | 5,212 | 5,593 | *5,167* | *0.44* |
| Uzbekistan | | | | 1,000 | 1,000 | 1,200 | 2,000 | 3,500 | 3,945 | 3,744 | 3,999 | 4,242 | *3,886* | *0.33* |
| Lebanon | 10,842 | 13,000 | 10,000 | 8,500 | 17,800 | 6,900 | 9,259 | 3,953 | 2,109 | 3,999 | 3,974 | 4,048 | *3,617* | *0.31* |
| Montenegro | – | – | – | – | – | – | 4,003 | 3,935 | 3,931 | 3,972 | 3,947 | 3,921 | *3,941* | *0.33* |
| Portugal | 137,900 | 51,000 | 37,000 | 6,313 | 3,496 | 1,967 | 2,957 | 3,039 | 3,161 | 3,402 | 3,740 | 3,380 | *3,344* | *0.28* |

| | | | | | | | | | | | | | | |
|---|---|---|---|---|---|---|---|---|---|---|---|---|---|---|
| Southern Hemisphere | | | | | | | | | | | | | | |
| Brazil | 22,099 | 38,007 | 24,482 | 15,924 | 17,207 | 23,697 | 25,727 | 29,071 | 26,910 | 25,891 | 23,674 | 22,526 | *25,614* | *2.16* |
| Peru | 6,584 | 5,008 | 3,663 | 3,924 | 3,716 | 2,751 | 3,082 | 2,570 | 2,674 | 2,743 | 2,947 | 3,095 | *2,806* | *0.24* |
| Colombia | – | – | 100 | – | 3,030 | 1,480 | 1,580 | 2,606 | 2,360 | 2,473 | 2,574 | 2,413 | *2,485* | *0.21* |
| South Africa | 1,492 | 1,303 | 1,271 | 1,173 | 1,956 | 1,800 | 1,797 | 1,998 | 2,148 | 2,146 | 2,197 | 1,972 | *2,092* | *0.18* |
| Argentina | 5,400 | 3,100 | 1,000 | 898 | 887 | 873 | 873 | 868 | 869 | 869 | 868 | 868 | *868* | *0.07* |
| Bolivia (Plurinational State of) | 450 | 3,185 | 3,010 | 3,250 | 3,600 | 1,686 | 1,524 | 602 | 617 | 657 | 667 | 686 | *646* | *0.05* |
| Chile | – | – | – | 22 | 46 | 69 | 130 | 164 | 163 | 180 | 140 | 104 | *150* | *0.01* |
| Australia | 597 | 116 | 112 | 88 | 76 | 79 | 88 | 62 | 46 | 40 | 40 | 41 | *46* | *0.00* |
| **World (Total)** | **1,173,936** | **928,805** | **1,080,549** | **1,116,953** | **997,044** | **1,090,791** | **1,065,429** | **1,164,849** | **1,062,450** | **1,161,101** | **1,224,852** | **1,315,588** | ***1,185,768*** | ***100*** |

**Table 3.3.** Countries with the largest import of fresh figs as quantity (t) and value ($) (FAOSTAT, 2021).

| | Import (t) | | | | | | | | | | Mean | Percentage | | Import (1000 $) | | | | | | | | | | Mean | Percentage |
|---|---|---|---|---|---|---|---|---|---|---|---|---|---|---|---|---|---|---|---|---|---|---|---|---|---|
| Country | 1990 | 1995 | 2000 | 2005 | 2010 | 2015 | 2016 | 2017 | 2018 | 2019 | (2015–2019) | | Country | 1990 | 1995 | 2000 | 2005 | 2010 | 2015 | 2016 | 2017 | 2018 | 2019 | (2015–2019) | |
| Germany | 1,827 | 2,887 | 2,837 | 3,201 | 5,445 | 7,827 | 6,678 | 8,671 | 8,954 | 9,952 | *8,416* | *13.30* | Germany | 4,079 | 5,431 | 4,734 | 8,165 | 16,557 | 21,180 | 17,055 | 26,222 | 25,438 | 27,604 | *23,500* | *12.44* |
| France | 1,508 | 3,873 | 5,091 | 6,412 | 8,112 | 7,536 | 8,277 | 7,240 | 8,143 | 8,040 | *7,847* | *12.40* | Austria | 391 | 1,124 | 3,471 | 8,910 | 16,184 | 18,122 | 17,080 | 20,715 | 16,663 | 23,021 | *19,120* | *10.12* |
| Austria | 201 | 552 | 2,067 | 5,563 | 6,167 | 5,767 | 6,054 | 6,300 | 5,751 | 7,745 | *6,323* | *9.99* | United States of America | 245 | 1,461 | 802 | 1,569 | 3,713 | 7,813 | 21,130 | 21,065 | 23,270 | 20,662 | *18,788* | *9.94* |
| Russian Federation | – | – | 3,050 | 328 | 1,923 | 750 | 943 | 1,640 | 5,428 | 5,915 | *2,935* | *4.64* | France | 2,640 | 6,637 | 6,736 | 12,294 | 18,736 | 16,283 | 17,572 | 17,057 | 19,794 | 19,233 | *17,988* | *9.52* |
| United Arab Emirates | 560 | 1,145 | 13 | – | 29 | 2,720 | 2,997 | 3,100 | 710 | 3,876 | *2,681* | *4.24* | Switzerland | 1,519 | 1,883 | 2,176 | 2,828 | 5,056 | 17,321 | 17,812 | 19,309 | 19,189 | 18,062 | *18,339* | *9.71* |
| Switzerland | 468 | 663 | 841 | 925 | 1,504 | 3,688 | 3,820 | 3,682 | 3,554 | 3,590 | *3,667* | *5.79* | United Kingdom | 1,222 | 1,257 | 2,779 | 4,623 | 7,270 | 16,381 | 16,951 | 16,142 | 13,734 | 14,550 | *15,552* | *8.23* |
| United Kingdom | 449 | 648 | 1,205 | 1,426 | 1,757 | 3,884 | 5,450 | 4,115 | 2,944 | 3,471 | *3,973* | *6.28* | Netherlands | 704 | 2,038 | 3,222 | 5,325 | 6,078 | 9,021 | 7,378 | 11,463 | 12,058 | 12,566 | *10,497* | *5.56* |
| United States of America | 103 | 1,060 | 427 | 571 | 846 | 1,501 | 3,961 | 3,550 | 3,645 | 3,416 | *3,215* | *5.08* | United Arab Emirates | 470 | 770 | 15 | – | – | 6,400 | 7,074 | 8,590 | 2,243 | 10,170 | *6,895* | *3.65* |
| China | – | 83 | 4,551 | 996 | 60 | 2,464 | 1,924 | 656 | 1,012 | 2,768 | *1,765* | *2.79* | Russian Federation | – | 0 | 628 | 424 | 5,510 | 1,422 | 1,746 | 3,688 | 8,521 | 9,832 | *5,042* | *2.67* |
| Netherlands | 386 | 976 | 1,510 | 1,777 | 1,708 | 2,370 | 1,612 | 2,491 | 2,482 | 2,679 | *2,327* | *3.68* | Canada | 2,069 | 1,847 | 2,953 | 4,476 | 7,518 | 7,506 | 8,469 | 8,413 | 7,838 | 9,058 | *8,257* | *4.37* |
| Canada | 1,069 | 720 | 1,239 | 1,430 | 1,850 | 1,749 | 1,987 | 2,088 | 1,546 | 2,008 | *1,876* | *2.96* | Belgium | 982 | 1,992 | 1,848 | 3,291 | 5,750 | 5,401 | 6,196 | 5,872 | 6,812 | 5,462 | *5,949* | *3.15* |
| Ethiopia | | | | 1 | 1 | 1 | – | – | – | 1,643 | *329* | *0.52* | China | 1 | 34 | 1,993 | 915 | 248 | 3,993 | 3,759 | 1,954 | 3,173 | 4,279 | *3,432* | *1.82* |
| Belgium | 529 | 1,141 | 1,133 | 1,470 | 1,836 | 1,724 | 2,005 | 1,935 | 1,797 | 1,611 | *1,814* | *2.87* | Bulgaria | – | – | – | – | 83 | 135 | 794 | 501 | 3,097 | 2,757 | *1,457* | *0.77* |
| Djibouti | – | – | – | – | – | 4 | 1,437 | 2,513 | 1,790 | 1,542 | *1,457* | *2.30* | Italy | 25 | 166 | 383 | 1,017 | 1,029 | 1,579 | 1,532 | 2,505 | 1,794 | 2,747 | *2,031* | *1.08* |
| Kuwait | – | 711 | 552 | 67 | 43 | 672 | 978 | 1,311 | 1,444 | 1,479 | *1,177* | *1.86* | Poland | – | 7 | – | 378 | 306 | 1,200 | 1,732 | 1,355 | 2,503 | 2,613 | *1,881* | *1.00* |
| Syrian Arab Republic | – | – | – | – | – | 255 | – | 28 | 23 | 1,113 | *284* | *0.45* | Republic of Korea | 66 | 612 | 799 | 1,097 | 1,004 | 1,705 | 1,684 | 1,964 | 2,158 | 2,495 | *2,001* | *1.06* |
| Bulgaria | – | – | – | – | 43 | 59 | 320 | 247 | 916 | 931 | *495* | *0.78* | Ethiopia | – | – | – | 2 | 5 | 1 | – | – | – | 1,981 | *396* | *0.21* |
| Italy | 6 | 62 | 171 | 585 | 403 | 526 | 533 | 808 | 640 | 904 | *682* | *1.08* | Croatia | – | – | – | 164 | 22 | 1,416 | 1,938 | 2,181 | 2,139 | 1,936 | *1,922* | *1.02* |
| Slovakia | – | – | – | 1 | 130 | 53 | 78 | 52 | 119 | 817 | *224* | *0.35* | Qatar | 96 | 63 | 27 | 152 | 127 | 2,359 | 2,065 | 2,676 | 2,120 | 1,755 | *2,195* | *1.16* |
| Poland | – | 11 | – | 98 | 92 | 453 | 627 | 347 | 653 | 752 | *566* | *0.89* | Djibouti | – | – | – | – | – | 2 | 1,006 | 2,690 | 1,517 | 1,743 | *1,392* | *0.74* |
| Sweden | – | 200 | 177 | 257 | 316 | 425 | 547 | 670 | 649 | 671 | *592* | *0.94* | Sweden | – | 471 | 366 | 907 | 1,462 | 1,647 | 1,680 | 1,589 | 1,416 | 1,510 | *1,568* | *0.83* |
| Croatia | – | – | – | – | 10 | 546 | 715 | 764 | 687 | 603 | *663* | *1.05* | Kuwait | – | 296 | 361 | 56 | 12 | 318 | 536 | 1,549 | 2,820 | 1,477 | *1,340* | *0.71* |
| Hungary | 0 | 1 | 0 | 196 | 234 | 691 | 61 | 701 | 809 | 540 | *560* | *0.89* | Ireland | 22 | 7 | 109 | 100 | 254 | 427 | 310 | 668 | 387 | 1,429 | *644* | *0.34* |
| Somalia | – | – | – | – | – | – | 1,815 | 2,136 | 1,373 | 535 | *1,172* | *1.85* | Mexico | – | – | 22 | – | – | 4,568 | 4,194 | 5,955 | 4,316 | 1,406 | *4,088* | *2.16* |
| Qatar | 80 | 269 | 65 | 360 | 345 | 990 | 1,178 | 928 | 588 | 513 | *839* | *1.33* | Czechia | – | 878 | 16 | 127 | 590 | 1,315 | 1,052 | 1,369 | 1,475 | 1,355 | *1,313* | *0.70* |
| Republic of Korea | 35 | 454 | 829 | 709 | 281 | 397 | 407 | 457 | 445 | 510 | *443* | *0.70* | Hungary | – | 2 | – | 329 | 534 | 1572 | 134 | 1,439 | 1,563 | 1,277 | *1,197* | *0.63* |
| Ireland | 6 | 1 | 40 | 22 | 71 | 139 | 85 | 242 | 82 | 498 | *209* | *0.33* | Oman | – | – | – | – | 4 | 119 | 171 | 128 | 131 | 1,177 | *345* | *0.18* |
| Armenia | – | – | – | 2 | – | – | – | 118 | 363 | 497 | *196* | *0.31* | Spain | 10 | 6 | 74 | 175 | 182 | 563 | 418 | 862 | 961 | 1,051 | *771* | *0.41* |
| Czechia | – | 381 | 6 | 50 | 189 | 490 | 461 | 568 | 482 | 492 | *499* | *0.79* | Denmark | 75 | 192 | 448 | 842 | 1,367 | 1,019 | 1,035 | 921 | 1,395 | 960 | *1,066* | *0.56* |
| Mexico | – | – | 13 | – | – | 1,699 | 1,720 | 2,305 | 1,535 | – | *1,452* | 2.29 | Syrian Arab Republic | – | – | – | – | – | 202 | – | 108 | 82 | 883 | *255* | *0.13* |
| **World** | **9,299** | **18,137** | **28,206** | **29,838** | **36,004** | **52,870** | **62,074** | **64,650** | **63,389** | **73,485** | ***63,294*** | ***100*** | **World** | **15,333** | **29,121** | **35,441** | **61,936** | **104,793** | **158,244** | **172,096** | **200,798** | **199,312** | **214,160** | ***188,922*** | ***100*** |

**Table 3.4.** Countries with the largest export of fresh figs as quantity (t) and value ($) (FAOSTAT, 2021).

| Country | Export (t) | | | | | | | | | | Mean | Percentage |
|---|---|---|---|---|---|---|---|---|---|---|---|---|
| | 1990 | 1995 | 2000 | 2005 | 2010 | 2015 | 2016 | 2017 | 2018 | 2019 | (2015–2019) | |
| Saudi Arabia | 996 | 1,106 | – | – | 274 | 76 | 5,324 | 10,057 | 6,564 | 6,699 | 5,744 | 18.60 |
| Austria | 33 | 71 | 1,870 | 4,795 | 5,740 | 4,959 | 4,819 | 5,192 | 4,787 | 5,588 | 5,069 | 16.41 |
| United Arab Emirates | – | – | 17 | 1 | 12 | 1,245 | 657 | 255 | 372 | 3,551 | 1,216 | 3.94 |
| Spain | 136 | 1,303 | 1,818 | 1,525 | 2,091 | 3,288 | 2,545 | 2,839 | 3,448 | 3,309 | 3,086 | 9.99 |
| Netherlands | 131 | 848 | 1,066 | 2,313 | 2,367 | 2,313 | 1,872 | 2,870 | 2,614 | 2,977 | 2,529 | 8.19 |
| Italy | 1,113 | 1,950 | 1,390 | 1,448 | 2,463 | 3,730 | 3,102 | 3,024 | 3,089 | 1,612 | 2,911 | 9.43 |
| Iran (Islamic Republic of) | – | – | – | 180 | – | 785 | – | – | 109 | 1,577 | 494 | 1.60 |
| Brazil | 543 | 601 | 707 | 837 | 817 | 1,365 | 1,191 | 1,307 | 1,394 | 1,386 | 1,329 | 4.30 |
| Germany | 429 | 54 | 284 | 545 | 1,195 | 2,097 | 1,415 | 1,241 | 1,510 | 1,173 | 1,487 | 4.81 |
| Mexico | 219 | 405 | 272 | – | – | 905 | 1,125 | 1,252 | 1,267 | 1,044 | 1,119 | 3.62 |
| Afghanistan | – | – | – | 7 | 5 | 246 | 554 | 477 | 9 | 901 | 437 | 1.42 |
| Armenia | – | – | – | – | – | – | – | – | 371 | 893 | 253 | 0.82 |
| Bulgaria | – | – | – | – | 10 | 25 | 105 | 96 | 795 | 857 | 376 | 1.22 |
| Peru | – | – | 97 | 55 | 67 | 59 | 267 | 201 | 723 | 801 | 410 | 1.33 |
| France | 189 | 140 | 481 | 745 | 946 | 1,079 | 804 | 765 | 607 | 742 | 799 | 2.59 |
| Hungary | – | 1 | – | 193 | 166 | 625 | 0 | 1,136 | 801 | 488 | 610 | 1.97 |
| Greece | 217 | 121 | 220 | 169 | 102 | 290 | 313 | 527 | 441 | 453 | 405 | 1.31 |
| Belgium | 82 | 339 | 968 | 1,573 | 401 | 681 | 782 | 643 | 443 | 410 | 592 | 1.92 |
| Syrian Arab Republic | 100 | 94 | – | – | 63 | 42 | 223 | 223 | 77 | 377 | 188 | 0.61 |
| Lebanon | 460 | 200 | 122 | 170 | 112 | – | 127 | – | – | 287 | 83 | 0.27 |
| Uzbekistan | – | – | 64 | – | – | – | 2 | 56 | 236 | 216 | 102 | 0.33 |
| Egypt | – | 20 | 9 | 43 | – | 51 | 82 | 47 | 25 | 201 | 81 | 0.26 |
| Switzerland | | | | | | 104 | 167 | 56 | 29 | 171 | 105 | 0.34 |
| Argentina | – | 52 | – | 109 | 112 | 107 | 91 | 122 | 66 | 118 | 101 | 0.33 |
| Albania | – | – | – | – | 14 | 259 | 252 | 108 | 20 | 115 | 151 | 0.49 |
| China | 18 | – | 39 | 123 | 3 | 37 | 51 | 69 | 86 | 107 | 70 | 0.23 |
| Pakistan | 18 | – | – | 33 | 126 | 242 | 241 | 131 | 91 | 87 | 158 | 0.51 |
| Portugal | 4 | 2 | – | – | 64 | 42 | 78 | 85 | 9 | 83 | 59 | 0.19 |
| Chile | 139 | 207 | 125 | 56 | 111 | 123 | 89 | 97 | 94 | 82 | 97 | 0.31 |
| India | – | – | – | – | – | 6 | 54 | 63 | 55 | 59 | 47 | 0.15 |
| **World** | **8,237** | **13,337** | **11,050** | **25,016** | **17,857** | **25,564** | **27,173** | **33,923** | **30,881** | **36,904** | **30,889** | **100** |

| Country | Export (1000 $) | | | | | | | | | | Mean | Percentage |
|---|---|---|---|---|---|---|---|---|---|---|---|---|
| | 1990 | 1995 | 2000 | 2005 | 2010 | 2015 | 2016 | 2017 | 2018 | 2019 | (2015–2019) | |
| Austria | 79 | 82 | 2,426 | 7,333 | 14,907 | 14,990 | 13,205 | 17,496 | 13,649 | 16,194 | 15,107 | 17.44 |
| Netherlands | 531 | 2,727 | 3,113 | 7,441 | 12,591 | 12,323 | 10,838 | 15,770 | 15,608 | 15,342 | 13,976 | 16.14 |
| Spain | 263 | 2,555 | 2,637 | 2,973 | 3,886 | 7,092 | 6,130 | 6,097 | 7,862 | 8,475 | 7,137 | 8.24 |
| Brazil | 596 | 1,253 | 1,269 | 2,215 | 4,473 | 6,940 | 6,304 | 6,608 | 6,940 | 6,709 | 6,700 | 7.74 |
| Saudi Arabia | 480 | 793 | – | – | 312 | 233 | 3,813 | 7,625 | 5,548 | 6,371 | 4,718 | 5.45 |
| Mexico | 316 | 1,473 | 656 | – | – | 3,777 | 4,458 | 5,410 | 6,047 | 4,955 | 4,929 | 5.69 |
| Peru | – | – | 363 | 215 | 362 | 431 | 1,596 | 1,567 | 5,033 | 4,850 | 2,695 | 3.11 |
| Germany | 1,346 | 168 | 574 | 1,255 | 3,643 | 6,207 | 4,347 | 4,160 | 5,460 | 4,164 | 4,868 | 5.62 |
| United Arab Emirates | 150 | 76 | – | – | – | 2,869 | 2,097 | 1,318 | 403 | 4,046 | 2,147 | 2.48 |
| Italy | 1,590 | 2,906 | 1,927 | 2,322 | 4,836 | 6,736 | 6,173 | 6,945 | 6,737 | 3,931 | 6,104 | 7.05 |
| France | 412 | 458 | 1,052 | 2,472 | 4,428 | 4,365 | 3,556 | 3,421 | 3,178 | 3,547 | 3,613 | 4.17 |
| Bulgaria | – | – | – | – | 29 | 78 | 273 | 288 | 2,844 | 2,792 | 1,255 | 1.45 |
| Afghanistan | – | – | – | 2 | 13 | 420 | 570 | 862 | 67 | 2,207 | 825 | 0.95 |
| Belgium | 204 | 654 | 1,305 | 3,342 | 2,815 | 3,252 | 3,222 | 2,554 | 2,668 | 2,156 | 2,770 | 3.20 |
| Greece | 733 | 434 | 638 | 567 | 395 | 1,003 | 1,121 | 1,809 | 1,815 | 1,789 | 1,507 | 1.74 |
| Iran (Islamic Republic of) | 0 | 560 | – | – | – | 1,210 | – | – | 422 | 1,247 | 576 | 0.66 |
| Hungary | – | 1 | – | 315 | 343 | 1,371 | – | 2,283 | 1,466 | 1,007 | 1,225 | 1.41 |
| Switzerland | 1 | 0 | 30 | 6 | 21 | 528 | 794 | 319 | 212 | 985 | 568 | 0.66 |
| Armenia | – | – | – | – | – | – | – | – | 205 | 906 | 222 | 0.26 |
| Chile | 101 | 180 | 119 | 154 | 393 | 932 | 682 | 648 | 595 | 692 | 710 | 0.82 |
| China | 5 | 1 | 138 | 164 | 9 | 280 | 393 | 636 | 586 | 631 | 505 | 0.58 |
| Lebanon | 100 | 70 | 28 | 49 | 100 | – | 109 | – | – | 505 | 123 | 0.14 |
| Argentina | – | 70 | – | 201 | 470 | 347 | 394 | 700 | 358 | 475 | 455 | 0.53 |
| Syrian Arab Republic | – | – | – | 126 | – | 52 | 292 | 230 | 148 | 442 | 233 | 0.27 |
| Pakistan | 16 | – | – | 54 | 210 | 559 | 668 | 503 | 469 | 356 | 511 | 0.59 |
| India | – | – | – | – | – | 44 | 363 | 448 | 428 | 350 | 327 | 0.38 |
| Egypt | – | 6 | 4 | 21 | – | 73 | 109 | 205 | 146 | 301 | 167 | 0.19 |
| Czechia | – | – | – | 3 | 84 | 133 | 212 | 208 | 105 | 225 | 177 | 0.20 |
| Croatia | – | – | – | 57 | – | 256 | 296 | 508 | 501 | 223 | 357 | 0.41 |
| Uzbekistan | – | – | 64 | 1 | – | – | 2 | 49 | 161 | 211 | 85 | 0.10 |
| **World** | **10,578** | **20,108** | **17,509** | **44,751** | **56,190** | **78,492** | **73,904** | **91,058** | **91,627** | **98,007** | **86,618** | **100** |

**Table 3.5.** Countries with the largest import of dried figs as quantity (t) and value ($) (FAOSTAT, 2021).

| Country | Import (t) 1990 | 1995 | 2000 | 2005 | 2010 | 2015 | 2016 | 2017 | 2018 | 2019 | Mean (2015–2019) | Percentage (2015–2019) |
|---|---|---|---|---|---|---|---|---|---|---|---|---|
| India | 463 | 930 | 1,816 | 2,651 | 4,561 | 10,927 | 8,411 | 12,726 | 9,725 | 11,935 | *10,745* | *12.61* |
| France | 6,512 | 6,738 | 7,375 | 9,746 | 8,553 | 9,340 | 10,564 | 9,654 | 10,107 | 9,517 | *9,836* | *11.54* |
| United States of America | 4,833 | 8,782 | 3,817 | 6,177 | 6,127 | 7,292 | 7,240 | 6,977 | 7,750 | 8,846 | *7,621* | *8.94* |
| Germany | 8,340 | 8,740 | 9,531 | 9,993 | 8,290 | 7,890 | 8,328 | 8,341 | 7,996 | 7,965 | *8,104* | *9.51* |
| Italy | 5,480 | 6 292 | 6,089 | 5,641 | 4,838 | 5,206 | 4,749 | 3,977 | 3,659 | 4,333 | *4,385* | *5.15* |
| United Kingdom | 3,756 | 2,172 | 2,295 | 3,529 | 2,264 | 5,377 | 4,185 | 3,871 | 2,856 | 3,927 | *4,043* | *4.74* |
| Saudi Arabia | 1,166 | 0 | 503 | 1,157 | 823 | 1,715 | 2,915 | 4,032 | 3,060 | 3,352 | *3,015* | *3.54* |
| Belarus | – | – | – | – | – | 525 | 2,283 | 2,730 | 2,882 | 3,272 | *2,338* | *2.74* |
| China | 449 | 2,925 | 6,186 | 5,354 | 922 | 2,149 | 2,298 | 2,439 | 2,427 | 2,269 | *2,316* | *2.72* |
| Turkey | – | 757 | 1,015 | 1,037 | 713 | 876 | 503 | 824 | 951 | 2,205 | *1,072* | *1.26* |
| Austria | 2,212 | 1,597 | 1,405 | 1,801 | 1,482 | 1,839 | 1,878 | 1,588 | 1,923 | 2,186 | *1,883* | *2.21* |
| Canada | 1,111 | 1,287 | 1,389 | 1,925 | 1,748 | 2,115 | 2,006 | 2,280 | 2,589 | 2,113 | *2,221* | *2.61* |
| Netherlands | 454 | 1,005 | 1,281 | 2,582 | 1,655 | 2,281 | 2,288 | 2,662 | 2,496 | 2,046 | *2,355* | *2.76* |
| Australia | 1,203 | 980 | 1,075 | 1,413 | 1,159 | 2,433 | 2,481 | 1,777 | 1,879 | 1,773 | *2,069* | *2.43* |
| Iraq | – | – | – | – | – | 589 | 1,332 | 1,053 | 1,112 | 1,675 | *1,152* | *1.35* |
| Japan | 196 | 678 | 1,894 | 1,383 | 1,130 | 1,756 | 1,456 | 1,681 | 1,535 | 1,598 | *1,605* | *1.88* |
| Belgium | 689 | 885 | 1,220 | 1,069 | 1,137 | 873 | 1,159 | 1,023 | 1,116 | 1,357 | *1,106* | *1.30* |
| Portugal | 640 | 1,047 | 1,640 | 1,983 | 1,308 | 1,224 | 1,257 | 992 | 1,124 | 1,233 | *1,166* | *1.37* |
| Ukraine | – | – | – | 968 | 700 | – | – | – | – | 1,192 | *238* | *0.28* |
| Denmark | 795 | 891 | 983 | 1,196 | 1,146 | 1,325 | 1,009 | 1,161 | 1,139 | 1,036 | *1,134* | *1.33* |
| Spain | 1,943 | 1,670 | 2,373 | 2,514 | 1,324 | 1,204 | 1,019 | 817 | 981 | 989 | *1,002* | *1.18* |
| Israel | 1,053 | 750 | 1,120 | 1,423 | 954 | 797 | 1,005 | 927 | 856 | 912 | *899* | *1.06* |
| Slovakia | – | 141 | 329 | 1,320 | 638 | 595 | 791 | 710 | 872 | 892 | *772* | *0.91* |
| Poland | – | 494 | 538 | 1,116 | 785 | 1,149 | 939 | 932 | 915 | 890 | *965* | *1.13* |
| Egypt | 19 | 302 | 1,539 | 911 | 1,030 | – | 639 | 442 | 691 | 857 | *526* | *0.62* |
| Serbia | 1,362 | – | 77 | 711 | 760 | 810 | 1,048 | 790 | 801 | 849 | *860* | *1.01* |
| Sweden | 1,274 | 948 | 991 | 1,254 | 746 | 913 | 799 | 879 | 873 | 647 | *822* | *0.96* |
| Romania | – | 103 | 338 | 503 | 420 | 617 | 569 | 603 | 568 | 493 | *570* | *0.67* |
| Bosnia and Herzegovina | – | – | – | 474 | 350 | 364 | 350 | 437 | 389 | 428 | *394* | *0.46* |
| Czechia | – | – | 697 | 832 | 609 | 532 | 692 | 732 | 513 | 403 | *574* | *0.67* |
| World | **50,609** | **58,218** | **66,333** | **87,818** | **72,152** | **84,862** | **83,174** | **86,165** | **84,272** | **87,632** | ***85,221*** | ***100*** |

| Country | Import (1000 $) 1990 | 1995 | 2000 | 2005 | 2010 | 2015 | 2016 | 2017 | 2018 | 2019 | Mean (2015–2019) | Percentage (2015–2019) |
|---|---|---|---|---|---|---|---|---|---|---|---|---|
| India | 554 | 1,309 | 2,639 | 13,420 | 32,737 | 77,742 | 59,678 | 101,277 | 81,219 | 96,123 | *83,208* | *21.81* |
| France | 12,922 | 13,989 | 14,010 | 22,006 | 32,763 | 39,528 | 42,520 | 41,301 | 47,198 | 40,987 | *42,307* | *11.09* |
| Germany | 16,810 | 17,082 | 14,966 | 22,120 | 3,2260 | 36,489 | 35,517 | 38,621 | 39,231 | 35,169 | *37,005* | *9.70* |
| United States of America | 6,807 | 11,943 | 6,029 | 8,750 | 11,029 | 21,757 | 19,906 | 20,688 | 23,796 | 23,818 | *21,993* | *5.76* |
| Italy | 10,753 | 13,025 | 10,997 | 13,106 | 19,002 | 21,217 | 20,530 | 19,366 | 17,959 | 19,570 | *19,728* | *5.17* |
| China | 1,547 | 6,975 | 7,938 | 8,878 | 6,048 | 14,085 | 16,218 | 17,549 | 16,841 | 14,412 | *15,821* | *4.15* |
| Canada | 2,624 | 2,884 | 3,100 | 5,057 | 8,258 | 10,622 | 11,922 | 13,310 | 16,239 | 14,327 | *13,284* | *3.48* |
| United Kingdom | 5,121 | 4,401 | 4,585 | 76,56 | 8,001 | 14,762 | 12,301 | 13,638 | 12,521 | 13,202 | *13,285* | *3.48* |
| Japan | 494 | 2,281 | 5,515 | 4,985 | 6,537 | 12,626 | 9,614 | 11,511 | 11,626 | 11,870 | *11,449* | *3.00* |
| Austria | 2,568 | 2,286 | 1,738 | 4,421 | 6,475 | 9,041 | 9,426 | 8,650 | 11,223 | 11,482 | *9,964* | *2.61* |
| Turkey | – | 260 | 1,818 | 2,075 | 3,027 | 4248 | 2,305 | 3,473 | 4,832 | 10,577 | *5,087* | *1.33* |
| Belarus | – | – | – | – | – | 1,619 | 4,132 | 4,215 | 5,444 | 9,239 | *4,930* | *1.29* |
| Netherlands | 867 | 2,095 | 2,073 | 5,401 | 6,005 | 9,278 | 9,837 | 12,172 | 11,521 | 9,011 | *10,364* | *2.72* |
| Australia | 1,808 | 2,164 | 2,087 | 3,444 | 4,897 | 8,765 | 8,249 | 8,370 | 8,366 | 7,847 | *8,319* | *2.18* |
| Saudi Arabia | 509 | 0 | 365 | 264 | 981 | 3,245 | 5,022 | 7,315 | 8,024 | 7,528 | *6,227* | *1.63* |
| Belgium | 1,454 | 1,889 | 2,195 | 2,631 | 4,966 | 4,053 | 5,753 | 5,593 | 5,789 | 5,697 | *5,377* | *1.41* |
| Egypt | 31 | 444 | 1,861 | 1,028 | 3,558 | – | 2,327 | 1,836 | 4,497 | 4,579 | *2,648* | *0.69* |
| Denmark | 1,511 | 1,747 | 1,684 | 2,824 | 4,872 | 6,436 | 4,591 | 5,762 | 5,396 | 4,133 | *5,264* | *1.38* |
| Spain | 3,817 | 3,232 | 3,854 | 5,232 | 4,933 | 4,576 | 4,150 | 3,322 | 3,939 | 3,967 | *3,991* | *1.05* |
| Portugal | 997 | 1,830 | 2,298 | 3,887 | 3,639 | 3,165 | 3,232 | 3,178 | 3,498 | 3,779 | *3,370* | *0.88* |
| Israel | 2,372 | 1,888 | 2,504 | 3,088 | 3,900 | 3,829 | 4,419 | 4,667 | 4,264 | 3,659 | *4,168* | *1.09* |
| Sweden | 2,822 | 2,214 | 1,905 | 3,127 | 3,421 | 5,171 | 4,323 | 5,241 | 5,593 | 3,646 | *4,795* | *1.26* |
| Slovakia | – | 266 | 466 | 3,013 | 2,256 | 2,163 | 3,350 | 2,944 | 3,870 | 3,592 | *3,184* | *0.83* |
| Poland | 0 | 747 | 710 | 1,954 | 2,726 | 4,332 | 3,195 | 3,495 | 3,769 | 3,292 | *3,617* | *0.95* |
| Iraq | – | – | – | – | – | 1,997 | 3,705 | 3,207 | 2,177 | 3,158 | *2,849* | *0.75* |
| Singapore | – | 251 | 516 | 554 | 1,834 | 1,824 | 2,488 | 2,868 | 2,756 | 2,648 | *2,517* | *0.66* |
| Serbia | – | – | – | – | 945 | 1,649 | 2,292 | 2,028 | 1,764 | 2,041 | *1,955* | *0.51* |
| Ukraine | – | – | – | 451 | 1,365 | – | – | – | – | 1,981 | *396* | *0.10* |
| Czechia | – | – | 988 | 1,653 | 2,611 | 2,031 | 2,679 | 3,172 | 2,403 | 1,703 | *2,398* | *0.63* |
| Romania | – | 116 | 66 | 735 | 901 | 2,175 | 1,794 | 2,200 | 2,152 | 1,670 | *1,998* | *0.52* |
| **World** | **90,773** | **109,081** | **110,345** | **180,410** | **271,324** | **362,431** | **342,246** | **400,802** | **403,722** | **398,769** | ***381,594*** | ***100*** |

**Table 3.6.** Countries with the largest export of dried figs as quantity (t) and value ($) (FAOSTAT, 2021).

| Country | Export (t) 1990 | 1995 | 2000 | 2005 | 2010 | 2015 | 2016 | 2017 | 2018 | 2019 | Mean (2015–2019) | Percentage (2015–2019) |
|---|---|---|---|---|---|---|---|---|---|---|---|---|
| Turkey | 33,532 | 35,657 | 43,066 | 52,595 | 58,252 | 67,186 | 69,683 | 79,236 | 78,370 | 84,923 | 75,880 | 61.69 |
| Iran (Islamic Republic of) | 50 | 2,610 | 10,184 | 8,933 | 5,072 | 6,967 | 8,886 | 11,093 | 13,307 | | 10,063 | 8.18 |
| Afghanistan | 205 | 600 | 1,750 | 2,393 | 1,671 | 6,246 | 7,585 | 11,945 | 9,593 | 11,778 | 9,429 | 7.67 |
| Spain | 883 | 2,593 | 2,856 | 3,851 | 2,699 | 3,523 | 3,819 | 4,734 | 4,476 | 4,037 | 4,118 | 3.35 |
| Syrian Arab Republic | 5,220 | 3,162 | 2,635 | 2,090 | 4,286 | 1,436 | 2,460 | 3,436 | 3,435 | 3,938 | 2,941 | 2.39 |
| Greece | 4,095 | 6,569 | 4,210 | 2,527 | 1,454 | 3,491 | 2,855 | 3,958 | 2,694 | 3,158 | 3,231 | 2.63 |
| Belarus | – | – | – | – | – | 438 | 2,177 | 2,597 | 2,708 | 3,124 | 2,209 | 1.80 |
| Germany | 601 | 1,008 | 1,597 | 1,651 | 1,453 | 1,768 | 1,936 | 1,937 | 1,997 | 2,641 | 2,056 | 1.67 |
| United States of America | 2,323 | 7,651 | 2,649 | 4,733 | 5,960 | 4,539 | 4,721 | 4,520 | 3,639 | 2,301 | 3,944 | 3.21 |
| Netherlands | 45 | 457 | 631 | 1,528 | 1,930 | 1,362 | 1,742 | 1,576 | 1,441 | 1,196 | 1,463 | 1.19 |
| Austria | 31 | 189 | 163 | 631 | 741 | 858 | 1,784 | 798 | 816 | 1,015 | 1,054 | 0.86 |
| France | 125 | 626 | 857 | 1,495 | 721 | 1,061 | 1,102 | 884 | 866 | 882 | 959 | 0.78 |
| Slovakia | – | – | 25 | 584 | 366 | 441 | 375 | 538 | 668 | 704 | 545 | 0.44 |
| Georgia | – | – | – | – | – | 1 | 67 | 248 | 180 | 562 | 212 | 0.17 |
| South Africa | 12 | 6 | 99 | 66 | 73 | 547 | 536 | 591 | 558 | 539 | 554 | 0.45 |
| India | – | 38 | 81 | 50 | 112 | 126 | 51 | 167 | 91 | 535 | 194 | 0.16 |
| United Kingdom | 208 | 254 | 154 | 55 | 93 | 279 | 288 | 289 | 930 | 440 | 445 | 0.36 |
| Israel | – | 10 | 41 | 546 | 443 | 823 | 736 | 581 | 497 | 417 | 611 | 0.50 |
| Italy | 521 | 246 | 382 | 521 | 670 | 402 | 358 | 360 | 334 | 311 | 353 | 0.29 |
| Belgium | 29 | 140 | 615 | 261 | 436 | 221 | 371 | 460 | 279 | 300 | 326 | 0.27 |
| Mexico | – | – | – | 659 | 616 | 724 | – | – | – | 286 | 202 | 0.16 |
| Tunisia | – | – | – | – | 64 | 1,858 | 723 | 411 | 330 | 283 | 721 | 0.59 |
| Denmark | 75 | 30 | 91 | 71 | 192 | 522 | 443 | 378 | 371 | 272 | 397 | 0.32 |
| Canada | 125 | 2 | 2 | 61 | 5 | 9 | 89 | 78 | 259 | 184 | 124 | 0.10 |
| Albania | 445 | – | – | – | 155 | – | – | – | – | 138 | 28 | 0.02 |
| China | 30 | 1,565 | 4,885 | 3,563 | 176 | 299 | 194 | 271 | 143 | 108 | 203 | 0.17 |
| Singapore | – | 18 | 115 | 43 | 58 | 77 | 108 | 66 | 55 | 99 | 81 | 0.07 |
| Portugal | 243 | 57 | 308 | 36 | 91 | 99 | 116 | 124 | 105 | 80 | 105 | 0.09 |
| Iraq | – | – | – | – | – | – | – | 6 | 3 | 75 | 17 | 0.01 |
| Slovenia | – | – | 25 | 22 | 480 | 4 | 16 | 11 | 84 | 67 | 36 | 0.03 |
| **World** | **49,599** | **64,167** | **78,797** | **94,387** | **92,351** | **108,667** | **116,952** | **133,192** | **131,269** | **124,928** | ***123,002*** | ***100*** |

| Country | Export (1000 $) 1990 | 1995 | 2000 | 2005 | 2010 | 2015 | 2016 | 2017 | 2018 | 2019 | Mean (2015–2019) | Percentage (2015–2019) |
|---|---|---|---|---|---|---|---|---|---|---|---|---|
| Turkey | 62,387 | 72,601 | 68,038 | 105,077 | 184,678 | 252,180 | 249,501 | 298,505 | 288,950 | 286,348 | 275,097 | 59.62 |
| Afghanistan | 319 | 800 | 2,400 | 11,734 | 6,847 | 24,607 | 33,861 | 81,552 | 67,716 | 85734 | 58,694 | 12.72 |
| Iran (Islamic Republic of) | 65 | 65 | 7,156 | 9,119 | 19,181 | 26,836 | 36,063 | 27,632 | 33,612 | | 31,036 | 6.73 |
| Germany | 1,467 | 2,039 | 2,967 | 5,339 | 7,813 | 10,364 | 10,701 | 11,180 | 12,875 | 14,612 | 11,946 | 2.59 |
| Greece | 6,751 | 13,005 | 7,684 | 5,858 | 5,422 | 11,513 | 10,156 | 13,957 | 10,798 | 14,523 | 12,189 | 2.64 |
| United States of America | 5,358 | 9,949 | 7,720 | 12,376 | 21,618 | 17,527 | 19,225 | 16,859 | 17,477 | 13,472 | 16,912 | 3.67 |
| Spain | 1,089 | 2,984 | 2,799 | 4,800 | 5,094 | 8,552 | 9,871 | 10,806 | 12,210 | 12,017 | 10,691 | 2.32 |
| Syrian Arab Republic | 11,295 | 3,406 | 2,431 | 1,560 | 5,244 | 1,649 | 2,945 | 5,321 | 8,995 | 8,947 | 5,571 | 1.21 |
| Austria | 66 | 459 | 330 | 1,927 | 3,628 | 5,089 | 10,095 | 4,731 | 5,451 | 6,021 | 6,277 | 1.36 |
| Netherlands | 108 | 959 | 1,179 | 3,233 | 4,120 | 4,072 | 6,639 | 7,329 | 7,207 | 5,432 | 6,136 | 1.33 |
| France | 351 | 1,970 | 2,836 | 6,528 | 3,896 | 5,875 | 6,057 | 5,550 | 5,650 | 5,422 | 5,711 | 1.24 |
| South Africa | 17 | 5 | 36 | 50 | 443 | 2,851 | 3,188 | 3,172 | 3,260 | 3,586 | 3,211 | 0.70 |
| Slovakia | – | – | 29 | 1,311 | 1,707 | 1,950 | 1,924 | 2,569 | 3,411 | 3,304 | 2,632 | 0.57 |
| Belarus | – | – | – | – | – | 260 | 1,495 | 1,805 | 1,813 | 2,418 | 1,558 | 0.34 |
| Italy | 891 | 685 | 930 | 1,604 | 3,484 | 2,944 | 2,576 | 2,666 | 2,597 | 2,156 | 2,588 | 0.56 |
| Israel | 3 | 46 | 197 | 1,205 | 2,214 | 3,341 | 3,085 | 2,676 | 2,341 | 1,782 | 2,645 | 0.57 |
| Denmark | 199 | 54 | 160 | 324 | 1,106 | 3,289 | 2,619 | 2,543 | 2,588 | 1,754 | 2,559 | 0.55 |
| Belgium | 74 | 458 | 1,250 | 743 | 1,927 | 1,207 | 1,449 | 2,515 | 1,586 | 1,627 | 1,677 | 0.36 |
| United Kingdom | 386 | 288 | 336 | 272 | 491 | 1,481 | 1,352 | 1,167 | 2,555 | 1,458 | 1,603 | 0.35 |
| Mexico | – | – | – | 2,134 | 4,454 | 3,071 | – | – | – | 1,440 | 902 | 0.20 |
| China | 102 | 2,797 | 4,468 | 3,608 | 900 | 1,400 | 835 | 1,520 | 1,009 | 737 | 1,100 | 0.24 |
| Singapore | – | 67 | 207 | 94 | 371 | 462 | 734 | 453 | 344 | 633 | 525 | 0.11 |
| Georgia | – | – | – | – | – | – | 35 | 441 | 197 | 609 | 256 | 0.06 |
| India | – | 35 | 45 | 27 | 64 | 121 | 108 | 337 | 325 | 572 | 293 | 0.06 |
| Tunisia | – | – | – | – | 120 | 1,862 | 1,335 | 1,292 | 650 | 480 | 1,124 | 0.24 |
| Canada | 83 | 7 | 5 | 173 | 17 | 51 | 185 | 200 | 601 | 364 | 280 | 0.06 |
| Slovenia | – | – | 29 | 36 | 2,011 | 20 | 33 | 72 | 539 | 355 | 204 | 0.04 |
| Portugal | 372 | 140 | 490 | 165 | 528 | 385 | 444 | 454 | 468 | 295 | 409 | 0.09 |
| Hungary | – | 2 | – | 17 | 129 | 73 | 196 | 153 | 179 | 277 | 176 | 0.04 |
| Czechia | – | – | 54 | 118 | 432 | 549 | 641 | 484 | 505 | 241 | 484 | 0.10 |
| **World** | **92,719** | **113,879** | **114,791** | **184,268** | **293,564** | **396,902** | **419,611** | **510,674** | **501,094** | **478,703** | ***461,397*** | ***100*** |

thus indicating an interest by consumers for such processed fruit. With regards to the world export of dried figs, data from 2019 report 124,928 t worth almost 480 million dollars (Table 3.6). Turkey is the great leader in this market (~68% in 2019) followed by countries such as Iran, Afghanistan, Spain, Syria and Greece for quantity, together with Germany, USA, Austria, the Netherlands and France for value because of the processing activities in such non-producing countries (except USA).

## 3.2 The Fig in the Northern Hemisphere

The fig was among the first tree fruits domesticated by humans. Fig trees were cultivated in the Mediterranean basin for more than 6000 years. Fresh and dried figs are a staple of Middle Eastern and Mediterranean cuisine. Fig cultivars are well adapted to dry and warm climates, but some cold-hardy cultivars are known to exist in colder climates. During fruit development and ripening, higher temperatures are required for adequate sugar accumulation and proper maturation. Rain or high humidity during summer exerts a negative impact on fruit quality. For commercial sun-drying (very common in many Mediterranean areas), a long, dry and hot summer is a prerequisite. From the Mediterranean area, figs moved towards the New World, in particular to California (USA) with the Spanish colonization and successively with other emigrants (Italians, Greeks, etc.). Almost all fig production (>95%) is located in the Northern Hemisphere (Table 3.2), primarily in Mediterranean countries.

### Turkey

Turkey is the world leader in fig production and consumption with an annual production in 2019 of 310,000 t on an area of 52,116 ha (ranks #2 in the world for harvested area) and a yield of 5.9 t/ha (FAOSTAT, 2021). Turkey produced 23.56% of the global fig output in 2019 (25.78% on a 5 year average) on 17.98% of world area (17.76% on a 5 year average) (FAOSTAT, 2021) and over half of the world's dried figs (Yilmaz *et al.*, 2017). Both harvested area and fig production significantly increased in the last 50 years (Tables 3.1 and 3.2). Turkey is the world's largest fig exporter (~90% of its production) with more than 70% sent to EU markets (Çalişkan and Polat, 2008). Around 30% of figs are consumed as fresh fruit, and the rest (70%) are eaten as dried figs either locally or worldwide. The world dried fig export was 124,928 t in 2019 (Table 3.6) with 84,923 t exported from Turkey. This constitutes the greatest amount of Turkish dried fig production and 68.0% and 61.7% of the world's exports in 2019 and as a 5 year average value, respectively (FAOSTAT, 2021). Turkey contributes almost half of the whole world dried fig production (FAOSTAT, 2021). The amount of dried fig exports from Turkey doubled in 20 years and the pattern seems to be still growing, with a strong increase in the value in the last 10 years (Table 3.6). No recent data are available for the export of fresh figs (Table 3.4) (FAOSTAT, 2021).

Fig production (from growing to processing) employs about 250–300,000 people, making it an important sector in the Turkish agricultural economy. Fig production is concentrated in the coastal belt of Turkey facing the Mediterranean Sea and often fig trees are interplanted with other fruit species such as olive. In particular, the most important areas are the large and small Menderes valleys, where dried fig production is a common and traditional activity. The Marmara, Mediterranean, Black Sea and Southeast Anatolia regions produce fresh figs (Yilmaz *et al.*, 2017). In particular, fig cultivation is located in Aydin, Bursa, Izmir, Mersin, Hatay, Balikesir and Gaziantep provinces (Fig. 3.3). 60–70% of the harvested area is located in Aydin province with 62% and 75% production of fresh and dried figs, respectively (Arpaci, 2017, Uzundumlu *et al.*, 2018). Dried figs are mainly produced in the Aegean region, especially in Aydin and Izmir provinces, while fresh figs are mainly farmed in the Southern Marmara region, especially in Bursa province (Uzundumlu *et al.*, 2018). Nowadays 80% of the fig orchards are established on slope areas (hilly and mountainous lands), moving from more favorable traditional flat areas, now used for other crops (olive, pomegranate, citrus, plum), due to the increase in irrigation potentiality and the establishment of geothermal energy plants in fig production areas (Arpaci, 2017).

Thirty percent of the fig production in Turkey is consumed fresh in the domestic market

**Fig. 3.3.** Production regions in Turkey. (Author's own figure.)

and 70% is consumed dried either in domestic or foreign markets. The primary cultivar is Sarilop (Calymirna), grown in the Küçük and Büyük Menderes river basins in the Aegean region of Turkey, mainly used for the production of dried figs since this cultivar accounts for 90% of Turkish fig production; only a small amount of Sarilop is consumed as fresh fruit. Other fig cultivars used for the production of dried figs are Keten Köyneği (also as fresh figs) in the Sanlurfa region, Halebi and Sultani in the Gaziantep region, Melli (also as fresh figs) and Mut. Fresh figs are produced in the areas of Bursa, Içel (Mut), Manisa (Turgutlu-Salihli), Burdur, Adana and Hatay. The Bursa Siyahi (Black Bursa) is the fresh fig most requested in global markets followed by Yediveren, Goklop, Morguz, Sari Zeybek (fresh and dried), and Yesilguz (Polat and Çalışkan, 2017). Several local cultivars do exist including Patlican, Siyah and Beyaz in the Black Sea region (Karadeniz, 2003).

## Egypt

In Egypt, figs are the major fruit produced for local consumption. Egypt was the largest producing country in Africa in 2019 producing 225,295 t on 31,674 ha and a yield of 7.1 t/ha (FAOSTAT, 2021). Egypt produced 17.13% of global fig output on a 10.93% area in 2019 (16.82% and 10.10%, respectively, on a 5 year average) and ranked second for world fig production and fourth for harvested area in 2019. Both harvested area and fig production have more than doubled in the last 30 years; however the pattern of production has been stable in the last few years (Tables 3.1 and 3.2). Egypt's export (81 t, as 2015–2019 average) and import of fresh figs and export of dried figs is very low, whereas there is a certain amount (526 t, as 2015–2019 average) of imported dried figs used particularly during Ramadan (FAOSTAT, 2021).

The main fig production area is on the northwestern coast, from Alexandria to Mersa Matruh and Sidi Barani (Fig. 3.4). Fig is also cultivated in the Sinai governorate and Siwa oasis, but the Matrouh governorate produces around 90% of Egyptian fresh figs. These areas are characterized by a dry climate with 180 mm average annual precipitation (Weather and Climate, 2020). The most important Egyptian fig cultivar is Soltani (Sultani), also known as Fayoumi, Ramadi, Barshoumy, Sidi Gaber and Hegazi. Soltani is considered the local standard cultivar. Other cultivars are Kommathri, Adasi-Abiad, Bakori, Adasi-Ahmer, Hammori, Asuani (Assouany), Koummassi, Abboudi and Kahramani (El Rayes, 1995). All these cultivars are common-type and produce figs parthenocarpically. These cultivars are used primarily for fresh consumption and much less as a dried product, since dried figs are also imported from other countries.

## Morocco

Morocco reported 153,472 t of figs produced on 62,969 ha and a yield of 2.4 t/ha in 2019 (FAOSTAT, 2021). Morocco produced 11.67% of

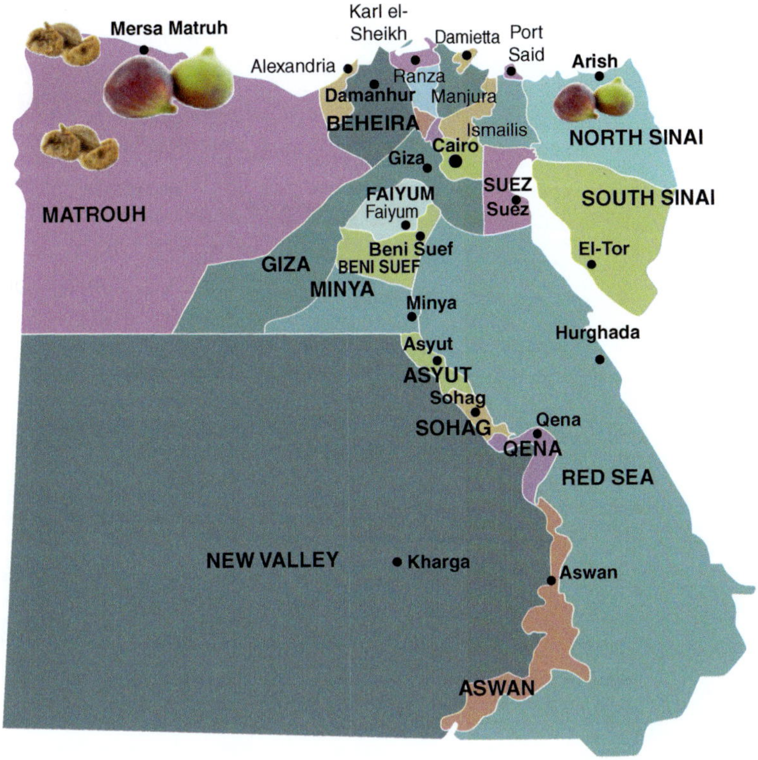


**Fig. 3.4.** Production regions in Egypt. (Author's own figure.)

global fig output on 21.73% of area in 2019 (10.62% and 20.91%, respectively, on a 5 year average) and ranked third for world fig production and first for harvested area in 2019 (Tables 3.1 and 3.2). Since 1990, the harvested area has increased by more than 20,000 hectares whereas fig production more than doubled; however, fig production shows very large fluctuations from year to year (Tables 3.1 and 3.2). Morocco's export of fresh figs in the most recent years was between 20 and 80 t (FAOSTAT, 2021), whereas export of dried figs, although very low in amount and value, showed a significant increase in the years 2015–2017 followed by a sharp decrease in the years 2018–2019 (FAOSTAT, 2021).

The Moroccan fig culture is ancient, and has particular importance in the traditional agroecosystems of the heterogeneous northern Riffian Mountains and oases where ~90% of fig orchards are planted (Fig. 3.5). The Riffian region is a very heterogeneous environment (climate, soil, slopes, crops, etc.) where fig is one of the most important agricultural products (Gharbaoui, 1981). Within this northern region, 40% of the fig tree area is in the Taounate region, with 1.7 million trees on 21,820 ha and an annual production of around 30,000 t, around 20–25% of Moroccan production (Table 3.2). Taounate is the heart of the fig sector, where the fig cultivation is essential for the economy of the area. In Morocco, there are many local cultivars but some cultivars have been introduced, such as Marseilles, Brunswick, Brown Turkey and Celeste. The main cultivars in Morocco are Ghouddane (Rond and Oblong), Ournakssi, Filalia and minor cultivars include Bioudi, El-Quoti Lezregs, Boussbatti Hamra, Hafer Jemel, El Quoti Lebied Fassi, Aboucherchaou, Hafer el Baeo, Homrana, Mounouacha, Lamdar Labiad, Lamdar Lakhal, Sbaa Ourkoud, Qoti Labiad, Qoti Lakhal and Sebi (El Rayes, 1995).

## Iran

Iran produced 130,328 t of figs on 18,655 ha and a yield of 7.0 t/ha in 2019 (FAOSTAT,

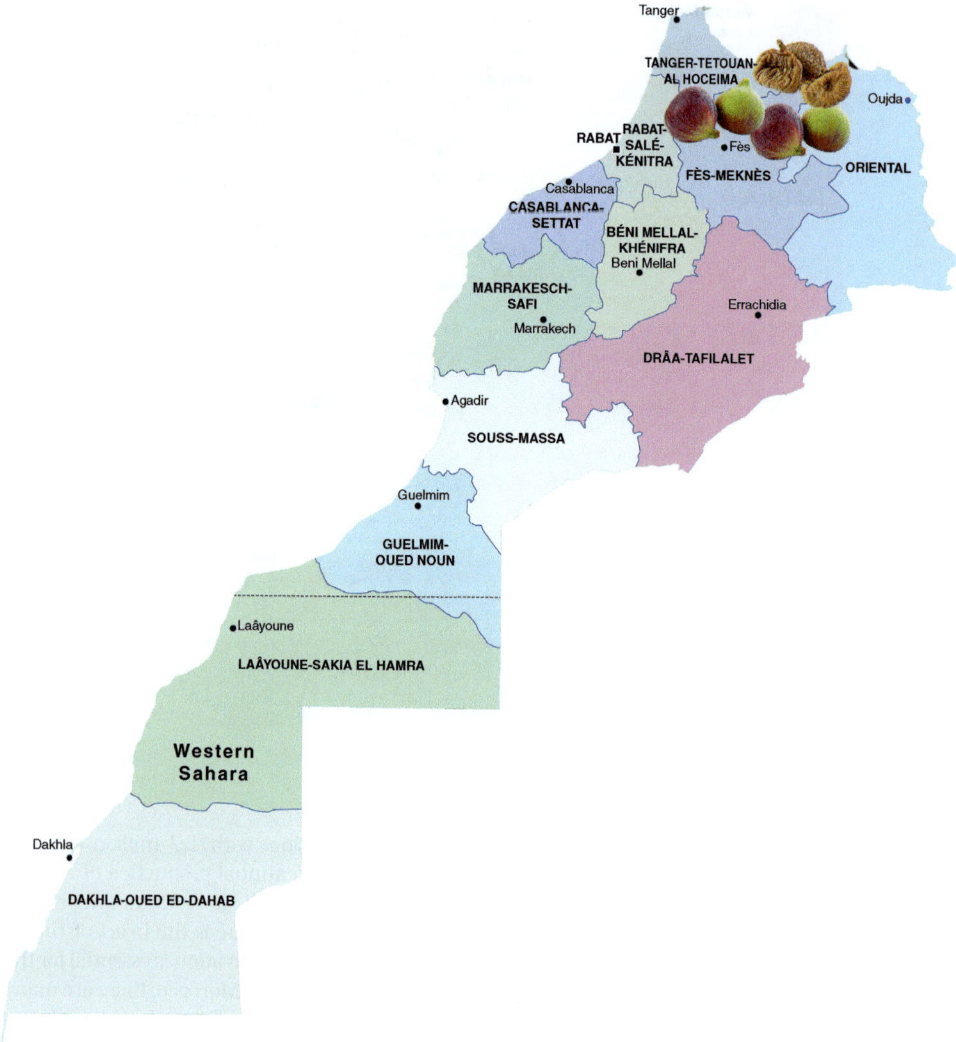


**Fig. 3.5.** Production regions in Morocco. (Author's own figure.)

2021). Iran produced 9.91% of global fig output on 6.44% of area in 2019 (7.05% and 8.88%, respectively, on a 5 year average) and ranks fourth for world fig production and fifth for harvested area in 2019 (Tables 3.1 and 3.2). Since 1990, the harvested area has reduced by almost 10,000 hectares but fig production has increased significantly in the last few years (Tables 3.1 and 3.2). Despite the significant amount of fig production, fresh fig export is very limited and fluctuates from year to year, but with a noteworthy increase in 2019 (1577 t, ranking seventh for fresh fig export). The export of dried figs has grown in the last few years with more than 13,000 t in 2018 ranking Iran second for dried fig exports in the world for quantity and third for value (Tables 3.4 and 3.6).

Figs are grown all over the country except for the Hamadan province (Fig. 3.6) because of the very cold winters. About half of Iran's fig production is from Fars province, followed by Lorestan, Kermanshah and Semnan provinces (Ministry of Agriculture Jihad, 2019). Most of the orchards in Fars province are under rain-fed

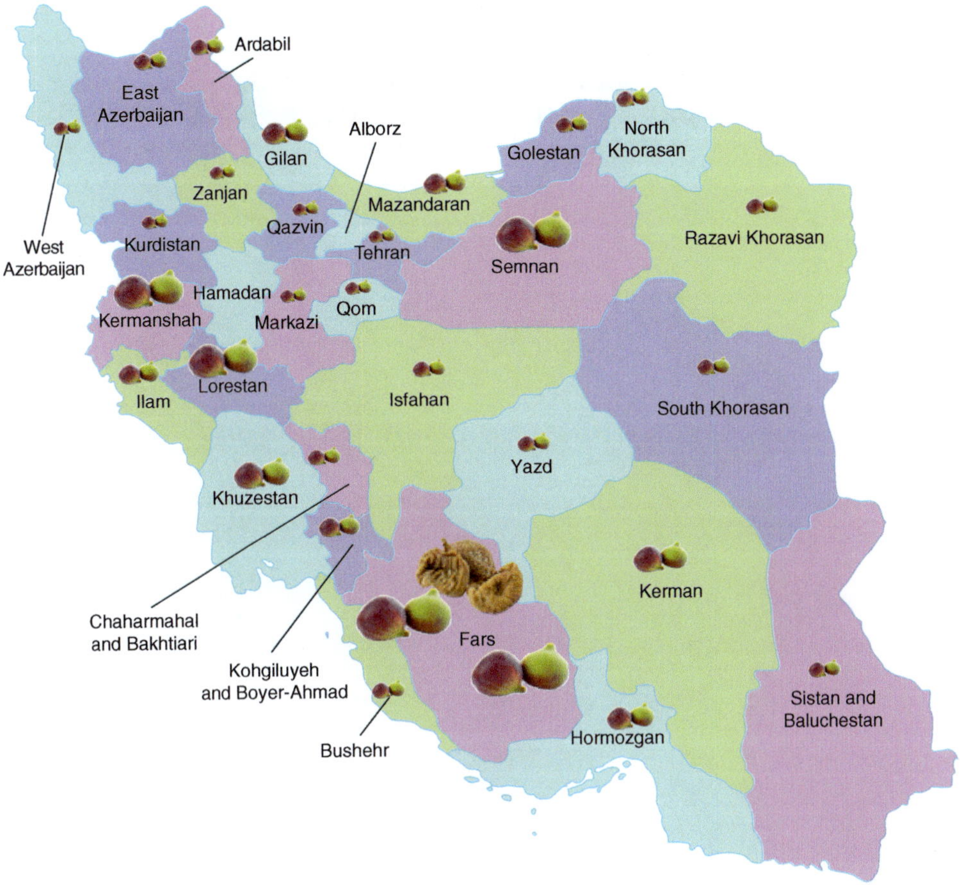


**Fig. 3.6.** Production regions in Iran. (Author's own figure.)

management. Estahban city in Fars province is important area for dried fig production in Iran. The most important fig cultivars in Iran are Sabz, Siah, Shah Anjir, Meshki, Rono, Matti, Kashki, Siah-e-Dorosht, Peyves and Manjifi (Fars province), Zard-e-Piazi (Markazi province), Meshki (Lorestan province) and Manjifi (Kermanshah province).

## Algeria

Algeria produced 114,092 t of figs on 39,438 ha and a yield of 2.8 t/ha in 2019 (FAOSTAT, 2021). Algeria produced 8.67% of global fig output on a 13.61% area in 2019 (10.31% and 14.40%, respectively, on a 5 year average) and ranked fifth for world fig production and third for harvested area in 2019 (Tables 3.1 and 3.2). In the last 50 years, the harvested area has followed a stable pattern (small increase in 1970–1980) whereas fig production has reached 120,000 t in the last 10 years compared to 50,000–70,000 t for the previous decades (Tables 3.1 and 3.2). Although this is important fig production, fresh and dried figs are mostly used for domestic consumption since export is inconsistent (FAOSTAT, 2021).

Algeria has a rich genetic diversity of fig cultivars, mostly poorly studied. The fig area is in the northern mountain regions of Bejaia, Tizi Ouzou and Sétif, where fig is the leading fruit crop after olive (Fig. 3.7). The Algerian Ministry

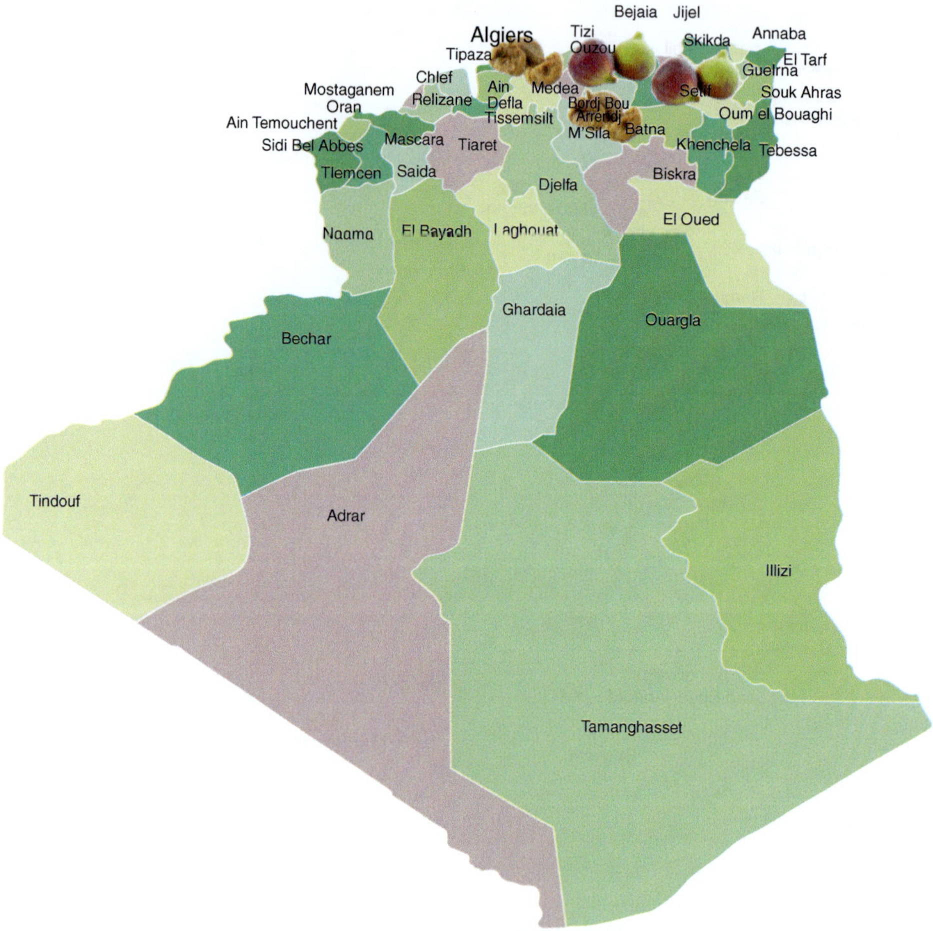


**Fig. 3.7.** Production regions in Algeria. (Author's own figure.)

of Agriculture has undertaken several fruit crop development programmes to increase interest in improving fig production, including dried figs for export. Recently, the Technical Institute of Fruit-bearing Arboriculture in Algeria characterized 40 cultivars (Meziant *et al.*, 2015; Mahmoudi *et al.*, 2018). However, cultivar numbers in Algeria are likely to be much greater, as many different cultivars have local names. The main local cultivars in Algeria are represented by both uniferous types (Abiarous SK, Bidha, Hamra, Onk Elhamam, Zarrouk, Chatwi, Boughandjo, Safra, El fessi, Alekak SK, Hafer lebghal, Abiarous Be, Chetoui, Abgaiti, Avouhvoul, Taarlit, Taranimt and Tameriout) and biferous types (Bakkor, Bither, Roudane and Bejaia).

## Spain

Spain produced 3.92% of global fig output on a 5.04% area with a yield of 3.5 t/ha in 2019 (3.5% and 4.73%, respectively, on a 5 year average) and ranked sixth in world fig production and seventh for harvested area in 2019 (Tables 3.1 and 3.2). Spanish fig production has gradually decreased from ~30,000 ha and ~100,000 t per year in the 1970s to 41,585 t on 13,502 ha in 2015–2019 but there has been a growing interest in this fruit crop in recent years (FAOSTAT, 2021). In the last few years the harvested area for fig cultivation has been almost stable (12,000–13,000 ha), reaching 14,600 ha in 2019 and with a remarkable increase (from 25,000 to 30,000 t up to 50,000 t) in fig production in just a few years (Tables 3.1 and 3.2). Fresh fig production is either used for domestic consumption or for export, since Spain is amongst the most important fresh fig exporters (3309 t and ~8.5 million dollars in 2019, ranked fourth and third for quantity and value, respectively) (Table 3.4). The export of Spanish dried figs is even more consistent both for quantity and value, with more than 4000 t and 12 million dollars in 2019 (Table 3.6). In the last few years, the pattern of fig production has shown a slight increase, with positive effects on export of both fresh and dried figs. Small quantities of dried figs are also imported; less than 1000 t in 2019 (Table 3.5).

The past importance of the fig is evident in the Spanish literature, which mentions the 'famous figs' from Turón, black and white types (Perez-Jimenez *et al.*, 2012). The fig was very important in the past in Spain for human consumption due to the nutritional properties of the fruit. Limited information on commercial cultivars, little research on cultivation techniques, lower production costs in other countries, and limited harvest and postharvest fruit handling time all contributed to this decline (Pereira *et al.*, 2017) but things have changed recently.

The main fig fruit-producing areas in Spain (Fig. 3.8) are Extremadura–Andalusia and neighboring regions (54%), the Levantine area, including the Balearic Islands (26%), and Galicia (14%). In these regions, all the fig fruits

**Fig. 3.8.** Production regions in Spain. (Author's own figure.)

(fresh/dry consumption and industrial use) are harvested by hand at the firm-mature stage from May to October (Pérez-Sánchez *et al.*, 2016). The two main fig gene banks established in Spain are Finca La Orden-Valdesequera (Badajoz, Extremadura) and Campo Experimental de Son Mut Nou (Mallorca, Balearic Islands). The current research on the fig carried out in Spain focuses on both improving the production (yield) and harvesting of high-quality dried figs. The objective of the harvesting research is to develop a system, with nets or meshes, that eliminates both hand harvesting and contact with soil and its mycotoxin-producing fungi (CICYTEX, 2018). Superintensive training systems, such as for olive and almonds, are also being investigated. In Spain, the cultivars destined for fresh consumption are Albacor, San Antonio, Banane, Cuello Dama Blanco, Cuello Dama Negro, Tiberio, Tres Voltas L'Any, Colar Elche, Brown Turkey, Blanca Bétera and De Rey. Albacor, Cuello Dama Blanco, De Rey and Colar Elche are dual-use, fresh and dried. Minor cultivars are Lampaga, Moscatel Negra, Napolitana Negra, Negra Cabezuela, Picholetera and Calabacita. New cultivars are also in field evaluation. All cultivars, except Tres Voltas L'Any, are in the Official Register of Commercial Fig Varieties, although only Cuello Dama Blanco, Colar Elche and Brown Turkey are produced commercially (Pereira *et al.*, 2017).

## Syrian Arab Republic

Syria produced 43,015 t of figs on 9,435 ha and a yield of 4.6 t/ha in 2019 (FAOSTAT, 2021). Syria produced 3.27% of global fig output on 3.26% of area in 2019 (3.27% and 3.30%, respectively, on a 5 year average) and ranked seventh for world fig production and eight for harvested area in 2019 (Tables 3.1 and 3.2). In the last 50 years the harvested area for fig cultivation has reduced by half (more than 10,000 ha) but with a limited decrease in fig production (Tables 3.1 and 3.2). Fresh fig production is used for domestic consumption and 1113 t of fresh figs were imported and 377 t exported in 2019 (Tables 3.3 and 3.4). The export of dried figs is more consistent with almost 4000 t in 2019 (ranked fifth), although this has significantly reduced in the last 30 years (~1,000 t); in the 1990s Syria was the second for dried fig exports in the world both for quantity and value (Table 3.6).

Syria has a fig gene bank maintained by the Arab Center for the Studies of Arid Zones and Dry Lands (ASCAD). The Syrian trade of dried figs has recently shown a slight increase with export of dried figs mainly to Egypt, Jordan, Lebanon, Malaysia and the Gulf countries. The dried fig industry is located in northern Ariha and al-ZawIya Mountain of Idlib Province (SANA, 2020).

Fig trees grow wild in the coastal mountains near Latakia and in areas close to Damascus, Aleppo and Idlib (Fig. 3.9). Many landraces distributed in Syria are adapted to the local climate, some of them for fresh consumption and others for dried fig production, accounting for at least 150 cultivars. Some local varieties are Alzabaly, Sultani, Shenshary, Kalze, Khany, Halaby, Sukkari, Bshari, Ghozlani, Khoudairi, Baiadi, Ahmar, Aswad jalladi, Zerkani, Shebli, Shami, Riadi, Oussely, Giziany, Sammaky, Bourkati, Batraky and Zoaibli (El Rayes, 1995).

Cultivars are also classified according to the skin color of fruit: green-yellow (Khudeiri, Sultani, Khani, Halabi, Sefrawi, Esseli, Anzouki, Boukrati, Birtati, Sumacki, Shtawi, Asfar and White Shami), brown-red (Zeibily-Sammaki, Afani and Shincher) and black-violet (Oubeidi, Habachi, Zamouhi, Assouad, Zaraki and Shiblawi) (El Rayes, 1995).

## United States

The United States produced 28,174 t of figs on 2,488 ha and a yield of 11.3 t/ha in 2019 (FAOSTAT, 2021). It produced 2.14% of the global fig output on 0.86% of area in 2019 (2.42% and 0.91%, respectively, on a 5 year average) and ranked eighth for world fig production and only 15th for harvested area in 2019 (Tables 3.1 and 3.2). In USA the harvested area for fig cultivation has reduced more than half in the last 20 years (since 2000) with a similar decrease in fig production which accounted for more than 50,000 t in the year 2000 (Tables 3.1 and 3.2). Fresh figs are also imported, with an increasing pattern in the last few years, 3416 t and more than 20 million dollars in 2019, ranked eighth and third for quantity and value, respectively (Table 3.3). Dried figs are either imported or

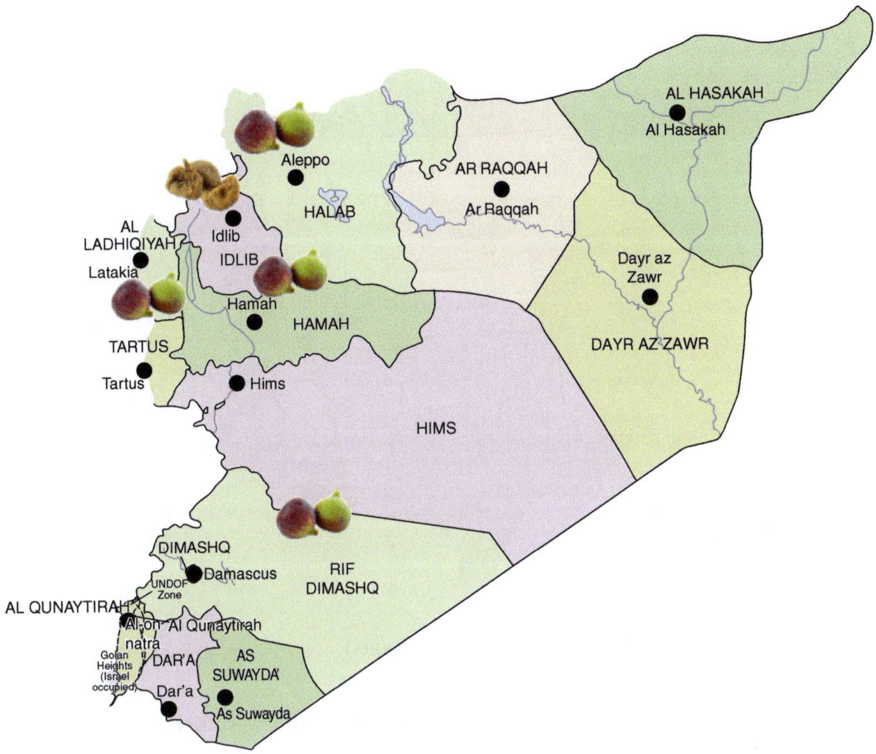


**Fig. 3.9.** Production regions in Syria. (Author's own figure.)

exported, and the USA ranked third for quantity (8846 t) and fourth for value (~23.8 million dollars) in dried fig imports and ranked ninth for quantity (2301 t) and sixth for value (~13.5 million dollars) in dried fig exports in 2019 (Tables 3.5 and 3.6).

The increased interest in figs (mainly dried) by growers and processing factories is the result of better pre- and post-harvest management and selection of cultivars with higher fruit quality such as weight, width, shape, and color of the skin and pulp, taste and acidity (Flaishman *et al.*, 2008; Pereira *et al.*, 2017). Nearly all US figs are grown in California (Fig. 3.10), introduced by Spanish missionaries in the 1760s and cultivated in missions along the California coast. Subsequently, primarily Mediterranean emigrants introduced other fig cultivars from their country of origin. Central California has an ideal climate for fig cultivation, particularly along the 38° latitude line. California produces six commercial cultivars, for fresh and dried use, and are often known by other names in their country of origin: Calimyrna/Sarilop (dried/fresh), Kadota/Dottato (dried/fresh), Black Mission (dried/fresh), Sierras (dried/fresh), Brown Turkey (fresh), and Tiger Fig/Panachee (fresh).

## Tunisia

Tunisia produced 24,619 t of figs on 16,863 ha and a yield of 1.5 t/ha in 2019 (FAOSTAT, 2021), equivalent to 1.87% of global fig output on a 5.82% area in 2019 (2.16% and 6.09%, respectively, on a 5 year average) and ranked ninth for world fig production but sixth for harvested area in 2019 (Tables 3.1. and 3.2). The harvested area in Tunisia increased around 20–30 years ago, and the same happened for fig production (Tables 3.1 and 3.2). Figs are generally used for local consumption but in the last 10 years a small market for the export of dried figs has

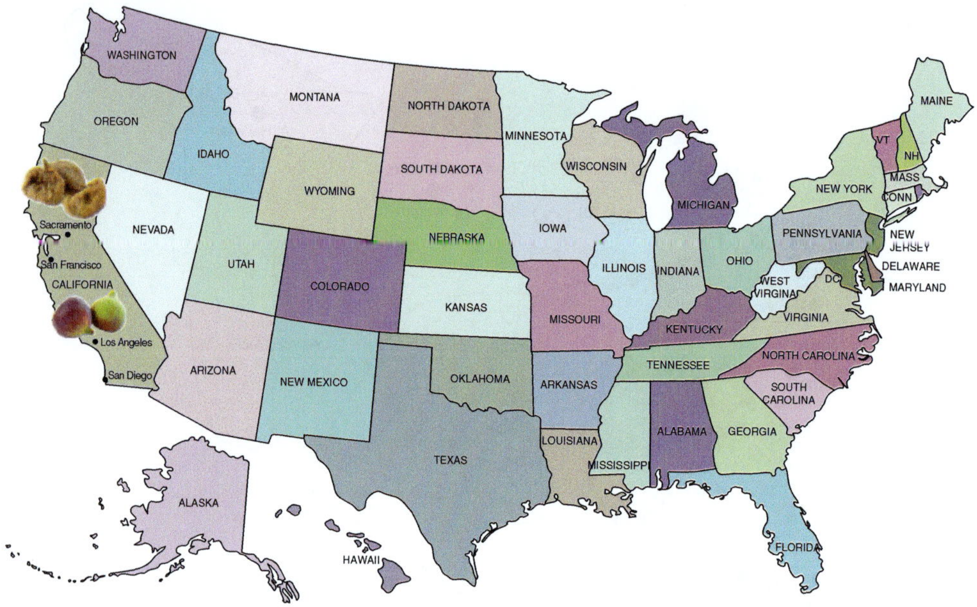


**Fig. 3.10.** Productions regions in the USA. (Author's own figure.)

developed (Table 3.6). Import and export of fresh figs and import of dried figs are negligible both as quantity and value (FAOSTAT, 2021).

Like other Mediterranean countries, Tunisia has cultivated figs since ancient times, often with date palms in the south and olive trees elsewhere. Fig is a secondary crop, after the most important, olive. Despite their socio-economic and historical importance, dried figs are commonly eaten during Ramadan. Fig orchards occur as small rainfed plantations, very often associated with other species such as olive and grape or as scattered trees in home gardens. Few regular plantations are located in the central and north parts of the country. Fig represents the principal tree species of several agro-ecosystems such as the Jessours and oases of south Tunisia. Frequent cultivar exchanges among regions produced the current multiple local landraces, high biodiversity and many cultivars well suited for fig production. There are five major fig producing areas in Tunisia (Fig. 3.11): south-east, south-west, center-east, north-east and north-west. In the south-east (Gabès, Médenine and Tataouine), figs are cultivated in the Jessours (Matmata, Beni Kheddache and Douiret) for dried production of excellent quality; coastal areas (Gabès, Jerba and Zarzis) produce precocious figs. In the south-west (Kébili, Tozeur and Gafsa) figs are cultivated in oases under palm trees in the typical three leveled system. In the center-east (Sousse, Monastir, Mahdia and Sfax) fig is cultivated mainly with olive but also in small regular plantations; in coastal areas mainly brebas are produced. In the north-east (Cap-Bon, Tunis and Bizerte) fig is cultivated with grape and citrus. In the north-west (Béja and Siliana) in mountainous villages (Djebba and Kesra) figs ripen late (Chatti, 2007; Mars *et al.*, 2008). Regional distribution of fig orchards has changed in the last few years due to the growing interest in south-east and west-center Tunisia. The center-west (Sidi Bouzid and Kairouan) may be considered an emerging fig producing area (Mars *et al.*, 2008).

However, severe genetic erosion is now threatening the local germplasm (Mars *et al.*, 2008). Since the late 1980s, after a survey to preserve the local fig germplasm and collect regional cultivars, five *ex situ* collections were established (Lahbib, 1984; Mars *et al.*, 1994; Rhouma, 1996; Mars *et al.*, 2008). Fig nomenclature is confused due to the denominations based on the color, size and time of fruit ripening or geographical origin. Fig cultivars are numerous and well adapted to the local pedo-climatic

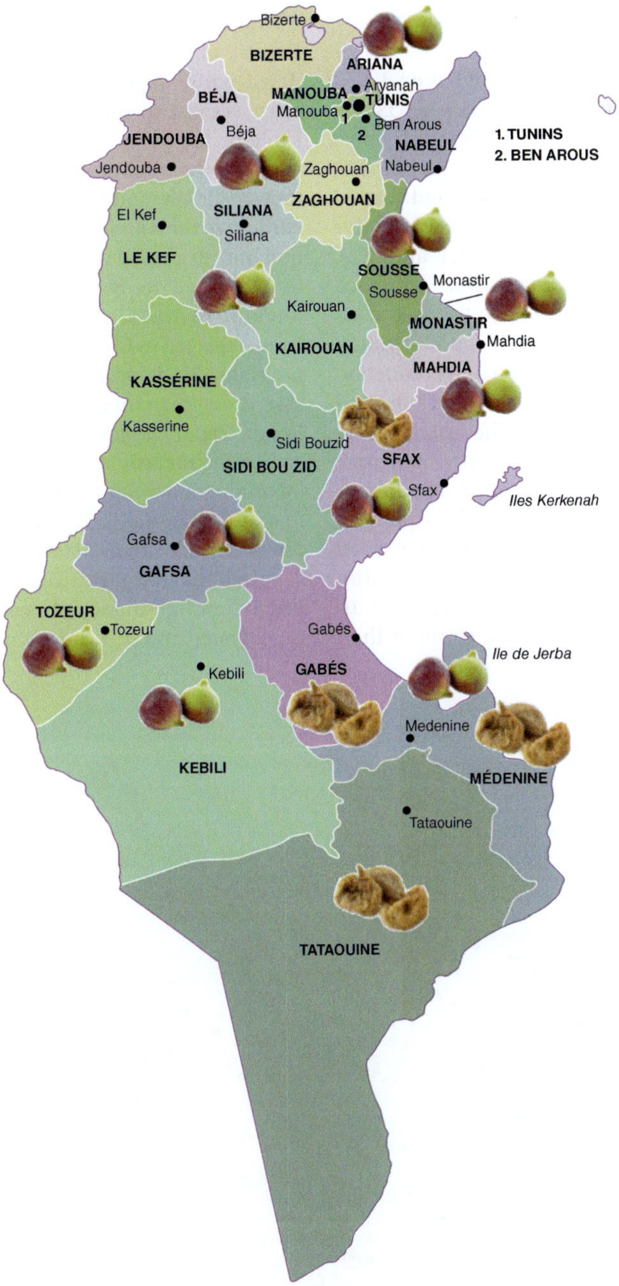


**Fig. 3.11.** Production regions in Tunisia. (Author's own figure.)

conditions and mainly belong to the Smyrna type. Lahbib (Lahbib, 1984) reported 25 local cultivars in the central parts of Tunisia of which 7 were of San Pedro type and 18 of Smyrna or common type. The main commercial Tunisian cultivars are Zidi, Soltani, Bither, Abiadh and Bither Akhal but there are many others with a limited diffusion such as Bousrari, Fellioui, Derji, Boukhobza, Nemri, Thamari, Deri and Black Birchi (Mars *et al.*, 2008; El Rayes, 1995).

## Afghanistan

Afghanistan produced 24,319 t of figs on 3,529 ha and a yield of 6.9 t/ha in 2019 (FAOSTAT, 2021). Afghanistan produced 1.85% of global fig output on a 1.22% area (0.84% and 0.64%, respectively, on a 5 year average) and ranked 10th for world fig production but 12th for harvested area in 2019 (Tables 3.1 and 3.2). The harvested area in Afghanistan decreased around 20 years ago, but in the last few years small signs of more interest in fig cultivation emerged both as cultivated area and production (Tables 3.1 and 3.2). Figs are generally used for local consumption but in the last 10 years the export of fresh figs significantly increased, reaching 901 t in 2019 (Table 3.4) whereas import of fresh figs is negligible both as quantity and value (FAOSTAT, 2021). The export of dried figs is much more important for Afghanistan with 11,778 t (ranked third) exported in 2019 corresponding to a value of more than 85 million dollars (ranked third) (Table 3.6). Export of dried figs accounted for 20% of dried fruits export in 2017 for Afghanistan with respect to only 6% in 2013 and in the same period the export value increased by 29% reaching the value of the more traditional dried almonds (Rasoly and Chandrashekar, 2018). According to the Kandahar Chamber of Commerce and Industries (www.acci.org.af/en/), 3% of figs are exported to Pakistan and the remaining amount to India; more lately figs have also been exported to China, Russia and United Arab Emirates. Fig is produced in most of Afghanistan's big provinces such as Herat, Balkh, Farah, Zabul and Jowzjan (Fig. 3.12). But southern Kandahar province is known as the center of fig production as almost 6000 t of dried figs (almost 34 million dollars value) produced in this province were exported to India.

In Kandahar province, a large local cultivar (the yellow fig), that can be eaten fresh or dried, is cultivated and successively processed through

Fig. 3.12. Production regions in Afghanistan. (Author's own figure.)

drying in a traditional manner. The yellow fig trees produce two crops per year, and after ripening, the fruits are picked by hand, flattened while still fresh, and then a hole is made in the middle to hang the figs in the sun to dry. This fig is known for its excellent aroma and sweet flavor. This traditional processed fruit is produced in limited quantities. There are an estimated 50–60 farmers growing this cultivar, with an annual production of no more than 5000 kg per year. The cultivar is propagated by cuttings, which are planted in orchards after rooting. This fig is particularly well adapted to the local climate. The dried yellow fig is often consumed at parties and celebrations such as the New Year and Muslim holidays. Today, the local yellow fig and associated drying method are at risk of being lost, because many of the fig orchards were destroyed or damaged during Afghanistan's civil wars. Siah, Spin, Tashkhordani, Kandahari, Zard Kandahari, and Zaed Kohi are known fig cultivars in the country.

### Albania

Albania produced 22,024 t of figs on 1,489 ha and a surprising yield of 14.8 t/ha in 2019 (FAOSTAT, 2021). Albania produced 1.67% of global fig output on only 0.51% of area in 2019 (1.83% and 0.52%, respectively, on a 5 year average) and ranked 12th for world fig production but 19th for harvested area in 2019 (Tables 3.1 and 3.2). The harvested area in Albania has increased (+500 ha) at the beginning of 2000, whereas fig production almost doubled (Tables 3.1 and 3.2). Figs are generally used for local consumption but in the last 10 years there has been growing interest by growers to export fresh figs (Table 3.4). Import of fresh and dried figs is very low (FAOSTAT, 2021), whereas the export of dried figs accounted for 138 t in 2019 (Table 3.6).

The fig production in Albania is now increasing, primarily in the Berat area followed by Fier, Shkodra, Vlora, Permet, Elbasan and Gramsh. The cultivation area also includes the western coastal regions from Koplik to Saranda and the internal regions (Fig. 3.13) extending to the borders of hilly areas characterized with Mediterranean climates (Koka, 2010).

Fig production was primarily used for export of dried figs, but in recent years the local production of fresh fruits or processed products (cooked figs, jam, etc.) has notably increased. Over 80 local fig cultivars have been identified, but only 40 cultivars have been fully described. The most important cultivars are Roshnik (green), Perdhikuli (yellow), Melacak (brown), Bishtgjati (purple), Rotllar (black), Kraps I zi, Cingell, Cipull, Allaxhir, Draz, Patllixhan, Morait, Peshtanak, Kallamata and Shkronjes (Kokaj, private communication).

### Greece

Greece produced 19,730 t of figs on 3,990 ha and a yield of 4.9 t/ha in 2019 (FAOSTAT, 2021). Greece produced 1.50% of global fig output on 1.38% of area in 2019 (1.73% and 1.43%, respectively, on a 5 year average) and ranked 13th for world fig production and 11th for harvested area in 2019 (Tables 3.1 and 3.2). The harvested area in Greece dramatically reduced from 20,000 ha in 1990 to the current value, whereas fig production reduced by almost 10-fold from 120,000 t produced in 1990 (Tables 3.1 and 3.2). After this large reduction in fig production in Greece, in the last 10 years data have been more stable with only some year-to-year obvious fluctuations. Greece's import of fresh figs is low with 432 t in 2019 (FAOSTAT, 2021), and export of fresh figs shows similar quantities with respect to import with average exports of 400 t in the last few years (Table 3.4). Greece ranked sixth and fifth in the world dried fig exports for quantity and value in 2019, respectively, with a continuously growing pattern since 2000 (Table 3.6); dried fig imports are almost negligible (FAOSTAT, 2021).

Greek fig production has been declining for the last 20 years and dried fig production has declined 75%; in 2018 Greece produced only 4% of the world dried fig production (FAOSTAT, 2021). However, most recently 5550–6500 t were produced on about 2100 ha contributing to the increased export of dried figs (Vasilis Demopoulos, private communication). Figs were a valuable fruit in ancient Greece for winter food, nutritional content and medical uses. The fig tree is ubiquitous in Greece, from the

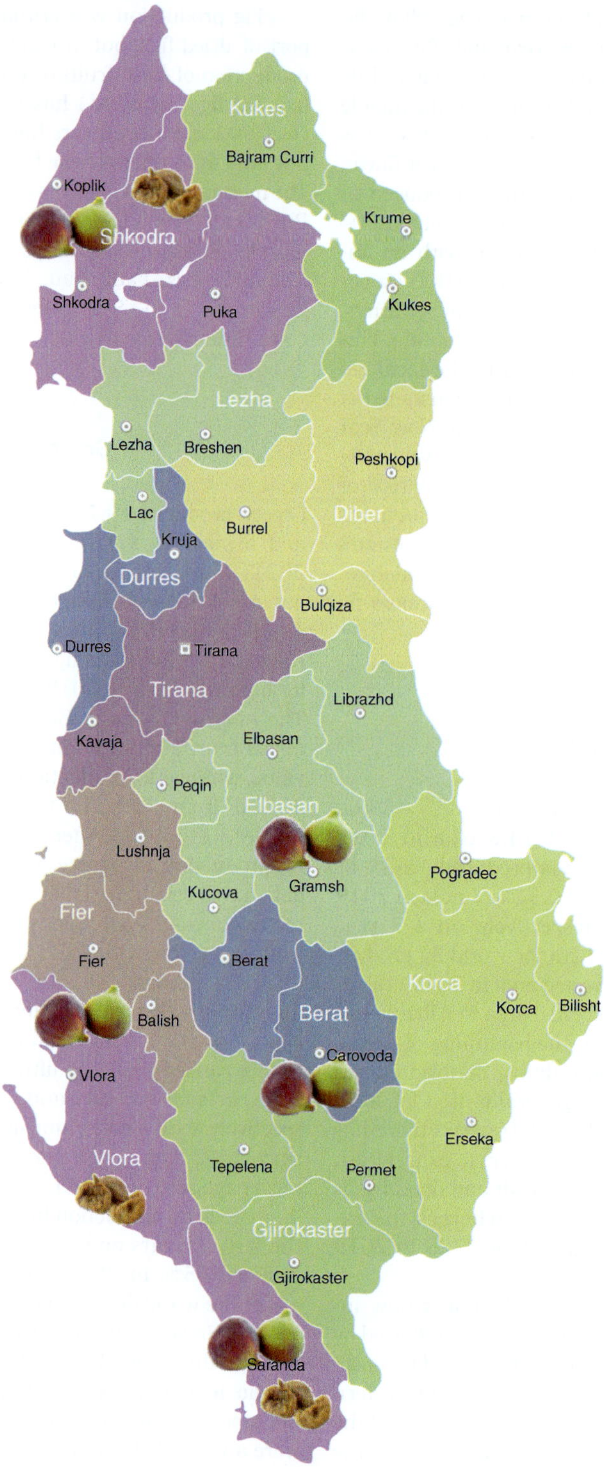


**Fig. 3.13.** Production regions in Albania. (Author's own figure.)

northern regions of Macedonia and Thraki to the southern regions of Crete (Fig. 3.14). However, most commercial cultivation is in the Peloponnese, Evia and the Aegean islands, half in the southern Peloponnese, Messenia and Laconia, and the rest in northern Evia Island. More than 50% is exported to the European Union countries and the USA, Australia and Canada. Fig orchards are typically family-owned, small, fragmented holdings. Fresh fig production is dispersed across southern Greece, in the Evia and Lesbos Islands, and Attica; it is not commercially important and few data on area and production are available. The Attica region produced 1500 t, and another 2500 t is produced in the rest of Greece. The main cultivars are Kalamon or Tsapela (Messenia and Laconia – dried figs), Basilica black or white (Attica – fresh figs), Kymi (Evia – fresh and dried figs), Smyrneiki, Politika (Lesbos – fresh figs). There are also many other local fig cultivars with regional names and limited local interest (Fragasana, Livano, Kanates, etc.).

## China

China produced 17,225 t of figs on 2,937 ha and a yield of 5.9 t/ha in 2019 (FAOSTAT, 202021). It produced 1.31% of global fig output on 1.01% of area in 2019 (1.33% and 0.97%,

**Fig. 3.14.** Production regions in Greece. (Author's own figure.)

respectively, on a 5 year average) and ranked 14th for world fig production and 13th for harvested area in 2019 (Tables 3.1 and 3.2). The harvested area in China increased five-fold from ~600 ha in 1995 to the current values whereas fig production increased by more than eight-fold from only 2000 t in the same period (Tables 3.1 and 3.2). Although there was a significant increase in fig production at the beginning of 2000s, the pattern has been more stable both as production and harvested area in the last few years. China's imports of fresh figs is increasing but values have been very variable since 1990, with the highest quantity imported in 2000 and with 2768 t imported in 2019 (Table 3.3); export of fresh figs is much smaller with respect to import accounting only for 107 t in 2019 (Table 3.4). China ranks ninth and sixth for world dried figs imports for quantity and value, respectively, with a generally stable pattern since 2015 (Table 3.5). China's exports of dried figs were small both as quantity (108 t) and value (0.74 million dollars) in 2019 and with fluctuations from year to year (Table 3.6).

The fig was introduced in China about one thousand years ago, but its commercial cultivation began only in the last decades. A large number of *Ficus* species are used as food and for medicinal properties in ayurvedic and traditional Chinese medicine, especially among local people (Lansky *et al.*, 2008) and before the early 1980s, when Chinese people seldom planted fig plants, and even had not heard of fig plants (Lianju *et al.*, 2003). Several provinces in eastern China such as Jiangsu, Shanghai, Zhejiang, Shandong, Fujian, Guangdong and others (Sichuan) have begun to grow fig trees commercially since 1990 and a well-known growing area is close to Atushi City in Xinjiang (Fig. 3.15), known as the 'county of figs' (Ren and Abudu, 2000; Lianju *et al.*, 2003; Jun *et al.*, 2011).

The fig cultivars grown in China are divided into two types: for fresh consumption only and

**Fig. 3.15.** Production regions in China. (Author's own figure.)

for both fresh and dried consumption. The main fresh fig cultivars in China are Masui Dauphine, Nixuan #1, Lukang #1 and Tanikawa. The cultivars that could be used both as fresh and for processing are Brunswick and Xinjiang Yellow, the former planted mainly in the south of China, and the latter in the south-west of Xinjiang. A production region for figs with green skin is the Jiaodong peninsula in Shandong Province. The Qingpi cultivar, with a green skin, is one of the most important Chinese cultivars produced in the Shandong Province. Other fig cultivars are Bojihong, Conadria, Armenian and Orphan. Red-purple skin figs are also produced in China and one of the most popular cultivars is Zibao (mutation of Qingpi). In northern China, fig cultivation is often done in greenhouses and high tunnels for protection from the winter cold. In the south, cultivation is mostly in open fields. As living standards have increased, the popularity of figs has also increased, favoring fig cultivation, production and consumption.

## India

India produced 14,914 t of figs on 5,734 ha and a yield of 2.6 t/ha in 2019 (FAOSTAT, 2021). India produced 1.13% of global fig output on 1.98% of area in 2019 (1.23% and 1.99%, respectively, on a 5 year average) and was ranked 15th for world fig production and 9th for harvested area in 2019 (Tables 3.1 and 3.2). The harvested area in India significantly increased from ~900 ha in 1980 to the actual value whereas fig production increased seven-fold from only 2000 t in 1970–1980 (Tables 3.1 and 3.2). Fresh figs are used for local consumption but small amounts (50–60 t) are also exported (Table 3.4). India ranked first in the world in 2019 as an importer of dried figs, both for quantity and value, with a significant growing pattern in the last few years (Table 3.5); export of dried figs is much less consistent with 535 t reached in 2019 (Table 3.6).

India is now an important player in the global fig market and the edible fig is locally known as 'anjeer'. The main cultivars grown in the different states of India are Black Ischia, Brown Turkey, Turkish White, Kabul and Marseilles, together with other cultivars with local names (Dinkar, Daulatabab) and the most important cultivated is Poona used for fresh consumption. There are also hybrids between *F. carica* (common fig) and indigenous species of *Ficus*. Currently cultivation is primarily in the western parts of Maharashtra (Pune), Karnataka and Gujarat, Uttar Pradesh (Lucknow and Saharanpur), and small areas of Tamil Nadu (Coimbatore) (Fig. 3.16). In Maharashtra, Poona and Dinkar cultivars are commercially cultivated in an area of around 2000 ha, whereas cultivars named Bellary and Krishna are grown in Karnataka (area >3000 ha). At Saharanpur, Brown Turkey, Bangalore, Black Ischia and Lucknow are successfully grown. Around Mumbai, there is only the local cultivar Poona. In Southern India, Poona and Marseilles are the most popular cultivars. The figs commonly grown in India belong to the Common type and grow parthenocarpically, but recently other cultivars (Conadria, Deana, Excel, King, Panachee, Texas, etc.) have been introduced to be evaluated for possible cultivation in the Punjab area (Rattanpal *et al.*, 2017).

## Japan

Japan produced 14,175 t of figs on 1,019 ha and a yield of 13.9 t/ha in 2019 (FAOSTAT, 2021). Japan produced 1.08% of global fig output on only 0.35% of area in 2019 (1.20% and 0.36%, respectively, on a 5 year average) and ranked 16th for world fig production and 23rd for harvested area in 2019 (Tables 3.1 and 3.2). Both the harvested area and fig production in Japan have been stable in the last 15 years (Tables 3.1 and 3.2). Import of fresh figs is of very low commercial interest both for quantity (a few tonnes) and value, and export is absent (FAOSTAT, 2021); the import of dried figs shows a growing interest in Japan with a doubled value in the last 10 years to almost 12 million dollars in 2019 (Table 3.5).

The common fig, 'ichijiku' in Japanese, was probably introduced from China to Japan in the 17th century, but the details of this introduction remain unknown (Ikegami *et al.*, 2009). The cultivation is mainly concentrated around cities (Fig. 3.17) in intensive training systems and often in greenhouses. The oldest cultivar is Houraihi, probably cultivated for centuries but Masui

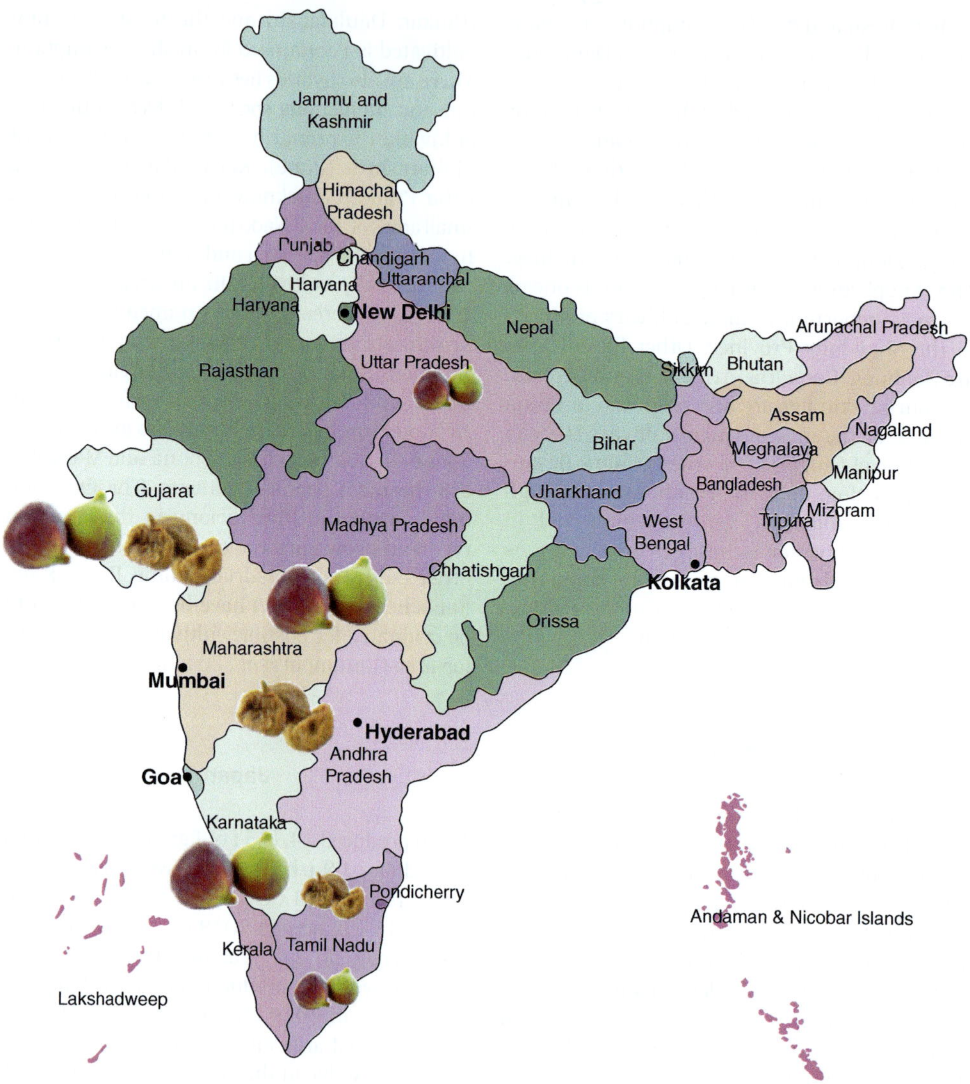


**Fig. 3.16.** Production regions in India. (Author's own figure.)

Dauphine is currently the major cultivar. Other local fig cultivars are Common-type figs, such as Toyomitsuhime and Himehourai and a San Pedro type cultivar named Tanikawa.

## Azerbaijan

Azerbaijan produced 12,074 t of figs on 1880 ha with a yield of 6.4 t/ha in 2019 (FAOSTAT, 2021). Azerbaijan produced 0.92% of global fig output on 0.65% of area in 2019 (0.86% and 0.64%, respectively, on a 5 year average) and ranked 17th for world fig production and 18th for harvested area in 2019 (Tables 3.1 and 3.2). The harvested area in Azerbaijan has increased to around 700 ha in the last 20 years but fig production increased by 50% in the same period (Tables 3.1 and 3.2). Import and export of fresh figs is insignificant both as quantity and value, and also the import of dried figs accounted for less than 100 t and export for only 1 t in 2019 (FAOSTAT, 2021).

**Fig. 3.17.** Production regions in Japan. (Author's own figure.)

Figs are mostly cultivated in Absheron peninsula, Lankaran, Aran and less in inner areas (Shaki-Zaqatala, Guba-Khachmaz) (Fig. 3.18). The fig cultivars have local names Sary-Gyanja, Bozov Burnu, Sary-Lob and Sary-Absheron. The Talysh region also has many indigenous cultivars.

## Italy

Italy produced 11,830 t of figs on 2150 ha with a yield of 5.5 t/ha in 2019 (FAOSTAT, 2021). Italy produced 0.90% of global fig output on 0.74% of area in 2019 (0.98% and 0.80%, respectively, on a 5 year average) and ranked 18th

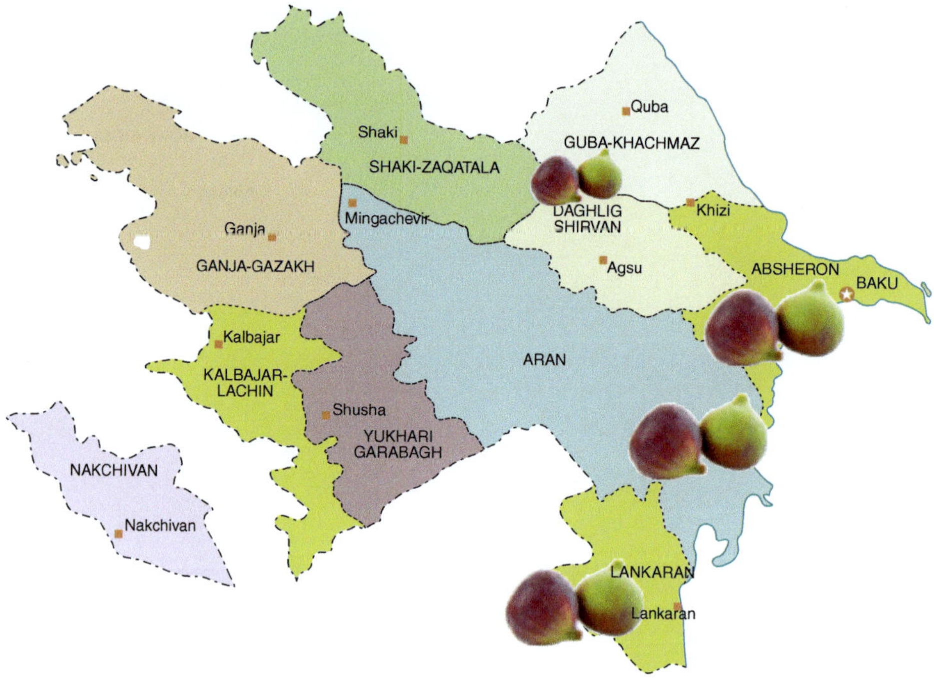


**Fig. 3.18.** Production regions in Azerbaijan. (Author's own figure.)

for world fig production and 17th for harvested area in 2019 (Tables 3.1 and 3.2). The harvested area in Italy constantly decreased from 39,000 ha in 1970 (the second harvested area in the world) to the recent value and fig production reduced from the impressive quantity of 198,000 t in 1970 (second world producer) to only ~12,000 t nowadays (Tables 3.1 and 3.2). Production has increased in the last few years because of the greater interest of growers in brebas and dried figs. Italy's imports of fresh figs have been almost stable in the last few years both as quantity and value with a sharp increase in 2019 up to 904 t and 2.7 million dollars (Table 3.3). With regards to export of fresh figs, Italy ranks 6th and 10th for quantity and value, respectively, with a clear growing pattern since the beginning of 2000 but with a sudden drop in 2019 (Table 3.4). Export of fresh figs (primarily brebas) is mainly directed to France and other European countries. Italy ranks fifth in the world dried figs imports for both quantity and value, with stable values in the last few years (Table 3.5) and around 80% of imported dried figs are from Turkey. The export of dried figs shows low and stable quantities in these years (~300 t) with an export value of ~2.5 million dollars (Table 3.6). Fig cultivation in Italy is located in southern regions (Fig. 3.19) such as Puglia (3328 t), Calabria (2659 t), Campania (2463 t), Sicily (1142 t), and Basilicata (1001 t) (ISTAT, 2021). In the Puglia region, fig cultivation was spread over 30,000 ha 50–60 years ago but was subsequently substituted by other fruit crops, i.e., table grape, sweet cherry, pomegranate, olive.

The fig tree has been in Italy since ancient times, and it was the main fruit crop in some areas of southern Italy with olive, grape and almond. Grassi (1982) identified 442 landraces from different Italian regions, but the morphological, phenological, productive and carpological observations identified many as synonyms. In Italy, fig cultivars are named for their place of origin, visual and organoleptic characteristics or ripening time. Regions often have only a few trees of a local cultivar as they were often removed to plant other fruit tree species. The most important cultivar is Dottato (Kadota), which can be used fresh, but it is generally dried. Other cultivars are Petrelli, Domenico Tauro, San

**Fig. 3.19.** Production regions in Italy. (Author's own figure).

Giovanni and Gentile, primarily for fresh breba production. Cultivars for fresh production (main crop) are numerous but with a limited cultivation (Albanera, Brogiotto bianco e nero, Callara, Citrullara, Della Monaca, Natalino, Melanzana, Rigato, Troiano, etc.) (Ferrara *et al.*, 2016; 2017; Marcotuli *et al.*, 2019). The high nutritional content, newer processing techniques, and better orchard management resulted in the European Community awarding the Protected Designation of Origin (DOP) certificate to the cultivar Dottato as 'Fico Bianco' cultivated in the Cilento area (Campania region) and 'Fico di Cosenza' in the Calabria region.

## Libya

Libya produced 9673 t of figs on 2554 ha and a yield of 3.8 t/ha in 2019 (FAOSTAT, 2021). Libya produced 0.74% of global fig output on 0.88% of area in 2019 (0.82% and 0.92%, respectively, on a 5 year average) and ranked 19th for world fig production and 14th for harvested area in 2019 (Tables 3.1 and 3.2). The harvested area in Libya has reduced by 500 ha in the last 10 years with only a small reduction in fig production in the same period (Tables 3.1 and 3.2). Imported fresh figs account for a few tonnes and export is almost nil; the import of dried figs was 222 t and 0.35 million dollars in 2019 whereas export was close to zero (FAOSTAT, 2021). In Libya, figs are grown in many areas of the country and are produced mostly for fresh consumption with a small portion being sun dried (Geddeda and Gumaty, 2015). Figs are mainly cultivated in regions such as Tripolitania (Tripoli and Tarhuna districts) and Cyrenaica (Benghazi district) (Fig. 3.20) where cultivars Baiodhi and Hammori are grown (Geddeda and Gumaty, 2015).

## Mexico

Mexico produced 9466 t of figs on 1322 ha and had a yield of 7.2 t/ha in 2019 (FAOSTAT, 2021). Mexico produced 0.72% of global fig output on only 0.46% of area in 2019 (0.64% and 0.45%, respectively, on a 5 year average) and ranked 20th both for world fig production and harvested area in 2019 (Tables 3.1 and 3.2). The harvested area in Mexico, after a significant

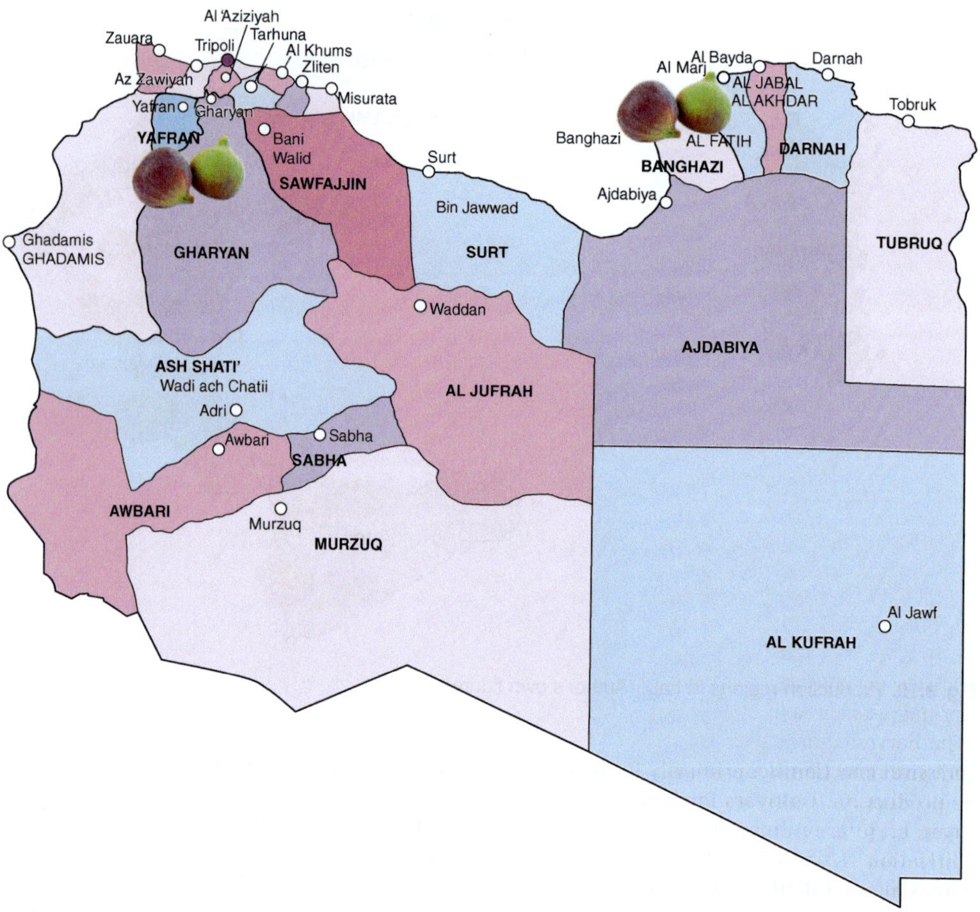


**Fig. 3.20.** Production regions in Libya. (Author's own figure.)

decrease between 1980–1990, slightly increased in the successive years reaching a stable level of 1200–1300 ha; fig production has increased in the last years from ~3600 t in 2010 up to ~9500 in 2019 (Tables 3.1 and 3.2). Import of fresh figs in Mexico was ~1500 t and ~4.0 million dollars, but there was a sharp drop in 2019 (Table 3.3). Mexico ranked 10th in 2019 for export of fresh figs for quantity and 6th for value with a positive trend (Table 3.4). Import and export of dried figs are of lower commercial quantity and value for Mexico, with ~2600 and ~600 t of import and export, respectively, between 2005 and 2010 and export of less than 300 t in 2019 (FAOSTAT, 2021).

The new Mexican orchards are planted with productive cultivars selected for fruit quality parameters such as weight, width, shape, and color of the skin and pulp, taste, aroma, soluble solids content and acidity, and are managed with modern agricultural techniques, irrigation, training systems and fertilization. Figs are cultivated (Fig. 3.21) in Morelos (58%), Hidalgo (14%), Veracruz (10%), Baja California Sur (6.5%), Distrito Federal (3.5%), Puebla (2.6%), Durango (2.4%), San Luis Potosí (1.5%), Sonora (1.1%), and Baja California (0.4%) (Sagarpa, 2010). Only in Baja California Sur and Sonora fig orchards are irrigated (Botti and Muchnik, 1998). The cultivars most appreciated and grown are Colar (black figs), Goina (similar to Colar), Ñoral (green figs), and Verdal and Blanca (used for processing).

### Portugal

Portugal produced 3380 t of figs on 4130 ha and had a yield of only 0.8 t/ha in 2019 (FAOSTAT, 2021). Portugal produced only 0.26% of global fig output on 1.43% of area in 2019 (0.28% and 1.46%, respectively, on a 5 year average) and ranked only 26th for world fig production but 10th for harvested area in 2019 (Tables 3.1 and 3.2). The harvested area in Portugal showed an impressive reduction from 200,000 ha in 1970 to ~4,000 in 2019 and also the production reduced from 137,400 in 1970 to the very low value of 2019 (Tables 3.1 and 3.2). A small but increasing export of fresh figs (~80 t) can be observed in the last few years (Table 3.4), whereas import of dried figs (~1,200 t and 3.3 million dollars) has been almost constant in the last few years (Table 3.5) and export of dried figs is of low commercial value, <0.5 million dollars (FAOSTAT, 2021).

The traditional fig tree cultivation regions are the Algarve (47% of fig area), Ribatejo e Oeste (30%), Alentejo (16%), and Trás-os Montes (6%) (Fig. 3.22). In these regions, a consistent reduction in fig cultivation has been registered in the last 30 years (Martins, 2008). The fig tree was introduced to Portugal by Phoenicians, Romans, and later, by the Arabs. Figs are used for fresh and dried consumption and also for alcohol production. The modern Algarve orchards in Torres Novas produce the Lampa Preta (breba and main crop), Princesa (breba and main crop), and Bebera Branka cultivars for fresh fig markets. Since fig production was very important in the past, in Portugal there are still many local cultivars with a less limited presence nowadays used as fresh fruit (Cachopeira, Bacela, Da Ponte, Moscatel Branco, Brava, Carvalhal), dried fruit (Côtea, Belmaldil, Olho de Passarinho) or for both uses (Urjal, Algarve, Pingo de Mel, Castelhana Branca).

## 3.3 The Fig in the Southern Hemisphere

The fig is grown between 20° and 40° latitude in the Southern Hemisphere but in fewer countries than in the Northern Hemisphere. As fig requires little or almost no chilling to break dormancy, in warm climates it grows like an evergreen species. In the tropics, fig cultivation can be found at altitudes up to 3000 m above sea level (a.s.l.). Temperatures below −10°C are the main limiting factors in site selection; new growth in early spring or secondary growth in late autumn, mainly on young trees, maybe damaged at −1°C. Low temperatures are a cold risk for fig wasps over-wintering in caprifig mamme fruits. The leading fig producing country in the Southern Hemisphere is Brazil whereas production is of lower value in the other countries.

### Brazil

Brazil produced 22,526 t of figs on 2208 ha and a yield of 10.2 t/ha in 2019 (FAOSTAT, 2021).

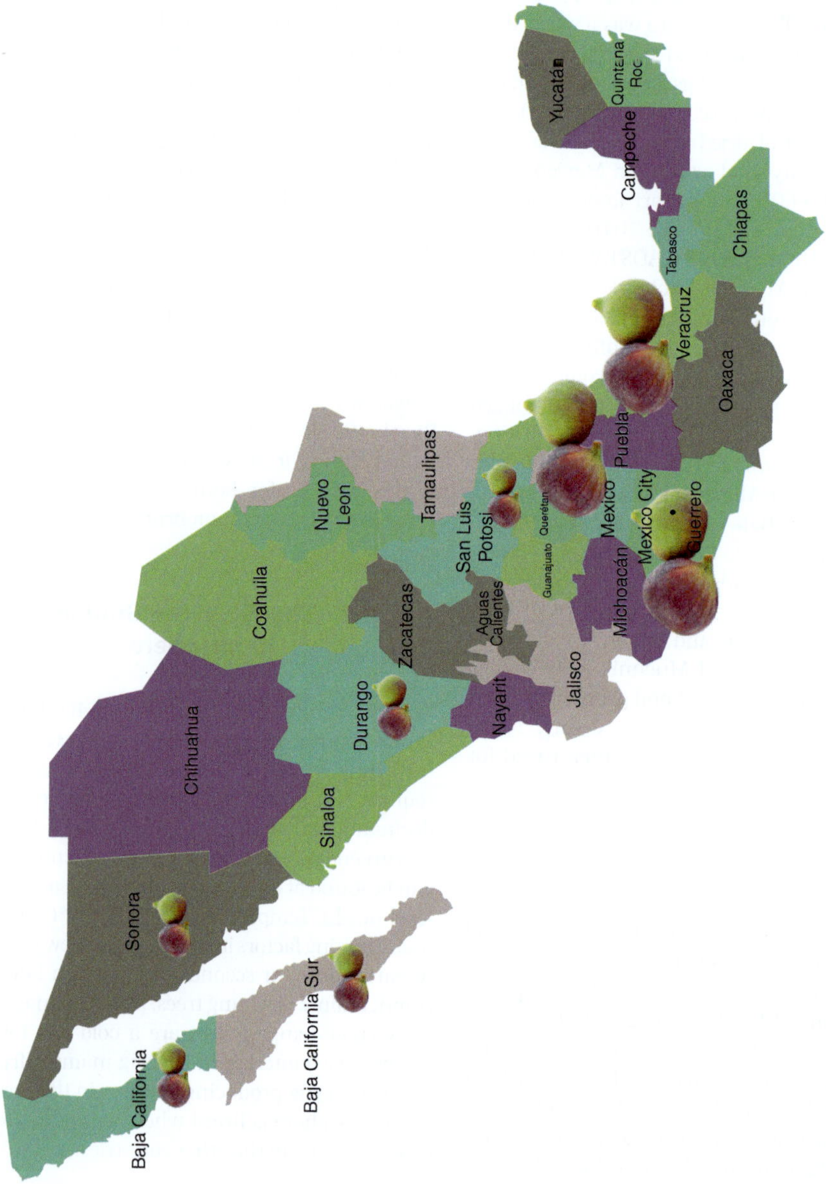


**Fig. 3.21.** Production regions in Mexico. (Author's own figure.)

Brazil produced 1.71% of global fig output on 0.76% of area (2.16% and 0.90%, respectively, on a 5 year average) and ranked 11th for world fig production and 16th for harvested area in 2019 (Tables 3.1 and 3.2). The harvested area in Brazil has reduced by 600 ha in the last 5 years, and even more with respect to the highest values reached between 1970 and 1990, and fig production shows a corresponding slight reduction (Tables 3.1 and 3.2). Export of fresh figs in Brazil has seen a constant slight growth both for quantity and value, and Brazil ranked 8th and 4th in 2019 for export of fresh figs for quantity and value, respectively (Table 3.4). Import of fresh figs and export of dried figs are of no commercial value for Brazil and import of dried figs accounted for ~300 t and almost 1.2 million dollars in 2019 (FAOSTAT, 2021).

**Fig. 3.22.** Production regions in Portugal. (Author's own figure).

Commercial fig production in Brazil started in 1910 in the Valinhos region of São Paulo state with the Roxo de Valinhos (Brown Turkey) imported from the Italian Adriatic coast (Leonel and Damatto Junior, 2008). Roxo de Valinhos is the only cultivar with economic value due to its rusticity, vigor and productivity, and produces both breba and main crop. It is sensitive to handling and highly perishable. Other cultivars planted on small areas are Kadota, Verdona Longa, Brunswick, Nobile and Corfu (Botti and Muchnik, 1998). Brazil has a growing export market for fresh figs, since Northern and Southern Hemisphere figs have complimentary harvest seasons. Fig is cultivated in the Brazilian states of Rio Grande do Sul (39.42%), São Paulo (35.15%) and Minas Gerais (18.75%) (Fig. 3.23). More than 80% of the fig production in the State of São Paulo is concentrated in Valinhos County, Campinas region. In the State of São Paulo, production is mainly destined for the fresh market, whereas in the other states, fig fruits are also used for processing, jam, marmalade, cookies and dried.

## Peru

Peru produced 3095 t of figs on 490 ha and a yield of 6.3 t/ha in 2019 (FAOSTAT, 2019). Peru produced only 0.24% of global fig output on only 0.17% of area in 2019 (0.24% and 0.19%, respectively, on a 5 year average) and ranked 30th for world fig producer and 29th for harvested area in 2019 (Tables 3.1 and 3.2). The harvested area follows a very stable pattern whereas fig production has slightly increased in the last few years (Tables 3.1 and 3.2). Export of fresh figs in Peru shows a very significant growth both for quantity (from ~60 t in 2015 to ~800 t in 2019) and value (from ~0.4 million dollars up to ~5 million dollars in 2019) (Table 3.4). Import of fresh figs and export and import of dried figs are of no commercial value for Peru (FAOSTAT, 2021).

Peru's fig production has increased rapidly in the last 30 years. The production region is in the coastal and mountain regions of Tacna, Moquegua, Arequipa, Lima and Ancash (Fig. 3.24). Total fig production is still low, but if it follows the pattern of the of table grape industry, Peru will become an important Southern Hemisphere exporter. The main fresh export cultivar is Toro sentado.

## Colombia

Colombia produced 2413 t of figs on 102 ha and a noteworthy yield of 23.7 t/ha in 2019 (FAOSTAT, 2021). Colombia produced only 0.18% of global fig output on only 0.04% of area in 2019 (0.21% and 0.04%, respectively, on a 5 year average) and ranks 31st for world fig production and 41st for harvested area in 2019 (Tables 3.1 and 3.2). Both the harvested area and fig production are very stable in Colombia (Tables 3.1 and 3.2). Export of fresh figs in Colombia is very low (~60 t) and import is even much less, whereas import and export of dried figs has no commercial value (FAOSTAT, 2021). Fig cultivation (parthenocarpic cultivars) in Colombia is mainly concentrated in the regions of Boyaca and Valle del Cauca (Fig. 3.25), but there are smaller plantations also in Antioquia, Caldas, Risaralda and Quindio (Peña and Fischer, 1993).

## South Africa

South Africa produced 1972 t of figs on 610 ha and a yield of 3.2 t/ha in 2019 (FAOSTAT, 2021). South Africa produced only 0.15% of global fig output on only 0.21% of area in 2019 (0.18% and 0.23%, respectively, on a 5 year average) and ranked 33rd for world fig production and 26th for harvested area in 2019 (Tables 3.1 and 3.2). The harvested area is very stable in South Africa whereas the fig production slightly increased in the last few years (Tables 3.1 and 3.2). Export of dried figs in South Africa shows a significant growth both for quantity (from 73 t in 2010 up to ~500 t) and value (from 0.4 million dollars in 2010 up to ~3.2 million dollars) in the last 10 years (Table 3.6). Import and export of fresh figs are of no commercial value for South Africa and import of dried figs accounted for 121 t and less than 0.5 million dollars in 2019 (FAOSTAT, 2019).

Production is concentrated in the Klein-Karoo district of the Western Cape and a little in the north-west close to Johannesburg (Fig. 3.26).

**Fig. 3.23.** Production regions in Brazil. (Author's own figure.)

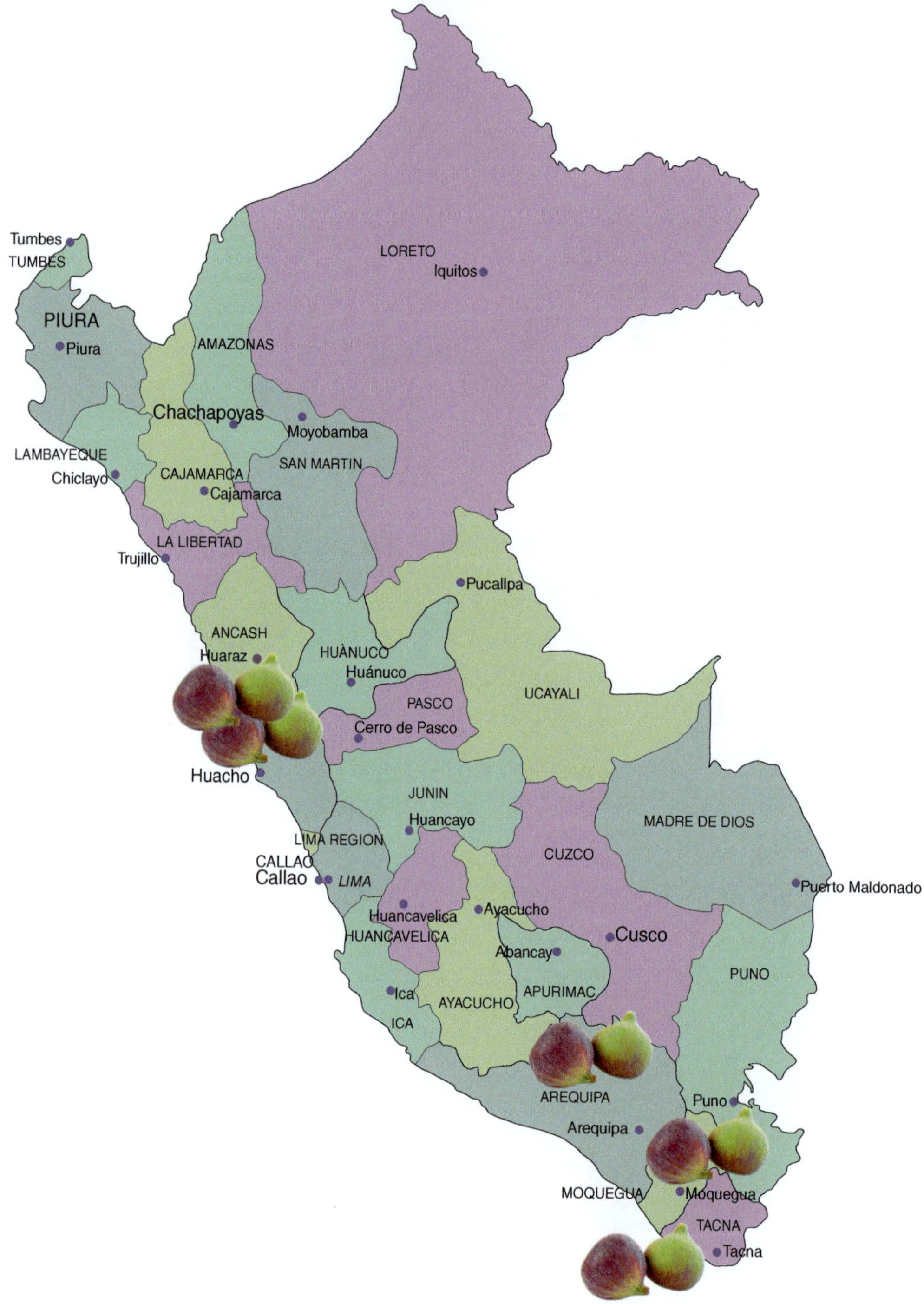


**Fig. 3.24.** Production regions in Peru. (Author's own figure.)

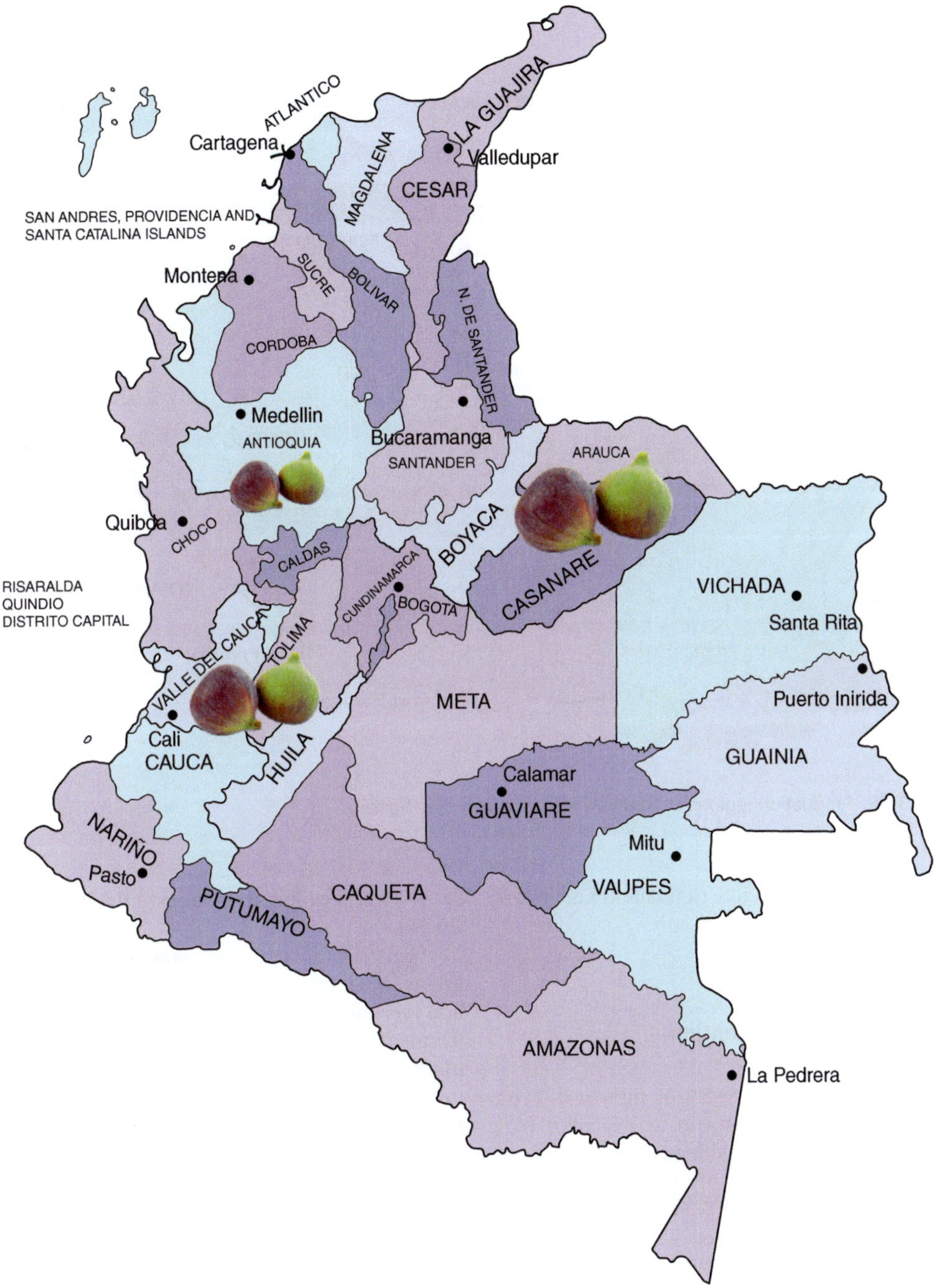


**Fig. 3.25.** Production regions in Colombia. (Author's own figure.)

While fig production is not increasing, the fig export industry and domestic demand are increasing due to changing trends in dried and fresh fruit consumption. The main South African cultivars are Ronde de Bordeaux and Bourjassotte Noire. Tangier, Deanna and King are less planted and supplemented by the older cultivars; Old Cape White and Adams are used for local

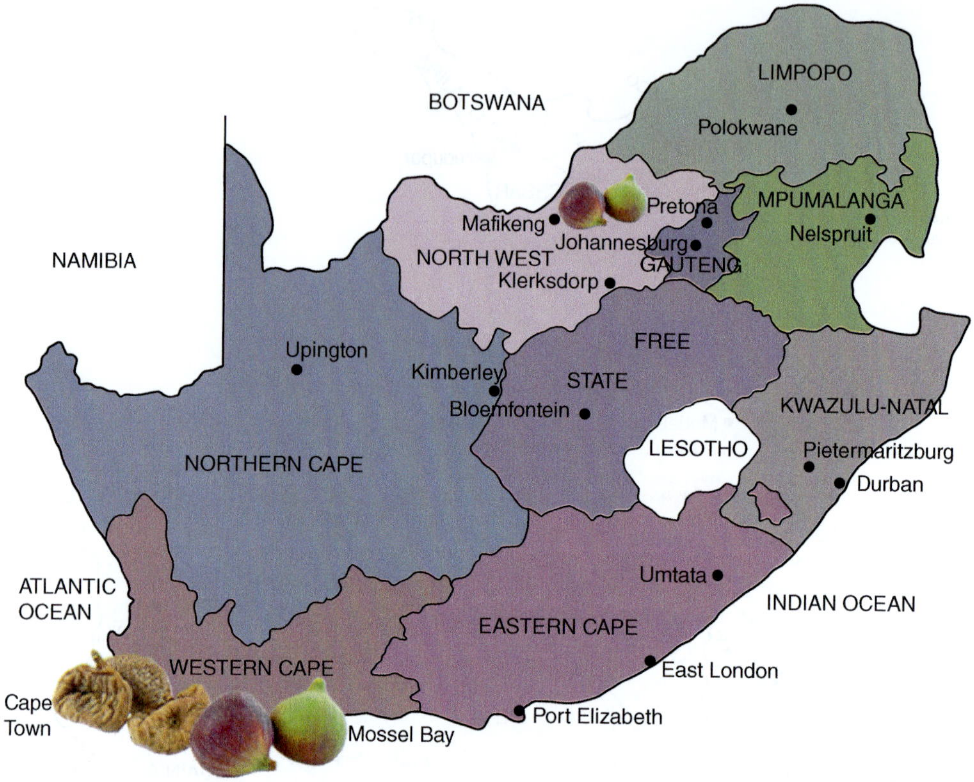


**Fig. 3.26.** Production regions in South Africa. (Author's own figure.)

consumption, and White Genoa and Kadota are used for fig preserves.

## Argentina

Argentina produced 868 t of figs on 276 ha and a yield of 3.1 t/ha in 2019 (FAOSTAT, 2021). Argentina produced only 0.07% of global fig output on only 0.10% of area in 2019 (0.07% and 0.10%, respectively, on a 5 year average) and ranked 38th for world fig production and 34th for harvested area in 2019 (Tables 3.1 and 3.2). Both the harvested area and fig production are almost stable in Argentina (Tables 3.1 and 3.2). Export of fresh figs accounted for around 100 t (less than 0.5 million dollars) in 2019 whereas import had no commercial value; 59 t of dried figs were also imported (in a reducing trend) in 2019 and export was almost nil (FAOSTAT, 2021).

Better cultivars and production innovations, like increased tree density, drip irrigation and mechanical pruning, have facilitated establishment of new orchards for European export markets. However, the local production still cannot supply the domestic market, and dried figs are imported annually from Turkey and Chile (Prataviera, 2003). The major Argentinian cultivars are Brown Turkey, Guarinta, Brogiotto Bianco, Servantine, Kadota and Genoa (Prataviera, 2003). Figs can be grown commercially throughout Argentina, with the exception of the three southern provinces, Chubut, Santa Cruz and Tierra del Fuego. The main producing areas are in the north-west (70% of production in San Juan, Santa Fé and San Luis provinces), the Cuyo region, and in the province of Buenos Aires (Fig. 3.27). In the central area of the province of Santa Fé fig cultivation is relatively recent. Fig is

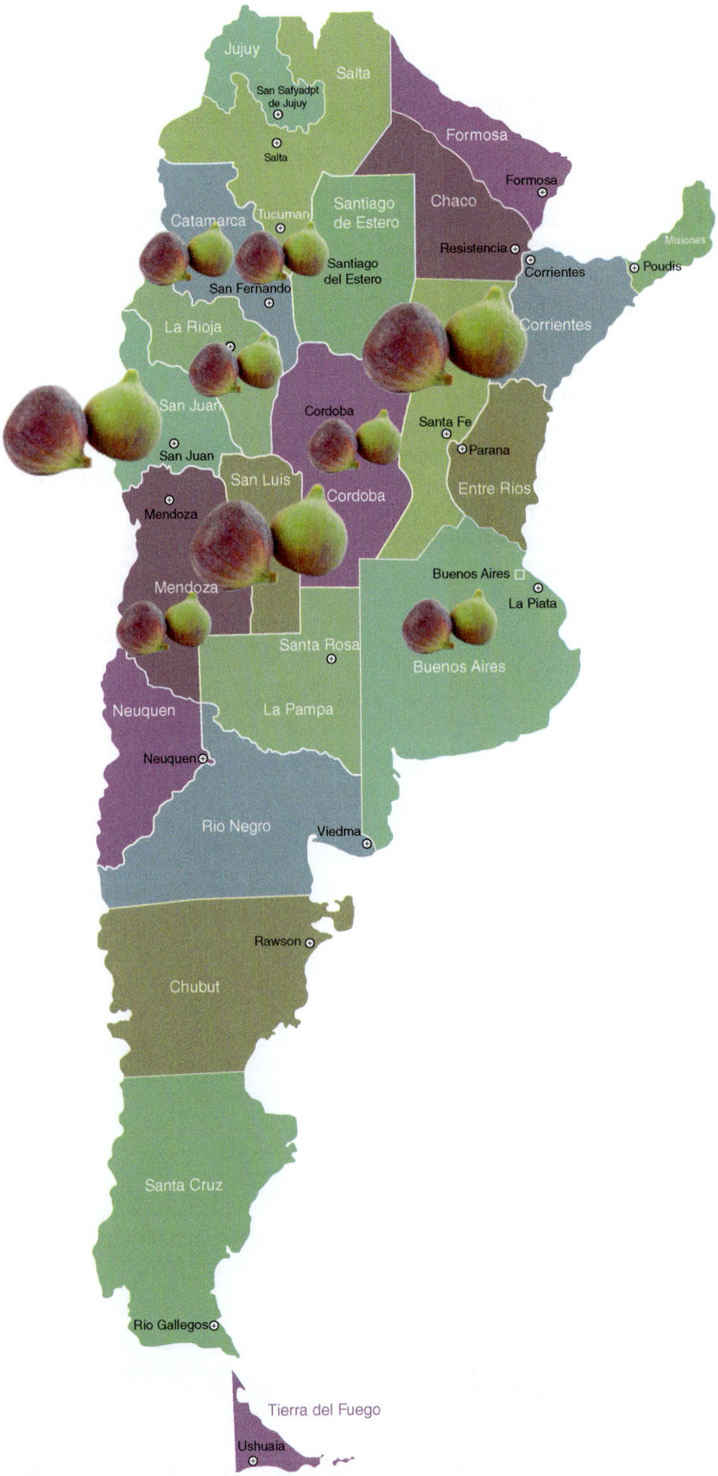


**Fig. 3.27.** Production regions in Argentina. (Author's own figure.)

also cultivated in Tucumán, Córdoba, Catamarca and La Rioja (Botti and Muchnik, 1998). In the last few years, growers have planted small fig orchards (0.1–2.0 ha) for diversification of fruit tree production. Currently, their production meets the demand of Santa Fé, Esperanza and nearby towns, as well as small sweet industries. The cultivars of fig most cultivated in this area are Turco (Brown Turkey), Guarinta and Kadota due to good adaptation to climate, and high yield and quality (Catraro, 2014). The most important cultivars grown in Argentina are Negra Común (Black Mission), King, Turco (Brown Turkey), Guarinta, Uñigal and Kadota (Botti and Muchnik, 1998). In Argentina fig is also curative and in the central region of Córdoba province dried fruits of the fig are used as a topical medication (skin pustules and furuncles) and the latex oozing from freshly harvested figs is used to treat warts (Morton, 1987).

## Chile

Chile produced 104 t of figs on only 65 ha and had a yield of 1.6 t/ha in 2019 (FAOSTAT, 2021). Chile produced only 0.01% of global fig output on only 0.02% of area in 2019 (0.01% and 0.03%, respectively, on a 5 year average) and ranked 45th for world fig production and 44th for harvested area in 2019 (Tables 3.1 and 3.2). Both the harvested area and fig production are significantly decreasing (Tables 3.1 and 3.2). Export of fresh figs accounted for around 82 t in 2019 (79% of domestic production) whereas import is not relevant (a few tonnes); import and export of dried figs is nil (FAOSTAT, 2021).

The fig industry has developed slowly in Chile and production is mainly devoted to the export of fresh figs. Fig is cultivated in small orchards in the north and central regions of Chile but is mostly concentrated in the Metropolitan area of Santiago where climatic conditions are more favorable (Fig. 3.28) (Botti and Muchnik, 1998). There is little interest in the crop, a lack of information on production and cultivation techniques, and few well adapted fig cultivars. However, the fig has the potential to replace traditional fruit crops with higher water demands in dry areas (Cancino *et al.*, 2000a, b).

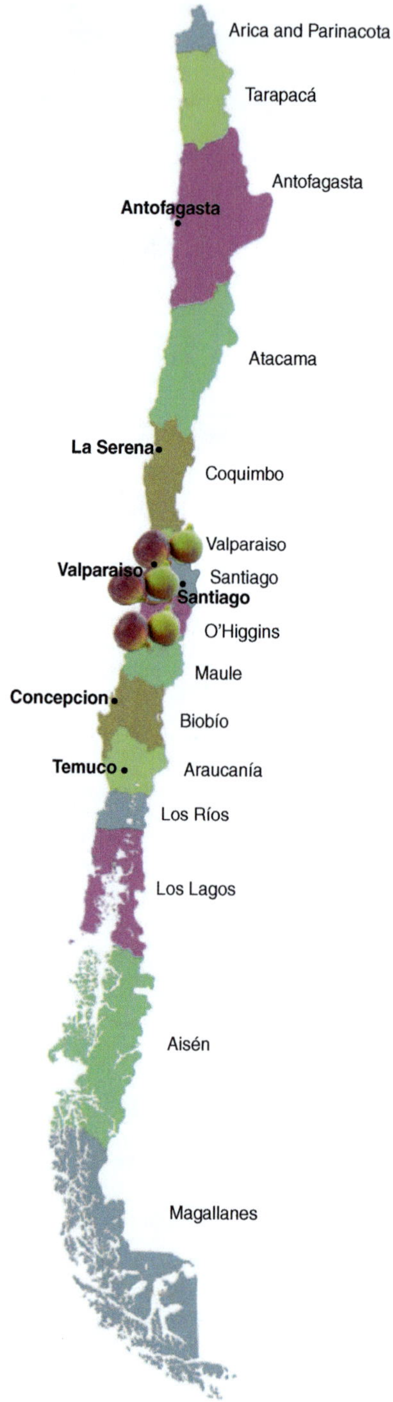


**Fig. 3.28.** Production regions in Chile. (Author's own figure.)

The main fig cultivar is Negra Común (Black Mission), followed by Kadota, White Adriatic, Kennedy, Brown Turkey, Brown Ischia, Royal Vineyard, Magdalena and Larga de Burdeos.

### Australia

Australia produced 41 t of figs on only 67 ha and had a yield of 0.6 t/ha in 2019 (FAOSTAT, 2021). Australia produced only 0.003% of global fig output on only 0.02% of area (0.004% and 0.02%, respectively, on a 5 year average) and ranked only 48th for world fig production and 42nd for harvested area in 2019 (Tables 3.1 and 3.2). The harvested area is almost stable but fig production has dramatically reduced in Australia because of decreasing interest in this fruit crop (Tables 3.1 and 3.2). There is no import and export of fresh figs in Australia; the import of dried figs has shown a general reduction both in quantity and value (to a lesser extent) in the last few years (Table 3.5) and only a few tonnes of dried figs are destined for export (FAOSTAT, 2021).

The Smyrna type cultivars (Smyrna, Calimyrna) and their caprifigs are grown in the Riverland in South Australia. Common type cultivars are also grown; Black Genoa, Brown Turkey, Preston Prolific, White Adriatic and Cape White together with a San Pedro type cultivar, Adam. Dried fig production was an important horticultural enterprise in the Murrumbidgee irrigation area during the 1920s (thanks to immigrants), but this industry declined in the 1950s, when cheaper imported figs entered the domestic market. Commercial fig production in NSW today is for the fresh fruit market (NSW-Agriculture, 2002). However, the fig in Australia has a double face. In the temperate regions of the south and some areas of the Northern Territory, South-eastern Queensland and Norfolk Island, the fig is considered almost an environmental weed. In fact, in the wetlands around Perth and the riparian zones in the Wooroloo Brook catchment fig ranks in the Environmental Weed Strategy of Western Australia, due to its perceived invasiveness and impact. In South Australia, especially in several conservation areas of Belair National Park, Kaiserstuhl Conservation Park, Cleland Conservation Park, Sturt Gorge Recreation Park, Onkaparinga River Recreation Park, Horsnell Gully Conservation Park and Morialta Conservation Park, and in Yarra Bend Park, in suburban Melbourne, the fig has already become a naturalized species but with limited commercial interest (Fig. 3.29).

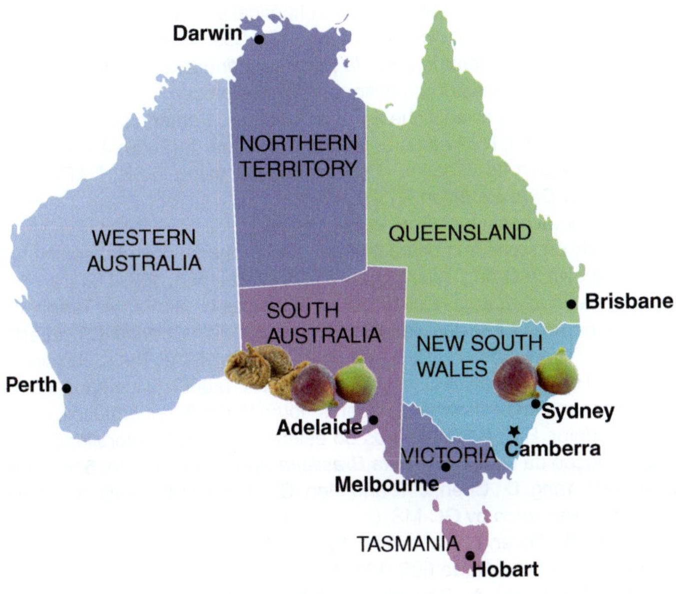


**Fig. 3.29.** Production regions in Australia. (Author's own figure.)

## References


Aksoy, M.A. and Beghin, J.C. (2005) *Global Agricultural Trade and Developing Countries*. The World Bank, Washington, DC.

Arpaci, S. (2017) An overview on fig production and research and development in Turkey. *Acta Horticulturae* 1173, 57–62.

Botti, C. and Muchnik, E. (1998) *Fig Production In Latin America*. International Society for Horticultural Science (ISHS), Leuven, Belgium, pp. 37–42.

Çalişkan, O. and Polat, A.A. (2008) Fruit characteristics of fig cultivars and genotypes grown in Turkey. *Scientia Horticulturae* 115, 360–367.

Cancino, J., Lewin, P. and Botti, C. (2000a) Comercio de higos registra permanente expansión (primera parte). *Revista Fedefruta* 85, 9–12.

Cancino, J., Lewin, P. and Botti, C. (2000b) Tendencia al alza permite proyectar panorama favorable para el higo (parte final). *Revista Fedefruta* 86, 9–11.

Catraro, M.A. (2014) *El Cultivo de la Higuera: Producción de higos y su deshidratación como método para el agregado de valor del producto.* Graduation thesis, Universidad Nacional del Litoral, Facultad de Ciencias Agrarias. Esperanza, Santa Fe, Argentina.

Chatti, K., Saddoud, O., Salhi Hannachi, A., Mars, M., Marrakchi, M., and Trifi, M. (2007) Inferring of genetic diversity and relationships in a Tunisian fig (*Ficus carica* L.) germplasm collection by random amplified microsatellite polymorphisms. *Journal of Integrative Plant Biology* 49, 386–391.

CICYTEX, CDICYTDE (2018) www.cicytex.juntaex.es/en/ (accessed 29 April 2020).

El Rayes, R. (1995) The fig tree in the Mediterranean region and in Syria. *Cahiers Options Méditerranéennes* 13, 79–83.

FAOSTAT (2021) *World Fig Production*. www.fao.org/faostat/en/#data/QC (accessed 11 April 2021).

Ferrara, G., Mazzeo, A., Gallotta, A., Pacucci, C., Matarrese, A.M.S., Tarantino, A., Incerti, O., Marcotuli, I., Nigro, D., Blanco, A. and Gadaleta, A. (2017) Fruit-set and SSR markers of fig cultivars from Puglia region, Southeastern Italy. *Acta Horticulturae* 1173, 39–44.

Ferrara, G., Mazzeo, A., Pacucci, C., Matarrese, A.M.S., Tarantino, A., Crisosto, C., Incerti, O., Marcotuli, I., Nigro, D., Blanco, A. and Gadaleta, A. (2016) Characterization of edible fig germplasm from Puglia, southeastern Italy: Is the distinction of three fig types (Smyrna, San Pedro and Common) still valid? *Scientia Horticulturae* 205, 52–58.

Flaishman, M., Rodov, V. and Stover, E. (2008) The fig: botany, horticulture and breeding. *Horticultural Reviews* 34, 113–196.

Gededda, Y.I. and Gumaty, K. (2015) Fruit physical and chemical characteristics of two fig cultivars grown in different regions of Libya. *Acta Horticulturae* 1089, 357–359.

Gharbaoui, A.E. (1981) *La Terre et l'homme dans la péninsule tingitane: étude sur l'homme et le milieu naturel dans le Rif occidental*. Institut scientifique, Rabat, Morocco.

Grassi, G. (1982) Il fico: Acquisizioni scientifiche italiane dell'ultimo trentennio. *Frutticoltura* 44, 7–16.

Ikegami, H., Nogata, H., Hirashima, K., Awamura, M. and Nakahara, T. (2009) Analysis of genetic diversity among European and Asian fig varieties (*Ficus carica* L.) using ISSR, RAPD, and SSR markers. *Genetic Resources and Crop Evolution* 56, 201–209.

ISTAT (2021) www.istat.it/en/ (accessed 11 April 2021).

Karadeniz, T. (2003) A study on some fruit characteristics and propagations of these by hardwood cuttings of local fig cultivars grown in Ordu (Turkey). *Acta Horticulturae* 605, 107–112.

Koka, T. (2010) *Fig germplasm in Albania (Ficus carica)*. Monography. albPaper, Tirana, Albania.

Lahbib, T. (1984) Etude pomologique des variétés de figuier (*Ficus carica* L.) répertoriées au Sahel tunisien. *Mémoire de Fin d'Études du Cycle de Spécialisation* INAT, Tunis, Tunisie.

Lansky, E.P., Paavilainen, H.M., Pawlus, A.D. and Newman, R.A. (2008) Ficus spp. (fig): Ethnobotany and potential as anticancer and anti-inflammatory agents. *Journal of Ethnopharmacology* 119, 195–213.

Leonel, S. and Damatto Junior, E.R. (2008) Efeitos do esterco de curral na fertilidade do solo, no estado nutricional e na produção da figueira. *Revista Brasileira de Fruticultura* 30, 534–539.

Li, J., Tian, Y.-Z., Sun, B.-Y., Yang, D., Chen, J.-P. and Men, Q.-M. (2011) Analysis on volatile constituents in leaves and fruits of *Ficus carica* by GC-MS. *Chinese Herbal Medicines* 4, 63–69.

Lianju, W., Weibin, J., Kai, M., Zhifeng, L. and Yelin, W. (2003) The production and research of fig (*Ficus carica* L.) in China. *Acta Horticulturae* 605, 191–196.

Mahmoudi, S., Khali, M., Benkhaled, A., Boucetta, I., Dahmani, Y., Attallah, Z. and Belbraouet, S. (2018) Fresh figs (*Ficus carica* L.): Pomological characteristics, nutritional value, and phytochemical properties. *European Journal of Horticultural Science* 83, 104–113.

Marcotuli, I., Mazzeo, A., Nigro, D., Giove, S., Giancaspro, A., Colasuonno, P., Prgomet, Ž., Prgomet, I., Tarantino, A., Ferrara, G. and Gadaleta, A. (2019) Analysis of genetic diversity of *Ficus carica* L. (Moraceae) collection using simple sequence repeat (SSR) markers. *Acta Scientiarum Polonorum Hortorum Cultus* 18, 93–109.

Mars, M., Carraut, A., Marrakchi, M., Gouiaa, M. and Gaaliche, F. (1994) Ressources génétiques fruitières en Tunisie (poirier, oranger, figuier, grenadier). *Plant Genetic Resources Newsletter* 100, 14–17.

Mars, M., Chatti, K., Saddoud, O., Salhi-Hannachi, A., Trifi, M. and Marrakchi, M. (2008) Fig cultivation and genetic resources in Tunisia, an overview. Acta *Horticulturae* 798, 27–32.

Martins, M.B. (2008) Structural issues on the Portuguese fig production: recent developments. *Acta Horticulturae* 798, 19–26.

Meziant, L., Saci, F., Bey, M.B. and Louaileche, H. (2015) Varietal influence on biological properties of Algerian light figs (*Ficus carica* L.). *International Journal of Bioinformatics and Biomedical Engineering* 1, 237–243.

Ministry of Agriculture Jihad (2019) Statistics of Agricultural Productions. https://www.maj.ir/page-NewEnMain/en/0 (accessed 29 April 2020).

Morton, J. 1987. Fig (*Ficus carica*). In: Fruits of warm climates. Julia F. Morton, Miami, FL, pp. 47–50.

NSW-Agriculture (2002) *Fig growing in NSW*. Agfact H3.1.19, Orange, New South Wales, Australia. www.dpi.nsw.gov.au/__data/assets/pdf_file/0017/119501/fig-growing-nsw.pdf (accessed 5 August 2021).

Peña, A. and Fischer, G. (1993) El cultivo del brevo *Ficus carica* en la zona alta de la Colombia. *Agro Desarollo* 4 (1–2), 231–243.

Pereira, C., Serradilla, M.J., Pérez-Gragera, F., Martín, A., Villalobos, M.C. and López-Corrales, M. (2017) Evaluation of agronomic and fruit quality traits of fig tree varieties (*Ficus carica* L.) grown in Mediterranean conditions. *Spanish Journal of Agricultural Research* 15, e0903.

Perez-Jimenez, M., Lopez, B., Dorado, G., Pujadas-Salva, A., Guzman, G. and Hernandez, P. (2012) Analysis of genetic diversity of southern Spain fig tree (*Ficus carica* L.) and reference materials as a tool for breeding and conservation. *Hereditas* 149, 108–113.

Pérez-Sánchez, R., Morales-Corts, M.R. and Gómez-Sánchez, M.A. (2016) Agromorphological diversity of traditional fig cultivars grown in central-western Spain. *Genetika* 48, 533–546.

Polat, A.A. and Çalışkan, O. (2017) Correlations among important fruit quality and plant characteristics of some fig genotypes. *Journal of Life Sciences* 11, 141–144.

Prataviera, A.G. (2003) El cultivo de la Higuera: una producción alternativa en marcha. In: Revista de Información sobre Investigación y Desarrollo Agropecuario. Producciones Alternativas: Idia XXI. Instituto Nacional de Tecnología Agropecuaria, Buenos Aires.

Rasoly, M. and Chandrashelar, H.M. (2018) Export performance of dried fruits from Afghanistan - A study in Afghanistan. *International Journal of Research in Business Studies and Management* 5, 16–22.

Rattanpal, H., Sidhu, G. and Bons, H. (2017) Tree and fruit characteristics of fig cultivars grown in Punjab. *Acta Horticulturae* 1173, 117–120.

Ren, D.S., Duan, X.L. and Abudu, K.D. (2000) Development and utilization of figs in Xinjiang. *Quarterly Forest By-Product Spec China* 53, 45–46.

Rhouma, A. (1996) Les ressources phytogénétiques oasiennes: le figuier (*Ficus carica* L.). *Actes des 3èmes journées nationales sur les acquis de la recherche agronomique, vétérinaire et halieutique*, Nabeul, Tunisie.

SANA (2020) https://www.sana.sy/en/?p=203649 (accessed 5 August 2021).

Uzundumlu, A., Oksuz, M.E. and Kurtoglu, S. (2018) Future of fig production in Turkey. *Journal of Tekirdag Agricultural Faculty* 15, 138–146.

Weather and Climate (2020) *World Weather and Climate Information* www.weather-and-climate.com/average-monthly-Rainfall-Temperature-Sunshine-in-Egypt (accessed 29 April 2020).

Yilmaz, S., Gozlekci, S. and Ersoy, N. (2017) A review of fig sector in Turkey. *Acta Horticulturae* 1173, 409–414.

# 4 Biodiversity, Germplasm Resources and Breeding Methods

**H. Ikegami[1]*, J.I. Hormaza[2], F.C. Balas[3], M.J. Serradilla[3] and M. López-Corrales[4]**

[1]*Fukuoka Agriculture and Forestry Research Center, Yukuhashi, Japan;* [2]*IHSM La Mayora – CSIC – UMA, Malaga, Spain;* [3]*Junta de Extremadura, Finca La Orden-Valdesequera Research Centre (CICYTEX), Badajoz, Spain;* [4]*Junta de Extremadura, Agri-Food Technological Institute of Extremadura (INTAEX-CICYTEX), Badajoz, Spain*

## 4.1 Introduction

The common fig (*Ficus carica* L.) is considered to be the oldest domesticated plant (Kislev *et al.*, 2006) and has been an important food source in human history (Flaishman *et al.*, 2008b). The fig is also a historic fruit mentioned frequently in the Old Testament. Fig trees produce nutritious fruit rich in calcium and fiber (Arvaniti *et al.*, 2019), and the perceived nutritional value has been boosted on account of people's increased health consciousness in recent years. Management of genetic resources and breeding of the fig are essential approaches to meet market demands. The earliest attempts to develop improved cultivars were conducted in the late nineteenth century and subsequent artificial crossing and selection led to the development of commercial cultivars in the mid-twentieth century. Fig is usually vegetatively propagated for commercial cultivation, but is propagated from seeds for breeding purposes.

This chapter focuses on basic matters pertaining to fig breeding. Specifically, we discuss fig genetic resources that form the foundation of breeding and genetic diversity, breeding history, target traits of the breeding process and the characteristics and application of conventional and contemporary breeding methods.

## 4.2 Biodiversity and Germplasm Resources

*Ficus carica* is a diploid species ($2n = 2x = 26$) that belongs to the Moraceae, a family with over 1400 species distributed in about 40 genera. The genus *Ficus* L. contains about 700 species, mainly in the tropics, currently divided into six subgenera (Berg, 2003). The common fig is thought to have originated in the western Asia (Stover *et al.*, 2007) and, together with the grapevine and the olive, is considered one of the three classical fruit trees associated with the beginning of horticulture in the Mediterranean Basin (Zohary and Spiegel-Roy, 1975; Zohary and Hopf, 2000; Janick, 2005; Khadari *et al.*, 2005) where it is currently present in wild and feral forms in most countries of this region.

*Corresponding author: ikegami@farc.pref.fukuoka.jp


DOI: 10.1079/9781789242881.0004

The long history of fig cultivation has resulted in hundreds of cultivars worldwide and, consequently, germplasm characterization and diversity studies are valuable for efficient management of fig genetic resources. Condit (1955) described more than 600 fig cultivars. Synonymies and homonymies are common in vegetatively propagated fruit tree species, such as figs. Fig cultivar identification has usually been carried out using traditional methods based on phenotypic traits (International Plant Genetic Resources Institute (IPGRI) and International Centre for Advanced Mediterranean Studies (CIHEAM), 2003). However, traditional characterization using morphologic and agronomic parameters is expensive and often ineffective in distinguishing genotypes due to the influence of the environment, the limited number of discriminating traits, and the need to study a high number of characters in organs that are not always present.

Various studies have reported the use of morphological traits (Papadopoulou *et al.* 2002; Chatti *et al.*, 2004; Çalişkan and Polat, 2008; Aljane and Ferchichi, 2010; Giraldo *et al.*, 2010; Podgornik *et al.*, 2010; Aljane *et al.*, 2012; Almajali *et al.*, 2012) and isozyme markers (Cabrita *et al.*, 2001; Chessa and Nieddu, 2005) for fig characterization. However, these parameters are influenced by environmental conditions and the phenological status of the trees. To overcome these difficulties, different molecular tools such as RAPDs (Khadari *et al.*, 1995; Galderisi *et al.*, 1999; Cabrita *et al.*, 2001; Papadopoulou *et al.*, 2002; Chessa and Nieddu, 2005; Salhi-Hannachi *et al.*, 2005; Akbulut *et al.*, 2009; Ikegami *et al.*, 2009; Chatti *et al.*, 2010; Dalkiliç *et al.*, 2011), ISSRs (Salhi-Hannachi *et al.*, 2004a,b, 2005; Khadari *et al.*, 2005; Ikegami *et al.*, 2009; Chatti *et al.*, 2010; Almajali *et al.*, 2012), AFLPs (Cabrita *et al.*, 2001; Baraket *et al.*, 2009, 2011), RFLPs (Khadari *et al.*, 2005) or SSRs (Khadari *et al.*, 2001, 2005; Giraldo *et al.*, 2005; Saddoud *et al.*, 2005, 2011; Bandelj *et al.*, 2007; Achtak *et al.*, 2009, 2010; Ikegami *et al.*, 2009; Aradhya *et al.*, 2010; Chatti *et al.*, 2010; Baraket *et al.*, 2011; Perez-Jimenez *et al.*, 2012; do Val *et al.*, 2013; Balas *et al.*, 2014) have been used for fig germplasm characterization and diversity analyses.

Different fig germplasm collections have been reported in several countries (Flaishman *et al.*, 2008b; Aljane *et al.*, 2018) (Fig. 4.1). In Europe, the most important collections are in Italy with over 400 accessions, Porquerolles in France with over 150 accessions, Badajoz in Spain with over 230 accessions and other minor collections in Greece (Aljane *et al.*, 2018). There are also important collections in Elbreyli in Turkey with over 300 accessions, Algeria, Morocco, Tunisia, Syria, Georgia, Turkmenistan and the USA (Aljane *et al.*, 2018).

**Fig. 4.1.** Fruit morphological attributes among different pomegranate accessions. (Figure used with permission from Cenap Yilmaz).

## 4.3 Reproductive Biology of *Ficus carica*

The premise of gynodioecy, ecotypes, fruit set diversity and symbiotic relation with insects complicates the understanding of the reproductive biology of *F. carica*. Therefore, knowledge of the background would be helpful when making a breeding plan. The common fig is a gynodioecious species with two sexual forms: male trees (caprifigs) that produce syconia with separate male and female flowers, and female trees that produce syconia with only female flowers that will develop into edible seeded figs if pollinated. Since only male trees produce pollen, the common fig is functionally a dioecious species. Three types of female figs can be distinguished (Flaishman *et al.*, 2008b): the common type that develops fruits parthenocarpically without pollination and can produce one (unifera varieties) or two (bifera varieties) crops, the Smyrna type that requires pollination with pollen from caprifigs, and the San Pedro type that produces parthenocarpically a first crop (brebas) and a second main crop (figs) after pollination with pollen from caprifigs. Pollination in the genus *Ficus* is dependent on the coevolution of the different species of the genus with pollinating wasps of the family Agaonidae (for reviews see Kjellberg *et al.*, 1987; Cook and Rasplus, 2003; Jackson, 2004; Machado *et al.*, 2005; Herre *et al.*, 2008). In the case of the common fig, pollination (caprification) is performed by *Blastophaga psenes* L. The caprifigs produce fruits in three crop cycles during each growing season, each harboring the larvae, pupae and, temporarily, the adults of the pollinating *Blastophaga* (Flaishman *et al.*, 2008b): profichi that ripen in early summer, mammoni that ripen in autumn and mamme that overwinter on the tree and mature in spring. Only the profichi carry pollen and are used for pollination.

## 4.4 Breeding History

For a thorough understanding of fig breeding, there is much to be gained from looking back on its history. The 'dawn' period of fig breeding, almost entirely using a hybridization and selection approach, is summarized in detail by Condit (1928) and Storey and Condit (1969).

Prior to the initiation of controlled breeding, 'fig seedlings sprang up in all sorts of unexpected places, the seeds having been scattered by man, bird and beast. The varieties of figs grown in gardens or in commercial orchards seem to be the result of chance pollinations rather than any controlled or prearranged crosses' (Condit, 1928). This perspective has been supported by many authors.

The following breeding attempts were conducted in Mediterranean coastal regions, which is the original area of fig cultivation. Trabut (1922) in Algeria performed crosses between *F. carica* and two closely related species, *F. palmata* and *F. pseudocarica* L. du Sablon found approximately 100 fig trees undoubtedly derived from seeds and observed progeny segregation in France. Pelicano and Guglielmi (1908) mention the need for breeding alternative cultivars to 'Dottato' in Italy (Condit, 1928).

In the USA, numerous breeders have aimed to select superior lines derived from natural or artificial hybrid seeds since the latter half of the nineteenth century. The earliest attempt was initiated in 1875 by Rev. Robert Jope (Condit, 1928). Subsequently, Swingle (1908, 1912), Hunt (1911, 1912), Burbank (1914), Rixford (1918, 1926), Noble (1922), and Condit (1928) engaged in breeding efforts. The main goal was to obtain parthenocarpic seedlings, but many of the early attempts obtained only male individuals and Smyrna figs in advanced generations, as the seed parent was derived from a Smyrna cultivar. The seedlings raised did not represent an improvement on the existing cultivars. Following this period, the most significant and large-scale breeding programme in the 20th century was implemented in the USA. Ralph E. Smith of the University of California initiated a breeding project in 1922, which was inherited by Condit in 1928 and subsequently by Storey in 1969 and continued until the 1980s (Storey, 1975; Stover *et al.*, 2007). The four cultivars 'Conadria', 'Di Redo', 'Flanders', and 'Excel' were the first cultivars selected in this breeding programme. 'Conadria' was the first cultivar in history to be developed in a planned breeding programme. The cultivation area of 'Conadria' had attained 267 ha (Storey, 1975) and it is among the most successful fig cultivars on the market. Its name is a contraction of 'Condit's Adriatic'. The cultivar 'Tena' was also selected in this programme and released in the mid-1970s (Doyle and Ferguson, 1998). The research findings derived from the programme have been systematically summarized, and the genetic model proposed became the basis of current molecular research (Storey and Condit, 1969; Storey, 1975).

An additional breeding programme to develop better-adapted cultivars through seedling selection was instituted at Louisiana State University in 1954 (O'Rourke, 1966). 'O'Rourke' is a cultivar selected from seedlings raised in the programme by E. N. O'Rourke in 1960 (Johonson *et al.*, 2010).

The University of California established a new fig breeding programme in 1991 aimed at developing cultivars that did not require caprification (see Chapter 5) and showed other fruit characteristics comparable to those of Calimyrna figs. In this programme, female lines of Calimyrna figs are backcrossed with caprifigs that have persistent traits and a certain percentage of the

Calimyrna genome. The resultant cultivar 'Sequoia' which is a common-type and has advantages over Calimyrna figs was released for use in the fresh market (Doyle and Ferguson, 1998; Doyle *et al.*, 2003; Stover *et al.*, 2007).

In addition to those in the USA, breeding programmes have been implemented in several other countries. There is a record that G. N. Slykov of the All-Union Institute of Plant Industry, Leningrad in the former USSR was a fig breeder (Storey, 1975; IBPGR, 1986). The Horticultural Experiment Station of the Ministry of Agriculture and Commerce in Japan selected the SanPedro-type cultivar 'Tanikawa' in 1921. However, it was not widely cultivated because it can set only the first crop (the summer crop in Japan) and the fruit quality traits are poor. About 70 years later, a full-scale breeding programme was established in Fukuoka Prefecture in 1989, from which the cultivars 'Himehorai' (Awamura *et al.*, 1998) and 'Toyomitsuhime' (Nogata and Awamura, 2005) were registered in 2000 and 2006, respectively. The former produces low yields and was not commercially viable, but the latter is characterized by strong sweetness and the cultivation area had expanded to 53 ha as of 2018. In Israel, a breeding programme to develop fig cultivars for fresh and dry consumption was initiated in 2005 and targeted fruit quality and shelf-life extension as the major objectives. Diverse genetic resources were used for breeding, including 300 genotypes and hybrids of common fig and other *Ficus* species. Annual assessment of about 8000 hybrid seedlings reveals the best parents and cross combinations, and ~60 promising hybrids are selected. One early- and one late-ripening cultivar derived from the programme are in commercial production (Flaishman *et al.*, 2017).

## 4.5 Breeding Programmes

In spite of the economic and ecological importance of the common fig in different regions of the world, there has been little genetic improvement of figs. Most of the current production still relies on old accessions, locally grown farmer selections. As more than 80% of fig cultivation depends on rain availability, the sustainability of fig production is jeopardized by climate change.

### Breeding criteria

In recent years, the commercial fruit sector has focused more on the quality and shelf-life demanded by retailers than on the organoleptic quality demanded by consumers. However, this trend is gradually evolving, mainly due to changes in consumer purchasing patterns that prioritize flavor over other organoleptic characteristics such as color, size, etc. These consumers' requests are increasingly being adopted by the industry and becoming the main goals of modern breeding programmes. The future will lie in developing new cultivars that satisfy the industry demands, reduce postharvest losses, and satisfy the consumers' flavor expectations.

Current fig breeding is aimed at meeting different requirements by all stakeholders in the agri-food chain. In terms of agronomy and production, the main breeding goals are:

Higher yielding varieties.
Short juvenile period/rapid entry into production.
Low vigorous trees in order to facilitate harvesting.
Adequate tree architecture that allows modern formation systems such as espalier, super-intensive frames, net undercover or greenhouse production.
Parthenogenesis/avoid caprification.
Extended ripening season, early and late ripening varieties.
Extended on-tree fruit useful life.
Pest and disease tolerance.
Soil salinity tolerance.
Drought resistance.
Frost resistance.
Low chilling requirements.

Industry requirements are more related to fruit quality through postharvest suitability in order to achieve better marketing standards, especially for fresh product, and satisfy consumer tastes:

Large fruits.
Firm fruits.
Hard and elastic skin.
Absence of skin cracking.
Small ostiole.
Fruit fungal resistance.
Low fruit respiratory metabolism.
Extended postharvest fruit useful life.
High sugar content, especially for dried figs.

High acidity.
High anthocyanin (dark-skinned figs)/hydroxycinnamic acid (light-skinned figs) content.
Homogeneity of fruit size and shape.

In the case of both figs and brebas, breeding programmes should aim to increase the resistance of the skin to cracking, and, thus, minimize the impact of skin damage and reduce the risk of infection by microorganisms. Besides, another great limitation of brebas and figs is the size of the ostiole. To ensure a longer postharvest life, the size of the ostiole needs to be as small as possible (Crisosto *et al.*, 2011). Therefore, within a fig breeding programme, it is essential to develop cultivars that bear fruits with resistant skin damage and closed ostiole to ensure better adaptation to the processes of harvesting, trading and distribution. Along with these prerequisites mentioned above, another fundamental trait is firmness. The search for firm cultivars is a priority in order to guarantee postharvest life, together with the search for cultivars with low respiratory metabolism (Crisosto *et al.*, 2011).

Regarding organoleptic characteristics, these depend mainly on the ratio of soluble solids to titratable acidity. This maturation index shows a positive correlation with consumer acceptance (Crisosto *et al.*, 2010). Organic acids are the necessary substrate to maintain the respiratory activity of the fruit during its postharvest life, so it is essential to develop high titratable acidity cultivars (Valero and Serrano, 2010). To ensure good consumer acceptance, high acidity has to be compensated with high soluble solids content.

Brebas and figs are considered high functional value foods due to presence of bioactive compounds such as anthocyanins among others, as well as for their high fiber, vitamins and minerals content (Veberic and Mikulic-Petkovsek, 2016). It would be interesting to establish high anthocyanin content for dark-skinned and red-fleshed cultivars, or hydroxycinnamic acid content for light-skinned cultivars as selection criteria for new progenies. The increase of the functional power of these fruits will have a positive impact in consumer acceptance.

Another interesting aim is to identify caprifig accessions with better performance as male parents in edible fig breeding programmes. Caprifig traits such as amount of pollen production, pollen viability and germination ratios, ripening period or disease and pest tolerance and resistance have an effect on fruit quality and phytochemical characteristics of edible figs (Caliskan *et al.*, 2017). Stover and Aradhya (2008) have identified two persistent caprifigs that could be very valuable to breed high-quality figs that would not require caprification to set heavy crops.

### Criteria for rootstock improvement

Fig trees are easy to propagate with soft or hardwood cuttings and, therefore, there has been little effort to develop fig rootstocks. However, in some regions, particular problems could be solved using appropriate rootstocks. An example is salinity problems in arid and semiarid areas from Iran where annual precipitation is less than annual evapotranspiration. This triggered Zarei *et al.* (2016) to evaluate the offspring of the variety Sabz and four caprifig varieties for salinity tolerance. They identified a salt-tolerant genotype with potential as a fig rootstock as it has the ability to effectively exclude $Na^+$ and $Cl^-$ from leaves in soils with up to 6 dS × $m^{-1}$ salinity.

### Characters related to the tree

Architectural traits in fig tree breeding programmes have been limited to early work by Lauri and Caraglio (1993). However, this is a promising way to improve vegetative development and improve regular bearing (Costes *et al.*, 2004). Gaaliche *et al.* (2016) quantified branch development in fig trees and the pattern of lateral branch distribution. These authors described an architectural typology characterizing all cultivars and established three distinct groups based on branch direction (orthotropic or plagiotropic), position (basitonic, mesotonic and acrotonic), and ramification distribution (continuous or diffuse).

The fig tree is reportedly moderately saline tolerant. Caruso *et al.* (2017) reported fig leaves remained healthy and green, and gas exchange parameters relatively high, after a few weeks of root exposure to 100 mM sodium chloride (NaCl). However, concentrations of ≥200 mM NaCl resulted in extensive leaf necrosis, leaf abscission and a dramatic decrease in leaf photosynthesis. Recently, a transcriptome analysis of leaves of fig plants exposed to salt stress identified a repertoire of salt-regulated genes (genes involved in hormone biosynthesis, proline accumulation and redox system) and identified new genes not previously associated with salinity responses (Vangelisti *et al.*, 2019).

## 4.6 Breeding Techniques

Most of the world fig cultivars are wild seedling selections and open-pollinated seedlings of cultivated trees, identified by anonymous breeders (Flaishman *et al.*, 2008b). However, since the beginning of controlled cross-breeding in the USA (Swingle, 1908), major breeding programmes have been implemented by several groups (O'Rourke, 1966; Storey, 1975; Doyle and Ferguson, 1998; Nogata and Awamura, 2005; Flaishman *et al.*, 2008b, 2017). Most of these programmes are based on traditional cross-breeding. In this section, we review both classical and newer breeding techniques, with a focus on the latest genomic approaches.

### Cross-breeding

Traditional controlled crosses are one of the most commonly used methods for expanding genetic variation. There are many reviews of classical breeding approaches to cross-breeding (Condit, 1947; Storey, 1975, 1976, 1977; Storey and Condit, 1969; Obenauf *et al.*, 1978; Ferguson *et al.*, 1990; Mars, 2003; Flaishman *et al.*, 2008b). An overview of the works accomplished in the past in this area are as described in the Breeding History section.

When developing a breeding plan, it is worth considering the genetic and genomic background of figs. Fig (*F. carica*) is a diploid ($2n = 26$), gynodioecious or dioecious, and highly heterogeneous species. Being diploid means that the inheritance pattern is simple compared with homogeneous and polyploid species. Gynodioecy requires female and male lines for mating. Breeders can use superior varieties or lines as female parents. In contrast, male lines are less readily available. Breeders should select and save the male lines from breeding populations for use as male parents.

Storey's cross-pollination method (Storey, 1975) is the most common used in fig trees. In this method, two holes are drilled in the female fruit and pollen is blown into the syconium cavity with a puffer. However, this method is technically challenging. An alternative method is to cut and pollinate the fruit directly (O'Rourke, 1966); the details are shown in Fig. 4.2. The pollination should be done in the first fruit development stage (Crane and Brown, 1950; Crane and Baker, 1953). When exposed to wind and rain after pollination, seed set decreases, so it is beneficial to cover the fruit with a paper bag after pollination, especially in rainy areas. The pollinated fruits mature on the tree. After fruit harvesting, the seeds should be separated from the flesh in water, then washed and air dried (Fig. 4.2).

Vermiculite or sandy soil are the best substrates for germinating the seeds (Storey, 1975). Sterilizing the soil improves the germination rate. The beds should be maintained at 25–30°C and at high relative humidities. Germination will start within 2 weeks to 2 months (Fig. 4.3). Pretreating the seeds with concentrated sulphuric acid for about 1 min and removing the seed coat will improve germination. Depending on the parents, a syconium can produce few hundred to a maximum of about 2000 seeds. Backcrossing is often used to maintain the target trait(s) during the breeding process, but as the homogeneity of the parents' genome increases, the seed fertility decreases.

When the plant is about 5–10 cm tall, after the juvenile stage, the seedlings can be transplanted into a fine structured soil. Subsequent growth depends on the temperature, but under ideal growing conditions, plants will bear fruit in 1–2 years. It requires 2–3 years to produce the initial fruits for evaluation; tree traits require longer to evaluate. Since the fig is a perennial tree crop, when the scale of a breeding population grows and becomes more complex, establishing a breeding plan that takes advantage of additional techniques such as marker-assisted selection (MAS) and genomic selection (GS), as described later, is desirable.

### Interspecific hybridization

Interspecific crosses are used to develop new species. Some desirable traits not present in *F. carica*

**Fig. 4.2.** Fig pollination methods. (A) Various caprifig syconia. Fruits are broken and pollen is dropped onto medicine wrapping paper. (B) Medicine wrapping paper with pollen and a disposable dropper made of resin. Pollen can be stored for a long time in a sealed vial with silica gel at −20°C. (C) Advanced pollination method (Storey, 1975): holes are made to insert pollen at two positions, in the ostiole of syconia of about 15–25 mm and at the equator. (D) Simple pollination method: syconia is cut at the top one-third line (O'Rourke, 1966). (E) After pollination, the syconia is covered with a bag to protect and keep the structure moist. The bag may be made of paper or waterproofed to prevent intrusion of rainwater. (Author's own figure.)

are found in closely-related species and it is possible to transfer those traits between species using conventional breeding methods. In the genus *Ficus* around 800 different species are estimated (Corner, 1965; Hortorum, 1976).

In fig species with syconia, such as *F. erecta*, the method of pollination is the same as that of *F. carica*. It has long been recognized that *F. carica* can be crossed with other closely-related species (Storey, 1975). In contrast, crossing between

**Fig. 4.3.** Extraction and sowing of seeds from pollinated fruit. (A) After crushing a mature pollinated fruit, the fruit is sunk in a beaker containing water and stirred with a magnetic stirrer to separate seeds and flesh. (B) Fertile seeds will sink in water. (C) Seeds that sink are dried on filter paper. Remove any remaining flesh completely to avoid a decrease in the germination rate. Seeds should be stored under dry and refrigerated conditions. (D) Seeds on 1 mm grid paper. (E) Seeds germinate in 2–8 weeks when sown in vermiculite or sandy soil. Germination time varies depending on temperature and mating combinations. (F) Germination also may be performed on filter paper under saturated humidity. (Author's own figure.)

distantly related species often does not yield fertile seeds, or seeds that germinate and plants that grow poorly. This is described as an interspecific incompatibility (reproductive barrier), and the detailed mechanism in fig has not been addressed. To obtain interspecific hybrids, the parent materials should be as closely related to *F carica* as possible. Another way to avoid the incompatibility is to cross genetically different individuals within the species.

The oldest scientific report on interspecific crossing within the genus *Ficus* was a hybrid between *F. carica* and *F. palmata* and between *F. carica* and *F. pseudo-carica* (Trabut, 1922). Condit (1947) made the same cross later and reported its behavior. He further hybridized *F. pumila* with *F. carica*, using pollen of the latter species (Condit, 1950). Yakushiji *et al.* (2012) first reported the hybridization between *F. carica* and *F. erecta*. *F. erecta* is resistant to *Ceratocystis* canker, caused by *Ceratocystis fimbriata* (Kato *et al.*, 1982), a serious disease of *F. carica*. The breeding goal was to introduce this resistance trait into *F. carica*. Subsequently, the hybrid F1 was backcrossed to *F. carica*, and individuals closer to *F. carica* (BCF1) were obtained (Yakushiji *et al.*, 2019). Flaishman *et al.* (2017) reported interspecific hybrids from three combinations of *F. carica* × *F. pumila*, *F. carica* × *F. palmata* and *F. carica* × *F. auriculata*. Some of the interspecific hybrids showed common fig characteristics, while other hybrids produced a new type of fruit (Flaishman *et al.*, 2017).

## Mutation breeding

Mutagenesis is another effective approach to expanding the available genetic variation in crops. The technical advantages of inducing mutations include the ability to artificially create completely new mutations absent in extant populations, and the ability to adjust the mutation pressure. The latter feature makes it possible to improve only some of the desired traits by applying minimal mutation pressure to a large number of clones of desirable cultivars. Artificial mutagenesis uses two approaches: physical (radiation) and chemical (compounds). In addition, mutation can also be induced during *in vitro* culture.

Radiation mutations have been obtained using gamma irradiation of cuttings, seeds, pollen and budsticks (Akhund-Zade, 1981; Özen *et al.*, 2017). For cuttings, seeds and pollen, dosage levels from 50 to 100 Gy were used. With budsticks, five radiation doses were tested (0, 10.0, 16.9, 25.3, 50.7, 60.7 Gy); the $LD_{50}$ was 50.7 Gy for Sarilop and 25.3 Gy for Bursa Siyahi (Özen *et al.*, 2017). Regardless of the mutation induction method, in order to obtain a desired mutant, it is beneficial to develop a survival curve in advance to determine how much mutation pressure can be applied. Generally, low lethality increases the probability of obtaining an individual with desired mutations. Alkylating agents such as EMS and ENU are often used as chemical mutagens, but chemical mutagen use for figs has not been reported.

*In vitro* culture is also a way of inducing somatic mutagenesis (Jona and Gribaudo, 1991; Haelterman and Docampo, 1994; Lopez Corrales *et al.*, 1998; Miguel and Lima-Costa, 1998; Nobre and Romano, 1998; Yakushiji *et al.*, 2003; Kim *et al.*, 2007). In many cases, however, the mutation rate during culture is unclear or tends to be lower relative to other methods. Until now no useful mutants have been identified in fig using *in vitro* culture.

In addition to artificial mutations, natural mutations occur under normal conditions. Natural mutations are usually observed as individuals or organs that behave differently from the population in commercial orchards. Several natural skin color fig mutants have been reported. In Israel, 'Red Kadota' was obtained as a skin-colored mutant of uncolored 'Kadota' (Flaishman *et al.*, 2007). In China, 'Purple Peel' (Zibao) was identified as a skin color mutant of the uncolored variety 'Green Peel' (Qingpi).

## Ploidy manipulation

Ploidy manipulation has been used in the improvement of many crops (Mason, 2016). The well known advantages of chromosome doubling include the elimination of sterility and sterilization (fruit seedlessness). In the case of fruit trees, the purpose is often to increase fruit size. The most common method of polyploidy induction is colchicine treatment. Storey (1975) reported using colchicine treatment in a few fig cultivars in 1970, but no cytological information guaranteed that the resulting genotypes were polyploid.

More recently, fig polyploids have been obtained by maintaining the axillary buds in sterile culture for 1–4 days in a medium containing 0.01–0.05% colchicine (Furukawa, 1997). Polyploidy acquisition efficiency was 24.1% at most (7/29). It was also possible to obtain polyploids by treating hybrid seeds with 0.01% colchicine solution for 8–16 h (Awamura and Nogata, 2003). All these artificial polyploids were created for the purpose of increasing fruit size (Fig. 4.4). However, all the individuals obtained had defects, weak growth or small fruits, and were not suitable for further development as cultivars or breeding lines.

Only diploid figs had been thought to exist under natural conditions. However, a natural triploid was discovered in Italy (Falistocco, 2009). Later, tetraploids were also discovered among wild figs (Falistocco, 2016). The author presented two possibilities for the origin of 3x plants: sexual polyploidization by means of the formation of 2n gametes and hybridization between 2x and 4x genotypes (Rhoades 1936; Müntzing 1937). The frequencies of polyploidy genotypes in this study (8.1-8.5%) support the first hypothesis as the most probable (Falistocco, 2016). Such triploids produced non-reduced gametes and could be used to enhance the flexibility of fig ploidy manipulation.

## Marker-assisted selection (MAS)

Perennial crops such as fruit trees are characterized by a long juvenile period before full production, and therefore, a long period between generations. These characteristics are obstacles to efficient breeding. Marker-assisted selection (MAS) can overcome the problems of extended juvenility and long generation periods through selection using genetic markers linked to causative genes or quantitative trait loci (QTL) associated with traits of interest.

**Fig. 4.4.** Increasing fruit size in figs has been one of the breeding priorities. (Figure used with permission from Alimohammad Yavari.)

In fig breeding, the most significant trait that might be the first target for MAS is the sex phenotype. Sex determination is considered to be controlled by a single locus with two closely linked pairs of alleles (Storey, 1975). The males and females segregate in the progeny of the most frequent crossing combination, producing Aa (male) and aa (female) genotypes. Only female lines are suitable for commercial fig production, while only male strains can supply the pollen necessary for mating. Thus, the utility value of sex-linked DNA markers that can discriminate the sex phenotype is extremely high. The candidate sex-determining gene, *RAN1* (*RESPONSIVE TO ANTAGONIST 1*), was identified through genomic analysis. A CAPS marker that was closely linked to male and female traits was also identified (Mori *et al.*, 2017; Ikten and Yilmaz, 2019). In addition, convenient PCR-based *RAN1*-linked markers have been developed (Tsugane *et al.*, 2019).

Another trait currently available for MAS is *Ceratocystis* canker disease resistance. This is a soil-borne disease (Kato *et al.*, 1982) that can lead to the death of the whole tree. The disease is most likely to occur in hot humid conditions (Kajitani and Masuya, 2011). Resistance against *Ceratocystis* canker is controlled by a single gene (Yakushiji *et al.*, 2019). This true resistance gene was not found in *F. carica*, but was found to be present in *F. erecta*. Hence, the resistance gene was identified from analysis using the two species and their hybrids. A CAPS marker for this gene that can be applied in MAS has been developed (Shirasawa *et al.*, 2020).

Syconia parthenocarpy (or persistence) is another desirable trait candidate for MAS marker conversion; a genetic model with one set of alleles (one gene) to explain parthenocarpy trait inheritance has been identified (Storey, 1975). The progenies from the crosses between caduceus and persistent including a reciprocal cross produced 1:1 segregation (Saleeb, 1965). For skin color traits, transcriptome and gene expression analyses have been conducted in recent years (Wang *et al.*, 2017; Wang *et al.*, 2019). Several MYB genes have been proposed as key regulators of skin color. Thus far no marker has been identified that can be effectively applied to fruit color selection.

## Genomic selection

Genomic selection (GS) is, in a broad sense, a form of MAS (Singh and Singh, 2015), and is based on the genotype information for marker alleles covering the entire genome. The effects of many marker loci across the entire genome are evaluated, and the estimated value as a marker is used to calculate the genomically-enhanced estimated breeding value (GEBV) for each individual or line. The GS system accesses two types of population: a training population and a breeding population. The training population is used to train the GS model and to estimate the marker effects required to assess individual GEBVs. Breeding populations are subjected to GS to achieve the desired improvements and to select desirable lines to be used as new varieties or new lines.

In the case of the common fig, applying GS appears to have great potential due to the diploid nature of the genome's composition (Condit, 1964). However, no successful cases of GS applications have been reported in fig. In general, GS is more versatile than MAS, because it can predict many traits including quantitative traits simultaneously. The training population does not always require a segregating population, and an existing genetic resource population can be used, and, therefore, it is easy to start GS modeling without any dedicated material preparations.

GS has many advantages as an unprecedented selection method and is expected to be applied to many species. However, it should be noted that there are also some drawbacks to applying GS. First, there is a limit in the phenotype measurement resolution. The prediction accuracy depends on the phenotypic measurement accuracy. The genome-wide marker typing for correctly predicting many traits simultaneously is likely to be high or the accuracy of phenotype prediction is reduced under variable environmental conditions. GS may not be an invincible method for many situations (Hanley, 2018).

## Genetics and genomics

The objective of plant breeding is to genetically improve organisms. Breeding methods are rapidly connected to developments in genetics and genomics. Genome assembly and genome-wide polymorphism information (SNPs, etc.) based on next-generation sequencing (NGS) technologies enable the construction of high-density genetic linkage maps and pseudo-chromosomes that allow approaches such as linkage analysis and genome-wide association studies (GWASs). The availability of GWASs is ground-breaking in the sense that it is no longer necessary to consider the cost of preparing dedicated populations for analysis. Furthermore, because of the low cost of acquiring polymorphism (sequence) information, genetic analyses using various genetic resources can be carried out on many traits. This section summarizes the current status of *Ficus* whole-genome sequences and its application to marker development.

Decoding of the first draft genome sequence of *F. carica* was reported by Mori *et al.* (2017). The target cultivar was 'Horaishi', a cultivar introduced to Japan about 400 years ago. The number of scaffolds obtained by this assembly was 27,995 with 36,138 predicted genes, which reached 248 Mb length and 90% of the complete single-copy genes in BUSCO but covered only 70% of the entire fig genome. Linkage analysis based on polymorphisms obtained from the assembly confirmed that the fig has 13 chromosomes (Condit, 1928; Mori *et al*, 2017).

Usai *et al.* (2020) reported a second common fig genome sequence for the 'Dottato' cultivar. They collected a total of 26.4 Gb long read sequences from the PacBio sequel system (PacBio, Menlo Park, CA, USA) for the assembly. The assembly had 333 Mb, larger than the first assembly (Mori *et al.*, 2017) and close to the estimated genome size. This data set covers 93.3% of a complete and single copy of the BUSCO gene and a total of 37,840 predicted genes (Usai *et al.*, 2020). A total of 407 sequences (266 Mb) were anchored to fig linkage groups to establish 13 pseudomolecule sequences (Usai *et al.*, 2020).

The genome sequence of a wild fig relative, *F. erecta*, was also reported (Shirasawa *et al.*, 2020). This species is known as a donor of resistance for *Ceratocystis* canker. A total of 21.3 Gb long read sequences (65X genome coverage) from the PacBio Sequel system were assembled into 538 contigs covering 331 Mb. The complete, single copy BUSCO gene coverage is 92.3% and a total of 51,806 genes were predicted in the sequences. In the process, 13 pseudomolecule sequences consisting of 204 sequences (273 Mb) were established in accordance with a genetic map for a BC1F1 population (Shirasawa *et al.*, 2020).

In a study by Mori *et al.* (2017), 64 SSRs and 8180 SNPs were mapped on the linkage map, and the isolation of a sex determination gene is the first target of analysis using this information. Storey (1975) assumed two closely-linked genes, the A and G genes, to explain sex determination of fig. In this model, the A gene determines the presence or absence of male flowers while the G gene determines the length of female flowers. The A gene, essentially responsible for the sex phenotype, *RAN1* (*RESPONSIVE TO ANTAGONIST 1*) was identified as the most likely candidate gene. *RAN1* encodes a copper-transporting ATPase (Hirayama *et al.*, 1999) and mediates copper transport into the cell, which is presumed to be involved in the functional ability of ethylene receptors. The GWAS described above confirmed that the SNP in the *RAN1* exon completely matched the sex expression of 122 genetic resources. The genetic resources used in the analysis included *F. palmata*, a very close relative of *F. carica*, which supported the hypothesis that *RAN1* is a causative gene for fig sex determination because its cross-species association has been confirmed. However, since the mutation sites targeted in the analysis are based on a draft genome with a coverage of 70%, reanalysis with an updated genome should be done.

Another case is the *F. erecta Ceratocystis* canker resistance locus (Yakushiji *et al.*, 2012). To map the resistance locus on the genome sequence, Shirasawa *et al.* (2020) employed a BC1F1 population generated from *F. erecta* and *F. carica* (Yakushiji *et al.*, 2019), and genome-wide SNP genotypes of restriction-site associated DNA sequencing (ddRAD-Seq) technology. A total of 13,320 SNPs detected a GWAS peak in chromosome 2 of the *F. erecta* genome. Graphical genotyping of the BC1F1 population refined the locus to a 66.3 kb region harboring nine candidate genes, of which only one candidate gene (Fer_r1.1chr02_g012310.1) shows sequence similarity with resistance genes. Transcriptome analysis confirmed the stable expression pattern of this gene in stems, regardless of pathogen infection. The counterpart sequence is conserved in susceptible fig genomes, but the corresponding gene has a frameshift mutation in the first exon (Shirasawa *et al.*, 2020). It is unknown whether this gene is the final answer. However, many deleterious mutations are found in the gene sequences, suggesting that there may be multiple susceptible alleles in fig lines.

## Genetic transformation

Genetic transformation may be defined as introducing a foreign gene that does not occur endogenously in the recipient into its genome, with the intention of modifying the feature(s) of the recipient organism. Until the emergence of transformation technology, the introduction of novel traits needed the occurrence of mutations linked to the target trait. Using genetic transformation, it is possible to develop modified plants in a shorter period than using conventional breeding approaches. Plants created by transformation do not arise under natural conditions and are, therefore, classified as genetically modified organisms (GMOs), and are regulated by the relevant administrative authorities in each country.

Critical factors in genetic modification experiments are the overall efficiency of the transformation techniques used to stably incorporate the desired genetic material into plant cells or tissues, and the regeneration techniques used to produce viable whole plants from transformed cells (Flaishmann *et al.*, 2008a). Since the efficiency of the culture system varies depending on the plant species, it was necessary to establish appropriate culture conditions for figs. Yakushiji *et al.* (2003) first reported organ formation and plant regeneration from the vegetative organs of fig. In this culture system, adventitious buds were induced by culturing in MS medium containing phloroglucinol and the growth regulator 2,4-D and thidiazuron (TDZ). The first demonstration of *F. carica* transformation was reported by Yancheva *et al.* (2005). For growth regulators they adopted IBA and TDZ as auxin and cytokinin, respectively, and increased the culture efficiency to obtain regenerated plants by keeping the explants aligned in a certain direction. A transformation efficiency of 7.8% was achieved.

The 'Kadota' variety has also been successfully transformed with this system (Flaishmann *et al.*, 2008a). The efficacy of combining IBA and TDZ was confirmed by a Korean group (Kim *et al.*, 2007). Subsequently, Soliman *et al.* (2010) established a culture method for efficient callus induction and shoot formation, and carried out transformation with *Agrobacterium* or gene gun bombardment using these explants. The transformation efficiency was 17.5% when using *Agrobacterium* and 12% when using a gene gun method.

Transformation is not only used directly as a means of improving varieties, but also contributes to advancing breeding programmes by improving the accuracy of MAS through gene function analysis experiments. The tool of choice for breeding has shifted from genetic transformation to genome editing, and it is assumed that in the future, transformation will be adopted mainly for experimental purposes.

## Genome editing

Genome editing is a technology that can introduce a specific mutation into a target endogenous gene in a host genome. Whereas transformation is based on the use of foreign genes, genome editing allows the editing of targeted DNA sequences based only on the host genome. Thus, genome editing can produce organisms that are indistinguishable from those arising by natural mutation. Recently, for this reason, there have been increasing numbers of cases where genome-edited plants are not classified as GMOs in different countries and, consequently, genome editing is one of the most promising genetic modification tools.

Currently, there are three systems commonly used in genome editing experiments: transcription activator-like effector nucleases (TALEN) (Christian *et al.*, 2010; Miller *et al.*, 2011), zinc finger nucleases (ZFNs) (Kim *et al.*, 1996; Carroll, 2011) and clustered regularly interspaced short palindromic repeats-associated proteins (CRISPR-Cas) (Jinek *et al.*, 2012; Mali *et al.*, 2013a,b; Cong *et al.*, 2013; Kim and Kim, 2014). The most popular system is CRISPR-Cas9. This system uses two molecules, called Cas9 protein and a guide RNA that is specifically designed for the desired mutation. The guide RNA binds to the target DNA site, then the Cas9 protein serves to cleave the DNA at that target site. An error may occur when the DNA sequence of the cut target site is repaired. By selecting an individual in which a repair error has occurred from a plurality of candidates, an individual whose target site has been edited can be selected.

ZFN was first applied for targeted mutagenesis in *F. carica* (Peer *et al.*, 2015). In this experiment, whole plants with a repaired uidA gene due to deletion of a stop codon were regenerated and the ZFN-mediated gene modifications were stable. Application of the CRISPR-Cas system in *F. carica* was first performed in *FcPDS* (phytoene desaturase) gene (Flaishman *et al.*, 2020). The *FcPDS* gene encodes an important enzyme in the carotenoid biosynthetic pathway, and loss of function of this gene results in the *Arabidopsis* albino and dwarf phenotypes (Qin *et al.*, 2007). It was confirmed that an albino individual was obtained in the same way as in the known *Arabidopsis thaliana* mutant when *FcPDS* lost its function by genome editing (Flaishman *et al.*, 2020).

These experiments have shown that genome editing technologies can function in *F. carica*. However, genome editing based on the current CRISPR-Cas9 methods usually creates genetically modified plants because foreign genes carried on vectors are also introduced. As a means of avoiding the problem of having these plants classified as GMOs, it has been reported that genome-edited individuals were obtained by directly introducing cleavage enzymes and guide molecules to the growth points of shoot tips and protoplasts, without using the vectors that introduce foreign DNA (Sauer *et al.*, 2016; Hamada *et al.*, 2018). However, when culturing cells, somatic cell mutations are likely to occur and, therefore, it is necessary to be careful of the possibility of individuals appearing with unwanted mutations. With regard to the social aspects of genome editing, intellectual property related to the technology and the regulatory status of genome-edited organisms in each country are factors that make it difficult to apply this technology and use the products generated with it (Friedrichs *et al.*, 2019; Jansing *et al.*, 2019). There are currently no problem-free approaches to genetic modification and it is unclear whether genome manipulation will be used more in future research and development in view of social and regulatory attitudes.

## 4.7 Future Perspectives

The economic importance of fig production has grown and will continue rising in the coming years since the worldwide demand for fresh and dried figs is increasing. In the last two decades, fig cultivation has shifted from being a traditional, under-utilized, rainfed crop, to a modern, irrigated crop. Interest is growing in intensive and super-intensive, greenhouse and hydroponic cultivation. Trait diversity maintained in germplasm banks will facilitate facing the challenges of technological innovations in plant improvement, improving organoleptic quality and postharvest management, and the abiotic stresses drought and salinity generated by climate change, and new pests and diseases.

Fruit quality, shelf-life, harvest time and disease resistance have remained essential breeding goals. Figs are fruits with strong palatability, and, therefore, taste and food functionality are important traits to be improved. In implementing efficient breeding, it is vital to establish standard evaluation methods for traits such as fruit quality, shelf-life, disease resistance, etc. Traditional cross-breeding will continue to be a fundamental breeding method that is not expected to change in its essential features. Interspecific crossing will be another promising choice to expand the variation in fig genetic resources and to provide cultivated figs with new traits such as intrinsic disease resistance. Key achievements have been obtained in molecular biology, biotechnology and breeding with this species. The application of MAS is the first step towards molecular breeding, to be expanded to MAS with GS based on multiple markers. Genome editing technologies will contribute to increase useful mutations and to enable precise genetic modifications to complement MAS. Future genome analyses will identify genes related to many economically important traits and lead to the development and use of DNA markers linked to traits of agronomic interest. All this implies that, as in other fruit species, improved varieties will eventually replace most of the landraces and traditional varieties currently grown. However, the genetic diversity represented by these landraces and traditional varieties is an essential tool that should be maintained as a resource in germplasm collections.

## References


Achtak, H., Oukabli, A., Ater, M., Santoni, S., Kjellberg, F. and Khadari, B. (2009) Microsatellite markers as reliable tools for fig cultivar identification. *Journal of the American Society of Horticultural Science* 134, 624–631.

Achtak, H., Ater, M., Oukabli, A., Santoni, S., Kjellberg, F. and Khadari, B. (2010) Traditional agroecosystems as conservatories and incubators of cultivated plant varietal diversity: the case of fig (*Ficus carica* L.) in Morocco. *BMC Plant Biology* 10, 28.

Akbulut, M., Ercisli, S. and Karlidag, H. (2009) RAPD-based study of genetic variation and relationships among wild fig genotypes in Turkey. *Genetics and Molecular Research* 8, 1109–1115.

Akhund-Zade, I.M. (1981) Radiation mutagenesis in subtropical crops. L-ya Vses Konf.po *Prikl. Radiobiol.* 1981, 50–51 (in Russian, English abstract).

Aljane, F. and Ferchichi, A. (2010) Assessment of genetic diversity among some southern Tunisian fig (*Ficus carica* L.) cultivars based on morphological and chemical characters. *Acta Botanica Gallica* 157, 171–182.

Aljane, F., Nahdi, S. and Essid, A. (2012) Genetic diversity of some accessions of Tunisian fig tree (*Ficus carica* L.) based in morphological and chemical traits. *Journal of Natural Product and Plant Resources* 2, 350–359.

Aljane, F., Essid, A. and Nahdi, S. (2018) Improvement of fig (*Ficus carica* L.) by conventional breeding and biotechnology. In: Al-Khayri, J., Jain, S. and Johnson, D. (eds) *Advances in Plant Breeding Strategies: Fruits*. Springer.

Almajali, D., Abdel-Ghani, A.H. and Migdadi, H. (2012) Evaluation of genetic diversity among Jordanian fig germplasm accessions by morphological traits and ISSR markers. *Scientia Horticulturae* 147, 8–19.

Aradhya, M.K., Stover, E., Velasco, D. and Koehmstedt, A. (2010) Genetic structure and differentiation in cultivated fig (*Ficus carica* L.). *Genetica* 138, 681–694.

Arvaniti, O.S., Samaras, Y., Gatidou, G., Thomaidis, N.S. and Stasinakis, A.S. (2019) Review on fresh and dried figs: Chemical analysis and occurrence of phytochemical compounds, antioxidant capacity and health effects. *Food Research International* 119, 244–267.

Awamura, M. and Nogata, H. (2003) Induced tetraploid Fig in vitro by colchicine treatment. In: Kashimura, Y. (ed.) *Horticulture Research (Japan)* 72 (Suppl.2), 273.

Awamura, M., Yahata, D., Nogata, H., Shoda, K. and Kanafusa, K. (1998) A new fig cultivar 'HIME-HOURAI'. *Bulletin Fukuoka Agriculture Research Centre* 17, 115–118.

Balas, F.C., Osuna, M.D., Domínguez, G., Pérez-Gragera, F. and López-Corrales, M. (2014) Ex situ conservation of underutilised fruit tree species: establishment of a core collection for *Ficus carica* L. using microsatellite markers (SSRs). *Tree Genetics and Genomes* 10, 703–710.

Bandelj, D., Javornik, B. and Jakse, J. (2007) Development of microsatellite markers in the common fig, *Ficus carica* L. *Molecular Ecology Notes* 7, 1311–1314.

Baraket, G., Chatti, K., Saddoud, O., Ben Abdelkarim, A., Mars, M., Trifi, M., Salhi Hannachi, A. (2011) Comparative assessment of SSR and AFLP markers for evaluation of genetic diversity and conservation of fig, *Ficus carica* L., genetic resources in Tunisia. *Plant Molecular Biology Reporter* 29, 171–184.

Baraket, G., Chatti, K., Saddoud, O., Mars, M., Marrakchi, M., Trifi, M. and Salhi-Hannachi, A. (2009) Genetic analysis of Tunisian fig (*Ficus carica* L.) cultivars using amplified fragment length polymorphism (AFLP) markers. *Scientia Horticulturae* 120, 487–492.

Barghini, E., Mascagni, F., Giordani, T., Solorzano Zambrano, L.J., Natali, L. and Cavallini, A. (2017) An insight into structure and composition of the fig genome. *Acta Horticulturae* 1173, 69–74.

Berg, C.C. (2003) Flora malesiana precursor for the treatment of Moraceae 1: The main subdivision of *Ficus*: the subgenera. *Blumea* 48, 167–78.

Burbank, L. (1914) *Luther Burbank*. Burbank Press, New York, pp. 297.

Cabrita, L.F., Aksoy, U., Hepaksoy, S. and Leitao, J.M. (2001) Suitability of isozyme, RAPD and AFLP markers to assess genetic differences and relatedness among fig (*Ficus carica* L.) clones. *Scientia Horticulturae* 87, 261–273.

Çalişkan, O. and Polat, A.A. (2008) Fruit characteristics of fig cultivars and genotypes grown in Turkey. *Scientia Horticulturae* 115, 360–367.

Caliskan, O., Bayazit, S., Ilgin, M. and Karatas, N. (2017) Morphological diversity of caprifig (*Ficus carica var. caprificus*) accessions in the eastern Mediterranean region of Turkey: Potential utility for caprification. *Scientia Horticulturae* 222, 46–56.

Carroll D. (2011) Genome engineering with zinc-finger nucleases. *Genetics* 188, 773–782.

Caruso, G., Gennai, C., Ugolini, F., Marchini, F., Quartacci, M.F. and Gucci, R. (2017) Tolerance and physiological response of young *Ficus carica* L. plants irrigated with saline water. *Acta Horticulturae* 1173, 137–142.

Chatti, K., Salhi-Hannachi, A., Mars, M. *et al.* (2004) Genetic diversity and phylogenic relationships in Tunisian fig (*Ficus carica* L.) cultivars mediated by RAPD. *Biol. Tunisie* 1, 14.

Chatti, K., Baraket, G., Ben Abdelkrim, A., Saddoud, O., Mars, M., Trifi, M. and Salhi Hannachi, A. (2010) Development of molecular tools for characterization and genetic diversity analysis in Tunisian fig (*Ficus carica*) cultivars. *Biochemical Genetics* 48, 789–806.

Chessa, I. and Nieddu, G. (2005) Analysis of diversity in the fruit tree genetic resources from a Mediterranean island. *Genetic Resources and Crop Evolution* 52, 267–276.

Christian, M., Cermak, T., Doyle, E.L., Schmidt, C., Zhang, F., Hummel, A., Bogdanove, A.J. and Voytas, D.F. (2010) Targeting DNA double-strand breaks with TAL effector nucleases. *Genetics* 186, 757–761.

Condit, I.J. (1928) Fig breeding. *Journal of Heredity* 19, 417424.

Condit, I.J. (1947) *The Fig*. Chronica Botanica, Waltham, Massachusetts.

Condit I.J. (1950) Fig characters as affected by climate. *Proceedings of the American Society of Horticultural Science* 55, 114–118.

Condit, I.J. 1955. Fig varieties: A monograph. *Hilgardia* 23, 323–538.

Condit I.J. (1964) Cytological studies in the genus *Ficus* III. Chromosome numbers in sixty-two species. *Madrono* 17, 153–154.

Cong, L., Ran, F.A, Cox, D., Lin, S., Barretto, R., Habib, N., Hsu, P.D., Wu, X., Jiang, W., Marraffini, L.A. and Zhang, F. (2013) Multiplex genome engineering using CRISPR/Cas systems. *Science* 339, 819–23.

Cook, J.M. and Rasplus, J.Y. (2003) Mutualists with attitude: coevolving fig wasps and figs. *Trends in Ecology and Evolution* 18, 241–248.

Corner, E.J.H. (1965) Check-list of Ficus in Asia and Australia, with keys to identification. *Gardens' Bul. Singapore* 21, 1–186.

Costes, E., Lauri, P.E., Laurens, F., Moutier, N., Belouin, A., Delort, F., Legave, J.M. and Regnard, J.L. (2004) Morphological and architectural traits on fruit trees which could be relevant for genetic studies: A review. *Acta Horticulturae* 663, 349–356.

Crane, J.C. and Brown, J.G. (1950) Growth of the fig fruit, *Ficus carica* var. Mission. *Proceedings of the American Society for Horticultural Science* 56, 93–97.

Crane, J.C. and Baker, R.E. (1953) Growth comparisons of the fruits and fruitlets of figs and strawberries. *Proceedings of the American Society for Horticultural Science* 62,142–153.

Crisosto, C.H., Bremer, V., Ferguson, L. and Crisosto, G.M. (2010) Evaluating quality attributes of four fresh fig (*Ficus carica* L.) cultivars harvested at two maturity stages. *Hortscience* 45, 707–710.

Crisosto, C.H., Ferguson, L., Bremer, V., Stover, E. and Colelli, G. (2011) Fig (*Ficus carica* L.), in postharvest biology and technology of tropical and subtropical fruits. In: Yahia, E.M. (ed.) *Cocona to Mango* Vol 3. Woodhead Publishing Ltd, Cambridge, UK, pp. 134–158.

Crossa, J., Pérez-Rodríguez, P., Cuevas, J., Montesinos-López, O., Jarquín, D., De los Campos, G., Burgueño, J., González-Camacho, J.M., Pérez-Elizalde, S. and Beyene, Y. (2017) Genomic selection in plant breeding: methods, models, and perspectives. *Trends in Plant Sciences* 22, 961–975. DOI:10.1016/j.tplants.2017.08.011

Dalkiliç, Z., Mestav, H.O., Günver-Dalkiliç, G. and Kocatas, H. (2011) Genetic diversity of male fig (*Ficus carica caprificus* L.) genotypes with random amplified polymorphic DNA (RAPD) markers. *African Journal of Biotechnology* 10, 519–526.

do Val, A.D.B., Souza, C.S., Ferreira, E.A., Salgado, S.M.L., Pasqual, M. and Cançado, G.M.A. (2013) Evaluation of genetic diversity in fig accessions by using microsatellite markers. *Genetics and Molecular Research* 12, 1383–1391.

Doyle, J. and Ferguson, L. (1998) Breeding persistent figs with Calimyrna quality. *Acta Horticulturae* 480, 259–261.

Doyle, J.F., Ferguson, L. and Herman, K. (2003) Fig cultivar development and evaluation. *Acta Horticulturae* 605, 29–32.

Falistocco, E. (2009) Presence of triploid cytotypes in the common fig (*Ficus carica* L.). *Genome* 52(11), 919–25. DOI:10.1139/g09-068.

Falistocco, E. (2016) Recurrent events of polyploidy in *Ficus carica* L. (Moraceae). *International Journal of Plant Sciences* 177(4), 319–325.

Ferguson, L., Michailides, T.J. and Shorey, H.H. (1990) The California fig industry. *Horticulture Reviews* 12, 409–490.

Flaishman, M.A., Goluhowicz, S., Yablowicz, Z., Erez, A., Solomon, A., Kerem, A. and Rotcheld, G. ( 2007) Red Kadota-new registered fig variety (in Hebrew). *Alon Hanotea* 61, 684–686.

Flaishman, M.A., Yablowich, Z., Golohowich, S., Solomon, A., Cohen, Y., Perl, A., Yancheva, S.D., Kerem, A. and Haklay, E. (2008a) Molecular breeding in Fig (*Ficus carica* L.) by the use of genetic transformation. *Acta Horticulturae* 798, 151–158.

Flaishman, M.A., Rodov, V. and Stover, E. (2008b) The fig: botany, horticulture, and breeding. In: Janick, J. (ed) *Horticultural Reviews*, Wiley, New York, pp. 113–196.

Flaishman, M.A., Peer, R., Freiman, Z.E., Izhaki, Y. and Yablovitz, Z. (2017) Conventional and molecular breeding systems in fig (*Ficus carica* L.). *Acta Horticulturae* 1173, 1–10.

Flaishman, M.A., Peer, R., Raz, A., Cohen, O., Izhaki, K., Bocobza, S., Lama, K., Pliner, M. and Levy, A. (2020) Advanced molecular tools for breeding in Mediterranean fruit trees: genome editing approach of Ficus. *Acta Horticulturae*. 1280, 1–10. DOI:10.17660/ActaHortic.2020.1280.1.

Friedrichs, S., Takasu, Y., Kearns, P., Dagallier, B., Oshima, R., Schofield, J. and Moreddu, C. (2019) Meeting report of the OECD conference on "Genome Editing: Applications in Agriculture - Implications for Health, Environment and Regulation". *Transgenic Research* 28, 419–463.

Furukawa, M. (1997) Production of tetraploid by colchicine treatment in fig (*Ficus carica* L.). In: Murakami, Y. (ed.) *Horticulture Research (Japan)* 66 (Suppl.2), 238–239.

Gaaliche, B., Aïachi-Mezghani, M., Trada, M., Costesc, E., Lauric, P.E. and Mars, M. (2016) Shoot architecture and morphology of different branch orders in fig tree (*Ficus carica* L.). *International Journal of Fruit Science* 16 (4), 378– 394.

Galderisi, U., Cipollaro, M., Di Bernardo, G., De Masi, L., Galano, G. and Cascino, A. (1999) Identification of the edible fig "Bianco del Cilento" by random amplified polymorphic DNA analysis. *HortScience* 34, 1263–1265.

Giraldo, E., Viruel, M.A., Lopez-Corrales, M. and Hormaza, J.I. (2005) Characterisation and cross-species transferability of microsatellites in the common fig (*Ficus carica* L.). *Journal of Horticultural Science and Biotechnology* 80, 217–224.

Giraldo, E., López-Corrales, M. and Hormaza, J.I. (2010) Selection of the most discriminating morphological qualitative variables for characterization of fig germplasm. *Journal of the American Society for Horticultural Science* 135, 240–249.

Haelterman, R.M. and Docampo, D.M. (1994) *In vitro* propagation of mosaic-free fig (*Ficus carica* L.) cultivars using thermotherapy and shoot-tip cultures. *Revista de Investigaciones Agrarias* 25 (3), 15– 22.

Hamada, H., Liu, Y., Nagira, Y., Miki, R., Taoka, N. and Imai, R. (2018) Biolistic-delivery-based transient CRISPR/Cas9 expression enables in planta genome editing in wheat. *Scientific Reports* 8, 14422.

Hanley, Z. (2018) Kiwifruit (*Actinidia* spp.) breeding. In: Al-Khayri, J., Jain, S. and Johnson, D. (eds) *Advances in Plant Breeding Strategies: Fruits*. Springer.

Hepaksoy, S. and Aksoy, U. (2006) Propagation of *Ficus carica* L. clones by *in vitro* culture. *Biologia Plantarum* 50, 433–436.

Herre, E.A., Jander, K.C. and Machado, C.A. (2008) Evolutionary ecology of figs and their associates: Recent progress and outstanding puzzles. *Annual Review of Ecology, Evolution and Systematics* 39, 439–458.

Hirayama, T., Kieber, J.J., Hirayama, N., Kogan, M., Guzman, P., Nourizadeh, S., Alonso, J.M., Dailey, W.P., Dancis, A. and Ecker, J.R. (1999) Responsive-to-antagonist, a Menkes/ Wilson disease-related copper transporter, is required for ethylene signaling in Arabidopsis. *Cell* 97(3), 383–93.

Horii, T., Morita, S., Kimura, M., Kobayashi, R., Tamura, D., Takahashi, R.U., Kimura, H., Suetake, I., Ohata, H., Okamoto, K., Tajima, S., Ochiya, T., Abe, Y. and Hatada, I. (2013) Genome engineering of mammalian haploid embryonic stem cells using the Cas9/RNA system. *PeerJ* 1, e230. DOI:10.7717/peerj.230

Hortorum, L.H.B. (1976) *Hortus third: A concise dictionary of plants cultivated in the United States and Canada*. Macmillan.

Hunt, W.B. (1911) Fig breeding. *University of Georgia Bulletin* 11, 146–148.

Hunt, W.B. (1912) Fig breeding. *University of Georgia Bulletin* 12, 107–110.

Ikegami, H., Nogata, H., Hirashima, K., Awamura, M. and Nakahara, T. (2009) Analysis of genetic diversity among European and Asian fig varieties (*Ficus carica* L.) using ISSR, RAPD, and SSR markers. *Genetic Resources and Crop Evolution* 56, 201–209.

Ikten, H. and Yilmaz, Y. (2019) Validation of gender specific CAPS marker in Turkish fig (*Ficus carica* L.) collection and F1 progenies. *Notulae Botanicae Horti Agrobotanici Cluj-Napoca* 47(3), 867–871. DOI:10.15835/nbha47311499

IPGRI and CIHEAM (2003) *Descriptors for Figs*. International Plant Genetic Resources Institute, Rome, Italy, and International Centre for Advanced Mediterranean Agronomic Studies Paris, France.

Iwata, H., Hayashi, T., Terakami, S., Takada, N., Saito, T. and Yamamoto, T. (2013) Genomic prediction of trait segregation in a progeny population: a case study of Japanese pear (*Pyrus pyrifolia*). *BMC Genetics* 14, 81.

Iwata, H., Minamikawa, M.F., Kajiya-Kanegae, H., Ishimori, M. and Hayashi, T. (2016) Genomics-assisted breeding in fruit trees. *Breeding Science* 66, 100–115.

Jackson, A.P. (2004) Cophylogeny of the *Ficus* microcosm. *Biological Reviews* 79, 751–768.

Janick, J. (2005) The origin of fruits, fruit growing, and fruit breeding. *Plant Breeding Reviews* 25, 255–320.

Jansing, J., Schiermeyer, A., Schillberg, S., Fischer, R. and Bortesi, L. (2019) Genome editing in agriculture: Technical and practical considerations. *International Journal of Molecular Sciences* 20(12), 2888. DOI:10.3390/ijms20122888

Jinek, M., Chylinski, K., Fonfara, I., Hauer, M., Doudna, J.A. and Charpentier, E. (2012). A programmable dual-RNA-Guided DNA endonuclease in adaptive bacterial immunity. *Science* 337, 816–21.

Johnson, C.E., O'Rourke, E. and Boudreaux, J.E. (2010) 'O'Rourke' fig. *HortScience* 45, 826–827.

Jona, R. and Gribaudo I. (1991) *Ficus* spp. In: Bajaj, Y.P.S. (eds) *Trees III. Biotechnology in Agriculture and Forestry, vol* 16. Springer.

Kajitani, Y. and Masuya, H. (2011) *Ceratocystis ficicola* sp. nov., a causal fungus of fig canker in Japan. *Mycoscience* 52, 349–353.

Kato, K., Hirota, K. and Miyagawa, T. (1982) A new disease, ceratocystis canker of fig caused by *Ceratocystis fimbriata* Ellis et Halsted. *Shokubutuboeki* 36, 55–59 (In Japanese).

Khadari, B., Hochu, I., Santoni, S. and Kjellberg, F. (2001) Identification and characterisation of microsatellite loci in the common fig (*Ficus carica* L.) and representative species of genus *Ficus*. *Molecular Ecology Notes* 1, 191–193.

Khadari, B., Oukabli, A., Ater, M., Mamouni, A., Roger, J.P. and Kjellberg, F. (2005) Molecular characterization of Moroccan fig germplasm using intersimple sequence repeat and simple sequence repeat markers to establish a reference collection. *HortScience* 40, 29–32.

Khadari, B., Lashermes, P. and Kjellberg, F. (1995) RAPD fingerprints for identification and genetic characterization of fig (*Ficus carica* L.) genotypes. *Journal of Genetics and Breeding* 49, 77–86.
Kim, H. and Kim, J.S. (2014) A guide to genome engineering with programmable nucleases. *Nature Reviews Genetics* 15, 321–334.
Kim, Y.G., Cha, J. and Chandrasegaran, S. (1996) Hybrid restriction enzymes: zinc finger fusions to Fok I cleavage domain. *Proceedings of the National Academy of Sciences* 93, 1156–1160.
Kim, K.M., Kim, M.Y., Yun, P.Y., Chandrasekhar, T., Lee, H.Y. and Song, P.S. (2007) Production of multiple shoots and plant regeneration from leaf segments of fig tree (*Ficus carica* L.). *Journal of Plant Biology* 50(4), 440–446.
Kislev, M.E., Hartmann, A. and Bar-Yosef, O. (2006) Early domesticated fig in the Jordan Valley. *Science* 312, 1372–1374.
Kjellberg, F., Gouyon, P.H., Ibrahim, M., Raymond, M. and Valdeyron, G. (1987) The stability of the symbiosis between dioecious figs and their pollinator: a study of *Ficus carica* L. and *Blastophaga psenes* L. *Evolution* 41, 693–704.
Lauri, P.E. and Caraglio, Y. (1993) Architectural analysis of fig trees as a tool to study genetic diversity in cultivars. *Acta Horticulturae* 349, 265–267.
López-Corrales, M., Gella, R., Marin, J.A. and Toribio, F. (1998) Elimination of fig mosaic from fig shoot-tip cultures by thermotherapy. *Acta Horticulturae* 480, 173–178.
Machado, C.A., Robbins, N., Gilbert, M.T.P. and Herre, E.A. (2005) Critical review of host specificity and its coevolutionary implications in the fig/fig-wasp mutualism. *Proceedings of the National Academy of Sciences USA* 102, 6558–6565.
Mali, P., Yang, L., Esvelt, K.M., Aach, J., Guell, M., DiCarlo, J.E., Norville, J.E. and Church, G.M. (2013a) RNA-guided human genome engineering via Cas9. *Science* 339, 823–826.
Mali, P., Esvelt, K.M. and Church, G.M. (2013b) Cas9 as a versatile tool for engineering biology. *Nature Methods* 10, 957–963.
Mars, M. (2003) Fig (*Ficus carica* L.) genetic resources and breeding. *Acta Horticulturae* 605, 19–27.
Mason, A.S. (ed.) (2016) *Polyploidy and Hybridization for Crop Improvement.* Taylor & Francis Group, Boca Raton, FL. DOI:10.1201/9781315369259
Mawa, S., Husain, K. and Jantan I. (2013) *Ficus carica* L. (Moraceae). *Phytochemistry, Traditional Uses and Biological Activities Evidence-based Complementary and Alternative Medicine* 3, 974256.
Miguel, M.G. and Lima-Costa, M.E. (1998) Proteolytic and clotting activities of cell suspension cultures of *Ficus carica* under stress saline conditions. *Acta Horticulturae* 480, 155–160.
Miller, J.C., Tan, S., Qiao, G., Barlow, K.A., Wang, J., Xia, D.F., Meng, X., Paschon, D.E., Leeb, M. and Wutz, A. (2011) Derivation of haploid embryonic stem cells from mouse embryos. *Nature* 479, 131–134.
Mori, K., Shirasawa, K., Nogata, H., Hirata, C., Tashiro, K., Habu, T., Kim, S., Himeno, S., Kuhara, S. and Ikegami, H. (2017) Identification of RAN1 orthologue associated with sex determination through whole genome sequencing analysis in fig (*Ficus carica* L.). *Scientific Reports* 7, 41124. DOI:10.1038/srep41124
Müntzing, A. (1937) The effects of chromosomal variation in Dactylis. *Hereditas* 23, 113–235.
Muriithi, L.M., Rangan, T.S. and Waite, B.H. (1982) *In vitro* propagation of fig through shoot tip culture. *Horticulture Science* 17, 86–87.
Noble, E.G. (1922) The work of the Yuma reclamation project experiment farm in 1919 and 1920. *USDA Circular* 221, 3–37.
Nobre, J. and Romano, A. (1998) *In vitro* cloning of *Ficus carica* L. adult trees. *Acta Horticulturae* 480, 161–164.
Nogata, H. and Awamura, M. (2005) A New Fig Cultivar 'H156-70'. *Bulletin Fukuoka Agriculture Research Centre* 24, 104 –107.
Obenauf, G., Gerdts, M., Leavitt, G., Crane, J. (1978) *Commercial Dried Fig Production in California*. Div. Agr. Sci., Univ. of California, Leaflet 21051.
O'Rourke, E.N., Jr. (1966) Breeding new figs for Louisiana. *Louisiana Society for Horticulture Research* 3, 35–41.
Özen, M., Kocataş, H., Çobanoğlu, F., Ertan, B., Kösoğlu, İ., Tan, N., Şahin, B., Belge, A., Konak, R., Aksoy, U. and Gülşen, O. (2017) Mutation breeding studies on fig. *Acta Horticulturae* 1173, 93–98.
Papadopoulou, K., Ehaliotis, C., Tourna, M., Kastanis, P., Karydis, I. and Zervakis, G. (2002) Genetic relatedness among dioecious *Ficus carica* L. cultivars by random amplified polymorphic DNA analysis, and evaluation of agronomic and morphological characters. *Genetics* 114, 183–194.
Peer, R., Rivlin, G., Golobovitch, S., Lapidot, M., Gal-On, A., Vainstein, A., Tzfira, T. and Flaishman, M.A. (2015) Targeted mutagenesis using zinc-finger nucleases in perennial fruit trees. *Planta* 241, 941–951.

Perez-Jiménez, M., López, B., Dorado, G.M., Pujadas-Salvá, A., Guzmán, G., Hernandez, P. (2012) Analysis of genetic diversity of southern Spain fig tree (*Ficus carica* L.) and reference materials as a tool for breeding and conservation. *Hereditas* 149, 108–113.
Podgornik, M., Vul, I., Vrhovnik, I., Mavsar, D.B. (2010) A survey and morphological evaluation of fig (*Ficus carica* L.) genetic resources from Slovenia. *Scientia Horticulturae* 125, 380–389.
Pontikis, C.A. and Melas, P. (1986) Micropropagation of *Ficus carica* L. *HortScience* 21, 153.
Qin, G., Gu, H., Ma, L., Peng, Y., Deng, X.W., Chen, Z. and Qu, L.J. (2007) Disruption of phytoene desaturase gene results in albino and dwarf phenotypes in Arabidopsis by impairing chlorophyll, carotenoid, and gibberellin biosynthesis. *Cell Research* 17, 471–482.
Rhoades, M.M. (1936) Note on the origin of triploidy in Maize. *Journal of Genetics* 33, 355–357.
Rixford, G.P. (1918) Smyrna fig culture. *USDA Bulletin* 7321–43.
Rixford, G.P. (1926) Work with seedling figs. *Calif. Ann. Fig. Inst. Proc.* 10, 6–9.
Saddoud, O., Baraket, G., Chatti, K., Trifi, M., Marrakchi, M., Mars, M. and Salhi-Hannach, M. (2011) Using morphological characters and simple sequence repeat (SSR) markers to characterize Tunisian fig (*Ficus carica* L.) cultivars. *Acta Biologica Cracoviensia* 53, 7–14.
Saddoud, O., Salhi-Hannachi, A., Chatti, K., Mars, M., Rhouma, A., Marrakchi, M. and Trifi, M. (2005) Tunisian fig (*Ficus carica* L.) genetic diversity and cultivar characterization using microsatellite markers. *Fruits* 60, 143–153.
Saleeb, W.F. (1965) Genetics and cytology of syconium persistence in *Ficus carica*. *Unpublished PhD thesis*, University of California.
Salhi-Hannachi, A., Chatti, K., Mars, M., Marrakchi, M. and Trifi, M. (2004a) Comparative analysis of genetic diversity on two collections fig cultivars based on random amplified DNA and inter repeats fingerprints. *Genetic Resources and Crop Evolution* 52, 563–573.
Salhi-Hannachi, A., Trifi, M., Zehdi, S., Hadfi, J., Mars, M., Rhouma, A. and Marrakchi, M. (2004b) Inter simple sequence repeat fingerprints to assess genetic diversity in Tunisian fig (*Ficus carica L.*). *Genetic Resources and Crop Evolution* 51, 269–275.
Salhi-Hannachi, A., Chatti, K., Mars, M., Marrakchi, M., Trifi, M. (2005) Comparative analysis of genetic diversity in two Tunisian collections of fig cultivars based on random amplified polymorphic DNA and inter simple sequence repeats fingerprints. *Genetic Resources and Crop Evolution* 52, 563–573.
Sauer, N.J., Narvaez-Vasquez, J., Mozoruk, J., Miller, R.B., Warburg, Z.J., Woodward, M.J., Mihiret, Y.A., Lincoln, T.A., Segami, R.E., Sanders, S.L., Walker, K.A., Beetham, P.R., Schopke, C.R. and Gocal, G.F. (2016) Oligonucleotide-mediated genome editing provides precision and function to engineered nucleases and antibiotics in plants. *Plant Physiology* 170, 1917–1928.
Shirasawa, K., Yakushiji, H., Nishimura, R., Morita, T., Jikumaru, S., Ikegami, H., Toyoda, A., Hirakawa, H. and Isobe, S. (2020) The *Ficus erecta* genome to identify the Ceratocystis canker resistance gene for breeding programs in common fig (*F. carica*). *Plant Journal* www.onlinelibrary.wiley.com/doi/abs/10.1111/tpj.14703 (accessed 3 September 2021).
Singh, B.D. and Singh, A.K. (2015) *Marker-Assisted Plant Breeding: Principles and Practices*. Springer, New Delhi.
Soliman, H.I., Gabr, M. and Abdallah, N.A. (2010) Efficient transformation and regeneration of fig (*Ficus carica* L.) via somatic embryogenesis. *GM Crops* 1, 40–51.
Storey, W.B. (1975) Figs. In: *Advances in Fruit Breeding*, Purdue University Press, pp. 568–589.
Storey, W.B. (1976) Fig *Ficus carica* (Moraceae). In: Simmonds, N.W. (ed.) *Evolution of Crop Plants*. Longman, London, pp. 205–208.
Storey, W.B. (1977) *The Fig (Ficus carica* Linnaeus*): Its Biology, History, Culture, and Utilization*. Jurupa Mountains Cultural Center, Riverside, California.
Storey, W.B. and Condit, I.J. (1969) Fig (*Ficus carica* L.). In: Ferwerda, F.P. and Wit, F. (eds) *Outlines of Perennial Crop Breeding in the Tropics. Misc. Pap.* 4. Wageningen, The Netherlands, pp. 259–267.
Stover, E., and Aradhya, M. (2008) Fig genetic resources and research at the US national clonal germplasm repository in Davis, California. *Acta Horticulturae* 798, 57–68.
Stover, E., Aradhya, M., Ferguson, L. and Crisosto, C.H. (2007) The Fig: Overview of an ancient fruit. *HortScience* 42, 1083–1087.
Swingle, W.T. (1908) Some points in the history of caprification in the life history of the fig. *Off. Rpt. 34th Fruit-Growers' Conv. Cal.* 178–187.
Swingle, W.T. (1912) Cooperative distribution of new varieties of Smyrna figs and Caprifigs. *U.S. Dept. Agr. Bur. Pl. 1nd.* 438, 1–6.
Trabut, L. (1922) On the origins of the fig (in French). *Rev. Bot. Appl. Colon.* 2, 393–396.

Tsugane, T., Watanabe, M., Toya, T., Fukami, M. (2019) Development of an SNP marker for sex determination of figs using PCR. (in Japanese) *CAFRC Research Bulletin* 11, 73–77.

Usai, G., Mascagni, F., Giordani, T., Vangelisti, A., Bosi, E., Zuccolo, A., Ceccarelli, M., King, R., Hassani-Pak, K., Zambrano, L.S., Cavallini, A. and Natali, L. (2020) Epigenetic patterns within the haplotype phased fig (*Ficus carica* L.) genome. *The Plant Journal*. www.onlinelibrary.wiley.com/doi/abs/10.1111/tpj.14635 (accessed 3 September 2021).

Valero, D. and Serrano, M. (2010) Fruit ripening. *Postharvest Biology and Technology for Preserving Fruit Quality*. CRC Press, Boca Raton, FL, pp. 7–47.

Vangelisti, A., Zambrano, L.S., Caruso, G., Macheda, D., Bernard, R., Usai, G., Mascagni, F., Giordani, T., Gucci, R., Cavallini, A. and Natali, L. (2019) How an ancient, salt-tolerant fruit crop, *Ficus carica* L., copes with salinity: a transcriptome analysis. *Scientific Reports* 9, 2561. DOI:10.1038/s41598-019-39114-4

Veberic, R. and Mikulic-Petkovsek, M. (2016) Phytochemical Composition of Common Fig (*Ficus carica L.*) Cultivars. In: Simmonds, M.S.J. and Preedy, V.R. (eds) *Nutritional Composition of Fruit Cultivars*. Academic Press, London, pp. 235–255.

Woeste, K.E. and Kieber, J.J. (2000) A strong loss-of-function mutation in RAN1 results in constitutive activation of the ethylene response pathway as well as a rosette-lethal phenotype. *Plant Cell* 12(3), 443–455.

Yakushiji, H., Mase, S. and Sato, Y. (2003) Adventitious bud formation and plantlet regeneration from leaves of fig (*Ficus carica* L.). *Journal of Horticultural Science and Biotechnology.* 78, 874–878.

Yakushiji, H., Morita, T., Jikumaru, S., Ikegami, H., Azuma, A. and Koshita, Y. (2012) Interspecific hybridization of fig (*Ficus carica* L.) and *Ficus erecta* Thunb. a source of Ceratocystis canker resistance. *Euphytica* 183(1), 39–47. DOI:10.1007/s10681-011-0459-1

Yakushiji, H., Morita, T. and Jikumaru, S. (2019) Ceratocystis canker resistance in BC1 populations of interspecific hybridization of fig (*Ficus carica*) and *F. erecta*. *Scientia Horticulturae* 252, 71–76. DOI:10.1016/j.scienta.2019.03.039

Yancheva, S.D., Goluhowicz, S., Yablowicz, Z., Perl, A. and Flaishman, M.A. (2005) Efficient Agrobacterium-mediated transformation and recovery of transgenic plants from fig (*Ficus carica* L.). *Plant Science* 168, 1433–1441.

Zarei, M., Azizi, M., Rahemi, M. and Tehranifar, A. (2016) Evaluation of NaCl salinity tolerance of four fig genotypes based on vegetative growth and ion content in leaves, shoots, and roots. *HortScience* 51, 1427–1434.

Zohary, D. and Hopf, M. (2000) *Domestication of Plants in the Old World*. 3rd edn, Oxford University Press, Oxford, UK.

Zohary, D. and Spiegel-Roy, P. (1975) Beginnings of fruit growing in the old world. *Science* 187, 319–327.

# 5 World Fig Cultivars

**F. Aljane[1]*, M.H. Neily[1], M.G.F. Rodrigues[2], C. Mazri[3], V. Stournaras[4], T. Kokaj[5], A. Yavari[6], L. Ferguson[7] and Ali Sarkhosh[8]**
[1]*Laboratoire d'Aridoculture et Cultures Oasiennes, Institut des Régions Arides (IRA), Université de Gabès, Tunisia;* [2]*Fruticulture and Floriculture Applied Laboratory, Faculty of Agricultural and Technological Sciences (FCAT - Dracena), São Paulo State University (UNESP), São Paulo, Brazil;* [3]*Département des Sciences Agronomiques, Université Le Colonel M'hand Akli Oulhade, Bouira, Algéria;* [4]*Department of Olive and Horticultural Crops, Hellenic Agricultural Organization (DEMETER) Institute of Olive Tree, Subtropical Crops and Viticulture, Kalamata, Greece;* [5]*Institute of Plant Genetic Resource, Agriculture University of Tirana, Albania;* [6]*Pomologist and Consultant, Tehran, Iran;* [7]*Department of Plant Sciences, College of Agricultural and Environmental Sciences, University of California Davis, California, USA;* [8]*Horticultural Sciences Department, University of Florida, Gainesville, Florida, USA*

## 5.1 Introduction

Among fruits, fig is noted for having numerous cultivars. The same genotypes are often given the different names or different genotypes given the same name; synonymy and homonymy (Aljane and Ferchichi, 2009; Aljane, 2011; Aljane *et al.*, 2012; Aljane and Nahdi, 2014). Figs were often named for their fruit color, shape, local region or harvest date (Caliskan and Polat, 2008; Aljane and Ferchichi, 2009). Natural populations and spontaneous subforms have also been reported in several Mediterranean regions and Central Asia.

Figs have been produced with traditional methods in multiple countries for centuries in many countries (Minangoin, 1931). Few cultivars are commercially propagated for new plantations (Aljane, 2016). Cultivars can be categorized by their wild or semi-wild form, as cultivated accessions, including commercial cultivars, local traditional cultivars, selected cultivars, seedlings, landraces and cultivars improved through mutation, hybridization and genetic methods. This chapter is a broad review of local figs, commercial and improved cultivars.

## 5.2 Fig Types

There are four distinct types of figs based on cropping and pollination characteristics (Fig. 5.1). They are caprifig fig (*Ficus carica* var. *sylvestris* Shinn.), Common-type figs (*Ficus carica* var. *hortensis* Shinn.), Smyrna-type figs (*Ficus carica* var. *smyrna* Shinn.), and San Pedro-type figs (*Ficus carica* var. *intermedia* Shinn.) (Flaishman *et al.*, 2008). There are multiple cultivars within each type. Among the types described above, 75% of the cultivars are Common-type, 18% Smyrna-type, 7% San Pedro-type and caprifigs (IBPGR, 1986).

*Corresponding author: fateh_aljane@yahoo.fr


DOI: 10.1079/9781789242881.0005

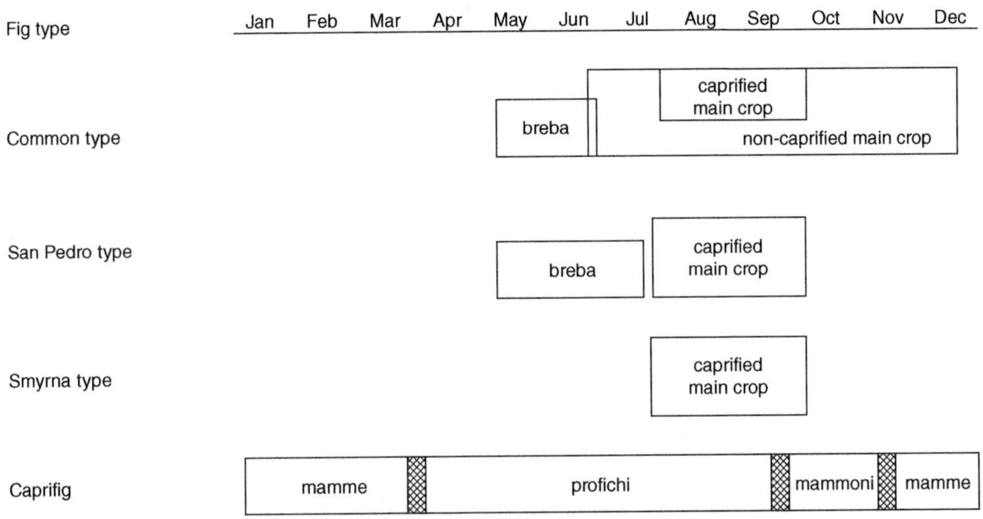


**Fig. 5.1.** Different fig types, caprification and cropping. (Figure used from Flaishman *et al.*, 2008 with permission from RightsLink.)

***Caprifig-type:*** Caprifigs are characterized by the short-styled pistillate flowers lining the syconium. The ovaries encased in the syconium may be inhabited by the larvae of the fig insect *Blastophaga psenes* Cav. In one or more crops, staminate flowers, which produce pollen, may be present as well. Three crops of a caprifig tree are generally recognized, and, as Eisen (1901) predicted, their Italian names have become household words in various fig-growing countries. These names are mamme, the first or winter crop, maturing in early April; profichi, the second or spring crop, maturing in June; and mammoni, the third crop, maturing in late summer (Singh *et al.*, 2015).

***Smyrna-type:*** Smyrna-type figs require pollination with pollen from caprifigs and mature into fruit with viable seeds. However, without pollination, fruits (both of the breba crop (first crop) and of the main crop) will fall from the tree before maturation. The commercial fig industries of Asia Minor, Iran, Turkey, Greece, Algeria and Portugal are based upon this type of fig. California previously was, but has now transitioned to the Common-type fig. 'Sarilop' (Calimyrna) and 'Sabz' are the most common and widely grown cultivars of this type (Condit,1941; Singh *et al.*, 2015; Jafari *et al.*, 2018).

***San Pedro-type:*** San Pedro is regarded as an intermediate type. It produces the first crop, known as a breba crop, that persists to maturity as a parthenocarpic fig that does not require fertilization. The second crop (main crop) requires pollination to develop to maturity. This type and Common-type figs both produce a breba. The 'Zidi' (Tunisia), 'King', and 'Gentile' are San Pedro-type figs (Singh *et al.*, 2015; Aljane, 2016).

***Common-type:*** The Common-type has individual pistillate flowers and produces figs without pollination or fertilization. The fruit contains infertile seeds. Some cultivars produce a low yield of large fruit with the first, breba, crop and then produce a good larger second, main, crop of smaller fruit. 'Bayoudhi' ('Tunisia'), 'Kadota' ('Dottato'), 'Mission', 'Adriatic', 'Brown Turkey', 'Celeste', and 'Conadria' are Common-type figs (Singh *et al.*, 2015; Aljane, 2016).

## 5.3 Fig Cultivars in the Northern Hemisphere

### Turkey

'Sarilop' is widely cultivated for the fresh market in the Aegean, Aydin and Izmir regions, 'Bursa Siyahi' and 'Bardacik' in the İzmir region, and 'Akça' in the Germencik region. 'Keten Köyneği', 'Melli' and 'Sarı Zeybek' are cultivated in the Nazilli area for both dried and fresh figs. 'Halebi' and 'Sultani' are grown in the Gaziantep region for dried figs (Arpaci *et al.*, 2018). 'Bursa

Siyahi' accounts for most of Turkey's fresh fig exports and is characterized by a high yield, big fruit size and long shelf life. 'Sari Zeybek' has high fruit quality, natural color, soft texture, honey-like flavor and a pleasant smell when dried. Two Turkish dried fig cultivars, 'Sarı Zeybek' and 'Divrek Kara', were patented in 2017. 'Sari Zeybek' is distinguished by a light external color and a small ostiole. 'Divrek Kara' has a dark external color, low sugar content and higher antioxidant activity (Arpaci *et al.*, 2018).

### Egypt

The Egyptian cultivars include 'Sultani', 'Aboudi', 'White Adcy', 'Gizi' and 'Black Mission' (Koyuncu, 1998; Koyuncu *et al.*, 1998; Nabil Sabet and Abou-Ellail, 2013; Abou-Ellail *et al.*, 2014). 'Sultani', the most widely grown cultivar has multiple names; 'Fayoumi', 'Ramadi', 'Barshoumy', 'Sidi Gaber' and 'Hegazi'. 'Asuani', 'Kommathri', 'Adasi-Abiad', 'Adasi-Ahmer', 'Abboudi' and 'Kahramani' are Common-type fresh market cultivars. Six additional cultivars, 'Assali', 'Akhardi', 'Abiadi', 'Azraki', 'Amouri' and 'Ahmadi' were introduced from Palestine and another four from Cyprus, 'Pyrgos', 'Lefkara', 'Vartika' and 'Vasanata'.

### Morocco

There are 277 cultivars cultivated throughout Morocco. The most important commercial cultivars are 'El Messari', which is also called 'Homrame' or 'Johri', 'Lembdar Labiad', 'Lembdar Lakhal', 'Rhouddane', 'El Koté' and 'Aounq Hmam' (Bibi and Afechtal, 2017). In the El Jadida province 'Mtioui' is 68% of the production followed by 'Kahouli' at 32%, and the 'Hamouri bifère' cultivar is less than 1% (El Hajjam *et al.*, 2018). In the Ouazzane area 'Homran' is the most abundantly grown cultivar at 20.2%, followed by the 'El masfah' or 'Koté Lazrak' (19.0%) and 'Ghouddane' cultivars (17.4%) (El Hajjam *et al.*, 2018).

### Algeria

More than 40 cultivars including 'Alelake', 'Tabouharchaout' and 'Taranimt', including caprifigs, are cultivated in Algeria (Aid-Houchi and Mazri, 2001; Mazri-Kartout and Aid-Houchi, 2001; Meziant *et al.*, 2015). The light colored 'Taamriwthe', and dark 'Azenjar' are the dominant cultivars in Algeria (Bachir Bey *et al.*, 2013; Bachir Bey and Louaileche, 2015).

### Iran

The most important Iranian fig cultivars are 'Sabz', 'Peyves', 'Siah', 'Shah Anjir', 'Matti', 'Kashki', 'Rono', 'Zard-e-Piyazi', 'Meshki', 'Manjifi', 'Sabzak', 'Se Fasle', 'Kolmi' and 'Keshani'. The 'Sabz' cultivar comprises 95% percent of the orchards. Smyrna-type cultivars are produced for the dried fig markets (Jafari *et al.*, 2016; Baziar *et al.*, 2018; Khadivi *et al.*, 2018). The primary caprifigs are 'Pouzdonbali', 'Dane Sefid', 'Shah Anjiri', 'Aatashi', 'Khakestouni', 'Aowgizi', 'Ghad Shanei', 'Sibi', 'Kouhaki', 'Maharloui' and 'Fel-e-Sarbast'. The 'Pouzdonbali' is the earliest caprifig (Rahemi and Jafari, 2008).

### Syria

As with most fig producing countries, Syria also classifies its fig cultivars by color. The green-yellowish cultivars are 'Khedeiri', 'Sultani', 'Khani', 'Halabi', 'Sefraouli', 'Esseli', 'Anzuki' and 'Boukrati'. The brown-reddish cultivars are 'Zeilbily', 'Afani' and 'Shincheri'. The black-purple cultivars are 'Oubeidi', 'Habachi', 'Zamouchi' and 'Assouad' (Al Ibrahim, 2000; Al Ibrahim and Bari, 2006).

### Spain

The most popular Spanish cultivars are are 'Blanco', 'San Antonio', 'De Rey', 'Napolitana Negra', 'Lampaga', 'Moscatel Negra', 'Negra Cabezuela', 'Picholetera', 'Albacor', 'Calabacita' and 'Cuello Dama' (Lopez-Corrales and Balas, 2014). Other fresh market cultivars are 'San Antonio', 'Banane', 'Negro', 'Tiberio', 'Tres Voltas L'Any', 'Colar Elche', 'Brown Turkey' and 'Blanca Bétera' (López-Corrales *et al.*, 2011). Only 'Cuello Dama Blanco', 'Colar Elche' and 'Brown Turkey' are produced commercially. All the other cultivars are Common-type figs and the 'Tiberio', a San Pedro-type. 'Cuello Dama Blanco' and 'Colar Elche' have dual-use, fresh consumption and dried fig fruit (Stover *et al.*, 2007). All cultivars except 'Tres Voltas L'Any' are included in the Official Register of Commercial Fig Cultivars (BOE, 2011).

## United States of America

In California, the main Common-type fig cultivars are 'Kadota', 'Mission', 'Adriatic', 'Conadria' and 'Brown Turkey'. The main Smyrna-type cultivar was previously 'Calimyrna' caprified by 'Brawley', 'Croisic', 'Roeding 3' and 'Stanford'. San Pedro-types are not commercially important in California. Some of the new and underutilized fig cultivars for the fresh and dried market are 'LSU Purple', 'Italian 376', 'Aklo Lalo', 'Italian 358', 'Patlican', 'Georgian Black', 'Early Violet', 'Black Fig #1' and 'Vedral Lounge' (Gordon *et al.*, 2018).

## Tunisia

Tunisian cultivars often have multiple figs in different regions (Aljane and Ferchichi, 2009; Aljane *et al.*, 2012). The Smyrna-types were dominant in the southeast of Tunisia. In the northeast, the Common and Smyrna-types are equally represented. A few San Pedro-types are cultivated; 'Bither Abiadh', 'Bither Akhal', 'Besbessi' and 'Bouholi' (Mars *et al.*, 2008). A large number of cultivars 'Zidi', 'Bither Abiadh', 'Bither Akhal', 'Soltani', 'Wedlani', 'Ragoubi', 'Bayoudhi', 'Safouri', 'Bouholi', 'Zidi', 'Wahchi' and 'Soltani' are grown in different parts of the country (Mars *et al.*, 2008).

## Albania

Many fig cultivars are grown in Albania, such as 'Roshniks', 'Kallamata', 'Perdhikuli' and 'Stambolli'. Other cultivars are 'Melacak', 'Bradashesh', 'Kraps', 'Shengjinas', 'Dersak', 'Cingell', 'Rotllar' and 'Bishtagjati'. The 'Roshniks' fig cultivar is particularly suitable for drying as the fruits have high sugar content. The 'Roshniks' and 'Perdhikuli' cultivars are characterized by their small size, light color, soft texture and red flesh. The 'Melacak', 'Bishtagjati' and 'Rotllar' are medium sized, dark colored and have firm skin (Hoxha *et al.*, 2015).

## India

India has many commercial fig cultivars. 'Poona' is the most popular cultivar. 'Bangalore', 'Bellary', 'Coimbatore', 'Daulatahad', 'Dindigual', 'Ganjam', 'Hindupur', 'Lucknow' and 'Saharanpur' cultivars resemble 'Poona'. 'Dinker' is an improved 'Daultabad' with better yield and fruit quality and is gaining in commercial importance. The 'Excel' and 'Conardia' cultivars are suitable for high-density planting and have little fruit splitting. 'Conardia' figs are good for drying. 'Deanna' fruits are generally consumed fresh (Singh *et al.*, 2015).

## China

The main fresh figs in China are 'Bojihong', 'Masui Dauphine', 'Nixuan No1', 'Lukan No 1' and 'Tanikawa' (Jianye *et al.*, 1997; Lianju, 2013). The cultivars 'Brunswick' and 'Xin jing Yellow' are used for fresh consumption and processing. Wang *et al.* (2017) reported two additional cultivars: Green Peel ('Qingpi'), a highly prized green cultivar and a bud mutation of 'Green Peel' with dark purple skin, 'Zibao'.

## Greece

Years of breeding have produced the figs cultivated in Greece (Greek Figs, 2020). Among the dried figs, recognized with a PDO (Protected Designation of Origin) is 'Xera Syka Kymis' or dried fig 'Kymis' (OJEC, 1996) and the 'Xira Syka Taxiarchi', or dried figs 'Taxiarchi' (OJEC, 2013). The 'Kalamon' and 'Kymis' cultivars are two of the important cultivars for dried fig but can be consumed fresh. 'Vasilika Melissi' and 'Vasilika Mavra' are two of the most important fresh market cultivars.

### Portugal

Portugal has more than 64 fig cultivars grown in various regions. 'Lampa Preta', 'Pingo de Mel', 'Maia', 'Dauphine' and 'Grise de Tarason' are easier to produce, have better fruit quality and ripen earlier. 'Pingo de Mel', 'Bêbera', 'Blanca' and 'Palmares' are late-season fig cultivars (Maia de Sousa, 2003). 'Lampa Preta', 'Princesa' and 'Maia' are grown for the breba crop and 'Pingo de Mel', 'Ponte', 'Princesa' and 'Berbera Branca' produce main crops in the autumn (Sousa and Gomez Pereira, 1995).

### Japan

Japan produces the 'Masui Dauphine', a large fruit cultivar named for the man who introduced them in the early 1900s (Japan Guide, 2019). The few other minor fig cultivars grown in Japan have many unique characteristics, compact to spreading growth habits, fruit color, shape, taste, size and plant hardiness (Alfrey, 2016).

### Italy

Italy grows multiple cultivars. Some produced exclusively for the brebas, 'fiorone' in Italian; 'Petrell' (syn. S. Giovanni, S. Antonio, Columbro, etc.), 'Tauro', 'Fiorone della Regina', 'Fiorone Nero', 'Nero di Terlizzi', 'Gentile Nero Lungo', 'Paradiso', 'S. Pietro', 'Fiorone Precocissimo', 'Fiorone Testa di Gatto' and 'Fiorone Gigante'. Others are produced primarily for the main crop figs, 'fico' in Italian; 'Dottato', 'Abat', 'Albanera', 'Brogiotto' (white and black), 'Callara', 'Citrullara', 'Della Monaca', 'Frecazzano' (white and black), 'Lattarola', 'Melanzana' (green and black), 'Natalino', 'Reale', 'Regina', 'Rigato', 'Troiano' and 'Zingarello' (white and black) (Ciarmielloa *et al.*, 2015). The most important dried cultivar is 'Dottato' ('Kadota'). The primary fresh cultivars, produced for the breba crop, are 'Brogiotto Nero', 'Petrelli', 'Domenico Tauro', 'San Giovanni' and 'Gentile'.

### Mexico

The 'Black Mission' cultivar has adapted to local environmental conditions in Central Mexico, yielding several distinct landraces with valuable and previously unseen genetic diversity (García-Ruiz *et al.*, 2013).

### France

The most important French commercial biferous fig cultivars are 'Grise de Saint-Jean', the earliest breba producer 'Brunswick', 'Dauphine', (syns. 'Grise De Tarascon' and 'Boule d'Or'), 'Breva', and 'Noire De Caromb'. The uniferous cultivars are 'Ronde De Bordeaux', 'Violette De Solliès' (syns. 'Bourjassotte Noire' and 'Barnissotte Noire'), 'Col De Dame', and 'Tardive Des Vannes', which ripens later than 'Col De Dame' (Baud, 2008). The best dried cultivars are 'Ronde de Bordeaux,' 'Figue de Marseille', 'Marseillaise', 'Negronne', 'Tena', 'Sucrette' and 'Nefiach'.

## 5.4 Fig Cultivars in the Southern Hemisphere

### Brazil

The main cultivars grown are: 'Brown Turkey', 'Roxo de Valinhos', 'Kadota', 'Pingo de Mel', 'Verdona Longa', 'Brunswick', 'Nobile' 'Korfu' and 'Figo Rojo' (Botti and Muchnik, 1998; Rodrigues *et al.*, 2018).

### Peru

'Black Mission', 'Red House', 'Red Queen' and 'Turkey AC' are grown in Peru (García and Sinova, 2019).

### South Africa

The most popular fig cultivar in South Africa is the 'Evita', a small to medium-sized fruit with purple skin, red flesh, excellent quality and long shelf life. Other dark-skinned cultivars are 'Adam', 'Black Genoa', 'Tangiers', 'King', 'Black Mission', 'Cape Black', 'Cape Brown' and 'Brown Turkey'. The light-skinned cultivars are 'Dauphine', 'Cape White', 'Kadota', 'White Genoa' and 'Deanna' (Louw, 2019).

### Argentina

The most commonly grown fig in Argentina is the 'Negra Comun' ('Black Mission'), 70% of the total production. The crops is purple-skinned with a light red pulp. Other cultivars include the 'King', cultivated only for the brebas. It is a yellow skinned fruit with an amber pulp. The biferous 'Turco' is dark skinned with red pulp, other cultivars produced for both the breba and main crop are the 'Guarinta', a yellow-skinned fruit with red pulp, 'Unigal', a dark-skinned fruit with dark red pulp, and the 'Kadota', a light-skinned fruit with an amber pulp (Botti and Muchnik, 1998).

### Chile

The cultivars grown in Chile include 'Negra Comun', 'Kadota' and 'White Adriatic' (ODEPA, 1997; Botti and Muchnik, 1998). There is no published country specific information about the characteristics of these cultivars. However, there is no reason to think these cultivars are different in Chile except for climatic effects.

### Australia

Many dark-skinned 'Brown Turkey', 'Black Genoa', and light-skinned 'White Adriatic', 'Excel' and 'Prestons Prolific' cultivars are produced in Australia (Brien and Hardy, 2002; Daleys, 2002).

## 5.5 Characteristics of Important Fig Cultivars

### 'Sarilop'; syns. 'Lob Injir' or 'Ingir', 'Erbeyli', 'Calimyrna' and 'Aidin'

The name is derived from sari, 'yellow', and lop, 'delicious'. This cultivar has been grown in Turkey's Meander Valley for several centuries. When imported into California in 1886 it was renamed 'Calimyrna' (Condit, 1955; 1956; Obenauf *et al.*, 1978; Papadopoulou *et al.*, 2002). It accounts for 90% of fig production in Turkey (Arpaci *et al.*, 2018) and once was California's most important commercial dried fig cultivar but is now no longer produced commercially in California. The trees are vigorous and have an upright growth habit. 'Sarilop' is a Smyrna-type. It is grown for both fresh and dried fruit consumption. The fruit is large and golden-yellow (Fig. 5.2). It has a honey-like taste, with notes of nuttiness from the numerous seeds. This cultivar is characterized by a long harvest period (July to September in the Northern Hemisphere), light-yellow skin color in fresh fruit, pinkish pulp color and a high fruit weight (65–70 g). The dried fruit has a whitish-yellow skin color, a medium moisture ratio 22–14% and high total soluble solids (TSS or °Brix) 22 to 23% (Nakilcioğlu and Hişil, 2013; Arpaci *et al.*, 2018).

### 'Bursa Siyahi', 'Bursa Black'

This cultivar is the most common fig in the Marmara and Bursa regions of northwest Turkey. It is Smyrna-type. The fruit ripens from August to mid-October in the Aegean region and from the beginning of September to the middle of November in the Bursa region. The fruits are large and round with dark purple to purplish-black skin, dark red pulp, a short neck and are easily peeled (Fig. 5.3). For a fresh fig, it has extremely long shelf life and good transportability. The fresh fruit pH is 4.55, total acidity (TA) is 0.18–0.19% and TSS is 18–22%. Bursa Black fruits are suitable for fresh consumption or processing (Arpaci *et al.*, 2018).

### 'Beyaz Orak'

This is a San Pedro-type with early production. The tree is a semi-dwarf to medium height. In

**Fig. 5.2.** 'Sarilop' ('Calimyrna'). (Figure used with permission from Megan Crivelli.)

the Northern Hemisphere, the first crop ripens from mid-June to the first week of July. The second, main crop ripens at the end of July to first week of September. The fruits are large, 78–100 g, peel easily, and have a distinctive odor. The peel is yellowish-green. The pulp is light pink (Fig. 5.4). The seeds are small to medium-sized. The ostiole is generally slightly open. The fresh fruit has poor storage life and should be consumed within 1–2 days postharvest ambient temperatures. The second, main crop, fruits have a milder flavor than the first crop, breba fruits.

## 'El Messari'

The 'El Messari', also known as 'Homrame', 'Lamtel', 'Johri' and 'Mtioui' is a San Pedro-type grown in Morocco for fresh first crop and late second crop, or autumn, figs. This cultivar is vigorous with upright growth habit and big leaves (Walali, 1995). It has moderate yield, 27 kg/tree, a long harvest period from May to June for the breba crop and July to September for fig autumn production. The yellowish-green fresh fruit (Fig. 5.5), weighs approximately 44 g, and is round. It is a sweet fruit, 20.5% TSS (El Hajjam *et al.*, 2018).

**Fig. 5.3.** 'Bursa Siyahi'. (Figure used with permission from Birgül Ertan.)

**Fig. 5.4.** 'Beyaz Orak'. (Figure used with permission from Birgül Ertan.)

## 'Lambdar Labiadh' and 'Lambdar Lakhal'

These are San Pedro-type cultivars long grown commercially in northern Morocco for fresh first crop figs (Tayou, 1985; Walali, 1995; Ait Haddou *et al.*, 2013). 'Lambdar Labiadh' and 'Lambdar Lakhal' are morphologically similar, except for the peel color. 'Lambdar Labiadh' is yellowish-green while 'Lambdar Lakhal' is blackish-purple. These trees have an upright growth habit and moderate sized leaves. The fruit has a flattened shape (Walali, 1995).

## 'El Koté'

'El Koté' is a Smyrna-type cultivated in Morocco to produce autumn dried figs (Walali, 1995; El Hajjam *et al.*, 2018). The trees have a spreading growth habit and large leaves (Walali, 1995). The yield is moderate, 30 kg/tree and the second harvest extends into mid-September. The fruit is yellowish-green, small, 34 g and spherical. It has a high TSS of 23% and produces a very sweet drying dried fruit (El Hajjam *et al.*, 2018).

**Fig. 5.5.** 'El Massari'. (Photo: El Hajjam *et al.*, 2018.)

### 'Azendjar'

This Smyrna-type is the second major cultivar produced in Algeria for fresh and dried consumption. 'Azendjar' is a large, 55–60 g, fruit with dark smooth flesh, small crunchy seeds, and sweet honey and tangy flavor. It is prone to skin cracking at the ostiole. It has a long harvest period from July to September (Fig. 5.6). The dried fruit is a whitish-black color with a medium moisture ratio, 15–25%, and TSS 50–60% sugar ratio (Mazri-Kartout and Aid-Houchi, 2001).

### 'Thaâmriwt'

This Smyrna-type fig is Algeria's most important commercial cultivar for both fresh and dried consumption. It is a large greenish-yellow fruit, 45–70 g (Fig. 5.7) that often splits as it ripens. The pulp has a gelatinous consistency and numerous seeds. This cultivar is characterized by an extended harvest from July to September. The dried fruit has a whitish-brown skin color, a medium moisture ratio 12–20%, and TSS 55–65% (Mazri-Kartout and Aid-Houchi, 2001).

**Fig. 5.6.** 'Azendjar'. (Figure courtesy of C. Mazri.)

**Fig. 5.7.** 'Thaâmriwt'. (Figure courtesy of C. Mazri.)

### 'Bakkor Biadh' (syns. 'Amazigh' and 'Abakour Amellal')

This is a Common-type cultivar. 'Bakkor' in Arabic means early, and 'Biadh' or 'Ammellal' in Amazigh language means white. This cultivar produces a heavy breba crop of large, 68–114 g, fruit with greenish-yellow skin, a pink and succulent pulp, and turbinate fruit shape (Fig. 5.8). It has thick tough skin with numerous white lenticels, is easy to peel, matures at the beginning of June, has a wide ostiole of 7–10 mm, and few seeds. The second crop fruits are bell-shaped with a short stalk, also large, 40–80 g, yellowish-green skin with sparse white lenticels, has a pink and juicy pulp, are moderately sweet, do not crack easily and mature in July. Both crops are consumed fresh. The second-crop figs are sometimes dried but are a mediocre product (Mahmoudi *et al.*, 2018).

### 'Bakkor Khal'

This vigorous productive Common-type cultivar is favored by Algeria for its abundant production of flavorful brebas and high quality second-crop dried figs. The first-crop fruits are ovoid, 35–67 g, with no neck and a short stalk, a thin, black, easily peeled skin with numerous lenticels, a 3–6 mm ostiole and ripens in June. The dark red (Fig. 5.9) aromatic pulp is less sweet than most cultivars and has a slightly bitter aftertaste. The smaller second crop is small with ovoid fruits weighing 20–38g, an aromatic, sweet slightly juicy pulp containing numerous seeds and a black skin with sparse pink lenticels. This cultivar has an ostiole

**Fig. 5.8.** 'Bakkor Biadh'. (Figure used with permission from Souhila Mahmoudi.)

width of 2–6 mm and is prone to longitudinal cracking. It matures from mid-August to September. The brebas are suitable only for fresh consumption, but the second crop make excellent dried figs (Mahmoudi *et al.*, 2018).

### 'Sabz'

'Sabz' is Iran's most important commercial Smyrna-type fig cultivar, accounting for approximately half of Iran's production. 'Sabz' means green in Farsi. It is a moderately vigorous tree with a slightly spreading growth habit, and figs can be consumed either fresh or dried. The medium sized, 30–35 g, fruits have a greenish-yellow skin that ripens to a golden-yellow and numerous seeds which produce a sweet nutty flavor (Fig. 5.10). It ripens from July to October and can be dried on the tree. The dried fruit has a whitish-yellow skin color, a low, 8–14% moisture (Zare and Jafari, 2018) and TSS 48–62% (Marzoughi Sotoudeh and Zare, 2018).

### 'Shah Anjir'

'Shah Anjir' means big fig in Farsi. It is an important Smyrna-type fresh fig cultivar in Iran. It is a vigorous tree with a semi-upright growth habit. It is large, 50–60 g, with a yellow-amber pulp, greenish-yellow skin and open ostiole when ripe (Zare and Jafari, 2018) (Fig. 5.11). The numerous seeds produce a sweet nutty taste. It ripens from July to August. Fresh fruit has a sugar ratio of 12–14% (Marzoughi Sotoudeh and Zare, 2018). 'Shah Anjir' is drought-sensitive (Jafari *et al.*, 2018).

### 'Siah'

'Siah' means black in Farsi. It is another major fresh Smyrna-type cultivar that, when dried, has medicinal purposes. The tree has a vigorous spreading growth habit. The fruit is medium

**Fig. 5.9.** 'Bakkor Khal'. (Figure used with permission from Souhila Mahmoudi.)

**Fig. 5.10.** 'Sabz'. (Figure used with permission from Moslem Jafari.)

**Fig. 5.11.** 'Shah Anjir'. (Figure used with permission from Moslem Jafari.)

sized, 30–40 g, has a red pulp with numerous seeds, dark purple skin (Fig. 5.12) and a sour and sweet nutty flavor when ripe (Zare and Jafari, 2018). It ripens from July through August. The sugar ratio is 22–28% (Marzoughi Sotoudeh and Zare, 2018). 'Siah' is drought-tolerant and can be used as a rootstock for rain-fed cultivation (Jafari *et al.*, 2018).

## 'Meshki'

'Meshki' means dark black in Farsi. This Common-type cultivar is grown primarily for fresh consumption, but can also be dried. The tree is moderately vigorous with an upright growth habit. When fully ripe the fruit is large, 60 g, with a black skin, pink pulp and sweet taste (Fig. 5.13).

**Fig. 5.12.** 'Siah'. (Figure used with permission from Moslem Jafari.)

**Fig. 5.13.** 'Meshki'. (Figure used with permission from Bardia Moradi.)

This cultivar has very early fruit set and ripens from June to September. It has TSS 16–18% and short postharvest life. In a well-irrigated orchard, the average yield is 60 kg per tree. The breba crop is smaller than the main crop (Zare and Jafari, 2018).

## 'Al-Osaili'

'Al- Osaili' means honey in Arabic. This Common-type fig is highly prized in, and cultivated throughout, Syria for fresh and dried consumption, jams and jellies. The tree is vigorous with a spreading growth habit. The leaf has five deeply spatulate lobes. The oblate fruit is medium sized, 26–30 g, has a medium thick, greenish-yellow skin that peels easily (Fig. 5.14) and a creamy white pulp with a smooth texture. The stalk is separated easily from the shoot. It ripens in early August and has an extended harvest season. It is a sweet fruit with a TSS of 19.36–27.9% and TA 0.18%.

## 'Tamouzi'

'Tamouzi' means July in Arabic and refers the early ripening of this Common-type fig consumed and cultivated throughout Syria under multiple names. It is consumed both fresh and dried. The tree is vigorous with a semi-erect growth habit. The leaves have five deeply spatulate lobes. The fruit is very large, 69–76 g, has a moderately thick, easily peeled yellowish-green skin and red pulp (Fig. 5.15). The fruit stalk is moderately long and thick and easily separates from the shoot. The fruit is moderately sweet having TSS 19%, and TA 0.24%.

## 'Khudeiri'

'Khudeiri' is a Common-type commercial cultivar grown throughout Syria for fresh or dried consumption. The tree is vigorous with a semi-erect growth habit. The spatulate leaves have five moderately deep lobes. The oblate fruit is small to medium, 43–46 g, has a medium thick, easily peeled, green skin with yellow spots and a red to purple juicy pulp (Fig. 5.16). It ripens in

**Fig. 5.14.** 'Al- Osaili'. (Figure used with permission from Ghada Kattmah.)

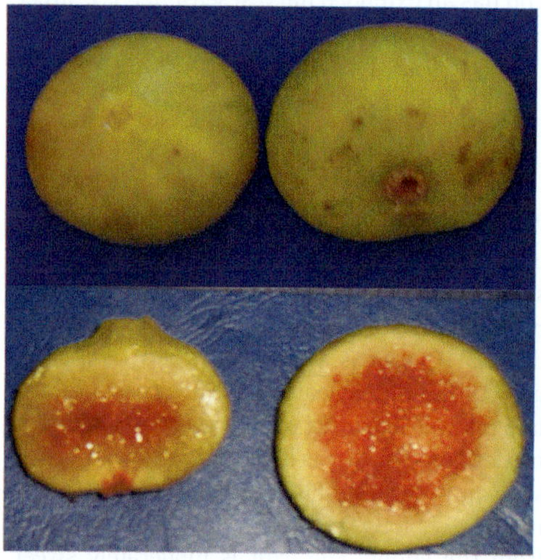

**Fig. 5.15.** 'Tamouzi'. (Figure used with permission from Ghada Kattmah.)

**Fig. 5.16.** 'Khudeiri'. (Figure used with permission from Ghada Kattmah.)

August. The fruit is sweet having TSS 2–25.82% and TA 16% total acidity.

## 'Calabacita'

'Calabacita' is a biferous, Common-type cultivar native to the Extremadura region of Southwest Spain. It is the major cultivar grown for dried consumption in Spain. The tree has low vigor and a semi-erect growth habit. The fruit skin is thin and greenish-yellow with ribbing (Fig. 5.17). They produce a small breba crop and a medium to large main crop. The two crops ripen from spring through late September. The dried figs average 8 g and TSS 80–85%. Originally cultivated exclusively on dry land for dried fig production, orchards are now being irrigated for fresh and dried fruit production. Irrigation considerably increases both fruit yield and average fruit size.

### 'Albac' (syns.: 'Colar', 'Cuello Dama Negro', 'Mission', 'Franciscana')

'Albaco' is an ancient biferous Common-type cultivar and the most widely grown in Spain for fresh consumption; primarily the brebas. The tree has moderate vigor and a semi-erect growth habit. The fruits are piriform with a very short or absent neck, black skin and whitish to purple pulp (Fig. 5.18). The variability in pulp color is due to the presence of different clones in the same orchards. It produces a moderate breba and main crop. The brebas ripen in late June to mid-July. The main crop ripens from early August, peaking in late August, through late September. Postharvest, 'Albaco' tolerate handling and transport well and is, therefore, a good cultivar for fresh marketing. 'Albaco' is well adapted to rainfed conditions but with irrigation produces breba and main crop figs with similar average weights ranging from 50 to 70 g.

**Fig. 5.17.** 'Calabacita'. (Figure used with permission from Margarita López-Corrales.)

**Fig. 5.18.** 'Albacor'. (Figure used with permission from Margarita López-Corrales.)

### 'Cuello Dama Blanco' (syns.: 'Kadota', 'Dottato', 'Ottato')

'Cuello Dama Blanco' is an ancient biferous, Common-type cultivar cultivated in the higher altitude Extremadura and Castilla y León regions of western Spain for both fresh and dried consumption. The tree is moderately vigorous with a semi-erect growth habit. The spherical fruits have an amber pulp and greenish-yellow thick, elastic skin that peels easily and is pest resistant (Fig. 5.19). They produce a small breba crop and very large main crop with fruit sizes decreasing as the season progresses. Fruits ripen the first week of August through to the end of September. The average fresh fruit size is 40–45 g with a TSS of 20–22%. The dried fruits are also large, 9–10 g, and excellent quality. While 'Cuello Dama Blanco' is adapted to humid rainfed soils in mountainous areas at 800–900 m, the first autumn rains frequently coincide with the late dried fig harvests.

**Fig. 5.19.** 'Cuello Dama Blanco'. (Figure used with permission from Margarita López-Corrales.)

### 'Mission' (syns. 'Black Mission', 'California Black', 'Franciscana' and 'Negra')

The 'Mission' fig was the first Common-type cultivar planted in California and is still the primary cultivar for fresh and dried fig production as well as juice and concentrates (Ferguson, *et al.* 1990; California Dried Fig Advisory Board and California Fig Institute, 2020). Planted by the Catholic missionary fathers in their 'Mission' orchards (Condit, 1947) it is similar to the cultivars grown in southern Spain under different names (Eisen, 1901). The trees are large and vigorous with a spreading growth habit (Ferguson, *et al.* 1990). Both breba and main crops have a purplish-black skin, a light strawberry colored pulp, numerous hollow seeds (Fig. 5.20) and a sweet, rich flavor. However, the breba figs vary in size and shape and are often larger and longer, oblong or pear-shaped in cool coastal climates (Eisen, 1901; Condit, 1955). Skin cracking is common in ripening fruits. 'Mission' is quite sensitive to frost. The breba harvest is in late June and the main crop ripens in mid-August through September. The fruits have a TSS of 17–23% and TA of 0.30–0.52%. 'Mission' fig trees are almost always infected with fig mosaic virus, which can affect the color and shape of leaves, but usually does not affect fruit yield or quality.

### 'Brown Turkey'

The name 'Brown Turkey' is used for two distinct Common-type fig cultivars; the 'English Brown Turkey' and 'California Brown Turkey'. The latter is more properly referred to as 'San Piero'. The tree has a dwarfed growth habit (Condit, 1955). The breba fruits are small to medium sized with an oblique-pyriform shape and a thick curved neck. The breba ostiole is medium sized and open with violet-brown scales. Breba crop fruits have mahogany-brown skin, tinged with violet and a light strawberry colored pulp (Fig. 5.21). Breba fruit flavor is fair to mild in quality and richness. Second-crop figs are typically small to medium, turbinate or oblately shaped, lack a neck and average 28 g per fruit. The ostiole is medium sized and open with violet-brown scales. Second-crop fruits typically have large conspicuous white flecks scattered on the brown to burnt umber skin and amber to light strawberry pulp. The second crop fruits are also fair quality, seedless and have a sweet but not rich flavor with an average TSS of 18% and TA of 0.02% (Condit, 1955). This cultivar is characterized by a long ripening season compared to the other cultivars, early June through late September and can produce a third crop in desert climates.

### 'Sierra'

'Sierra' is an unpatented Common-type cultivar released by Doyle and Ferguson (2005). The Sierra breba is suitable for fresh markets and the main crop is suitable for dried fig production. The breba and main crop fruits are large and round with a yellow-green skin, reddish-amber pulp (Fig. 5.22) and a tightly closed ostiole. The breba crop ripens in late June producing a low yield of

**Fig. 5.20.** 'Mission'. (Figure used with permission from Megan Crivelli.)

**Fig. 5.21.** 'Brown Turkey'. (Figure used with permission from Megan Crivelli.)

large fruit. The main crop ripens mid-August through September producing a large crop of medium size fruit that maintains its fruit size well into the fall. The fruit flavor and quality equal that of 'Mission' but are less than that of 'Calimyrna'.

### 'King' ('Desert King')

The origin and identity of the 'Desert King', a San Pedro-type cultivar are unknown (Condit, 1955). The breba crop is especially abundant, with large, 30 g, fruits with a green, glossy skin and violet-red pulp (Fig. 5.23), TSS 18–22% and TA 0.51–0.64%. It is sold fresh in local markets. In cool coastal climates the brebas mature in late July. This cold-tolerant fig is grown on small farms in the cooler Pacific Northwest as it tolerates rains well.

**Fig. 5.22.** 'Sierra'. (Figure used with permission from Megan Crivelli.)

**Fig. 5.23.** 'King'. (Photo: www.groworganic.com). (Figure used with permission from Louise Ferguson.)

### 'Excel'

'Excel' is a Common-type fig produced for fresh markets, drying and/or processing. It is a small, bell-shaped fruit, 24.72 g, with a greenish-yellow skin color and yellow-white pulp (Fig. 5.24). The TSS is 15.48%, and TA 0.19% (Singh *et al.*, 2015). Harvest time is from mid to late September.

### 'Panachee' (Tiger Fig)

'Panache' is a Common-type, which originated from France, with striking green and yellow stripes. It is a color chimera or sport of the type described by Collins (1919) and Condit (1928). The fruit is pyriform, medium sized with a prominent neck, a yellow and green striped skin, an amber-pink to dark red pulp (Fig. 5.25) and an intense flavor for processing. It ripens in late August and has TSS 16.8–23.1% and TA 0.61–0.66%.

### 'Zidi'

'Zidi' is a commercially important fresh market Smyrna-type fig grown in multiple areas of Tunisia (Gaaliche *et al.*, 2012). There is production in the northwest regions of Kesra and Djebba (Mars *et al.*, 2009; Aljane and Ferchichi, 2010), in the central-east regions of Sahel, Bekalta

**Fig. 5.24.** 'Excel'. (Photo: https://www.naturehills.com). (Figure used with permission from Louise Ferguson.)

(Lahbib, 1984; Trad *et al.*, 2013); and in the south-eastern mountains of Matmata (Ferchichi and Aljane, 2007, Aljane, 2011), Beni Kheddache (Ancillotti *et al.*, 1988; Ben Salah *et al.*, 1995), Tataouine and Bir Amir (Aljane and Ferchichi, 2009). The fruits are large, 60–100 g per fruit, pyriform with a long curved neck, a purplish-black skin and deep red-purple pulp (Fig. 5.26) with TSS 17.2–24% and TA 0.40–0.61% (Aljane and Ferchichi, 2010; Trad *et al.*, 2013). 'Zidi' is a high yielding cultivar, producing from 60 to 80 kg/tree, with better yields in coastal zones (Trad *et al.*, 2013). The harvest period extends from July to August (Aljane and Ferchichi, 2010; Trad *et al.*, 2013).

## 'Bouholi'

'Bouholi' is a San Pedro-type cultivar. It is the primary commercially important fig produced in Djebba, northwestern Tunisia's (Mars *et al.*, 2009; Aljane and Ferchichi, 2010), major fig production region (ONUDI, 2019). It is the first Tunisian fig to have a registered designation of origin (AOC) (JORT, 2012; ONUDI, 2019). It has conical buds, light green cordate leaves and a large, 64–135 g, lightly striped oblate fruit with purplish-black skin and light red pulp (Fig. 5.27) (Aljane and Ferchichi, 2010; Gaaliche *et al.*, 2012). The TSS is 18% (Aljane and Ferchichi, 2010). The breba ripens in June and the main crop ripens in August (Mars *et al.*, 2009; Aljane and Ferchichi, 2010).

**Fig. 5.25.** 'Panachee'. (Figure used with permission from Megan Crivelli.)

## 'Soltani'

'Soltani' means king or noble in Farsi. 'Soltani' is a Smyrna-type with a wide geographical distribution. It is cultivated in the oasis of Gafsa (Ben Salah and Lejri, 1992), the region of Djebba (Aljane and Ferchichi, 2010; Gaaliche *et al.*, 2012), and the region of Sahel (Lahbib, 1984). The fruits are pyriform with a long, curved neck, green to purple skin and red pulp (Fig. 5.28), and have a TSS of 16–20%. It ripens from mid-August to September (Aljane and Ferchichi, 2010; Aljane *et al.*, 2018).

## 'Bither Abiadh' and 'Bither Akhal'

In Tunisia, the term 'Bither' designates biferous cultivars. 'Bither Abiadh' and 'Bither Akhal' are San Pedro-type cultivars grown throughout Tunisia. They are cultivated in the mountainous

**Fig. 5.26.** 'Zidi'. (Figure courtesy of F. Aljane.)

**Fig. 5.27.** 'Bouholi'. (Figure courtesy of F. Aljane.)

Matmata region, Tataouine and Beni Khed-dache, in Sahel, Ghadhbna, in the Kerkennah Islands, along the coast in Zarizs and Djerba, and as garden trees everywhere (Ben Salah *et al.*, 2010; Aljane, 2011; 2016). 'Bither Abiadh' and 'Bither Akhal' are vigorous trees with large leaves and are very productive (Aljane, 2011).

'Bither Abiadh' produces moderate size, 47–58 g, oblate fruits with a short neck, yellowish-green skin and a reddish-white pulp (Fig. 5.29). The fruit ostiole is open, 1.9–2.5 mm (Aljane *et al.*, 2012). The breba crop is consumed fresh and the main crop is dried. The dried fruit is whitish-yellow in color with abundant seeds (Ben Salah *et al.*, 2010; Aljane, 2011). The breba crop ripens from June to mid-July and the main crop ripens in August (Aljane, 2011). Fruit has TSS of 18–21% and TA, 0.04–0.1% (Aljane *et al.*, 2012).

'Bither Akhal' is oblately shaped with a protruding neck. The 30–50 g fruits have a reddish-yellow skin and a reddish-white pulp with many seeds and a large ostiole, >3 mm (Fig. 5.30) (Aljane, 2016). Fruits have TSS in the range of 20% (Ben Salah *et al.*, 2010). Crops are better suited for fresh consumption (Aljane, 2016).

**Fig. 5.28.** 'Soltani'. (Figure courtesy of F. Aljane.)

**Fig. 5.29.** 'Bither Abiadh'. (Figure courtesy of F. Aljane.)

## 'Bayoudhi'

The 'Bayoudhi' is a Common-type cultivar described by Guillochon (1927). It is commercially cultivated in the southeastern mountain regions of Matmata with olive and almond trees and in the coastal zones of Zarizs and Djerba (Aljane, 2004, 2011; Ben Salah *et al.* 2010). The small spheroid fruits, 20–39 g, lack a neck, have greenish-yellow skin and pink-red pulp (Fig. 5.31), and a small ostiole opening, <1 mm. It ripens in June (Ferchichi and Aljane, 2007; Aljane and Ferchichi; 2018). While it has good postharvest handling characteristics it is also an excellent quality dried fig therefore much of the main crop production is processed (Aljane, 2010; 2011; Ben Salah *et al.* 2010).

**Fig. 5.30.** 'Bither Akhal'. (Figure courtesy of F. Aljane.)

## 'Kraps i Zl'

This San Pedro-type fig cultivar is grown in the Tirana zone of Albania for fresh consumption (Kokaj, 2019). The fruit has black skin and red pulp (Fig. 5.32). The breba crop ripens in June and the main crop in August. It has TSS 16% (Kokaj, 2019; T. Kokaj, unpublished data).

## 'Shqau'

This Common-type fig cultivar found in the Tirana zone of Albania produces a single crop that ripens in August and is destined for fresh consumption (Koka, 2003; Kokaj, 2018). The fruit skin is green and the pulp is brown to pink (Fig. 5.33). It is very sweet; TSS is 26% (T. Kokaj, unpublished data).

**Fig. 5.31.** 'Bayoudhi'. (Figure courtesy of F. Aljane.)

**Fig. 5.32.** 'Kraps i Zl'. (Figure courtesy of T. Kokaj.)

## 'Rotllar'

'Rotllar' is a Common-type cultivar produced in the Tirana zone of Albania, primarily the main crop for fresh markets (Koka, 2003; Kokaj, 2018). The skin is black and the pulp is red (Fig. 5.34). The main crop ripens in August. TSS is 24% (T. Kokaj, unpublished data).

**Fig. 5.33.** 'Shqau'. (Figure courtesy of T. Kokaj.)

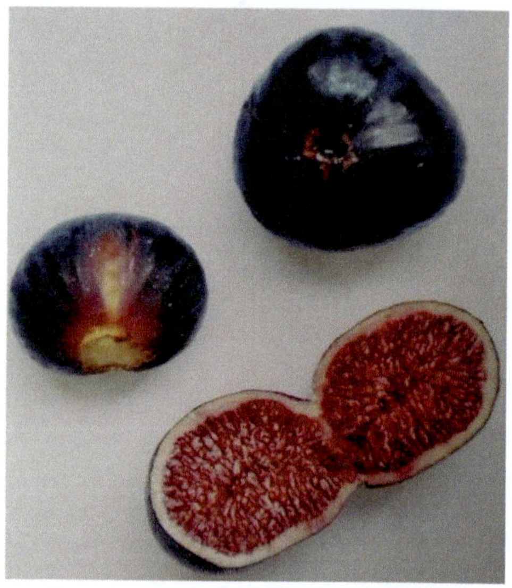

**Fig. 5.34.** 'Rotllar'. (Figure courtesy of T. Kokaj.)

### 'Kallamata'

'Kallamata' is a Smyrna-type grown on the Albanian coast of Himara near the Ionian Sea. It is consumed both as a fresh and dried fruit (Koka, 2003; Kokaj, 2018). The skin is green with a pink pulp (Fig. 5.35). It ripens from August to September. It has a TSS of 27% (T. Kokaj, unpublished data).

### 'Poona'

'Poona' is a Common-type fig consumed fresh in India and used in the Indian deserts Khatta and Mittha bahar. The small, 25 g, average length 5.58 cm, diameter 12.0 cm, pear-shaped fruits have a reddish-green skin and pink pulp (Fig. 5.36). The fruits ripen in summer from March to May and during the rainy season from August to October. It has a TSS of about 15%, and TA 0.21% (Singh *et al.*, 2015).

### 'Bojihong'

'Bojihong' is a Common-type, high quality fig for fresh consumption. In Chinese 'boji' means female on water and 'hong' red. Designated as A132 when introduced from the US in 1998 it is commercially grown in almost every province, accounting for 40% of China's fresh fig production. Trees are vigorous with relatively soft branches that require support for the larger, late season main crop. The fruits are obovate, large, 70–100g with deep red skin and a strawberry pulp (Fig. 5.37). It ripens from July to November in South China. In the greenhouses near Beijing and subtropical regions, production can be extended through January.

### 'Qingpi'

'Qingpi' means green, 'qing', peel, 'pi', in Chinese. It is a Common-type cultivar with outstanding fresh-eating and freeze-dried processing quality.

**Fig. 5.36.** 'Poona'. (Figure used with permission from V.S. Supe.)

**Fig. 5.35.** 'Kallamata'. (Figure courtesy of T. Kokaj.)

**Fig. 5.37.** 'Bojihong'. (Figure used with permission from Huiqin Ma.)

It was introduced to Weihai City in Handong Province by British missionaries in the late 18th century. It is the most important commercial cultivar in Weihai and other Jiaodong Peninsula regions of Shandong Province. It accounts for almost 40% of China's fresh fig production. The vigorous, cold hardy, and drought resistant trees are large with a round canopy and strong trunk. The large, 80–100 g breba and 40–50 g main crop fruits have a yellowish-green skin, obvious ribbing, and a deep red pulp (Fig. 5.38). When ripe, sap and sugars spill from the small ostiole. The breba crop ripens from late June to mid-July and the main crop from early August to late October. Ripe main crop fruits have TSS 22–24%. When freeze-dried the peels are light green and the flesh becomes pink.

### 'Xinjiang Huang'

'Xinjiang Huang' is named for 'Xinjiang', the region where this cultivar is grown, and 'huang', yellow in Chinese. It is a Common-type fig, used for fresh and dried consumption and jams. Introduced from Central Asia via the 'Silk road' several centuries ago, 'Xinjiang Huang' now accounts for more than 90% of Xinjiang's fig production. The primary production regions are south of Xinjiang in Atushi and Kashgar City. It has moderate vigor and produces medium to high yields in dry climates. In China's central regions with heavy precipitation, it is an overly vigorous cultivar and poorly productive. The fruit is medium-sized, 50–70 g, is oblate with flat ends, a very short stalk and no obvious neck. When fully ripe, the fruit is golden-yellow with light yellow pulp, sometimes with a reddish hue (Fig. 5.39). It ripens July to October. The fresh fruit has TSS 18–24%. The dried fruit is whitish-yellow with 21% moisture and TSS 60%.

**Fig. 5.38.** 'Qingpi'. (Figure used with permission from Huiqin Ma.)

### 'Kalamon'

'Kalamon' (synonymy 'Tsapela') is among the important commercial Smyrna-type dried fig cultivars in the southern Peloponnese, Messinia and Lakonia, in Greece. 'Kalamon' trees have a semi-erect growth habit. The medium sized fruit is oblate and green-skinned with red pulp (Fig. 5.40). It ripens from August to mid-September.

### 'Kymis'

'Kymis' (synonymy of 'Kimis') is another commercially important Smyrna-type cultivated for dried fig production in Greece (Fig. 5.41). Produced primarily in the Evia Island's Kymi region 'Kymis' trees have a spreading growth habit. The firm, oblate medium sized fruit, width 35–49 mm, length 28–54 mm (Papadopoulou *et al.*, 2002), ripens from August to mid-September.

**Fig. 5.39.** 'Xinjiang Huang'. (Figure used with permission from Huiqin Ma.)

## 'Vasilika Mellisi'

'Vasilika Melissi' (syn. 'Vasilikali White') is among the commercially important Greek Smyrna-type fresh fig cultivars. The major production areas are in Attiki and the Peloponnese. The trees have a compact growth habit. The firm, globose, medium sized, width 35–49 mm, length 28-54 mm, fruit has a green, easy to peel skin with irregular purple patches and a dark red pulp (Fig. 5.42). It ripens from August to mid-September. The stalk does not easily separate from the stem (Papadopoulou *et al.*, 2002).

## 'Vasilika Mavra'

'Vasilika Mavra' (syn. 'Vasilika Black') is another commercially important Smyrna-type fig cultivated for fresh fig production in Greece, primarily in Attiki and the Peloponnese. 'Vasilika Mavra' trees have a semi-upright growth habit. The medium sized fruit, width 35–49 mm, length 28–54 mm has a turbinate/urceolate shape, a black, easily peeled skin and red pulp (Fig. 5.43) (V. Stournaras, unpublished data). It ripens from August until September.

## 'Masui Dauphine'

'Masui Dauphine' (c.v. San Piero *sensu*; Condit, 1955) is a Common-type fig introduced from California to Japan in 1908. It is Japan's major fresh fig cultivar, accounting for 70% of Japan's production. It is cultivated primarily in the southern Tokai, Kinki and Chugoku regions. The tree's poor cold tolerance precludes its cultivation

**Fig. 5.40.** 'Kalamon'. (Figure courtesy of V. Stournaras.)

**Fig. 5.41.** 'Kymis'. (Figure courtesy of V. Stournaras.)

**Fig. 5.42.** 'Vasilika Meliss'. (Figure courtesy of V. Stournaras.)

**Fig. 5.43.** 'Vasilika Mavra'. (Figure courtesy of V. Stournaras.)

in northern Japan. Trees have intermediate vigor and a spreading growth habit. The oblong to pyriform shaped fruits are very large, 70–150 g, have purple skin and pink pulp (Fig. 5.44). The second-crop, harvested from the sucker shoots of heavily pruned trees, ripens from late July through autumn. The fruit is of moderate quality with 14–18%, total soluble solids and low titratable acidity. The large fruit size and high yields, 3–4 t/1000 m$^2$, are the reasons it is cultivated in Japan. 'Summer Red' is a bud mutation of 'Masui Dauphine' found in the Aichi Prefecture of Japan. It was registered as No. 3351 under the Plant Cultivar Protection and Seed Act of Japan on January 18, 1993. The characteristics are almost the same as 'Masui Dauphine', but it ripens a week earlier and fruits have less cracking.

### 'Houraishi'

'Houraishi' (c.v. Horaigaki *sensu*; Condit, 1955) is the oldest Common-type fig cultivar in Japan. It was probably introduced in the 17th century and is traditionally produced in the western regions of Kyushu, Shikoku and Chugoku. The trees are more cold tolerant (cold hardiness) than the 'Masui Dauphine' cultivar. The trees are vigorous with a semi-erect growth habit. The oblong-pyriform or globose-ovoid fruits are medium to large, 60–80 g, have brown skin, and pink pulp (Fig. 5.45). The prolific second crop is harvested from dense shorter shoots of thinned-out pruned trees and sold as fresh fruit from late August through autumn; it has a late, short harvest period. This cultivar has TSS 15–18% and moderate titratable acidity. The ostiole cracks easily.

**Fig. 5.44.** 'Masui Dauphine'. (Figure used with permission from Akihiro Hosomi.)

### 'Brunswick'

'Brunswick' is another Common-type fig introduced from America to Japan in 1871, before 'Masui Dauphine'. The tree's cold tolerance makes cultivation possible in the northern Tohoku region. The trees are low vigor with a spreading growth habit. The oblong-pyriform shaped fruits are medium to large, 50–70 g, have no neck and a brown skin and amber pulp (Fig. 5.46). The large second crop ripens from mid-August and is marketed as fresh fruit or processed into fig syrup. The apical and basal fruits on the same branch differ in maturity.

### 'Toyomitsuhime'

'Toyomitsuhime' is a Common-type hybrid cultivar developed by the Fukuoka Agricultural

**Fig. 5.45.** 'Houraishi'. (Figure used with permission from Akihiro Hosomi.)

**Fig. 5.46.** 'Brunswick'. (Figure used with permission from Akihiro Hosomi.)

Research Center, Fukuoka, Japan. It was registered as No. 14414 under the Plant Cultivar Protection and Seed Act of Japan on August 22, 2006. Its production is expanding as a local and special cultivar in Fukuoka Prefecture. The trees are moderately vigorous with a semi-erect growth habit. The oblong-pyriform shaped fruits are large, average 80 g, with purple skin and pink pulp (Fig. 5.47). The large second crop is produced on sucker shoots of heavily pruned trees, ripens from summer through autumn and is sold as fresh fruit. This cultivar has a high TSS, above 17%, low TA and an average fruit yield of 3 t/1000 m$^2$. 'Toyomitsuhime' is juicier and sweeter than 'Masui Dauphine' and is resistant to thrip. However, the fruits are smaller than those of 'Masui Dauphine' and are frequently scratched by its leaves, which have long and slender stalks.

## 'Dottato' ('Kadota')

'Dottato', believed to have been introduced by the Greeks, is Italy's primary Common-type cultivar for dried fig production. The Greek name, 'optào/opteteòn' means to dry. It is produced in the Puglia, Campania and Calabria regions. The EU awarded 'Dottato' the Protected Designation of Origin (DOP) certificate for the 'Fico Bianco' cultivated in Cilento area of Campania and for the 'Fico di Cosenza' in Calabria. The trees are moderately vigorous with an open growth habit. The light green leaf is three-lobed with a medium length stalk. 'Dottato' belongs to the Common-type figs; wasp pollination can affect fruit TSS, color and weight. The slightly oblong fruit is small–medium sized, 35–55 g with a yellowish-green skin and amber-red pulp (Fig. 5.48). The small breba crop ripens early but often drops prior to ripening. The large main crop ripens from early to late August. The juicy fruit has a fine texture, sweet, honey-like flavor, and low to medium number of druplets. Its hard, rubbery flesh is ideal for canning (Condit, 1956) and 'Kadota' once was the primary canned fig in California.

**Fig. 5.47.** 'Toyomitsuhime'. (Figure used with permission from Akihiro Hosomi.)

## 'Petrelli' (S. Giovanni, S. Antonio, Columbro)

'Petrelli', grown in southern Italy, is a San Pedro-type fig with multiple names in other regions. It has been grown since ancient times in the Bari and Brindisi provinces for breba production. The trees are moderately to highly vigorous with an open-upright growth habit. The light green leaf is generally five-lobed with a few three-lobed leaves and a medium-long length stalk. The oblong breba fruit is large, 50–90 g, with an easily peeled, thin to moderately thick green skin that does not crack, a medium to coarse textured pink-amber pulp, little juice, and only moderately sweet, often described as having a 'neutral' flavor (Fig. 5.49). The main crop fruit is smaller and more globose than the breba fruit, 30–60 g, with a thin, easily peeled greenish-yellow skin, a medium to coarse textured, fragrant, amber and very sweet pulp. In commercial production the breba yield is very high and the main crop yield is very low if not caprificated. The brebas ripen midseason and early season for the main crop. The postharvest life is limited, so the fruits are marketed domestically; a small percentage is exported.

## 'Tauro' (Minghtaur, Domenico Tauro)

'Tauro', a 'San Pedro-type cultivar is Italy's most important fresh breba fig for export markets. It is grown in Italy's Puglia region in northern Bari Province, in Ruvo di Puglia, Corato and Terlizzi. Its origin is obscure; it has been suggested it was introduced before 1700, or was a local wild tree. The trees are moderately vigorous with an open-upright growth habit. The green leaves are

**Fig. 5.48.** 'Dottato' ('Kadota'). (Figure used with permission from Giuseppe Ferrara.)

**Fig. 5.49.** 'Petrelli'. (Figure used with permission from Giuseppe Ferrara.)

three- or five-lobed with a medium length stalk. The oblong breba fruit is large to very large, 50–100 g, with a medium thickness, easily peeled, green skin with dark purple over color (Fig. 5.50), and no cracks. The medium-coarse textured pulp is pink-amber, slightly juicy and of mild flavor. The breba crop yield is very high and the main crop yield is very low if not caprificated. The brebas fruits have a relatively long post-harvest life and tolerate handling and shipping well for a fresh fig.

### 'Violette de Solliès' ('Parisienne', 'Bourjasotte Noire', 'Barnisotte')

'Violette de Solliès' is a uniferous Common-type cultivar named for Solliès Pont southeast of Provence in the Vallée du Gapeau, 'the garden of Provence'. This region, due to its microclimate, has been the main French fig production area since the 19th century (Figues du Monde, 2019). The 'Violette de Solliès', among France's best fresh fig cultivars, obtained the AOC label in 2006 and the AOP label in 2011 (Figues du Monde, 2019). Today, the Solliès fig makes up 75% of France's fig production with yields around 2000 t/yr (Figues du Monde, 2019). Established trees with a developed root system prefer humid soils and adequate irrigation to produce high quality fruit (Fig Trees of the Catalan Countries and my Roof, 2020). The turbine form, spinning top, flattened shape fruits are large, 70 g, with a visibly ribbed purple to black skin, a dense, firm, supple texture, amber-raspberry to garnet-red pulp with many fine beige seeds, and a slightly open ostiole at maturity (Fig. 5.51) (Baud, 2008). The single crop ripens from mid-August to mid-November. Its thick, solid epidermis facilitates post harvest handling and fresh marketing (Baud, 2008). When fully ripe the Solliès fig has a minimum sugar content of 14 °Brix and is characterized by a good sugar acid balance and rich flavor. Before maturity it can be insipid (Baud, 2008). This cultivar is very sensitive to the Black Fig Fly, *Silba adipata* McAlpine, Diptera: Lonchaeidae (Figues du Monde, 2019; Fig Trees of the Catalan Countries and my Roof, 2020).

### 'Col de Dame Noire' ('Coll de Senyora Negra', 'Cuello de Dama Negra', 'Figues des Dames', 'Pera', 'Fraga')

This old Common-type cultivar is cultivated for fresh consumption in many regions of Spain (including Aragon and Catalonia, particularly Lérida), in France (Pyrénées, Languedoc, Roussillon, Ariège) and North Africa (Kabilie) (Figues du Monde, 2019). It is grown for fresh consumption

**Fig. 5.50.** 'Tauro'. (Figure used with permission from Giuseppe Ferrara.)

**Fig. 5.51.** 'Violette De Solliès'. (www.frenchfigfarm.com/). (Author's own figure.)

in France. The medium size and vigorous trees have a semi-erect canopy. They produce best in cool soils. In sandy soils they must be grafted to a nematode-resistant rootstock. The medium sized, 55–60 g, pyriform shaped fruits with an elongated neck, short peduncle, and small ostiole have a thick purplish-black skin, dense flesh and dark red pulp (Fig. 5.52) (Baud, 2008). It ripens at the end of the hot season from September through to the first frost, and despite its thick skin, will crack if it rains (Baud, 2008; Fig Trees of the Catalan Countries and my Roof, 2020). 'Col de Dame Noire' is a prolific producer and tolerates storage and transportation well. This cultivar is also very sensitive to the Black Fig Fly, *Silba adipata* McAlpine, Diptera: Lonchaeidae (Figues du Monde, 2019; Fig Trees of the Catalan Countries and my Roof, 2020).

**Fig. 5.52.** 'Col de Dame Noire'. (Figure used with permission from Margarita López-Corrales).

## 'Grise de Saint-Jean' ('Cotignane', 'Observantine', 'Grisette', 'Cordelière', 'Grisette de Montpellier', 'Coucourelle grise', 'Célestine', 'Capoutchin' and 'Mestissegrise')

The 'Grise de Saint-Jean' cultivar is among the most important San Pedro-type biferous commercial fresh fig cultivars in France, particularly southern France (Drouet, 2015). The trees are vigorous and large, from 3.5 to 6 m tall and 6–10 m in diameter, and cold-sensitive (Baud, 2008; Pepinieres Baud, 2019). The large, 40–45 round breba fruits have a thin, green skin streaked with purple and dark red pulp (Fig. 5.53). The smaller breba crop ripens in July. The larger, but smaller fruited, 35 g, main crop ripens in stages from late August to October (Baud, 2008). This fig is best consumed fresh. Cold storage diminishes the fig's aroma and damages the fruit epidermis (Drouet, 2015).

## 'Ronde de Bordeaux'

'Ronde de Bordeaux' is a uniferous, Common-type fig cultivated for dried fig processing. It produces early brebas and is one of the earliest main crops (Baud, 2008). The small, 25–35 g, round fruit lacks a neck, has a thin black skin and pink pulp and a slightly open ostiole (Fig. 5.54). It ripens mid-August to September. (Baud, 2008; Fig Trees of the Catalan Countries and my Roof, 2020). Its thin skin and sensitivity to humidity make it ideal for drying (Baud, 2008; Fig Trees of the Catalan Countries and my Roof, 2020).

### 'Roxo-de-Valinhos'

Brazil's single commercial fig cultivar is the vigorous high-yielding Common-type 'Roxo-de-Valinhos' (Rodrigues *et al.*, 2019b). It is grown in the subtropical regions where harvesting can be staggered throughout the year with the irrigation and pruning practices that produce larger, sweet fruits the domestic market prefers (Santos, 2017). The vigorous tree is severely pruned to the size of a shrub so that fruit is produced on the current year's shoot growth (Correa and Boliani, 1999). The large, 60–90 g and 7.5 cm long, oblong-pyriform shaped fruits have a short, thick neck, typical reddish-purple bracts and large ostiole. The dark purple-violet skin is slightly furrowed and the pulp is pinkish-reddish and juicy (Fig. 5.55) Pereira and Nachtigal (1999). The flavor is pleasant and bittersweet (Pereira, 1981). 'Roxo-de-Valinhos' fruits can be processed when green or ripe but are primarily consumed fresh when ripe.

**Fig. 5.53.** 'Grise de Saint-Jean'. (Author's own figure).

**Fig. 5.54.** 'Ronde de Bordeaux'. (Figure used with permission from Keith Wilson.)

### 'Negrito'

'Negrito', a sweeter variation of the Common-type 'Roxo-de-Valinhos' cultivar with a darker peel and pulp is excellent for fresh consumption and processing into jam (UFRGS, 2020). The 'Negrito' is produced in Brazil's southern temperate regions that permit the severe dormant pruning that produces fresh figs during the Northern Hemisphere's off season; a significant market advantage (Dalastra *et al.*, 2009; Parra Lerma, 1996). The dormant pruning eliminates most of the previous season's growth; heading branches to 5–10 cm, leaving two outward facing buds to form the new canopy (Santos *et al.*, 2001). The resulting in-season shoot growth produces

**Fig. 5.55.** 'Roxo-de-Valinhos'. (Figure courtesy of M.G.F. Rodrigues.)

high quality figs with an elongated shape, 5–8 cm long, and dark purple when ripe (Fig. 5.56) (Medeiros, 2002).

### 'Pingo de Mel'

'Pingo de Mel' is another Common-type fig produced in Brazil. When mature, a drop of exudate at the ostiole attracts bees; hence its name 'Pingo de Mel', honey drop (Sousa, 2010). This small to medium sized, 35 g, round fruit with a medium peduncle and small ostiole has a smooth greenish-yellow tough peel that resists cracking and amber pulp (Fig. 5.57). Like the 'Roxo-de-Valinhos' cultivar, it is pruned to produce crop on the current year's growth. This cultivar is harvested late to increase the sweetness for processing and fresh consumption. The fruit's tough peel and small ostiole make it resistant to cracking and insect infestation resulting in better postharvest life (Bisi, 2015; Aljane *et al.*, 2018; Rodrigues *et al.*, 2019b).

**Fig. 5.56.** 'Negrito'. (Author's own figure.)

**Fig. 5.57.** 'Pingo de Mel'. (Figure courtesy of M.G.F. Rodrigues.)

### 'Parisian' or 'Evita' ('Violette de Solliès', 'Parisienne', 'Bourjasotte Noir', 'Barnisotte Noir', 'Brogiotto Nero', 'Negro Largo')

'Parisian' or 'Evita' is a Common-type uniferous cultivar that accounts for 75% of South Africa's fresh fig exports. It is cultivated in 5–50 ha orchards throughout the Western Cape. Larger fruit are produced along the coastal strip, averaging 50 g, whereas in the dry, hotter interior areas the figs average 35 g. The trees are very large and vigorous with a spreading growth habit. The turbinate-pyriform medium sized fruits, 35–50 g, have a tiny 2–3 mm ostiole, and a tough, thin purplish-black skin and light strawberry pulp with a moderate number of seeds (Fig. 5.58). It ripens from late January through mid-May in the Southern Hemisphere. The flavor is exceptionally rich, balanced and berry-like. While it has good postharvest life the Western Cape's high temperatures and wind cause fruit scarring, decreasing export pack-outs.

### 'Tangier'

'Tangier' is a Smyrna-type that now accounts for 10% of South Africa's exports. Its popularity is increasing as it produces large figs in all the Western Cape's microclimates. It is a medium size erect tree kept compact in high density trellised plantings of up to 1000 trees per hectare.

**Fig. 5.58.** 'Parisian' or 'Evita' (Figure used with permission from Keith Wilson.)

The large to very large, 65 g, pyriform fruits have prominent ribs and a green tip at the neck, a purple to black tough skin resistant to wind scarring, a dense, moderately seeded strawberry pulp, and a medium sized, 2–4 mm, ostiole (Fig. 5.59). In the Southern Hemisphere the figs ripen from mid-January through to mid-March. The flavor is rich and fruity with a balanced sweetness. The tough peel resists robust wind scarring but is vulnerable to cracking with irregular or over irrigation. Pack-outs often exceed 90%.

## 'Cosmic'

'Cosmic' is a Smyrna-type now dominating the new plantings in South Africa's Western Cape. Currently it is 2% of exports but is expected to grow exponentially. The tree is compact on its own roots but vigorous on a rootstock. The figs are similar to 'Bursa Black'. The turbinate-pyriform fruit is very large, 72–140 g, light purple to black with prominent ribbing and has a tiny ostiole, 1–2 mm. The pulp is strawberry colored with few seeds and juicy (Fig. 5.60). The fruit shape is more turbinate when grown on rootstock and more turbinate-pyriform when own-rooted. In the Southern Hemisphere they ripen from mid-February through May. If harvested when the neck is light green, they will develop their full black color within a day. It has good postharvest shelf life and transportability.

## 'Spen'

'Spen' is originally from Kandahar province of Afghanistan and widely cultivated in the Shah Wali Kot district of Kandahar (Fig. 5.61). This cultivar is also cultivated in Herat, Helmand and Zabul provinces. 'Spen' means white in the Pashto language of Afghanistan, which refers to the light color of the fruit. It is the most important commercial cultivar used to produce dried figs in Afghanistan. The fruit skin color is light green and pulp color is light pink. The fruit is medium in size and has TSS of 19%. 'Spen' is a Common-type of fig, does not need wasps for pollination and produces a single crop. The cultivar ripens from July 10 until the end of August. Due to differences in the climate in different villages the

**Fig. 5.60.** 'Cosmic'. (Figure used with permission from Keith Wilson.)

**Fig. 5.59.** 'Tangier'. (Figure used with permission from Keith Wilson.)

**Fig. 5.61.** 'Spen'. (Figure courtesy of A. Yavari.)

harvesting period is more than two months. In villages with a warmer climate the first harvest starts in mid-June, but in other villages it starts in mid-July and lasts until the end of August.

## Conclusion

As this chapter only begins to demonstrate, the fig is a global fruit, its numerous cultivars adapted to multiple microclimates, production methods, markets and uses. It has been a part of multiple cultures for millennia and its adaptability, and new plant improvement, production and postharvest technologies ensure it will only increase in variability, production and popularity.

## Acknowledgments

The authors would like to acknowledge the valuable help of N. Habib (CFPPA Comminges Ariège, France), Z.E. Benettayeb (Université Hassiba Benbouali, Chlef, Algeria), A. Lehad (Ecole Nationale Supérieure Agronomique d'Alger, El Harrach, Algeria), T. Demarquest ('Pépinière Figues Du Monde', Douelle, France), A. Baud ('Pépinière Baud', le Palis Vaison la Romaine, France), A. Hosomi (Toyo Institute of Food Technology, Kawanishi, Japan), H. Ma (Department of Fruit Tree Sciences, College of Horticulture, China Agricultural University, Beijing, China), M. López-Corrales (Junta de Extremadura, Agri-Food Technological Institute of Extremadura (INTAEX-CICYTEX), Badajoz, Spain) and B. Ertan (Fig Research Institute, Aydın, Turkey).

Also G. Ferrara (Department of Soil, Plant and Food Sciences, University of Bari 'Aldo Moro', Bari, Italy), M. Jafari (Estahban Fig Research Station, Fars Agricultural and Natural Resources Research Center, Shiraz, Iran), S. Mahmoudi (Department of Agronomic Sciences, University 'Med. BOUDIAF', M'sila, Algeria), Keith Wilson (Exotic fruits breeder and private consultant, Cape Town, South Africa), Ghada Kattmah (General Commission for Scientific Agricultural Research, Damascus, Syria), V.S. Supe (Associate Director of Research, NARP, Ganeshkhind, Pune (MS), India) and N.A. Awd (Horticultural Research Institute, Agricultural Research Center, Egypt), who generously provided us with photos. We would like to thank all the authors of this chapter and apologize to the many other colleagues, whose original publications are not cited due to space limitations.

## References

Abou-Ellail, M., Mahfouze, S.A., Magda, A.M. and Mustafa, N.S.A. (2014) Using biochemical and simple sequence repeats (SSR) markers to characterize (*Ficus carica* L.) cultivars. *World Applied Sciences Journal* 29 (3), 313–321. DOI.10.5829/idosi.wasj.2014.29.03.13835

Aid-Houchi, A. and Mazri, C. (2001) Evaluation de l'Eat a des Figuieraies de Kabylie. *Proceeding des Troisiemes Journées Scientifiques de l'INRAA*. INRAA, Bejaia, Algeria, p. 9.

Ait Haddou, L., Blenzar, A., Messaoudi, Z., Van Damme, P., Zinedine, F. and Sakhaoui A. (2013) Caractérisation pomologique de 22 cultivars locaux du figuier (*Ficus carica* L.) au Maroc. *European Journal of Scientific Research*, 416–428.

Alfrey, P. (2016) *Dig the Fig – The Essential Guide to All You Need to Know About Figs.* www.permaculturenews.org/2016/09/30/dig-fig-essential-guide-need-know-figs-ficus-carica (accessed 16 October 2019).

Al Ibrahim, A. (2000) *Arbre du Figuier: Culture- Variétés- Multiplication- Conduite*. Ministère d'Agriculture, Damas, Syria.

Al Ibrahim, A. and Bari, A. (2006) Fig genetic resources in Syria: varietal diversity, agronomic and economic importance. *Proceeding of the Conference on Sustainable Agricultural Development and Food Security*. Tishreen University, Lattakia, Syria, pp. 1–14.

Aljane, F. (2004) Prospection et caractérisation des variétés locales de figuier (*Ficus carica* L.) dans les Jessours méridionaux des djebels Matmata. *MSc Thesis*, Faculté des Sciences de Sfax, Tunisie.

Aljane, F. (2011) Caractérisation et évaluation des accessions locales de figuier (*Ficus carica* L.) en Tunisie et sélection des plus performantes. *MSc Thesis*, Faculté des Sciences de Tunis, Tunis, Tunisie.

Aljane, F. (2016) Analysis of genetic diversity in Tunisian fig (*Ficus carica* L.) germplasm bank revealed by RAPD markers and morphological characters. *European Journal of Sciences Research* 142(2), 172–192.

Aljane, F. and Ferchichi, A. (2009) Assessment of genetic diversity among some southern Tunisian fig (*Ficus carica* L.) cultivars based on morphological descriptors. *Jordan Journal of Agricultural Sciences* 5 (1), 1–15.

Aljane, F. and Ferchichi, A. (2010) Assessment of genetic diversity of Tunisian fig (*Ficus carica* L.) cultivars using morphological and chemical characters. *Acta Botanica Gallica* 157(1), 171–182.

Aljane, F., Nahdi, S. and Essid, A. (2012) Genetic diversity of some accessions of Tunisian fig tree (*Ficus carica* L.) based in morphological and chemical traits. *Journal of Natural Production Plants Resources* 2(3), 350–359.

Aljane, F., Essid, A. and Nahdi, S. (2018) Improvement of fig (*Ficus carica* L.) by conventional breeding and biotechnology. In: Al-khayri, J.M., Mohan, J.S. and Johnson, D.V. (eds) *Advances in Plant Breeding Strategies: Fruits* Vol 3. Springer, Al-Hassa, Saudi Arabia; Helsinki, Finland and Ohio, USA, pp. 343–375. DOI: 10.1007/978-3-319-91944-7_9.

Ancillotti, M., Ben Salah, M. and Loumerem, M. (1988) Contribution à l'étude de la culture du figuier (*Ficus carica* L.) et de certaines de ses populations cultivées dans la région de Beni Kheddache (Gouvernorat de Médenine). Rapport du projet pour la mise en valeur du Sahara tunisien, Institut des Régions Arides de Médenine, Médenine, Tunisia.

Arpaci, S., Konak, R. and Çiçek, E. (2018) A national value: Turkish figs. *Proceedings of the 30th International Horticultural Congress 12 - 16 August 2018*, Istanbul, Turkey, pp. 15.

Bachir Bey, M., Louaileche, H. and Zemouri, S. (2013) Optimization of phenolic compound recovery and antioxidant activity of light and dark dried fig (*Ficus carica* L.) cultivars. *Food Sciences Biotechnologies* 22(6), 1613–1619. DOI: 10.1007/s10068-013-0258-7.

Bachir Bey, M. and Louaileche, H. (2015) A comparative study of phytochemical profile and in vitro antioxidant activities of dark and light dried fig (*Ficus carica* L.) cultivars. *Journal of Phytopharmacy* 4(1), 41–48.

Baud, P. (2008) *Le figuier Pas à Pas*. Eds. EDISUD. Un Editeur de la Compagne des Editions de la Lesse, Aix-En-Provence, France.

Baziar, G., Jafari, M., Sharifi-Noori, M.S. and Samarfard, S. (2018). Evaluation of genetic diversity among Persian fig cultivars by morphological traits and RAPD markers. *Hortscience*, 53(5), 613–619.

Ben Salah, M. and Lejri, M.H. (1992) Description pomologique de quatre variétés de figuier (*Ficus carica* L.) des oasis de Gafsa. *Revue des Régions Arides* 16–20.

Ben Salah, M., Ancilotti, M. and Loumerem, M. (1995) Etude pomologique de six variétés de figuier (*Ficus carica* L.) typiques de Beni Kheddache. *Plant Genetic Resources Newsletter* 104, 16–20.

Ben Salah, M., Mosbah, A., Rhihane, M. and Hamid, H. (2010) Répertoire et caractérisation des variétés population de figuier dans la zone montagneuse de Matmata. Procceding du 3ème Meeting International Aridoculture et Cultures Oasiennes : Gestion et Valorisation des Ressources et Applications Biotechnologiques dans les Agrosystèmes Arides et Sahariens Institut des Régions Arides de Médenine, Décembre 2009, Djerba, Tunisia. *Revue des Régions Arides, Numéro spécial* 24, 135–141.

Bibi, I. and Afechtal, M. (2017) Programme de certification du figuier (*Ficus carica* L.) au Morocco. *Revue Marocaines des Sciences Agronomiques et Véterinaires* 5(4), 400–404.

Birk, V. (2015) *Figo é o motivo da maior festa do interior de Nova Petrópolis*. www.displaycard.com.br/noticias/figo-e-o-motivo-da-maior-festa-do-interior-de-nova-petropolis (accessed 8 May 2020).

Bisi, R.B. (2015) Rooting of cuttings from 15 cultivars for diversification of ficiculture. 61f. Dissertation. *MSC Thesis*. Federal University of Lavras, Lavras, Brazil.

BOE, 2011. Order ARM/2935/2011, that established the inclusion of several varieties of different species in the Register of Commercial Varieties. Boletin oficial del Estado no 262, 31/10/ 2011

Botti, C. and Muchnik, E. (1998) Fig production in Latin America. *Acta Horticulturae* 480, 37–42. DOI: 10.17660/ActaHortic.1998.480.4

Brien, J. and Hardy, T. S. (2002) Fig Growing in NSW. www.agric.nsw.gov.au. (accessed 22 October 2019).

California Dried Fig Advisory Board and California Fig Institute (2020) Statistical Review of the California Fig Industry: 2019–2020 Fiscal Year. Califorina Dried Fig Advisory Board and California Fig Institute, Madera, CA, USA. www.californiafigs.com

Caliskan, O. and Polat, A.A. (2008) Fruit characteristics of fig cultivars and genotypes grown in Turkey. *Scientia Horticulturae* 115, 360–367.

Ciarmielloa, L.F., Piccirilloa, P., Carillob, P., De Lucaa, A. and Woodrowb, P. (2015) Determination of the genetic relatedness of fig (*Ficus carica* L.) accessions using RAPD fingerprint and their agro-morphological characterization. *South African Journal of Botany* 97, 40–47. DOI:10.1016/j.sajb.2014.11.012.

Collins, J.L. (1919) A case of chimera in a fig. *Journal of Heredity* 10(1), 6–7.
Condit, I.J. (1928) Other fig chimeras. *Journal of Heredity* 19(2), 49–53.
Condit, I.J. (1941) Fig characteristics useful in the identification of cultivars. *Hilgardia* 14(1), 1–68.
Condit, I.J. (1947) *The Fig*. Chronica Botanica Co., Waltham, Massachusetts.
Condit, I.J. (1955) Fig cultivars: A monograph. *Hilgardia* 11, 323–538.
Condit, I.J. (1956) Promising new seedling fig. *California Agriculture* 10, 4–14.
Correa, L.S. and Boliani, A.C. (1999) *Fig Tree Culture, From Planting to Marketing*. Ilha Solteira, FAPESP, Brazil.
Dalastra, I.M., Pio, R., Campagnolo, M.A., Dalastra, G.M., Chagas, E.A. and Guimarães, V.F. (2009) Pruning time in the production of 'Roxo de Valinhos' green fig in organic sistem in the west region of Paraná State. *Revista Brasileira de Fruticultura* 31(2), 447–453.
Daleys (2002) Fabulous Figs. Newsletter. www.daleysfruit.com.au/newsletter/january2002.htm (acessed 22 October 2019).
Doyle, J.F. and Ferguson, L. (2005) Sierra: A new non-caprifying calimyrna, p. 60. In *Third Int. Symp. Fig. 16–20 May*, Vilamoura, Portugal.
Drouet, F. (2015) *Culture de Figuier: Quelques observations et conseils*. www.fruitiers-rares.info/articles75a80/article76-culture-Figuier-Ficus-carica.html (accessed 10 May 2019).
Eisen, G. (1901) The fig: its history, culture, and curing. *USDA Bulletin* Government Printing Office.
El Hajjam, A., Ezzahouani, A. and Sehhar, E.A. (2018) Conduite technique et inventaire des variétés Moroccoaines locales de figuier (*Ficus carica* L.) dans quatre principaux sites de production, provinces de Chefchaouen, El Jadida, Ouezzane, et Taounate. *Revue Marocaine des Sciences Agronomiques & Vétérinaires* 4, 494–504.
Ferguson, L., Michailides, J.T. and Shorey, H.H. (1990) The California fig industry. *Horticulture Revue* 12, 409–490.
Figues du Monde (2019) www.figuesdumonde.wordpress.com/type/link (accessed 22 October 2019).
Fig Trees of the Catalan Countries and my Roof (2020) www.galgoni.com/ENG/Index_EN.php (accessed 13 May 2020).
Flaishman, M.A., Rodov, V. and Stover, E. (2008) The fig: botany, horticulture, and breeding. *Horticultural Reviews* 34, 113–195. DOI: 10.1002/9780470380147.ch2.
Gaaliche, B., Saddoud, O. and Mars, M. (2012) Morphological and pomological diversity of fig (*Ficus carica* L.) cultivars in northwest of Tunisia. *ISRN Agronomy*, 1–9. DOI. org/10.5402/2012/326461.
García-Ruiz, M.T., Mendoza-Castillo, V.M., Valadez-Moctezuma, E. and Muratalla-Lúa, A. (2013) Initial assessment of natural diversity in Mexican fig landraces. *Genetics and Molecular Research* 12(2), 3931–3943. DOI:10.4238/2013.September.23.12.
Gil, M., Pérez, F., Giraldo, E., Cortésand, J. and López-Corrales, M. (2011) Classification and synonymy of Albacor fig in the genebank at Finca La Orden, Spain. *Acta Horticulturae* 918, 961–963. DOI: 10.17660/ActaHortic.2011.918.125.
Gordon, P., Preece, J., Ferguson, L., Crisosto, C., Aradhya, M. and Norton, M. (2018) Field evaluation of new and underutilized fig cultivars for fresh and dried markets. *Executive Research Reports*. California Fig Institute, California.
Guillochon, L. (1927) La culture du Figuier en Tunisie. *Revue de Botanique Appliquée et d'Agriculture Coloniale* 7(65), 18–28. DOI:10.3406/jatba.1927.4483.
Hoxha, L., Kongoli, R. and Hoxha, M. (2015) Antioxidant activity of some dried autochthonous Albanian fig (*Ficus carica* L.) cultivars. *International Journal of Crop Science and Technology* 1(2), 20–25.
IBPGR (1986) *Ficus carica* L. In: *Genetic Resources of Tropical and Sub-tropical Fruits and Nuts.* International Board Plant Genetic Resources (IBPGR), Rome, Italy.
Jafari, M., Rahemi, M. and Hashighi, A.A. (2018) Role of fig rootstock on changes of water status and nutrient concentrations in 'Sabz' cultivar under drought stress condition. *Scientia Horticulturae* 230, 56–61.
Jafari, M., Zare, H., Golkar, G., Jokar, L. and Tabatabaei, Z. (2016) Evaluation of morphological characteristics in some fig (*Ficus carica* L.) genotypes. *Seed and Plant Improvement Journal* 32(2), Pe147–Pe163.
Japan Guide (2019) Japanese fruits. www.japanguide.com/e/e2347.html (accessed 16 October 2019).
Jianye, C., Yuxia, N. and Zilan, Z. (1997) Observation on biological characteristics of fig (*Ficus carica* L.). *Journal of Fruit Sciences* 14(1), 16–20.
JORT (2012) *Journal Officiel de la Republique Tunisienne* 47, 1461–1463.
Khadivi, A., Anjam, R. and Anjam, K. (2018) Morphological and pomological characterization of edible fig (*Ficus carica* L.) to select the superior trees. *Scientia Horticulturae* 238, 66–74.
Koka, T. (2003) Studies on local fig (*Ficus carica* L.) germplasm in Albania. *Acta Horticulturae* 605, 87–89. DOI:10.17660/ActaHortic.2003.605.11.

Kokaj, T. (2018) Genetic diversity of fig cultivars in ex situ conservation. *International Journal of Agriculture, Environment and Bioresearch* 3(03), 50–59.

Kokaj, T. (2019) Clonal selection in Cipull cv, Kraps i zi cv, Melacak cv at Tirana. *International Journal of Agriculture, Environment and Bioresearch* 4(02), 41–46.

Koyuncu, M. (1998) A study on some fruit characteristics in local fig cultivars grown in Hilvan (Uraf, southern Turkey). *Acta Horticulturae* 480, 83–85.

Koyuncu, M.A., Bostan, S.Z. and Koyuncu, F. (1998) Investigation on some physical and chemical characteristics in fig cultivars growth in ORDU. *Acta Horticulturae* 480, 87–88.

Lahbib, T. (1984) Etude pomologique des variétés de figuier (*Ficus carica* L.) répertoriées dans le Sahel tunisien. MSc Thesis, Institut National d'Agronomie de Tunis, Tunis, Tunisie).

López-Corrales, M., Gil, M., Pérez, F., Cortés, J., Serradilla, M.J. and Chome, P.M. (2011) Variedades de higuera: descripción y registro de variedades. In: *Variedades de Higuera Descripción y Registro de Variedades*. Ministerio de Medio Ambiente y Medio rural y Marino, Madrid. ISBN 978-84-491-1103-7

López-Corrales, M. and Balas, F. (2014) In: Hueso-Martín, J.J. and Cuevas-González, J. (eds) *La Fruticultura del Siglo XXI en España*. Cajamar Caja Rural. pp. 241–260.

Louw, M. (2019) Fig cultivars. Fig farming in South Africa. Available at: http://southafrica.co.za/fig-cultivars.html (accessed 21 October 2019).

Mahmoudi, S., Khali, M., Benkhaled, A., Boucetta, I., Dahmani, Y., Attallah, Z. and Belbraouet, S. (2018) Fresh figs (*Ficus carica* L.): Pomological characteristics, nutritional value, and phytochemical properties. *European Journal of Horticultural Science* 83(2), 104–113.

Maia de Sousa, R.M. (2003) Fig culture techniques. *Acta Horticulturae* 605, 99–101. DOI: 10.17660/ActaHortic.2003.605.14.

Mars, M. (2003) Fig (*Ficus carica* L.) genetic resources and breeding. *Acta Horticulturae* 605, 19–25.

Mars, M., Chatti, K. and Saddoud, O. (2008) Fig cultivation and genetic resources in Tunisia. An overview. *Acta Horticulturae* 798, 27–32.

Mars, M., Gaaliche, B., Ouerfelli, I. and Chouat, S. (2009) Systèmes de production et ressources génétiques du figuier (*Ficus carica* L.) a Djébba et Kesra, deux villages de montagne au nord-ouest de la Tunisie. *Revue des Régions Arides* 22, 33–45.

Mazri-Kartout, C. and Aid-Houchi, A. (2001) Contribution à la caractérisation de trois variétés de figuier dans la commune de Fréha Wilaya de tizi-Ouzou. d'Algérie. *Revue semestrielle de la Recherche Agronomique de l' INRA* 14, 53–63.

Marzoughi Sotoudeh, M. and Zare, H. (2018) Effect of different harvesting stages and storage temperature on quality indices and postharvest life of fresh fruit of edible fig cultivars. *Iranian Journal of Horticultural Science and Technology* 19(3), 277–292.

Medeiros, A.R.M. (2002) Fig (*Ficus carica* L.) from planting to home processing. *Pelotas: Embrapa. Technical Circular* 35.

Meziant, L., Saci, F., Bachir Bey, M. and Louaileche, H. (2015) Varietal influence on biological properties of Algerian light figs (*Ficus carica* L.) *International Journal of Bioinformatics and Biomedical Engineering* 1(3), 237–243. DOI.aiscience.org/journal/ijbbe

Minangoin, N. (1931) Monographie des variétés de figues tunisiennes. *Congres d'Agronomie du cinquantenaire. Tan. Ed. Imprimerie Baconnier, Algerie*, pp. 336–364.

Nabil Sabet, A.M. and Abou-Ellail, M. (2013) Genetic discrimination of Egyptian fig cultivars revealed by RAPD fingerprints. *International Journal of Agricultural Research* 8, 17–25. DOI:10.3923/ijar.2013.17.25.

Nakilcioğlu, E. and Hışıl, Y. (2013) Research on the phenolic compounds in sarilop (*Ficus carica* L.) fig cultivar. *Gida Bilimi ve Teknolojisi Dergisi (GIDA)* 38(5), 267–274. DOI:10.5505/gida.2013.08208.

Obenauf, G., Gerdts, M., Leavitt, G. and Crane, J. (1978) Commercial dried fig production. University of California. *Agriculture Sciences Leaflet*. California, USA.

ODEPA (1997) *Centro Documentacion*. Santiago, Chile.

OJEC (1996) Official Journal of the European Communities No L148/1 (21.6.1996). Commission Regulation (EC) No 1107/96.

OJEC (2013) Official Journal of the European Union L133/11 (17.5.2013). Commission Implementing Regulation (EU) No 455/2013.

ONUDI (2019) *Guide des bonnes pratiques de production des figues à Djébba (Béjà)*. Organisation des Nations Unies pour le Développement Industriel, Ministère de l'Agriculture, des Ressources Hydrauliques et de la Pêche and Ministère de l'Industrie et des PME, Tunis, Tunisie.

Papadopoulou, K., Ehaliotis, C., Tourna, M., Kastanis, P., Karydis, I. and Zervakis, G. (2002) Genetic relatedness among dioecious *Ficus carica* L. Cultivars by random amplified polymorphic DNA analysis, and evaluation of agronomic and morphological characters. *Genetica* 114, 183–194.
Parra Lerma, J. (1996) Estudio Fenológico de un Huerto de Higuera (*Ficus carica* L.) en el Ejido Francisco Villa, Municipio de Comundu, Baja California Sur. *MSc thesis*. Universidad Autónoma de Nayarit, Spain.
Pepinieres Baud (2019) *Figuiers and Grenadiers*. www.figbaud.com/choixfiguiers.html (accessed 2 July 2019).
Pereira, F.M. (1981) *Fig Tree Culture*. Livroceres, São Paulo, Brazil.
Pereira, F.M. and Nachtigal, J.C. (1999) Botany, biology and fig cultivars. In: Corrêa, L. de S. and Boliani, A.C. (eds) *Fig Tree Culture,* from *Planting* to *Marketing*. Ilha Solteira, FAPESP, Brazil, pp. 25–36.
Rahemi, M. and Jafari, M. (2008) Effect of Caprifig-type on quantity and quality of estahban dried fig *Ficus carica* cv. sabz. *Acta Horticulturae* 798, 249–252.
Rodrigues, M.G.F., Santos, T.P., Ferreira, A.F.A., Monteiro, L.N.H., Nakanishi, E.S. and Boliani, A.C. (2019) Morphological characterization of active germoplasm bank fig tree accessions. *Revista Brasileira de Fruticultura* 41(5), e-074.
Rodrigues, M.G.F., Monteiro, L.N.H., Ferreira, A.F.A, Dos Santos, T.P., Pavan, B.E., Neves, V.A.B. and Boliani, A.C. (2018) Genetic variability in morphological characters among fig tree accessions fig genetic conservation. *Genetics and Molecular Research* 17(4), 1–9. gmr16039931. DOI:10.4238/gmr16039931.
Santos, D.F., Martins, F.B. and Torres, R.R. (2017) Impacts of climate projections on water balance and implications on olive tree in Minas Gerais. *Revista Brasileira de Engenharia Agrícola e Ambiental* 21(2), 77–82.
Santos, S.C., Tarsiano, M.A.A. and Corrêa, L.S. (2001) Economic analysis of the fig tree subjected to variations in pruning time and number of branches, in the region of Selvíria-MS. *Revista Brasileira de Fruticultura* 23, 64–68.
Singh, A., Prakash, J., Meghawal, P.R. and Ranpise, S.A. (2015) The fig (*Ficus carica*). In: Ghosh, S.N. (ed.) *Breeding of Underutilized Fruit Crops Part I*. Jaya Publishing House, New Delhi, India, pp. 149–179.
Sousa, R.M. and Gomes Pereira, J. (1995) Notes sur quelques espèces fruitières sous-utilisées au Portugal (néflier japonais, figuier, kaki et grenadier). *Cahiers Options Méditerranéennes* 13, 63–67.
Sousa, R.M. (2010) Fig cultivars, driving techniques and potential for Brazil. In: Ferraz, A.C.O., Souza, F.C. and Honório, S.L. 2° *Brazilian Symposium on the Fig Culture. Campinas, Feagri, Brazil*, pp. 35–40.
Souza, O.P., Melo, B. and Mancin, C.A. (2020) *The Culture of the Fig Tree*. www.fruticultura.iciag.ufu.br (accessed 8 May 2020).
Specialty Produce (2020) Greek Figs www.specialtyproduce.com/produce/Greek_Figs_18086.php (accessed 10 May 2020).
Stover, E., Aradhya, M., Ferguson, L. and Crisosto, C.H. (2007) The fig: overview of an ancient fruit. *HortScience* 42, 1083–1087.
Tayou, A. (1985) Etude technique et économique de la culture du figuier dans région de Chefchaouen. *MSC thesis*. Institut Agronomique et Vétérinaire Hassan II, Rabat, Morocco.
Trad, M., Le Bourvellec, C., Gaaliche, B., Ginies, C., Renard, C.M.G.C. and Mars, M. (2013) Caprification modifies polyphenols but not cell wall concentrations in ripe figs. *Scientia Horticulturae (Amsterdam)* 160, 115–122. DOI:10.1016/j.scienta.2013.05.037.
UFRGS (2020) *Cultivars*. Universidade Federal Do Rio Grande Do Sul www.ufrgs.br/afeira/materias-primas/frutas/figo/variedades (accessed 8 May 2020).
Walali, L.D. (1995) Quelques espèces fruitières d'intérêt secondaire cultivées au Maroc. In: Llácer, G., Aksoy, U. and Mars M. (eds) Underutilized fruit crops in the Mediterranean region CIHEAM, Zaragoza, Spain. *Cahiers Options Méditerranéennes* 13, 47–51.
Wang, Z., Cui, Y., Vainstein, A., Chen, S. and Mal, H. (2017) Regulation of fig (*Ficus carica* L.) fruit colour: Metabolomic and transcriptomic analyses of the flavonoid biosynthetic pathway. *Frontiers in Plant Science* 8, 1–15. DOI:10.3389/fpls.2017.01990.
Weber, A.A. (2020) *Cultura da Figueira*. Universidade Federal do Pampa Campus Itaqui. docplayer.com.br/50476865-Universidade-federal-do-pampa-campus-itaqui-cultura-da-figueira-prof-dr-anderson-weber.html (accessed 8 May 2020).
Zare, H. and Jafari, M. (2018) *Morphological Characteristics of Some Iranian Fig Cultivars*. Marjaeelm, Shiraz, Iran.

# 6 Propagation Techniques and Nursery Management

**H. Yakushiji[1]*, E. Ertan[2], B. Ertan[3] and G. Günver Dalkılıç[2]**
[1]*Division of Fruit Tree Production Research, Institute of Fruit Tree and Tea Science, NARO, Hiroshima, Japan;* [2]*Department of Horticulture, Faculty of Agriculture, University of Aydin Adnan Menderes, Aydın, Turkey;*
[3]*Fig Research Institute, Aydın, Turkey.*

## 6.1 Introduction

Fig (*Ficus carica*) can be propagated in several ways, such as cutting, layering and grafting (Condit, 1947; Morton, 1987). The simplest method of propagating a tree vegetatively is by cuttings. A cutting (usually a piece of stem of the parent plant) is cut off and stuck into soil. Fig is most commonly propagated by cuttings to ensure that new trees have the same genotype as the parent tree. Many fig cultivars have a high ability to root from both green shoots and dormant branches without the need for hormone rooting promoters. However, rooting agents are effective for improving nursery efficiency and for cultivars with low rooting ability. Plant tissue culture can also be used for the mass production of clones under artificially controlled conditions (Pontikis and Melas, 1986), and for the genetic improvement of fig cultivars through selection (Flaishman *et al.*, 2017). Since there is no graft incompatibility between cultivars of *F. carica*, various grafting techniques can also be used to top-work existing trees (Morton, 1987). Although no dedicated rootstock cultivars are in common use, some rootstocks can enhance scion growth, resistance to soil diseases and pests such as nematodes, and drought tolerance. In nursery production, nursery management is greatly influenced by natural conditions. This makes it necessary to optimize many factors simultaneously: seedling propagation, nursery site, soil preparation, cutting and grafting, transplanting, and fertilization and irrigation for efficient and stable nursery propagation. This chapter discusses recent research on fig propagation techniques and nursery management.

## 6.2 Propagation Techniques

### Seedlings

Pollinated fig syconia contain many small fruits and seeds. According to Storey (1975), viable fig seeds can be separated from unviable seeds by floating them in water, a viable seed will sink and the seed selected for germination. The seed which floats is discarded. Viable seeds can be germinated by spreading on a relatively fine grade of vermiculite and covered lightly with the

*Corresponding author: aoe@affrc.go.jp


DOI: 10.1079/9781789242881.0006

same material (Storey, 1975). Seed germination can be promoted by the application of gibberellic acid ($GA_3$): $GA_3$ at 500 or 1000 mg $L^{-1}$ achieved the highest rates of germination and emergence of 'Bursa Siyahi' and 'Sarilop' (Çalişkan *et al.*, 2012).

Seedlings are not guaranteed to have the same traits as either parent. Thus, fig seedlings are important only for propagating seedling rootstocks (Rattanpal and Sidhu, 2017) and for breeding new cultivars (Flaishman *et al.*, 2008) or interspecific hybrids (Storey, 1975; Yakushiji *et al.*, 2012: Flaishman *et al.*, 2017, Yakushiji *et al.*, 2019). In India, 'Brown Turkey' seedlings are used as rootstock as they promote scion vigor, even under adverse conditions (Rattanpal and Sidhu, 2017). Figs cannot be grown from seed in Japan owing to the absence of the fig pollinator wasp *Blastophaga psenes*, therefore fruit set is parthenocarpic, and no viable seed is produced.

### Cutting

Propagation from cuttings, cloning, is the most common technique for large-scale nursery plant production. This method depends greatly on the rooting ability of the genotype. Generally, *F. carica* is easy to root from both dormant shoots (Tekintas and Seferoglu, 1998; Karadeniz, 2003; Zerhoune, 2003; Dolgun and Tekintas, 2009; Aljane and Nahdi, 2014; Nahdi and Aljane, 2014; Kuntagol *et al.*, 2017) and softwood from current shoots (Sousa *et al.*, 2013; Bisi *et al.*, 2016). Therefore, cutting is the most commonly used clonal method. This technique has the advantage of rapid, large-scale production of superior genotypes.

#### Hardwood cutting

Dormant one- to three-year-old cuttings, or the basal portion of vigorous first-year shoots with the heel of a two-year-old branch at the base are taken during dormant winter pruning (Morton, 1987). Hardwood cuttings need to be prepared in early spring before bud break. The rooted cuttings should be ready for transplanting the following winter.

It might take several months from collecting hardwood cuttings of mother plants then storage in winter to rooted plants. The storage conditions of cuttings affect rooting and growth. Takagaki *et al.* (1997) found a strong correlation between the water content of cuttings and the dry weights of the roots and shoots, regardless of the storage period. They proposed storing hardwood cuttings in sand with high water content (30%) for 10–20 days at 5–10°C. However, Antunes *et al.* (2003) reported the stratification duration was negatively correlated with rooting due to secondary necrosis and pathogens. Hosomi *et al.* (2015) found that growth of hardwood cuttings was influenced by the stem portion used but not by the time of collection. Conversly, Nava *et al.* (2014) reported that rooting of woody cuttings of 'Roxo de Valinhos' did not depend on position, depth or coverage in medium grain sand. In tropical areas figs thrive at high elevations of 800–1800 m (Morton, 1987) such as the tropical highland areas of northern Thailand with its year-round cool climate. To improve fig propagation from hardwood cuttings at elevations of 1300 and 720 m in northern Thailand, Pipattanawong *et al.* (2008) used plastic film covers to increase the temperature, promote early callus formation and the emergence of buds and roots, allowing fig propagation in cool areas.

The anatomical and histological studies of hardwood cuttings (Seferoğlu and Tekintaş, 1998) demonstrate the cells in the proximal sides of cuttings differentiate into root primordia from near the cambium through the cortex. The root primordia cells then divide, broadening the sclerenchymatous cells approximately 50 days after planting even though the cutting developed callus xylem 40 days after planting. However, no tissue formation was observed to suppress the root formation of the cuttings because one-line sclerenchymatous cells were suppressed by ray parenchyma cells. Thus, Seferoğlu and Tekintaş (1998) indicated that sclerenchymatous cells did not affect root formation in figs.

#### Bud or tip cuttings

For bud or tip cuttings, 5–8 cm long sections with at least one bud, are collected from one-year-old shoots or two-year-old branches

**Fig. 6.1.** Bud or tip cutting. (Figure courtesy of H. Yakushiji.)

(Dolgun and Tekintas, 2009) (Fig. 6.1). Under optimum conditions, this method propagates plants with high efficiency. Dolgun and Tekintas (2009) showed that bud cuttings taken from the tip, middle and bottom of one- or more than 2-year-old shoots of 'Calimyrna' ('Sarilop') could both develop adventitious roots and elongate shoots in a mist house. Tip cuttings of one-year-old shoots had the highest rooting ratio. Bottom heating and misting are used when propagating of bud cuttings (Yildiz, 1999; Dolgun *et al.*, 2003; Dolgun and Tekintas, 2009).

### Apical cuttings in lopping management

To prolong the production period of fig production and yield, fig trees can be managed by the 'lopping' system, in which trees are pruned to grow 6 or 12 vigorously growing branches, which are lopped when they reach the 16th leaf, producing two vigorous shoots from the branch tip. These two branches are lopped when they reach the sixth leaf (Peche *et al.*, 2017). This sixth leaf lopping to produce two new branches is repeated another two or three times, for a total of four loppings until the end of the production season. Peche *et al.* (2017) evaluated the woody apical cuttings of 'Roxo de Valinhos' from trees subjected to different lopping treatments in Brazil. Apical cuttings collected from the lopping in midsummer increased the percentage of rooted and sprouting cuttings, and the application of indole 3-butyric acid (IBA) at 2000 mg $L^{-1}$ for 10 s increased the proliferation of roots and shoots.

### Herbaceous (softwood) cutting

Herbaceous (softwood) stem cuttings are best taken during the growing season, spring to summer (Rajeswara-Reddy *et al.*, 2008; Bisi *et al.*, 2016). They must be kept cool and moist by mist until placed in the propagating medium (Bisi *et al.*, 2016). Unlike hardwood cuttings, the top two leaves of cuttings are retained (Bisi *et al.*, 2016), or all leaves are removed (Rajeswara-Reddy *et al.*, 2008) (Fig. 6.2.a). If rooting powder (IBA) is used, the cutting is trimmed just below a node and dipped into the powder (Rajeswara-Reddy *et al.*, 2008). The cuttings are placed in the propagation medium with intermittent misting and bottom heat at 20–25°C. Adventitious roots normally appear in 3–4 weeks and elongate thereafter (Fig. 6.2.b). The rooted cuttings then grow in the greenhouse until being moved to nursery rows the following spring. The next season, they are planted in the orchard.

### Semi-hardwood cutting

A variation of the softwood cutting is the semi-hardwood cutting. Semi-hardwood cuttings are taken from the current season's growth. All leaves except the top leaf are removed, and the cuttings are trimmed to the required length by removing the terminal portion just above a bud (Bisi *et al.*, 2016). Cuttings 15–23 cm long are taken in summer to early autumn. The top leaf is cut in half along to the midrib. If necessary, they are treated with rooting powder (Sousa *et al.*, 2013) and inserted into a cutting bed with intermittent misting and optional bottom heat. Rooted plants are hardened off before bedding in nursery rows the following season.

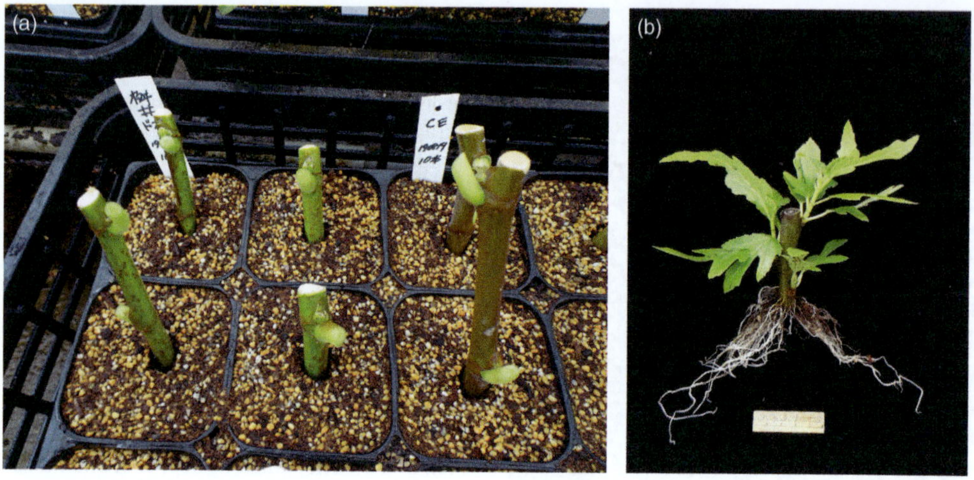


**Fig. 6.2.** Softwood cutting; (a) bedding just after cuttings, (b) formation of adventitious roots and new bud sprouts from scion 40 days after cutting. Bar is 5cm. (Figure courtesy of H. Yakushiji.)

## Aerial root-cutting

All species of the West African *Ficus* subgenus *Urostigma* can be propagated by aerial root cuttings (Danthu *et al.*, 2002). All have aerial roots characterized by a hemi-epiphytic development. Root cuttings must be planted upright. The highest rooting rate was observed at the beginning of the dry season (November) in Senegal, after the rainy season, and the lowest at the end of the cool season, when growth started. Root cuttings root better when collected while the parent plant is dormant, as the aerial roots are well supplied with stored carbohydrates at the end of the rainy season (March–May) (Danthu *et al.*, 2002). This method might support the conservation of some genetic resources.

## Rooting induction with artificial auxins

Although fig trees root well from cuttings, the ability depends on the cultivar. The rooting efficiency of cultivars with low rooting ability can be improved by rooting agents such as auxins. Artificial auxins, IBA or 1-naphthyl acetic acid (NAA), have been tested on fig cuttings. Two IBA treatment methods were tried: a high concentration for several seconds and a low concentration for 24 h. In the quick-dip method, the best results with Caprifig cultivars were obtained with 1000 mg $L^{-1}$ IBA for 20 s (Zarei and Aboutalebi, 2013) or at 2500 mg $L^{-1}$ for 5 s (Rajeswara-Reddy *et al.*, 2008). A mixture of IBA (2500 mg $L^{-1}$) + NAA (2500 mg $L^{-1}$) was more effective than either alone with both hardwood and semi-hardwood cuttings (Rajeswara-Reddy *et al.*, 2008). In the slow-dip method, the best results were obtained with IBA at 100 mg $L^{-1}$ for 24 h with hardwood cuttings (Antunes *et al.*, 2003; Chalfun *et al.*, 2003). Chalfun *et al.* (2003) reported that IBA treatment increased the dry matter weight of both roots and the aerial parts. Using air layering, Bisi *et al.* (2016) found that a pine bark substrate and IBA at 1000 mg $L^{-1}$ gave the best rooting results. These results all demonstrate auxins are an effective method for increasing the efficiency and stability of nursery propagation of cultivars with low rooting ability.

## Layering

Layering is the practice of inducing roots on a piece of stem still attached to its parent plant. The new plant is severed after it has grown roots. Although the use of nursery stock is preferable, propagation of figs by air or ground layering can be successful (Morton, 1987). Dolgun and Tekintas (2008) suggested stem layering as a cheaper method of producing fig nursery plants. Layering can be successful with fig cultivars that are difficult to propagate by other clonal methods.

## Air layering

Air layering is a method of propagating new plants from stems still attached to the parent plant. After the bark, cambium and phloem layer are removed from around the circumference of the shoot, the site is wrapped with damp moss to encourage root initiation. Due to the possibility of stem breakage, Bisi *et al.* (2016) recommended air layering in the middle of the branch without wounding: they successfully air-layered figs by brushing a solution of IBA at 1000 mg $L^{-1}$ on a 2 cm portion of the branch and wrapped the stem with pine bark and moistened vermiculite (2:1) secured within transparent plastic tied at both ends to prevent moisture loss. Fifteen cultivars had high callus formation (>50%), some cultivars did not root despite IBA application (Bisi *et al.*, 2016). These collective results suggest air layering can be used to propagate fig, but it is time-consuming, labor-intensive and best limited to small-scale use.

## Stem layering

Dolgun and Tekintas (2008) used stem layering to produce 'Sarilop' nursery plants in Turkey. Dormant mother plants were oriented horizontally at a spacing of 70 × 60 cm after the fall of the leaves in the nursery. In February, all the buds on the ground side of the stems were removed. The stems were bent over and fixed to the ground surface with ∩-shaped galvanized wire hoops. When the new shoots, which developed from the mother plants, reached 20 cm, the bases were covered by 10 cm of soil, and more soil added as the shoots grew and rooted. The bottom parts of the new plants formed from shoots was carefully taken away, and rooted plants were cut off from mother plants. Most of the plants (76.5%) were suitable for sale. The authors suggested that stem layering is a low-cost alternative for producing fig nursery plants. However, it poses a risk of transferring nematodes (*Meloidogyne* spp.) and soil diseases to new orchards. For this reason, this method needs to be carried out in a field free from nematodes and soil diseases, or in a field in which they have been controlled.

## Grafting and budding

The fig can be budded onto existing rootstock with nematode or soil disease resistance to change fruit varieties and achieve rapid propagation (Storey, 1975). Alternatively, an existing cultivar may be grafted and replaced with other cultivars. For breeding, seedlings can be encouraged to begin bearing within three years through grafting, as opposed to the 5–8 years required for a tree grown from seed (Storey, 1975). The simple bark graft is the easiest and fastest method. Fig can be top-worked by T-budding (Krezdorn and Glasgow, 1970; Rattanpal *et al.*, 2014; Rattanpal and Sidhu, 2017) into vigorous one-year-old shoots on heavily pruned trees, or patch-budding older shoots (Hartmann *et al.*, 2011). Cleft- or bark-grafting can also be used to top work existing varieties (Condit, 1947; Morton, 1987) (Fig. 6.3). Grafting is usually performed during the dormant season on trees in the orchard, while softwood grafting is done in the nursery. Cutting-grafting that carries out both grafting and cutting at the same time has been successful using dormant branches for rootstocks and scions (Hosomi, 2006; Ushijima *et al.*, 2013).

**Fig. 6.3.** Top working (hardwood grafting). (Figure courtesy of H. Yakushiji.)

## Budding

Fig cuttings are easily rooted, but sudden changes in air temperature and soil moisture content can affect rooting and shoot development. Field propagation has the risk of soil-based bacterial and fungal diseases that decrease propagation success and growth. These pathogens have been cited as among the reasons the efficiency of field propagation in Turkey is about 50–60% (Seferoğlu and Tekintaş, 1998; Cobanoglu *et al.*, 2004). Rattanpal *et al.* (2014) used inverted T-budding (Fig. 6.4) on seedlings of 'Brown Turkey' in the last week of August, at the end of rainy season, in Punjab, India, at 30–40 cm above the soil level. They reported no significant difference between eight cultivars for number of days from budding onto 'Brown Turkey' seedlings to sprouting, and no incompatibility was recorded with eight cultivars one year after budding. The budding success among the eight fig cultivars grafted onto 'Brown Turkey' seedlings ranged from 68.4% to 94.8% (Rattanpal and Sindhu, 2017). Thus, Rattanpal and Sidhu (2017) concluded that budding at the end of rainy season can be used to propagate superior varieties of fig under Punjab conditions in India. Similarly, Krezdorn and Glasgow (1970) reported successfully budding and grafting several *F. carica* cultivars onto *F. glomerata*, *F. cocculifolia*, *F. gnaphalocarpa* and *F. palmata* and did not observe any incompatibility with any of the rootstock and scion species combinations.

## Softwood grafting

Softwood grafting is a simple and rapid method of vegetatively propagating figs (Fig. 6.5). The

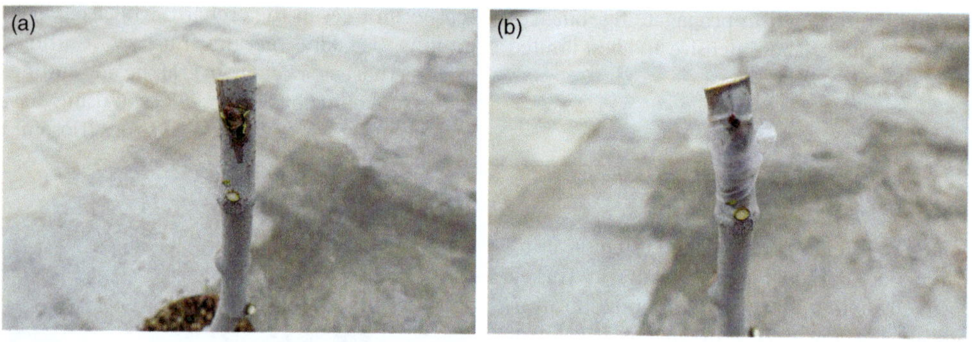


**Fig. 6.4.** Inverted T-budding: (a) T-budding before covering, (b) budding after covering with thin vinyl tape. (Figure courtesy of H. Yakushiji.)

**Fig. 6.5.** Softwood grafting: (a) before covering with thin vinyl tape, and (b) the grafted union of 'Masui Dauphine' grafted on rootstock obtained from $BC_1$ interspecific hybrid between *F. carica* and *F. erecta*. (Figure courtesy of H. Yakushiji.)

method is easy and requires no special skills. Scions are grafted onto 6- to 10-month-old rootstocks of a similar girth. A scion cut into a wedge shape is inserted into a vertical slit in the rootstock, and the joint is firmly wrapped with waterproof tape. Hosomi *et al.* (2015) tested combinations of rootstock and scion for raising nursery stock within the current year by softwood grafting. The side-grafting survival rate was higher with the immature, as opposed to the semi-mature portion of a 'Masui Dauphine' scion shoot. Side grafting is also easy, convenient, quick, inexpensive, and has a good success rate (33–80%).

## Cutting-grafting

Cutting-grafting involves grafting a scion onto a rootstock cutting to shorten the time by performing both tasks simultaneously (Hosomi, 2006; Ushijima *et al.*, 2013) (Fig. 6.6). Ushijima *et al.* (2013) grafted a single dormant node of

**Fig. 6.6.** Cutting-grafting. (a) Nursing of fig plants from graft cutting: grafted seedling before cutting, (b) bedding, and (c) plant derived from graft cuttings after 3 months. (Figure courtesy of H. Yakushiji.)

'Toyomituhime' or 'Horaishi' onto a 20–25 cm cutting of dormant wood of the Ceratocystis canker-resistant rootstock 'Kibaru'. The survival rate of both scions was 95%. The shoot length in the current year's growth reached 120–150 cm and the shoot had a basal diameter of 16 mm and produced nursery stock within 1 year.

### Tissue culture

Tissue culture of fig (*F. carica* L.) is an efficient method of producing true-to-type plantlets for mass multiplication (Ling *et al.*, 2018). Propagation by the conventional method of cuttings and grafting is limited and slow. Micropropagation techniques, using shoot tips and nodal explants, have been used for rapid and large-scale propagation of multiple fig cultivars (Kumar *et al.*, 1998). *In vitro* studies with multiple fig cultivars reported successful shoot regeneration using apical buds (Gella *et al.*, 1998; Kumar *et al.*, 1998; Toma and Tamer, 2016; Al-Shomali *et al.*, 2017), apical shoot buds (Nobre *et al.*, 1998; Hepaksoy and Aksoy, 2006; Mustafa and Taha, 2012; Mustafa *et al.*, 2013; Qrunfleh *et al.*, 2013; Taha *et al.*, 2013; Danial *et al.*, 2014; Darwesh *et al.*, 2014; Bayoudh *et al.*, 2015; Elazab and Shaaban, 2015; Dessoky *et al.*, 2016; Mustafa *et al.*, 2016; Hassan and Zayed, 2018; Shatnawi *et al.*, 2019), lateral shoot buds, axillary shoot tips and nodal explants (Fráguas *et al.*, 2004a, b; Dhage *et al.*, 2015; Shahcheraghi and Shekafandeh, 2016; Shekafandeh and Shahcheraghi, 2017; Ling *et al.*, 2018; Sen and Patel, 2018) and leaves (Yakushiji *et al.* 2003; Kim *et al.*, 2007; Soliman *et al.*, 2010; Dhage *et al.*, 2012). *In vitro* proliferation and rooting of *F. carica* are affected by many factors. These factors include culture medium strength, concentration and composition, particularly of growth regulators; medium composition has particularly strong effects on growth (Mustafa *et al.*, 2016). The Murashige and Skoog (MS) culture medium (Murashige and Skoog, 1962) is widely used for fig tissue culture. Its high salt concentrations produce significant tissue and cell growth. The Woody Plant Medium (WPM) (Lloyd and McCown, 1980) is the second most widely used medium for *in vitro* culturing of bushes and trees. The WPM has been shown to be an effective micropropagation medium for the common fig (Pasqual and Ferreira, 2007). Mustafa *et al.* (2013) reported that fig leaf number and shoot length was induced using a double strength Murashige and Skoog medium. However, the best shoot proliferation was obtained on half strength MS medium. A study by Shatnawi *et al.* (2019) on *F. carica* showed that benzyl amino purine (BAP) at 0.4 mg $L^{-1}$ in combination with 0.2 mg $L^{-1}$ IBA produced the maximum *in vitro* propagation rate, with 4.2 shoots per explant. The best root proliferation was obtained with 1.5 mg $L^{-1}$ IBA. Bayoudh *et al.* (2017) optimized a micropropagation and acclimatization protocol; they reported more rapid and higher rates of *ex vitro* healthy root production (90%) using a peat substrate. Yakushiji *et al.* (2003) developed a method for inducing organogenesis from leaf explants of *F. carica* (Fig. 6.7), but their frequency of adventitious bud differentiation was low. Leaf fragments of fig tree (*Ficus carica* L. 'Masui Dauphine') obtained from *in vitro* shoot cultures were cultured on MS medium supplemented with different combinations of 2,4-D, TDZ and 0.5 mM phloroglucinol. The addition of 2,4-D induced root formation and the presence of phloroglucinol increased root formation significantly. When 2,4-D and TDZ were added to MS medium containing phloroglucinol, the explants produced adventitious buds on the vein of leaf explants. The addition of phloroglucinol was also effective in inducing adventitious bud formation. Regenerated plantlets were successfully established in soil after a short period of acclimatization. Fig. 6.8 shows some tissue culture studies.

**Fig. 6.7.** Adventitious buds from fig leaf explants. (Figure courtesy of H. Yakushiji.)

**Fig. 6.8.** Some tissue culture studies in fig: (a) leaf growth in fig shoot tips, (b) callus formation in fig leaf explants, (c) shoot development on the fig shoot tips after application of thermotherapy for 3 weeks, (d) shoot development on fig shoot tips without thermotherapy treatment, (e) shoot development on fig shoot tips in medium containing active carbon, (f) adventitious root formation on shoot under low light conditions. (Figure courtesy of G. Günver Dalkılıç.)

### Meristem culture

Fig mosaic disease is ubiquitous in Mediterranean countries with temperate and subtropical climates (Sümer, 2014). The disease causes economic losses by defoliating plants and causing fruit drop. *In vitro* meristem culture and thermotherapy are the primary methods of producing fig mosaic virus-free plants by using single-shoot-tip elongation (Murithi *et al.*, 1982; Pontikis and Melas, 1986; Günver and Ertan, 1998). Thermotherapy can be applied in three different ways:

(1) thermotherapy is first applied to a plant then meristem culture is established; (2) thermotherapy is first applied to *in vitro* shoots, then meristem/shoot tip culture is established; (3) thermotherapy is applied to a cultured meristem/shoot tip (George, 1993). Gella *et al.* (1998) established a fig shoot culture infected with fig mosaic virus, then applied *in vitro* thermotherapy demonstrating that *in vitro* cultivated fig shoots are tolerant to prolonged heat treatment. The advantage of an *in vitro* thermotherapy application is it can produce a larger explant (>1 mm) than meristem culture. Günver Dalkılıç *et al.* (2017) reported that after thermotherapy application, shoot survival ratio and new shoot emergence ratios were lower compared to the control. The collection season of shoot tips does not affect the viability and proliferation but adding activated charcoal to the culture medium improved shoot development of virus-free fig plantlets (Demiralay *et al.*, 1998). Meristem culture, together with thermotherapy treatment, is recommended to obtain virus-free fig plant materials. Although cultured plants appear healthy, dsRNA virus tests are recommended for a definitive evaluation of the plant virus status (Comlekcioglu *et al.*, 2007). A meristem size of approximately 0.5 mm provides the best results, 70–90%, for eliminating the virus from caprifig and female fig varieties (Bayoudh *et al.*, 2017).

### Somatic embryogenesis

In somatic embryogenesis, organized embryos arise directly on the explant or indirectly from callus (Singh *et al.*, 2016). Soliman *et al.* (2010) used a regeneration and transformation system to propagate the 'Sultani' cultivar via somatic embryogenesis.

### Encapsulation of explants

A synthetic or artificial seed has been defined as a somatic embryo encapsulated inside a coating and is considered to be analogous to a zygotic seed. There are several different types of synthetic seed: somatic embryos encapsulated in a water gel; dried and coated somatic embryos; dried and uncoated somatic embryos; somatic embryos suspended in a fluid carrier; and shoot buds encapsulated in a water gel (Singh *et al.*, 2016). Nodal segments were more responsive than shoot tips for shoot multiplication. Recent studies suggest the encapsulation of nodal segments significantly improved the exchange and conservation of fig germplasm as well as being more easily adapted for large scale commercial cultivation (Sharma *et al.*, 2014).

## 6.3 Producing Rootstocks From Seeds

Because fig can easily be propagated and repeatedly repropagated through cuttings to maintain desirable cultivars and because there is also considerable opportunity for phenotypic variability, produced by natural mutations and environmental factors, within a cultivar (Flaishman *et al.*, 2008) figs are propagated primarily by cuttings. However, genetically different plants can be grown from seed produced by the female tree relatively easily. Furthermore, genetic variability in fig is enhanced by the obligatory outcrossing within a species, resulting in the production of new seedling individuals with potentially favorable, or unfavorable characteristics. Thus far the genes determining if a seed from a single syconium will produce a male caprifig or a fruit producing female fig have not been determined (Beck and Lord, 1988). In fig production, seeds are important primarily for propagation of seedling rootstocks, and for obtaining hybrid plants in breeding programmes (Hartmann *et al.*, 2002).

### Seed extraction

Figs contain many small fruits and seeds. To extract seeds from fresh figs, first grind the fruit to a pulp in the blender. After filtration, the seeds are washed with water and dried in a shade. Fig. 6.9 shows the seed extraction steps. According to the size of ripe fig fruit, 600–1200 seeds can be obtained. The average weight of a thousand dried fig seeds is 1.2–1.3 g. Viable seeds can be separated from inviable seeds by floating them in water. Viable seeds will sink, seeds that do not sink to the bottom are inviable and should be discarded.

**Fig. 6.9.** (a) Preparation stage for seed extraction, (b) fragmentation of fruit pulp in a blender, (c) addition of water into the flask, (d) empty the seeds from the flask to tray, (e) seeds before filtration, and (f) extraction of fig seeds. (Figure courtesy of B. Ertan.)

## Seed dormancy

Mature dry fig seeds will not germinate and grow until chilled above freezing under moist conditions (stratification). Such chilling releases the embryo from endodormancy. The optimum temperature and duration of chilling vary widely among species (Westwood, 1993). Çalışkan *et al.* (2012) reported fig seeds to have a physiological dormancy that requires stratification to satisfy the dormancy requirement. Also, fig seed germination of old seed was higher (60–70%) than fresh seeds. Seed dormancy can also be caused by poor embryo development (Karam and Al-Salem, 2001). Seed physiological dormancy in seeds is related to the endogenous growth regulator levels, the inhibitor abscisic acid (ABA) and promoters, the gibberellins: (GA) (Hartmann *et al.*, 2002). Powell (1987) reported stratification stimulated GA synthesis. Çalışkan *et al.* (2012), reported stratification was not sufficient to break dormancy in 'Sarılop' seeds. Treatments of 500 and 1000 mg $L^{-1}$ $GA_3$ has been successful in breaking dormancy (Furutani and Nagao, 1987). Also, 21 days for stratification and 3% $KNO_3$ treatments for 'Bursa Siyahı' gave higher germination (>80%) than 'Sarılop'; possibly due to cultivar variation.

## Seed germination

Seed germination success showed no significant differences between water priming for 24 h and the control treatment. Seed germination is influenced by internal factors that influence dormancy including the seed coat, an undeveloped embryo or endogenous inhibitors (Agrawal and Dadlani, 1995). There are different methods of overcoming dormancy. These may differ among species and include soaking in water (Çetinbaş and Koyuncu, 2006), stratification, scarification (Al Absi, 2010), $KNO_3$ (Furutani and Nagao, 1987) and gibberellin applications (Çetinbaş and Koyuncu, 2006). Fig seeds are generated primarily in breeding programmes. Another recent use was the production of virus-free seedlings for fig mosaic research (Çağlayan *et al.*, 2009). Detailed information on fig seed germination and seedling emergence is limited; Ellis *et al.* (1985) reported full germination is difficult to achieve in *Ficus carica*.

## Seedling propagation

The production of fig seedlings using seeds is exclusively for the development of research on genetic improvement of the species. In Brazil, seedlings produced by seeds are not used due to the absence of *Blastophaga psenes* (L) (Ferreira *et al.*, 2009), the natural pollinator the fig tree (Anjam *et al.*, 2017), which is not found in the country. Reinforcing the use of seeds for hybridization studies, Yakushiji *et al.* (2012) performed artificial pollination in *F. carica* plants using *F. erecta* pollen and observed that plants generated had high resistance to Ceratocystis canker, contributing to studies to obtain resistant rootstocks to this soilborne disease (Yakushiji *et al.*, 2019). Fig seeds should be mixed with 50% volume of wood ash or fine garden sand before sowing. Fertile seeds can be germinated by spreading them on a tray of moist soil mix (Hartmann, 2011), or 1:1 peat+perlite medium. The mixture should be evenly distributed over the surface of the growing medium in a 4–6 inch deep tray with bottom drainage. Fig seeds need a warm and humid environment to germinate. If fig seeds are kept at a temperature near 25°C, and 60–70% RH, the seeds should germinate within two weeks to three months. This can be accomplished by placing a plastic cover or bag over the top of the tray and placing the tray in a warm area with indirect sunlight. As long as condensation appears on the inside surface of the bag or lid, the seeds do not need to be watered. After small sprouts with leaves appear on the surface of the growing medium, the plants can be acclimated to the humidity of the surrounding environment by removing the lid or bag each day for a few more hours each time. Fig seedlings adapt to lower humidity as long as they receive regular watering.

## Transplanting

When the seedlings have reached 5–10 cm in height, they can be gently removed from the seedling tray and planted into 3 L volume regular potting soil and be watered well for the first two weeks after transplanting: a thorough watering once every two days is best. After this period, they should be watered only when the surface of the soil feels dry to the touch. Fig seedlings also perform better when kept in indirect sunlight for the first month.

### Grafting seedlings

The seeds of many fruit produce trees that are different from the parent because seeds themselves are produced by sexual reproduction, and they receive genes from a male and female to form. As they are a cross from two sets of genes, fig trees are not 'true to seed', that is, their seeds will produce a different variety of tree from the parent. Propagation methods that use material from the parent trees such as grafting are a form of vegetative, or asexual reproduction, as genes only come from one parent to produce identical genetic clones. According to Cato the Censor, budding and grafting of fig trees was done in the 3rd century BCE in much the same way it is done today (Condit, 1947). All budding requires the rootstock to be actively growing so that the bark slips easily. Three bud types are used when top-working fig rootstocks: T-budding for younger smaller shoots and shield or patch buds for older larger shoots (Hartmann *et al.*, 2011). The scion, with mature well-formed buds with a shield matching with the rootstocks that are actively growing can be cut anytime. After insertion, the bud can be wrapped with cloth, tape or rubber. After budding, the stock must be girdled or cut back in order to force out the bud. Generally, bud wrapping stays on no longer than three weeks. When the budding is done determines when the bud should be forced. Buds placed in the stock during late summer will remain dormant until spring. Figs can also be top worked in the dormant season with a cleft or bark graft. Also, shield- or patch-budding, or cleft- or bark-grafting techniques can be used to topwork an existing orchard to produce the same result; a fruit-bearing tree identical to the source of the bud (Morton, 1987).

## 6.4 Nursery Management

A plant nursery is a place where any kind of plant is grown for the sake of being moved or transplanted later. It can occupy a field, garden, greenhouse or other forms of growing space. The nursery site should be located in nutrient-rich/medium soil, near to a water source, free from soil pathogens and insects, with availability of cheap and skilled labor, and good access to the main road for easy transportation. The site should be on a gently sloping area and away from other tall crops; this is important for good drainage as well as to encourage air circulation. An appropriate site must be selected for the most effective, efficient and economical design of a nursery. The purpose and target of plants to be produced will decide the site selection and its improvement. Careful observation of site conditions and an assessment of past and present climatic records are important. As mentioned above, fig trees (*F. carica*) can be propagated successfully through cutting, grafting, layering and tissue culture (Valio, 1986; Kabasakal, 1990; Brum, 2001; Chalfun *et al.*, 2003; Fráguas *et al.*, 2004a, b). The primary fig tree propagation method is cuttings. Mother plant cuttings in the appropriate conditions will produce roots and shoots, becoming a new plant identical to the mother plant. Cuttings from mature, well-grown, and one-year-old shoots with little pith are best for this for propagation method. The cuttings are placed in sand for a stratification period, then planted directly into a nursery bed or pot. This is the traditional method of growing nursery fig trees (Kabasakal, 1990), and ensures that new trees are female and 'true to type', identical to the tree from which cuttings were collected.

### Production of nursery trees by cuttings

Commercial nursery fig tree production is essentially done in the same way as field production; cuttings from mother trees are rooted in a growing medium, again forming a new plant identical to the mother plant (Valio, 1986). Hardwood cuttings from one-year-old shoots are stratified in sand for 45–50 days at 4°C before planting in containers or nursery beds. With this traditional method of producing nursery trees, rooting is not a problem. However, soil-borne nematodes can be major problem in nursery production (Kılınç *et al.*, 2007). Little information is available on damage to figs resulting from a nematode infestation. The root-knot (*Meloidogyne incognita, Meloidogyne javanica*), dagger (*Xiphinema index*), and lesion (*Pratylenchus vulnus*) nematodes are known to cause a reduction in growth and yield. The trees are likely to be more susceptible to temperature and water stress. The symptoms described below are indicative of a nematode problem but are not diagnostic as they could result from other causes as well. Lesion nematode infested trees may appear stunted with very few feeder roots. The roots may have reddish-brown lesions that eventually turn dark. Root-knot nematode infested trees are also likely to have reduced growth and appear stunted. The roots have distinctive swellings, called galls. Dagger nematode causes gall formation on root tips. These pests are controlled with either steam or chemical soil sterilization. Steam sterilization is not always economically viable. Previously, methyl bromide was the most common chemical soil fumigant. However, under the Montreal Protocol, it is banned in many countries.

## Soilless cultivation

Soilless cultivation (Burrage, 1999; Van Os, 2000) is an option for producing disease-free high-quality nursery fig trees. Soilless culture techniques have been used for several years in Mediterranean floriculture and vegetable production (Jensen and Collins, 1985; Resh, 1987; Maloupa *et al.*, 1992; Molitor, 1998; Jensen, 1999). Generally, this culture technique protects plants from most soil-borne pathogens, if proper culture practice is followed, and root disease control in soilless culture focuses only on disinfection methods (Runia, 1995). Although disinfection has been effective, the cost is significantly higher than in the conventional method (Koohakan *et al.*, 2004). From the late 1970s, as an alternative to expensive chemical soil disinfection methods and possible plant residues, growers switched to soilless culture (Rijck and Schrevens, 1998). Multiple studies investigated alternative substrates for nursery production (Bugbee *et al.*, 1991; Tyler *et al.*, 1993; Evans *et al.*, 1996; Chong and Lumis, 2000). Most of the components of the alternative substrate were non-toxic to plants and were successfully used to amend conventional substrates. However, regional availability and a limited supply of uniform and consistent quality product reduced their widespread usage (Wright *et al.*, 2006). Producing trees from cuttings with traditional methods resulted in losses due to root damage during the transplant process. Substrate culture, defined as the cultivation of plants in substrates enriched with nutrient solutions, is a new technology for producing fig trees in the nursery. Trough culture is one soilless culture method. Concrete, asbestos concrete or brick concrete troughs, 30–120 cm wide and 15–20 cm deep can be constructed within the nursery building. Bed length is only limited by the greenhouse size and the number of plants to be cultivated. However, troughs can also be prepared much less expensively by digging the trough directly into the soil (Fig. 6.10). It is recommended to give a slope of 0.1% every 15 m (Sevgican, 2003). Drainage pipes are laid in the bottom of the troughs, then covered with a 5 cm layer of stone or pebbles. The pipes connect with a central drainage channel which must be opened at the lower end of the trough. This not only prevents saturation of the soil medium but facilitates collecting the nutrient solution for reuse.

Şirin *et al.* (2010) examined the effect of three substrates, perlite (100%), peat (50%) + perlite (50%), and fine sawdust (100%) versus a soil control for 8 months with *F. carica* cv. 'Sarilop' ('Calimyrna') cuttings in high-tunnel trough culture. The peat + perlite and perlite substrates produced the best growth and plant quality, in that order. Perlite is very porous, has intense capillary action, retaining 3–4 times its weight in water. Perlite provides roots with a well-aerated but adequately irrigated environment (Olympios, 1992; Kreen *et al.*, 2002). Combining perlite with peat further enhances the substrate by increasing the cation exchange capacity (CEC), an important advantage of peat (Verdonck, 1991). Peat effectively increases nutrient uptake because it acts as a reservoir, slowly releasing nutrients into the rhizosphere. Hardwood cuttings for trough culture are taken from one-year-old shoots of 25 cm length or longer, and stratified in the sand for 45–50 days at 4°C before planting. The stratified fig cuttings are planted at 25 × 25 cm (Fig. 6.11). A drip irrigation system is used to supply water and nutrients.

In the soilless substrate that has culture systems, nutrient management is a key factor in

**Fig. 6.10.** (a) General view of troughs covered with plastic sheeting, (b) trough filled with substrate. (Figure courtesy of E. Ertan.)

**Fig. 6.11.** The appearance of the cuttings planted in the troughs after burst (peat medium). (Figure courtesy of E. Ertan.)

producing quality plants (Jung *et al.*, 1994; Roh *et al.*, 1995). Each plant species has specific requirements. However, there are few reports in the literature of suitable nutrient solutions for nursery fig trees. Kılınç *et al.* (2007) tested five different nutrient solution formulations in high-tunnel and open-field conditions in substrate culture. They reported that Hewitt's and

Hoagland's nutrient solutions produced the best quality nursery fig trees in both high-tunnel and open-field conditions, respectively. Table 6.1 presents the Nutrient Solution Formulations (NSF) used (Kılınç *et al.*, 2007).

These studies (Kılınç *et al.*, 2007; Şirin *et al.*, 2010) demonstrate fig nursery trees can be successfully propagated with soilless substrate culture with nutrient solution formulations in both troughs and containers (Fig. 6.12).

**Table 6.1.** Nutrient Solution Formulations (NSFs) used in the research.

| | Nutrient solution formulations (NSFs) | | | | |
|---|---|---|---|---|---|
| | I (Hoagland and Arnon) | II (Hewitt) | III (Çolakoğlu, a) | IV (Çolakoğlu, b) | V (Steiner) |
| Nutrient elements | (mg $L^{-1}$) | (mg $L^{-1}$) | (mg $L^{-1}$) | (mg $L^{-1}$) | (mg $L^{-1}$) |
| N | 210 | 168 | 50 | 150 | 167 |
| P | 31 | 41 | 26 | 31 | 31 |
| K | 234 | 156 | 66 | 234 | 277 |
| Mg | 48 | 36 | 10 | 30 | 49 |
| Ca | 160 | 160 | 33 | 100 | 183 |
| S | 64 | 48 | 5 | 15 | 111 |
| Fe | 2.5 | 2.8 | 2.6 | 8 | 1.33 |
| Mn | 0.5 | 0.55 | 1.6 | 5 | 0.62 |
| B | 0.5 | 0.54 | 0.50 | 1.5 | 0.44 |
| Cu | 0.02 | 0.064 | 0.66 | 2 | 0.02 |
| Zn | 0.05 | 0.065 | 1.0 | 3 | 0.11 |
| Mo | 0.01 | 0.048 | 0.066 | 0.2 | 0.048 |

**Fig. 6.12.** Fig nursery trees grown in trough culture. (Figure courtesy of E. Ertan.)

## 6.5 Rootstock

As the fig is easy to propagate from cuttings, own-rooted nursery trees are widely used for planting, and rootstocks are not commonly used. Research on the effect of the rootstock-scion relationship on resistance to environmental stresses, soil disease or nematodes is limited. Storey (1975) evaluated both cultivated and wild fig species such as *F. cocculifolia*, *F. pumila* and *F. racemosa* for resistance to root-knot nematode (*M. incognita*). Hosomi *et al.* (2002) investigated the soil-borne disease known as fig orchard transplant problem, which is one of the soil problems that can cause poor growth and low yields, by planting the same plants in the previous fig fields. Condit and Horne (1933) demonstrated that fig leaf mosaic diseases (fig mosaic virus) reduce yield. Flock and Wallace (1955) demonstrated the only way to produce fig trees free of mosaic virus is to propagate from virus-free mother plants as the pathogen is transmitted by grafting and by an insect vector (the eriophyid mite, *Aceria ficus*).

### Graft compatibility of wild *Ficus* species

Some wild fig species are highly resistant to *Ceratocystis* canker (Hosomi *et al.*, 2012; Yakushiji *et al.*, 2012), root-knot nematode (Condit, 1947; Hosomi, 1993), and soil sickness (Hosomi, 2002) and could be used as rootstocks. Therefore evaluating graft compatibility between rootstocks of wild *Ficus* species and scions of *F. carica* would be of great value. Hosomi (2017) evaluated the graft compatibility of 'Masui Dauphine' (*F. carica*) on 21 species of *Ficus*. The scion was most compatible with *F. palmata* (subgenus *Ficus*), followed by *F. septica* (subgenus *Sycomorus*) and *F. stipulata* (subgenus *Synoecia*). The scion had low compatibility with *F. hirta* and *F. erecta* (also subgenus *Ficus*). Scion growth on all 21 species of the subgenus *Urostigma* (e.g., *F. lyrata*) was very weak. Members of the subgenus *Sycomorus* (including *F. racemosa* and *F. gnaphalocarpa*) showed a good grafting compatibility (Condit, 1969; Krezdorn and Glasgow, 1970). Although 'Masui Dauphine' grafted on *F. palmata* grew well for 3 years, it was weaker than the own-rooted 'Masui Dauphine', indicating that *F. palmata* was an unsuitable rootstock. Since wild species are difficult to use directly as rootstocks, interspecific hybrids with *F. carica* could also have potential as rootstocks. Interspecific hybrids of *F. carica* × *F. pumila* have been produced (Condit, 1950; Flaishman *et al.*, 2017), and hybrids of *F. carica* × *F. erecta* have been tried (Yakushiji *et al.*, 2012; 2019). Although the graft affinity of the $F_1$ rootstocks was improved, physiological disorders such as withering roots prevented their use as practical rootstocks. Nevertheless, as the graft compatibility between $BC_1$ and fig is high (Fig. 6.3b; Yakushiji *et al.*, 2019), $BC_1$ rootstocks resistant to *Ceratocystis* canker may be practical in fields contaminated with spores of this pathogen.

### Rootstocks for drought stress

Drought stress reduces fig tree vigor, yield and survival. Fig trees are characterized as drought-tolerant, but growth can be limited depending upon drought severity (Tapia *et al.*, 2003). Some commercial Moroccan genotypes tolerate water potential as low as −4.0 MPa at midday (Oukabli *et al.*, 2008). Rootstocks can alter scion leaf water potential, relative water content, stomatal conductance and nutrient content. Jafari *et al.* (2018) reported the P and K concentrations in leaves, stomatal conductance in the scion leaves, and root volume of rootstocks were positively correlated with the relative water content of the scion under drought stress. For example, 'Siah' and 'Dehdz' rootstocks transferred more water to 'Sabz' scion. These results indicate that the selection of drought resistant rootstock is associated with the potential of water and nutrient uptake under low soil fertility. This suggests stomatal conductance of the scion under drought conditions can be used to screen fig rootstocks for drought tolerance.

### Rootstocks with resistance to Ceratocystis canker

In hot, humid climates such as southern Japan or Brazil, the fungus *Ceratocystis ficicola*, is a major fig disease (Kato *et al.*, 1982; Johnson *et al.*, 2005; Morita *et al.*, 2016; Kajitani and Masuya, 2011). Formerly called *C. fimbriata* Ellis et Halsted (Kato *et al.*, 1982), it is one of the most severe diseases of common fig in Japan. A similar disease has been found in Brazil (Johnson *et al.*, 2005). The disease invades the roots and trunks of figs, eventually killing the tree. Fungicide applications after the appearance of canker stain are ineffective. The most promising approach is to develop a resistant rootstock. The 'Masui Dauphine' is particularly susceptible, 'Celeste' is relatively resistant (Hosomi and Kawaradani, 2004). A new rootstock cultivar, 'Kibaru', bred from 'Celeste' × 'H238-107', with field resistance to Ceratocystis canker, was recently released in Japan (Himeno *et al.*, 2015). However, *F. carica* rootstock remains at risk of infection in contaminated fields. *Ficus erecta*, a wild fig species indigenous to Japan, is extremely resistant to Ceratocystis canker (Shimizu and Miyoshi, 1999; Morita *et al.*, 2011), but its extremely low graft compatibility with *F. carica* eliminates it as a rootstock (Hosomi, 1993; 2017). Yakushiji *et al.* (2012) obtained $F_1$ hybrids of *F. carica* × *F. erecta* with resistance equal to that of *F. erecta*, but the hybrids are unusable as rootstock because of physiological disorder in which roots gradually die. Yakushiji *et al.* (2019) performed backcross hybridizations ($BC_1$) between four common figs (seed parent) and an $F_1$ caprifig type (pollen parent) through artificial pollination. The segregation ratios of resistance and susceptibility (Fig. 6.13) were consistent with a 1:1 Mendelian model, suggesting that Ceratocystis canker resistance in *F. erecta* is controlled by a single dominant gene and the $BC_1$ population possesses true resistance. The healthy growth of $BC_1$ seedlings, propagability of cuttings, and graft compatibility indicate that the $BC_1$ population is a practical source for creating rootstocks resistant to Ceratocystis canker (Table 6.2).

## Rootstocks for root-knot nematode and soil sickness

Nematodes are microscopic, eel-like roundworms. Root-knot nematodes (*Meloidogyne incognita*) cause distinctive swellings, called

**Fig. 6.13.** Ceratocystis canker lesions of $BC_1$ seedlings after stem inoculation with *Ceratocystis ficicola*. The whole plants of resistant (left) and susceptible (right). (Photo: Yakushiji *et al.* (2019), author's own figure.)

**Table 6.2.** Graft compatibility and nursery growth of 'Houraishi' scions grafted to $BC_1$ rootstocks (4 months after hardwood grafting).

| $BC_1$ lines | No. of rootstocks tested | No. of grafts | Successful unions (%) | Nursery height (cm) | Rootstock diameter (mm) | Shoot diameter (mm) |
|---|---|---|---|---|---|---|
| NLBN7-2 | 6 | 4 | 66.7 | 86.8 $^{ab}$ | 12.3 $^{b}$ | 9.9 $^{b}$ |
| NLBN7-5 | 10 | 8 | 80.0 | 105.3 $^{b}$ | 10.9 $^{ab}$ | 10.0 $^{b}$ |
| BNBN7-3 | 6 | 4 | 66.7 | 105.0 $^{b}$ | 15.0 $^{b}$ | 11.3 $^{b}$ |
| BNBN7-5 | 5 | 5 | 100.0 | 46.4 $^{a}$ | 9.9 $^{a}$ | 6.6 $^{a}$ |

Values in a column followed by different letters differ significantly (Tukey-Kramer HSD, $p = 0.05$). Source: Yakushiji *et al.* (2019)

galls, on the roots of affected fig trees. The formation of these galls damages the water- and nutrient-conducting abilities of the roots. Trees infected with root-knot nematodes show a gradual decline, have chronically poor health and have visibly weak shoot, leaf and fruit growth. Nematodes are challenging to control, spread easily from field to field in soil on tools, boots or infested plants. The best way to manage nematodes is to use resistant rootstocks.

Some *Ficus* species such as *F. racemosa* are resistant to nematodes but are not practical as rootstocks due to low graft compatibility or poor tolerance of low temperatures in the northern USA (Condit, 1947). Some *F. carica* cultivars and many other *Ficus* species are tolerant or even immune to some nematodes that compromise commercial fig plantings (Storey, 1975). Even though *F. erecta* is more resistant than 'Masui Dauphine', it also has low graft compatibility (Hosomi, 1993). Among 20 commercial fig cultivars tested as nematode-tolerant rootstock, 'Zidi' is a vigorous and suitable rootstock for 'Masui Dauphine' trees in orchards with soil sickness (Fig. 6.14; Hosomi *et al.*, 2002).

## 6.6 Future Perspectives

Nursery production of fig trees is an integral part of the world's commercial fig industries. Sufficient vigorous and healthy fig nursery trees are the first requirement for establishing a commercial fig industry. However, fig nurseries can also play a valuable role in improving the world fig industries. Well-run commercial nurseries can easily cooperate with researchers in applied trials evaluating substrates and nutrient solutions, graft compatibility and early tree growth. As figs can easily be propagated by cutting, own-rooted plants have been widely used. However, it is also important to study and select rootstocks that can cope with soil diseases or environmental stresses such as drought. In the future, it will be necessary to develop technologies for early selection by elucidating these mechanisms and developing DNA markers.

**Fig. 6.14.** Comparison of 'Masui Dauphine' growth in an orchard infected by soil sickness between own-rooted (left square) and grafted on 'Zidi' rootstock (right square). (Figure used with permission from Akihiro Hosomi.)

## References

Agrawal, P.K. and Dadlani, M. (1995) *Techniques in Seed Science and Technology*. 2nd edn. South Asian Publishers Limited, India.

Al Absi, K.M. (2010) The effects of different pre-sowing seed treatments on breaking the dormancy of mahaleb cherries, *Prunus mahaleb* L. seeds. *Seed Science and Technology* 38, 332–340.

Aljane, F. and Nahdi, S. (2014) Propagation of some local fig (*Ficus carica* L.) cultivars by hardwood cuttings under the field conditions in Tunisia. *International Secularly Research Notices* 2014, 1–5. DOI: 10.1155/2014/809450

Al-Shomali, I., Sadder, M.T. and Atehyeha, A. (2017) Culture media comparative assessment of common fig (*Ficus carica* L.) and carryover effect. *Jordan Journal of Biological Sciences* 10, 13–18.

Anjam, K., Khadıvı-Khub, A. and Sarkhosh, A. (2017). The potential of caprifig genotypes for sheltering *Blastophaga psenes* L. for caprification of edible figs. *ErwerbsObstbau, Heidelberg* 59, 45–49.

Antunes, L.E.C., Chalfun, N.N.J., Pasqual, M., Dutra, L.F. and Cavalcante-Alves, J.M. (2003) Factors affecting on rooting of figs (*Ficus Carica* L.) cuttings. *Acta Horticulturae* 605, 141–146. DOI: 10.17660/ActaHortic.2003.605.22

Bayoudh, C., Labidi, R., Majdoub, A. and Mars, M. (2015) *In vitro* propagation of caprifig and female fig varieties (*Ficus carica* L.) from shoot-tips. *Journal of Agricultural Science and Technology* 17, 1597–1608.

Bayoudh, C., Elair, M., Labidi, R., Majdoub, A., Mahfoudhi, N. and Mars, M. (2017) Efficacy of tissue culture in virus elimination from caprifig and female fig varieties (*Ficus carica* L.). *The Plant Pathology Journal* 33, 288–295. DOI: 10. 5423/ PPJ.OA.10.2016.0205

Beck, N.G. and Lord, E.M. (1988) Breeding system in *Ficus carica*, the common fig. II. Pollination events. *American Journal of Botany* 75, 1913–1922.

Bisi, R.B., Locatelli, G., Barbosa, C.M.A., Pio, R. and Balbi, R.V. (2016). Rooting of stem segments from fig tree cultivar. *Acta Scientiarum.Agronomy* 28, 370–385. DOI: 10, 4025/actasciagron. V38i3.28116.

Brum, G.R. (2001) Micropropagação da figueira (*Ficus carica* L.) 'Roxo de valinhos'. 41f. *Dissertação Mestrado em Agronomia*. Universidade Federal de Lavras, Lavras, Brazil.

Bugbee, G.J., Frink, C.R. and Migneault, D. (1991). Growth of perennials and leaching of heavy metals amended with a municipal leaf sewage sludge and street sand compost. *Journal of Environmental Horticulture* 9, 47–50.

Burrage, S.W. (1999) The nutrient film technique (NFT) for crop production in the Mediterranean region. *Acta Horticulturae* 491, 301–306. DOI: 10.17660/ActaHortic.1999.491.45

Çağlayan, K., Medina, V., Gazel, M., Serçe, Ç.U., Serrano, L., Achon, A., Soylu, S., Çalişkan, O. and Gumuş, M. (2009) Putative agents of fig mosaic disease in Turkey. *Turkish Journal of Agriculture and Forestry* 33, 469–476. DOI: 10.3906/tar-0807-20

Çalişkan, O., Mavi, K. and Polat, A. (2012) Influences of presowing treatments on the germination and emergence of fig seeds (*Ficus carica* L.). *Acta Scientiarum.Agronomy* 34, 293–296.

Çetinbaş, M. and Koyuncu, F. (2006) Improving germination of *Prunus avium* L. seeds by gibberellic acid, potassium nitrate and thiourea. *Horticultural Science. Prague* 33, 119–123.

Chalfun, N.N.J., Pasqual, M., Norberto, P.M., Dutra, L.F. and Cavalcante-Alves, J.M. (2003) Rooting of fig (*Ficus carica* L.) cuttings: cutting time and IBA. *Acta Horticulturae* 605, 137–140. DOI: 10.17660/ActaHortic.2003.605.21

Chong, C. and Lumis, G.P. (2000) Mixtures of paper mill sludge, wood chips, bark, and peat in substrates for pot-in-pot shade tree production. *Canadian Journal of Plant Science* 80, 669–675. DOI: 10.4141/P99-083

Cobanoglu, F., Sahin, B., Kocatas, H. and Ozen, M. (2004) Productivity and quality parameters on fig young plant in Tube. *Gazi OsmanPasa University, Journal of Agriculture of Faculty* 21, 1–8.

Comlekcioglu, S., Kuden, A.B., Aka Kacar, Y. and Kamberoglu, M.A. (2007) Meristem culture of two fig cultivars in Turkey. *Fruits* 62(2), 125–131. www.edpsciences.org/fruits.

Condit, I.J. (1947) *The Fig*. Chronica Botanica Co., Waltham, Massachusetts.

Condit, I.J. (1950) An interspecific hybrid in *Ficus*. *Journal of Heredity* 41, 165–168.

Condit, I.J. (1969) *Ficus, The Exotic Species*. University of California, Division of Agricultural Sciences, Berkeley, California.

Condit, I.J. and Horne, W.T. (1933) A mosaic of the fig in California. *Phytopathology* 23, 887–896.

Danial, G.H., Ibrahim, D.A., Brkat, S.A. and Khalil, B.M. (2014) Multiple shoots production from shoot tips of fig tree (*Ficus carica* L.) and callus induction from leaf segments. *International Journal of Pure and Applied Sciences and Technology* 20, 117–124.

Danthu, P., Soloviev, P., Gaye, A., Sarr, M., Seck, M. and Thomas, I. (2002) Vegetative propagation of some West African *Ficus* species by cuttings. *Agroforestry Systems* 55, 57–63.
Darwesh, H.Y., Bazaid, S.A. and Abu Samra, B.N. (2014) *In vitro* propagation method of *Ficus carica* at Taif governorate using tissue culture technique. *International Journal of Advanced Research* 2, 756–761.
Demiralay, A., Yalcin-Mendi, Y., Aka-Kacar, Y. and Cetiner, S. (1998) *In vitro* propagation of *Ficus carica* L. var. Bursa Siyahi through meristem culture. *Acta Horticulturae* 480, 165–166. DOI: 10.17660/ActaHortic.1998.480.28
Dessoky, D.S., Attia, A.O. and Mohamed, A.A.M. (2016) An efficient protocol for *in vitro* propagation of fig (*Ficus carica* sp) and evaluation of genetic fidelity using RAPD and ISSR markers. *Journal of Applied Biology & Biotechnology* 4, 57–63. DOI: 10.7324/JABB.2016.40406
Dhage, S.S., Pawar, B.D., Chimote, V.P., Jadhav, A.S. and Kale, A.A. (2012) *In vitro* callus induction and plantlet regeneration in fig (*Ficus carica* L.). *Journal of Cell and Tissue Research* 12, 3395–3400.
Dhage, S.S., Chimote, V.P., Pawar, B.D., Kale, A.A., Pawar, S.V. and Jadhav, A.S. (2015) Development of an efficient *in vitro* regeneration protocol in fig (*Ficus carica* L.). *Journal of Applied Horticulture* 17, 160–164.
Dolgun, O. and Tekintas, F.E. (2008) Production of fig (*Ficus carica* L.) nursery plants by stem layering method. *Agriculturae Conspectus Scientificus* 73, 157–160.
Dolgun, O. and Tekintas, F.E. (2009) Effective use of vegetative material in fig (*Ficus carica* L.) nursery plant production. *African Journal of Agricultural Research* 4, 701–706.
Dolgun, O., Tekintas, F.E., Seferoglu, G. and Sahin, N. (2003) Organization for nursery plant of Sarilop and Bursa Siyahi fig cultivars. *Tubitak Project*. 2574–1, Aydin, Turkey.
Elazab, D.S. and Shaaban, M.M. (2015) The impact of sucrose concentration on root growth and development in fig (*Ficus carica* L.) *in vitro*. *Assiut Journal of Agricultural Sciences* 46, 67–75.
Ellis, R.H., Hong, T.D., Roberts, E.H. (1985) *Handbook of Seed Technology for Gene Banks, Volume II. Compendium of Specific Germination Information and Test Recommendations*. IBPGR, Rome.
Evans, M.R., Konduru, S. and Stamps, R.H. (1996) Source variation in physical and chemical properties of coconut coir dust. *HortScience* 31, 965–966.
Ferreira, E.A., Pasqual, M. and Tulmann Neto, A. (2009) *In vitro* sensitivity of fig plantlets to gamma ray. *Scientia Agricola* 66, 540–542. DOI: 10.1590/S0103-90162009000400017
Flaishman, M., Rodov, V. and Stover, E. (2008) The fig: botany, horticulture, and breeding. In: Janik, J. (ed.) *Horticultural Reviews 34*, John Wiley & Sons, Inc., USA, pp.113–196.
Flaishman, M.A., Peer, R., Freiman, Z.E., Izhaki, Y. and Yablovizt, Z. (2017) Conventional and molecular breeding systems in fig (*Ficus carica* L.). *Acta Horticulturae* 1173, 1–9. DOI: 10.17660/ActaHortic.2016.1173.1
Flock, R.A. and Wallace, J.M. (1955) Transmission of fig mosaic by the eriophyid mite *Aceria ficus*. *Phytopathology* 45, 52–54.
Fráguas, C.B., Pasqual, M., Dutra, L.F. and Cazetta, J.O. (2004a) Micropropagation of fig (*Ficus carica* L.) 'Roxo De Valinhos' plants. *In Vitro Cellular & Developmental Biology-Plant* 40, 471–474. DOI: 10.1079/IVP2004562
Fráguas, C.B., Pasqual, M. and Pereira, A.R. (2004b) Multiplicação in vitro de *Ficus carica* L. Efeito da Cinetina e do Ácido Giberélico. *Ciência Agrotecnologia, Lavras* 28, 49–55.
Furutani, S.C. and Nagao, M.A. (1987) Influence of temperature, $KNO_3$, $GA_3$ and seed drying on emergence of papaya seedlings. *Scientia Horticulturae* 32, 67–72. DOI: 10.1016/0304-4238(87)90017-3
Gella, R., Marin, J.A., Corrales, M.L. and Toribio, F. (1998) Elimination of fig mosaic from fig shoot tip cultures by thermotherapy. *Acta Horticulturae* 480, 173–176. DOI: 10.17660/ActaHortic.1998.480.30
George, E.F. (1993) *Plant Propagation by Tissue Culture, Part 1, The Technology*. Exegetics Ltd. pp. 575.
Günver, G. and Ertan, E. (1998) A study on the propagation of figs by the tissue culture techniques. *Acta Horticulturae* 480, 169–172. DOI: 10.17660/ActaHortic.1998.480.29
Günver Dalkılıç, G., Ertan, B., Dağ, S. and Görücüoğlu, P. (2017) Effect of *in vitro* thermotherapy application on shoot development in some fig cultivars. *Acta Horticulturae* 1173, 195–196. DOI: 10.17660/ActaHortic.2016.1173.33
Hartmann, H.T., Kester, D.E., Davies, F.T. and Geneve, R.L. (2002) *Plant Propagation: Principles and Practices*, 7th edn. Prentice Hall, USA.
Hartmann, H.T., Hudson, T., Kester, D.E., Dale, E.K., Davies Junior, F.T. and Geneve, R.L. (2011) *Plant Propagation: Principles and Practices*. 8th edn. Prentice Hall, Boston, pp. 928.
Hassan, S.A.M. and Zayed, N.S. (2018) Factor controlling micropropagation of fruit trees: a review. *Science International*, ISSN 2305-1884, Open Access 1–10.

Hepaksoy, S. and Aksoy, U. (2006) Propagation of *Ficus carica* L. clones by *in vitro* culture. *Biologica Plantarum* 50, 433–436. DOI: 10.1007/s10535-006-0063-8

Himeno, S., Awamura, M., Nogata, H., Ishibashi, M. and Inoue, Y. (2015) A new fig rootstock cultivar 'Kibaru' resistant to Ceratocystis canker. *Bulletin of the Fukuoka Agriculture and Forestry Research Center* 1, 76–81.

Hosomi, A. (1993) Availability of some tree species (Moraceae) for nematode resistant rootstock in the fig cultivation. *Proceeding of The Kansai Plant Protection Society* 35, 39–40. DOI: 10.4165/kapps1958.35.0_39

Hosomi A. (2006) Trial rearing of 'Masui Dauphine' fig culture with vigorating rootstocks. *Kinki Chugoku Shikoku Agriculture Research* 8, 93–96.

Hosomi, A. (2017) Variation in graft compatibility of wild *Ficus* species as rootstock for common fig trees (*Ficus carica*). *Acta Horticulturae* 1173, 199–206. DOI: 10.17660/ActaHortic.2016.1173.34

Hosomi, A. and Kawaradani, M. (2004) Degree of resistance to ceratocystis canker by *Ceratocystis fimbriata* in fig varieties 'Zidi' and 'King'. *Annual Report of The Kansai Plant Protection Society*, 46, 29–32. DOI: 10.4165/kapps.46.29

Hosomi, A., Dan, M. and Kato, A. (2002) Screening of fig varieties for rootstocks resistant to soil sickness. *Journal of the Japanese Society for Horticulture Science* 71, 171–176. DOI: 10.2503/jjshs.71.171

Hosomi, A., Miwa, Y., Furukawa, M. and Kawaradani, M. (2012) Growth of fig varieties resistant to Ceratocystis canker following infection with *Ceratocystis fimbriata*. *Journal of the Japanese Society for Horticulture Science* 81, 159–165. DOI: 10.2503/jjshs1.81.159

Hosomi, A., Miwa, Y. and Isobe, T. (2015) Choice of cuttings and scions for raising fig (*Ficus carica* L.) nursery stocks by softwood grafting. *Horticultural Research (Japan)* 14, 197–203. DOI: 10.2503/hrj.14.197

Jafari, M., Raheml, M. and Haghighi, A.A.K. (2018) Role of fig rootstock on changes of water status and nutrient concentrations in 'Sabz' cultivars under drought stress condition. *Scientia Horticulturae* 230, 56–61. DOI: 10.1016/j_scienta.2016.10.014

Jensen, M.H. (1999) Hydroponic worldwide. *Acta Horticulturae* 481, 719–729. DOI: 10.17660/ActaHortic.1999.481.87

Jensen, M.H. and Collins, W.L. (1985) Hydroponic vegetable production. *Horticultural Review* 7, 483–558. DOI: 10.1002/9781118060735.ch10

Johnson, J.A., Harrington, T.C. and Engelbrecht, G.J.B. (2005) Phylogeny and taxonomy of the North American clade of the *Ceratocystis fimbriata* complex. *Mycologia* 97, 1067–1092. DOI: 10.3852/mycologia.96.5.1067

Jung, H.B., Ito, T. and Maruo, T. (1994) Effects shading and $NO_3$:$NH_4$ ratios in the nutrient solution on the growth and yield of pepper plants in nutrient film technique culture. *Journal of the Japanese Society for Horticulture Science* 63, 371–376. DOI: 10.2503/jjshs.63.371

Kabasakal, A. (1990) İncir Yetiştiriciliği. *Tarımsal Araştırmaları Destekleme ve Geliştirme Vakfı.* Yayın No: 20, Yalova.

Kajitani, Y. and Masuya, H. (2011) *Ceratocystis ficicora* sp. nov., a causal fungus of fig canker in Japan. *Mycoscience* 52, 349–353. DOI: 10.1007/S10267-011-0116-5

Karadeniz, T. (2003) A study on some fruit characteristics and propagations of these by hardwood cuttings of local fig cultivars grown in Ordu (Turkey). *Acta Horticulturae* 605, 107–112. DOI:10.17660/ActaHortic.2003.605.16

Karam, N.S. and Al-Salem, M.M. (2001) Breaking dormancy in *Arbutus andrachne* L. seeds by stratification and gibberellic acid. *Seed Science and Technology* 29, 51–56. DOI: 10.3923/ajps.2006.695.699

Kato, K., Yokota, K. and Miyagawa, T. (1982) A new disease, Ceratocystis canker of fig caused by *Ceratocystis fimbriata* Ellis et Halsted. *Plant Protection* 36, 55–59.

Kılınç, S.S., Ertan, E. and Seferoğlu, S. (2007) Effects of different nutrient solution formulations on morphological and biochemical characteristics of nursery fig trees grown in substrate culture. *Scientia Horticulturae* 113, 20–26. DOI: 10.1016/j.scienta.2006.01.032

Kim, K.M., Kim, Y., Yum, P.Y.T., Chandrasekha, T., Lee, H.Y. and Song, P.S. (2007) Production of multiple shoots and plant regeneration from leaf segments of fig tree (*Ficus carica* L.). *Journal of Plant Biology* 50, 440–446. DOI: 10.1007/BF03030680

Koohakan, P., Ikeda, H., Jeanaksorn, T., Tojo, M., Kusakari, S., Okada, K. and Sato, S. (2004) Evaluation of the indigenous microorganisms in soilless culture: occurrence and quantitative characteristics in the different growing systems. *Scientia Horticulturae* 101, 179–188. DOI: 10.1016/j.scienta.2003.09.012

Kreen, S., Svensson, M. and Rumpunen, K. (2002) Rooting of Clematis microshoots and stem cuttings in different substrates. *Scientia Horticulturae* 96, 351–356. DOI: 10.1016/S0304-4238(02)00126-7

Krezdorn, A.H. and Glasgow, S.K. (1970) Propagation of *Ficus carica* on tropical species of *Ficus*. *Proceedings of the Tropical Region American Society for Horticultural Science* 14, 156–164.

Kumar, V., Radha, A. and Chitta, S.C. (1998) *In vitro* plant regeneration of fig (*Ficus carica* L. cv. gular) using apical buds from mature trees. *Plant Cell Reports* 17, 717–720. DOI: 10.1007/s002990050471

Kuntagol, P., Patil, D.R., Venka teshalu, Biradar, I.B., Nagaraj, G., Manjunath, G. and Prakash, D.P. (2017) Potentiality of different varieties for rooting of cuttings under open and shade house conditions in northern dry zone of Karnataka, India. *International of Journal Microbiology and Applied Sciences* 11, 1763–1768. DOI: 10.20546/ijcmas.2016.611.212

Ling, W.T., Liew, F.C., Lim, W.Y., Subramaniam, S. and Chew, B.L. (2018) Shoot induction from axillary shoot tips explants of fig (*Ficus carica*) cv. Japanese BTM 6. *Tropical Life Sciences Research* 22, 165–174. DOI: 10.21315/tlsr2018.29.2.11

Lloyd, G. and McCown, B.H. (1980) Commercially-feasible micropropagation of Mountain Laurel, *Kalmia latifolia*, by shoot tip culture. *Combined Proceedings, International Plant Propagators' Society* 30, 421–426.

Maloupa, E., Mitsios, I., Martinez, P.F. and Bladenopoulou, S. (1992) Study of substrates used in gerbera soilless culture grown in plastic greenhouse. *Acta Horticulturae* 323, 139–144. DOI: 10.17660/ActaHortic.1993.323.12

Molitor, H.D. (1998) Environmentally sound production in horticulture – the European way. *Journal of the Japanese Society for Horticulture Science* 67, 1224–1228. DOI: 10.2503/jjshs.66.1224

Morita, T., Jikumaru, S. and Kuroda, S. (2016) Disease development in *Ficus carica* plants after inoculation with *Ceratocystis ficicola*. (1) Relationship between xylem dysfunction and wilt symptoms. *Japanese Journal of Phytopathology* 82, 301–309. DOI: 10.3186/jjphytopath.82.301

Morita, T., Mise, D. and Jikumaru, S. (2011) Relative resistance of *Ficus erecta* and cultivars of *F. carica* to *Ceratocystis fimbriata* using their seedlings. *Annual Report of The Kansai Plant Protection Society*, 53, 51–52. DOI: 10.4165/kapps.53.51

Morton, J. (1987) Fig. In: *Fruits of Warm Climates*. Julia F. Morton, Miami, Florida, pp. 47–50. www.hort.purdue.edu/newcrop/morton/fig.html (accessed 22 October 2019).

Murashige, T. and Skoog, F. (1962) A revised medium for rapid growth and bioassays with tobacco tissue cultures. *Physiologia Plantarum* 15, 473–496. DOI: 10.1111/j.1399-3054.1962.tb08052.x

Murithi, L.M., Ranganand, T.S. and Waite, B.H. (1982) *In vitro* propagation of fig through shoot tip culture. *HortScience* 17, 86–86.

Mustafa, N.S. and Taha, R.A. (2012) Influence of plant growth regulators and subculturing on *in vitro* multiplication of some fig (*Ficus carica* L.) cultivars. *Journal of Applied Sciences Research* 8, 4038–4044.

Mustafa, N.S., Hassan, S.A.M., Taha, R.A. and Zayed, N.S. (2016) Studies on the behavior of proliferated shoots and roots of two fig cultivars *in vitro*. *International Journal of ChemTech Research* 9, 1–6.

Mustafa, N.S., Taha, R.A., Hassan, S.A.M. and Zaid, N.S.M. (2013) Effect of medium strength and carbon source on *in vitro* shoot multiplication of two *Ficus carica* cultivars. *Journal of Applied Sciences -Research* 9, 3068–3074.

Nahdi, S. and Aljane, F. (2014) Propagation of some local fig (*Ficus carica* L.) cultivars by hardwood cuttings under the field conditions in Tunisia. *International Scholarly Research Notices* 2014, 1–5. DOI: 10.1155/2014/809450

Nava, G.A., Wagner, J.A., Mezalira, E.J., Cassol, D.A. and Alegretti, A.L. (2014) Rooting of hardwood cuttings of Roxo de Valinhos fig (*Ficus carica* L.) with different propagation strategies. *Revista Ceres* 61, 989–996. DOI: 10.1590/0034-737X201461060015

Nobre, J., Romano, A., Aksoy, U., Ferguson, L. and Hepaksoy, S. (1998) *In vitro* cloning of *Ficus carica* L. adult trees. *Acta Horticulturae* 480, 161–164. DOI: 10.17660/ActaHortic.1998.480.27

Olympios, C.M. (1992) Soilless media under protected cultivation. Rockwool, peat, perlite and other substrates. *Acta Horticulturae* 323, 215–234. DOI: 10.17660/ActaHortic.1993.323.20

Oukabli, A., Mekaoui, A., Ibnouali-El-Aloui, M. and Bari, A. (2008) Contribution to identification of fig (*Ficus carica*) genotypes tolerant to drought. *Acta Horticulturae* 798, 87–93. DOI: 10.17660/ActaHortic.2008.798.10

Pasqual, M. and Ferreira, E.A. (2007) Micropropagation of fig tree (*Ficus carica* sp). In: Mohan, J. and Häggman, H (eds) *Protocols for Micropropagation of Woody Trees and Fruits*. Springer, Dordrecht, The Netherlands, pp. 409–416.

Peche, P.M., Moura, P.H.A., Curi, P.N., Melo, E.T., Figueiredo, A.L., Zambon, C.R., Coutinho, G. and Pio, R. (2017) Rooting of apical cuttings of fig trees with a lopping system. *Acta Horticulturae* 1173, 217–220. DOI: 10.17660/ActaHortic.2016.1173.36.

Pipattanawong, N., Tiwong, S., Thongyean, B., Darak, R., Thamin, P. and Techa, W. (2008) Improvement of propagation by hardwood cuttings with and without using plastic pavilions in fig (*Ficus carica* L.). *Kasetsart Journal (Nat. Sci.)* 42, 207–214.

Pontikis, C.A. and Melas, P. (1986) Micropropagation of *Ficus carica* L. *HortScience* 21, 153–154.

Powell, L.B. (1987) Hormonal aspects of bud and seed dormancy in temperate-zone woody plants. *HortScience* 22, 845–850.

Qrunfleh, I.M., Shatnawi, M.A. and Al-Ajlouni, Z.I. (2013) Effect of different concentrations of carbon source, salinity and gelling agent on *in vitro* growth of fig (*Ficus carica* L.). *African Journal of Biotechnology* 12, 936–940.

Rajeswara-Reddy, K.V., Pulla-Reddy, C. and Goud. P.V. (2008) Effect of auxins on the rooting of fig (*Ficus carica* L.) hardwood and semi hardwood cuttings. *Indian Journal of Agricultural Research* 42, 75–78.

Rattanpal, H.S. and Sidhu, G.S. (2017) Studies on propagation of fig (*Ficus carica* L.) cultivars by budding. *Acta Horticulturae* 1173, 173–176. DOI: 10.17660/ActaHortic.2016.1173.29

Rattanpal, H.S., Singh, G. and Singh, G. (2014) Standardization of rapid propagation technique for fig (*Ficus carica* L.) cultivars by budding. *HortFlora Research Spectrum* 3, 94–96.

Resh, H.M. (1987) *Hydroponic Food Production: A Definitive Guidebook of Soilless Food Growing Methods*, 3rd edn. Woodbridge Press Publishing Co., Santa Barbara, California.

Rijck, G.D. and Schrevens, E. (1998) Distribution of nutrients and water in rockwool slabs. *Scientia Horticulturae* 72, 277–285. DOI: 10.1016/S0304-4238(97)00144-1

Roh, M.Y., Bae, J.H., Lee, Y.B., Park, K.W. and Kwon, Y.S. (1995) Effect of concentration of nutrient solution on early yield and fruit quality of tomato (*Lycopersicon esculentum* Mill.) in substrate culture. *Journal of Biological Production Facilities & Environmental Control* 4, 68–73.

Runia, W.T. (1995) A review of possibilities for disinfection of recirculation water from soilless cultures. *Acta Horticulturae* 382, 221–229. DOI: 10.17660/ActaHortic.1995.382.25

Seferoğlu, G. and Tekintaş, F.E. (1998) Anatomical and histological development of rooting on the fig hardwood cuttings. *Acta Horticulturae* 480, 115–116. DOI: 10.17660/ActaHortic.1998.480.17

Sen, D.K. and Patel, R.M. (2018) *In vitro* mass multiplication of fig (*Ficus carica* L.) through nodal segment explants. *Journal of Cell and Tissue Research* 18, 6435–6440.

Sevgican, A. (2003) Örtüaltı Sebzeciliği (Topraksız Tarım) Cilt: II. *Sayı: 4*, İzmir, Turkey.

Shahcheraghi, S.T. and Shekafandeh, A. (2016) Micropropagation of three endemic and endangered fig (*Ficus carica* L.) genotypes. *Advances in Horticultural Science* 30, 129–134. DOI: 10.13128/ahs-20248.

Sharma, S., Shahzad, A., Mahmood, S. and Saeed, T. (2014) High-frequency clonal propagation, encapsulation of nodal segments for short-term storage and germplasm exchange of *Ficus carica* L. *Trees* 29, 345–353.

Shatnawi, M.A., Shibli, R.A., Shahrour, W.G., Al-Qudah, T.S. and Abu-Zahra, T. (2019) Micropropagation and conservation of fig (*Ficus carica* L.). *Journal of Advances in Agriculture* 10, 1669–1679. DOI: 10.24297/jaa.v10i0.8160

Shekafandeh, A. and Shahcheraghi, S.T. (2017) *Ex vitro* rooting and survival of regenerated shoots from three fig (*Ficus carica* L.) genotypes. *Agriculturae Conspectus Scientificus* 82, 383–386.

Shimizu, S. and Miyoshi, T. (1999) Occurrence of control of fig ceratocystis canker caused by *Ceratocystis fimbriata*. *Plant Protection* 53, 25–26.

Singh, B.M., Rajoriya, C.M., Wani, I.A., Rawat, R.S. and Jat, B.L. (2016) *In vitro* studies of *Ficus carica* and its application in crop improvement. *International Journal for Research in Applied Science & Engineering Technology* 4, 135–148.

Şirin, U., Ertan, E. and Ertan, B. (2010) Growth substrates and fig nursery tree production. *Scientia Agricola* 67, 633–638. DOI: 10.1590/S0103-90162010000600003

Soliman, H.I., Gabr, M. and Abdallah, N.A. (2010) Efficient transformation and regeneration of fig (*Ficus carica* L.) via somatic embryogenesis. *GM Crops* 1, 40–51.

Sousa, C.M., Busquet, R.N. and Miranda, M.A.S.V.R. (2013) Effects of auxin and misting on the rooting of herbaceous and hardwood cuttings from the fig tree. *Revista Ciência Agronômica* 44, 334–338. DOI: 10.1590/S1806-66902013000200016

Storey, W.B. (1975) Figs. In: Janick, J. and Moore, J. N. (eds) *Advances in Fruit Breeding*. Purdue University Press, West Lafayette, Indiana. pp.568–589.

Sümer, G. (2014) Elemination of fig mosaic disease agents by the methods of shoot tip and thermotherapy and testing of Sarilop and Bursa Siyahi fig production material by one step RT-PCR. *MSc thesis*, (inTurkish), Aydın Adnan Menderes University, Aydın, Turkey.

Taha, R.A., Mustafa, N.S. and Hassan, S.A. (2013) Protocol for micropropagation of two *Ficus carica* cultivars. *World Journal of Agricultural Sciences* 9, 383–388. DOI: 10.5829/idosi.wjas.2013.9.5.1802

Takagaki, M., Udagawa, Y. and Takahashi, E. (1997) Effect of pretreatment on rooting and shoot growth of *Ficus carica* L. cuttings. *The Technical Bulletin of Faculty of Horticulture, Chiba University* 51, 227–230.

Tapia, R., Botti, C., Carrasco, O., Prat, L. and Franck, N. (2003) Effect of four irrigation rates on growth of six fig tree varieties. *Acta Horticulturae* 605, 113–118. DOI: 10.17660/ActaHortic.2003.605.17

Tekintas, F.E. and Seferoglu, G. (1998) Propagation of fig by hardwood cuttings in the field conditions (*Ficus carica* L.). *Acta Horticulturae* 480, 119–120. DOI: 10.17660/ActaHortic.1998.480.18

Toma, R.S. and Tamer, Y.S. (2016) Direct *in vitro* regeneration of fig (*Ficus carica* L. cv. Benaty) from apical and lateral buds. *Journal of Zankoy Sulaimani* 18–1, (Part-A) 77–86.

Tyler, H.H., Warren, S.L., Bilderback, T. and Perry, K.B. (1993) Composted turkey litter: II. Effect on plant growth. *Journal of Environmental Horticulture* 11, 137–141.

Ushijima, K., Asakuma, H., Muramoto, K. and Kusano, N. (2013) Shortening of grafted nursery period by novel method of cutting graft of fig (*Ficus carica* L.) rootstock cultivar 'Kibaru'. *Bullen of the Fukuoka Agriculture and Forestry Research Center* 32, 76–80.

Valio, I.F.M. (1986) Auxinas. (2) In: Feri, M.G. (ed.). *Fisiologia Vegetal.* EPV, Sao Paulo Brazil, pp. 39–72.

Van Os, E.A. (2000) New developments in recirculation systems and disinfection methods for greenhouse crops. In: *Proceedings of the 15th Workshop on Agricultural Structures and ACESYS (Japan)*, pp. 81–91.

Verdonck, O. (1991) Horticultural substrates. *21st Int. Course on Vegetable Production. Wageningen*, 95.

Westwood, M.N. (1993) *Temperate-zone Pomology Physiology and Culture*. Timber Press, Portland, Oregon.

Wright, R.D., Browder, J.F. and Jackson, B.E. (2006) Ground pine chips as a substrate for container-grown woody nursery crops. *Journal of Environmental Horticulture* 24, 181–184.

Yakushiji, H., Mase, N. and Sato, Y. (2003) Adventitious bud formation and plantlet regeneration from leaves of fig (*Ficus carica* L.). *The Journal of Horticultural Science Biotechnology* 78, 874–878. DOI: 10.1080/14620316.2003.11511712

Yakushiji, H., Morita, T., Jikumaru, S., Ikegami, A., Azuma, A. and Koshita, Y. (2012) Interspecific hybridization of fig (*Ficus carica* L.) and *Ficus erecta* Thunb., a source of ceratocystis canker resistance. *Euphytica* 183, 39–46. DOI: 10.1007/s10681-011-0459-1

Yakushiji, H., Morita, T. and Jikumaru, S. (2019) Ceratocystis canker resistance in $BC_1$ populations of interspecific hybridization of fig (*Ficus carica*) and *F. erecta*. *Scientia Horticulturae* 252, 71–76. DOI: 10.1016/j.scienta.2019.03.039

Yildiz, H. (1999) Investigation on increasing of nursery stock production in Bursa Siyahi fig cultivar. *Master Thesis*. Institute of National and Applied Sciences, Adnan Menderes University, Turkey.

Zarei, N. and Aboutalebi, A. (2013) Evaluation the rooting behavior of caprifig cultivars cuttings in Fars province. *International Journal of Agriculture Innovations and Research* 2, 397–400.

Zerhoune, M. (2003) Propagation of fig (*Ficus carica* L.) varieties under field conditions. *Acta Horticulturae* 605, 103–106. DOI:10.17660/ActaHortic.2003.605.15

# 7 Environmental Requirements and Site Selection

**A. Aytekin Polat[1]* and A. Sarkhosh[2]**
[1]*Department of Horticulture, Faculty of Agriculture, Mustafa Kemal University, Hatay, Turkey;* [2]*Horticultural Sciences Department, University of Florida, Gainesville, Florida, USA*

## 7.1 Introduction

Although the cultivated fig (*Ficus carica* L.) is a subtropical climate plant, it is grown successfully in many areas of the world that have cool winters and hot dry summers. Fig trees are especially well adapted to Mediterranean climate conditions. However, they can be grown in more humid regions, including the tropics and subtropics, where fruit splitting and disease pressure are commercial production issues (Stover *et al.*, 2007). Fig trees can also grow well under low humidity and high light and require a dry climate and full sunlight for maximum fruit production (Ferguson *et al.*, 1990; Flores, 1990; Melgarejo, 1996). It has been reported that figs can survive under temperatures as high as 45°C (Botti *et al.*, 2003; Jalikop, 2009).

The climatic conditions have an important effect on productivity and fruit quality; the climate requirements of fresh fig varieties are different from those destined for drying. The best quality fig fruits are produced in the Mediterranean and the drier warm-temperate climates. A comparative study on the effects of climatic conditions in the production of fresh and dry figs revealed that dry fig production is strongly dependent on climatic conditions and is successful mostly under dry and warm temperate climates (Botti *et al.*, 2003)

The fig tree is tolerant of a wide range of environmental conditions, has a low chilling requirement, can withstand some frost and is tolerant of drought. Figs for drying are best produced in regions with long sunny days and relatively low humidity, while figs for preserving and canning are best produced in regions with hot summers and higher humidity. Wind and rain during the caprification season may interfere with fig wasps flying. However, gentle breezes during the drying season are desired as they favor fruit maturation on the tree (Childers *et al.*, 1995). The climate and soil requirements of the fig are given in detail below.

## 7.2 Temperature

The fig tree has characteristics of both tropical and temperate zone trees. However, its culture is limited more by low winter temperatures than by high summer heat. The fig is predominantly grown in countries with a Mediterranean climate. It is a deciduous subtropical tree, which loses its leaves for only a short period in California (Childers *et al.*, 1995). With extra care, figs can grow in cooler and wetter areas. However, diseases limit utility in tropical climates. Depending on the cultivar, mature and fully

*Corresponding author: aapolat@mku.edu.tr


DOI: 10.1079/9781789242881.0007

dormant trees can be cold hardy from −6 to −9°C (Fruit & Nut Research & Information, 2019), while actively growing plants can be damaged at −1°C (Fig. 7.1).

## Growing season temperatures

Growth and production of fig trees is strongly dependent on climatic conditions. The low temperatures during autumn, cold winter conditions, and the growing season temperatures and rainfall all affect tree growth, crop production and quality (Flaishman *et al.*, 2008). Fig is hardy to −9°C in winter and tolerant of 40°C in summer. The annual average temperature is 18–20°C, with 30–32°C commonly experienced in July–September during ripening and dry periods, which are important for dry figs (European Commission, 2016). High summer temperatures favor vegetative growth. Daytime temperatures much above 38°C cause premature ripening and skin toughness. However, higher temperatures produce smaller fruit, and cause sunburn damage in fruit and exposed branches of the tree, particularly if the tree is water stressed. Therefore, the highest temperature should not exceed 40°C in fig cultivation areas.

Temperature decreases in autumn slow shoot growth and precipitate terminal bud development. This process affects late autumn crop production. In areas with night temperatures above 12°C, such as in the coastal areas in Israel and in the Imperial Valley in California, fig trees produce fruits in November and December and fruit maturation will continue until leaf fall (Flaishman and Al Hadi, 2002). Generally, autumn production will reduce the dormant bud numbers and, therefore, next year's breba crop production. Breba crop production can be successful in relatively mild winters. In addition to affecting the pollination success in figs, climate conditions also significantly affect the size, shape, skin and pulp color of fig fruits (Condit, 1947). Cooler climates cause fruits to have greener as opposed to yellow skin, more vivid pulp colors, and larger, more elongated fruits.

## Dormant season temperatures

Low temperatures are as impactful on figs as high temperatures. Fig trees can withstand varying degrees of subfreezing temperatures. The degree to which they can avoid cold injury depends on the variety, the amount of moisture in the soil, and the tree's dormancy state. Fig trees are often injured by early or late frosts, when they are not fully dormant or have begun spring growth, killing the younger growth. Fig cultivation cannot be carried out successfully in places where the lowest temperature falls below −9°C. In places where the minimum temperature is −13 or −14°C, fig trees will suffer severe freeze damage and death. The degree of low temperature, duration and the origin of the fig are essential factors. Temperatures of less than −9°C, which persist for a very short period are not harmful, while temperatures above −9°C with a long duration can be harmful (Fig. 7.2). Young trees can be damaged by early frosts resulting from temperatures falling to −3 or −4°C in

**Fig. 7.1.** Spring frost damage is one of the major problems in fig orchards during late winter or early spring. (Figures used with permission from Margarita López-Corrales and Cenap Yilmaz.)

**Fig. 7.2.** Freezing injury in a fig tree. (Figures used with permission from Walter Reeves and Akihiro Hosomi.)

October–November, and depending on variety, may die at temperatures of −6 to −9°C in winter. Also, they are very susceptible to frost damage in spring, especially if frosts are severe. Late spring frosts resulting from temperatures that fall below −1°C at the end of March and April may damage new shoot growth (Fig. 7.1) and reduce the crop. Temperatures of −4 to −7°C at the end of winter lead to *mammoni* crop damage and male wasps in male figs. In this case, the *profichi* caprifig fruit subsequently has insufficient wasps. For this reason, the placement of male figs in orchard establishments should be in a south-facing location, close to the north side, and in a frost-free location. During frosts in fig orchards and nurseries smoke from burning tires, straw or oil, and overhead sprinkler irrigation or fogging, especially in nurseries have been used for frost protection. In addition, to preventing frost damage, wind machines can be used to mix heavier cold and lighter warm air in the orchard.

### Chilling requirement

The chilling requirement of fig trees is still not well understood. Fig requires some winter cold to complete the chilling requirement, but grow better with mild winters. In tropical climates, figs may not drop their leaves in winter (Mendes-Pereira, 1981; Ferguson *et al.*, 1990), cease growth, defoliate or go dormant (Flaishman *et al.*, 2008). Munroe (2002) indicated that figs do not require winter chilling to break dormancy, while Andersen and Crocker (2016) reported fig cultivars to require 100 hours or less below 7.2°C during the dormant season to promote normal vegetative and reproductive bud development. However, others report the chilling requirement for figs is less than 300 hours (California Rare Fruit Growers, 1996; Stover *et al.*, 2007), and that good quality figs can be produced in regions with slight winter chill (Erez and Yablowitz, 2000). The fig dormant period length is dependent upon the local climate (Erez and Shulman, 1982). Oukabli and Mekaoui (2012) indicated successful fig cultivation is possible if a warm growing season is followed by a relatively cold winter.

## 7.3 Light

High solar irradiation is a key factor in maximizing fruit production. Fig requires full sun to produce good quality fruit, at least 8 hours per day (California Rare Fruit Growers, 1996; Givan, 2008; Ernst, 2018). Zhang *et al.* (2018) reported that fig trees require full sunlight to produce strong plants, fruit set on lower nodes, and fruit with high sugar content, shiny skin and excellent quality. If the light conditions are poor or trees are improperly pruned, the branches will be thin and set less fruit, only in higher nodes, and the fruit will be smaller, and of poor quality. Early morning sun is particularly important to dry dew reducing, relative humidity and fungal diseases. However, high temperatures and arid conditions in spring and summer can cause sunburn damage (Fig. 7.3) and early leaf fall if severe (Cutajar and Mifsud, 2017). In these cases, the orchard can be covered with any shade material such as net.

## 7.4 Topography and Elevation

Although fig thrives in the Taurus mountains of Asiatic Turkey at 1463 m, it is generally grown below 900 m in Turkey. In southern India, 'Marseilles' flourishes above 1525 m. In tropical areas, fig thrives between 800 and 1800 m. In California, fig is successfully grown in depressions 300 m below sea level and up to 900 m in the mountains. In Arizona, figs are produced at 1372 m and in Nairobi, East Africa, up to 2743 m elevation but produce best at 1829 m (Condit, 1947). However, most commercial fig production is at lower elevations in interior valleys or coastal locations where winters are not severe. Choosing the correct location to establish an orchard is the primary factor for successful fruit production. Orchards should be in an area free of late spring frost.

Air drainage is important for regulating temperature and avoiding pooling of cold air in the orchard. The orchard site should be a place where the cold air will not accumulate and can flow to a lower place. Buildings, wooded areas and fences can cause pooling of cold air resulting in spring frost damage. Fig trees should not be planted too close to these obstructions. Ideally, fig orchards should be established in the middle of the slope. Planting in valleys or bottomlands where the air will pool and settle will result in frost-damaged trees (Basinger *et al.*, 2016).

Almost all the dry fig cultivation in Turkey is now in the Büyük and Küçük Menderes basin. Fifty to sixty years ago, figs were produced primarily in the lowlands. However, as lowlands were developed, fig production shifted to the mountainous slopes (Fig. 7.4) (Arpacı *et al.*, 2018). About 83.4% of total dried fig plantations in Turkey are located on the mountainous slopes.

Poor soil structure and erosion are the major problems in mountainous areas, requiring soil and water management. On flat land, fig is planted at 7 × 7 or 8 × 8 m spacing for dried fig production or at a 6 × 4 m spacing for fresh fig production. On slopes and in highland regions, trees are usually planted at a 6 × 6 m spacing because the trees develop smaller canopies (Arpacı *et al.*, 2018). About 75% of the total sloping lands of Fars province in the southwest of Iran is under rainfed fig cultivation, where rainwater harvesting is a traditional practice for irrigating fig trees in Estahban (Fig. 7.5). The Estahban region produces 6.5% of the world fig

**Fig. 7.3.** Sunburn damage to fig fruit (left) and tree trunk and limbs (right) can significantly reduce fruit quality and tree performance. (Figures used with permission from Margarita López-Corrales and Cenap Yilmaz.)

**Fig. 7.4.** Fig orchards in lowland (left) and mountainous slope (right) conditions in Turkey. (Figures used with permission from Cenap Yilmaz.)

**Fig. 7.5.** A view of rainfed fig orchards in Estahban region of Fars province, Iran (left). Small dams are made around trees to harvest rainwater (right). (Figures used with permission from Moslem Jafari and Alimohammad Yavari.)

production. Fig production in the Estahban region has been located mostly on steep slopes, since time immemorial, keeping the system alive, and is a sustainable heritage for future generations. Estahban figs orchards are rainfed crops with 2–3 times supplementary irrigation.

Although traditional rainfed orchards have a low yield, these orchards are mostly established on sloping lands and have a key role in reclaiming lands and preventing soil erosion. Water harvesting in sloping lands is a valuable method that provides suitable conditions for establishment of commercial rainfed orchards. However, for establishment and management of rainfed fruit orchards, scientific and engineering principles such as macro- and micro-catchment, land topography, supplementary irrigation, straw and organic matter, sand filters, mulch and cover crops should be considered. Sloping lands are more suitable for fig cultivation, if provided with enough supplementary irrigation (Yadollahi *et al.*, 2013).

## 7.5 Water Availability and Quality

The fig is suited to arid zones due to its drought tolerance. However, there is little data-based information on fig water requirements (Tapia *et al.*, 2003). Fig roots are well developed, which increases tolerance to drought, but the fig's large leaves easily transpire moisture, especially when new shoots are rapidly developing and the fruit is growing. During these stages the water demand is high. If the soil is too dry, plant development will be affected, the fruit skin will thicken, and fruit size and quality will decrease. Severe water stress can cause fruit drop; the fig tree does not tolerate saturated soils. If standing in water for over a week, fruit and leaf drop, and even tree death will result (Zhang *et al.*, 2018). Fig trees are extremely drought tolerant after planting but require regular irrigation during orchard establishment, and produce better yields if regularly irrigated; most commercial orchards are irrigated. In dried fig production the irrigation is reduced in the weeks before maturation. For fresh production irrigation is maintained through harvest (Morton, 1987; Stover *et al.*, 2007).

For the highest yields, fig trees need a continuous supply of water during the summer. Rainfall is usually sufficient to produce a reasonable crop, even when lack of water in late summer and fall results in cessation of shoot elongation and fruiting. The frequency and the amount of irrigation vary according to the soil type, and soils that are shallow or retain less moisture need more frequent irrigation (Krezdorn and Adriance, 1961).

Soil moisture must be managed carefully because most fig tree roots are shallow in soils that dry quickly. Regular irrigation and mulching maintain tree health and vigor and reduces fruit drop (McEachern, 1996). There are no standards specific to figs for irrigation water quality. However, irrigation water should be low in sodium (<3 mg $L^{-1}$), chloride (<4 mg $L^{-1}$), and boron (< 0.67 mg $L^{-1}$) salts (Ayers and Westcot, 1989). Most cultivars require 2.5 acre-feet (76 cm-ha) of irrigation water at maturity, in addition to winter rains (Ferguson *et al.*, 1990).

## 7.5 Wind

Wind can affect fruit quality. Strong winds in late spring and early summer can scar young fruit, decreasing fruit quality. Windy weather and rains during the caprification season may interfere with fig wasp flight disrupting caprification (Childers *et al.*, 1995). However, wind during ripening and drying is among the more significant factors affecting the quality. In Turkey, during August and September, the northeast wind and westerly sea breezes create a suitable climate for drying the best dried cultivar, Aydın İncir.

In Turkey's Menderes plains in Aydin province northern and southern winds produce high-quality figs. Fig quality cannot be achieved when one wind direction is dominant. North and east winds are dry as they originate in the interior of Anatolia, and are suitable for producing dried figs. When the east winds dominate during the entire growing season the figs are not sweet, and are small and dry, with thick skins. West winds increase fruit size and splitting, decreasing fruit quality. Therefore, east and west winds should be approximately equal in intensity (Duyar, 1998).

For example, the east winds in Turkey called Gümüşkanat, begin in the early morning continuing until 12 or 1 pm. These winds accelerate the drying of fresh figs on the tree. The west winds, called Denizyeli, begin at midday, continuing until 8 or 9 pm. If these humid west winds dominate the fresh fig fruit skins become wet. If continually wet the fruit size increases easily, often resulting in the fruit splitting radially from the ostiole, decreasing fruit quality. A high relative humidity also exacerbates fungal disease development in the ostiole. In addition to dominant humid Denizyeli winds, a high water level in soil will also increase the relative humidity in the orchard resulting in not sweet enough and watery fruits.

## 7.6 Rain and Humidity

Traditionally, fig cultivation was rainfed, without supplemental irrigation. The optimal rainfall requirement for good production is 625–675 mm per year. If rainfall is less than 550 mm, supplemental irrigation is necessary to produce a good crop. If the precipitation is too high, the trees will be excessively vegetative and set fewer, large and tasteless fruit. For dried fig production, precipitation through November–June is optimal. Cloudless days without rainfall are optimal from July–September. Rain during fruit development and ripening can cause the fruit to split and spoilage from fungal pathogens (Morton, 1987). Rain late in the summer and during harvest is particularly damaging.

Avoid sites and soils with poor infiltration and water remaining 24 hours after a rain or irrigation. In areas with poor drainage, roots suffer from hypoxia, stunted growth and eventually may die (McEachern, 1996). If it rains during fruit ripening the high temperatures and humidity will exacerbate disease, decrease fruit sugar content and quality, and increase ostiole cracking

(Zhang *et al.*, 2018) (Fig. 7.6). When selecting cultivars for a specific region care should be taken to investigate that the historical precipitation patterns do not overlap with the selected cultivar's final ripening stage.

Dried figs are best produced in arid regions with low relative humidity. Figs for preserving and canning are produced in regions with high humidity (Childers *et al.*, 1995). The relative humidity is among the major factors affecting dried fig quality during maturation. During the drying season (August–October) a relative humidity between 40 and 50% is ideal. High relative humidity during the maturation season will also cause splitting at the ostiole (Fig 7.6).

## 7.7 Soil Properties

The chemical (such as pH) and physical properties of the orchard soil can influence the uptake of plant nutrients and consequently resistance to stress conditions. Thus soil properties should be fully evaluated before orchard establishment (Cutajar and Mifsud, 2017).

Fig trees can be grown on a wide range of soils, including heavy clays, loams and light sands, but the top 1.0–1.5 m should be well-drained. Figs do not tolerate poorly drained sites. Good drainage is a more important consideration than soil fertility. The best fig soils are at least 1–1.5 m deep, well-drained, sandy-clay, loam soils with sufficient organic matter and lime. Alluvial soils rich in lime and potassium are considered acceptable. In very calcareous soils, the fruits remain small. The optimum pH range for fig soils is 6.0–7.8 (Aksoy *et al.* 2001). However, figs can tolerate a pH ranging from 5.5 to 8.0 (Flaishman *et al.*, 2008). Acidic soils are unsuitable. Fig trees grow well in lime-rich soils but are not salinity tolerant.

As would be expected for a tree that requires a well-drained soil the ground water level is an important site selection consideration (Fig. 7.7). Soils with ground water less than 2 m from the surface with annual fluctuations are not suitable for fig cultivation. In heavily saturated soils, the roots will suffer from hypoxia and the fruit will develop ostiole splits and fungal diseases (Brotóns Gallardo *et al.*, 2003).

### Salinity

There are different opinions on the salinity resistance of the fig tree in the literature. Morton (1987) reported that the fig tree is fairly tolerant of moderate salinity. Others report it is moderately tolerant of high salinity (IBPGR 1986; Golombek and Ludder 1993; Tous and Ferguson, 1996; Caruso *et al.*, 2017). On the other hand, Zhang *et al.* (2018) suggest that the fig has high resistance to salinity; however, when the salt content is more than 0.3%, fig survival rate is low and it displays slow growth. Figs can only be planted in soil with salinity this high after preplant modification (Zhang *et al.* 2018). Although there are very different opinions on this subject, it should not be forgotten that the fig tree is a plant that is less resistant to soil

**Fig. 7.6.** Soil moisture fluctuation or sudden rain before or during harvesting results in fruit splitting injury. (Figures used with permission from Moslem Jafari.)

**Fig. 7.7.** Raised beds (left) are recommended for planting fig trees in regions with heavy soil or waterlogging problems (right). (Figures used with permission from Margarita López-Corrales and Rosa Hernandorena.)

salinity. The type of salt is as important as the amount of salt in the soil. Fig trees are sensitive to sodium and boron in the soil. In fig trees the limit of soil salinity at which yield loss begins is ECe = 3–6 dS/m (Maas, 1993).

## 7.8 Greenhouse and High-tunnel Production

Although fig trees have traditionally been cultivated in orchards, excellent results are now being obtained with greenhouse production (Fig. 7.8). Figs are grown in greenhouses and high-tunnels for following reasons: frost and heavy rain protection, extending production window, and higher and more profitable earlier yields in normal climates. Site selection for greenhouse and high-tunnel production considers most of the same factors as used in siting an orchard location: topography, soil and water quantity and quality, and marketing factors.

### Location

Site selection is crucial for profitable and sustainable production in protected climates. Heating in cool climates and cooling in hot climates are prohibitively expensive. Greenhouse production is more manageable in temperate climates. Countries with average annual temperatures between 10 and 20°C are most suitable: Spain, France, Japan, Turkey, Italy, Greece and Israel. Greenhouse production is increasing rapidly in these countries as the average winter temperatures are moderate, reducing heating costs, the biggest input in greenhouses. Total solar radiation and year-round change is the most important climate factor affecting greenhouse production; radiation between 380 and 760 nm is the main energy source for plant growth in greenhouse cultivation. Researchers have stated that in order to provide optimum conditions in terms of radiation in greenhouse cultivation, the radiation values reaching the earth should be 65–70 kWh $m^{-2}$ per month (Krug *et al.*, 2007; Zabeltitz, 2011). If production is to be throughout the year when at times solar radiation and duration are insufficient, artificial lighting can be used.

### Topography

The greenhouse should be built on level ground, ideally with a 1.5% slope in all directions to promote cold air drainage. When orienting the greenhouses, always default to what is best for drainage and to avoid shading from other structures. In determining the direction of the greenhouse, care should be taken to ensure that the plants within will not shade one another. Generally, the plant rows should be in the south–north direction to receive east–west sun throughout the day. In the north, orienting the side of the greenhouse with the most glazing to true solar south is the best practice. In order to increase the light transmittance in the winter, the greenhouses should be directed in the east–west direction. This should be adjusted according to the region where the cultivation is made. For

**Fig. 7.8.** Fig production in greenhouse. (Figures used with permission from Tecnología Hortícola.)

example, in Turkey (25°40' to 44°49' E longitude, and 35°51' to 42°06' N latitude), the greenhouse roof slope angle should be 27–26° in order to provide the maximum level of light transmission. It should be away from streams and canals prone to flooding.

## Soil

Within the greenhouse the figs can be grown in bags, pots or raised beds or directly in the amended soil. With containers or directly in the ground the soil is generally treated for pests and pathogens, and amended for nutrition and drainage as standing water vectors diseases. Adequate content of humus is desirable in greenhouse soils as well as in open fields, but there are no standards of organic matter content. On the other hand, in some greenhouse cultivation, artificial media that contain no humus at all, such as rock wool, are used. Also, volcanic soil or perlite sacks in the pots can be used as substrate. Volcanic soil mixed with other materials lightens the weight and increases the filtration of the mixtures. The soil can be covered with a polypropylene layer to prevent weeds from emerging. Ideally a greenhouse soil for fig production should be at least 2 m deep, a sandy-loam, have 2.1–4.5% organic matter (Kotsiras *et al.*, 2017), pH 6.0–7.8, an ECe below 6 dS $m^{-1}$ and free of pests and diseases. Also, care must be taken to avoid salt accumulation in the greenhouses due to low rainfall runoff.

In addition, fig trees can grow in soilless culture under greenhouse cultivation to get rid of all the inconveniences of traditional farming. So, growers could benefit from it by improving yields. In a study by Melgarejo *et al.* (2007) the chosen growing media were perlite sacks of 60 and 40 L, and open polypropylene furrows (18 cm high and 25 cm wide) filled with perlite (Fig. 7.8). This type of soil-free culture allowed irrigated farms to boost their fig production from 4500 kg $ha^{-1}$ per year up to 81,000 kg $ha^{-1}$ per year; that is an 18-fold yield increase compared to traditional farming. Likewise, water efficiency would also be maximized. A 90% water reduction was achieved by applying this growing technique. Furthermore, fertilizers and pesticide applications, as well as farming costs (hand labor) may be reduced by growing the appropriate fig cultivars.

## Water availability and quality

Greenhouse irrigation supplies should free from diseases and pests; the most reliable sources are deep wells as there is less potential for pathogen contamination versus surface water. However, well water may be saline. Supplemental rainwater can be collected from the greenhouse's gutters to fill water storage containers. However, stored water can develop algae which can clog greenhouse irrigation systems.

## Greenhouse factors

The most important greenhouse factors in protected cultivation of fig are light, temperature, relative humidity, precipitation and ventilation. Greenhouse temperatures vary according to cultivated varieties, growth stages and climatic regions. In general, the temperature is maintained at approximately 15–30°C with heating and ventilation. The average temperature, should be 11–20°C during vegetative growth stages and 30–32°C during fruit ripening. But if it gets too much hotter, a misting system should kick in to cool the trees. One of the most important elements of successful cultivation under cover is not to neglect the ventilation in greenhouses. Greenhouse ventilation can be automatically controlled by lateral and upper openings. Also, shading or cooling of the greenhouse is required. During the intense summer heat, it may be necessary to throw a shade cloth on top of the greenhouse to keep it cool and reduce the solar radiation. To avoid temperature stress during summer (because of solar radiation), polycarbonate surfaces can be whitened with calcium hydroxide sprays.

Fig trees need plenty of sun (8 hours at least) which helps ripen the fruits. Although greenhouses are suitable for plant development in the winter season, plant development slows down if sufficient light is not provided. Sunlight loss with plastic film should be limited to 30% or less. Supplemental lighting is expensive because of electricity costs, so mulching the ground with reflective sheets is practical. If the light intensity is above the desired level, shortening in the shoot lengths, hardening and feathering of the leaves and also sunburn occur in connection with the temperature. Therefore, in greenhouses, shading material should be used in a very hot summer. To protect the plants from the hot weather in the summer months, the greenhouses can be covered with a 40% light mesh.

Relative humidity higher than desired level results in reduced plant resistance, closure of stomata and sweating, slowed growth and development, and increased disease and pests, resulting in reduced product quality. The relative humidity rate being lower than desired causes the yield and quality of the fruits to decrease and the plant to fade. Lower relative humidity is better except for bud burst stage, but it is difficult to control. Precipitation such as rain, snow and hail can be challenging in greenhouse production.

## Economic factors

Due to their perishability and high production costs, fresh greenhouse fig production should be established close to major markets. The distance to markets, especially in export-focused production, can be a limiting factor in profitable greenhouse cultivation (Castilla and Baeza, 2013). In order to reduce transportation costs, greenhouses need to be close to fuel. No work can be done in any greenhouse without electricity; the cheapest and easiest available are preferred to reduce the cost. There must be an energy source that can be used continuously and cheaply in the place where the greenhouse will be installed. It is very necessary for fertilizer and disinfectant for the greenhouse. It is desirable that these be located close to the greenhouse to reduce costs. Finding qualified workers is one of the most important problems in greenhouse cultivation. Knowing plant cultivation in the open is not enough to grow plants in the greenhouse. It is absolutely necessary that the worker can read the greenhouse temperature and humidity from the tools and adjust the greenhouse ventilation according to these values.

## Conclusions

The edible fig (*F. carica*) is considered to be a subtropical plant, but can be grown successfully in many areas of the world which have cool winters and hot dry summers. Figs are especially well adapted to Mediterranean climate conditions in which the annual average temperature is 18–20°C and annual rainfall is 625–675 mm. Therefore, figs are predominantly grown in countries that are dominated by a Mediterranean climate. As a matter of fact, 70% of the world's figs are produced in Mediterranean countries. Depending on the cultivar, mature and fully dormant trees can be cold hardy to −6 to −10°C while actively growing plants can be damaged at −1°C. Fig trees can also grow well under low humidity and high light, and require a dry climate and full sunlight for maximum fruit production. Temperatures reaching 30–32°C in ripening and drying periods are highly significant to dry figs. For drying, the fruit ripening and drying season must be arid and cloudless. Rainfall during the drying period causes a decrease in fruit quality and therefore is not desired. The relative humidity of the weather is among the major factors affecting dried fig quality during maturation. During the drying season relative humidity between 40 and 50% is ideal. While high humidity causes darkening of the fruit color and split of fruit, low humidity results in thick fruit skin, which results in loss of quality. Also, the prevailing winds in the ripening and drying season of figs are amongst the most significant factors affecting the quality of figs. In Turkey, which is the main producer country, in the August and September months, the northeast wind and westerly sea breeze creates a suitable climate for drying figs. Figs are tolerant of most types of soil, but perform best on well-drained, reasonably fertile soils.

The ideal conditions for dried fig production are: temperate and rainy winters, 30–40°C summer temperatures especially in July–September, 45–50% relative humidity, alluvial, clayish-loam soil, and average annual precipitation of around 650 mm. When the famous Turkish fig cultivar, 'Sarılop', is grown in the Büyük and Küçük Menderes basins of the Aegean region, where the climate and soil conditions meet these desired criteria, some of the best quality dried figs in the world are produced. Since table fresh figs are not as demanding as dried figs in terms of climate and soils, they can be grown economically wherever the temperature and precipitation demands are met. However, production in open field shows some problems in production such as: high incidence of pests and diseases, fruit decay during the rainy season, mishandling of plantations and damage caused by frost. Hence, lower yields are obtained and in some cases the plantations are abandoned. The search for higher yields in fig has led to implementing high-planting densities under greenhouse conditions. Another important aspect to be considered is that the fig requires lots of heat units to achieve good quality of fruit, which limits the expansion of cultivation in regions with low temperatures. In this case, production systems in controlled environments or protected agriculture are a good production alternative.

Growing fig trees in intensive production systems under hydroponic and greenhouse conditions is an excellent alternative production system since it allows increasing leaf area index, harvest index and fruit yield in relation to the total biomass of the plant. With this, up to 20 more times of fresh fruit yield can be obtained, compared with open field plantations.

In site selection for greenhouse production, most of the same factors as used in siting an orchard location should be considered: topography, soil and water quantity and quality, and marketing factors. In addition, the most important greenhouse factors such as light, temperature, relative humidity, precipitation and ventilation should be considered, as well as the most important marketing or economic factors such as road transportation, market accessibility (proximity to settlements, market demands), availability of services (fuel, fertilizer, disinfectant and electricity), and availability of labor.

## References

Aksoy, U., Balci, B., Can, H.Z. and Hepaksoy, S. (2001) Some significant results of the research work in Turkey on fig. *Acta Horticulturae* 605, 173–181.

Andersen, P.C. and Crocker, T.E. (2016) *The Fig*. U.S. Department of Agriculture, UF/IFAS Extension Service, University of Florida, IFAS, Florida A & M University Cooperative Extension Program, and Boards of County Commissioners Cooperating. www.edis.ifas.ufl.edu (accessed 12 October 2019).

Arpacı, S., Konak, R. and Çiçek, E. (2018) A national value: Turkish figs. *Chronica Horticulturae* 58(2), 85–91.

Ayers, R.S. and Westcot, D.W. (1989) *Water Quality for Agriculture*. FAO Irrigation And Drainage. Paper No.29. Rome.

Basinger, N.T., Owle, J.M., Piner, A. and Parker, M.L. (2016) *North Carolina Production Guide for Smaller Orchard Plantings*. AG-814. North Carolina Cooperative Extension Service, Raleigh, NC. www.content.ces.ncsu.edu/north-carolina-production-guide-for-smaller-orchard-plantings (accessed 15 October 2019).

Botti, C., Franck, N., Prat, L., Ioannidis, D. and Morales, B. (2003) The effect of climatic conditions on fresh fig fruit yield, quality and type of crop *Acta Horticulturae* 605, 37–42.

Brotóns Gallardo, P., Martínez Cano, C.N., Faz Cano, A. and Marín Sanleandro, P. (2003) Pedological characterization of fig plantation in the southeast of Elche (Alicante, S.E. Spain). *Acta Horticulturae* 605, 295–301.

California Rare Fruit Growers (1996) Fig Fruit Fact. www.crfg.org/pubs/ff/fig.html (accessed 15 October 2019).

Caruso, G., Gennai, C., Ugolini, F., Marchini, F., Quartacci, M.F. and Gucci, R. (2017). Tolerance and physiological response of young *Ficus carica* L. plants irrigated with saline water. *Acta Horticulturae* 1173, 137–142.

Castilla, N. and Baeza, E. (2013) Greenhouse site selection. In: GAPs for greenhouse vegetable crops: Principles for Mediterranean climate areas. *FAO Plant Production and Protection Paper* 217, 21–33. www.fao.org/3/i3284e/i3284e.pdf (accessed 3 February 2019).

Childers, N.F., Morris, J.R. and Sibbett, G.S. (1995) *Modern Fruit Science - Orchard and Small Fruit Management*. 10th edn. Horticultural Publications, Gainesville, Fl, p. 632.

Condit, I.J. (1947) *The Fig*. Chronica Botanica Co. Waltham, Massachusetts.

Cutajar, S. and Mifsud, D. (2017) *Good Agricultural Practice (GAP) for Fig Tree Cultivation*. Malta University Consulting, Plant Protection Directorate, Lija, Malta, p. 31.

Duyar, E. (1998) Impact of ecological changes on fig plantations in Big Menderes (meander) basin. *Acta Horticulturae* 480, 311–315.

Erez, A. and Shulman, Y. (1982) Growing breva figs in a meadow orchard; a possibility for commercial fig growing. *Hassadeh* 63, 969–971.

Erez, A. and Yablowitz, Y. (2000) Técnicas de producción de brevas en Israel, In: Botti, C. (ed.) *Cultivos Frutales para Zonas Aridas*. Escuela de Agronomia, Facultad de Ciencias Agronómicas, Univ. Chile, pp. 1–13.

Ernst, M. (2018) *Figs*. CCD-CP-135. Center for Crop Diversification, University of Kentucky College of Agriculture, Food and Environment, Lexington, KY. www.uky.edu/ccd/sites/www.uky.edu.ccd/files/figs.pdf (accessed 1 October 2019).

European Commission (2016) *Specification, Aydın 'Inciri*. European Commission, Agriculture and Rural Development, Agriculture and Food. www.ec.europa.eu/agriculture/quality/door/registeredName.html?denominationId=8900#top (accessed 15 March 2019).

Ferguson, L., Michalides, T.J. and Shorey, H.H. (1990) The California fig industry. *Horticultural Review* 12, 409–490.

Flaishman, M.A. and Al Hadi, F.A. (2002) Fig growth in Israel (in Hebrew). *Alon Hanotea* 56, 56–57.

Flaishman, M.A., Rodov, V. and Stover, E. (2008) The fig: Botany, horticulture, and breeding. *Horticultural Review* 34, 97–113.

Flores, A. (1990) *La Higuera*. Ed. Mundi-Prensa, Madrid. p. 190.

Fruit & Nut Research & Information (2019) *Fig in California*. University of California, Division of Agriculture and Natural Resources. www.fruitandnuteducation.ucdavis.edu/fruitnutproduction/Fig/ (accessed 15 October 2019).

Givan, R.U. (2008) The NAFEX FIG Page. Growing Figs. NAFEX Fig Interest Group Coordinator. www.figdatabase.com/uploads/The%20NAFEX%20FIG%20Page.pdf (accessed March 2019).

Golombek, S.D. and Ludders, P. (1993) Effects of short-term salinity on leaf gas exchange of the fig (*Ficus carica* L.). *Plant and Soil* 148, 21–27.

IBPGR (1986) *Genetic resources of tropical and sub-tropical fruits and nuts*. International Board Plant Genetic Resources (IBPGR), Rome.

Jalikop, S.H. (2009) *Fig*. Division of Fruit Crops, Indian Institute of Horticultural Research, Bangalore, India.

Kotsiras, A., Amorgianiotis, T., Tsiavtari, E., Antonopoulos, D.F. and Demopoulos, V. (2017) GIS based soil suitability evaluation for fig orchards in Southern Peloponnese, Greece. *Acta Horticulturae* 1173, 207–212. DOI: 10.17660/ActaHortic.2017.1173.35

Krezdorn, A.H. and Adriance, G.W. (1961) *Fig Growing in the South*. Agricultural Research Service, United States Department of Agriculture, Agriculture Handbook No. 196.

Krug, H., Romey, A. and Rath, T. (2007) Decision support for climate dependent greenhouse production planning and climate control by modelling. I. Modelling climate. *European Journal of Horticultural Science* 72(3), 97–103.

Maas, E.V. (1993) Testing crops for salinity tolerance. In: Maranville, J.W., Baligar, B.V., Duncan, R.R., Yohe, J.M. (eds) *Proceedings of Workshop on Adaptation of Plants to Soil Stresses. INTSORMIL*. Pub. No. 94-2, Univ of Nebraska, Lincoln, NE, August 1-4-1993. pp. 234–247.

McEachern, G.R. (1996) *Figs*. Texas A & M University College Station, TX. December 9, 1996.
Melgarejo, P. (1996) *La Higuera* (*Ficus carica* L.). Universidad Politécnica de Valencia, Orihuela, España.
Melgarejo, P., Martínez, J.J., Hernández, F., Salazar, D.M. and Martínez, R. (2007) Preliminary results on fig soil-less culture. *Scientia Horticulturae* 111, 255–259.
Mendes-Pereira, F. (1981) *Cultura da Figueira*. Série Fruticultura Boletim N° 2, Livroceres Ltda. Piracicaba, Brasil. 73 pp.
Morton, J. (1987) Fig. In: *Fruits of Warm Climates*. Morton, J. F., Miami, FL, pp. 47–50.
Munroe, A. (2002) *Fig Growing in NSW*. NSW Agriculture, Orange, New South Wales, September 2002.
Oukabli, A. and Mekaoui, A. (2012) Dormancy of fig cultivated under Moroccan conditions. *American Journal of Plant Sciences* 3, 473–479.
Stover, E., Aradhya, M., Ferguson, L. and Crisosto, C.H. (2007) The fig: Overview of an ancient fruit. *HortScience* 42(5), 1083–1087.
Tapia, R., Botti, C., Carrasco, O., Prat, L. and Franck, N. (2003) Effect of four irrigation rates on growth of six fig tree varieties. *Acta Horticulturae* 605, 113–118.
Tous, J. and Ferguson, L. (1996) Mediterranean fruits. In: Janick, J. (ed.), *Progress in New Crops*. ASHS Press, Arlington, VA. pp. 416–430.
Yadollahi, A., Sarikhani Khorami, S. and Kohandel, A. (2013). *Practical Methods of Water Harvesting for Horticultural Crops Cultivation in Rainfed Condition* (in Persian). Jahad Daneshgahi Press. pp.197.
Zabeltitz, C.V. (2011) *Integrated Greenhouse Systems for Mild Climates*. Springer, pp. 137–143.
Zhang, Y., He, J., Chen, J. and Ling, P.H. (2018) *Site Selection of Fig Orchard, Vol.* 3, *No. 1.* Research in Agriculture, ISSN 2740-4431. www.scholink.org/ojs/index.php/ra (accessed 6 September 2021).

# 8 Orchard Establishment and Management

**M. Jafari[1], M. López-Corrales[2], A.J. Galán[3], A.I. Galván[3], A. Hosomi[4], H. Nogata[5], H. Ikegami[5], F. Balas[3], M.J. Serradilla[3], C.A. Garza-Alonso[6], A. Yavari[7], L. Ferguson[8], and A. Sarkhosh[9]***

*[1]Fig Research Station, Fars Agricultural and Natural Resources Research and Education Center, Agricultural Research, Education and Extension Organization (AREEO), Estahban, Fars, Iran; [2]Junta de Extremadura, Agri-Food Technological Institute of Extremadura (INTAEX-CICYTEX), Badajoz, Spain; [3]Junta de Extremadura, Finca La Orden-Valdesequera Research Centre (CICYTEX), Badajoz, Spain; [4]Toyo Institute of Food Technology, Kawanishi, Japan; [5]Fukuoka Agriculture and Forestry Research Center, Yukuhashi, Japan; [6]Facultad de Agronomía – Universidad Autónoma de Nuevo León, Escobedo, Mexico; [7]Private pomologist and consultant, Tehran, Iran; [8]Department of Plant Sciences, College of Agricultural and Environmental Sciences University of California Davis, California, USA; [9]Horticulural Sciences Department, University of Florida, Gainsville, Florida, USA*

## 8.1 Introduction

Traditionally fig was cultivated under rainfed conditions at low densities and the fruit processed primarily into dried figs. Modern fig orchards are now irrigated and planted at moderate to high densities to produce both fresh and dried figs. This chapter summarizes how fig growers can take advantage of the wide range of edible fig cultivars, their collective climatic and edaphic adaptability and the multiple methods of manipulating fig production to achieve the global fig production of today.

## 8.2 Site Selection

Prior to establishing a modern fig plantation, microclimate, irrigation water availability and quality, and the soil physical and chemical characteristics should be assessed. These topics are covered in detail in Chapter 7: *Environmental Requirements and Site Selection*, which summarizes site selection as follows 'Although the cultivated fig (*Ficus carica* L.) is a subtropical climate plant, it is grown successfully in many areas of the world that have cool winters and hot dry summers.' Fig trees are especially well adapted to Mediterranean climate conditions. However,

*Corresponding author: sarkhosha@ufl.edu

 *The Fig: Botany, Production and Uses* (eds. Sarkhosh, A., Yavari, A. and Ferguson, L.)
DOI: 10.1079/9781789242881.0008

they can be grown in more humid regions, including the tropics and subtropics. In addition to climate, phytopathogen analyses for root-knot nematodes, *Meloidogyne* spp., particularly on light and sandy soils should be conducted (Casadomet *et al.*, 2016; Condit, 1947; Kawase, 1990; Hosomi *et al.*, 2002; Casadomet *et al.*, 2016). The soil should also be analysed for the fungi *Armillaria mellea*, *Rosellinia necatrix* and *Roesleria hypogea* in heavy, saturated soils (Costa, 2019). Soil diagnostics in the laboratory must take place for identification. Based on chemical rules and regulations in the region, chemical fumigation or non-chemical alternatives can be considered as management strategies. These topics are covered in detail in Chapter 12: *Fungal, Bacterial, and Nematode Problems in Figs.*

## 8.3 Cultivar and Rootstock Selection

Variety selection is based upon climate, particularly the danger of fall, winter and spring cold damage, fruiting and ripening dates, soil and water availability and quality, fruit yield and quality, production infrastructure, consumer preferences, fresh or dried production, and accessible markets. Figs for fresh consumption should be large fruits with a small ostiole, good color, a skin resistant to handling damage, detach easily from the peduncles, and good post-harvest transport, storage and shelf life. Figs for dried consumption should have a high soluble solids content, ripen early, dry well on the tree, detach easily and have smooth skin. Chapter 5: *World Fig Cultivars* gives a complete discussion of fig cultivars, their climatic and production requirements. Also, an updated list of registered cultivars is generally available in each country. This information certifies the identity of commercially acquired trees (López-Corrales *et al.*, 2011).

### Rootstocks

Most fig cultivars are grown on their own roots. However, rootstocks can be used in problematic soils or to provide vigor or tolerance to abiotic factors such as drought or salinity. In Japan, Hosomi *et al.*, (2002) evaluated different commercial fig cultivars as rootstocks for the 'Masui Dauphine' variety in root-knot nematode infested soils. They concluded the 'Zidi' variety's high vigor conferred some tolerance in these soils. In Spain, the vigorous variety 'Bourjassotte Blanche' has been used to promote the vegetative development of the 'Bourjassotte Noire' cultivar (Condit, 1947). In Iran, the 'Siah' and 'Dehdez' cultivars have been evaluated for improving the amounts of drought tolerance of the 'Sabz' cultivar (Jafari *et al.*, 2018).

## 8.4 Orchard Design

Orchard design is determined by the site topography, cultivar, fresh or dried production and orchard operations, particularly harvesting. These parameters all influence the planting design and the between and in-row spacing. For example, 'Calimyrna' ('Lob Injir') cultivar has a naturally spreading growth habit and is allowed to grow tall when producing dried figs. Therefore, it is planted at higher spacings and lower densities (Fig. 8.1). Fresh fig cultivars like 'Dottato' ('Kadota') and 'San Piero' are planted at lower spacings and higher densities, and the trees are pruned low to facilitate harvesting (Fig. 8.2).

Ideally, orchards are planted with the rows oriented from north to south to take advantage the sun exposure from east to west throughout the day. Tree spacings range from 4 × 4 m to 12 × 12 m, 625 to 69 trees per hectare respectively (Pereira *et al.*, 2015). Three orchard designs predominate; the square, the triangular or offset, and the contour designs.

The square and rectangular orchard designs are the most popular and easiest to develop; spacing between trees and rows is the same or spacing within the row is slightly less than between rows. Cultivar and final product, fresh or dried, soil fertility, and training system determine the spacing, generally 6–12 m. Every four trees form a square, or rectangle, with trees at right angles to one another (Fig. 8.3). If the orchard is rainfed with no irrigation lines to be disturbed, cultivation can be done in both directions. This orchard design is used in both rainfed and irrigated orchards.

In the triangular, off-set square design, the trees are also planted in a square design but the trees in alternate rows are planted midway, or

**Fig. 8.1.** A typical dried fig orchard planted at higher spacings and lower densities with wide row middles to ensure good light penetration for on tree drying before dropping. The trees are allowed to grow tall because most figs drop or are blown off to the ground before winnowing and collecting from the ground. (Left figure courtesy of M. López-Corrales and right figure used with permission from Birgül Ertan.)

**Fig. 8.2.** A typical fresh fig orchard planted at lower spacings and higher densities with narrow row middles to maximize the use of orchard space. The trees are pruned low to facilitate harvesting from the ground. (Left figure used with permission from Rosa Hernandorena www.hernandorena.com/en/ and right figure courtesy of M. López-Corrales.)

off-set, from trees in the adjoining rows. This design is less popular than the square design but in sloping, steep terrain and rainfed orchards this design improves water catchment and controls runoff better than a square or rectangular design.

The contour design is used in hilly areas with sloping terrain. As with the square design the trees are planted in rows but follow the contour of the slope. This design prevents the soil erosion that cultivation and irrigation would produce if trees were planted in straight lines. With this design this system layout is done the same as in the square design. The lower contour line is established first and marking for successive rows is done from the base to the top. Bench terraces should be built when the slope is greater than 10% (Fig. 8.4).

## High-density fig orchards

A high-density orchard is defined as any orchard with more than 1000 trees per hectare. The objective is increasing net return by maximizing orchard area use, increasing early production and facilitating orchard operations, particularly pruning, spraying and harvesting by limiting tree size. High density systems can be established as free standing trees or trained on an espalier.

**Fig. 8.3.** A rainfed fig orchard in Estahban (Iran), with a square orchard design. This design facilitates cultivation in both directions in rainfed orchards without irrigation systems. (Figure courtesy of A. Yavari.)

**Fig. 8.4.** Contour design in rainfed fig orchard in Kazeroun, Iran. (Figure courtesy of M. Jafari.)

The high light exposure with an espalier produces excellent fresh fruit color. The major disadvantage of both methods is high establishment costs of the additional trees, irrigation system, the trellis if the trees are espaliered and hand labor for planting and pruning (Majid *et al.*, 2020). One interesting innovation with potential for both types of high-density dried fig orchards, free standing and espalier, is installing mesh nets beneath the tree canopies to collect dropped dried figs, facilitate harvest collection and prevent the soil contact that proliferates the postharvest fungal pathogens (Gilbert and Senyuva, 2008) (Fig. 8.5). Alternatively, the mesh can be placed on the ground (Figs. 8.6 and 8.7). The Centre for Scientific and Technological Research of Extremadura, CICYTEX, is investigating super high-density fig orchards, 1000 trees per hectare using the dried fig cultivars 'Cuello Dama Blanco', 'Calabacita' and 'Picholetera' using the below-tree mesh nets. 'Calabacita' was the most productive variety with 12.87 kg/tree, 19.970 kg/ha in the 9th green year, followed by 'Cuello Dama Blanco' with 10.38 kg/tree, 10,380 kg/ha, and 'Picholetera' with 10.18 kg/tree, 10,180 kg/ha (Galván *et al.*, 2021).

## 8.5 Land Preparation Before Planting (Site Preparation)

As with all tree crops, the first step in site selection, after determining climate, is analyzing the soil by either the pit or auger method (FAO, 2021a; Globe Canada, 2021b). Fig does not tolerate saturated soil conditions; eliminating cemented soil layers and hardpans that impede drainage is necessary. The fig tree's fibrous root system can spread 15 m laterally and 2 m vertically, but most root biomass is within the top 60–70 cm of the soil. Therefore, subsoiling to 50–80 cm depth allows soil aeration without inverting the horizon (Fig. 8.8). This is expensive but improves permeability and encourages root exploration. Cross-ripping is optimal, but, if it cannot be done, at least one pass down the tree row should be done. To effectively break stratification in the soil profile, this should be done when the soil contains 5–15% water.

After subsoiling, the soil is cultivated, levelled and floated (Fig. 8.9). If the soil is heavy with a potential for waterlogging, tree row berms 50–70 cm high and 100–120 cm wide should be formed. During soil preparation, preplant amendments, particularly limestone as fig trees

**Fig. 8.5.** Below-tree mesh nets in a high-density fig orchard. The nets prevent contact with the soil, decreasing fungal pathogen contamination and facilitate harvest collection. (Figure courtesy of M. López-Corrales.)

**Fig. 8.6.** Below-tree mesh nets on the ground in a high-density fig orchard. (Figure courtesy of M. López-Corrales.)

**Fig. 8.7.** Mesh nets below fig trees trained on an espalier for fresh and dried fig production 'Cuello Dama Blanco' ('Kadota'). The earlier larger figs can be hand harvested for the fresh market and the later smaller fruits can drop, to be collected later for drying. Note the above tree nets for bird protection. (Figure courtesy of M. López-Corrales.)

prefer soils between pH 6.0 and 8.0, can be incorporated (Flaishman *et al.*, 2008). In regions with soilborne disease and nematode problems, fumigation must be done before staking, hole digging and planting. Soil texture, preparation method, temperature and moisture amount are very important in fumigation efficiency (Wilhelm *et al.*, 1974; ISHS, 2004). In regions with restrictions on fumigant use or in organic production systems, non-fumigant alternative options for

**Fig. 8.8.** Subsoiling preplant to improve permeability and increase root exploration. (Figure courtesy of M. López-Corrales.)

**Fig. 8.9.** Cultivating after subsoiling. (Figure courtesy of M. López-Corrales.)

soil disinfestation such as cultivar/rootstock selections, soil solarization or stem application should be considered (Shennan *et al.*, 2014).

## 8.6 Irrigation System Selection and Installation

Traditional fig production was rainfed, supplemented with furrow or flood irrigation. Now most figs are produced with drip irrigation modified for fresh or dried fig production. The specific irrigation method, system design and operation are a function of microclimate, terrain soil structure and chemistry, initial cost, availability and local experience. The irrigation system infrastructure: hydrants, pipes, solenoid valves and lines, are installed before planting. The drippers are usually placed closed to the young tree after planting.

Low humidity and a dry orchard floor are important for dried fig production. Typically, as the tree grows the irrigation line is moved further from the canopy to avoid both increasing humidity within the canopy and a wet soil surface under the tree where the ripening figs will fall. Irrigation intervals are longer and the volume is greater than for fresh fig production to encourage deep root growth and a dry soil surface.

For fresh fig production, lower humidity within the tree canopy is also important for disease and pest control but a dry orchard floor is not necessary. The typical drip irrigation system starts with a single dripper close to the tree for the first 3 years, increasing to two drippers/tree in the fourth year and moved 20–30 cm further from the trunk each successive year to produce a larger wetted soil volume. Irrigation of fresh figs is generally based on orchard evapotranspiration (ETo) measurements and is generally more frequent and smaller volume than dried fig irrigation. Depending upon tree size, spacing, climate and crop, fresh fig production can require 2500 $m^3$ water/ha/year from mid-May to late September (Pereira *et al.*, 2015).

## 8.7 Planting and Young Tree Care

Traditionally fig plantations in the Mediterranean basin were established by partially burying variable lengths of curved shoots in the ground with the final five or six buds remaining above the soil. This traditional method has been replaced by planting either bare root trees raised in a nursery bed or container-grown trees, both propagated from hard or softwood cuttings. For both methods, the hardwood cuttings are collected in the fall or winter, 12–16 months before the orchard planting. The cuttings are rooted then grown in beds or containers through a full growing season. At planting, the young trees should have a trunk diameter of 12–25 mm at 150 mm above the ground. Trees less than 10 mm or over 25 mm do not establish as well, growing more slowly. Producing young fig trees for orchard establishment in is covered in Chapter 6: *Propagation Techniques and Nursery Management*. Obtaining good quality healthy fig trees free of fig mosaic virus (FMV) is among the major problems in fig orchard establishment.

The time of planting depends on whether the trees are bare root or container trees. Bare-root plants are maintained in a soil trench, dormant until danger of damaging cold is past but planted within the winter dormant period. Container-grown trees can be planted in fall (autumn) so that the plants acclimatize to the final site, or in late spring after the danger of frost is past but before temperatures rise. Fig trees planted at the beginning of the dormant season often develop a root system before leaf out in the spring. This can be advantageous, but young trees are more susceptible to cold injury. In areas where freeze damage potential is high, planting should be delayed until just before dormancy is broken in late winter or early spring. With spring planting, there is less danger of young tree loss to frosts. However, planting should be as early in the spring as possible, after temperatures have begun to rise above 10°C at night.

If bare root, trees are removed from the trench, transported to the field and planted the same day. Care should be taken to avoid desiccation from wind during transport or planting. Before planting trees, marker stakes are inserted using a slack line or GPS to maintain the tree line. The container-grown rooted trees are delivered to the field, placed next to the stakes, and planted the same day. The holes can be dug with a shovel or mechanically with a tractor-mounted auger. With both, but particularly with an auger, the smooth sides of the holes make it difficult for the roots to explore the adjacent soil. If the soil is wet, using an auger may make the walls of the hole seal off, impeding root growth. If the soil is wet, it is better to delay planting or use a shovel and scratch the insides of the hole with a hoe or shovel. The planting hole should be just slightly deeper and wider than the root mass and potting soil. Typical minimum dimensions of the hole are 50 cm in diameter and 50 cm deep. If the hole is too deep, the young tree will settle with irrigation. Just before planting the fig roots can be trimmed. Fig roots on rooted cuttings are wide angled, 90° from the cutting, and brittle – breaking easily and sensitive to being exposed to air during planting. Any broken or damaged roots should be cut off. The roots should be trimmed slightly if they are curling around the pot soil mass to avoid curling in the hole. Trees should be set in the hole at the same depth they were in the pot. Once the root ball is in the hole, the sides should be backfilled with soil to maintain the root crown level, the tree firmly tamped down into the hole but topsoil should not be placed on top of the potting soil. This could impede irrigation water entering the root ball. If the trees are to be staked it should be done at planting with the stake inserted at the edge of the root ball, on the downwind side. If the site is windy use two stakes perpendicular to the wind. Fertilizer should not be put in the planting hole.

At planting the trees are 'headed', cut back with a sharply slanted cut at 60–75 cm. The tree should be oriented in the hole with this sloping cut facing east, away from the afternoon sun. This will encourage branching as it breaks dormancy, and reduces water loss by leaves until the roots are more developed. Any side shoots should be trimmed back to above the first basal bud. Irrigate the tree lightly to settle the soil and remove air pockets. Do not irrigate heavily within the first two months after planting as later rains could then flood the planting hole, saturating the roots. Also, water uptake by the roots is reduced for a short time after planting because the root systems may be damaged during planting. After planting, the trunks can be wrapped in white cardboard or painted with white latex paint to protect the tree from sunburn. A typical paint mixture is 2 L of white latex paint mixed with 1 kg of copper in 16 L of water. The paint should be at least 15 cm below the young treetop to allow the new shoots to grow unobstructed. Weeds compete for water and nutrients and should be controlled, especially near the base of the tree by hoeing and mulch.

After planting young fig trees, care in the first year consists primarily of monitoring the soil moisture in the area immediately around the base of the young tree for deficits and excess; both equally harmful to young trees. Light, monthly irrigations should be sufficient through late spring. By summer, regular irrigation should begin. The amount of irrigation will depend upon soil texture (sandy soils may require more frequent irrigation than loams or soils with clay). A typical irrigation schedule for the first five years is given in Table 8.1.

Young trees do not require much fertilization. Some sources recommend providing 40–100 g of nitrogen for young trees after the first year and until the trees reach maturity. Yearly suggested fertilization rates range from 100 to 200 g of nitrogen per tree (Vachamidis and Vemmos, 2010).

The amounts given in Table 8.2 can be divided by the number of trees per hectare and applied below the tree above the root zone and before irrigation. Do not let the root zone get very dry before delivering irrigation with fertilizer.

Any inexpensive granular nitrogen fertilizer is suitable. If the pH is high, UN32 is cheap and soluble. Calcium ammonium nitrate, CAN, a mixture of 20–30% $CaCO_3$ and 70–80% ammonium nitrate containing 27% N and 20% of ground limestone is also a good fertilizer as fig fruits have a high percentage of Ca, and CAN fertilizer supports good fruit quality.

**Table 8.1.** A young tree irrigation schedule.

| Year | Applied m³/ha | L/ha/year |
|---|---|---|
| 1 | 0.04 | 411,160 |
| 2 | 0.09 | 935,111 |
| 3 | 0.16 | 1,644,642 |
| 4 | 0.18 | 1,850,222 |
| 5 + | 0.24 | 2,466,963 |

**Table 8.2.** A young tree fertilization schedule.

| Year | Applied N kg/ha/year |
|---|---|
| 1 | 11 |
| 2 | 22 |
| 3 | 34 |
| 4 | 45 |
| 5+ | 56 |

As with all fruit trees, the training method is dictated by the harvest method. Most fig trees, for fresh or dried production, if trained, are trained to a vase shape. The tree is headed with a slanting cut at 60–75 cm and a framework of branches at approximately 45° angles spaced along the upper 45–60 cm of the trunk are selected. This spacing and angle avoids narrow crotches or flat scaffolds that may break with heavy fruit loads in the future. The trunk sprouts should be eliminated to avoid competition with the main trunk.

Before they produce fruit, fig trees have few pests. The leaf miner, *Simaethis pariana* and *Choreutis nemorana*, the fig-tree skeletonizer moth or fig leaf roller, a species of moth of the family Choreutidae, are the major pests of young fig trees. For more information about fig tree insect pests visit Chapter 14: *Arthropod Pests of Fig and Their Management*.

## High-density precision orchard planting

Precision agriculture for mechanically establishing fig orchards is starting in Spain. Automated tractors guided by a GPS system with Real Time Kinematic (RTK) correction are achieving a horizontal precision of ±3cm and a yield of 5–10 ha in a 6 hour shift, depending on the planting density. The tractor has an automatic planter attached that opens a 40 cm deep trench, inserts the fig tree and stake at the specified coordinate and closes the trench (Fig. 8.10). The unit requires three workers, two supply and one who directs the tractor from the end of the row.

The unit can carry 500–1000 trees, 35–45 cm tall with a root ball volume of at least 1 L, the size required to avoid desiccation (Fig. 8.11). This system is now being used to plant intensive or super-intensive orchards 3 ha or larger. It is not economically viable for more widely spaced or smaller orchards. Nurseries are beginning to offer this planting system along with tree and super-intensive orchard design to the field setting of the plants.

## Young tree training

Young tree training is done before a fruit tree reaches maturity and bears fruit. The purpose is to build a strong tree that occupies maximum orchard floor space and will reliably produce maximum annual crops of good quality figs. Training delays and decreases early crop but is necessary for efficient orchard operations in maturity; particularly harvesting. Producing fresh and dried figs requires different training methods as fresh figs are picked from the tree while dried figs are allowed to drop to the ground, nets or tarps below the tree.

Fig trees for both dried and fresh production are trained to an open vase shape. The major difference is trees trained for dried fig production are trained to provide an open space under the trees to facilitate drying and collection of dropped fruit. During the first dormant season the initial topping cut for dried fig production trees is at ~1 m from the ground, versus the 60 cm topping for a fresh fig tree. This cut eliminates the apical dominance and generates the side shoots that will become the primary scaffolds (Fig 8.12, left). The following winter 2–5 primary scaffolds are selected, separated by at least 10 cm vertically on the trunk and distributed around the horizontal trunk axis. These primary scaffolds should be topped at 45–60 cm, the length increasing annually. Select buds that will not only generate a vase shape that fills the tree's allocated space but also shades the upper surface of the interior scaffolds as fig wood sunburns easily (Fig 8.12, right). Remove the first fruits within the first two years to avoid infesting the wood with botrytis (*Botrytis cinerea*). The objective is to produce a vase shaped tree that shades 80% of the orchard floor at 12 noon, is easily harvestable without a ladder, 2.5 m tall, and produces regular annual crops (Fig. 8.13).

## Mature tree pruning

Mature tree pruning is done annually during the dormant season. The objective is to maintain the tree's height, to keep it within its allotted orchard floor space, and annually produce regular, good quality crops. To renew bearing wood within a harvestable height, 2–2.5 m, generally requires pruning the current season's growth to the best developed 3–4 basal buds with heading cuts. Flat, weak, low, crossing and unproductive branches should be removed with thinning cuts

**Fig. 8.10.** RTK-GPS planting system. (Figure used with permission from DALIA nurseries.)

(Fig. 8.14). Trunk suckers are removed during winter pruning and during the summer.

Young 'Adriatic' trees trained for dried fig production require little pruning; only removal of low flat branches and thinning the interior branches. 'Adriatic' tree branches droop if heavily cropped, exposing the upper branches, resulting in sunburn.

Like the 'Adriatic', the 'Mission' tree branches freely producing many laterals. Selection and spacing of the scaffolds for 'Mission' trees for fresh production is important because 'Mission' trees form acute angles that split with heavy crops. Central wire bracing is sometimes used to prevent breakage. Heavy pruning of 'Mission' trees should be avoided as it proliferates unproductive shoots.

**Fig. 8.11.** Young fig tree with root ball for precision planting. (Figure used with permission from DALIA nurseries.)

**Fig. 8.12.** Formation pruning of fig trees for fresh consumption. (Photo courtesy of M. López-Corrales.)

**Fig. 8.13.** Young trees at the end of tree training entering bearing. (Figure courtesy of M. López-Corrales.)

**Fig. 8.14.** Well-trained mature bearing fresh fig tree before and after dormant pruning. (Figure courtesy of M. López-Corrales.)

### Spring tipping to advance maturity

In Spain, where summer drought and winter frost do not occur, removing the terminal buds of the fruiting shoots in late spring advances fruit growth and maturation of figs, particularly the brebas, and also stimulates new fruiting buds adjacent to the brebas. De La and Tinie (1693) stated that the 'removal of terminal buds in early spring makes the fig shoot out the sooner and consequently to ripen fruits sooner, since the first that come out of the tree are always the first ripe upon that tree.' Columella and Pliny also described this method. The latter stated that if the tip of each branch was removed just as it began to put forth leaves, the strength and productivity of the tree were greatly increased. The practice is common in France, especially at Argenteuil, and is described and illustrated by Auguste and Gustave Rivière (1928).

Argenteuil pruning also advances breba maturity. However, it decreases production and therefore is only performed on trees with heavy breba crops. The terminal bud is removed early, followed by removal of all but the lowest axillary buds. Most of the potential fruit is lost on these branches. The two remaining buds produce breba buds. This is done on 30% of the branches. After brebas harvest those shoots are pruned above the replacement shoots at its base.

### Pruning caprifig trees

Caprifig trees are trained at early ages and maintained low to facilitate collecting. The only training is the removal of suckers and interfering branches. When mature the mamme crop is valuable so the trees are not pruned heavily in winter. If necessary, thinning is done midsummer after harvesting the profichi crop. In Spain, non-caprifying varieties are grown with San Pedro types such as 'Nazaret' or 'Tiberio'. Therefore, the establishment of caprifigs in commercial plantations is not necessary.

### Rejuvenation pruning

Rejuvenation pruning is severe pruning done to neglected, damaged, poorly yielding or very large trees that can no longer be efficiently farmed. The tree is topped about 0.5 m from the ground during the dormant season (Fig. 8.15). Pruning compound with fungicide is often applied. New scaffolds will be selected from the shoots on the trunk and trained for 3–4 years. Rejuvenation pruning takes advantage of the existing root system's ability to produce a new tree rapidly.

### Girdling

Girdling is done to inhibit phloem conduction of non-structural carbohydrates to the roots, making them available for fruit growth. A discontinuous circular, Z-shaped cut in the bark, down to the cambium, of fruit bearing branches is done before final fruit ripening. Girdling is done primarily in fresh fig production to increase fruit size but is not common. Girdling has also been

**Fig. 8.15.** Initial severe topping cut to rejuvenate (left) and a rejuvenated tree three years later (right). (Figures courtesy of M. López-Corrales.)

done to dwarf trees (Gawankar *et al.*, 2019; Hossain and Boyce, 2009; Gawankar *et al.*, 2019). Mano *et al.* (2011b) reported that girdling ameliorated freeze injury during the dormant season and at bud break by enhancing the nutrient reserves in the dormant branches and delaying bud break.

## 8.8 Orchard Floor Management

### Mulching

Reimer (1910) suggested mulching under fig trees (Fig. 8.16). Experiments with fresh 'Lemon' and 'Celeste' fig production in a humid climate demonstrated a straw mulch produced 95% more fruit and 50% more foliage versus a clean orchard for better cold resistance (Cooley, 1930; Anderson, 1931). However, mulching is not feasible for dried fig production with naturally dropped fruit.

### Tillage

Fig orchard floors for dried fig production are generally maintained weed free with smooth surfaces to facilitate fruit drying and collection. Orchard floors for fresh fig production can have a mowed cover crop (Starnes, 1907) or be maintained as a clean orchard floor with shallow, 20–40 cm deep tillage to prevent cutting shallow roots (Stansel and Wyche, 1932). In rainfed orchards, the orchards are tilled after harvest to better capture winter rain, before fruit set to eliminate weeds and incorporate fertilizer, and summer to break up the surface crust and decrease evaporation. In modern intensive and super-intensive irrigated plantations, both for fresh consumption and for drying, no tillage is generally carried out.

### Irrigation

Dried and fresh figs can be grown in rainfed or irrigated orchards and greenhouses. In all situations, for both fresh and dried fig production irrigation is the production practice with the greatest effect on tree growth, fruit quality and yield. Individual fig orchard water requirements and irrigation schedules can be developed by estimating the water requirements based on historical records or by monitoring water use with meteorological tools and calculating replacement volumes: budget irrigation. Fig can be irrigated using all irrigation methods: rainfed, water catchment, flood, micro sprinkler and drip. Irrigation efficiency can be tested by monitoring the soil water status by direct hand examination, using soil moisture tension or conductivity technology tree water status methods such as stem water potentials. However, the latter, tree water status measurements, are rarely used, perhaps because the milky sap exudate from the leaf petiole would interfere with stem water potential measurement. How to irrigate a mature orchard is dependent upon microclimate, soil and water quality and quantity, cultivar, fresh or dried fig production and available irrigation technology. Verifying irrigation efficiency is dependent upon available technology to monitor the climate, and the soil and tree water status. All this is covered completely in Chapter 11: *Irrigation and Water Requirements*.

**Fig. 8.16.** Plastic in-row mulch under for fresh fig production. (Left figure courtesy of M. López-Corrales and right figure used with permission from Rosa Hernandorena www.hernandorena.com/en/)

## 8.9 Fertilization

As stated in Chapter 10: *Soil and Nutritional Requirements* the fertilization and soil amendment practices vary by fig production area and there is not enough research to suggest generic fertilization guidelines, aside from replacing nutrients exported during harvest. However, nitrogen is the most widely applied nutrient and traditionally the only recommended nutrient. In areas that have been under long-term production, have low CECs, or are low in potassium, additional potassium is likely to increase yield and fruit quality. Fertilization practices for other orchard crops recommend replacing what was removed with yield and taking leaf tissue concentrations, uptake timing, and tree health into account: nutrients should be available at the time of greatest uptake, and trees that are not actively growing or not in good health or have low yield will require less, if any fertilization. However, aside from Turkey, there are no leaf tissue standards developed for most nutrients, and it is unknown when the rate of uptake for nitrogen and potassium are highest, though nutrient demand during syconia development should be higher than the vegetative growth periods or after fruit maturation.

Nitrogen applications should occur in-season and start when leaves are large enough to begin transpiration, which is when deciduous trees typically begin taking up nitrogen. Previous publications have suggested 18–36 kg of nitrogen per hectare per year for 'Calimyrna' ('Sarilop') figs (Obenauf, 1978), skipping a year if the soil is especially fertile. It is not recommended to provide nitrogen to figs planted in their first year. Other sources recommend providing 40–100 g of nitrogen for young trees after the first year until they reach maturity. Afterward yearly fertilization rates range from 100 to 200 g of nitrogen per tree (Vachamidis and Vemmos, 2010). We speculate nitrogen applications should end around August or September, or at the latest by the last harvest of the season due to research on other crops, which has shown that the majority of nitrogen is taken up during fruit development (Rosecrance *et al.*, 1996; Muhammad *et al.*, 2015).

## 8.10 Caprification Management

Dried fig production requires caprification, also known as pollination. Caprification is the process of transferring caprifig fruit pollen to the edible fig syconia by the fig wasp, *Blastophaga psenes* L. Edible fig cultivars are separated into three horticultural types based on the caprification requirement of the breba and main crops: Smyrna, San Pedro and Common. Caprifig trees produce three annual syconia: mamme, mammoni, and the profichi. Profichi figs contain the fig wasp (Fig. 8.17) and active pollen grains; they are 'insectiferous'. Those without wasps are 'polleniferous' and are useless for caprification. Although only profichi directly pollinate edible figs, the three are required for the fig wasp's life cycle (Obenauf *et al.*, 1978). For more information about caprification read Chapter 9: *Fig Wasps and Pollination.*

Smyrna-type figs do not produce a breba crop but require caprification for main crop production. San Pedro-type figs generate parthenocarpic breba crops. The second, main, crop requires caprification. Common-type fig cultivars may or may not produce a breba crop, which persists without caprification. Most of the common fig cultivars produce moderate to large main crops without caprification. Thus, common-type figs are often referred to as parthenocarpic.

Caprification is done from May to July when the edible female fig fruits are the size of a hazelnut. The caprification process should be repeated every few days as the female syconia gradually become receptive from the base to apex of the shoot over three to four weeks (Rahemi and

**Fig. 8.17.** Profichi fruit of 'Pouzdonbali' caprifig with *Blastophaga* pollinator. (Figure courtesy of M. Jafari.)

Jafari, 2008). Because the wasps emerge from the caprifigs at sunrise, producers collect them early in the morning before they emerge. The number of caprifigs, typically 25–100, frequency of placement in the tree and length of time in the tree, all depend on the female cultivar, tree age and size, and weather. The caprifigs are placed in small nets (Fig. 8.18), slotted paper bags (Fig. 8.19), buckets with bottom holes (Fig. 8.20) or strung on strings, hung in multiple shaded areas of the tree in the morning when pollination is most effective as frequently as twice a week or as infrequently as once a season (Jafari, 1999; Stover *et al.*, 2007). Distribution within the tree improves caprification efficiency. Chemical pest and weed control, hanging pheromone traps, and tillage should not be done during caprification. Temperatures above 38°C reduce the *Blastophaga* wasps life span to less than an hour (Jevanandam and Corlett, 2013); this significantly decreases caprification efficiency. The caprifigs are collected and destroyed after caprification.

Non-caprified fruits do not develop, turn yellowish in color and drop (Ferguson *et al.*, 1990).

## 8.11 Harvesting

### Expansion of fruit harvesting season in tropical and subtropical climate (case of Mexico)

Chapter 15: *Fig Fruit Maturity, Harvest Methods and Technologies* covers all aspects of harvesting fresh and dried fig fruits. As Chapter 15 details, the major differences in fig fruit harvesting are dictated by the same factor that dictates differences in most fig fruit production, mainly whether the final product is fresh or dried figs. However, countries like Mexico in tropical and subtropical climates have an advantage, as can be seen in Table 8.3. Mexico has a wide range of harvest dates in different regions. However, all the states want to extend their harvest to have figs in the off season, when there's little fruit in the market.

### Techniques to extend the harvest period

Figs behave like evergreens in tropical and subtropical climates. This allows harvesting

**Fig. 8.18.** Putting the caprifigs into the nets for caprification. (Figure used with permission from Birgül.)

**Fig. 8.19.** Caprifigs placed in slotted paper bags in 'Calimyrna' fig orchards. (Author's own photo.)

**Fig. 8.20.** Different methods of caprifig usage in caprification. (Figure courtesy of A. Yavari.)

throughout the year. The harvest time of figs can be extended by pruning before leaf senescence. To extend the harvest season, pruning must be staggered. There are two ways to do this. The first is the usual method of producing fruits. The entire orchard is harvested and pruned simultaneously, forcing the vegetative and fruiting cycle of all the trees into the same timeline.

**Table 8.3.** Harvest months in the main fig producing states in Mexico.

| Harvest months of principal producer states of Mexico | | | | | | | | | | | |
|---|---|---|---|---|---|---|---|---|---|---|---|
| J | F | M | A | M | J | J | A | S | O | N | D |
| | | Baja California, Sonora, Chihuahua | | | | | | | | | |
| | | | | | Durango, Coahuila | | | | | | |
| | | | Michoacán, Jalisco, Querétaro, Aguascalientes, Zacatecas | | | | | | | | |
| Morelos | | | | | | | | | | Morelos | |
| | | | | | CDMX, Hidalgo, Puebla, Veracruz | | | | | | |

Or, the orchard can be divided into sections in which the pruning is done at different times so that the fruit is produced, ripened and harvested at different times. To facilitate management, the orchard can be divided into quadrants, and pruning will be at 4–6 week intervals (Fig. 8.21).

Another technique for extending the harvest periods is the so-called 'stem continuity', method in which the apical bud is eliminated to force productive new growth. The technique also produces high quality fruits.

The method consists of removing the apical bud and disbudding from shoots with growing fruits (Fig. 8.22A and B). This favors the accumulation of photoassimilates in that section of the shoot, producing higher quality fruit. A new vigorous vegetative shoot will form lower on the disbudded stem (Fig. 8.22C). This new stem will also produce fruit, and soon the two stems will be equal in size (Fig. 8.22D). The section of the stem that has already produced fruit must be left free of leaves to favor air flow, lower humidity and deter insects and diseases (Fig. 8.22E, red circle), while the new stem will continue to produce fruit (Fig. 8.22E, yellow circle).

## 8.12 Fig Cultivation in Japan

### Cultivation history and commercial cultivars

Figs arrived in Japan in the 17th century (Kaibara, 1709) and were a dooryard plant until commercial cultivation began in the late 19th century in western Japan. Approximately 12,000 t are now produced on 950 ha. All commercial fig varieties are Common-type as there are no wasps. The second fresh crop is harvested from summer to fall. The dominant variety, 'Masui Dauphine' (c.v. San Piero sensu; Condit, 1955), makes up 70% of Japan's production (Fig. 8.23). The other main fig varieties are 'Houraishi (Horaishi)' (c.v. Horaigaki sensu; Condit, 1955), the oldest introduced, and currently produced in western Japan;

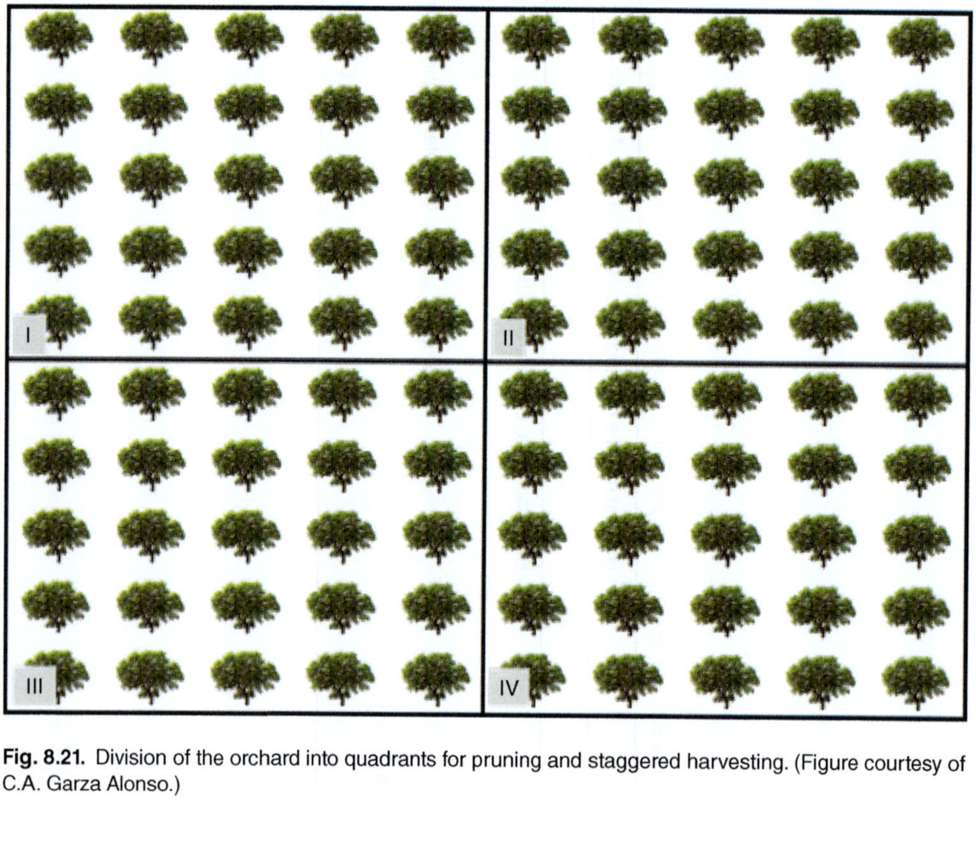


**Fig. 8.21.** Division of the orchard into quadrants for pruning and staggered harvesting. (Figure courtesy of C.A. Garza Alonso.)

**Fig. 8.22.** Stem continuity technique to lengthen the fig harvest. (A, B) removal of the apical bud (pinch), (C) appearance of the new vegetative shoot, (D) growth of the new productive stem, (E) new stem with developing figs (yellow circle) and stem section with previously harvested figs (red circle). (Figure courtesy of C.A. Garza Alonso.)

**Fig. 8.23.** Fruits of fig variety 'Masui Dauphine'. (Figure courtesy of A. Hosomi.)

'Brunswick', which has a high resistance to cold and is cultivated mainly in northern Japan; and 'Toyomitsuhime', a local hybrid variety produced in Fukuoka Prefecture (Nogata and Awamura, 2005). These cultivars were selected for their large fruit size and high fruit yield. Recent interest has focused on better tasting varieties, even if smaller fruited. (For more information, read Chapter 5: *World Fig Cultivars*).

## Propagation

In Japan, own-rooted stocks are used. However, soil sickness (probably caused by unknown microorganisms and root-knot nematodes) and Ceratocystis canker are serious soilborne threats. Therefore, fig rootstocks are used. For example, vigorous varieties such as 'Zidi' are used for 'Masui Dauphine' trees to compensate for the dwarfing caused by soil sickness (Hosomi *et al.*, 2002). The 'Negronne' variety (Hosomi *et al.*, 2012), a hybrid 'Kibaru' (Himeno *et al.*, 2015) and an interspecific hybrid 'Reikoudai-ichigou' (Yakushiji *et al.*, 2019) are used for their resistance to Ceratocystis canker; hard wood grafting has been used to raise nursery stocks. Recently, cutting grafting (Hosomi, 2006; Ushijima *et al.*, 2013) or soft wood grafting (Hosomi *et al.*, 2015b) have been recommended to promote early growth of nursery stocks (Fig. 8.24). (For more information, read Chapter 7: *Environmental Requirements and Site Selection*).

**Fig. 8.24.** Propagation of the nursery stocks by cutting grafting. (Figure courtesy of A. Hosomi.)

## Establishing the field

Flat land is preferred for fig cultivation everywhere. However, Japan's flat land is poorly drained. Therefore, drainage mitigation is the most important preparation measure for fig orchard establishment in Japan. Land adjoining rivers with relatively good drainage is often chosen for fig orchards. Deep plowing and underdrains are common drainage measures but are expensive and inconvenient if the fields are to be reused for rice production in the future (Kabumoto, 1985). Therefore, open channeling and riding are done prior to orchard planting. Significant composting on former rice producing ground is done preplant (Fig. 8.25).

## Planting

Nursery stocks are planted in early spring to avoid the severe winter cold season but sometimes in the fall. They are sometimes planted on a mound to enhance drainage. Planting density is dictated by variety vigor, regardless of the training method. 'Masui Dauphine', intermediate in tree vigor, are planted at a high density, 100 trees per 1000 $m^2$ (Mano, 2015) (Fig. 8.26). Extremely dense planting (e.g. 692 trees per 1000 $m^2$), is recommended for early production (Mano and Hamada, 2005; Mano *et al.*, 2011a; Kadowaki *et al.*, 2012). For 'Toyomitsuhime', intermediate in tree vigor, planting densities similar to 'Masui Dauphine' are recommended (Nogata and Awamura, 2005). For 'Houraishi', high in tree vigor, a very low planting density, 10–20 trees per 1000 $m^2$, is recommended (Awamura, 2015).

## Training and pruning

Tree training and pruning is specific to each variety. 'Masui Dauphine' trees are typically trained either by base-form training or straight-line training with short pruning (Fig. 8.27). Base-form training involves various branched structures and the distribution of the shoots in a planar space. After harvesting, the bearing shoots are shortened to a few nodes to act as the mother shoots for the next season, and similar pruning occurs annually. Like cordon spur pruning of grapes, straight-line training involves bilateral horizontal limbs (primary scaffold limbs). It greatly improves the cultivation efficiency because the shoots elongate linearly in a regular fashion (Kabumoto *et al.*, 1985). However, there is less fruit color development on the lower shoot and a higher risk of limb freezing injury (Horimoto *et al.*, 1994). Recent training and pruning methods have been developed. With 'high limb

**Fig. 8.25.** Amending a former ride paddy with wood compost before planting a fig orchard. (Figure courtesy of A. Hosomi.)

**Fig. 8.26.** 'Masui Dauphine' nursery stocks are planted approximately 100 trees per 1000 m². (Figure courtesy of A. Hosomi.)

**Fig. 8.27.** 'Masui Dauphine' fig trees with base-form (left) and straight-line (right) training, both immediately after short pruning. (Figure courtesy of A. Hosomi.)

straight-line training' the shoots are elongated horizontally or downward from the limbs at 180 cm, versus conventional straight-line training shoots trained upward from 40 cm (Fig. 8.28). This prevents freezing injury by reducing the temperature differences on limb surfaces between day and night (Mano *et al.*, 2012b), and promotes fruit coloring (Hosomi *et al.*, 2013; Matsuura *et al.*, 2014). 'Overlap limb training' has mono-lateral limbs, the basal parts of which overlap with the apical parts of the neighboring tree's limbs (Souda *et al.*, 2016). The basal parts of the limbs that are vulnerable to frost damage are protected by the overlapping. 'Renewal long pruning' combines long pruning and continuous renewal pruning and produces a few long dormant shoots that serve as both mother shoots and as replaceable limbs (Fig. 8.29). This prevents excess shoot vigor and induces early maturation and better fruit size (Hosomi *et al.*, 2015a). It also revives limbs damaged by freezing injury or pests.

In 'Masui Dauphine' trees, regardless of the training and pruning method, the shoots are thinned to restrict the numbers to 2500–2800 shoots per 1000 m$^2$ of bearing shoots. The combination of a relatively denser planting and less shoots induces vigorous shoot growth. Although excessive shoot growth decreases fruit color development and produces more cracked fruit (Senda *et al.*, 1996), the increased shoot vigor produces higher yields (Hosomi, 2020). 'Houraishi' fig trees are typically trained to an open center with thinning cuts. The trees are often trained on a horizontal trellis to support and spread the canopy (Fig. 8.30). Compared to the 'Masui Dauphine', this training produces more, 5000 per 1000 m$^2$, and shorter, 0.2–0.4 m, shoots and a larger, 50–100 m$^2$, per tree, canopy in the 'Houraishi' variety (Awamura *et al.*, 1993; Awamura, 2015). More shorter shoots producing larger early season fruit (Hosomi, 2020), when market prices are higher, is the advantage.

## Irrigation and soil management

Although Japan has relatively high rainfall, summer irrigation is necessary. In mid-July, when the rainy season ends, leaves and young fruits may wither when roots cannot meet the transpiration demand with sudden solar radiation. In-row irrigation systems or flooding the tree row middles are both common (Fig. 8.31).

**Fig. 8.28.** Normal growth of 'Masui Dauphine' fig trees with high limb straight-line training (upper photo), and growth inhibited by frost damage on adjoining trees with conventional training (lower photo). (Figure courtesy of A. Hosomi.)

Common fertilization programs include 13–24 kg, 10–20 kg, and 14–24 kg per 1000 $m^2$ of N, $P_2O_5$, and $K_2O$, respectively, applied three times annually through soil (Kabumoto, 1985). However, with Japan's higher rainfall the soil becomes acidic, and phosphoric acid becomes insoluble (Hirai *et al.*, 1961). Because figs prefer relatively alkaline soils, neutralizing this acidity is done with winter lime applications. Large amounts of organic matter, such as bark compost, are also incorporated in the winter to improve the soil's physical properties, such as water and nutrient holding capacities.

## Pest control

Fig has 32 diseases (PSJ, 2020) and 57 insect pests (JSAEZ, 2006). Ceratocystis canker, caused by *Ceratocystis ficicola* (Kajitani and Masuya, 2011), is among the most serious diseases (Figueiredo and Pinheiro, 1969; Kato *et al.*, 1982). The disease invades the roots and trunk, eventually killing the tree (Fig. 8.32). The major fig varieties, 'Masui Dauphine' and 'Houraishi' are susceptible. Fungicides are ineffective. The most common but not complete control is the disease-resistant rootstock varieties described earlier (Hosomi *et al.*, 2012; Himeno *et al.*, 2015). Recently, a true resistant rootstock was bred by interspecific hybridization (Yakushiji *et al.*, 2012; Yakushiji *et al.*, 2019).

Soil sickness (SS) is the rise of negative conditions for plant vegetative and reproductive performances induced into the soil. Three main hypotheses have been proposed to explain SS: (i) soil nutrient depletion or imbalance; (ii) build-up of soilborne pathogen and parasite populations, coupled with a shift in soil microbial community composition; and (iii) release of phytotoxic and autotoxic compounds during decomposition of crop residues. Evidence from both agroecosystems and natural plant communities undoubtedly ruled out the nutrient

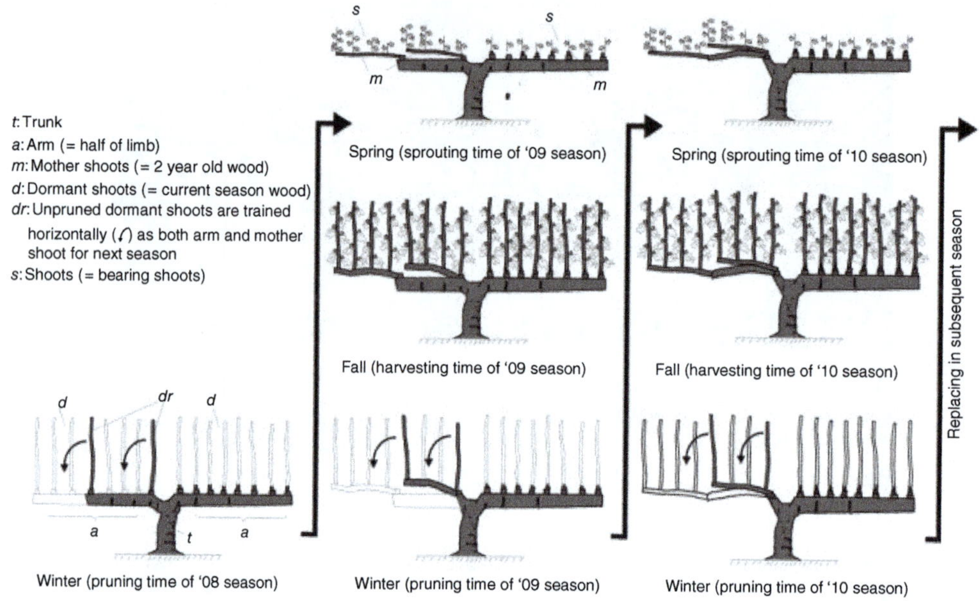


**Fig. 8.29.** An arm (horizontal shoot, left limb in figure) received 'renewal long pruning'. Two dormant shoots (*dr*) were retained and other woody parts (blank parts in the figure) were removed. These two shoots were trained horizontally and act as both arm and mother shoot for the next season. Similar pruning is done the following season on the same arm so that the wood age of the limb was reset completely. The opposite arm (right limb in figure) acted as usual short pruning (Hosomi *et al.*, 2015a).

**Fig. 8.30.** 'Houraish' trees with thinned-out pruned shoots spread on the horizontal trellis. (Figure courtesy of A. Hosomi.)

**Fig. 8.31.** Flood irrigation in the row middles. (Figure courtesy of A. Hosomi.)

**Fig. 8.32.** 'Masui Dauphine' tree dead from Ceratocystis canker. (Figure courtesy of A. Hosomi.)

deficiency as a primary causal factor (Cesarano *et al.*, 2017), caused by root-knot nematodes (Taylor, 1943; Sato and Shichijo, 1953) and other unknown microorganisms (Hosomi and Uchiyama, 1998) that seriously inhibits fig tree growth (Fig. 8.33) (Hirai and Hirano, 1949; Hosomi, 2011). There is no definitive control to soil sickness, other than rotating production

**Fig. 8.33.** 'Masui Dauphine' trees dwarfed by soil sickness (front) and healthy trees (back). (Figure courtesy of A. Hosomi.)

with other crops, often rice. However, vigorous rootstocks as discussed earlier (Hosomi *et al.*, 2002), and soil improvement such as soil solarization before planting in order to eliminate soilborne plant pathogen, increasing the amounts of soil organic matter (SOM) can mitigate the dwarfing of 'Masui Dauphine' fig trees.

Rust caused by *Phakopsora nishidana* (Ito and Honnma, 1938) is also an important disease. The symptoms occur on the leaves, sometimes on fruits, and induce defoliation. The fungicides hexaconazole, Azoxystrobin, triflumizole or cupric oxysulfateare, applied through August and September can control rust (OPPPO, 2020).

Rhizopus rot, caused by *Rhizopus stolonifer* rapidly degrades the fruit just before harvest (Edgerton, 1911). It often manifests after shipping (Nojima, 1934), harming sales (Fig. 8.34). The pathogen is insect transmitted, primarily by Drosophila (Hosomi and Kusakari, 1995) during the fall rainy season. Prevention includes sanitation, removing mature fruit and rain protection. The fungicide thiophanate-methyl is an effective chemical control (OPPPO, 2020).

**Fig. 8.34.** 'Masui Dauphine' fruit infested by Rhizopus rot. (Figure courtesy of A. Hosomi.)

Thrips such as *Thrips tabaci* and *Frankliniella sp.* (Hansen, 1929) are the most serious insect problems for Japanese fig production. The thrips enter the fruit cavity through the ostiole, precipitating pulp discoloration without external symptoms (Fig. 8.35). The inability to identify the damage by external appearance makes this problem serious. Various insecticides (spinosad, acetamiprid, thiamethoxam, acephate, chlorfenapyr, and permethrin, among others) can control thrips (OPPPO, 2020). However, insecticide resistance is increasing. A labor-intensive method to seal the eyes, preventing insect entrance, has been developed (Ichikawa and Naganawa, 2004). Other control methods, laying reflective sheets on the ground under the trees (Morishita, 2002) and covering the field perimeter with red netting (Shirotsuka *et al.*, 2018), are expensive.

Longicorn beetles (*Apriona japonica* and *Psacothea hilaris*) feed on the shoots and leaves and produce larvae that damage branches and trunks with burrowing. Acetamiprid and clothianidin during the adult flight and injecting permethrin into the larval feeder pores (OPPPO, 2020) are effective controls. Non-woven fabric coated with the microbial pesticide, *Beauveria brongniartii* hung in the trees is another control. Conidia of this entomopathogenic fungus prevented the proliferation of beetles, in addition to killing adult insects (Higuchi *et al.* 1997).

**Fig. 8.35.** 'Masui Dauphine' fruit infested by thrips. (Figure courtesy of A. Hosomi.)

Ficus mottle mite (*Aceria ficus*) occurs from spring to fall, peaking in summer, causing mosaic-like symptoms on the leaves, and sometimes on the fruit (Nakasone and Kusakari, 1991). It can be controlled by spraying the fenpyroximate, pyridaben and tebufenpyrad in July when the adults are present (OPPPO, 2020). *A. ficus*, a mite, overwinters in dormant buds on the shoot tips (Shibao and Tanaka, 1997). Short pruning is effective in reducing the mite density.

Vertebrate pests damage figs primarily by feeding on the fruit. Birds, (crow, gray and starling), and raccoon dog, common raccoon, masked palm civet and wild boar all eat figs. Nets, customized for each animal type, are the easiest most effective control method. Sounds and lights are less effective, but electric fences, which are more assuredly effective, are becoming widespread.

## Weed control

Some farmers use herbicides, glyphosate and glufosinate three times annually for weed control. However, many farmers worry that herbicides may damage the shallow roots produced by high ground water. Therefore mulching the berm with rice straw has long been used to control weeds. Recently, it has become challenging to obtain rice straw, therefore farmers are using geotextile plastic sheets (Fig. 8.36), to control weeds as well as thrips and improve fruit color development with reflected light (Mano, 2004).

## Meteorological damage

Freezing injury is among the most serious threats to Japanese fig production. The major cultivar 'Masui Dauphine' is particularly susceptible to cold stress (Mano *et al.*, 2010). It can withstand −10°C when dormant but will be damaged at −4°C in early spring (Horimoto *et al.*, 1994). Spring frost damage destroys buds and causes limb necrosis. Wrapping the branches with rice straw in early spring provides some protection. As mentioned above, rice straw has been difficult to obtain in recent years, and

**Fig. 8.36.** White sheet mulch controls weeds, thrips and improves fruit color with reflected light. (Figure courtesy of A. Hosomi.)

therefore non-woven fabric or aluminum sheets are increasingly used instead (Fig. 8.37). Training and pruning, as mentioned above, is also a method for preventing frost damage. Mano *et al.*, (2011b) reported branch girdling promoted nutrition accumulation and increased freezing tolerance.

Mild winds causing leaves to scratch fruits are more serious problems than rare tree felling in strong winds. Enclosing orchards with windproof nets and tying the shoots with wire in straight-line training prevents leaves from touching fruits.

## Protected fig cultivation in Japan

Most protected fig cultivation techniques are based on open field cultivation. Fig cultivation in Japan is usually a family enterprise with family labor, generally a couple. Harvesting and shipping is 60% of the labor. Therefore, a couple cannot manage more than 20–30 acres. Greenhouse cultivation is essential for expanding the production capacity. Greenhouse cultivation also has the advantages of improving profitability through early shipment, reducing harvest losses due to rainfall, and producing better fruit quality. Fig greenhouse cultivation began in 1947–1948 with the development of a straight line training shape 'Ichimonji' for the 'Masui Dauphine' cultivar. Greenhouse cultivation of 'Houraishi' began in 1985. By 1999, the total area of greenhouse fig cultivation was 4942 acre (heated 1479 acre, unheated 3463 acre) (Table 8.4).

### *Heated greenhouse*

Round roof steel pipe houses are the most common greenhouses for protected fig cultivation (Fig. 8.38). The height of a 'Masui Dauphine' tree with straight line training is 1.8 m. Therefore, the greenhouse eaves height should be 1.8–2.0 m high and the building 3.0–3.6 m and

**Fig. 8.37.** Early spring 'Masui Dauphine' trees wrapped in nonwoven fabric to prevent frost damage. (Figure courtesy of A. Hosomi.)

approximately 4–6 m wide. Heat is generally supplied by an oil-fired boiler. However, while capital investment for oil fired boilers is low, operating costs are high. Therefore, the oil-fired boiler is often supplemented with a heat pump to reduce the total cost. Farmers can use the heat pumps, which have a low operating cost, first and supplement with oil-fired boilers.

### *Unheated greenhouse*

An open-center tree shape is approximately 1.8 m tall, therefore the eaves should be 2 m high, the ridge 3.6 m, and the end walls approximately 7 m wide. Because figs need high night temperatures, vertical lining curtains are used as insulation to increase efficiency. The greenhouse should be sufficiently tall to prevent summer high-temperature damage. To maintain daytime temperatures below 30°C, ventilation fans are installed at each end. The fans alone cannot totally regulate high temperatures, so side and end panels can be opened and closed. Fully open houses have also been introduced to help plants cope with higher temperatures (Fig. 8.39).

### *Rain-shelter house: pipe frame structure*

An unheated greenhouse made of arched pipes set at 1.5 m intervals with the upper part covered with plastic film, as shown in Fig. 8.40. The objective is to prevent rainfall from degrading fruit quality. Because greenhouse cultivation produces less-colored fruit than field production, light reflecting sheets are often spread on the greenhouse floor and the roof film is rolled up in sunny weather.

## Site selection and orientation for greenhouses and tunnels

All greenhouses or tunnels should be constructed on flat ground, in full sun, not in windy or storm prone areas and not on easily saturated soils. The structure's long axis should be north–south with the tree rows parallel for maximum sun exposure.

**Table 8.4.** The area of greenhouse cultivation by cultivar in Japan (unit: acre).

| Masui Dauphine | | | Summer Red | | | Houraishi | | | Toyomitsuhime | | |
|---|---|---|---|---|---|---|---|---|---|---|---|
| Heated | Non-heated | Total | Heated | Non-heated | Total | Heated | Non-heated | Total | Heated | Non-heated | Total |
| 1112 | 1345 | 2457 | 82 | 30 | 112 | 125 | 88 | 213 | 160 | 2000 | 2160 |

**Fig. 8.38.** Round roof type steel pipe house heated cultivation with the side vinyl covers opened in summer. (Figure used with permission from Syuichi Himeno.)

**Fig. 8.39.** Full open type greenhouse. The ceiling film can be opened, preventing excessive summer temperature. (Figure used with permission from Syuichi Himeno.)

## Cultivar and rootstock selection and planting in greenhouses and tunnels

Generally, the varieties, rootstocks and planting, training, and irrigation are the same for field, greenhouse and tunnel cultivation. Varieties for both greenhouse and open-field cultivation are 'Masui Dauphine' and 'Houraishi'; 'Summer Red', a spontaneous color mutant of 'Masui Dauphine' registered in 1999 and 'Toyomitsu-hime' a 2006 hybrid. (For more information, read Chapter 5: *World Fig Cultivars*.)

**Fig. 8.40.** Tunnel covering; the tunnel on the left has ceiling and side vinyl. The tunnel on the right has ceiling vinyl only. Note the reflective mulch on the tunnel floor. (Figure used with permission from Syuichi Himeno.)

Figs are usually self-rooted trees regardless of cropping type. However, when Ceratocystis canker disease pressure is severe, grafting onto the resistant rootstocks 'Negronne' and 'Kibaru' is recommended. When root-knot nematodes pressure is high grafting onto 'Zidi', a vigorous rootstock gives some protection. (For more information, read Chapter 7).

### *Propagation techniques and nursery management*

Planting holes are prepared in the fall of the previous year when compost and lime carbonate are mixed with topsoil. Planting is completed by early spring. The cultivar spacings and trees/10 acre are shown in Table 8.5. When planting, do not allow the roots to dry. Cut the damaged roots back to healthy tissue. In the planting hole, spread the roots in all directions without overlap and plant with the root tips facing down. Ensure the roots are in firm contact with the soil, and tamp the down tree firmly but do not change the soil level at the tree crown or cover with ground soil. After planting, irrigate and stake at the edge of the root ball. Secure the tree to the stake with ties.

To produce a straight-line training shape, the planted nursery trees are cut back at 40–60 cm and also where the two buds that will produce the primary scaffold branches are at a 20–30° angle to the tree's trunk axis. The cuts should be covered with a pruning compound.

## Irrigation

Fig trees have large leaves with a high transpiration rate, the growth of the above-ground parts is negatively affected when the soil moisture content is 20% or lower. Elongation of the above-ground parts stops when the soil water content is 15% or lower, and leaves wilt when the soil water content is 13% or lower (Morita *et al.*, 1955). Because fig trees are shallow-rooted, irrigation should be frequent and limited.

**1.** From film covering to budding stage: As a general guide, irrigate when the soil moisture is 1.8–2.0 bar (Fig. 8.41). Spraying the trees with water and increasing the humidity in the greenhouse are effective ways to control tree water status until budding.

**2.** From leaf development stage to pre-harvest stage: Maintain the soil moisture content at 1.8 and 2.5 pF. Irrigate on sunny days when the leaf transpiration and soil evaporation rates are high.

**3.** Maturity: Fig fruits rapidly absorb water as they ripen and grow (Yahata and Nogata, 2000). If the soil moistures content is maintained at 2.4 pF during maturation and the soil dries, fruit water content is lowered, the sugar content increases and final fruit weight is lower than that of fruits grown in a soil moisture content of 2.0 pF or lower. Generally, the soil moisture content should be maintained at 1.8–2.2 pF during fruit maturation. Irrigation should be done in the morning when the pF is 1.8–2.0 so that it does not exceed 2.2.

**Table 8.5.** Interval and number of plants in fig facility cultivation.

| Cultivar | Interval | | | Number of plants |
|---|---|---|---|---|
| | Between plants (m) | × | Between columns (m) | (plants/10 acre) |
| Masui Dauphine | 4.0–5.0 | | 2.0–2.5 | 80–125 |
| Houraishi | 7.0 | | 7.0 | 20 |

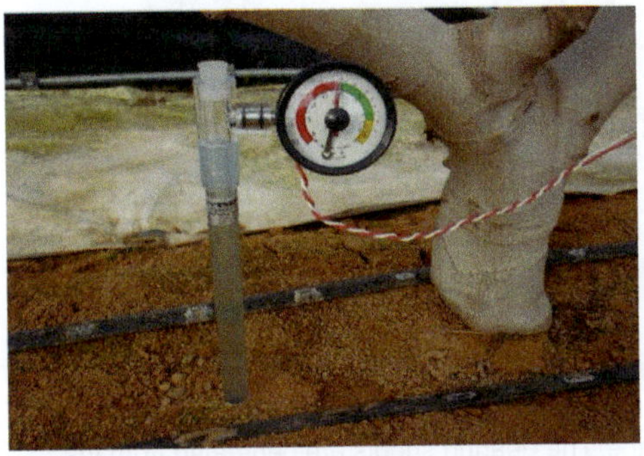

**Fig. 8.41.** Measurement of soil water tension by pF meter. (Figure used with permission from Syuichi Himeno.)

**Fig. 8.42.** Pruning of straight-line training tree shape. In this training, the bearing shoots generated from the maternal branch are cut, leaving 1 or 2 buds. (Figure used with permission from Syuichi Himeno.)

Water management also differs depending on the cultivar. For example, in the high-sugar cultivar 'Toyomitsuhime', fruit quality deteriorates because of abnormal ripening and the achenes remain immature under dry conditions (soil pF 2.2 or higher) (Fig. 8.41). In the case of 'Masui Dauphine', it was suggested that the amount of irrigation during maturation should be adjusted according to root zone depth and tree vigor (Mano *et al.*, 2012a). Because many factors related to the irrigation of fig trees remain unclear, further studies are required to optimize irrigation practices in greenhouses.

## Pruning and training 'Masui Dauphine' (straight-line training)

In greenhouse cultivation, pruning is performed at the same time as plastic film covering. The current year's shoot growth should be pruned above the first or second internode from the base (Fig. 8.42). Disbudding is done after 2–3 leaves have grown and should end by the 5–6 leaf stage. The shoot training should begin when approximately 10 leaves have grown (Fig. 8.43).

Because greenhouse grown figs have large leaves and poor light reception, it is best to use a wider branch spacing than in open-field cultivation (40 cm). In midsummer, the tip of the bearing shoot is pinched at approximately 20–25 nodes (Fig. 8.44). When several lateral shoots develop, the tip of the shoot is re-pinched with 2–3 leaves, and all the others are scraped off. Branch management is important in greenhouse cultivation because insufficient light can lead to poor bearing.

## 'Houraishi' (open-center training)

Because 'Houraishi' is vigorous, the pruning consists primarily of thinning cuts to maintain

**Fig. 8.43.** Training style of straight-line training tree shape. The bearing shoots are generated from the maternal branch at intervals of 40–50 cm. (Figure courtesy of H. Nogata.)

**Fig. 8.44.** Pinched shoot tip. (Figure used with permission from Syuichi Himeno.)

tree size and shape that will produce a good crop in a greenhouse. These thinning cuts will produce a first crop at the tip of the mother shoot. There are more, shorter, bearing shoots than in open-field cultivation (Fig. 8.45). However, if the bearing shoots are crowded, color development and sugar decrease; therefore, the bearing shoots should be evenly distributed within the canopy. For this reason, 'Houraishi' shoot thinning (Fig. 8.46) and training (Fig. 8.47) should be done at several different times to obtain adequate light interception throughout the canopy.

## Temperature control: unheated greenhouse

Daytime temperatures should be maintained at 20–25°C for about a week after covering, and 25–30°C subsequently. Night temperatures should be as warm as possible. Temperatures below 15°C during the bearing season produce malformed fruit. Therefore the side plastic film should be lowered when temperatures fall below 15°C.

## Temperature control: heated greenhouse

With heated greenhouses, temperatures can be manipulated to change harvest timing. Immediately after enclosing the greenhouse with film, minimum temperature should be raised with heaters from 10 to 15°C. To improve heat retention, additional layers of film and liners can be installed before heaters are turned on in fall. Fig bud endodormancy is deepest during late fall; buds begin breaking endodormancy in early to mid-winter. The liners can then be removed in mid-spring, at the end of March, and the side film can be removed at the end of May. As with unheated greenhouse cultivation, maximum

**Fig. 8.45.** Pruning of open natural tree shape (Houraishi). Focus is on thinning out and arranging the bearing shoots evenly. (Figure courtesy of H. Nogata.)

**Fig. 8.46.** Bud picking of open natural tree shape (Houraishi). Many buds sprout because the maternal branch is not cut back. (Figure courtesy of H. Nogata.)

greenhouse daytime temperature should not exceed 30°C as that will decrease fruit quality. High temperatures can be lowered by opening the front side of each greenhouse, the troughs of the continuous greenhouses, and opening the ceiling film of a single house, in addition to side ventilation.

### Light control

Fig is a species with relatively strong shade tolerance and a light saturation point of 40,000 lux (Amano *et al.*, 1972). In greenhouse conditions, light is often insufficient because fig leaves enlarge under the lower light levels, causing the upper leaves to shade the lower leaves. In Japan, the illuminance when sunny is about 100,000 lux and 30,000–40,000 lux when cloudy. The light-shielding rate of most films is about 20% and increases with age. Although film reuse is not recommended, they are often reused for 2 years. Additional light is supplied with electrical lights and light-reflecting materials. Artificial light during ripening is expensive but improves fruit color development and sugar accumulation (Kimura *et al.*, 1992). Light-reflecting sheets (Fig. 8.48)

**Fig. 8.47.** Arrangement of bearing shoots (Houraishi). Bearing shoots are evenly arranged on the plane of the shelf. (Photo courtesy of H. Nogata.)

**Fig. 8.48.** Light-reflecting sheet covering the ground. (Figure used with permission from Syuichi Himeno.)

installed on the greenhouse floor increase light intensity, improving fruit quality (Mano and Hamada, 2003), and prevented thrips (Thysanoptera) from flying and harming fruits (Takahashi, 1986; Shirotsuka *et al.*, 2018).

### Covering time and sprouting management in heated greenhouses

The best time to install the plastic film covering in unheated cultivation is from late winter to early spring. If the covering is applied before that the chilling requirement is insufficient for bud break and fruit set will be poor. The greenhouse should be covered after pruning and heated in early winter. Irrigate and maintain humidity in the house until bud break. Each tree needs 10 mm of water every 1–2 days. Irrigation should be done in the morning to prevent decreasing the soil temperature. Note that poor temperature and water management is likely to cause bud break failure and bud death, particularly when the temperature exceeds 35°C.

#### Period before harvest

Irrigate with 15–20 mm of water every other week during the shoot leaf growth period and before harvesting. Insufficient water during this period decreases new shoot growth and fruit set. Transpiration and soil moisture evaporation are reduced by cloudy weather and rain; irrigation amount and interval should be modified to account for the weather. The film lining should be removed by late spring, when the side film outside temperatures are 15°C.

#### During the harvesting season

During the harvesting season, irrigate approximately 10 mm every 2–3 days. Specifically, irrigate soil with a pF of 1.8–2.0 in the morning to ensure that it does not become over dried. Irrigation during this period affects fruit enlargement; when the amount of water is large, the change in soil moisture increases, and the number of fruits increases. In contrast, when the amount of irrigation is small, fruit enlargement decreases and the occurrence of high-temperature damage symptoms increases. The upper plastic film should be kept in place until the end of harvest.

#### Fertilization and soil management

Fig trees grown in greenhouses cultivation are more vigorous than trees grown under open-field cultivation. Fertilization should be decreased by 20–30%. In greenhouse cultivation, fertilizer runoff is decreased and the soil pH may increase. If the pH exceeds 7, there is no need to apply lime.

## 8.13 Rainfed Orchard Establishment and Management

Rainfed fig production is the traditional method of fig farming. Iran is among the semi-arid regions of the world with extensive rainfed fig orchards. The Estahban region in southeast Iran has about 26,000 ha of rainfed fig orchards. The primarily dried fig cultivar is 'Sabz'.

The primary difference between rainfed and irrigated fig cultivation is rainfed orchards are often located on northern slopes, using a contour design (Fig. 8.4) to cache spring rain or snowmelt as opposed to square or triangular, off-set, designs on flat land (Fig. 8.3). However, rainfed orchards are always at wider spacings; often 12 × 12 m, 69 trees per hectare (Pereira *et al*., 2015).

The trees can also be propagated and planted in the same way and time as irrigated orchards. However, planting in the fall can take better advantage of the winter rains. And traditional direct planting methods are often used.

Direct planting uses a dormant 3–4 year old, 60–80 cm long, 3–4 cm diameter branch. The cutting is laid diagonally at 1–1.5 m deep, with the top extending out of the covering soil by 1–2 cm. The pit surface is covered with stones and a final layer of soil. This is method is called "Koushk" (Fig. 8.49). In early June, a few stones are removed to allow sunlight penetration to the cutting. Cuttings form roots within three months and new shoot growth within a month.

The rooted cuttings need daily irrigation from summer through fall for the first three years in regions with less than 250 mm of rain annually. The trees need irrigation every 7–10, 15–20, and 20–30 days per a month for the first, second and third year, respectively. At three years, rain or snowmelt supplies 200–250 mm and by April, the daily irrigation can be stopped, and the trees are irrigated with rain and micro-catchment water harvesting (Fig. 8.50).

The young tree can also be protected from sunburn on the western side of the young tree with white water-based paint or palm fronds, paper, or aluminum foil tree guards (Fig. 8.51). Plastic bags and high potash N–P–K fertilizers, 12–12–36 or 10–10–40 can offer freeze protection.

**Fig. 8.49.** New growth from hard wood cutting planted using the 'Koushk' method. (Figure courtesy of M. Jafari.)

**Fig. 8.50.** Eyebrows microcatchments in Estahban rainfed fig orchards. (Figure courtesy of M. Jafari.)

**Fig. 8.51.** Protection of fig plant from freezing (left) and sunburn (right) using date palm leaves. (Figure courtesy of M. Jafari.)

### Rainwater harvesting

Microcatchments for water harvesting are formed in the first fall. A 20 cm deep half-moon loop is dug on the lower slope side of the 'Koushk'. The size of the microcatchment increases as the tree grows and needs more water. Eventually, a low wall of soil or stone is added to increase capacity. Channels may be dug to guide water toward the microcatchment (Fig. 8.50).

## Training and pruning rainfed trees

Most rainfed fig trees are allowed to develop as multi-trunked bushes. The young tree can also be headed back to 50% of the total height at planting, allowed to grow one season, and thinned the following dormant season to three to eight strong, upright, well-spaced vertical and horizontal shoots. The leaders must be spaced far enough apart to allow them to increase to 10 cm in diameter without crowding one another. If the leaders are crowded at the base, high winds may split them. Training of rainfed trees is done for the same reasons irrigated trees are pruned, to produce bearing and facilitate orchard operations. However, rainfed trees are also trained and pruned to remain smaller so they need less water.

## Rejuvenation pruning and thinning

The major abiotic stress in rainfed fig cultivation is drought that results in slow growth and a high carbon to nitrogen ratio in stressed trees. In stressed and old rainfed fig trees, rejuvenation pruning is necessary to improve water and nutrient uptake, balance C/N ratio, increase bearing surface, and improve fruit quality. In rejuvenation pruning, all trunks are cut at 1 m, allowed to grow for a season, and thinned, much as in tree training, the following dormant season (Fig. 8.52).

## Management of physiological disorders

### *Sunburn*

Fig trees, particularly young or water-stressed trees, and rainfed trees, are sensitive to sunburn, primarily on the southwest tree quadrant or trunks. Heavy pruning, which exposes previously shaded wood to strong sunlight on the top surface of exposed branches, also causes sunburn. Sunburn appears as cracked and dead bark. The best control is good irrigation and painting after heavy pruning with a white water-based paint (Fig. 8.53) (Ferguson *et al.*, 1990).

Sunburned wood is easily confused with bacterial canker (Fig. 8.54). The sunburned wood is on sun-exposed surfaces and remains intact and dry, while a bacterial canker infection manifests as discolored destroyed wood.

Fruits and trees sunburn for the same reasons, particularly as they ripen, become heavier and droop out of the canopy cover (Fig. 8.55). The typical symptom is bleached patches on the exposed fruit skin, which later become brownish. Fruit sunburn can also be managed partially with good irrigation and moderate pruning. Leaves in water-stressed trees can sunburn. The leaves develop round faded patches and necrosis. The only prevention is sufficient irrigation, shading or painting with white water-based paint. Kaolin sprays can reduce leaf sunburn in

**Fig. 8.52.** Two same-age fig trees with (left) and without (right) rejuvenation pruning under a drought year. (Figure courtesy of M. Jafari.)

**Fig. 8.53.** Barks are painted white to reflect light and avoid sunburn. (Figure courtesy of M. Jafari.)

**Fig. 8.54.** Bark cracking in rainfed fig tree by direct sunlight (left picture) and canker (right picture). (Figure courtesy of L. Ferguson (left) and M. Jafari (right).)

**Fig. 8.55.** Fig fruits exposed to direct sunlight. (Figure courtesy of L. Ferguson.)

rainfed fig trees before the fruit ripening season. Spraying 5% kaolin in mid-May after full leaf out reduces sunburn in rainfed fig orchards (Zare *et al.*, 2017).

### *Fruit splitting*

Hepaksoy *et al.* (2007) reported variation of soil moisture content and relative humidity, dry

**Fig. 8.56.** Fruit splitting scores from none to severe on 'Sabz' cultivar. (Figure courtesy of M. Jafari.)

**Fig. 8.57.** Immature fruit dropping in non-caprified fig trees of Sabz cultivar. (Figure courtesy of M. Jafari).

winds, or heavy irrigation after a dry period all caused fruit splitting at the ostiole (eye) (Fig. 8.56). Splitting in 'Calimyrna' and other varieties may also result from excessive pollination, 'over-caprification' with the growth of too many developing seeds during fruit ripening resulting in fruit cracking. The caprifig cultivar can also have a significant effect on ostiole cracking. The caprification of 'Sabz' cultivar by 'Pouzdonbali' caprifig creates fig fruits with a large crack in their ostioles (Rahemi and Jafari, 2008). Aydın and Kaptan (2015) reported a positive correlation of cracking with B (boron) content in soil, leaf and peel and negative correlation of soil potassium (K), leaf Ca and peel Ca content and soil Ca/B, leaf Ca/B, soil Ca/(K+Mg+Na) and fruit Ca/(K+Mg+Na) ratios.

### *Fruit drop*

Non-caprified fruits do not develop; they turn yellowish in color and drop (Ferguson *et al.*, 1990). Interaction between reproductive and vegetative growth in fig trees is another factor in immature fruit drop. Rapidly growing, highly vegetative trees may cause fruit drop. However, the heavy fruit drop seen in Fig. 8.57 is generally a function of non-caprification. The over growth in fig trees appears indirectly to mediate fruit dropping (Gaaliche *et al.*, 2011).

## References

Amano, S., Hino, A., Daito H. and Kuraoka, T. (1972) Studies on the photosynthetic activity in several kinds of fruit trees. I. Effect of some environmental factors on the rate of photosynthesis (In Japanese with English abstract). *Journal of the Japanese Society for Horticultural Science* 41, 144–150. DOI:10.2503/jjshs.41.144

Awamura, M. (2015). Cultivation of 'Houraishi' fig trees. In: Mano, T. (ed.) *Ichijiku-no-Sagyoubennrichou* (In Japanese). Noubunkyo, Tokyo. p. 28–30, 66–68, 97–98.

Awamura, M., Shoda, K., Nakayama, T.H., Nogata, H. and Takashima, E. (1993) Establishment of techniques for stabilization of fig production. (4) Effects of canopy structure on fruit production of fig 'Houraishi'. (In Japanese with English abstract). *Bulletin Fukuoka Agriculture Research Center* B-12, 65–68. https://agriknowledge.affrc.go.jp/RN/2010490742.pdf

Aydın, M. and Kaptan, M.A. (2015) Effect of nutritional status on fruit cracking of fig (*Ficus carica* L. cv. Sarılop) grown in high level boron contained soils. *Scientific Papers - Series A, Agronomy* 58, 20–25.

Casadomet, E., López-Corrales, M., Pérez Gragera, F., Senero, M., Pérez Ross, J. and Del Moral, J. (2016) *Plagas y enfermedades del cultivo de la higuera. Hojas divulgadoras.* Ministerio de Agricultura, Alimentación y Medio Ambiente, Extremadura, Spain.

Cesarano, G., Zotti, M., Antignani, V., Marra, R., Scala, F. and Bonanomi, G. (2017) Soil sickness and negative plant-soil feedback: A reappraisal of hypotheses. *Journal of Plant Pathology* 99(3), 545–570.

Condit, I.J. (1947) *The Fig* Chronica Botanica Co., Waltham, Massachusetts, pp. 109–110.

Condit, I.J. (1955) Fig varieties: A monograph. *Hilgardia* 23, 323–538.

Costa, A. (2019) *El Cultivo de la Higuera en el Campo de Albatera*. Newton publicaciones, Spain.

Edgerton, C.W. (1911) Diseases of the fig tree and fruit. *Louisiana Agriculture Experiment Station Bulletin* 126, 1–20.

FAO (2021) www.fao.org/fishery/docs/CDrom/FAO_Training/FAO_Training/General/x6706e/x6706e02.htm (accessed 5 August 2021).

Ferguson, L., Michailides, T.J. and Shorey, H.H. (1990) The California fig industry *Horticulture Review* 12, 409–490.

Figueiredo, P. and Pinheiro, E.D. (1969) Uma nova doença da figueira (*Ficus carica* L.) na região de Valinhos, SP (In Portuguese). *Biologico* 35, 227–230.

Flaishman, M.A., Rodov, V. and Stover, E. (2008) *The Fig: Botany, Horticulture, and Breeding*. Westport, New York, p. 113.

Gaaliche, B., Lauri, P.E., Trad, M., Costes, E., and Mars, M. (2011). Interactions between vegetative and generative growth and between crop generations in fig tree (*Ficus carica* L). *Scientia Horticulturae* 131, 22–28.

Galván, A.I., Pérez-Gragera, F., Serradilla, M.J., Córdoba, M.G., Domínguez, G. and López-Corrales, M. (2021) Agronomic behavior and fruit quality attributes of three dried fig cultivars in super high density system. *Acta Horticulturae* 1310, 217–222. DOI: 10.17660/ActaHortic.2021.1310.33

Gawankar, M.S., Haldankar, P.M., Salvi, B.R., Parulekar, Y.R., Dalvi, N.V., Kulkarni, M.M., Saitwal, Y.S. and Nalage, N.A. (2019) Effect of girdling on induction of flowering and quality of fruits in horticultural crops - a review. *Advanced Agricultural Research & Technology Journal* 3(2), 201–215.

Gilbert, J. and Senyuva, H. (2008) Fungal and mycotoxin contamination of dried figs. A review. *Mycotoxins* 58(2). DOI: 10.2520/myco.58.73.

Globe Canada (2021) www.placebasedbasics.weebly.com/soil-analysis.html#:~:text=Pit%20Method%3A%20Students%20dig%20a,1.5%20x%201.5%20m%20wide (accessed 5 August 2021).

Hansen, H.N. (1929) Thrips as carriers of fig-decaying organisms. *Science* 69, 356–357. DOI:10.1126/science.69.1787.356.b

Hepaksoy, S. and Aksoy, U. (2006) Propagation of *Ficus carica* L. clones by in vitro culture. *Biologia Plantarum* 50(3), 433–436.

Hepaksoy, S., Aksoy, U., Can, H.Z. and Ul, M.A. (2007) Determination of relationship between fruit cracking and some physiological responses, leaf characteristics and nutritional status of some pomegranate varieties. *Chieam*, p. 87–92. www.resources.chiam.org.

Higuchi, T., Saika, T., Senda, S., Mizobata, T., Kawata, Y. and Nagai, J. (1997) Development of biorational pest control formulation against longicorn beetles using a fungus, Beauveria brongniartii (Sacc.) Petch. *Journal of Fermentation and Bioengineering* 84(3), 236–243. DOI:10.1016/S0922-338X(97)82061-2

Hirai, J. and Hirano, A. (1949) Studies on the fig sick soil. (I) Effect of sick soil upon the tree growth (In Japanese). *Studies from the Institute of Horticulture, Kyoto University* 4, 96–102.

Hirai, J., Nakagawa, S., Nanjo, Y. and Hirata, N. (1961) Studies on the amounts of the nutrient elements absorbed by the fig trees (In Japanese with English abstract). *Journal Japanese Society Horticultural Science* 30, 203–210. DOI: 10.2503/jjshs.30.203

Horimoto, M., Matsuura, K. and Araki, H. (1994) Freezing tolerance of fig trees and protective covering with aluminum coated film (In Japanese). *Bulletin Hyogo Pre. Agri. Inst. (Agriculture)* 42, 37–42. https://agriknowledge.affrc.go.jp/RN/2010561478.pdf (accessed 5 August 2021).

Hosomi, A. (2006) Trial rearing of 'Masui Dauphine' fig culture with vigorating rootstocks (In Japanese with English abstract). *Kinki Chugoku Shikoku Agriculture Research* 8, 93–97.

Hosomi, A. (2011) Incidence of soil sickness in areas producing 'Masui Dauphine' figs (In Japanese with English abstract). *Bulletin of Research Institute of Environment, Agriculture and Fisheries, Osaka* 4, 9–13. www.kannousuiken-osaka.or.jp/_files/00027946/h23-ichijiku.pdf (accessed 5 August 2021).

Hosomi, A. (2020) Relationship between shoot growth and fruit production of 'Masui Dauphine' and 'Houraishi' fig trees at various plant densities. *Horticulture Journal* 89, 375–383. DOI: 10.2503/hortj.UTD-164

Hosomi, A. and Kusakari, S. (1995) A yeast associated with soft rot of fig fruit in Japan (In Japanese with English abstract). *Annual Report Kansai Plant Protection* 37, 9–14. DOI: 10.4165/kapps1958.37.0_9

Hosomi, A. and Uchiyama, T. (1998) Growth inhibiting factors in sick soil of fig orchards (In Japanese with English abstract). *Journal Japanese Society Horticultural Science* 67, 44–50. DOI: 10.2503/jjshs.67.44

Hosomi A., Dan, M. and Kato, A. (2002) Screening of fig varieties for rootstocks resistant to soil sickness. *Journal Japanese Society Horticultural Science* 71, 171–176. DOI: 10.2503/jjshs.71.171

Hosomi, A., Miwa, Y., Furukawa, M. and Kawaradani, M. (2012) Growth of fig varieties resistant to Ceratocystis canker following infection with *Ceratocystis fimbriata*. *Journal Japanese Society Horticultural Science* 81(2), 159–165. DOI: 10.2503/jjshs1.81.159

Hosomi, A., Miwa, Y. and Mano, T. (2013) Shoot growth and fruit production of 'Masui Dauphine' fig trees having high limb position with downward shoots *Journal Japanese Society Horticultural Science* 82, 215–221. DOI: 10.2503/jjshs1.82.215

Hosomi, A., Isobe, T. and Miwa, Y. (2015a) Shoot growth and fruit production of the 'Masui Dauphine' variety of fig (*Ficus carica* L.) undergoing renewal long pruning. *JARQ* 49, 391–397. DOI: 10.6090/jarq.49.391

Hosomi, A., Miwa, Y. and Isobe, T. (2015b) Choice of cuttings and scions for raising fig (*Ficus carica* L.) nursery stocks by softwood grafting (In Japanese with English abstract). *Horticulture Research (Japan)* 14, 197–203. DOI: 10.2503/hrj.14.197

Hossain, A.B.M.S. and Boyce, A.N. (2009) Fig fruit growth and quality development as affected by phloem stress. *Bulgarian Journal of Agricultural Science* 15(3), 189–195.

Ichikawa, K. and Naganawa, M. (2004) Control of thrips by tape affixation to fig fruit apex (In Japanese with English abstract). *Research Bulletin Aichi Agricultural Research Center* 36, 35–39. https://agriknowledge.affrc.go.jp/RN/2010712333.pdf (accessed 5 August 2021).

Himeno, S., Awamura, M., Nogata, H., Ishibashi, M. and Inoue, Y., (2015) A new fig rootstock cultiver 'Kibaru' resistant to Ceratocystis canker (In Japanese with English abstract). *Bulletin Fukuoka Agriculture and Forestry Research Center* 1, 76–81. www.google.co.jp/url?sa=t&rct=j&q=&esrc=s&source=web&cd=&cad=rja&uact=8&ved=2ahUKEwiU96y91IDqAhWIL6YKHcRuD1YQFjAAegQIAhAB&url=http%3A%2F%2Ffarc.pref.fukuoka.jp%2Ffarc%2Fkenpo%2Fnourinkenpo-1%2F01-16.pdf&usg=AOvVaw2LtHFM9DIv2xTvhC7QQc_P (accessed 5 August 2021).

ISHS (2004) *Proceedings of the Sixth International Symposium on Chemical and Non-Chemical Soil and Substrate Disinfestation*. ISHS Working Group on Soil Borne Pathogens, Corfu, Greece.

Ito, S. and Homma, Y. (1938) Notae mycologicae Asiae orientalis III. *Transactions Sapporo Natural History Society* 15, 113–118. www.eprints.lib.hokudai.ac.jp/dspace/bitstream/2115/64235/1/Vol.15No.3_001.pdf (accessed 5 August 2021).

Jafari, M., Rahemi, M. and Kagkar-Haghighi, A.A. (2018) Role of fig rootstock on changes of water status and nutrient concentrations in 'Sabz' cultivar under drought stress condition. *Scientia Horticulturae* 230, 56–61.

Jevanandam, N. and Corlett, G.A.R. (2013) Climate warming and the potential extinction of fig wasps, the obligate pollinators of figs *Biology Letters* 9, 20130041.

JSAEZ (The Japanese Society of Applied Entomology & Zoology) (2006) *Major Insect and Other Pests of Economic Plants in Japan* (In Japanese). Japan Plant Protection Association, Tokyo.

Kabumoto, T. (1985) Fertilizing (Sehi), Orchard establishment and replanting (Kaien, Kaishoku). In: Kajuzensho, *Japanese Apricot, Fig and Loquat (Ume-Ichijiku-Biwa)* (In Japanese). Nobunkyo, Tokyo. p. 315–320, 341–348.

Kabumoto, T., Nakaoka, T., Taniguchi, T., Fujiwara, T. and Fujiwara, T. (1985) Studies on the training method of fig tree, 2 Developing new training method and on the tree growth, yield, quality and laboring hours for cultivation in straight line training (In Japanese). *Bull. Hyogo Pref. Agri. Center Exp., Ext. Educ.* 33, 74–78. https://agriknowledge.affrc.go.jp/RN/2010330147.pdf (accessed 5 August 2021)

Kadowaki, N., Sekine, N., Ebashi, K. and Tahira, K (2012) Effect of close planting on the growth, yield, and fruit quality of the fig tree (Ficus carica L.) after frost and cold damage. *Bulletin of the Horticultural Institute, Ibaraki Agricultural Cent.* 19, 1–8. https://agriknowledge.affrc.go.jp/RN/2010850023 (accessed 5 August 2021)

Kaibara, E. (1709) *Japanese Herbs (Yamato-Honzo) 9, 10 and 12* (In Japanese). Nagatachoubei. Kyoto. www.dl.ndl.go.jp/info:ndljp/pid/2605899

Kajitani, Y. and Masuya, H. (2011) *Ceratocystis ficicola* sp. nov., a causal fungus of fig canker in Japan *Mycoscience* 52, 349–353. DOI: 10.1007/s10267-011-0116-5

Kato, K., Hirota, K. and Miyagawa, T. (1982) A new disease, ceratocystis canker of fig caused by *Ceratocystis fimbriata* Ellis et Halsted (In Japanese). *Shokubutuboeki* 36, 55–59.

Kawase, K. (1990) Properties and utilization of rootstocks for fruit crops. *Crops*. Nohunyo, Tokyo. pp. 469–475.

Kimura, N., Sakakibara, M. and Kimbara, T. (1992) Lighting for coloring and quality in fig fruit in maturing period. *Research Bulletin Aichi Agriculture Research Center* 24, 195–204. http://agriknowledge.affrc.go.jp/RN/2010490085 (accessed 10 February 2022).

López-Corrales, M., Gil, M., Pérez, F., Cortés, J. and Serradilla, M.J. and Chome, P.M. (2011) Variedades de higuera: descripción y registro de variedades. In: *Ministerio de Medio Ambiente y Medio rural y Marino*. Madrid. ISBN 978-84-491-1103-7.

Majid, I., Khalil Khalil, A. and Nazir, N. (2020) Economic analysis of high-density orchards. *International Journal of Advance Research in Science and Engineering* 7 (4).

Mano, T. (2004) Improvement in quality of the fig fruits by impermeable white sheet mulching (In Japanese). *Kinki Chugoku Shikoku Niokeru Shingijutu* 4, 105–107.

Mano, T. (2015) Planting row spacing of 'Masui Dauphine' fig trees. In: Mano, T. (ed.) *Ichijiku-no-Sagyoubennrichou* (In Japanese). Noubunkyo, Tokyo. p. 27.

Mano, T. and Hamada, K. (2003) Improvement in quality of the fig fruits by impermeable while sheet mulching (In Japanese). *Journal Japanese Society Horticultural Science* 72 (Suppl. 2), 346.

Mano, T. and Hamada, K. (2005) Effect of close planting on growth, fruits quality and yield in young fig tree (In Japanese). *Kinki Chugoku Shikoku Agri. Res.* 6, 72–75.

Mano, T., Mizuta, Y. and Moriguchi, T. (2010) Difference of degree on low temperature injury in variety of fig (In Japanese). *Horticulturel Research (Japan)*. 9 (Suppl. 2), 404.

Mano, T., Mizuta, Y. and Moriguchi, T. (2011a) Super-high density planting of fig (*Ficus carica* L.) for early recovery from sick soil and low temperature injury (In Japanese with English abstract). *Horticulturel Research (Japan)* 10, 367–373. DOI: 10.2503/hrj.10.367

Mano, T., Sugiura, T., Moriguchi, T. and Kuroda, H. (2011b). Effects of girdling on freezing injury of fig (*Ficus carica* L.) (In Japanese with English abstract). *Horticulturel Research (Japan)* 10, 573–579. DOI: 10.2503/hrj.10.573

Mano, T., Mizuta, Y. and Moriguchi, T. (2012a) Effects of non-woven fabric mulching and irrigation on growth and fruit quality of fig with different tree vigor (In Japanese with English abstract). *Horticulture Research (Japan)* 11, 219–225. DOI:10.2503/hrj.11.219

Mano, T., Mizuta, Y., Ito, A., Isobe, T., Hosomi, A. and Moriguchi, T. (2012b) Effect of height of primary scaffold branch on freezing injury of fig (*Ficus carica* L.) (In Japanese with English abstract). *Horticulture Research (Japan)* 11, 351–356. DOI: 10.2503/hrj.11.351

Matsuura, K., Mizuta, Y. and Ozeki, H. (2014) Effect of different heights of branch on the growth, yield and fruit quality of 'Masui-Dauphine' fig (In Japanese with English abstract). *Bulletin Hyogo Prefectural Technology Center for Agriculture, Forestry and Fisheries* 62, 1–6. https://agriknowledge.affrc.go.jp/RN/2010891605.pdf (accessed 5 August 2021).

Morishita, M. (2002) Reflective plastic mulch as control measure for thrips in fig plant (*Ficus carica* L.) (In Japanese with English abstract). *Annual Report Kansai Plant Protection* 44, 21–26. DOI: 10.4165/kapps 1958.44.0_21

Morita, Y., Ishihara, M., Yoneyama, K., Oguro, E., Nishida, M., Itakura, T. and Usui, S. (1955) Kajyuendo-jyo-no-kenkyu (In Japanese). *Bulletin National Institute Agricultural Science Series E. Horticulture* 4, 1–144.

Muhammad, S., Sanden, B.L., Lampinen, B.D., Saa, S., Siddiqui, M.I., Smart, D.R., Olivos, A., Shackel, K.A., DeJong, T. and Brown, P.H. (2015) Seasonal changes in nutrient content and concentrations in a mature deciduous tree species: Studies in almond (*Prunus dulcis* (Mill.) DA Webb). *European Journal of Agronomy* 65, 52–68.

Nakasone, W. and Kusakari, S. (1991) Occurrence and control of figs mosaic disease caused by mites (In Japanese). *Annual Report Kansai Plant Protection* 33, 116. DOI: 10.4165/kapps1958.33.0_116

Nogata, H. and Awamura, M. (2005) New fig cultivar 'H156-70' (In Japanese with English abstract). *Bulletin Fukuoka Agriculturel Research Center.* 24, 104–107. https://agriknowledge.affrc.go.jp/RN/2010712403.pdf (accessed 5 August 2021).

Nojima, T. (1934) Pathogenicity of *Rhizopus* fungi isolated from figs (In Japanese). *Annals Phytopathology Society Japan* 3, 87–89. DOI:10.3186/jjphytopath.3.56

Obenauf, G. (1978) Commercial dried fig production in California. www.agris.fao.org/agris-search/search.do?recordID=US201300572983 (accessed 5 August 2021).

OPPPO (Osaka Prefectural Plant Protection Office) (2020) *Guidelines for Crop Pest Control 2020* (In Japanese). OPPPO, Osaka. www.jppn.ne.jp/osaka/shishin/body/2003/2003_10_01_itijiku.pdf (accessed 5 August 2021).

Pereira, C., Serradilla, M.J., Martín, A., Villalobos, M.C., Pérez-Gragera, F. and López-Corrales, M. (2015) Agronomic behaviour and quality of six cultivars for fresh consumption. *Scientia Horticulturel* 185, 121–128. DOI: 10.1016/j.scienta.2015.01.026

PSJ (The Phytopathological Society of Japan) (2020) *Common Names of Plant Diseases in Japan* (2020.1 edition) (In Japanese). PSJ, Tokyo www.ppsj.org/mokuroku.html (accessed 5 August 2021).

Rahemi, M. and Jafari, M. (2008) Effect of caprifig type on quantity and quality of Estahban dried fig (*Ficus carica* cv.Sabz). *Acta Horticulturae* 798, 249–253.

Rosecrance, R.C., Weinbaum, S.A. and Brown, P.H. (1996) Assessment of nitrogen, phosphorus, and potassium uptake capacity and root growth in mature alternate-bearing pistachio (*Pistacia vera*) trees. *Tree Physiology* 16(11-12), 949–956.

Sato, K. and Shichijo, T. (1953) Growth of young fig trees in old fig orchard soils (In Japanese with English abstract). *Journal Japanese Society Horticulturel Science* 22, 163–166. DOI: 10.2503/jjshs.22.163

Senda, T., Manago, N., Sakakibara, M., Takeuchi, M. and Banno, M. (1996) The relationship between the growth pattern and productivity of fig trees and its diagnosis (In Japanese with English abstract). *Research Bulletin Aichi Agriculturel Research Center* 28, 247–251. https://agriknowledge.affrc.go.jp/RN/2010550034.pdf (accessed 5 August 2021).

Shennan, C., Muramoto, J., Mazzola, M., Butler, D., Rosskoph, E., Kokalis-Burelle, N., Momma, K., Kobara, Y. and Lamers, J. (2014) Anaerobic soil disinfestation for soil borne disease control in strawberry and vegetable systems: Current knowledge and future directions. *Acta Horticulturae* 1044, 165–175.

Shibao, M. and Tanaka, H. (1997) Overwintering sites and stages of the fig bud mite, *Eriophyes ficus* COTTE (In Japanese with English abstract). *Annual Report Kansai Plant Protection* 39, 1–3. DOI: 10.4165/kapps1958.39.0_1

Shirotsuka, K., Mizuno, A. and Shibao, M. (2018) Control of thrips on fig crops using insect-proof netting and diffuse-reflection mulch sheets (In Japanese with English abstract). *Annual Report Kansai Plant Protection* 60 157–160. DOI: 10.4165/kapps.60.157

Souda, K., Matsuura, K. and Tanaka, T. (2016) Effects of training that primary scaffold branch overlap the trunk of next tree on the growth, the yield and the fruit quality of 'Masui-Dauphine' fig (In Japanese). *Horticulturel Research (Japan)* 15 (Suppl. 1), 299.

Stover, E., Aradhya, M., Ferguson, L. and Crisosto, C.H. (2007) The fig: Overview of an ancient fruit. *HortScience* 42, 1083–1087.

Takahashi, A. (1986) A physical method for controlling thrips which attack the inside of fig fruits (*Ficus carica* L.) (In Japanese). *Bulletin Shizuoka Citrus Experiment Station* 22, 33–40. http://agriknowledge.affrc.go.jp/RN/2010350197 (accessed 10 February 2022).

Taylor, A.L. (1943) The effect of root-knot on fig seedlings. *Plant Disease Reporter* 27, 224.

Ushijima, K., Asakuma, H., Muramoto, K. and Kusano, N. (2013) Shortening of grafted nursery period by nove1 method of cutting graft of fig (*Ficus carica* L.) rootstock cu1tivar 'Kibaru' (In Japanese with English abstract). *Bulletin Fukuoka Agriculturel Research Center* 32, 76–80. https://agriknowledge.affrc.go.jp/RN/2010852157.pdf (accessed 5 August 2021).

Vachamidis, P. and Vemmos, S. (2010) Fig tree cultivation. *Agriculture-Livestock* 1, 64–72.

Wilhelm, S., Storkan, R.C. and Wilhelm, J.M. (1974) Preplant soil fumigation with methyl bromide-chloropicrin mixtures for control of soil-borne diseases of strawberries — A summary of fifteen years of development. *Agriculture and Environment* 1, 227–236.

Yahata, D. and Nogata, H. (2000) Relationship between increase in water content and cell wall composition in fig syconia during enlargement and ripening (In Japanese with English abstract). *Bulletin Fukuoka Agriculturel Research Center* 19, 76–80. https://agriknowledge.affrc.go.jp/RN/2010610472 (accessed 10 February 2022).

Yakushiji, H., Yamazaki, A., Jikumaru, S., Morita, T., Hoshino S., Sugawa S., Shirakami N. and Hamana Y. (2019) Reikoudai-ichigou *Japan Plant Variety Registration* 34378.

Yakushiji, H., Morita, T., Jikumaru, S., Ikegami, H., Azuma, A. and Koshita, Y. (2012) Interspecific hybridization of fig (*Ficus carica* L.) and *Ficus erecta* Thunb., a source of Ceratocystis canker resistance. *Euphytica* 183, 39–47. DOI: 10.1007/s10681-011-0459-1

Zare, H., Mohammadi, D., Golkar, G. and Jokar, L. (2017) The effects of kaolin spraying on some physiological characteristics of fig (*Ficus carica* cv. Sabz) under rainfed conditions. *10th National Horticultural Science Congress of Iran*. Tehran, Iran.

# 9 Fig Wasps and Pollination

**F. Kjellberg[1], S. van Noort[2,3]* and J.Y. Rasplus[4]**
*[1]CEFE, CNRS, University of Montpellier, Montpellier, France; [2]Research and Exhibitions Department, South African Museum, Iziko Museums of South Africa, Cape Town, South Africa; [3]Department of Biological Sciences, University of Cape Town, South Africa; [4]INRAE, IRD, CIRAD, SupAgro, University Montpellier CBGP, Montferrier-sur-Lez, France*

## 9.1 Introduction

The following text is illustrated by the slide show '*Ficus carica* and its pollination' (Kjellberg and Lesne, 2020). Slide numbers in the text refer to slide numbers in the first version of the slide show.

The obligate mutualism between fig trees (*Ficus*, Moraceae) and their pollinating fig wasps (Agaonidae, Hymenoptera) has been in existence for some 70 million years (Cruaud *et al.*, 2012). It entails complex patterns of reciprocal evolution between the partners that has resulted in highly constrained relationships between the pollinators and their host *Ficus* species (Ramirez, 1970; 1974; 1977; Galil, 1977; Weiblen and Bush, 2002; Cook and Rasplus, 2003; Cook and Segar, 2010; Cruaud *et al.*, 2010; 2012). Each partner cannot reproduce without the other. *Ficus* species are entirely reliant on the specialized agaonid fig wasps (Agaonidae; Chalcidoidea; Hymenoptera; slide 26) for pollination of their florets, which are encased inside the fig, an enclosed inflorescence (slide 8). Fig wasps locate their host tree by homing in on host-specific volatiles (Barker, 1985; van Noort *et al.*, 1989; Ware *et al.*, 1993; Hossaert-McKey *et al.*, 1994; Gibernau *et al.*, 1997; Grison *et al.*, 1999, Soler *et al.*, 2012; Souto-Vilarós *et al.*, 2018) emitted by glands located in the fig ostiole (Souza *et al.*, 2015) when the florets and stigmas are receptive for oviposition and pollination (Fig. 9.1A) (slides 28, 43). On locating receptive figs, the wasps negotiate a tight bract-lined opening situated at the apex of the fig termed the ostiole (slides 8, 41), to gain access to the florets lining the inside of the central cavity (slide 8). Negotiating the ostiole is no easy task, with the female wasp having to squeeze and labor her way between the tightly closed bracts. She is, however, remarkably adapted to do so. Her body, in particular her head and thorax, is extremely flattened and elongate (Fig. 9.1A). She also has rows upon rows of backward pointing teeth on her mandibular appendage (Fig. 9.2B), situated on the underside of her head, as well as a few strong teeth on her legs (Fig. 9.2D). They then pollinate the stigmas while ovipositing (Fig. 9.1B). They oviposit by inserting their ovipositor down the style (slide 10) to deposit an egg within the flower ovule, between the inner integument and the nucellus (Jansen-González *et al.*, 2012). The wasp larva feeds on endosperm tissue within the galled flower. This endosperm may result from

*Corresponding author: svannoort@iziko.org.za

 *The Fig: Botany, Production and Uses* (eds. Sarkhosh, A., Yavari, A. and Ferguson, L.)
DOI: 10.1079/9781789242881.0009

**Fig. 9.1.** *Blastophaga psenes*. (A) A female attracted to a receptive fig by host-specific volatiles emitted by the fig ostiole, (B) female ovipositing in a receptive fig of a male tree, (C) male opening a hole into a gall containing a female, (D) female escaping the fig dusted with pollen. (Figures courtesy of F. Kjellberg and J.Y. Rasplus.)

the double fertilization of the ovule (Jansen-González *et al.*, 2012) or its development is parthenogenetic (Leclerc du Sablon, 1908; Corner, 1933), initiated by the wasp larva (Borges and Kjellberg, 2014). They are not able to develop anywhere else, except within the figs of their host *Ficus* species. When the wasps become adult, they emerge from their natal fig loaded with pollen, and leave in search of a fresh receptive fig (Fig. 9.1D). The relationship is generally highly specific although some wasps may end up in the wrong host (Compton, 1990; Ware and

**Fig. 9.2.** *Blastophaga psenes*. (A) Antenna (SEM), (B) Head ventral, showing the mandibles and their appendage bearing 6 ridges, (C, D) Habitus male and female, lateral view, (E) Habitus male dorsal view, (F) Mesosoma female ventral view, white arrow shows the scar in place of pollen pocket. Scale = 200 μm excepted for habitus = 1 mm. ( Figure courtesy of J.Y. Rasplus.)

Compton, 1992; Moe *et al.*, 2011; Yang *et al.*, 2012). However, there are numerous instances where more than one pollinator species regularly develops within a host fig species, either with parapatric distributions throughout the range of the host (Rasplus, 1996; Yu *et al.*, 2019), or pollinators may co-occur within the same location (Rasplus, 1996; Molbo *et al.* 2003; Herre *et al.*, 2008; Compton *et al.*, 2009a; Rodriguez *et al*,. 2017). Rarely a single wasp species may be the main pollinator of two or more *Ficus* species (Rasplus, 1996; Cornille *et al.*, 2012; Wang *et al.*, 2016).

In some *Ficus* species, the wasps, before leaving their natal fig, walk to the stamens and load pollen into specialized pollen pockets (Frank, 1984). Once the wasps have entered a receptive fig, they extract pollen from the pollen pockets and deposit it on the stigmas (Frank, 1984). In active pollination systems in general, and in *Ficus* in particular, limited amounts of pollen are produced (Pellmyr *et al.*, 2020).

In about one third of *Ficus* species, including *F. carica* (Galil and Neeman, 1977), pollen transport is passive (Kjellberg *et al.*, 2001; Cruaud *et al.*, 2012). The fig produces a large amount of pollen and the wasps become coated with pollen on emerging from their natal fig (Galil and Neeman, 1977; slides 37, 38) (Fig. 9.1D). When they enter a receptive fig, the body of the wasp will brush against the elongate stigmas ensuring their pollination (Galil and Neeman, 1977; slides 42, 43).

Two breeding systems have evolved in the mutualism. About half of the 850 fig species are monoecious (seeds and wasps develop within the same fig), and the other half are gynodioecious (functionally dioecious, where male plants produce pollen and pollen vectors and female plants produce seeds) (Cruaud *et al.*, 2012). In receptive figs of monoecious species, pistillate flowers are stacked in several layers due to variable pedicel length. The styles vary in length so that flowers located close to the fig wall have long stigmas while flowers located closer to the fig cavity have shorter stigmas. The wasps preferentially oviposit in shorter styled flowers, located closer to the fig cavity for several reasons. One is that the larvae located closest to the fig wall are more exposed to parasitic non-pollinating wasps ovipositing through the fig wall (Al-Beidh *et al.*, 2012). Another reason is that female wasps developing in galls located close to the fig cavity are more likely to be already mated when wasps begin to emerge from a fig, and they emerge more rapidly from the fig and are larger (Anstett, 2001). As a result, ovules located close to the fig wall mainly turn into seeds while ovules located closer to the fig cavity are mainly transformed into galls hosting wasp larvae (Nefdt and Compton, 1996). These wasps will leave the fig in search of a receptive fig to enter. They transfer pollen from their natal fig to the next fig, ensuring the male function of the tree.

In functionally dioecious species, half of the trees (Valdeyron and Lloyd, 1979) have figs presenting very long styles (Condit, 1932; Beck and Lord, 1988a), so that wasps cannot oviposit (Fig. 9.3A,B) (slides 12, 42, 43). The wasps enter these figs by mistake, effect pollination and die without producing offspring (Anstett *et al.*, 1997). As these trees produce seeds and no wasps, they are female. The figs of the other trees have short styled female flowers (Condit, 1932; Beck and Lord, 1988a; slides 10, 12, 28–30). The wasps oviposit in many of these flowers transforming them into galls (Fig. 9.3B,C) (slides 31, 32). Non-oviposited flowers generally produce no seeds because the ovules are not fertilized. This is due, depending on species, to very precise pollination of the flowers in which the wasps oviposit (Jousselin and Kjellberg, 2001), pollen stigma incompatibilities (Jousselin and Kjellberg, 2001), stigma traits (Beck and Lord, 1988b) or lack of pollen in the fig from which the wasp emerged (Leclerc du Sablon, 1908). As the figs of these trees produce pollen, pollen vectors (wasps), and no (or almost no) seeds, these trees are functionally male (Galil and Neeman, 1977; Neeman and Galil, 1978; Bronstein *et al.*, 1992; Kerdelhué and Rasplus, 1996a, b).

*Ficus carica* is a functionally dioecious species, which is hypothesized to have enabled the functional adaptation of the species to the seasonal Mediterranean climates, the area encompassing the original distribution of this species (Kjellberg *et al.*, 1987). The pollinator *Blastophaga psenes* Linnaeus, 1758 is not able to differentiate between the two sexes and enters figs of trees of both sexes. The volatile attractants released by receptive figs of male and female *F. carica* show no difference in chemical composition (Gibernau *et al.* 1997; Proffit *et al.*, 2020). Although the figs on female trees are traps for the wasps, as the females are unable to reproduce in these figs, they are fooled into entering the fig as a result of the exact mimicking of male figs (size, color), and most importantly the chemical volatiles that the pollinators home in on (Patel *et al.*, 1995).

Although *F. carica* has been in cultivation for thousands of years, an understanding of the role that the pollinating wasp, *B. psenes* (Figs. 9.1 and 9.2) plays in the reproduction of the figs (the process of caprification) is only a relatively recent elucidation (Baker, 1961). The earliest detailed accounts of the caprification process date back to the mid-19th century (Westwood, 1840). Subsequent investigations into the biology of *F. carica* (Solms-Laubach, 1882; Eisen 1896, 1901; Du Sablon, 1908; Longo, 1909; Ravasini, 1911; Rixford, 1918; Condit, 1920, 1932; Cotte and Reyner, 1923; Grandi, 1923; Crane, 1951) have ensured the ongoing commercial success of this crop.

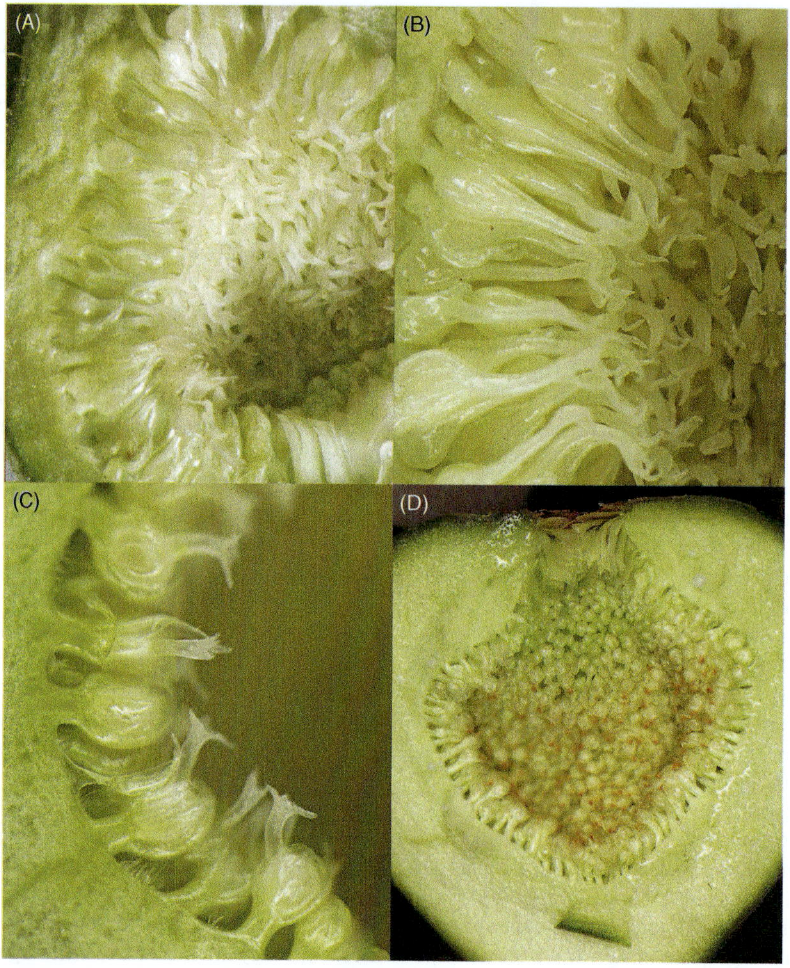


**Fig. 9.3.** *Ficus carica.* (A) Female tree, receptive fig; the elongate stigmas that will brush the body of the wasp, collecting pollen and ensuring seed set. (B) Detail of the elongate stigmas with papillae on which pollen will stick, the styles are long and do not allow wasp oviposition. (C) Male tree, inside a receptive fig the styles are short allowing wasp oviposition. (D) Male tree, fig a few days after oviposition; immature male flowers are visible in the upper part of the fig; in the lower part, female flowers show style with brown trace of ovipositor insertion, the pedicel elongates and the ovule begins to swell turning into a gall. (Figures courtesy of F. Kjellberg.)

## 9.2 Pollinator Wasp and Other Organisms Associated with *Ficus carica*

*Blastophaga psenes* is the only known pollinator of *F. carica* (Baker, 1961; Wiebes, 1994). The species has a large distribution and ranges from the Canary Islands to Uttarakhand in northern India (where it is associated with *F. palmata*), and from northern Iran to southern Algeria. However, no genetic analysis of the species has been attempted so far. Specimens from the Mediterranean area appear morphologically similar and may represent *B. psenes*. However, slight morphological differences between specimens from eastern localities (Iran, India), mostly associated with *F. palmata* (Fig. 9.4C,D), suggest that *B. psenes* as presently diagnosed may represent a species complex. Further studies are required to

**Fig. 9.4.** (A) *Ficus carica* (typical form), on rocks along river banks. (B) *F. carica* (form *Ficus colchica* Grossh), this is morphologically the most divergent form of *F. carica* with gracile branches and flexible leaves, Black Sea coast, Turkey. (C) *Ficus palmata*, Adiyaman Province, Turkey; *F. palmata* is distributed from Ethiopia to India and Nepal, and is the sole species closely related to *F. carica.* (D) *F. palmata* male tree with profichi. (Figures courtesy of F. Kjellberg.)

better define this supposedly well-known species. Morphologically, *B. psenes* has been thoroughly described by Grandi (1920, 1923, 1929), the female is winged and the male is apterous (Fig. 9.2C,D) like all other males of agaonids. The female is characterized by a third funicular segment subdivided into three segments (Fig. 9.2A); funiculars F5 to F8 bear one row of multiporous plate sensillae; only a residual scar is visible on the mesopleuron in place of a pollen pocket (Fig. 9.2F); a mesoscutum without a median sulcus; and metasomal spiracles with a small peritremata. In males, the metanotum is partly fused with the mesoscutum and propodeum, therefore the metapleurae are widely separated, but well defined through their internal septa (Fig. 9.2E), and the head is round bearing multiple spicules in its anterior third. Molecular phylogenies (Cruaud *et al.*, 2012) have placed *B. psenes* within a clade of agaonids (formerly described in the genus *Dolichoris*) associated with *Ficus* subgenus *Pharmacosycea* section *Oreosycea*. This result is corroborated by morphology and suggests that the association of *B. psenes* with *F. carica* (subgenus *Ficus*) may result from an ancient host shift. The venom gland reservoir (acid gland or poison bag) of *B. psenes*, and more generally in Agaonids, is exceptionally large compared to the venom gland of other chalcid wasps (Copland *et al.*, 1973). The secretion produced (possibly associated with larval secretion) is involved in the transformation of the flower (that has had an egg laid into the ovary) into a gall.

Females *B. psenes* have two large ovaries each containing from 130 to more than 200 ovarioles that are tightly packed (Grandi, 1929). Each ovariole contains a single egg ready for laying and no oocyte (Elias *et al.*, 2018). Indeed, all Agaonidae are pro-ovigenic. They mature all their eggs during pre-adult development (Grandi, 1929; Copland *et al.*, 1973; Dunn *et al.*, 2011; Jia *et al.*, 2014; Elias *et al.*, 2018) and no ovigeny occurs in adult females. Eggs of agaonids are composed of three parts, the ovum, a long peduncle and the ovariole, recognizable by the presence of a tunica propria. Because the ovariole is filled by a substance similar to the substance observed in the ovum, ovarioles were not detected by Grandi (1929) in *B. psenes*. The fate of the ovariole is still debated. In *B. psenes*, Grandi (1929) assumed the ovariole is incorporated with the peduncle into the ovum while observations in other genera suggest it may remain lodged in the flower style (Copland *et al.*, 1973) and may be used as a cue by other ovipositing females (Dunn *et al.*, 2011). The process of oviposition takes about two and a half minutes in *B. psenes*, and the style of the flowers in which an egg has been deposited develops a characteristic rusty coloration.

The egg of *B. psenes*, is deposited near the style insertion between the nucellus and the inner integument of the ovule. The egg hatches about 4–5 days after oviposition. The young larva is translucid, whitish and glabrous and exhibits 13 segments. The final larval instar is ventrally bent. The head bears two acute mandibles. The prothorax is longer than the two other thoracic segments, which bear ventro-laterally two rounded outgrowths. The number of larval instars is not known in *B. psenes* and is globally poorly known in Agaonidae, only having been recorded twice with four instars in *Pegoscapus* (Jansen *et al.*, 2012) and five instars in *Ceratosolen* (Jia *et al.*, 2014).

The post-embryonic development time of *B. psenes* depends on temperature, season and other parameters, but may take about two months (Grandi, 1929) for the spring generation. The post-embryonic development time until pupation is the same in males and females. However, males have a shorter pupation time and therefore emerge from their galls one day before females (protandry). Males copulate with multiple females while they are still enclosed in their galls (Fig. 9.1C). Females of the spring generation get abundantly dusted with pollen while in the process of emerging from their natal fig. The female lifespan of *B. psenes* is relatively short and varies from 5 to 8 days under laboratory conditions, but only two days in the natural environment (Kjellberg *et al.*, 1988).

Most *Ficus* species have a cohort of non-pollinating fig wasps also associated with their figs, which may be gallers (Compton and van Noort, 1992; Jansen-González *et al.*, 2014), cleptoparasites (Joseph, 1958) or true parasitoids (endo and exoparasitoids) of the galling wasp species and pollinators (Compton and van Noort, 1992; Tzeng *et al.* 2008; Yadav and Borges, 2018), or even hyperparasitoids (Compton *et al.*, 2009b). The galling non-pollinating wasps (at least the two species for which data is available) feed on galled nucellus and not, as the pollinators, on endosperm (Jansen-González *et al.*, 2014). This assemblage can vary from zero, or one or two species to well over 40 species (Compton and Hawkins, 1992; Hawkins and Compton, 1992; Chen *et al.*, 1999; Cook and Rasplus, 2003; Segar *et al.*, 2013; Wang *et al.*, 2015). The wasps are generally host-specific, but in Africa and South America some non-pollinating fig wasps are true generalists exploiting a series of local *Ficus* species (Mcleish *et al.*, 2012; Farache *et al.*, 2018). Within non-pollinating wasp communities, parapatric distributions of morphologically similar species suggest that competition is a major structuring factor (Darwell and Cook, 2017). The single known non-pollinating fig wasp associated with *F. carica* is *Philotrypesis caricae* L., 1762 (Pteromalidae, Sycoryctinae) (Figs. 9.5A–C and 9.5A–G). As do many of the non-pollinating fig wasps, *Philotrypesis* species oviposit through the fig wall into the ovules (slides 32–34) (Fig. 9.6A–C). This non-pollinating fig wasp is a cleptoparasite of the pollinator wasp larvae. The egg is laid as the pollinator egg between the inner integument and the nucellus (Joseph, 1958). It feeds initially, together with the host *Blastophaga psenes* larva, on the endosperm tissue (the development of which was initiated by the pollinator larva) inside the galled flower until the second larval instar, where after the *Philotrypesis* larvae dominates resulting in the death of the pollinator larvae (Joseph, 1958).

A diversity of other organisms develops inside figs and are transferred between figs by the pollinating wasps (Wohlfarter *et al.*, 2011). They include mites (Jauharlina *et al.*, 2012), nematodes (Martin *et al.*, 1973; Van Goor *et al.*, 2018), fungi (Caldis, 1927; Michailides *et al.*, 1996; Michailides and Morgan, 1998), and bacteria (Grimont *et al.*, 1979). The biology of the mites is poorly known. They have been recorded as feeding on ovule tissue, and may disperse autonomously between figs on a tree, while dispersal between trees is phoretic with the mites hitching a ride on the body of dispersing female pollinating wasps (Jauharlina *et al.*, 2012). Fig nematodes disperse inside the body of female pollinating fig wasps, and at least in the case of *Schistonchus africanus* do not colonize the wingless pollinator males (Vovlas *et al.*, 1998). They feed on wasp and plant tissue. The effect of nematodes on the wasps varies. *Parasitodiplogaster* nematodes appeared to have a strong effect on agaonid wasp reproductive success in Panama (Herre, 1993), while their effect is only marginal in Baja California (Van Goor *et al.*, 2018). It has been proposed that nematodes avoid entering the lethal trap constituted by dispersing non-pollinating fig wasps that oviposit through the fig wall, never entering a fig (Herre, 1993). However, *Parasitodiplogaster* nematodes in Baja California and the *Ficus carica* nematode *Schistonchus caprifici* (Gasparrini, 1864) colonize both the pollinator and the parasitoid fig wasps (Vovlas and Lariszza, 1996; Van Goor *et al.*, 2018). Pollinating wasps also carry fungi from ripe figs to receptive figs. This is the case for *Fusarium monoliforme*, which is carried from fig to fig by *Blastophaga psenes*, the pollinator of *Ficus carica*, and is responsible for a disease called endosepsis (Davey and Smith, 1933; Michailides *et al.*, 1996; Michailides and Morgan, 1998). The presence of *Fusarium moniliforme* strongly reduces the reproductive success of the wasp (Michailides *et al.*, 1996; Michailides and Morgan, 1998). *Blastophaga psenes* often exits the receptive fig it has entered, so that visited figs often contain no wasp corpse (Gibernau *et al.*, 1996). This can be seen as a wasp strategy to limit contamination by germs and parasites present in the wasp's body. Bacteria such as *Serratia ficaria* (Grimont *et al.*, 1979) have been found to be transferred between figs by the pollinators, though no negative effect on the wasps have been observed (Grimont *et al.* 1979; FK, personal observation).

## 9.3 Pollination Biology of Wild Common Fig, *Ficus carica* in its Native Habitat

*Ficus carica* L. 1753, is native to the whole of the Mediterranean basin (Khadari *et al.*, 2005b), where both cultivated plants and wild populations co-exist (Figs. 9.7A,B and 9.8A). Fossils, some of which date back 100,000 years, have been found in France (Planchon, 1864; Ambert *et al.*, 1995; slide 3). *Ficus carica* is also present along the south-eastern coast of the Black Sea and the Caspian Sea, where it presents somewhat different phenotypes and is genetically somewhat different (Khadari *et al.*, 2005a). These forms were described under the names *Ficus colchica* Grossh. (slides 4, 5) and *Ficus hyrcana* Grossh. They present more gracile shoots than typical *F. carica* (Fig. 9.4B). The arid-adapted species *Ficus palmata* (slides 6, 7) is the closest relative of *F. carica*, but is not conspecific (Berg and Corner, 2005; Khadari *et al.*, 2005b). It is distributed from Ethiopia to India and Nepal (Fig. 9.4C,D).

Descriptions of the year-round phenology of *Ficus carica* are confusing, because most are based on observations of male tree cultivars. The yearly number of wasp generations varies

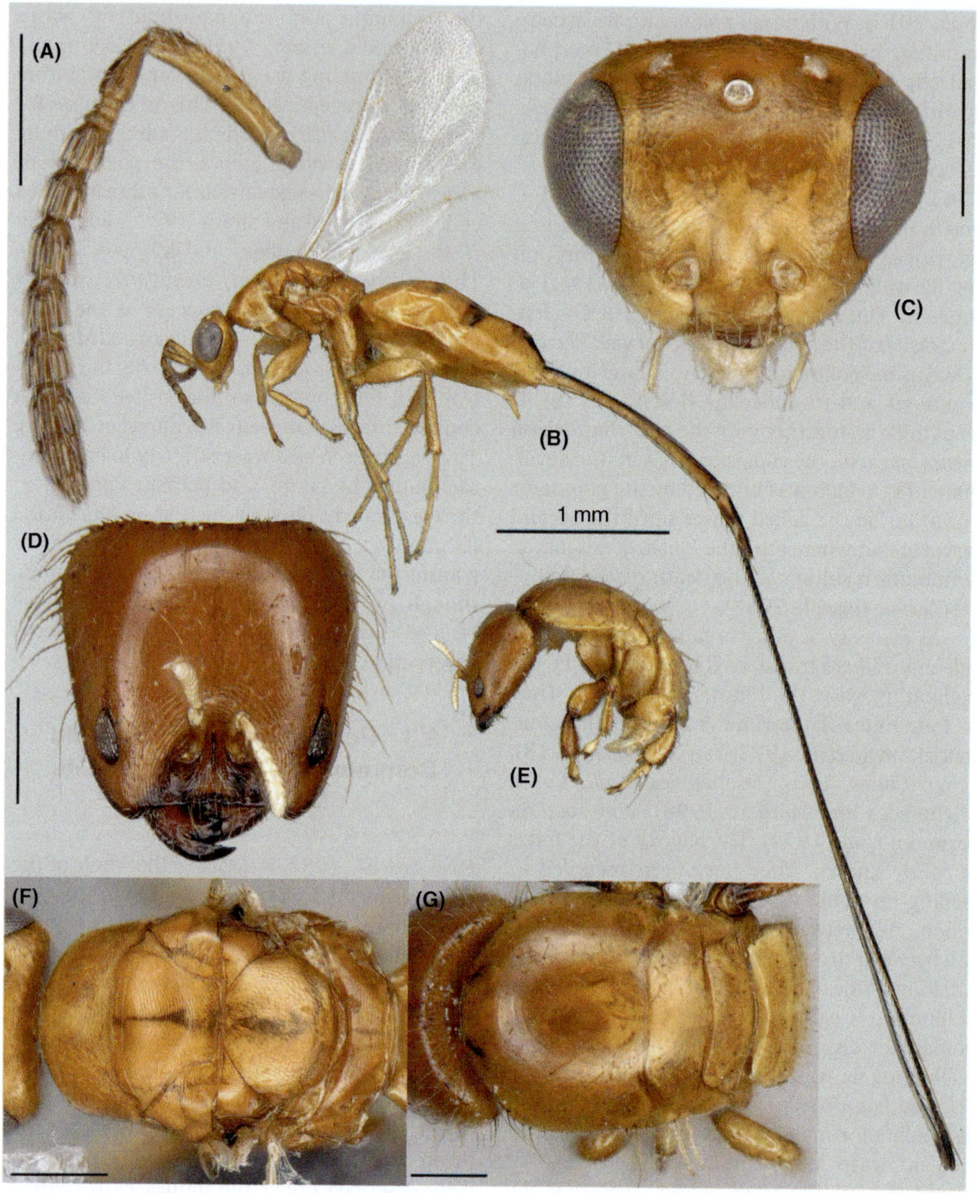


**Fig. 9.5.** *Philotrypesis caricae.* (A) Antenna. (B) Habitus female, lateral view. (C) Head female, frontal view. (D) Head male, frontal view. (E) Habitus male, lateral view. (F) Mesosoma female dorsal view. (G) Mesosoma male dorsal view. Scale = 200 μm excepted for habitus = 1 mm. (Figures courtesy of J.Y. Rasplus.)

geographically, but also depends on whether wild trees or some specific male tree cultivars are investigated. Here we first describe the year-round phenology of wild trees in the South of France, where *Ficus carica* is native, as presented by Valdeyron and Lloyd (1979) and improved by further observations in Kjellberg *et al.* (1987). This phenology is illustrated in slides 17–47.

Terminal buds flush in late March and the shoot grows until July. At each node, a fig bud may appear on the shoot that has developed in that year, at the axil of the leaf. Fig buds appear progressively along the shoot, from the base to

**Fig. 9.6.** *Philotrypesis caricae*. (A) Females ovipositing into a fig recently entered by *B. psenes*, on a male tree. (B, C) Female ovipositing and cleaning. (Figures courtesy of B. Schatz, J.Y. Rasplus and F. Kjellberg.)

the apex of the shoot. On female trees, fig buds develop rapidly into receptive figs that become receptive from mid-June to mid-July. Often the sub-terminal fig bud remains dormant until the next spring. However, the spring development of such sub-terminal fig buds stops rapidly after onset and the fig aborts. On male trees, fig buds appear later, from late July to the end of summer. The basal ones progressively become receptive before winter. The distal fig buds, closer to the apex of the branch and representing often about half of the fig buds, do not develop into receptive figs before winter. They overwinter as fig buds and, the next spring, they develop synchronously into receptive figs. These figs, which have overwintered as fig buds, are the sole figs producing abundant, functional pollen. These delayed polliniferous male figs are visited by wasps in April and release the next generation of wasps in June–July (Fig. 9.8B). These are the sole wasps in the yearly cycle that are abundantly dusted with pollen (Fig. 9.1D). When the wasps emerge from most of these figs, there are no receptive figs on male trees, but all the figs on female trees are receptive. Hence, these wasps visit female trees, ensure pollination and die without producing offspring. The latest wasps emerging from the latest polliniferous figs reach the first figs to become receptive on the shoot of that year of male trees, in late July–early August,

**Fig. 9.7.** *Ficus carica*. (A) Wild *F. carica* on the rocky bank of river Hérault, March 2020, St-Guilhem-le-désert, France. (B) Wild *F. carica* male in fields, July, Malemort, France; at that period, there are no receptive figs on male trees, the wasps can only go from male figs to figs on female trees. (Figures courtesy of F. Kjellberg and J.Y. Rasplus.)

enter them and oviposit. In Montpellier, in these figs, the wasp larvae reach the last larval stage in September–October and do not complete their metamorphosis before next spring. The figs will ripen the following April releasing the wasps that will ensure the colonization of the polliniferous figs. Hence, there are two yearly wasp generations.

In particularly warm years or in warmer climates, the wasps developing in figs that have been visited in late July become adult before winter. As fig buds develop progressively into receptive figs

**Fig. 9.8.** *Ficus carica*. (A) Wild *F. carica* female tree with ripe seed-containing figs. (B) Wild *F. carica* male, polliniferous figs at the stage of *Blastophaga psenes* emergence. (Figures courtesy of F. Kjellberg and J.Y. Rasplus.)

along the shoot of that year, the wasps find receptive figs on the same tree and visit them. These later figs are the ones that will overwinter containing last stage fig wasp larvae inside. Hence, depending on climatic conditions, there are two or three yearly generations of wasps. Detecting the yearly number of generations of wasps in winter is easy. If the figs hosting wasps in winter are at the base of the shoot that has developed in that year, then it is a two-yearly wasp generation cycle (slide 21). If the figs hosting the wasps are higher up on the shoot of that year, then it is a three-yearly wasp generation cycle. This conclusion can be confirmed by looking for fig seeds. In most years, in Montpellier, some seeds are found in male figs visited by wasps emerging from polliniferous figs that have overwintered as fig buds. Figs that develop directly on the shoot of that year do not produce functional pollen. Hence, there are no seeds in male figs visited by wasps emerging from figs developing on the shoot of that year.

However, fig and axillary vegetative bud development are under the control of apical dominance. Lopping off the terminal bud, or any factor reducing its vigor such as strong frost in winter, will result in early development of the sub-terminal vegetative bud and of the associated fig bud, making the reading of phenology somewhat more complex, but still feasible. For instance, early figs on male trees, that have been receptive at the same time as the main female tree crop, contain not only wasp offspring, but also many seeds and they ripen sweet (Condit, 1920), while the figs that are visited following the usual natural yearly cycle are visited by late emerging wasps and contain only few seeds. This is because the last wasps to emerge from a polliniferous fig carry reduced quantities of pollen (Kjellberg *et al.*, 2014). Another confusing factor is when fig trees grow in places allowing prolonged vegetative growth throughout summer, or when intensive pruning produces the same effect. In such cases, development of the fig bud becomes somewhat atypical (slide 81 for a cultivar in a moist climate).

Much confusion in the literature stems from the presence in cultivation of male tree genotypes presenting an atypical early onset of fig bud development (slides 65, 66). In some cultivars, fig buds begin to develop rapidly on the shoot of that year on male trees. As a result, figs become receptive in July, when wasps are emerging from the polliniferous figs. These figs are readily visited, and they release a new generation of wasps before winter. On leaving their fig, these wasps find receptive figs on the same male tree, visit them, and their offspring overwinter in these figs. Hence, figs on the shoot of that year on male trees may release wasps before winter for two different reasons, either (i) because of a long growth season extending late in autumn, or (ii) because they develop on particular cultivated genotypes. In the 19th and 20th century literature on *Ficus carica* cultivation, the figs developing on male trees and ripening before winter are called mammoni, whatever their origin. Figs overwintering with wasps inside are called mamme, whether visited by wasps emerging from polliniferous figs or from mammoni figs, and polliniferous figs are called profichi (Condit, 1947). Because the term mammoni represents two different biological phenomena, the classical literature is almost impossible to analyze.

On wild male trees, the colonization of the first receptive figs on the shoot of that year by wasps emerging from polliniferous figs is a sensitive point during which the continuity of wasp generations is difficult to observe. In some years the continuity of generations is poor, resulting in few visited figs on few male trees. Indeed, on average, only one in several thousand female wasps produced by polliniferous figs will breed successfully. The lack of receptive figs on male trees at the time when most figs are receptive on female trees is also observed in other *Ficus* species that grow in highly seasonal climates, and that have developed grouped fig receptivity on all female trees. This is the case for *Ficus exasperata* (Patel, 1996), and *Ficus erecta* (Tzeng *et al.*, 2006). The taxonomy and phylogeny of genus *Ficus* shows that these examples represent three independent instances of evolution of adaptation to highly seasonal climates (Cruaud *et al.*, 2012). Hence, the succession of generations on *Ficus carica* as described above corresponds to a general pattern that may emerge when *Ficus* species adapt to seasonal climates. The reason why male trees do not produce receptive figs at the time when figs are receptive on female trees is straightforward. When such figs are present, the wasps, which are guided to receptive figs by the volatile organic compounds they release (Soler *et al.*, 2012), move *en masse* to these nearby figs as described by Condit (1920) 'it is not at all uncommon to find mammoni figs with five or six females, either around the eye, or trying to force an entrance [slide 66]. If such figs are opened, the spaces between the bracts are literally packed full and black with the living and dead bodies of *Blastophaga*'. These wasps do not go to the female trees and hence, do not ensure the reproductive success of the male plant. Therefore, a male tree producing such figs is counter selected for under the premise of natural selection.

The matching of wasp emergence from overwintering figs and receptivity of the polliniferous figs lacks precision. The development of the overwintering wasps is stopped at the last larval stage by low temperatures. As a result, the overwintering wasps are all synchronized at the last larval stage. Their development resumes if sufficiently high temperatures are reached, and this sometimes happens in the middle of winter for trees growing in protected places (Joseph, 1958). If a fig-bearing branch is cut in winter and brought into a heated room in a water jar, after a few days, the wasps emerge. Conversely, the development of receptive figs depends on a sufficient period of vegetative growth, which depends on a sum of suitably warm degree-days.

Further, temperatures in April may drop below those required for wasp emergence so that emergences from figs stop for several days (Khadari *et al.*, 2005a,b), while the receptive figs continue to grow. Hence, receptive figs need to be and are capable of waiting for several weeks for the emergence of wasps (Khadari *et al.*, 2005a,b; slides 22, 23).

## 9.4 Fig Crops and Pollination Biology of Cultivated Common Fig, *Ficus carica*, in its Native Range

Wild female *F. carica* produce sweet tasty pollinated ripe figs in autumn (Fig. 9.8A) (slide 52), but they abort unpollinated figs. In places where *F. carica* thrives as wild trees, and where fig cultivation was limited, female trees germinating in ditches along fields were often preserved and even tended (slide 53). They were pollinated by wasps emerging from nearby male trees and produced large quantities of pollinated ripe figs. In the Montpellier region (South of France), these figs were often dried by farmers for family consumption during winter and could constitute about 5–10% of the caloric diet of a family during winter (Kjellberg and Valdeyron, 1984). Pollination was generally successful because of the presence of nearby wild male trees and because only few female trees were tended. Locally appreciated wild female genotypes were sometimes multiplied through cuttings within a village, benefiting from natural pollination (Kjellberg, 1995). This was originally the case for the variety Black Bursa (Bursa Siyahi) in the Bursa region of Turkey for the production of fresh figs. These figs require pollination, but as long as cultivation remained at a small scale, the peasants relied on natural pollination. When the number of cultivated trees increased to provide figs for the Istanbul market, a large fraction of the fig crop aborted and fig-cultivation experts from Izmir were called in. They introduced the controlled pollination techniques (see below) that were locally unknown in the Bursa region (U. Aksoy, personal communication). This easy incorporation of wild plants into cultivation systems may explain the strong genetic diversity of fig cultivars at the western limit of *F. carica*'s natural range (Achtak *et al.*, 2010) even though initial fig domestication is assumed to have occurred in the east Mediterranean. Indeed, there is no report of a genetic signature of introduction into the west Mediterranean of domesticated plants from the east Mediterranean. This is in stark contrast with the olive, another emblematic tree crop in the Mediterranean. Olive presents much more marked domestication traits and a clear signature of introductions of domesticates from the east Mediterranean to the west Mediterranean is demonstrated accompanied by some introgression in the west by local wild olive (Gros-Balthazard *et al.*, 2019).

Some of the most important fig cultivars, including the Turkish Sari Lop or Smyrna fig, require pollination for fruit development (Condit, 1947; Gaaliche *et al.*, 2011). The fruiting phenology of these cultivars is the same as that of wild female trees (Fig. 9.9B) (slide 57). However, they are mainly cultivated in locations that favor the natural drying of the ripe figs, i.e. where the climate at the end of summer and in autumn is warm and dry. Often the trees are planted in deep soils, in habitats where rocks are missing, i.e. where natural populations of *F. carica* are absent (Fig. 9.9A) (slides 55, 56, 58). Further, the trees are planted in large number, by far outnumbering wild local male fig trees if any are present. Therefore, the farmers need to manage pollination in a process called caprification. This is achieved by picking polliniferous figs (profichi in the agronomic literature) in June from male trees (caprifig trees in the agronomic literature) just before wasp emergence, and suspending them in female trees (slides 59, 61, 62). The profichi originate either locally from trees cultivated by the farmers (slides 60, 63, FK personal observation) or from wild growing local trees (Aumeeruddy-Thomas and Hmimsa, 2018) or the profichi may be provided by specialists who sell them on markets (slide 64). Often farmers use a mix of locally produced and commercial profichi figs to cover the period of fig receptivity (U. Aksoy, personal communication; FK personal observation). The farmers are conscious of the importance of the male figs that overwinter, sheltering the wasps (mamme figs) to ensure colonization of the profichi. In case of climatic accidents leading to a deficit of mamme figs on their cultivated caprifig trees, farmers may import mamme figs and suspend them in the male trees to ensure the visitation of the profichi (slide 67).

**Fig. 9.9.** (A). Fig plantation, Meander valley, Turkey. (B). Female tree, receptive figs of the second crop (= the crop present on wild female trees), requiring pollination; this phenotype is very similar to typical wild female trees, same locality. (Figure courtesy of F. Kjellberg.)

Caprification of caprifig trees with mamme figs is reported in Turkey with mamme figs from Greece (Eisen, 1901), for the Kerkhennah islands off Tunisia (G. Valdeyron, personal communication 1982, slide 67) and for the Greek Cyclades islands (Tournefort, 1717) with mamme figs from the continent. In Turkey, along the Meander valley, some villages are specialized in the production of profichi figs. In at least one of them, the caprifig cultivar in use produced very early receptive figs on the growth of the year facilitating local survival of the pollinating wasps despite the removal of most profichi from the trees (slides 64, 65).

A domestication trait in fig cultivars is the occurrence of parthenocarpy, i.e. fig development without pollination. Some such cultivars produce the same crop of figs that ripen in August–September as do wild female trees, and as do Smyrna figs. Such cultivars may be favored to avoid diseases transmitted by the pollinating wasps. Alternatively, male figs may be treated with fungicides (Michailides *et al.*, 1996, slides 68–76).

In some cultivars, the sub-terminal fig bud(s) that overwinters on the tree does not abort early when growth resumes in spring. At receptivity, these figs cannot be pollinated because mamme figs on male trees contain no pollen. These figs may continue to develop as parthenocarpic fruits and ripen at the same time as the profichi, in June–July in the South of France (slides 77, 78). They bear different names in different countries such as breba figs or figues-fleurs, or first crop figs. First crop figs have overwintered as fig buds (slide 78) and their development has been affected in the same way as figs that overwintered as fig buds on male trees. They include a thicker, spongier fig wall and larger size so that flowers are less packed within the fig cavity. Their styles are also somewhat shorter that those of second crop figs (compare slides 42, 43 with slides 79, 80). First crop figs are not a natural crop on female trees and, as such, are prone to early fruit dropping. Abortion is less marked in trees that have more resources available. This may be the reason for the cultivation of some cultivars in which the first crop is parthenocarpic but not the second one, as in such cultivars, in the absence of pollination, most second crop figs abort. When such trees are cultivated in low number in the Montpellier region, because of the availability of pollinating wasps stemming from wild trees, it is worthwhile removing by hand the second crop figs. In the Montpellier region we investigated a one hectare orchard of such a cultivar. The nursery man who provided the plants included a wild germinated tree for pollination. The tree turned out to be female. Nevertheless, we recorded that second fig crops were more visited on the adjacent fig trees than in the remainder of the orchard and as a result they aborted a larger fraction of their first fig crop (Kjellberg *et al.*, 1983). The simple explanation is that the second fig crop of the wild female tree was more attractive to the pollinators than the cultivar's figs, and hence attracted pollinating wasps that also visited neighboring trees. This trait associated with domestication has resulted in reduced attractivity of the second fig crop with an associated reduction in pollinator visits, a higher second fig crop abortion and hence more resources available for the first fig crop.

## 9.5 The History of Fig Cultivation and Pollination

Monkeys and apes relish eating figs produced by a diversity of *Ficus* species and early humans were probably no exception. Figs of *F. carica* were consumed by humans from early times. Indeed, charred fig seeds and even fig remains have been found in settlements along the Jordan valley dating back 11,000 BP (the occurrence of parthenocarpic figs in a site dating to 11,400 BP, in the Jordan Valley is doubtful) (Kislev *et al.*, 2006), and figs and fig trees are regularly included in old scriptures from the region. This is no surprise as the climate of Galilea is favorable for fig tree growth and wild *F. carica* is still thriving (personal observation). More unexpectedly, charred fig seeds appear in sites of the Rakka governorate along the Euphrates in the tenth millennium BP (Willcox, 2014). The dry climate in that region is not compatible with the presence of wild *F. carica* populations. Further Willcox (2014) stated: 'finds of figs are numerous on a number of early sites in the southern Levant where they were apparently used intensively at an earlier date than in the north. However they occur well away from their natural habitats, so even in the south it appears that figs and probably the cultivation of fig trees was spreading thanks to human contact over large distances.' Fuller and Stevens (2019) provide a map of early archaeological sites presenting evidence for figs. Several sites to the East are outside what we consider the natural range of *F. carica* suggesting fig cultivation outside its natural range. There is no early archaeological documentation of fig consumption in the Meander valley region. This contradicts the classical idea (Condit, 1947), included in the name *F. carica*, that Caria was the region of origin of the species. Indeed, the landscape in which the fig is cultivated south of Izmir is not compatible with the presence of large wild populations of fig trees.

The first known description of caprification is given by Theophrastus (391–288 BC). The term he used for caprification, erinasmos, is the same as used today in Greece and suffers no ambiguity as to its meaning. He states that the pollinators develop feeding on the seeds in the caprifigs. The classical roman word for male fig tree is *caprificus*, and the verb 'caprificare' translates as 'to caprify', so this term has also survived up to today. Such indigenous terminology for caprification is in stark contrast with old Hebraic scriptures in which there is no evocation of the male fig tree and no specific word for caprification (Goor, 1965). In the Turkish language, İncir is the edible fig, İncir Ağacı is the female fig tree and the caprifig tree is called Erkek İncir, i.e. male fig. İlek is the first male crop of the season. In Turkish, İlk is first and Erkek is male. However, in some regions, the names Tum İncir and Eşek İnciri are used for caprifig trees. Tum İncir means the unripe fig while Eşek İnciri means donkey figs, i.e. a fig that can only be eaten by donkeys (Oğuzhan Çalişkan, personal communication). Throughout the Arab speaking countries, the caprifig tree is called Dokkar (or dhakar), which means male. This is true from Portugal, where the caprifig tree is known as figo di toca (Eisen, 1901), to Syria. However, in the north of Syria, in the Idlib governorate, a specific word, 'tub' is used for caprifig tree, which could originate in Tum İncir (Oğuzhan Çalişkan, personal communication). These differences in terminology for caprifigs and caprification suggest independent origin of the practice of caprification in northern Syria, Turkey, Greece and Italy, while the practice of caprification may have been imported, or is of more recent origin in the south of the Mediterranean basin. A thorough comparative linguistic investigation of the terminology associated with caprification is required before firm conclusions can be drawn.

A series of historical accounts of caprification around the Mediterranean basin in the late 19th–early 20th century are provided by American authors that were involved in introducing *Blastophaga psenes* to America and in optimization of fig cultivation (Eisen, 1896, 1901; Condit, 1947). The agronomic literature describing caprification practice and the succession of fig crops in California demands some interpretation as they were conducted outside the natural range of the fig in the absence of natural populations. Eisen (1901) states that the caprifig trees should not be planted close to the female trees as they require sheltered places and shade. Indeed, female plants used for fig production are generally not planted in the natural habitat of *F. carica*. He also states that 'The various crops of the caprifig do not always succeed each other continuously. There is a frequent lapse of time between the falling of the profichi and the appearance of the mammoni.' Hence many of the caprifigs cultivated in California presented the typical phenology of wild caprifig trees presented above. Caprification of caprifig trees was also used early in California (Eisen, 1901). The difficult succession of crops on the male trees is still present in Condit's 1933 account of fig culture in California. He states, 'One weak point in the succession of crops, the break between the profichi and the mammoni crops, can in the case of some varieties be overcome by pinching back the tips of new shoots here and there over the tree, about the middle of May. This temporarily stops the terminal growth but stimulates development of lateral fruit buds which will be ready to receive blastophagas early in June.' Lopping off the terminal bud (in May in France) is an easy way to induce the early development of the subterminal fig bud, resulting in a fig that is receptive when wasps emerge from the profichi. Progressively caprifig cultivars, in which the phenological gap between profichi and mammoni was missing, were chosen. Such a cultivar is reported in Condit's (1955) compendium of fig cultivars 'On trees of the Milco at Fresno, the mammoni figs appeared in numbers a week or ten days before the profichi', and today only such cultivars are used.

*Ficus carica* has also been introduced for cultivation to various other parts of the world where there is a suitable climate, such as Australia (Morton, 1987), China (Morton, 1987), India (Rattanpal *et al.*, 2017), Japan (Morton, 1987), North America (including California (USA), Mexico, Costa Rica, and the West Indies (Condit, 1920, 1933, 1955; Morton, 1987; Armstrong, 1988, 2010), South America (including Argentina, Brazil, Chile, Colombia, Venezuela) (Morton, 1987), and South Africa (Tribolet, 1912; Wohlfarter *et al.*, 2011; Giliomee *et al.*, 2017). In some of the introduced areas, *F. carica* has become naturalized in suitable habitats. In southern California, trees establish in riverbeds and creeks among willow and cottonwood thickets, and have become invasive (Holmes *et al.*, 2014).

There are a huge number of varieties of *F. carica* in cultivation, in areas of origin (Achtak *et al.*, 2010) as well as in areas of introduction with some widely known introduced cultivars. Condit (1955) provides a key for almost 700 cultivars. Adriatic, Black Ischia, Brown Turkey, Brunswick, Genoa, Celeste, and Marseilles, Poona, and Purple Genoa are favored in warmer climates and do not require pollination (Morton, 1987). These varieties are usually those that are first introduced to new regions, because they do not require the more complicated horticultural practice of caprification. The Smyrna fig and its pollinator *B. psenes* is often subsequently introduced to various parts of the world presenting Mediterranean climates to facilitate the commercial cultivation of this variety, used for the dry fig market, which if not pollinated abort their figs before they reach maturity (Tribolet, 1912). Although the Smyrna fig (known as the Calimyrna fig in California) was introduced to California around 1880, the pollinator only followed around 1899 (Morton, 1987, Tribolet, 1912), after it was discovered that it was not present in the caprifig trees that had been established 20 years earlier. Experimental deduction through the process of artificial pollination elucidated the need for the pollinator wasp to perform this duty (Tribolet, 1912). Following the successful introduction and propagation in California, the Smyrna fig was imported into South Africa in 1902/1903 and the pollinator was introduced from California in 1908, and has been cycling successfully ever since then (Tribolet, 1912; Wohlfarter *et al.*, 2011).

In California and South Africa the caprifig trees (male trees) of *F. carica* produce three caprifig crops per year: the summer *profichi* crop (with galls and male flowers); an autumn *mammoni* crop (galls and abortive male flowers); and an overwintering *mamme* crop (galls and abortive male flowers) where the figs only mature the following spring (Armstrong, 1988; Giliomee *et al.*, 2007; Tribolate, 1912). Pollen laden female fig wasps leaving the *profichi* crop enter the *mammoni* crop and oviposit into the ovules; the resulting wasp progeny leaving this crop then oviposit in the *mamme* crop and those wasps then complete the cycle by ovipositing in the *profichi* crop again. Pollen from the *profichi* crop is used to pollinate the Smyrna (Calimyrna) orchards in midsummer, as the other two crops do not produce pollen in the figs (Armstrong, 2010). The wasps survive the cold winter months as larvae in the overwintering *mamme* crops. Female trees produce two fig crops per year: the *breba* crop, ripening in early summer; and the main second crop ripening in autumn. The process of caprification can aid the transmission of internal rot fungus or endosepsis (*Fusarium moniliforme* var. *fici* (Caldis, 1927) from the male caprifigs to the female Smyrna figs (Armstrong, 2010; Caldis, 1927; Davey and Smith, 1933; Michailides and Morgan, 1998; Wohlfarter *et al.*, 2011). Intervention is required to avoid this transmission. In California the overwintering *mamme* crop is harvested in spring and the figs are treated with a fungicide after being split open (Armstrong, 2010). They are then re-introduced to the orchards so that the emerging wasps can enter the *profichi* crop. This fungus can result in considerable crop losses, with hot and humid conditions being ideal for rapid transmission (Caldis, 1927; Michailides and Morgan, 1998; Wohlfarter *et al.*, 2011).

## Conclusion

The cultivated fig *Ficus carica* is a dioecious species that has been in cultivation for thousands of years, although an understanding of the role that the pollinating wasp *Blastophaga psenes* plays in the reproduction of the cultivated fig (the process of caprification) is only a relatively recent elucidation. The process of caprification where oviposition by the pollinator *B. psenes* is required to ensure successful ripening of the female figs still plays an important role in the cultivated fig industry. In natural wild populations of *F. carica*, figs suitable for consumption are produced on female trees only if they are pollinated by *Blastophaga psenes*, released from figs borne on male trees, which contain both wasps and pollen, and are usually not eaten. The male figs that produce the wasp pollinator were called 'caprifigs' because they were not eaten and were said to be 'only good for the goats'. Commercial cultivation of fresh figs for the consumer market usually centers on parthenocarpic cultivars, i.e. varieties that have been artificially selected for during the long association of cultivation by humans. Some of these varieties have been artificially selected to not need pollination to produce ripe figs and their propagation is via the process of cuttings (truncheons). Fig trees will normally abort their fig crop if the figs are not pollinated as the figs are then an energy cost with no benefit to the tree. This mode of vegetative propagation allowed for the anthropogenic expansion of the species' distribution to non-native regions across various parts of the world, and may have been one of the earliest forms of agricultural domestication of a plant species, predating cereal production. The best fig cultivars grown for dry fig production have mainly been cultivars that need pollination (caprification) such as the Turkish Sari Lop or Smyrna fig, which has a nutty flavor and contains seeds, in contrast to the sweeter fruit of the parthenocarpic varieties that are much softer without seeds (Armstrong, 1988; 2010; Tribolet, 1912). The pollinator has hence also been introduced to various parts of the world where *F. carica* is cultivated to enable the production and supply of figs to the dry fig market (Mars, 2003). In many regions, however, these cultivars are being progressively replaced by cultivars that do not require pollination in order to avoid sanitary problems and costs.

## References


Achtak, H., Ater, M., Oukabli, A., Santoni, S., Kjellberg, F. and Khadari, B. (2010) Traditional agroecosystems as conservatories and incubators of cultivated plant varietal diversity: the case of fig (*Ficus carica* L.) in Morocco. *BMC Plant Biology* 10, 28.

Al-Beidh, A., Dunn, D.W., Power, S.A. and Cook, J.M. (2012) Parasites and mutualism function: measuring enemy-free space in a fig–pollinator symbiosis. *Oikos* 121, 1833–1839. DOI: 10.1111/j.1600-0706.2011.20179.x

Ambert, P., Quinif, Y., Roiron, P. and Arthuis, R. (1995) Les travertins de la vallée du Lez (Montpellier, Sud de France). Datations $^{230}$Th/$^{234}$U et environnements pleistocènes. *Comptes Rendus de l'Académie des Sciences, Paris*, 32, série lia, 667–674.

Anstett, M.C. (2001) Unbeatable strategy, constraint and coevolution, or how to resolve evolutionary conflicts: the case of the fig/wasp mutualism. *Oikos* 95, 476–484.

Anstett, M.C., Hossaert-McKey, M. and Kjellberg, F. (1997) Figs and fig pollinators: evolutionary conflicts in a coevolved mutualism. *Trends in Ecology and Evolution* 12, 94–99.

Armstrong, W.P. (1988) The calimyrna fig and its wasp. *California Garden* 79, 135–138.

Armstrong, W.P. (2010) Calimyrna Figs in California. https://www2.palomar.edu/users/warmstrong/pljune99.htm (accessed 6 December 2019).

Aumeeruddy-Thomas, Y. and Hmimsa, Y. (2018) Fig and olive domestication in the Rif, Northern Morocco. Entangled human and tree lives and history. In: Stepanoff, C. and Vigne, J.D. (ed.) *Hybrid Communities: Biosocial Approaches to Domestication and Other Transspecies Relationships*, Routledge, London, pp. 179–198.

Baker, H.G. (1961) *Ficus* and *Blastophaga*. *Evolution* 15, 378–379.

Barker, N.P. (1985) Evidence of a volatile attractant in *Ficus ingens* (Moraceae). *Bothalia* 15, 607–611.

Beck, N.G. and Lord, E.M. (1988a) Breeding system in *Ficus carica*, the common Fig. I. Floral diversity. *American Journal of Botany* 75, 1904–1912.

Beck, N.G. and Lord, E.M. (1988b) Breeding system in *Ficus carica*, the common Fig. II. Pollination events. *American Journal of Botany* 75, 1913–1922.

Berg, C.C. and Corner, E.J.H. (2005) Moraceae - *Ficus*. Flora Malesiana Series I (Seed Plants) Volume 17/Part 2. National Herbarium of the Netherlands, Leiden.

Borges, R.M. and Kjellberg, F. (2014) New insights from the fig-fig wasp model interaction system. *Acta Oecologica* 57, 3–4.

Bronstein, J.L. (1992) Seed predators as mutualists: ecology and evolution of the fig/pollinator interaction. In: Bernays, E. (ed.) *Insect-Plant Interactions* Vol IV. CRC Press, London, pp. 1–44.

Caldis, P.D. (1927) Etiology and transmission of endosepsis (internal rot) of the fruit of the Fig. *Hilgardia* 2, 287–346.

Chen, Y.-R., Chuang, W.-C. and Wu, W.-J. (1999) Chalcid wasps on *Ficus microcarpa* L. in Taiwan (Hymenoptera: Chalcidoidea). *Journal of Taiwan Museum* 52, 39–79.

Compton, S.G. (1990) A collapse of host specificity in some African fig wasps. *South African Journal of Science* 86(1), 39–40.

Compton, S.G. and Hawkins, B.A. (1992) Determinants of species richness in southern African fig wasp assemblages. *Oecologia* 91, 68–74. DOI:10.1007/BF00317243

Compton, S.G. and van Noort, S. (1992) Southern African fig wasps (Hymenoptera: Chalcidoidea): resource utilization and host relationships. *Proceedings of the Koninklijke Nederlandse Akademie van Wetenschappen* 95, 423–435.

Compton, S.G., Grehan, K. and van Noort, S. (2009a) A fig crop pollinated by three or more species of agaonid fig wasps. *African Entomology* 17, 215–222.

Compton, S.G., van Noort, S., Mcleish, M., Deeble, M. and Stone, V. (2009b) Sneaky African fig wasps that oviposit through holes drilled by other species. *African Natural History* 5, 9–15.

Cook, J.M. and Segar, S.T. (2010) Speciation in fig wasps. *Ecological Entomology* 35, 54–66.

Condit, I.J. (1920) Caprifigs and Caprification. *California Agriculture Experimental Station Bulletin* 319, 341–377.

Condit, I.J. (1932) The structure and development of flowers in *Ficus carica* L. *Hilgardia* 6, 443–481.

Condit, I.J. (1933) Fig culture in California. California Agricultural Extension Service circular 77, The College of Agriculture, University of California, Berkeley, California.

Condit, I.J. (1947) *The Fig.* Chronica Botanica Co., Waltham, MA.

Condit, I.J. (1955) Fig varieties: A Monograph. *Hilgardia* 11, 323–538.

Cook, J.M. and Rasplus, J.Y. (2003) Mutualists with attitude: coevolving fig wasps and figs. *Trends in Ecology and Evolution* 18, 241–248.

Cook, J.M. and Segar, S.T. (2010) Speciation in fig wasps. *Ecological Entomology* 35, 54–66. DOI:10.1111/j.1365-2311.2009.01148.x

Copland, M.J.W., King, P.E. and Hill, D.S. (1973) The structure of the female reproductive system in the Agaonidae (Chacidoidea, Hymenoptera). *Journal of Entomology (A)* 48, 25–35.

Corner, E.J.H. (1933) A revision of the Malayan species of *Ficus*: *Covellia* and *Neomorphe. Journal of the Malayan Branch of the Royal Asiatic Society* 11(116), 1–65.

Cornille, A., Underhill, J.G., Cruaud, A, Hossaert-Mckey, M., Johnson, S.D., Tolley, K.A., Kjellberg, F., Van Noort, S., Proffit, M. (2012) Floral volatiles, pollinator sharing and diversification in the fig-wasp mutualism: insights from *Ficus natalensis* and its two wasp pollinators (South Africa). *Proceedings of the Royal Society B* 279, 1731–1739. DOI:10.1098/rspb.2011.1972

Cotte, J. and Reynier, A. (1923) La dioecie du figuier et *Blastophaga psenes*. *Comptes Rendus des Séances de la Société de Biologies* 88, 500–502.

Crane, J.C. (1951) Seasonal development of caprifig and seedless Calimyrna figs. *California Fig Institute American Research Conference Proceedings* 5, 8–10.

Cruaud, A., Jabbour-Zahab, R., Genson, G., Cruaud, C., Couloux, A., Kjellberg, F., Van Noort, S. and Rasplus, J.Y. (2010) Laying the foundations for a new classification of Agaonidae (Hymenoptera: Chalcidoidea), a multilocus phylogenetic approach. *Cladistics* 26, 359–387.

Cruaud, A., Ronsted, N., Chantarasuwan, B., Chou, L.S., Clement, W.L., Couloux, A., Cousins, B., Genson, G., Harrison, R.D., Hanson, P.E., Hossaert-McKey, M., Jabbourzahab, R., Jousselin, E., Kerdelhue, C., Kjellberg, F., Lopez-Vaamonde, C., Peebles, J., Peng, Y.Q., Pereira, R.A.S., Schramm, T., Ubaidillah, R., Van Noort, S., Weiblen, W.D., Yang, D.R., Yodpinyanee, A., Libeskind-Hadas, R., Cook, J.M., Rasplus J.Y. and Savolainen V. (2012) An extreme case of plant-insect co-diversification: figs and fig-pollinating wasps. *Systematic Biology* 61, 1029–1047.

Darwell, C.T. and Cook, J.M. (2017) Cryptic diversity in a fig wasp community – morphologically differentiated species are sympatric but cryptic species are allopatric. *Molecular Ecology* 26, 937–950.

Davey, A.E. and Smith, R.E. (1933) The epidemiology of fig spoilage. *Hilgardia* 7, 523–551.

Dunn, D.W., Jansen-Gonzalez, S., Cook, J.M., Yu, D.W. and Pereira, R.A.S. (2011) Measuring the discrepancy between fecundity and lifetime reproductive success in a pollinating fig wasp. *Entomologia Experimentalis et Applicata* 140(3), 218–225.

Du Sablon, L. (1908) Observations on the various forms of figs. *Revue Générale de Botanique* 20, 129–150.

Eisen, G. (1896) Biological studies on figs, caprifigs, and caprification. *Proceedings of the California Academy of Sciences* Ser. 2, 5, 897–1001.

Eisen, G. (1901) *The fig: its history, culture and curing with a descriptive catalog of the known varieties of Fig.* U.S. Department of Agriculture, Division of Pomology, Bulletin n° 9, Government Printing Office, Washington, USA.

Elias, L.G., Lino-Neto, J. and Pereira, R.A.S. (2018) Oogenesis and ovarian morphology in pollinating and non-pollinating fig wasps: evidence from adult and immature stages. *Invertebrate Reproduction and Development* 1–7. DOI:10.1080/07924259.2018.1469549

Farache, F.H.A., Cruaud, A., Rasplus, J.Y., Cerezini, M.T., Rattis, L., Kjellberg, F. and Pereira, R.A.S. (2018) Insights into the structure of plant-insect communities: Specialism and generalism in a regional set of non-pollinating fig wasp communities. *Acta Oecologica* 90, 49–59.

Frank, S. (1984) The behavior and morphology of the fig wasps *Pegoscapus assuetus* and *P. jimenezi*: Descriptions and suggested behavioral characters for phylogenetic studies. *Psyche* 91, 289–308.

Fuller, D.Q. and Stevens, C.J. (2019) Between domestication and civilization: the role of agriculture and arboriculture in the emergence of the first urban societies. *Vegetation History and Archaeobotany* 28, 263–282 DOI:10.1007/s00334-019-00727-4

Gaaliche, B., Trad, M. and Mars, M. (2011) Effect of pollination intensity, frequency and pollen source on fig (*Ficus carica* L.) productivity and fruit quality. *Scientia Horticulturae* 130, 737–742.

Galil, J. (1977) Fig biology. *Endeavour* 1, 52–56.

Galil, J. and Neeman, G. (1977) Pollen transfer and pollination in the common fig (*Ficus carica* L.). *New Phytologist* 79,163–171.

Gasparrini, G. (1863) Sulla maturazione e la qualit~t dei fichi dei contorni di Napoli. *Atti Accad. Pontaniana* 9, 99–118 (1864).

Gibernau, M., Buser, H.R., Frey, J.E. and Hossaert-McKey, M. (1997) Volatile compounds from extracts of figs of *Ficus carica*. *Phytochemistry* 46, 241–244.

Gibernau, M., Hossaert-McKey, M., Anstett, M.-C. and Kjellberg, F. (1996) Consequences of protecting flowers in a fig: a one-way trip for pollinators? *Journal of Biogeography* 23, 425–432.

Giliomee, J.H., Venter, E. and Wohlfarter M. (2007) Mediterranean black fig fly, *Silba adipata* McAlpine (Diptera: Lonchaeidae), recorded from South Africa. *African Entomology* 15, 383–384. DOI: 10.4001/1021-3589-15.2.383

Goor, A. (1965) The history of the fig in the Holy Land from ancient times to the present. *Economic Botany* 19, 124–135.

Grandi, G. (1920) Studio morphologico e biologico della *Blastophaga psenes* (L.). *Bollettino del Laboratorio di Zoologia Generale e Agraria Portici* 14, 63–204.

Grandi, G. (1923) Gli insetti dei caprifichi. *Rivista di Biologia* 5, 69–90.

Grandi, G. (1929) Studio morphologico e biologico della *Blastophaga psenes* (L.) (2a edizione riveduta). *Bollettino del Laboratorio di Entomologia Bologna* 2, 1–147.

Grimont, P.A.D., Grimont, F. and Starr, M.P. (1979) *Serratia ficaria* sp. nov., a bacterial species associated with Smyrna figs and the fig wasp *Blastophaga psenes*. *Current Microbiology* 2, 277–282.

Grison, L., Edwards, A.A. and Hossaert-McKey, M. (1999) Interspecies variation in floral fragrances emitted by tropical *Ficus* species. *Phytochemistry* 52, 1293–1299.

Gros-Balthazard, M., Besnard, G., Sarah, G., Holtz, Y., Leclercq, J., Santoni, S., Wegmann, D., Glémin, S. and Khadari, B. (2019) Evolutionary transcriptomics reveals the origins of olives and the genomic changes associated with their domestication. *The Plant Journal* 100, 143–157.

Hawkins, B. and Compton, S.G. (1992) African fig wasp communities: undersaturation and latitudinal gradients in species richness. *The Journal of Animal Ecology* 61, 361–372. DOI:10.2307/5328

Herre, E.A. (1993) Population structure and the evolution of virulence in nematode parasites of fig wasps. *Science* 259, 1442–1445.

Herre, E.A., Jander, K.C. and Machado, C.A. (2008) Evolutionary ecology of figs and their associates: recent progress and outstanding puzzles. *Annual Review of Ecology, Evolution and Systematics* 39, 439–458.

Holmes, K.A., Greco, S.E. and Berry, A.M. (2014) Pattern and process of fig (*Ficus carica*) invasion in a California riparian forest. *Invasive Plant Science and Management* 7, 46–58.

Hossaert-McKey, M., Gibernau, M. and Frey, J.E. (1994) Chemosensory attraction of fig wasps to substances produced by receptive figs. *Entomologia Experientia et Applicata*, 70, 185–191.

Jansen-González, S., Simone de Padua Teixeira, S.P. and Pereira, R.A.S. (2012) Mutualism from the inside: coordinated development of plant and insect in an active pollinating fig wasp. *Arthropod-Plant Interactions* 6, 601–609.

Jansen-González, S., Teixeira, S.P., Kjellberg, F. and Pereira R.A.S (2014) Same but different: larval development and gall-inducing process of a non-pollinating fig wasp compared to that of pollinating fig-wasps. *Acta Oecologica* 57, 44–50.

Jauharlina, J., Lindquist, E.E., Quinnell, R.J., Robertson, H.G. and Compton, S.G. (2012) Fig wasps as vectors of mites and nematodes. *African Entomology* 20, 101–110.

Jia, L.Y., Xiao, J.H., Niu, L.M., Ma, G.C., Fu, Y.G., Dunn, D.W. and Huang, D.W. (2014) Delimitation and description of the immature stages of a pollinating fig wasp, *Ceratosolen solmsi marchali* Mayr

(Hymenoptera: Agaonidae). *Bulletin of Entomological Research* 104(2), 164–175. DOI:10.1017/S0007485313000606

Joseph, K.J. (1958) Recherches sur les chalcidiens *Blastophaga psenes* (L.) et *Philotrypesis caricae* (L.) du figuier (*Ficus carica* L.). *Annales des Sciences Naturelles, Zoologie, 11ème série* 20, 197– 260.

Jousselin, E. and Kjellberg, F. (2001) The functional implications of active and passive pollination in dioecious figs. *Ecology Letters* 4, 151–158.

Kerdelhué, C. and Rasplus, J.Y. (1996a) Non-pollinating Afrotropical fig wasps affect the fig-pollinator mutualism in *Ficus* within the subgenus *Sycomorus*. *Oikos* 75(1), 3–14. DOI:10.2307/3546315

Kerdelhué, C. and Rasplus, J.Y. (1996b) The evolution of dioecy among Ficus (Moraceae): an alternative hypothesis involving non-pollinating fig wasp pressure on the fig-pollinator mutualism. *Oikos* 77(1), 163–166. DOI:10.2307/3545597

Khadari, B., Grout, C., Santoni, S., Hochu, I., Roger, J.-P., Ater, M., Aksoy, U. and Kjellberg, F. (2005a) Etude préliminaire des origines de *Ficus carica* L. et de sa domestication *Les Actes du BRG* 5, 53–65.

Khadari, B., Grout, C., Santoni, S. and Kjellberg, F. (2005b) Contrasted genetic diversity and differentiation among Mediterranean populations of *Ficus carica* L.: a study using mtDNA RFLP. *Genetic Resources and Crop Evolution* 52, 97–109.

Kislev, M.E., Hartmann, A. and Bar-Yosef, O. (2006) Early domesticated fig in the Jordan Valley. *Science* 312, 1372–1374.

Kjellberg, F. (1995) Le bon sauvage ou le figuier de mon village. *Fruits Oubliés* 1/95, 3–4.

Kjellberg, F. and Lesne, A. (2020) *Ficus carica* and its pollination. https://hal.archives-ouvertes.fr/hal-02516888v1 (accessed 12 April 2020).

Kjellberg, F. and Valdeyron, G. (1984) The pollination of the fig tree (*Ficus carica* L.) and its control in horticulture. *Acta Oecologica Oecologia Generalis* 5, 407–412.

Kjellberg, F., Aljibouri, A. and Valdeyron, G. (1983) Observations récentes sur la pollinisation du figuier. *Fruits* 38, 567–569.

Kjellberg, F., Gouyon, P.H., Ibrahim, M., Raymond, M., and Valdeyron, G. (1987) The stability of the symbiosis between dioecious figs and their pollinators: a study of *Ficus carica* L. and *Blastophaga psenes* L. *Evolution* 41, 693–704. DOI:10.1111/j.1558-5646.1987.tb05846.x.

Kjellberg, F., Doumesche, B. and Bronstein, J.L. (1988) Longevity of a fig wasp (*Blastophaga psenes*). *Proceedings Koninklijke Nederlandse Akademie van Wetschappen (C)* 91, 117–122.

Kjellberg, F., Jousselin, E., Bronstein, J.L., Patel, A., Yokoyama, J. and Rasplus, J.Y. (2001) Pollination mode in fig wasps: the predictive power of correlated traits. *Proceedings of the Royal Society of London Series B-Biological Sciences* 268, 1113–1121. DOI:10.1098/rspb.2001.1633

Kjellberg, F., Suleman, N., Raja, S., Tayou, A., Hossaert-McKey, M. and Compton, S.G. (2014) Some pollinators are more equal than others: factors influencing pollen loads and seed set capacity of two actively and passively pollinating fig wasps. *Acta Oecologica* 57, 73–79.

Leclerc du Sablon, M. (1908) Structure et développement de l'albumen du caprifiguier. *Revue Générale de Botanique* 20, 14–24, figs 1–6 planche 6.

Longo, B. (1909) Osservazioni e recerchi sul *Ficus carica*. *Annali di Botanica Genoa* 7, 235–256.

Mars, M. (2003) Fig (*Ficus carica* L.) genetic resources and breeding. *Acta Horticulturae* 605, 19–27. DOI:10.17660/ActaHortic.2003.605.1

Martin, G.C., Owen, A.M. and Way, J.I. (1973) Nematodes, figs and wasps. *Journal of Nematology* 5(1), 77–78.

Mcleish, M.J., Beukman, G., van Noort, S., Wossler, T.C. (2012) Host-plant species conservatism and ecology of a parasitoid fig wasp genus (Chalcidoidea; Sycoryctinae; Arachonia). *PLoS ONE* 7(9), e44804. DOI:10.1371/journal.pone.0044804

Michailides, T.J. and Morgan, D.P. (1998) Spread of *endosepsis* in Calimyrna fig orchards. *Phytopathology* 88, 637–647.

Michailides, T.J., Morgan, D.P. and Subbarao, K.V. (1996) Fig endosepsis. An old disease still a dilemma for California growers. *Plant Disease* 8, 828–841.

Moe, A.M., Rossi, D.R. and Weiblen, G.D. (2011) Pollinator sharing in dioecious figs (Moraceae). *Biological Journal of the Linnean Society* 103, 546–558.

Molbo, D., Machado, C.A., Sevenster, J.G., Keller, L. and Herre, E.A. (2003) Cryptic species of fig-pollinating wasps: implications for the evolution of the fig-wasp mutualism, sex allocation, and precision of adaptation. *Proceedings of the National Academy of Sciences of the United States of America* 100, 5867–5872.

Morton, J. (1987) Fig. In: Morton, J.F. and Dowling, C.F. (eds) *Fruits of Warm Climates*. Miami, Florida, USA, pp. 47–50.

Neeman, G. and Galil, J. (1978) Seed set in the 'male syconia' of the common fig *Ficus carica* L. (Caprificus). *New Phytologist* 81(2), 375–380.
Nefdt, R.J.C. and Compton, S.G. (1996) Regulation of seed and pollinator production in the fig-fig wasp mutualism. *The Journal of Animal Ecology* 65, 170–182.
Patel, A. (1996) Variation in a mutualism: phenology and the maintenance of gynodioecy in two Indian fig species. *Journal of Ecology* 84, 667–680.
Patel, A., Anstett, M.-C., Hossaert-McKey, M. and Kjellberg, F. (1995) Pollinators entering female dioecious figs: why commit suicide? *Journal of Evolutionary Biology* 8, 301–313.
Pellmyr, O., Kjellberg, F., Herre, E.A., Kawakita, A., Hembry, D., Holland, J.N., Terrazas, T., Clement, W., Segraves, K.A. and Althoff, D. (2020) Active pollination drives selection for reduced pollen-ovule ratios. *American Journal of Botany* 107, 164–170.
Planchon G. (1864) *Étude des tufs de Montpellier du point de vue géologique et paléontologique*. Thesis, Faculté des Sciences, Paris, 73 p.
Proffit, M., Lapeyre, B., Buatois, B., Deng, X., Arnal, P., Gouzerh, F., Carrasco, D. and Hossaert-McKey, M. (2020) Chemical signal is in the blend: bases of plant-pollinator encounter in a highly specialized interaction. *Scientific Reports* 10, 10071.
Ramirez, W.B. (1970) Host specificity of fig wasps (Agaonidae). *Evolution* 24, 680–691.
Ramirez, W.B. (1974) Coevolution of *Ficus* and Agaonidae. *Annals of the Missouri Botanical Garden* 61, 770–780.
Ramirez, W.B. (1977) Evolution of blastophagy. *Brenesia* 9, 1–13.
Rasplus, J.Y. (1996) The one-to-one species-specificity of the *Ficus*-Agaoninae mutualism : how casual? In: van der Maesen, L.J.G., van der Burgt, X.M. and van Medenbach de Rooy, J.M. (eds) *The Biodiversity of African Plants*. Kluwer Academic Publishers, Wageningen, The Netherlands, pp. 639–649.
Rattanpal, H.S., Sidhu, G.S. and Bons, H.K. (2017) Tree and fruit characteristics of fig cultivars grown in Punjab. *Acta Horticulturae* 1173, 117–120. DOI:10.17660/ActaHortic.2017.1173.20
Ravasini, R. (1911) *Die Feigenbaume Italiens und ihre Beziehungen zu Finander*. Bern, Germany.
Rixford, G.P. (1918) Smyrna fig culture. *USDA Bulletin (1915-23)* 732, 1–43.
Rodriguez, L.J., Bain, A., Chou, L.S., Conchou, L., Cruaud, A., Gonzales, R., Hossaert-McKey, M., Rasplus, J.Y., Tzeng, H.Y. and Kjellberg, F. (2017) Diversification and spatial structuring in the mutualism between *Ficus septica* and its pollinating wasps in insular South East Asia. *BMC Evolutionary Biology* 17, 207.
Segar, S.T., Pereira, R.A.S., Compton, S.G. and Cook, J.M. (2013) Convergent structure of multitrophic communities over three continents. *Ecology Letters* 16, 1436–1445.
Soler, C., Proffit, M., Bessière, J.M., Hossaert-McKey, M. and Schatz, B. (2012) When males change their scents in presence of females, the case of the plant *Ficus carica*. *Ecology Letters* 15, 978–985.
Solms-Laubach, H. Grafen zu. (1882) Die Herkunft, Domestikation und Verbreitung des gewoenlichen Feigenbaumes (*Ficus carica* L.). Dieterichsche Verlags Buchhandlung, Göttingen, Germany.
Souto-Vilarós, D., Proffit, M., Buatois, B., Rindos, M., Sisol, M., Kuyaiva, T., Michalek, J., Darwell, C.T., Hossaert-McKey, M., Weiblen, G.D., Novotny, V. and Segar, S.T. (2018) Pollination along an elevational gradient mediated both by floral scent and pollinator compatibility in the fig and fig-wasp mutualism. *Journal of Ecology* 106, 2256–2273.
Souza, C.D., Pereira, R.A.S., Marinho, C.R., Kjellberg, F. and Teixeira, S.P. (2015) Diversity of fig glands is associated with nursery mutualism in fig trees. *American Journal of Botany* 102, 1564–1577.
Tournefort, J.P. de (1717) *Relation d'un voyage du Levant. Vol. I, Imprimerie royale* Paris, France.
Tribolet, I. (1912) Caprification of smyrna figs. *Agricultural Journal of the Union of South Africa* 3, 247–256. https://journals.co.za/content/ajusa/3/2/AJA0000021_102 (accessed 5 August 2021).
Tzeng, H.Y., Lu, F.Y., Ou, C.H., Lu, K.C. and Tseng, L.J. (2006) Pollinational-mutualism strategy of *Ficus erecta* var. *beecheyana* and *Blastophaga nipponica* in seasonal Guandaushi Forest Ecosystem, Taiwan. *Botanical Studies* 47, 307–318.
Tzeng, H.Y., Tseng, L.J., Ou, C.H., Lu, K.C., Lu, F.Y. and Chou, L.S. (2008) Confirmation of the parasitoid feeding habit in *Sycoscapter*, and their impact on pollinator abundance in *Ficus formosana*. *Symbiosis* 45, 129–134.
Valdeyron, G. and Lloyd, D.G. (1979) Sex differences and flowering phenology in the common fig, *Ficus carica* L. *Evolution* 33, 673–685. DOI:10.1111/j.1558-5646.1979.tb04720.x
Van Goor, J., Piatscheck, F., Houston, D.D. and Nason, J.D. (2018) Figs, pollinators, and parasites: a longitudinal study of the effects of nematode infection on fig wasp fitness. *Acta Oecologica* 90, 140–150.
van Noort, S., Ware, A.B. and Compton, S.G. (1989) Pollinator - specific volatile attractants released from the figs of *Ficus burtt - davyi. South African Journal of Science* 85, 323–324.

Vovlas, N. and Larizza, A. (1996) Relationship of *Schistonchus caprifici* (Aphelenchoididae) with fig inflorescences, the fig pollinator *Blastophaga psenes*, and its cleptoparasite *Philotrypesis caricae*. *Fundamental and Applied Nematology* 19, 443–448.

Vovlas, N., Troccoli, A., Van Noort, S. and Van Den Berg, E. (1998) *Schistonchus africanus* n. sp. (Tylenchida: Aphelenchoididae) associated with *Ficus thonningii* (Moraceae) and its pollinator wasp *Elisabethiella stuckenbergi* (Agaonidae, Chalcidoidea, Hymenoptera). *Journal of Nematology* 30, 404–410.

Wang, G., Cannon, C.H., Chen, J. (2016) Pollinator sharing and gene flow among closely related sympatric dioecious fig taxa. *Proceedings of the Royal Society B* 283, 20152963.

Wang, R., Aylwin, R., Barwell, L., Chen, X.Y., Chen, Y., Chou, L.S., Cobb, J., Collette, D., Craine, L., Giblin-Davis, R.M., Ghana, S., Harper, M., Harrison, R.D., McPherson, J.R., Peng, Y.Q., Pereira, R.A.S., Reyes-Betancort, A., Rodriguez, L.J.V., Strange, E., van Noort, S., Yang, H.W., Yu, H. and Compton, S.G. (2015) The fig wasp followers and colonists of a widely introduced fig tree, *Ficus microcarpa*. *Insect Conservation and Diversity* 8, 322–336. DOI:10.1111/icad.12111

Ware, A.B. and Compton, S.G. (1992) Breakdown of pollinator specificity in an African fig tree. *Biotropica* 24, 544–549.

Ware, A.B., Kaye, P.T., Compton, S.G. and Van Noort, S. (1993) Fig volatiles: their role in attracting pollinators and maintaining pollinator specificity. *Plant Systematics and Evolution* 186, 147–156.

Westwood, J.O. (1840) On caprification as practised upon the Figs in the south of Europe and the Levant, with descriptions of the insects employed for that purpose; and observations upon the Agaon paradoxum Dalman. *Transactions of the Entomological Society of London* II, 214–224.

Weiblen, G.D. and Bush, G.L. (2002) Speciation in fig pollinators and parasites. *Molecular Ecology* 11, 1573–1578.

Wiebes, J.T. (1994) The Indo-Australian Agaoninae (pollinators of figs). *Verhandelingen der Koninklijke Nederlandsche (Nederlandse) Akademie van Wetenschappen* 92, 1–208.

Willcox, G. (2014) Food in the early neolithic of the Near East. In: Milano, L. (ed.) *Paleonutrition and Food Practices in the Ancient Near East.* History of the Ancient Near East: Monographs Vol. XIV 1–10. Padova, Italy.

Wohlfarter, M., Giliomee, J.H., Venter, E. and Storey, S. (2011) A survey of the arthropod pests and plant parasitic nematodes associated with commercial figs, *Ficus carica* (Moraceae), in South Africa. *African Entomology* 19, 165–172. DOI: 10.4001/003.019.0118

Yadav, P. and Borges, R.M. (2018) Host-parasitoid development and survival strategies in a non-pollinating fig wasp community. *Acta Oecologica* 90, 60–68. DOI: 10.1016/j.actao.2017.04.001

Yang, P., Li, Z., Peng, Y. and Yang, D. (2012) Exchange of hosts: can agaonid fig wasps reproduce successfully in the figs of non-host *Ficus*? *Naturwissenschaften* 99(3), 199–205.

Yu, H., Tian, E., Zheng, L., Deng, X., Cheng, Y., Chen, L., Wu, W., Tanming, W., Zhang, D., Compton, S.G. and Kjellberg, F. (2019) Multiple parapatric pollinators have radiated across a continental fig tree displaying clinal genetic variation. *Molecular Ecology* 28, 2391–2405.

# 10 Soil and Nutritional Requirements

**P. Gordon[1]*, L. Ferguson[2] and P. Brown[2]**
[1]*Cooperative Extension, Madera and Merced Counties, Division of Agriculture and Natural Resources, University of California, Madera, CA, USA;* [2]*Department of Plant Sciences, University of California, Davis, CA, USA*

## 10.1 Introduction

*Ficus carica* is tolerant of a wide range of soil textures ranging from sandy to clay loams (Ferguson *et al.*, 1990), and tolerant of high pH and carbonates in excess of 10% (Aydin, 1998; Flaishman *et al.*, 2008). Productive orchards can also exist in soils with a pH as low as 4.1–4.5 (Caetano *et al.*, 2006), though lime is required to increase productivity (Flaishman *et al.*, 2008). Unlike many deciduous tree crops, a fig tree's root system is shallow (Condit, 1947), and can be productive in shallow soils that would limit other orchard crops. However, rooting depth exceeding 1.2 m is preferable (Ferguson *et al.*, 1990). Like most crop plants, *F. carica* grows best on soils with high fertility (İrget *et al.*, 2008), however yield may decrease as growth becomes overly vegetative on the best soils, thus it may be best to avoid the most productive ground, for instance Class I soils as defined by the USDA NRCS. Traditionally both fresh and dried figs have been produced in arid Mediterranean climates with summer temperatures of 32°C or higher (Ferguson *et al.*, 1990). Limited local production has expanded into other temperate regions for the fresh fruit market or personal use.

## 10.2 Nutrition

Unlike many orchard crops, there is a dearth of fertilization research on *F. carica*. The bulk of the nutrition research has been done in Turkey, the world's leader in dried fig production. The Turkish leaf sampling method consists of sampling the third leaf from the shoot base (Kabasakal, 1983), typically also the location of the oldest fruit on the branch. Sampling methodologies in California and other areas recommend selecting the most recently developed leaf on sun exposed non-fruiting branches (Aksoy *et al.*, 1987a; Brown, 1994; Holstein *et al.*, 2015). These leaves will not be supplying nutrients to fruits, and therefore have higher levels of the plant mobile nutrients nitrogen, phosphorus, potassium and magnesium. This makes comparison of leaf standards between regions fraught at best. Additionally, production practices differ between the major fig production areas of the world; fig orchards in Turkey are typically rainfed and may or may not be fertilized, whereas orchards in California, Brazil and the Mediterranean countries in Europe are intensively irrigated and fertilized. Thus, it is more appropriate that fig production regions adapt their own leaf tissue standards. Because of this, this review will focus

*Corresponding author: pegordon@ucanr.edu


DOI: 10.1079/9781789242881.0010

on the role of these nutrients in fig growth and fruit quality and will not suggest critical leaf tissue levels but summarize the range of values from major surveys or research trials. We will also suggest future nutrition research needs.

While rootstocks are not widely used in fig production, there is some evidence that different genotypes may be more or less efficient at taking up nutrients, particularly ones critical for fruit quality such as potassium and calcium (Jafari *et al.*, 2018). There are also differences between rootstocks' ability to take up nutrients under drought stress conditions, suggesting it is beneficial for drought prone areas to adopt rootstocks better able to forage nutrients under low-water conditions. Jafari *et al.* (2018) examined several fig rootstocks and found that 'Matti' was best for nutrient uptake for well-irrigated figs, whereas 'Dehdez' was more appropriate for water stressed conditions.

Table 10.1 lists some of the leaf nutrient values found in four articles. Aksoy *et al.*, (1987c) reported the average values of 73 healthy, full-bearing orchards in the Big Menderes Valley. This report was part of a series establishing regional standards; only this one has been reported in the table. Aksoy and Anac (1993a) conducted a survey of the top five Sarilop orchards, as determined based on fruit quality, from both the Big and Small Meander Valleys in Turkey. Researchers sampled the third leaf from the base of each shoot at the onset of the fruit ripening period. Brown, 1996 conducted a survey of eight high vigor Calimyrna orchards in California. Brown sampled the most recently matured sun-exposed leaves on non-fruiting branches in July, and reported values, ±SD. Moreno *et al.* (1998) conducted a trial attempting to establish leaf tissue standards of a well fertilized, vigorous orchard. Researchers sampled mid-shoot leaves '12 times at 15-day intervals during four consecutive vegetative cycles'.

## 10.3 Macronutrient

### Nitrogen

Nitrogen is a critical component in amino acids, DNA and RNA, and is typically found in the highest amounts (Hawkesford *et al.*, 2012). Under the warm temperatures and high pH conditions typical of fig production, soil nitrifying bacteria eventually convert all forms of applied and native nitrogen into nitrate, which is negatively charged and therefore highly mobile in soils. Nitrate is an environmental contaminant in surface water and ground water, causing eutrophication in fresh and salt water ecosystems, causing health problems in animals and humans. Agricultural nitrogen applications have led to the deterioration of ground water and surface water quality (Schlesinger, 2009), therefore agricultural applications should be based on crop need, which in permanent crops is typically calculated by crop removal rates. Symptoms of nitrogen deficiency are chlorosis of the entire leaves (Fig. 10.1b) (Garza-Alonso *et al.* 2019),

**Table 10.1.** Compilation of leaf tissue nutrient levels gained via surveys of high vigor orchards, not including fertilization rate trials.

| Nutrient | Source | | | |
|---|---|---|---|---|
| | Aksoy *et al.*, 1987c | Aksoy and Anac, 1993a | Brown, 1994 | Moreno *et al.*, 1998 |
| N | 1.695–1.709% | 1.46–1.94% (av. 1.64) | 2.1% ± 0.3 | 26.4–26.6 mg $g^{-1}$ |
| P | 0.088% | 0.07–0.11% (av. 0.09) | 0.12% ± 0.01 | 1.44–1.51 mg $g^{-1}$ |
| K | 1.070–1.161% | 1.11–1.53% (av. 1.36) | 1.0% ± 0.3 | 1.08–1.13 mg $g^{-1}$ |
| Ca | 3.568–3.614% | 2.78–4.73% (av. 3.87) | 3.0% ± 0.3 | 3.40–3.45 mg $g^{-1}$ |
| Mg | 0.658–0.721% | 0.50–1.07% (av. 0.65) | 0.7% ± 0.1 | 0.45–0.49 mg $g^{-1}$ |
| S | . | . | . | 1.72–1.74 mg $g^{-1}$ |
| Zn | . | . | 12 ppm ± 2 | 30–35 µg $g^{-1}$ |
| Fe | . | . | 110 ppm ± 12 | 183–189 µg $g^{-1}$ |
| Mn | . | . | 90 ppm ± 25 | 86–92 µg $g^{-1}$ |
| Cu | . | . | 6 ppm ± 1 | 15–21 µg $g^{-1}$ |
| B | . | . | 100 ppm ± 45 | . |

**Fig. 10.1.** Images of macronutrient deficiencies in *Ficus carica* leaves. (a) No deficiencies, (b) nitrogen deficiency, (c) phosphorus deficiency, (d) potassium deficiency, (e) calcium deficiency, (f) magnesium deficiency. (Figures used with permission of Garza-Alonso *et al.*, 2019.)

which is common across plant species. Figs seem to have low nitrogen needs compared to other orchard crops, however there are few studies determining the crop removal rates. Nitrogen rates are also generally low since excess N results in a reduction in fruit quality (Aksoy *et al.*, 2003) delays in maturity (Proebsting and Warner, 1954; Aydın and Kaptan, 2015), and a possible increase in aflatoxin contamination (Özer and Derici, 1998). According to Hernandez *et al.* 1994, 10 t of fresh fruit, 1.3 t of immature fruit, and the associated prunings would export 65 kg of nitrogen per hectare. When only looking at the fresh fruit in this orchard, 10 t removed 20.5 kg of nitrogen. More work needs to be done examining the nutrient export rate of fresh and dried figs.

Nitrogen applications have been shown to increase yield, which may be due to nitrogen's effect on fig tree vigor. As fig syconia are produced at the node of a leaf, anything that increases the number of nodes produced in a season should result in increased yield. Proebsting and Warner (1954) applied 44.8 and 89.7 kg N $ha^{-1}$ to Adriatic, Calimyrna and Black Mission figs from 1946 to 1952. Both application rates significantly increased vigor and yield in contrast to unfertilized controls. In both varieties, there was a non-significant, erratic, but frequent reduction in the quality of harvested Adriatic and Calimyrna fruit at the highest N application rate, while Black Mission fig quality was unaffected. The difference among these cultivars may be due to the fact that Black Mission figs have a smaller ostiole that is fairly closed (Condit, 1955) and are less susceptible to insect damage. The applied nitrogen rates used in the studies of Proebsting and Warner (1954) are much higher than those recommended by Ferguson *et al.* (1990), which range from 22 to 45 kg N $ha^{-1}$. The winter application used by Proebsting and Warner (1954) likely reduced uptake efficiency thus requiring a higher rate to meet tree demand. Multiple in-season nitrogen applications result in higher uptake efficiency than dormant season applications.

In other studies, Holstein *et al.* (2015) found a positive relationship between leaf tissue nitrogen concentrations and yield ($R^2$ value = 0.54). Leaf nitrogen content has also been found to positively correlate with fig photosynthetic rate (Vemmos *et al.*, 2013). Hernandez *et al.* (1994) found that in Brazilian fig orchards there was a correlation between high nitrogen rates and sweeter fruit in the first harvest of the year, but no effect with later applications. The authors suggested the late season humid conditions exacerbated defoliation from fig leaf rust, resulting in fewer photosynthates and fruit with less sugar. Other studies did not examine fruit sugar levels in response to nitrogen applications.

*F. carica* is unusual among temperate tree crops in that it appears to convert nitrate into ammonium in its leaves (Proebsting and Tate, 1952) unlike other temperate tree species, in which this conversion is in the roots (Hawkesford *et al.*, 2012). Three studies have reported the levels of nitrate in fig leaves; Proebsting and Tate (1952) looked at nitrate levels in leaves as part of their fertilization trial (Proebsting and Warner, 1954) and found concentrations as high as approximately 5500 ppm in mid-April in Adriatic figs and 7000 ppm in late May in Calimyrna figs (Fig. 10.2). Levels declined as the season progressed. Interestingly, the authors found that early season nitrate levels in leaves corresponded better to nitrogen treatments than total nitrogen levels (Fig. 10.2). Moreno *et al.* (1998), reported leaf nitrate levels of 2.2–2.4 mg $g^{-1}$ (2200–2400 ppm); however, the sampling time was not reported. Caetano *et al.* (2006) examined the application of manure to orchards and reported leaf lamina nitrate levels of 1.3 g $kg^{-1}$ (1300 ppm), though in this study lamina nitrate levels did not respond to fertilization, possibly because manure has low levels of available nitrogen. More work needs to be done to examine the relationship between leaf nitrate and fertilization levels. Using petiole nitrate tests to determine crop nitrogen status is well established in many vegetable crops (Hartz and Hochmuth, 1996) and grapes (Christensen and Peacock, 2000). However, nitrate tests must be established using realistic standards and effective sample times; in tomatoes, even well fertilized plants have low nitrate levels after full bloom, making petiole nitrate tests ineffective at measuring plant nitrogen status (Hartz and Bottoms, 2009), and petiole nitrate tests are only effective at bloom in 'Thomson Seedless' grapes and established standards may not be applicable to all cultivars (Christensen and Peacock, 2000).

Reported leaf tissue analysis values range depending on the time of the year samples are taken, and can be as high as 5.45% (Ersoy *et al.*, 2001) in the spring, and as low as 1.04% mg $g^{-1}$ (Mordoğan *et al.*, 2013) in late summer once fruit has begun to ripen. Work done on 'Sarilop' (Calimyrna) figs found that leaf tissue levels above 1.75% selected from the third node from the base of the shoot at the onset of fruit maturation

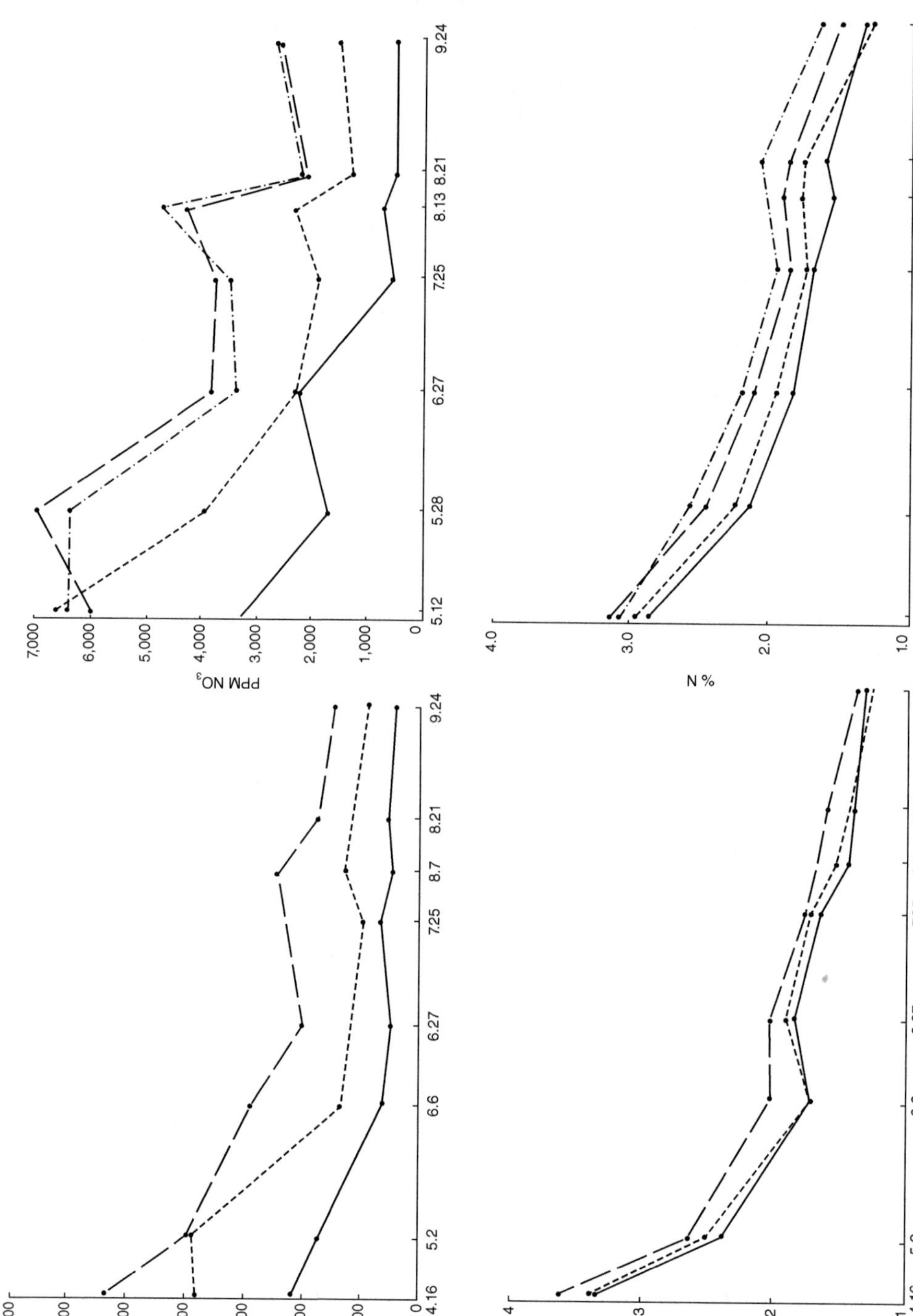


**Fig. 10.2.** Reprinted from Proebsting and Tate (1952). Adriatic leaf nitrate (top left) and Calimyrna leaf nitrate (top right) and total nitrogen (bottom left) and total nitrogen (bottom right) values gathered over eight and seven dates in 1951, respectively. Researchers applied ammonium sulfate to three plots at the rate of no nitrogen (solid line), 1.8 kg (short dash), and 3.6 kg (long dash), and in a single Calimyrna plot, 5.4 kg (alternate short and long dashes). Reprinted with permission from the American Society for Horticultural Science.

led to more, smaller, fruits with poor color (Aksoy *et al.*, 2003). This value is within range of regional reported values (Table 10.1) and is much lower than the 2.2–2.5% leaf tissue values reported by Ferguson *et al.* (1990) in Calimyrna figs. Leaf sampling procedures in Ferguson *et al.* (1990) were not stated but can be assumed to be standard for the region, which is to select the most recently matured sun-exposed leaf on non-fruiting branches in mid-summer. Sampling methodology could account for differences in recommended leaf tissue levels.

Moreno *et al.* (1998) conducted a multi-year fertilization trial using marketable yield in a high yielding 'Pellejo Duro' orchard to determine optimal levels of leaf nutrient levels in a high yielding, fertilized and irrigated orchard receiving 70 kg of nitrogen per hectare per year and determined the optimal nitrogen levels were 26.4–26.6 mg $g^{-1}$ (2.64–2.66%) in mid-shoot leaves selected in the middle of a 'vegetative cycle'.

Nitrogen applications have been looked at in great detail in *F.* carica, however there are some opportunities for future research. Many crop species have periods of higher or lower nitrogen uptake rates, and fertilization recommendations suggest targeting those periods for the heaviest nitrogen applications. However, the timing of the most rapid nitrogen uptake rates for *F. carica* are unknown. Additionally, as nitrogen is known to decrease fresh fruit quality, the critical leaf tissue values for nitrogen may differ between fresh and dried fruit.

## Phosphorus

Phosphorus is a critical component for some amino acids, phospholipids, and nucleic acids. Phosphorus availability in soils is highly dependent on pH, as it will precipitate into a variety of compounds, thus it is the least available of the plant essential elements. At low soil pH, phosphorus forms compounds with iron and aluminum, and at high soil pH it both forms compounds with calcium and magnesium and binds to calcium and magnesium carbonates. At any point in time, the amount of phosphorus available in the soil solution is far below crop needs, and a regular supply of phosphorus depends on the dissolution of phosphorus-containing compounds (Holford, 1997).

Requirements for tree crops are largely unknown (Brown, 1994) and the mycorrhizal relationships that many tree species form may facilitate phosphorus uptake. It is unknown whether *F. carica* forms mycorrhizal relationships (Brown, 1994). No studies the authors are aware of have examined phosphorus fertilization on its own, though several trials that have used complete fertilizers have reported leaf tissue levels. Garza-Alonso *et al.* (2019) artificially induced phosphorus deficiency in figs and observed a deep green coloration in the leaves, followed by patchy chlorosis which began in the center lobe, and then evolved into necrotic spots (Fig. 10.1c) Fig phosphorus removal rates are very low: Hernandez *et al.*, (1994) found that 10 t of fresh, mature fruit, 1.3 t of immature fruit, and prunings exported 4.3 kg of phosphorus (10 kg $P_2O_5$); approximately 1 kg of P (2.2 kg $P_2O_5$) was in the 10 t of fresh fruit.

Leaf phosphorus levels drop as the season progresses and have been reported to be as high as 0.4% in leaf tissues in early spring (Proebsting and Warner, 1954), and drop until they stabilize between May and June (Proebsting and Warner, 1954; Ersoy *et al.*, 2001). Moreno *et al.* (1998) reported that ideal leaf tissue levels range from 1.4 to 1.5 mg $g^{-1}$, which is higher than other summer leaf tissue levels in other studies of 0.12% in Brown (1994) and 0.07–0.11% in Aksoy and Anac (1993a). Aksoy *et al.* (1987c) report a regional standard of 0.088% in the Germencik region. These values are higher than leaf phosphorus levels reported in other trials that included phosphorus fertilization, such as Proebsting and Warner (1954), where leaves sampled from the basal portion of the shoot stabilized around 0.1% in July. While P availability is dependent on soil pH, the soil pH in these studies spanned a pH range of 4.8–6.8 (Brown, 1994), 7.3–8.3 (Proebsting and Warner, 1954), and 6.20–8.04 (Aksoy and Anac, 1993a).

## Potassium

After nitrogen, potassium is the element most needed by plants, though in figs calcium leaf tissue concentrations can be higher than either nutrient (İrget *et al.*, 2008). Fig fruit potassium export is slightly higher than nitrogen: Hernandez *et al.* (1994) found that 10 t of fresh fruit exported 22 kg of potassium (26.7 kg $K_2O$), whereas the total export that included 1.3 t of immature fruit and prunings increased the nutrient loss to 36 kg K (44 kg $K_2O$) per hectare. Aksoy and Akyuz (1993) found that the whole fig fruit is 1.5% potassium on a dry weight basis. Potassium is unusual in that a large portion of cellular potassium is not incorporated into tissues or enzymes; its most critical role in the plant is as an osmolyte, notably in the stomata. Because of its role in cell water balance, plants deficient in potassium have reduced photosynthetic rates and cell expansion. It is also an enzyme activator and is important in the phloem sugar loading and transport (Hawkesford *et al.*, 2012). Symptoms of potassium deficiency in fig are marginal chlorosis that expands toward the center of the leaf, followed by marginal necrosis (Fig. 10.1d).

In the soil, plant available potassium is bound to the cation exchange capacity, and is therefore immobile but labile. The vast majority of potassium in soils is in primary and secondary minerals, becoming available as minerals weather (Zörb *et al.*, 2014). Long-term agricultural production generally depletes available soil potassium, requiring potassium fertilization. Some soils that contain illite and vermiculite can fix potassium, making it unavailable in the short term, though it is also released over time. Potassium fixation occurs when the ion, which has a small hydrated radius, adsorbs between sheets of silicate clays, rendering it unavailable for uptake (Zörb *et al.*, 2014). This can severely reduce the efficiency and availability of potassium fertilizers in crop production. Application strategies to maximize potassium availability include heavy banding that saturates the fixation capacity of the root zone, or constant, small applications of soluble potassium that maintain levels of available potassium in the root zone. Some areas of fig production, such as the Central Valley of California, contain potassium fixing soils.

Potassium fertilization has been found to improve yield and fruit quality of *F. carica*. Soliman *et al.* (2018) applied high rates of potassium sulfate: 100, 200 and 400 g $K_2O$ per tree, to individual 'Brown Turkey' fig trees. The fruit weight and size, sugars, and moisture content increased with increasing potassium applications, while fruit acidity decreased. Though the researchers did not record their methodology for leaf tissue sampling, leaf potassium levels were low when compared to other surveys (Table 10.1) even after a second year of fertilization, and only reached 0.98% after two years of 400 g of $K_2O$ per tree. Trees in the control treatment had leaf tissue potassium levels of 0.64%. Yields were not reported; however, the trial's single tree replicates could not have resulted in reliable yield data. Holstein *et al.* (2015) fertigated 'Black Mission' and 'Sierra' fig trees with a combination of potassium and calcium thiosulfate. There was a positive marketable yield response to fertigation by both cultivars due to a reduction in cull fruit from cracking. The researchers found that leaf potassium levels were positively correlated with fruit size in 'Sierra' figs ($R^2$ value = 0.53), but fertilization did not affect 'Black Mission'. There was no response to fruit total soluble solids or titratable acidity in response to potassium and calcium fertigation. The applied potassium had no effect on fruit sugar content, perhaps because the potassium application was much lower, 19 g of $K_2O$ per tree, than in Soliman *et al.*'s 2018 trial. Additionally, the trees in Holstein *et al.*'s 2015 trial began the trial with higher leaf potassium levels: 'Black Mission' control trees ranged from 1.08 to 1.23% potassium and 'Sierra' fig trees ranged from 1.66 to 1.79%. Treated trees had 1.35–1.42% and 1.52–1.88% for Black Mission and Sierra, respectively. There may be genotypic differences in potassium demand; this should be explored in further studies.

Potassium levels in fig fruit have been reported to increase fruit cracking (Aksoy *et al.*, 1987b as cited in Aksoy and Akyuz, 1993). However, we located no supporting results in the literature and other evidence indicates potassium prevents fruit cracking. Aksoy and Akyuz (1993) examined potassium, calcium and magnesium concentrations in the fruit of three different varieties: Bursa Black, considered to be resistant to cracking, Sarilop, intermediate, and Goklop, highly susceptible. They reported whole fruit K/Ca ratios changed with maturity. The two cultivars susceptible to cracking remained relatively constant until Stage III of fruit development began around July 24 (Fig. 10.3). After that point, potassium concentrations increased relative to calcium, and the increase was greater in the highly susceptible Goklop. The Bursa Black K/Ca ratios steadily increased until Stage III, at which point they decreased as the fruit reached maturity (Fig. 10.3) (Aksoy and Akyuz, 1993). The relative differences suggest potassium plays a supportive role in fruit cracking, or more likely, that calcium plays a preventative role. The leaf tissue concentrations of potassium and calcium were not reported by the authors, thus it is unknown if the trees had adequate potassium and calcium levels.

Multiple studies have reported potassium applications decreased fruit cracking. Aydın and Kaptan (2015) examined fruit cracking in a boron contaminated soil and found that soil potassium levels were negatively correlated with cracking (correlation coefficient of −0.508). However, this correlation was weaker than that of other nutrients, such as soil and leaf calcium and calcium to boron ratios. Potassium sulfate sprays also significantly reduced cracking by 5.1% (Aydın and Kaptan, 2015) and calcium chloride sprays reduced cracking by 23.7%. Holstein *et al.* (2015) also found an increase in marketable yield due to a decrease in fruit cracking as a result of potassium applications, though calcium was applied with the potassium. Cracking in fig fruits is complex and reported to be influenced by tree water status in figs (Kong *et al.*, 2013). In other fruits, nitrogen status has also been identified as a possible nutritional cause of cracking (Opara *et al.*, 1997). The levels of these cations in the fruit can be antagonistic; Eryüce *et al.* (1993) found that fertilizing fig trees with potassium decreased both leaf and fruit calcium and fruit magnesium concentrations and increased the amount of nitrogen in fruits. More work needs to be done to elucidate the role and relative importance of potassium in fig fruit cracking.

Potassium leaf tissue levels change as the season progresses; leaf levels start out high in the spring, typically dropping rapidly as fruit develops and potassium is transported from the leaves to the fruit (Brown, 1994), although Proebsting and Warner (1954) reported a springtime increase in leaf potassium content between early May and late June before leaf tissue levels dropped again. Ideal tissue levels differ in the literature; Aksoy *et al.* (1987c) suggested leaf tissue levels from 1.1 to 1.2% for 'Sarilop' fruit in the Germincik region in Turkey (Table 10.1). This is lower than end-season values observed by Proebsting and Warner (1954) in a California fertilization trial with Adriatic figs. In the untreated control, summer leaf tissue levels were between 1.5 and 1.7%. Leaf tissue levels were lower in trees fertilized with nitrogen, closer to 1.05%, likely due to the increased growth and yield in nitrogen fertilized trees. Differences in sampling methodologies cannot explain these differences: Proebsting and Warner (1954) sampled basal leaves.

In Brown's survey (1994) of high and low vigor Calimyrna fig orchards in California, potassium was the one nutrient that could explain differences in vigor; high vigor orchards had July potassium leaf tissue levels of 1%, whereas low vigor orchards had leaf tissue levels approximately half of that. These orchards were not fertilized with potassium. Later studies on fresh figs by

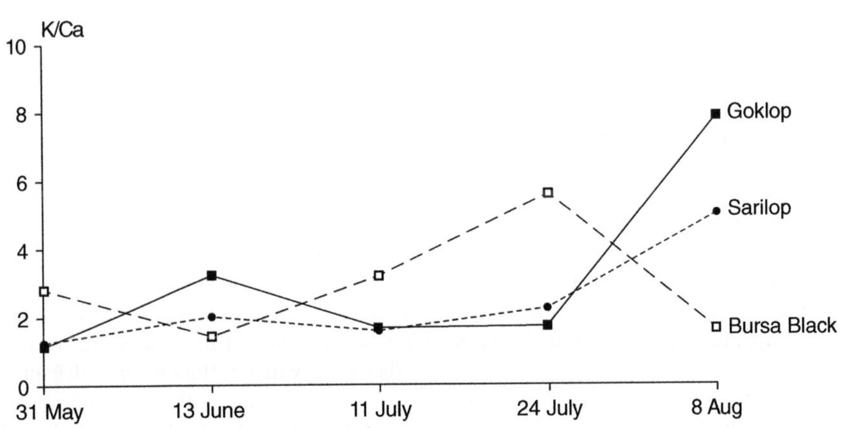


**Fig. 10.3.** Change of whole fruit K/Ca ratio during fruit development. (Figure used from Aksoy and Akyuz, 1993 with permission from RightsLink.)

Holstein *et al.* (2015) projected, based on a regression of potassium leaf concentrations and fruit size of the 'Sierra' cultivar, that leaf tissue levels should be above 2%. However, none of the leaf tissue values obtained in the trial reached 2% potassium. Additionally, the study included calcium in the fertilization program, which has been found to decrease fruit cracking. The reported levels are far higher than those reported by Moreno (1998), which suggested that leaf tissue levels should be between 1.08 and 1.13 mg $g^{-1}$ (0.108– 0.113%).

## Calcium

Calcium, bound to cell wall pectins, is necessary for cell wall stability and extension, and cellular membrane stability. It is also a second messenger in some signaling pathways. Due to its role in cell wall stability, it is critically important in fruit development and storage life (Hawkesford *et al.*, 2012). In fig, artificially induced calcium deficiency resulted in chlorosis and deformation of younger leaves (Fig. 10.1e) (Garza-Alonso *et al.*, 2019). The leaf deformation is due to calcium's role in cell walls. In the soil, calcium is bound to cation exchange sites and at high pH precipitates into calcium carbonate and is therefore immobile. The amount and availability tend to be high in the arid soils figs are typically grown in; in acidic soils it is less available and at lower concentrations. Calcium in soils can react with bicarbonates at soil pH above 7.0 and becomes less available for uptake. It can also bind with other nutrients in the soil directly, as with phosphorus, or indirectly, as with zinc binding to calcium carbonates in soils with high pH (Broadley *et al.*, 2012). In general, calcium in soil is available in quantities far above plant requirements, though soils with very low pH or high concentrations of aluminum, sodium or heavy metals can interfere with its uptake. Calcium is not phloem mobile, therefore calcium deficiencies can occur in young leaves or fruits as the tissues are partially supplied via the phloem (Hawkesford *et al.*, 2012).

In figs calcium has been found to be negatively correlated with fruit cracking (Aksoy and Akyuz, 1993), and negatively correlated with fruit sugar levels (Hakerlerler *et al.*, 1997), possibly in part due to its antagonistic role with potassium uptake. Calcium levels in the fruit have also been reported to be positively correlated with aflatoxin B1 contamination (Özer and Derici, 1998), however a later study found no connection between fruit nutrient levels and aflatoxin contamination (Demopoulous *et al.*, 2011a). Leaf calcium concentrations have been found to positively correlate with applied water due to calcium's increased availability in hydrated soils (Hernandez *et al.*, 1994; Jafari *et al.*, 2018), and decrease with applied nitrogen (Hernandez *et al.*, 1994), likely due to nitrogen's stimulation of growth.

The only leaf tissue standards set for calcium have been in Turkey, which changes on a regional basis but can range from 3.6 to 3.8% (Table 10.1). The reported concentrations in the literature can range from less than 1.5% (Caetano *et al.*, 2006) to higher than 6% (İrget *et al.*, 2008). This is likely due to the fact that calcium is taken up passively from the soil solution and stored in the leaves (Hawkesford *et al.*, 2012). Brown reported July leaf calcium levels of 3% in high vigor California orchards, similar to the values of 2.7–4.7% reported by Aksoy and Anac (1993a). However, all are an order of magnitude higher than standards set by Moreno *et al.* (1998) of 3.4–3.5 mg $g^{-1}$ dry weight (0.34–0.35%).

Multiple studies demonstrate a reduction in fruit cracking from applied calcium, even in soils with adequate levels. İrget *et al.* (2008) examined different rates of soil applied calcium on Sarilop fig trees ranging from 70 to 420 g of calcium per tree in addition to added nitrogen, phosphorus, and potassium (NPK), and one treatment with only nitrogen and calcium (Table 10.2a,b). The experiment was conducted at two sites with different fertilities as measured by CEC, soil organic matter, potassium and calcium. The addition of calcium containing fertilizers significantly decreased fruit cracking (Table 10.2b) at the Erbeyli site, which had lower soil fertility, and marketable yield increased when moderate rates of calcium were added with potassium over both the control and the NPK only treatment (Table 10.2a). The lowest measured cull ratio at the Erbeyli site, which is the ratio of cull fruit to total yield, was achieved with 420 g of calcium per tree plus NPK. At the higher fertility Kanlibahce site, any fertilizer increased marketable yield and decreased the cull ratio (Table 10.2a,b), and there

**Table 10.2.** From İrget *et al.* (2008). Researchers soil-applied increasing rates of calcium fertilizer with either nitrogen or nitrogen, phosphorus and potassium. The two sites differed in soil fertility: Kanlibahce was a higher fertility site than Erbeyli. (a) Marketable yield as a response to applied nutrients and (b) ostiole crack index as a response to applied nutrients. A rating of 0 has no cracking, 1 has cracking on less than 1/3 of the fruit, 2 has cracking between 1/3 and 2/3 of the length of the fruit, and 3 has cracking longer than 2/3 of the length of the fruit. Reprinted with permission from Elsevier.

(a) Marketable yield per tree with tested treatments

| | Marketable yield | |
|---|---|---|
| Treatment | Kanlibahce | Erbeyli |
| Control | 21.60 b | 10.94 d |
| NPK | 26.91 a | 13.41 c |
| NPK + 70 g Ca | 25.59 a | 14.65 abc |
| NPK + 140 g Ca | 25.49 a | 16.19 ab |
| NPK + 280 g Ca | 25.63 a | 16.57 a |
| NPK + 420 g Ca | 26.86 a | 15.29 abc |
| N + 420 g Ca | 25.92 a | 13.86 bc |
| Tukey | 3.12 | 2.36 |
| Year 1 | 30.83 a (2000) | 12.18 b (2001) |
| Year 2 | 20.13 b (2001) | 16.64 a (2002) |
| Tukey | 1.093 | 0.82 |
| Year *x* treatment ($p > F$) | ns | ns |

(b) Ostiole-end crack index calculated on a 0 to 3 scale

| | Ostiole-end crack/split index | | | |
|---|---|---|---|---|
| | Kanlibahce | Erbeyli | | |
| Treatment | Mean | 2001 | 2002 | Mean |
| Control | 1.50 a | 1.33 a | 1.78 a | 1.56 a |
| NPK | 1.40 ab | 1.17 a | 1.54 ab | 1.36 ab |
| NPK + 70 g Ca | 1.32 bc | 0.92 b | 1.25 bc | 1.09 bc |
| NPK + 140 g Ca | 1.37 bc | 0.83 b | 1.13 c | 0.98 c |
| NPK + 280 g Ca | 1.33 bc | 0.92 b | 1.13 c | 1.02 c |
| NPK + 420 g Ca | 1.29 bc | 1.17 ab | 1.05 c | 1.11 bc |
| N + 420 g Ca | 1.21 c | 1.21 ab | 1.00 c | 1.11 bc |
| Tukey | 0.15 | 0.41 | 0.41 | 0.31 |
| Year 1 | 1.36 b (2000) | | 1.08 b (2001) | |
| Year 2 | 1.58 c (2001) | | 1.27 a (2002) | |
| Tukey | 0.06 | | 0.13 | |
| Year *x* treatment ($p > F$) | ns | | 0.05* | |

were no significant differences between treatments. Sunscald was decreased at the Kanlibahce site with fertilization and significantly increased as applied calcium increased, though the sunscald severity at 420 g Ca per tree was never worse than the control. It is notable that leaf calcium levels in this trial were extremely high. At the Erbeyli site, which is the only site where leaf and fruit nutrient levels were reported, the control was at 5.62%. Both leaf and fruit calcium levels increased with applied calcium. Holstein *et al.* (2015) applied a combination of potassium and calcium thiosulfate via fertigation to Black Mission and Sierra trees and found a significant increase in marketable yield due to the reduction in the number of fruits culled from splitting. Black Mission culls dropped from 49% in the control to 43% in the treated trees, and Sierra dropped from 42 to 36%. The trial was not designed in a way that could determine whether the effect was due to calcium or potassium.

Foliar applications of potassium have also been reported to reduce cracking. Aksoy and Anac (1993b) applied a 1% $CaCl_2$ solution to three calcium deficient fig cultivars in Stage II of fruit development with differing susceptibilities to cracking. The varieties were Goklop, susceptible to cracking; Sarilop (Calimyrna), intermediate; and Bursa Black, considered resistant. The spray significantly decreased cracking in Goklop: mild cracking decreased from 46% of the control to 0% in treated fruit, and severe cracking dropped from 13.2 to 0.1%. Calcium sprays have also been found to counter the negative effects of excessive boron. Aydın and Kaptan (2015) applied 0.25% of $CaCl_2$ (w/v) to a fig orchard with high levels of soil boron and found that cracking in fruit decreased from 33.7% in the control to 10%. Leaf calcium levels were on average 1.26%, which is very low compared to regional standards. The researchers also examined different soil, leaf and fruit nutrient concentrations and found that cracking was negatively correlated with soil calcium levels ($r = -0.751$) and leaf calcium levels ($r = -0.596$).

Adequate calcium nutrition has also been linked to fruit postharvest storage quality. While studies on post-harvest calcium applications have been done, this chapter will focus on pre-harvest applications as the focus is on nutrition and fertilization and not postharvest physiology. Antunes *et al.* (2005) applied two 0.04% calcium oxide foliar sprays in the spring but found the sprays did not improve fresh fig postharvest quality; the authors speculated the solution was too weak.

## Magnesium

Magnesium is a critical component in cellular metabolism. It is the central atom in the chlorophyll heme, is important for some protein structures, and is required for many enzymatic reactions (Broadley *et al.*, 2012). Like calcium, deficiencies are typically only seen in acidic, heavily weathered soils, but deficiency symptoms can be induced if there are extreme cation ratios in the soil. In agricultural settings this can occur after heavy applications of calcium containing amendments or potassium fertilizer. In figs, magnesium deficiencies exhibit interveinal chlorosis (Garza-Alonso *et al.*, 2019) (Fig. 10.1f).

Magnesium deficiency has not been documented in figs, likely because they tend to be grown in high pH soils that are saturated with cations. Like calcium, magnesium leaf tissue levels steadily increase through the growing season (Proebsting and Warner, 1954; Ersoy *et al.*, 2001), with summer values varying by source. Brown (1994) reported fairly constant magnesium levels across high and low vigor sites, ranging from 0.7 to 0.9%. This is slightly higher than what was reported by Aksoy *et al.* (1987c), where ideal Mg values ranged from 0.658 to 0.721% in the Sarilop variety (Table 10.1). Aksoy and Anac (1993a) reported a range of values in orchards producing high quality fruits from 0.5 to 1.07%. All ranges are an order of magnitude higher than Moreno *et al.* (1998), which reported ideal values from 0.45 to 0.49 mg $g^{-1}$ (0.045–0.049%) (Table 10.1). The results may represent a conversion error, as results are much lower than the reported range for plants, which is 1.5–3.5 g $kg^{-1}$, or 0.15– 0.35% (Broadley *et al.*, 2012).

In fruit, magnesium levels can decrease or increase as fruit matures; it appears the concentration in fruit is variety dependent (Aksoy and Akyuz, 1993). The authors looked at levels of potassium, calcium and magnesium in fruit and found that in two cultivars, Goklop and Sarilop, magnesium levels decline as fruit matures. However, magnesium levels increase in Bursa Black as the fruit reaches maturity. Bursa Black is a variety that is resistant to cracking, whereas Goklop and Sarilop are sensitive or moderately sensitive to cracking, respectively. The authors concluded that magnesium may play a positive role in preventing fig fruit cracking. This does not agree with Aydın and Kaptan (2015), who found that fruit and soil Ca:(K+Mg+Na) were negatively correlated with fruit cracking. These conflicting reports may have to do with magnesium's antagonistic role with calcium uptake by the plant rather than a specific role in fruit cracking. In an experiment where a 1% calcium chloride solution was sprayed on three varieties of fig fruits in Stage II, magnesium levels in the fruit dropped in treated trees (Aksoy and Anac, 1993b). Similarly, soil fertilization with potassium significantly reduced fruit magnesium levels (Eryüce *et al.*, 1993). The interpretation of these results is difficult as Aydın and Kaptan (2015) were examining fig fruit cracking in a soil contaminated by boron, which may change plant response. Alternatively, magnesium may induce cracking only when calcium levels in the fruit are low (Aksoy and Akyuz, 1993).

### Sulfur

Sulfur is a component of some amino acids such as methionine, and thus a critical component of proteins. It is also important for some coenzymes and prosthetic groups (Broadley *et al.*, 2012). To the authors' knowledge, sulfur deficiencies in fig have never been investigated. Only Moreno (1998) suggested critical values for leaf sulfur: 0.17%. This is consistent with other reviews on plant nutrition, which list leaf tissue levels ranging from 0.1 to 0.5% (Broadley *et al.*, 2012).

## 10.4 Micronutrients

### Iron

Iron is involved in redox reactions and protein synthesis and is directly involved in photosynthesis (Broadley *et al.*, 2012). Iron deficiency is well documented in plants growing in high pH soils, and soils with large amounts of carbonates (Broadley *et al.*, 2012). However, iron deficiency has rarely been documented in figs, which may reflect this species' tolerance to soils with high pH and carbonates. Leaf iron concentration patterns through the season change depending on the source; Ersoy *et al.* (2001) reported that leaf values remain steady, whereas Brown (1994) reported that they decline as the season progresses. Moreno *et al.* (1998) suggested ideal fig leaf iron requirements are 183–189 $\mu g\ g^{-1}$ (ppm). This is higher than reported leaf levels of high yielding California fig orchards, which were 110 ppm in July (Brown, 1994). Iron deficiency symptoms were not reported in this study, and iron levels did not correlate with vigor in the Brown study.

### Zinc

In plants, zinc is an enzyme catalyst used by a wide range of enzymes encompassing all enzyme classes (Broadley *et al.*, 2012). It is deficient in weathered acidic soils and its availability is restricted at high pH and in soils high in carbonates or with large amounts of available phosphorus (Broadley *et al.*, 2012). Jafari *et al.* (2018) reported that zinc uptake in figs decreases under water stress. Moreno *et al.* (1998) reported ideal leaf tissue Zn levels to be 30–35 $\mu g\ g^{-1}$, which is much higher than levels observed by Brown (1994), who observed leaf tissue levels of 12 ppm in high vigor California Calimyrna orchards (Table 10.1). Leaf zinc concentrations decrease as the season progresses (Ersoy *et al.*, 2001).

To the authors' knowledge, only a single trial has examined zinc fertilization in *F. carica*. Hakerlerler *et al.* (1999) examined applying different rates of zinc sulfate to a soil with high pH (8.1–8.3) and low soil Zn (0.7–0.8 ppm) in the top foot of soil. Soil applications of 200 g $tree^{-1}$ and foliar applications of 0.1–0.15% resulted in the highest yields. Application of zinc fertilizer also significantly improved the fruit sugar content, rates as high as 600 g of zinc sulfate applied to the soil or a 0.2% solution of zinc sulfate applied to the foliage increased the fruit fructose content by 64.5% and 58.5% over the control, respectively. Texture and fruit coloration were negatively affected by zinc applications however, so the authors recommended the more moderate rates that also resulted in the highest yields. The trial was conducted with replicates containing only two trees and needs to be conducted on a wider scale to confirm field-scale responses. Additionally, leaf tissue concentrations of zinc in the soil application trial did not reflect yield, as the treatment with the highest yield had statistically similar lamina zinc levels to the control, which were 15.4 and 14.5 ppm, respectively. While Hakerlerler (1999) reported leaf lamina and petiole values for trees in the foliar zinc treatments, it is unlikely these values reflect true tissue concentrations. Foliar nutrient analyses are unreliable when micronutrient sprays have been applied to the foliage due to the complexing of metal micronutrients to leaf cuticles (Gordon, personal observation). Washing does not remove all the bound micronutrients.

## Copper

Copper is involved with redox reactions in plants and is complexed to proteins involved with cellular metabolism or mitigating oxidative stress. Copper deficiency is found in soils derived from parent material low in copper or with high organic matter (Broadley *et al.*, 2012). Leaf copper concentrations decrease as the season progresses (Ersoy *et al.*, 2001). Reported leaf tissue levels vary by source; Brown (1994) reported 6 ppm, much lower than the leaf tissue levels of 15–21 ppm leaf copper levels reported by Moreno (1998). Hakerlerler *et al.* (1997) found a negative correlation between copper in fruit tissues and sorbitol levels, however this nutrient has not been examined closely in fig nutrition research.

Özer and Derici (1998) reported a negative correlation between fruit copper concentrations and aflatoxins B1, B2 and G1. A second survey found no correlation between fruit nutrients and aflatoxin B1 levels (Demopoulos *et al.*, 2011a). A later examination of copper applied as Bordeaux mixture on figs followed by *Aspergillus flavus* inoculations found no effect on ripening figs, but increased aflatoxin contamination on figs when applied in the green stage. The authors speculated that aflatoxin concentrations increased due to the antibacterial activity of copper which allowed *A. flavus* to flourish unimpeded by competition (Demopoulos *et al.*, 2011b).

## Manganese

Manganese is a cofactor for many enzymes, and a component of superoxide dismutase, an essential enzyme that detoxifies free radical oxygen. It is also involved with cellular metabolism and involved in root extension. Manganese can be toxic to plants at low soil pH due to its increased availability. Its availability is reduced under high soil pH conditions with large concentrations of carbonates (Broadley *et al.*, 2012). In figs, leaf manganese has been positively correlated with fruit fructose, glucose and total sugar (Hakerlerler *et al.*, 1997). Fruit manganese levels have been found to negatively correlate with aflatoxin levels (Özer and Derici, 1998).

Leaf manganese levels in fig are typically higher at the end of the season (Brown, 1994), though Ersoy *et al.* (2001) reported a rapid increase in leaf Mn levels, then a decrease as the season progressed. Moreno (1998) reported the ideal leaf tissue levels as 86–92 ppm (Table 10.1). Brown (1994) reported leaf tissue levels of 90 ppm in high vigor Calimyrna orchards. However, in a single low vigor site surveyed by Brown (1994), July leaf levels were 250 ppm, and increased to 450 ppm by October. The soil pH at that site was 4.8–5.2, suggesting figs may be sensitive to manganese toxicity.

## Boron

Boron is immobile (Fig. 10.4) in *F. carica*, as this species does not transport photosynthates as sorbitol (Brown and Hu, 1996), and excess levels of it accumulate in the leaves (Brown, 1994). Boron is active in cell wall pectins, cell metabolism, membrane integrity, flower production and pollen tube growth (Broadley *et al.*, 2012). In soils, boron is weakly mobile, and its availability is dependent on soil factors as well as its solubility; adsorbed boron is not available to plants for uptake. Boron availability decreases as soil pH increases from 3 to 9 and as soil moisture content decreases. Due to the fact that it adsorbs to clays, its availability for uptake is higher in sandy soils, and deficiencies are most often seen in sandy soils due to the lack of adsorption sites (Goldberg, 1997). In irrigated agriculture, B

**Fig. 10.4.** Top: boron toxicity of fig leaves. Bottom: a Sarilop fig tree in the İncirliova province of Aydın with boron toxicity. (Figures used with permission from İzzet Zeki Atalay and M. Aydin.)

present in the irrigation water source is frequently the most important source of B for crops.

*F. carica* is considered to be sensitive to soil boron, which has a limit of 0.7 mg $kg^{-1}$ in a saturation paste extract (USDA ARS, 1954). *F. carica's* tolerance to poor quality soils and saline water make it likely that fig production could be pushed into soils with high boron as urban development has tended to push agricultural production into increasingly marginal lands. To date there has only been one study examining boron fertilization, done in an acidic tropical soil. The authors found no significant effect on yield despite increased boron uptake, however potassium, manganese and zinc leaf nutrient concentrations increased as a result of fertilization (Caetano *et al.*, 2006). In other work, high boron has been found to be associated with reduced fruit coloration (Aksoy *et al.*, 2003). The only survey of high performing fig orchards that has reported leaf boron concentrations is Brown (1994) which reported June levels of 100 ppm (Table 10.1).

Fruit cracking has been found to be associated with boron toxicity. Aydın and Kaptan (2015) examined fig fruit cracking in a soil with boron levels ranging from 1.60 to 3.94 mg $kg^{-1}$; fruit cracking was positively associated with soil ($r$ = 0.899) and leaf ($r$ = 0.577) tissue boron levels and negatively correlated with soil Ca:B ratios ($r$ = −0.932). Leaf tissue boron levels averaged 475 mg $kg^{-1}$ in fruit with less than 2% cracking and was as high as 563 mg $kg^{-1}$ in fruit with more than 20% cracking. The authors also reported some foliar toxicity symptoms at these soil boron levels. Aydın and Kaptan (2015) also evaluated foliar nutrient sprays applied three times from June and found spraying trees with a boric acid (0.25% B) solution significantly increased the cracked fruit percentage from 33.7 in the control to 42.9%. Applying potassium sulfate (0.25% K) or calcium chloride (0.25% Ca) solutions reduced the cracked fruit ratio in this high boron soil to 28.6 and 10.0, respectively.

## 10.5 Salinity and Sodicity

As competition for irrigation water increases, agricultural production is increasingly relegated to more marginal lands. The combination of droughts precipitated by climate change and salinization of soils from irrigation have sharply increased the global percentage of saline and sodic soils. Saline soils are defined as having a salinity above 4 dS $m^{-1}$ (Abrol *et al.*, 1988), and sodic soils have an exchangeable sodium percentage, or the amount of the cation exchange capacity that is taken up by sodium, greater than 15. Soil pH in sodic soils is often above 8.2 (Abrol *et al.*, 1988), soils can be saline as well as sodic. The studies cited in this review use two different methods for measuring salinity: the first is concentration, which is the measurement of the number of specific ions in solution, expressed here as millimoles. The second method is electrical conductivity of the soil saturation paste (ECe), which measures the ease of flow of electrons through a solution, and in soil solutions is a measurement of the total number of ions in solution. Since ECe is determined by both by the concentration of soluble Na and Cl in addition to other ions and molecules in solution (Ca, Mg, K, sulfate and bicarbonate) it is not a perfect measure of plant response to salinity as is the direct measurement of Na and Cl ion solution concentrations. Electrical conductivity, however, is widely used under field settings, as it can be measured easily with simple equipment and limited training.

The effect of salinity and sodicity on plants is two-fold: the increase of ions in the soil solution lowers the osmotic potential of the soil water, making it more difficult for plants to extract water. Additionally, the uptake of these ions, which are typically sulfates, sodium, chloride and boron, ultimately cause cell damage, called specific ion toxicity. Sodium has an additional effect on soils; its large hydrated radius and monovalent charge pushes clay particles apart, resulting in deflocculation of soil aggregates (Bronick and Lal, 2005). This produces soil surface crusting, reducing water infiltration and increasing surface pooling and runoff. The induced secondary anoxia in sodic soils increases the risk of soilborne fungal infections and impairs plant growth by limiting oxygen. Oxygen is needed for root processes such as nutrient uptake and growth.

Figs are moderately tolerant of soil salinity (Tanji and Kielen, 2003) and tolerate ECes as high as 6 dS $m^{-1}$. However, to the authors' knowledge, no in-field salinity-yield response trials have been conducted. Several studies using container seedlings have been reported. In multiple studies, growth was reduced or stopped at and above 50 mM sodium chloride (Okubo and Utsunomiya, 1996; Caruso *et al.*, 2015), though there was genetic variation (Zarei *et al.*, 2016). Zarei *et al.* (2016) reported the EC50 values of 'Sabz' fig hybrids ranged from 4.98 to greater than 8 dS $m^{-1}$ (Table 10.3).

**Table 10.3.** From Zarei *et al.* (2016). The EC50 values of 'Sabz' hybrid cultivar pollinated by different caprifig pollen parents exposed to four different concentrations of sodium chloride solution, 0.6 (S0), 4 (S1), 6 (S2), and 8 (S3) dS $m^{-1}$ for five weeks. Reprinted with permission from the American Society for Horticultural Science.

| | EC 50 | | | |
|---|---|---|---|---|
| Growth factor | S × P | S × K | S × Sh | S × D |
| Shoot height | >8 | 6.057 | 5.24 | 4.98 |
| Shoot diameter | 1.29 | 4.97 | 7.54 | 0.75 |
| Shoot fresh weight | >8 | >8 | >8 | >8 |
| Shoot dry weight | >8 | >8 | >8 | >8 |
| Root fresh weight | 6.71 | >8 | >8 | 6.06 |
| Root dry weight | >8 | >8 | >8 | 6.87 |

Generally, stomatal conductance continues after growth has stopped. Caruso *et al.*, (2015) found that stomatal conductance gradually declines until 200 mM, after which there was no significant decrease with any further increases in salinity. Golombek and Ludders (1990) found that photosynthetic rates of fig seedlings exposed to saline irrigation water ranging from 50 to 100 mM sodium chloride decreased as leaf sodium chloride concentrations increased and stomatal conductance decreased. The authors found that nonstomatal conductance rates declined to a lesser degree than stomatal conductance, which allowed photosynthesis to continue and increased the water use efficiency (Fig. 10.5). In a subsequent trial examining the effect of the ratio of root mass to leaf area in response to increasing salinity, Golombek and Ludders (1990) reported fig seedlings with intact root masses had an initial temporary increase in net photosynthesis due to nonstomatal conductance. This suggests that *F. carica* plants with larger root systems relative to leaf mass are better able to tolerate salinity stress. This characteristic could facilitate potential rootstock selection for salinity tolerance.

Figs, like other glycophytes, accumulate sodium and chloride and cannot completely prevent the movement of detrimental ions from the roots into shoots and leaves. The existing literature does not agree as to whether chloride ions can be partially excluded from the shoots and leaves. In an experiment where 'Masui Dauphine' fig seedlings were watered with a 50 mM solution of NaCl for four weeks, sodium and chloride initially accumulated in the stems, roots and leaves; however by week four, there were more sodium and chloride ions in the leaves and stems (Okubo and Utsunomiya, 1996). In contrast, Zarei *et al.* (2016) grew four seedling 'Sabz' hybrids from different caprifig pollen parents in salinity levels ranging from 0 to 8 dS $m^{-1}$. They found that while sodium and chloride ions accumulated in leaves, stems and roots, for the most part chloride accumulated to greater levels in root tissues. There were genotypic differences, however, as one genotype accumulated less chloride in the roots than the others, and leaf chloride levels actually exceeded root levels at 6 dS $m^{-1}$ and were similar at 8 dS $m^{-1}$.

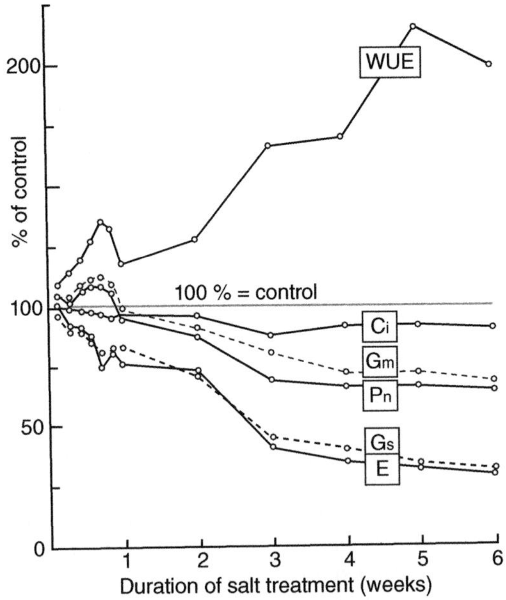


**Fig. 10.5.** Change in photosynthetic processes, expressed as a percentage of the control, over time. Ci is intercellular $CO_2$ concentration, Gm is nonstomatal conductance, Pn is photosynthesis, Gs is stomatal conductance, E is transpiration. (Figure used from Golombek and Ludders, 1990 with permission from RightsLink.)

Okubo and Utsunomiya (1996) found that 50 mM sodium chloride solutions decreased root potassium, stem magnesium, and both stem and leaf calcium. It is known that salt in the root system interferes with the xylem's ability to load calcium (Läuchli and Grattan, 2007). Leaf potassium levels in the salt treatment increased by week four. Zarei *et al.* (2016) also found that imposing salinity treatments on four fig genotypes increased leaf potassium levels and decreased root potassium levels. The authors speculated that increases in tissue potassium levels could be an osmotic response to maintain water levels in the cells (Okubo and Utsunomiya, 1996). A plant's ability to maintain high K:Na concentrations in the cytosol under saline conditions has been positively correlated with tolerance to salinity (Shabala and Pottosin, 2014), suggesting *F. carica's* salinity tolerance is a function of potassium accumulation rather than ion exclusion. Okubo and Utsunomiya (1996) found that the concentrations of sodium, chloride and potassium in fig latex increased under sodium chloride treatments, which resulted in the fig latex osmotic potential becoming more negative.

Generally, leaf symptoms of salinity, including phytotoxicity and defoliation, increase as irrigation water salinity increases. Caruso *et al.* (2015) reported minor symptoms of salt injury at 50 mM, while severe defoliation in *F. carica* did not occur until 200 mM sodium chloride. While other salinity studies in *F. carica* reporting defoliation or leaf symptoms did not use solution concentrations this high, they reported no defoliation at 100 mM (Golombek and Ludders, 1990, 1993; Okubo and Utsunomiya, 1996). Zarei *et al.* (2016) reported that among the four genotypes studied in their trial, leaf production slowed or halted at 6 dS $m^{-1}$ and defoliation occurred at 8 dS $m^{-1}$.

Zarei *et al.* (2016) and Caruso *et al.* (2017) found that stem water potential decreased as the solution salinity increased from 0 to 8 dS $m^{-1}$ and 0 to 250 mM sodium chloride solutions, respectively, and that they differed by genotype (Fig. 10.6). Caruso *et al.* (2017) separated stem water potential measures for each salinity concentration over several weeks, so no statistical conclusions could be drawn.

The cessation of growth around 50 mM due to osmotic effects, followed by a partial inability to prevent the buildup of toxic ions in above-ground portions of the plant classify *F. carica* as a partially tolerant species (Läuchli and Grattam, 2007), and it displays some traits that confer salt tolerance, such as the accumulation of potassium in plant tissues (Shabala and Pottosin, 2014). The cited tolerance of fig to salinities up to 6 dS $m^{-1}$ is somewhat supported by seedling trials. The seedling studies presented here are limited by the fact that they used sodium chloride as the salinizing agent, whereas in the field, salt affected soils and irrigation water may also have high concentrations of calcium, magnesium, sulfates, bicarbonates and carbonates, and sometimes boron (Läuchli and Grattan, 2007). Because of this, and because the behavior of mature plants or plants slowly exposed to salinity is different than acute salinity, in-field trials need to be established to fully assess *F. carica*'s tolerance to salinity. As shoot extension is critical for producing syconium, it is likely that the crop tolerance limit will be when shoot extension ceases.

## 10.6 Production Practices

Fertilization and soil amendment practices vary by fig production area and there is not enough research to suggest generic fertilization guidelines, aside from replacing nutrients exported during harvest. However, nitrogen is the most widely applied nutrient and traditionally the only recommended nutrient. In areas that have been under long-term production, have low CECs, or are low in potassium, additional potassium is likely to increase yield and fruit quality.

Fertilization practices for other orchard crops recommend replacing what was removed with yield and taking leaf tissue concentrations, uptake timing, and tree health into account; nutrients should be available at the time of greatest uptake, and trees that are not actively growing or not in good health or have low yield will require less, if any fertilization. However, aside from Turkey, there are no leaf tissue standards developed for most nutrients, and it is unknown when the rate of uptake for nitrogen and potassium are highest, though nutrient demand during syconia development should be higher than the vegetative growth periods or after fruit maturation.

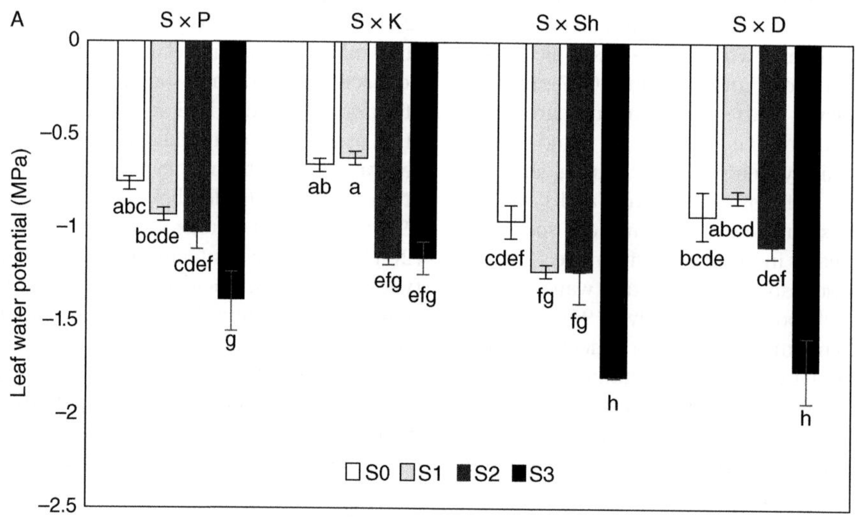


**Fig. 10.6.** From Zarei *et al.* (2016). Leaf water potentials of four different crosses of 'Sabz' cultivar with different caprifig pollen parents exposed to four different concentrations sodium chloride solutions, measuring 0.6 (S0), 4 (S1), 6 (S2), and 8 (S3) dS $m^{-1}$ for five weeks. Bars marked with the same letter are not significantly different at $P \leq 0.05$ level. (Figures used with permission from the American Society for Horticultural Science.)

Nitrogen applications should occur in-season and start when leaves are mature enough to transpire, which is when deciduous trees typically begin taking up nitrogen. Previous publications have suggested 18–36 kg of nitrogen per hectare per year for Calimyrna (Sarilop) figs (Obenauf, 1978), skipping a year if the soil is especially fertile. These recommendations were made before improvements in irrigation technology resulted in better yields, and California growers report applying much higher rates of nitrogen. It is not recommended to provide nitrogen to figs planted in their first year. Other sources recommend providing 40–100 g of nitrogen for young trees after the first year until they reach maturity, afterward yearly fertilization rates range from 100 to 200 g of nitrogen per tree (Vachamidis and Vemmos, 2010). We speculate nitrogen applications should end around August or September, or at the latest, by the last harvest of the season due to research on other orchard crops, which has shown that the majority of nitrogen is taken up during fruit development (Rosecrance *et al.*, 1996; Muhammad *et al.*, 2015) (Fig. 10.7).

Vachamidis and Vemmos (2010) recommend adding phosphorus as part of a complete fertilizer, which contains nitrogen, phosphorus and potassium. However, the literature review provides no evidence for its addition, and it is likely recommended as part of a standard practice. Potassium applications, as the nutrient is immobile in soils, can occur at any time. Rates can be many times higher than what is lost through harvest, however heavy applications in soils with low CECs can result in nutrient imbalances and temporary deficiencies of other cations such as magnesium, or most critically calcium. In other fresh fruit tree crops, recommended rates for remedying potassium deficient orchards are to shank in 0.5–1 kg $K_2O$ per tree (CDFA FREP, 2019). Surface application is also acceptable, however heavy irrigation or rain is needed to dissolve the fertilizer and move the potassium into the root zone. It is accepted in other fruit tree crops that potassium deficiencies take several years to remedy, so it is best to apply potassium fertilizer yearly, if possible.

Soil pH modification is required only in soils with pH below 5.5 (Flaishman *et al.* 2008). The lime requirement of a soil is determined by its current pH, the desired pH, and the buffer capacity of the soil, which is determined by acidic groups

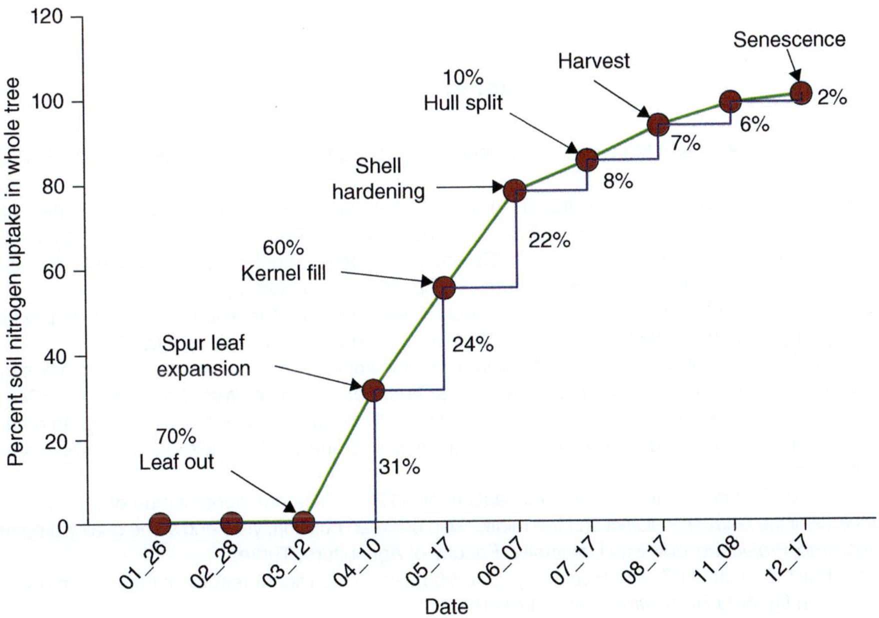


**Fig. 10.7.** From Muhammad *et al.* (2018). Nitrogen uptake rate in mature almond trees. Reprinted with permission from the University of California.

on soil particles and organic matter, carbonates, aluminum, ion exchange and weathering of minerals (Nelson and Su, 2010). Commercial agricultural laboratories can calculate the lime requirement of a soil. Liming can be done with a wide range of materials, including agricultural lime, ash and dolomitic limestone. It is best done before the orchard is planted, as the amendment only affects the pH of the immediately surrounding soil. Surface applications of lime may be applied to existing orchards, however it will only affect the upper portion of the soil and will require more time to take effect.

The application of organic matter has been found to increase the yield of *F. carica* in acidic tropical soils, likely due to the addition of available nutrients, an increase in soil pH from the manure, and complexation of aluminum by organic matter (Caetano *et al.*, 2006). In a study done by Mordoğan *et al.* (2013), mixtures of cow and sheep manure increased the vigor, size and soluble solids content of 'Sarilop' fig trees. Ostiole size was also decreased by fertilization, which has been observed in other fertilization trials (Proebsting and Warner, 1954). Leaf tissue analyses were also not affected in either trial. Uncomposted manure is a food safety risk, and the authors recommend fully composting any animal waste products before applying to orchards.

In order to fully maximize nutrient uptake from compost or composted manure, the product must be worked into the soil with shallow tillage. Surface applications provide some benefit, but nutrients, particularly nitrogen, can be lost by runoff, denitrification, and if the product has not been fully composted, volatilization of ammonia. Composts and particularly composted manures contain salts, and heavy applications may appreciably increase the ECe of a soil. Vachamidis and Vemmos (2010) recommend the addition of 20–30 t of digested manure per hectare before planting figs.

In order to prevent the buildup of salts in the soil, a regular leaching programme must be implemented. This requires water that has a lower salt load than the soil requiring reclamation, which can include rainfall, as well as a beneficial cation to push off sodium. There are many excellent reviews of the effects of salt on soil as well as leaching guidelines, including Abrol *et al.* (1988).

## Acknowledgements

Thank you to Thomas Sotiropoulos for providing a translation of 'Fig Tree Cultivation'.

## References

Abrol, I.P., Yadav, J.S.P. and Massoud, F.I. (1988) Salt-affected soils and their management *FAO Soils Bulletin* 39.

Aksoy, U., Anaç, D., Eryuce, N. and Yoltaş, T. (1987a) Determination and evaluation of the nutrient status of fig orchards in the Aegean region. *Journal of Ege University Faculty of Agriculture* 24(2), 21–35.

Aksoy, U. and Akyuz, D. (1993) Changes in K, Ca and Mg contents in different parts of the fig fruit during development. In: *Optimization of Plant Nutrition*. Springer, Dordrecht, pp. 309–312.

Aksoy, U. and Anaç, D. (1993a) Soil properties and mineral content of leaves in fig orchards producing high-quality fruits. In: *Optimization of Plant Nutrition*. Springer, Dordrecht, pp. 305–308.

Aksoy, U. and Anac, D. (1993b) The effect of calcium chloride application on fruit quality and mineral content of fig. In: *International Symposium on Postharvest Treatment of Horticultural Crops* 368 pp. 754–762.

Aksoy, U., Anaç, D., Hakerlerler, H. and Düzbastilar, M. (1987b) Nutrient status of Calimyrna fig orchards in Big Meander Valley and their relationships with yield and quality. *Taris Research and Development Center. Project* (006).

Aksoy, U., Anaç, D., Hakerlerler, H. and Düzbastilar, M. (1987c) The nutritional status of fig (cv Sarilop) orchards of the Germenik district and the relationship between nutrition, yield and some quality properties. *Taris Arge, Project no. 006* Ege University, Faculty of Agriculture, Turkey.

Aksoy, U., Balci, B., Can, H.Z. and Hepaksoy, S. (2003) Some significant results of the research-work in Turkey on fig. *Acta Horticulturae* 605, 173–181.

Antunes, M.D.C., Mendonca, V., Miguel, M.G., Neves, M.A., Martins, M.A., *et al.* (2005) The Effect of Irrigation Level and Pre-Harvest Treatments with Calcium Oxide on Storage Ability of Fig Fruits (Ficus carica'Lampa Preta'and'Bebera Branca'). In *III International Symposium on Fig* 798, 335–339.

Aydın, M. and Kaptan, M.A. (2015) Effect of nutritional status on fruit cracking of fig (*Ficus carica* L. cv. Sarilop) grown in high level boron contained soils. *Scientific Papers-Series A, Agronomy* 58, 20–25.

Aydin, S., Irget, M.E., Anac, D. and Oktay M. (1998) Nutrition of Fig. *Aegean Region 1st Agricultural Congress.*

Broadley M., Brown, P., Cakmak, I., Feng Ma, J., Rengel, Z., Zhao, F., (2012) Beneficial elements. In: Marschner P. (ed.) *Marschner's Mineral Nutrition of Higher Plants* (3rd edn), Academic Press, pp 249–269. DOI: 10.1016/B978-0-12-384905-2.00008-X

Bronick, C.J. and Lal, R. (2005) Soil structure and management: a review. *Geoderma* 124(1–2), 3–22.

Brown, P.H. (1994) Seasonal variations in fig (*Ficus carica* L.) leaf nutrient concentrations. *HortScience* 29(8), 871–873.

Brown, P.H. and Hu, H. (1996) Phloem mobility of boron is species dependent: evidence for phloem mobility in sorbitol-rich species. *Annals of Botany* 77(5), 497–506.

Caetano, L.C.S., de Carvalho, A.J.C. and Jasmim, J.M. (2006) Preliminary report on yield productivity and mineral composition of the fig tree as a function of boron and cattle manure fertilization in Brazil. *Fruits* 61(5), 341–349.

Caruso, G., Gennai, C., Ugolini, F., Marchini, F., Quartacci, M.F. *et al.* (2017) Tolerance and physiological response of young *Ficus carica* L. plants irrigated with saline water. *V International Symposium on Fig* 1173, 137–142.

CDFA FREP (2019) The California Department of Food and Agriculture. www.cdfa.ca.gov/is/ffldrs/frep/ (accessed 15 February 2022).

Christensen, L.P. and Peacock, W.L. (2000) Mineral nutrition and fertilization. In: *Raisin Production Manual* Vol. 3393. UCANR Publications, Oakland, California, pp. 102–114.

Condit, I.J. (1947) *The Fig*. Chronica Botanica Co., Waltham, Massachusetts.

Condit, I. (1955) Fig varieties: a monograph. *Hilgardia* 23(11), 323–538.

Demopoulos, V., Zokos, M., Tsiavtari, E., Varzakas, T., Kotsiras, A., *et al.* (2011a) Relationship between dried figs nutrition and aflatoxin B1 contamination. *I International Symposium on Mycotoxins in Nuts and Dried Fruits* 963, 79–82.

Demopoulos, V., Monas, A., Souliotis, E., Manolopoulou, E. and Varzakas, T. (2011b) Effect of copper on the aflatoxin contamination in dried figs. *I International Symposium on Mycotoxins in Nuts and Dried Fruits* 963, 183–188.

Ersoy, N., Gözlekci, S. and Kaynak, L. (2001) Seasonal variations in the content of nutrient elements in the leaves of fig (*Ficus carica* L. Yesilguz). *II International Symposium on Fig* 605, 269–275.

Eryüce, N., Colakoglu, H., Aydin, S. and Cokuysal, B. (1993) The effects of K and Mg fertilization on some quality characteristics and mineral nutrition of fig. *International Symposium on Quality of Fruit and Vegetables: Influence of Pre-and Post-Harvest Factors and Technology* 379, 199–204.

Ferguson, L., Michailides, T.J. and Shorey, H.H. (1990) The California fig industry. *Horticultural Reviews* 12, 409–490.

Flaishman, M.A., Rodov, V. and Stover, E. (2008) The fig: botany, horticulture. In: *Plant Nutrition—Physiology and Applications* Springer, Dordrecht, pp. 87–493.

Garza-Alonso, C.A., Olivares-Sáenz, E., Gutiérrez-Díez, A., Vázquez-Alvarado, R.E. and López-Jiménez, A. (2019) Visual symptoms, vegetative growth, and mineral concentration in fig tree (*Ficus carica* L.) under macronutrient deficiencies. *Agronomy* 9(12), 787–799.

Goldberg, S. (1997) Reactions of boron with soils. *Plant and Soil* 193(1–2), 35–48.

Golombek, S.D. and Lüdders, P. (1990) Gas exchange of Ficus carica in response to salinity. In: van Beusichem, M.L. (ed.) *Plant Nutrition—Physiology and Applications* Springer, Dordrecht, Netherlands, pp. 487–493.

Golombek, S.D. and Lüdders, P. (1993) Effects of short-term salinity on leaf gas exchange of the fig (*Ficus carica* L.). *Plant and Soil* 148(1), 21–27.

Hakerlerler, H., Saatci, N., Hepaksoy, S. and Aksoy, U. (1997) Fruit and leaf nutritional status of some fig clones and cultivars and relationships with some sugar fractions. *I International Symposium on Fig* 480, 247–252.

Hakerlerler, H., Aydın, Ş., İrget, M.E., Aksoy, U. and Tutam, M. (1999) The effect of soil and foliage applied zinc on yield and quality of fig/Ficus carica L. cv sarılop) for drying. *5th International Meeting on Soils with Mediterrraean Type of Climate (IM SMTC)* Barcelona (Catalonia), *Spain*.

Hartz, T.K. and Bottoms, T.G. (2009) Nitrogen requirements of drip-irrigated processing tomatoes. *HortScience* 44(7), 1988–1993.

Hartz, T.K. and Hochmuth, G.J. (1996) Fertility management of drip-irrigated vegetables. *HortTechnology* 6(3), 168–172.
Hawkesford, M., Horst, W., Kichey, T., Lambers, H., Schjoerring, J., *et al.* (2012) Functions of macronutrients. In: *Marschner's Mineral Nutrition of Higher Plants*. Academic Press, pp. 135–189.
Hernandez, F.B.T., Modesto, J.C., Suzuki, M.A., Correa, L.S. and Reichardt, K. (1994) Effects of irrigation and nitrogen levels on qualitative and nutritional aspects of fig-trees (*Ficus carica* L.). *Scientia Agricola* 51(2), 292–297.
Holford, I.C.R. (1997) Soil phosphorus: its measurement, and its uptake by plants. *Soil Research* 35(2), 227–240.
Holstein, H., Crisosto, G.M. and Crisosto, C.H. (2015) Fertigation with potassium increases size and yield in fresh figs growing in California. *V International Symposium on Fig* 1173, 177–182.
İrget, M.E., Aksoy, U., Okur, B., Ongun, A.R. and Tepecik, M. (2008) Effect of calcium based fertilization on dried fig (*Ficus carica* L. cv. Sarılop) yield and quality. *Scientia Horticulturae* 118(4), 308–313.
Jafari, M., Rahemi, M. and Haghighi, A.A.K. (2018) Role of fig rootstock on changes of water status and nutrient concentrations in 'Sabz' cultivar under drought stress condition. *Scientia Horticulturae* 230, 56–61.
Kabasakal, A. (1983) Seasonal variation of some mineral nutrients in 'Sarılop' Fig. *Doctoral Thesis* Soil Science Department, Ege University, Izmir, Turkey.
Kong, M., Lampinen, B., Shackel, K. and Crisosto, C.H. (2013) Fruit skin side cracking and ostiole-end splitting shorten postharvest life in fresh figs (*Ficus carica* L.), but are reduced by deficit irrigation. *Postharvest Biology and Technology* 85, 154–161.
Läuchli, A. and Grattan, S.R. (2007) Plant growth and development under salinity stress. In: *Advances in Molecular Breeding Toward Drought and Salt Tolerant Crops*. Springer, Dordrecht, pp. 1–32.
Mordoğan, N., Hakerlerler, H., Ceylan, Ş., Aydın, Ş., Yağmur, B., *et al.* (2013) Effect of organic fertilization on fig leaf nutrients and fruit quality. *Journal of Plant Nutrition* 36(7), 1128–1137.
Moreno, D.A., Pulgar, G., Víllora, G. and Romero, L. (1998) Nutritional diagnosis of fig tree leaves. *Journal of Plant Nutrition* 21(12), 2579–2588.
Muhammad, S., Khalsa, S.D.S. and Brown, P.H. (2018) *Nitrogen Management in Nut Crops*. University of California Agriculture and Natural Resources, Oakland, California.
Muhammad, S., Sanden, B.L., Lampinen, B.D., Saa, S., Siddiqui, M.I., *et al.* (2015) Seasonal changes in nutrient content and concentrations in a mature deciduous tree species: Studies in almond (*Prunus dulcis* (Mill.) DA Webb). *European Journal of Agronomy* 65, 52–68.
Nelson, P.N. and Su, N. (2010) Soil pH buffering capacity: a descriptive function and its application to some acidic tropical soils. *Australian Journal of Soil Research* 48, 201–207.
Obenauf, G. (1978) *Commercial Dried Fig Production in California*. Division of Agriculture and Sciences, University of California, California.
Okubo, M. and Utsunomiya, N. (1996) Effects of sodium choride on growth, gas exchange and ion concentration in latex of fig (*Ficus carica* L.). *Environment Control in Biology* 34(4), 259–265.
Opara, L.U., Studman, C.J. and Banks, N.H. (1997) Fruit skin splitting and cracking. *Horticultural Reviews* 19, 217–262.
Özer, K.B. and Derici, B. (1997) A research on the relationship between aflatoxin and ochratoxin a formation and plant nutrients. *I International Symposium on Fig* 480, 199–206.
Proebsting, E.L. and Tate, R. (1952) Seasonal changes in nitrate content of fig leaves. *Proceedings of the American Society for Horticultural Science* 60, 7–10.
Proebsting, E.L. and Warner, R.M. (1954) The effect of fertilizers on yield, quality and leaf composition of figs. *Proceedings of the American Society for Horticultural Science* 63, 10–18.
Richards, L.A. (1954) *Diagnosis and improvement of saline and alkali soils*, No. 2. US Department of Agriculture, p. 154.
Rosecrance, R.C., Weinbaum, S.A. and Brown, P.H. (1996) Assessment of nitrogen, phosphorus, and potassium uptake capacity and root growth in mature alternate-bearing pistachio (*Pistacia vera*) trees. *Tree Physiology* 16(11–12), 949–956.
Schlesinger, W.H. (2009) On the fate of anthropogenic nitrogen. *Proceedings of the National Academy of Sciences* 106(1), 203–208.
Shabala, S. and Pottosin, I. (2014) Regulation of potassium transport in plants under hostile conditions: implications for abiotic and biotic stress tolerance. *Physiologia Plantarum* 151(3), 257–279.

Soliman, S., Alebidi, A., Al-Obeed, R. and Al-Saif, A. (2018) Effect of potassium fertilizer on fruit quality and mineral composition of fig (*Ficus carica* L. cv. Brown Turky). *Pakistan Journal of Botany* 50(5), 1753–1758.

Tanji, K. and Kielen, N. (2003) *Agricultural drainage water management in arid and semi-arid areas.* FAO irrigation and drainage paper 61, annex 1: crop salt tolerance data. United Nations Food and Agriculture Organization, Rome.

USDA ARS (1954) Crop selection for saline soils: Web adaptation from Handbook 60. www.ars.usda.gov/pacific-west-area/riverside-ca/us-salinity-laboratory/docs/crop-selection-for-saline-soils/ (accessed May 2019).

Vachamidis, P. and Vemmos, S. (2010) Fig tree cultivation. *Agriculture-Livestock* 1, 64–72.

Vemmos, S.N., Petri, E. and Stournaras, V. (2013) Seasonal changes in photosynthetic activity and carbohydrate content in leaves and fruit of three fig cultivars (*Ficus carica* L.). *Scientia Horticulturae* 160, 198–207.

Zarei, M., Azizi, M., Rahemi, M. and Tehranifar, A. (2016) Evaluation of NaCl salinity tolerance of four fig genotypes based on vegetative growth and ion content in leaves, shoots, and roots. *HortScience* 51(11), 1427–1434.

Zörb, C., Senbayram, M. and Peiter, E. (2014) Potassium in agriculture–status and perspectives. *Journal of Plant Physiology* 171(9), 656–669.

# 11 Irrigation and Water Requirements

**M. Abdolahipour[1]*, A.A. Kamgar-Haghighi[2], A.R. Sepaskhah[3], N. Davatgar[4], and N. Dalir[5]**

[1]*Department of Water Engineering, College of Aburaihan, University of Tehran, Tehran, Iran;* [2]*Water Engineering Department, Shiraz University, Shiraz, Iran;* [3]*Water Engineering Department and Drought Research Institute, Shiraz University, Shiraz, Iran;* [4]*Soil and Water Research Institute, Agricultural Research, Education and Extension Organization (AREEO), Karaj, Iran;* [5]*Department of Soil Science, Tarbiat Modares University, Tehran, Iran*

## 11.1 Introduction

Research investigating fig tree irrigation is limited to the dominant varieties and industries. Although fig is considered drought tolerant, sustained water stress can affect tree and fruit growth. Mason (1911) investigated the effects of water stress on the growth of fig and olive trees planted in Palm Springs, California in 1891 (Condit, 1947). After twenty years, he reported that 'the contrast between the green of the olives on either side and the figs, which are dead save a few struggling sprouts, illustrates in a most marked way the comparative drought resistance of the two'. This strongly suggests in arid and desert climates figs require supplemental irrigation in addition to precipitation. Most fig orchards in the world are irrigated but rainfed orchards are cultivated in the arid and semi-arid Mediterranean climates of Iran, Egypt and Turkey. There is a lack of information about the fig's water needs and use (Flaishman *et al.*, 2007). This review will discuss water requirements and irrigation strategies of rainfed fig trees.

## 11.2 Irrigating Fig Orchards

There is limited international literature on the effects of irrigation on fig tree growth and productivity. The traditional Old World orchards in Europe, Africa and Asia were rarely irrigated. According to Grasovsky and Weitz (1930), irrigation, if used at all in Palestine, was during fruit development. In North America, Reimer (1910) reported eastern orchards were irrigated during the dry season: September and October when the fruit is ripening. In the western orchards of California, irrigation is essential except where the water table is high. In Fresno, California's Central Valley, with an average rainfall of 254 mm, most orchards need one or more flood irrigations a year (Condit, 1947). In Algiers, Algeria the heavy average rainfall of 719–1069 mm allows fig production without irrigation. The fig trees near Kalamata, Greece, are grown on terraced hillsides with an average annual rainfall of 867–1219 mm (Condit, 1947).

Poor growth and canopy development in young and mature fig trees is the best visual indicator of water status. In mature trees rapid increase in water supply during the ripening period will cause the fruit to split, resulting in fermentation and souring. Too much water in mid to late summer will lead to excessive vegetative growth, and larger trees, at the expense of fruit quality, and delay the defoliation that precipitates dormancy, making the trees

*Corresponding author: abdolahipour@ut.ac.ir


DOI: 10.1079/9781789242881.0011

vulnerable to early freezes. The Smyrna cultivar is particularly susceptible to fruit quality problems when vegetative growth is excessive (Brien and Hardy, 2002). Water stressed trees are particularly susceptible to attack by Pacific and red spider mite (Faghih and Sabet-Sarvestani, 2001).

## 11.3 Supplemental Irrigation of Rainfed Orchards

Fig trees can grow in rainfed conditions all over the world. However, under prolonged drought conditions, severe damage can occur in rainfed orchards normally tolerant to water shortage (Stover *et al.*, 2007; Hallaç-Türk and Aksoy, 2011; Gholami *et al.*, 2012; Karimi *et al.*, 2012). In this condition, supplemental irrigation (SI) can mitigate the negative effects of water stress. Supplemental irrigation (SI) is 'the addition of a limited amount of water to otherwise rainfed crops, when rainfall fails to provide essential moisture for normal plant growth, in order to improve and stabilize productivity' (Oweis *et al.*, 1999). Similar to other arid and semi-arid regions, Iranian fig growers apply SI in their orchards in drought conditions (Sharifzadeh *et al.*, 2012; Kamyab, 2015) (Fig. 11.1).

Supplemental irrigation at an inappropriate time or quantity may negatively affect fig trees. Abdolahipour *et al.* (2018) investigated the optimal time and amount of SI at different distances from trunks in a rainfed fig orchard (*Ficus carica* L. cv Sabz, Smyrna type) in Iran's Estahban region with an average annual rainfall of 354 mm. They designated three application positions, next to, 1–1.1 m from the trunk and outside the tree canopy, at three irrigation levels, none, 1000 and 2000 L, respectively (Fig. 11.2), and irrigations in early spring and mid-summer.

Compared to summer irrigation, early spring irrigation maintained the SWC for a longer time. Irrigating further from the tree trunks increased soil surface evaporation, especially in summer. This study suggests that using supplemental irrigation with 2000 L per tree of water in early spring near the tree trunk could provide higher SWC and lower surface evaporation for rainfed fig orchards during drought conditions in the study area.

## 11.4 Effects of Irrigation on Fig Tree Growth and Yield

Previous reports have confirmed the benefits of SI in improving the morphological traits and yield of fig trees under drought conditions. Abdolahipour et al. (2019c) investigated supplemental irrigation timing and amount of SI at different distances from tree trunks on morphological characteristics of rainfed fig trees under drought conditions. They reported irrigation outside the tree canopy appeared to be the best irrigation location but no significant differences in growth traits between irrigation amounts during the two study years. Similarly, Al-Desouki *et al.* (2009) reported SI had significant effects on plant growth of mature 'Sultani' fig trees in the western coastal zone of Egypt under rainfed conditions. El Kassas (1975) observed better vegetation. Leonel and Tecchio (2010) reported irrigation improved growth.

### Root length density

Fig trees have a relatively high root-to-shoot ratio of 1:1.44 (Abdel Razik and El Darier, 1991) with a root volume twice that of the canopy (Keleg *et al.*, 1981). The fig tree has a system of fibrous roots which spread widely both laterally and, in some soils, to a surprisingly great depth. Traub and Stansel (1930) found that a five-year-old Brunswick (Magnolia) fig tree in Texas had a root spread of 15.24 m, a single lateral reaching 10.7 m from the main trunk. Several researchers (Pandey *et al.*, 1984; Sponchiado *et al.*, 1989) have hypothesized that the ability of a plant to change its root distribution in the soil is an important mechanism for drought avoidance. Sponchiado *et al.* (1989) showed that the ability of a plant to change root distribution to avoid drought stress could vary by cultivar within a species.

An investigation (Abdolahipour *et al.*, 2020) of how SI affected root length density (RLD) of rainfed fig trees in Iran's Estahban region demonstrated the highest root concentrations at 15–45 cm depth late winter and late spring. However, by summer, the highest RLD was at 30–60 cm depth for both irrigated and non-irrigated trees. Differences were insignificant

**Fig. 11.1.** Supplemental irrigation of rainfed fig orchards, Estahban, Iran. (Figure courtesy of M. Abdolahipour.)

below 60 cm depth. They hypothesized the winter rains and the prior season's stored water significantly increased RLD. This root growth pattern was observed when SI was close to the trunk. However, application of SI 2.1–2.2 m from the tree trunks (outside of the tree canopy), developed the root horizontal extension to a greater distance.

## Yield

Drought, after climate per se, is the major factor limiting crop yield and quality worldwide. SI applied during critical stages of fig growth can improve fig tree production in drought conditions (Abdolahipour *et al.*, 2019b; Honar *et al.*, 2019). Dried fruit weight is a function of fruits number

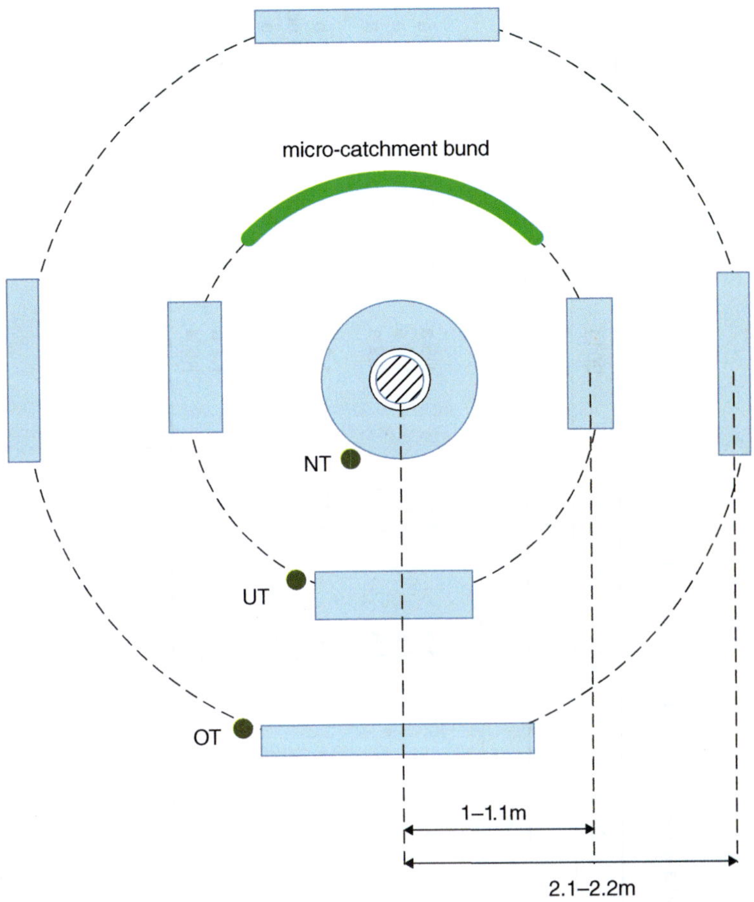


**Fig. 11.2.** Different irrigation application positions from tree trunk treatments in the experiment for tree with a canopy cover diameter about 3.2 m (shaded area: irrigation positions, hatched area: tree trunk, black points: access tube for measuring soil moisture, NT: near tree trunk, UT: under the tree canopy, and OT: out of tree canopy). (Figure used from Abdolahipour *et al.*, 2019c with permission from RightsLink.)

and fruit size (Aksoy *et al.*, 2001). The average fruit weight of an on-tree 'Sabz' sun dried fig fruit is 5.6 g, 40% of fresh ripe weight (Faghih and Sabet-Sarvestani, 2001). Abdolahipour *et al.* (2019b) determined 1000 and 2000 L SI per tree increased yield significantly equally (Table 11.1), demonstrating per tree per year, the lower 1000 L per tree per year is sufficient for economic production.

There was no significant difference in increasing the yield between early spring versus mid-summer SI timing (Table 11.1). However, the final fruit growth phase is highly influenced by irrigation, as a considerable fraction of dry and fresh weight is accumulated later in the season (Pavel and DeJong, 1993; Girona *et al.*, 2001). While there are very low simple correlation coefficients between rainfed crop yield and annual rainfall (Nix and Fitzpatrick, 1969), the rainfall occurrence during the different crop growth stages is more relevant (Sepaskhah *et al.*, 2003; Mishra *et al.*, 2008). Rainfall in winter is the vital parameter for fig yield (Bagheri and Sepaskhah, 2014). Theoretically, if there is sufficient soil water stored from the previous winter's rain, SI irrigation timing in rainfed conditions does not have a great effect on yield.

The relationship between seasonal rainfall distribution and rainfed fig production was investigated by Bagheri and Sepaskhah (2014) over 13 years in Iran's Estahban region to determine the critical rainfall periods for rainfed

**Table 11.1.** Fruit weight (kg/tree) based on mean annual total yield and diameter of fruit for different treatments in two experimental years (adopted from Abdolahipour *et al*., 2019b.).

| Year | | 2013 | 2014 | 2013 | 2014 | 2013 | 2014 | 2013 | 2014 |
|---|---|---|---|---|---|---|---|---|---|
| | | | | Diameter of fruit (mm) | | | | | |
| Treatment | No. | Total yield (kg/tree) | | <17 | | 17–22 | | >22 | |
| Distance | | | | | | | | | |
| AT | 24 | 8.3 a* | 7.2 ab | 3.9 a | 4.8 a | 2.5 a | 1.5 a | 1.8 ab | 0.9 a |
| UT | 24 | 5.8 c | 6.7 b | 2.1 b | 3.8 a | 2.4 a | 1.4 a | 1.3 b | 1.5 a |
| OT | 24 | 7.0 b | 8.6 a | 1.9 b | 4.1 a | 2.7 a | 2.2 a | 2.4 a | 2.3 a |
| Amount | | | | | | | | | |
| Rainfed | 24 | 5.8 c | 6.1 b | 2.4 a | 3.8 a | 1.9 b | 1.2 b | 1.5 b | 1.1 b |
| 1000 L | | 6.6 b | 8.0 a | 2.9 a | 4.4 a | 2.1 b | 2.0 a | 1.6 b | 1.6 ab |
| 2000 L | 24 | 8.7 a | 8.3 a | 2.7 a | 4.5 a | 3.6 a | 1.8 a | 2.4 a | 2.0 a |
| Time | | | | | | | | | |
| Early spring | 36 | 6.8 a | 7.9 a | 2.7 a | 4.7 a | 2.5 a | 1.7 a | 1.6 b | 1.4 a |
| Mid-summer | 36 | 7.3 a | 7.1 a | 2.6 a | 3.7 b | 2.6 a | 1.6 a | 2.1 a | 1.8 a |

*Means followed by the same letter are not statistically different according to Duncan multiple range test ($p < 0.05$). The difference between means is tested for each treatment, separately. (AT: near the tree trunk, UT: under the tree canopy, and OT: out of tree canopy).

fig production and supplementary irrigation water application. They found that the rainfall in spring, particularly in April, and in December has a negative influence on fig yield due to its interference with the life cycle of Blastophaga wasps for pollination, and supplementary irrigation in March is a good choice during droughts. On the other hand, an economic analysis of SI for Iran's Estahban region, with and without micro-catchment water harvesting systems (MCWHS) showed using MCWHS can lower the optimal amount of SI water (Khozaie and Sepaskhah, 2018). By using MCWHS, there was a 50% reduction in applied water, 14% lower fig production, and twice as high net income compared with maximum production. In order to obtain high net profit with a given amount of annual rainfall, the amount of optimal SI water decreased by increasing the unit price of water.

Abdolahipour *et al.* (2019a) investigated the interaction of SI and pruning on growth and yield in rainfed Sabz fig trees (Smyrna type) of Iran's Estahban region under drought conditions. The results regarding the growth characteristics showed a positive response of fig trees to the SI and the maximum yield achieved by using two irrigations in the last month of winter and in mid-spring. However, one irrigation in late winter showed the highest water productivity among irrigation treatments. Among pruning treatments, severe thinning out of one-year-old lateral branches resulted in appropriate vegetative and reproductive characteristics. Conclusively, a combination of severe thinning out of one-year-old lateral branches and one SI event with 750 L per tree (equal to 75 $m^3$ $ha^{-1}$) in late winter was recommended as the best choice in drought conditions.

Using a drag-hose sprinkler system for SI, Allam *et al.* (2007) found significantly higher fig yield in irrigated trees compared to those under rainfed conditions only.

## Fruit quality

The size and color are the major factors in marketing of dried figs (İrget *et al.*, 2008). Abdolahipour et al. (2019b) showed higher irrigation volume increased fruit size and consequently trees irrigated with 2000 L per tree had significantly higher fruit yields with larger fruit diameters (>22 mm) in both years (Table 11.1). Early spring irrigation produced significantly more small fruits. Results also indicated that trees under OT treatment produced significantly higher yields of larger fruit (>22 mm) diameter. Although irrigation treatment of 2000 L increased the weight of fruits with light color in two experimental years, it increased the dark brown colored fruits, significantly (Table 11.2).

Skin color is another important quality factor in dried fig marketing. Fruits with high total sugars, particularly reducing sugars interact with amino acids to give brown compounds which decrease dried fig quality (Abul-Fadl *et al.*, 2015). Other factors can also affect dried fig color: pollen source, droplet formation, air humidity, air temperature and sunlight, soil type, tree nutritional status, and soil water (Faghih and Sabet-Sarvestani, 2001; Flaishman *et al.*, 2007). Previous field observation and farmers' experiences in the different regions showed that excessive water can lead to dark skin color fruits. This might occur due to the sensitivity of fig trees to root rot as a result of extreme soil wetting (Dominguez, 1990).

## Physiological parameters

The physiological parameters could aid irrigation scheduling for figs. However, the gas exchange and photosynthesis parameters of fig trees are poorly documented (Higgins *et al.*, 1992; Clifford *et al.*, 1997; Can *et al.*, 1998; Can and Aksoy, 2007), for both irrigated and rainfed conditions.

The gas exchange characteristics are an indicator of the plant growth due to their direct relationship with the net productivity (Ashraf, 2004; Piao *et al.*, 2008; Ashraf and Harris, 2013). The decrease in net photosynthesis ($A_n$), is intensified by water stress, and it may result from increasing stomatal and non-stomatal limitations (Ben *et al.*, 1987; Ort *et al.*, 1994; Syvertsen and Lloyd, 1994; Tezara *et al.*, 1999). The reduction in the rate of photosynthesis under drought conditions is totally attributed to a suppression in the mesophyll conductance and the stomata closure at water stress (Flexas *et al.*, 2004, Chaves *et al.*, 2009; Ashraf and Harris, 2013).

The seasonal changes in the photosynthetic activity of the three fig cultivars of 'Fracasana'

**Table 11.2.** Mean fruit weight (kg/tree) based on skin color and TSS (Brix) for different treatments in two experimental years (adopted from Abdolahipour *et al.*, 2019b).

| Year | | 2013 | 2014 | 2013 | 2014 | 2013 | 2014 | 2013 | 2014 |
|---|---|---|---|---|---|---|---|---|---|
| | | Color of fruit skin | | | | | | | |
| Treatment | No. | Yellow | | Light brown | | Dark brown | | TSS (Brix) | |
| Distance | | | | | | | | | |
| AT | 24 | 1.7 b* | 1.0 b | 1.9 a | 1.9 b | 4.6 a | 4.3 a | 32.9 a | 32.5 a |
| UT | 24 | 1.7 b | 1.5 b | 1.9 a | 1.9 b | 2.2 b | 3.3 b | 32.7 a | 31.8 a |
| OT | 24 | 3.1 a | 2.7 a | 2.0 a | 2.8 a | 1.8 b | 3.1 b | 32.3 a | 31.0 a |
| Amount | | | | | | | | | |
| Rainfed | 24 | 1.8 b | 1.6 a | 1.6 b | 1.7 b | 2.4 b | 2.8 b | 34.2 a | 33.2 a |
| 1000 L | 24 | 1.7 b | 1.7 a | 1.8 b | 2.4 a | 3.1 a | 3.9 a | 32.7 b | 31.3 b |
| 2000 L | 24 | 3.1 a | 1.9 a | 2.4 a | 2.5 a | 3.2 a | 3.9 a | 31.0 c | 30.8 b |
| Time | | | | | | | | | |
| Early spring | 36 | 2.1 a | 1.7 a | 2.0 a | 2.4 a | 2.6 b | 3.8 a | 32.7 a | 31.7 a |
| Mid-summer | 36 | 2.2 a | 1.8 a | 1.9 a | 2.0 a | 3.1 a | 3.3 b | 32.5 a | 31.9 a |

*Means followed by the same letter are not statistically different according to Duncan multiple range test ($p < 0.05$). The difference between means is tested for each treatment, separately. (AT: near the tree trunk, UT: under the tree canopy, and OT: out of tree canopy).

(San Pedro type), 'Kalamon' and 'Mission' (Common type) was investigated by Vemmos *et al.* (2013) in a Mediterranean climate region in Greece. They reported a rapid increase in $A_n$ during leaf development, obtaining its highest value at full leaf expansion (in May). However, in the same climate and for a different fig cultivar ('De a Libra'), González-Rodríguez and Peters (2010) observed the maximum values of gas exchange parameters 20 days after full leaf expansion.

Abdolahipour et al. (2019c) reported the positive effects of supplemental irrigation on $A_n$. They also showed the influence of different SI on leaf water potential (LWP) during plant growth. Increasing the amount of applied water to 1000 and 2000 L of water, significantly increased LWP compared to that in non-irrigated trees. The highest $A_n$ and $g_s$ occurred in LWP around −1 MPa and air temperature of 20°C. An LWP equal to −2 MPa could be considered as a threshold for stomatal closure of rainfed fig trees, which put this plant among drought-tolerant trees. The highest leaf scale water use efficiency for trees irrigated with 2000 liters of water, out of the canopy, indicated higher increase of $A_n$ compared with $T_r$ after irrigation.

Some other physiological and biochemical characteristics of four of Estahban's fig cultivars, Sabz, Siah, Matti and Shah-anjir (Smyrna type), were evaluated under water stress treatments of 25, 50, 75 and 100% of field capacity (Shirbani *et al.*, 2013). The results showed drought stress decreases the chlorophyll and leaf relative water content and increases the proline and electrolyte leakage.

## 11.5 Irrigation Water Requirements of Figs

Tapia *et al.* (2003) evaluated the response of six varieties (planted at 6 × 4 m), of the 4th leaf fig trees, at four irrigation rates relative to estimated crop evapotranspiration ($ET_c$), adapted for young trees and based on 'Class A' pan evaporation data in a dry area in Chile with annual rainfall of 37 mm. The trial was 39 weeks long with weekly irrigation intervals. All treatments demonstrated a strong positive correlation between irrigation rate and shoot length. They concluded that it is possible to irrigate 3-year-old fig trees with 220 mm per year (2200 $m^3$ $ha^{-1}$ per year).

However, higher water requirements are reported by authors in other research. Hernandez *et al.* (1994) using a micro sprinkler system, evaluated the effects of six irrigation (0, 25, 50, 75, 100 and 125% of 'Class A' pan evaporation) and six nitrogen level (0, 150, 300, 450, 600 and 750 g per tree) treatments on 'Roxo de Valinhos' fig trees (Common type) in the state of São Paulo, Brazil. A micro sprinkler was installed for each two plants and irrigation was done three times a week. Irrigation showed positive effects on total productivity, length of branches and diameter of fruits. They recommended application of 75% of 'Class A' pan evaporation.

An irrigation experiment in southern California from 1928 to 1938 investigated flood irrigating four cultivars, Lob Injir (Smyrna type), San Piero, Mission and Dottato (Common type) twice or once a month, and once or not at all during the growing season. The trees irrigated twice a month had an average trunk circumference of 793 mm; trees irrigated once a month averaged 714 mm; those irrigated only once during the season 656 mm, and unirrigated trees averaged only 520 mm. Among the four varieties in the experiment, Lob Injir and San Piero produced the lowest and highest amount of fruit, respectively, and the Mission cultivar and Dottato had similar yields (Condit, 1947).

Some research has been done on drip irrigation of 'Roxo de Valinhos' fig trees (Common type) in Brazil. Sampaio *et al.* (1981) reported drip irrigation and pruning extended the fresh fig production season. The experiment was conducted with three periods of pruning during 1980 and 1981 and the irrigations were operated with a frequency of two irrigations per week. However, the fruit size and yields were affected by the lower winter temperatures and they recommended their practice for places with less severe winter temperature.

Leonel and Tecchio (2008; 2009; 2010) investigated the effects of pruning time and irrigation on 'Roxo de Valinhos' fig trees (Common type) in the state of São Paulo, Brazil. They reported August pruning and irrigation enhanced the growth and yield. Without irrigation, the fig tree could be pruned in the months of July and August, showing similar results.

The $K_c$ value of 0.47 developed by Olitta *et al.* (1979) is used in nearly all Brazilian research reports to calculate ET in a 'two-step method' (Andrade *et al.*, 2014). Also in Brazil, Andrade *et al.* (2014) evaluated the effect of drip irrigation on growth and yield of three year old 'Roxo de Valinhos' fig trees (Common type) as well as determine crop coefficients ($K_c$) and the cultivar's water requirement ($ET_c$) (Fig. 11.3). The fig tree crop coefficients ($K_c$) obtained through the crop cycle are shown in Fig. 11.3 for both irrigated treatments. The highest $K_c$ values corresponded to the early growing season, when rainfall was the lowest.

As can be seen in Fig. 11.3, the two-day irrigation interval treatment (I1) had consistently higher $K_c$ values versus four-day irrigation interval treatment (I2) due to higher evapotranspiration potential as a function of higher soil water content. This is consistent with Allen's (1998) theory that in situations with sufficient soil water, $K_c$ values increase significantly, approaching 1.0–1.2 for all crops. Compared with non-irrigated trees, irrigation produced a 27% yield increase. Interestingly, the irrigated trees produced more shoot internodes but not significantly longer growth than the unirrigated trees.

Goldhamer and Salinas (1999) analyzed the tree water relations and production data of commercial 20-year-old Black Mission fig trees in California's major Central Valley production region (Table 11.3). The amount of applied water in each treatment during summer is shown in Fig. 11.4. Based on the results, 67.8 cm of summer applied water was optimal for fruit yield and quality. As can be seen in Table 11.3, the greater fruit size was the primary yield component responsible for the improved production and profit. The disadvantage of using more water increased the amount of cull fruit due to increased fungal contamination from fruit dropped on the wet soil, especially if the soil has poor infiltration. Theoretically subsurface drip irrigation may eliminate the surface wetting that increases cull fruit.

Based on historical ETo and the determined crop coefficients, summer-applied water of 90.4 cm to Black Mission figs in California's San Joaquin Valley achieved the highest fruit production and grower profit in the Madera area (Table 11.4). In Guadajira, Spain, Pereira (2017) reported 2300 $m^3$ $ha^{-1}$ per year applied with drip irrigation to multiple varieties of 'common type' fig trees grown in a sandy loam with an average depth of 1.5 m produced no water stress. The

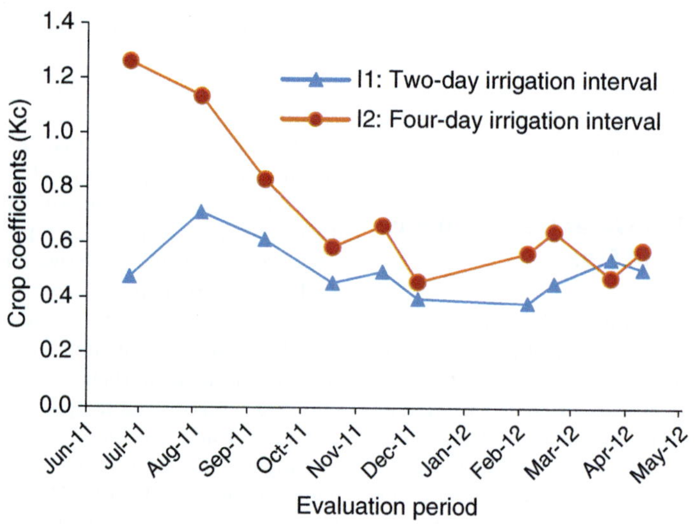


**Fig. 11.3.** Monthly crop coefficients ($K_c$) of the irrigated treatments. (Figure used from Andrade *et al.*, 2014 with permission from SciELO.)

**Table 11.3.** Mean 1994–1995 summer-applied water and yield components for seven irrigation regimes (adopted from Goldhamer and Salinas,1999).

| Irrigation treatment | Summer-applied water (cm) | 'Product' fruit value ($ $t^{-1}$) | Total fruit load (no. $tree^{-1}$) | 'Product' fruit yield (kg $ha^{-1}$) | Cull fruit yield (kg $ha^{-1}$) | Total fruit yield (kg $ha^{-1}$) |
|---|---|---|---|---|---|---|
| T1 | 27.2 | 1130 a | 4245 ns | 5195 ns | 310 a | 5507 a |
| T2 | 38.9 | 1160 abc | 4314 ns | 5601 ns | 321 a | 5920 ab |
| T3 | 50.3 | 1154 abc | 5236 ns | 6206 ns | 368 a | 6575 ab |
| T4 | 60.0 | 1151 ab | 4958 ns | 5977 ns | 498 ab | 6476 ab |
| T5 | 67.8 | 1183 c | 4625 ns | 6520 ns | 571 ab | 7091 ab |
| T6 | 77.2 | 1178 bc | 4632 ns | 6603 ns | 779 b | 7382 ab |
| T7 | 84.6 | 1182 bc | 4666 ns | 6677 ns | 746 b | 7423 b |

Values not followed by the same letter are statistically different using Fisher's protected least squares difference at the 95% confidence level; ns = no significant difference.

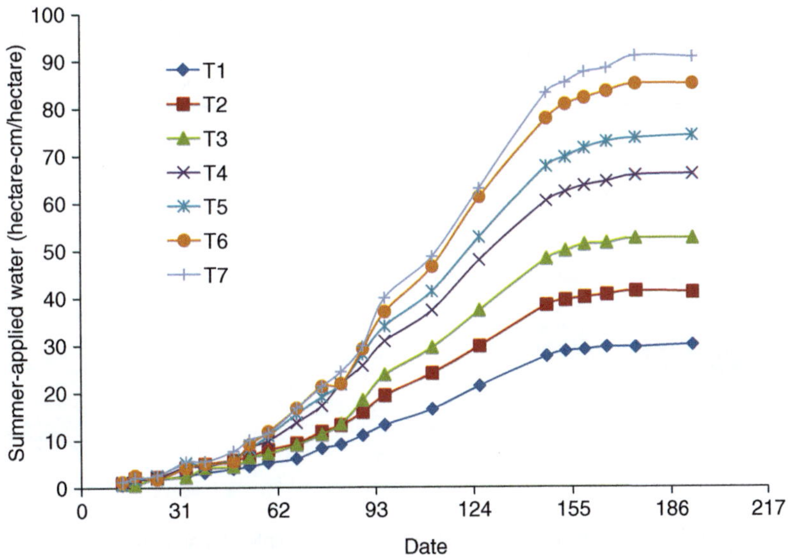


**Fig. 11.4.** Summer-applied water with time during 1995 for seven irrigation regimes. (Figure used from Goldhamer and Salinas, 1999 with permission from University of California, Agriculture and Natural Resources.)

climate was semi-temperate, with a mean rainfall of 550 mm.

## 11.6 Selection of Irrigation Method

Fig-orchard irrigation methods are no different than those practiced in other deciduous orchards. Furrow, flood, sprinkler and drip irrigation are all used in fig orchards worldwide (Fig. 11.5).

### Irrigation water quality

Compared to other types of tree fruits, fig trees are considered to be moderately tolerant to salt stress (Golombek and Lüdders, 1990). While the highest level of EC tolerance in irrigation water for most fruit trees is lower than 2 dS $m^{-1}$

**Table 11.4.** Summer water use of mature clean-cultivated Mission figs for an average weather year near Madera. Values based on crop coefficients developed in this study and long-term historical reference crop water use (ETo). (Adopted from Goldhamer and Salinas,1999). This resulted in 90.4 cm of summer-applied water and a maximum orchard water use value of 185 L/tree/day for mature Black Mission figs in a 4.3 × 6.1 m spacing.

| Date | Daily ETo (cm/day) | Crop coefficient ($K_c$) | Daily $ET_c$(cm/day) | Cumulative $ET_c$(cm) | Daily $ET_c$(L/tree/day) |
|---|---|---|---|---|---|
| Apr 1–15 | 0.43 | 0.18 | 0.08 | 1.3 | 20.1 |
| Apr 16–30 | 0.53 | 0.34 | 0.15 | 3.3 | 38.2 |
| May 1–15 | 0.58 | 0.49 | 0.28 | 7.6 | 74.2 |
| May 16–31 | 0.64 | 0.65 | 0.38 | 13.7 | 98.4 |
| June 1–15 | 0.66 | 0.82 | 0.53 | 21.8 | 140.4 |
| June 16–30 | 0.71 | 0.97 | 0.64 | 31.5 | 166.2 |
| Jul 1–15 | 0.71 | 1.02 | 0.71 | 42.2 | 185.1 |
| Jul 16–30 | 0.69 | 1.02 | 0.71 | 53.6 | 185.1 |
| Aug 1–15 | 0.66 | 1.02 | 0.69 | 63.8 | 174.9 |
| Aug 16–30 | 0.58 | 1.02 | 0.69 | 74.4 | 174.9 |
| Sept 1–15 | 0.51 | 0.88 | 0.46 | 81.3 | 116.2 |
| Sept 16–30 | 0.43 | 0.77 | 0.38 | 87.1 | 101.4 |
| Oct 1–15 | 0.36 | 0.64 | 0.23 | 90.4 | 59.1 |

**Fig. 11.5.** Surface drip irrigation in irrigated orchards (right) and rainfed fig orchards (left), Estahban, Iran. (Figure courtesy of E. Sheikhali.)

(Maas, 1993), irrigation water salinity of 5.5–6 dS m$^{-1}$ has been reported (Dominguez, 1990) for figs. Little information is available so it is assumed that the irrigation water quality for edible fig trees is also acceptable for caprifig orchards.

## 11.7 Fertigation

Under normal circumstances, fig trees require little or no fertilization; the natural fertility of native soils is often enough for a fig tree to grow and thrive. However, Proebsting and Tate (1952) observed that foliar concentration of total nitrogen decreased during the growing season.

Honar *et al.* (2019) investigated the effects of SI timing and potassium application on yield and yield quality of rainfed fig trees in Iran's Estahban region. Treatments included SI timing (in mid-March, mid-May and early August), potassium (K) fertigation application (no-fertilizer, 150 and 250 g fertilizer per tree) and foliar K application (no-fertilizer, 1.5 and 3.0 kg 1000 L$^{-1}$ water). Using 150 g per tree K by fertigation method or 3.0 kg 1000 L$^{-1}$ foliar K fertilizer showed significant positive effect on the production (Table 11.5) and yield quality and can be used for fertilization of fig trees under drought conditions.

**Table 11.5.** Mean yield (kg/tree) in different irrigation, fertilizer levels in different years (adopted from Honar *et al.*, 2021).

| | First year | | | | | | | |
|---|---|---|---|---|---|---|---|---|
| | Foliar K fertilizer (kg 1000 L$^{-1}$) | | | | K application by fertigation (g/tree) | | | |
| Irrigation treatment | 0 | 1.5 | 3.0 | Mean | 0 | 150 | 250 | Mean |
| Rain-fed | 6.75a* | 6.90a | 7.86a | **7.18C** | 6.75a* | 6.92a | 7.89a | **7.19C** |
| March | 9.62a | 9.98a | 12.32a | **10.64AB** | 9.62a | 11.33a | 11.87a | **10.94AB** |
| May | 8.71a | 10.38a | 11.27a | **10.12B** | 8.71a | 11.55a | 10.26a | **10.16B** |
| August | ---- | ----- | ---- | **----** | 9.41a | 11.26a | 12.34a | **11.00AB** |
| March and May | 10.42a | 10.84a | 13.58a | **11.61A** | 10.42a | 11.91a | 13.17a | **11.84A** |
| **Mean** | **8.88B** | **9.53B** | **11.26A** | | **8.98B** | **10.60A** | **11.11A** | |
| | Second and third year | | | | | | | |
| Rain-fed | 4.36f | 9.28bc | 6.51de | **6.72B** | 4.358g | 7.22c-f | 9.45ab | **7.01C** |
| March | 7.06cd | 7.15cd | 9.50b | **7.91AB** | 7.06d-f | 8.217b-e | 8.29d-e | **7.86BC** |
| May | 4.89ef | 7.99b-d | 7.19cd | **6.69B** | 4.89g | 5.77fg | 6.91ef | **5.85D** |
| August | ----- | ----- | ----- | **-----** | 7.81b-f | 9.10a-d | 9.24a-c | **8.72AB** |
| March and May | 7.96b-d | 6.70de | 12.52a | **9.06A** | 7.96d-e | 10.68a | 8.51d-e | **9.05A** |
| **Mean** | **6.07C** | **7.78B** | **8.93A** | | **6.42B** | **8.20A** | **8.48A** | |

*Mean followed by the same letters in each column and each row are not significantly different at 5% level of probability by Duncan multiple range test.

## References

Abdel Razik, M.S. and El Darier, S. (1991) Functional adaptations of fig trees (*Ficus carica*, L.) in agroecosystems of the western Mediterranean desert of Egypt. *Qatar University Scientific Bulletin* 11, 183–199.

Abdolahipour, M., Kamgar-Haghighi, A.A. and Sepaskhah, A.R. (2018) Time and amount of supplemental irrigation at different distances from tree trunks influence on soil water distribution, evaporation and evapotranspiration in rainfed fig orchards. *Agricultural Water Management* 203, 322–332.

Abdolahipour, M., Kamgar-Haghighi, A.A., Sepaskhah, A.R., Dalir, N., Shabani, A., Honar, T. and Jafari, M. (2019a) Supplemental irrigation and pruning influence on growth characteristics and yield of rain-fed fig trees under drought conditions. *Fruits* 74, 282–293.

Abdolahipour, M., Kamgar-Haghighi, A.A., Sepaskhah, A.R., Zand-Parsa, S. and Honar, T. (2019b) Effect of time and amount of supplemental irrigation at different distances from tree trunks on quantity and quality of rain-fed fig production. *Iran Agriculture Research* 38, 35–46.

Abdolahipour, M., Kamgar-Haghighi, A.A., Sepaskhah, A.R., Zand-Parsa, S., Honar, T. and Razzaghi, F. (2019c) Time and amount of supplemental irrigation at different distances from tree trunks influence on morphological characteristics and physiological responses of rainfed fig trees under drought conditions. *Scientia Horticulturae* 253, 241–254.

Abdolahipour M., Kamgar-Haghighi, A.A., Sepaskhah, A.R., Zand-Parsa, S. and Honar, T. (2020) Root length density of rainfed fig trees under different times, amounts, and positions of supplemental irrigation. *Journal of Agricultural Science and Technology* 22(4), 1137–1150.

Abul-Fadl, M.M., Ghanem, T.H., N., E.-B. and Nasr, A. (2015) Effect of some different drying methods on quality criteria of dried fig fruits. *Current Science International* 4, 548–566.

Aksoy, U., Can, H., Hepaksoy, S. and Sahin, N. (2001) *Fig Cultivation*. TARP Turkey Agricultural Research Project Press, Izmir, Turkey, p. 45.

Al-Desouki, M., Abd El-Rahman, I. and Sahar, A. (2009) Effect of some antitranspirants and supplementary irrigation on growth, yield and fruit quality of Sultani fig (*Ficus carica*) grown in the Egyptian western coastal zone under rainfed conditions. *Research Journal of Agriculture and Biological Sciences* 5, 899–908.

Allam, K., Adly, M. and Mourad, M. (2007) Effect of supplemental irrigation and intercropping treatments on the productivity of fig trees and lentil crop in the Northwest Coast. *Misr Journal Agricultural Engineering* 24, 88–102.

Allen, R.G. (1998) Crop evapotranspiration-Guidelines for computing crop water requirements. *FAO Irrigation and drainage paper 56*, FAO, Rome.

Andrade, I.P.d.S., Carvalho, D.F.d., Almeida, W.S.d., Silva, J.B.G. and da Silva, L.D. (2014) Water requirement and yield of fig trees under different drip irrigation management. *Engenharia Agrícola* 34, 17–27.

Ashraf, M. (2004) Some important physiological selection criteria for salt tolerance in plants. *Flora-Morphology, Distribution, Functional Ecology of Plants* 199, 361–376.

Ashraf, M. and Harris, P. (2013) Photosynthesis under stressful environments: an overview. *Photosynthetica* 51, 163–190.

Bagheri, E. and Sepaskhah, A.R. (2014) Rain-fed fig yield as affected by rainfall distribution. *Theoretical and Applied Climatology* 117, 433–439.

Ben, G.-Y., Osmond, C.B. and Sharkey, T.D. (1987) Comparisons of photosynthetic responses of *Xanthium strumarium* and *Helianthus annuus* to chronic and acute water stress in sun and shade. *Plant Physiology* 84, 476–482.

Brien, J. and Hardy, S. (2002) Fig growing in NSW. *Agfact* H3 *1*.

Can, Z.H. and Aksoy, U. (2007) Seasonal and diurnal photosynthetic behaviour of fig (*Ficus carica* L.) under semi-arid climatic conditions. *Acta Agriculturae Scandinavica Section B-Soil and Plant Science* 57, 297–306.

Can, H., Hepaksoy, S., Aksoy, U. and Kutlu, E. (1998) Leaf characteristics and net gas exchange of fig cultigens adapted to different climatic zones. *XXV International Horticultural Congress, Part 6: Culture Techniques with Special Emphasis on Environmental Implications* 516, 131–138.

Chaves, M.M., Flexas, J. and Pinheiro, C. (2009) Photosynthesis under drought and salt stress: regulation mechanisms from whole plant to cell. *Annals of Botany* 103, 551–560.

Clifford, S.C., Kadzere, I., Jones, H.G. and Jackson, J.E. (1997) Field comparisons of photosynthesis and leaf conductance in *Ziziphus mauritiana* and other fruit tree species in Zimbabwe. *Trees* 11, 449–454.

Condit, I.J. (1947) *The Fig*. Chronica Botanica Co. Waltham, MA, USA.

Dominguez, A.F. (1990) *La Higuera: Frutal Mediterraneo Para Climas Calidos*. Mundi-Prensa, Madrid, Spain.

El Kassas, S. (1975) Effect of soil moisture levels and nitrogen fertilization on yield, shoot growth and leaf composition of fig trees [Egypt]. *Assiut Journal of Agricultural Sciences* 4, 155–166.
Faghih, H. and Sabet-Sarvestani, J. (2001) *Fig: Planting, Cultivation and Harvesting.* Rahgosha Press, Shiraz, Iran.
Flaishman, M.A., Rodov, V. and Stover, E. (2007) The fig: botany, horticulture, and breeding. In: Janick, J. (ed.) *Horticultural Reviews*. John Wiley & Sons, Inc., Hoboken, NJ, USA, p. 113.
Flexas, J., Bota, J., Loreto, F., Cornic, G. and Sharkey, T. (2004) Diffusive and metabolic limitations to photosynthesis under drought and salinity in C3 plants. *Plant Biology* 6, 269–279.
Gholami, M., Rahemi, M. and Rastegar, S. (2012) Use of rapid screening methods for detecting drought tolerant cultivars of fig (*Ficus carica* L.). *Scientia Horticulturae* 143, 7–14.
Girona, J., Marsal, J., Mata, M., Arbonés, A. and Mata, A. (2001) The combined effect of fruit load and water stress in different peach fruit growth stages (*Prunus persica* L.). *VI International Symposium on Computer Modelling in Fruit Research and Orchard Management* 584, pp. 149–152.
Goldhamer, D.A. and Salinas, M. (1999) Black Mission fig production improved by heavier irrigation. *California Agriculture* 53, 30–34.
Golombek, S. and Lüdders, P. (1990) *Gas Exchange of Ficus carica in Response to Salinity, Plant Nutrition—Physiology and Applications*. Springer, pp. 487–493.
González-Rodríguez, A. and Peters, J. (2010) Strategies of leaf expansion in *Ficus carica* under semiarid conditions. *Plant Biology* 12, 469–474.
Grasovsky, A. and Weitz, J. (1930) *Fig Growing in Palestine.* CAB International, Wallingford, UK.
Hallaç-Türk, F. and Aksoy, U. (2011) Comparison of organic, biodynamic and conventional fig farms under rain-fed conditions in Turkey. *Cell Plant Science* 2, 22–33.
Hernandez, F.B.T., Suzuki, M.A., Buzetti, S. and Correa, L.S. (1994) Resposta da figueira (*Ficus carica* L.) ao uso da irrigação e nitrogênio na região de Ilha Solteira. *Scientia Agriculturae* 51, 99–104.
Higgins, S., Larsen, F., Bendel, R., Radamaker, G., Bassman, J., Bidlake, W. and Al Wir, A. (1992) Comparative gas exchange characteristics of potted, glasshouse-grown almond, apple, fig, grape, olive, peach and Asian pear. *Scientia Horticulturae* 52, 313–329.
Honar, T., Shabani, A., Abdolahipour, M., Dalir, N., Sepaskhah, A.R., Haghighi, A.A. and Jafari, M. (2021) Rain-fed fig trees response to supplemental irrigation timing and potassium fertiliser in micro-catchment. *Journal of Horticultural Science and Biotechnology* 96, 738–749.
İrget, M.E., Aksoy, U., Okur, B., Ongun, A.R. and Tepecik, M. (2008) Effect of calcium based fertilization on dried fig (*Ficus carica* L. cv. Sarılop) yield and quality. *Scientia Horticulturae* 118, 308–313.
Kamyab, S. (2015) *Evaluation of Possibility of Supplemental Irrigation Application in Rainfed Fig Orchards of Fars Province*. Shiraz University, Shiraz, Iran, p. 95 (in Persian).
Karimi, S., Hojati, S., Eshghi, S., Nazary Moghaddam, R. and Jandoust, S. (2012) Magnetic exposure improves tolerance of fig 'Sabz' explants to drought stress induced in vitro. *Scientia Horticulturae* 137, 95–99.
Keleg, F., El-Gazzar, A. and Zahran, A. (1981) Studies on root distribution of Jourdan almond and Sultani fig. *Alexandria Journal Agricultural Research* 29, 219–224.
Khozaie, M. and Sepaskhah, A.R. (2018) Economic analysis of the optimal level of supplemental irrigation for rain-fed figs. *Iran Agricultural Research* 37, 17–26.
Leonel, S. and Tecchio, M.A. (2008) Produção da figueira submetida a diferentes épocas de poda e irrigação. *Revista Brasiliera Fruticola* 30, 1015–1021.
Leonel, S. and Tecchio, M.A. (2009) Teores nutricionais em folhas e frutos de figueira, submetida a épocas de poda e irrigação. *Semina: Ciências Agrárias* 30.
Leonel, S. and Tecchio, M.A. (2010) Épocas de poda e uso da irrigação em figueira 'Roxo de Valinhos' na região de Botucatu, SP. *Bragantia* 69, 571–580.
Maas, E. (1993) Testing crops for salinity tolerance. *Proc. Workshop on Adaptation of Plants to Soil Stresses.* In: Maranville, J.W., Baligar, B.V., Duncan, R.R. and Yohe, J.M. (eds) *INTSORMIL. Pub. No. 94-2*, 1–4 August 1993, University of Nebraska, Lincoln, NE, USA, pp. 234–247.
Mason, S.C. (1911) *Drought Resistance of the Olive in the Southwestern States*. US Government Printing Office.
Mishra, A., Hansen, J.W., Dingkuhn, M., Baron, C., Traoré, S.B., Ndiaye, O. and Ward, M.N. (2008) Sorghum yield prediction from seasonal rainfall forecasts in Burkina Faso. *Agriculture and Forest Meteorology* 148, 1798–1814.
Nix, H.A. and Fitzpatrick, E.A. (1969) An index of crop water stress related to wheat and grain sorghum yields. *Agricultural Meteorology* 6, 321–337.

Olitta, A., Sampaio, V. and Barbin, D. (1979) Estudo da lâmina e freqüência de irrigação por gotejo na cultura do figo. *O Solo, Piracicaba* 71, 9–22.
Ort, D., Oxborough, K. and Wise, R. (1994) Depressions of photosynthesis in crops with water deficits. In: Baker, N.R. and Bowyer, J.R. (eds) *Photoinhibition of Photosynthesis: From Molecular Mechanisms to the Field*. Bios Scientific Publishers, Oxford, UK, pp. 315–329.
Oweis, T., Hachum, A. and Kijne, J. (1999) *Water Harvesting and Supplemental Irrigation for Improved Water Use Efficiency in Dry Areas, SWIM.* International Water Management Institute, Colombo, Sri Lanka.
Pandey, R., Herrera, W., Villegas, A. and Pendleton, J. (1984) Drought response of grain legumes under irrigation gradient: III. Plant growth. *Agronomy Journal* 76, 557–560.
Pavel, E. and DeJong, T. (1993) Source-and sink-limited growth periods of developing peach fruits indicated by relative growth rate analysis. *Journal of the American Society for Horticultural Science* 118, 820–824.
Pereira, L.S. (2017) Water, agriculture and food: challenges and issues. *Water Resource Management* 31(10), 2985–2999.
Piao, S., Ciais, P., Friedlingstein, P., Peylin, P., Reichstein, M., Luyssaert, S., Margolis, H., Fang, J., Barr, A. and Chen, A. (2008) Net carbon dioxide losses of northern ecosystems in response to autumn warming. *Nature* 451, 49.
Proebsting, E. and Tate, R. (1952) Seasonal changes in nitrate content of fig leaves. *Proceedings of the American Society for Horticultural Science. Alexandria*, pp. 7–10.
Reimer, F.C. (1910) *Fig culture in North Carolina*. North Carolina Agricultural Experiment Station of the College of Agriculture, North Carolina.
Sampaio, V.R., Olitta, A.F. and Oliveira, A.S. (1981) Efeitos das épocas de poda na produção do figo irrigado por gotejamento. *Anais da Escola Superior de Agricultura Luiz de Queiroz* 38, 847–856.
Sepaskhah, A.R., Honar, T. and Rezaee, A.R. (2003) Effects of rainfall distribution on the dryland wheat yield with an economic analysis in southern provinces of Iran. *Iran Agricultural Research* 22, 89–104.
Sharifzadeh, M., Kamgar-Haghighi, A.A., Sepaskhah, A.R., Honar, T., Abdolahipour, M., Kamyab, S. and Khosrozadeh, M. (2012) *Factors Contributing to Application of Supplemental Irrigation Technique in Fig Production: Evidence from a Survey in Iran*. National Drought Research Institute, Shiraz, Iran, p. 73 (in Persian).
Shirbani, S., Abdolahipour, M., Haghighi, J., Jafari, M. and Davarynejad, G.H. (2013) Physiological and biochemical responses of four edible fig cultivars to water stress condition. *Scholarly Journal of Agricultural Science* 3, 473–479.
Sponchiado, B., White, J., Castillo, J. and Jones, P. (1989) Root growth of four common bean cultivars in relation to drought tolerance in environments with contrasting soil types. *Experimental Agriculture* 25, 249–257.
Stover, E., Aradhya, M., Ferguson, L., Crisosto, C.H. (2007) The fig: overview of an ancient fruit. *HortScience* 42, 1083–1087.
Syvertsen, J. and Lloyd, J. (1994) Citrus. *Handbook of Environmental Physiology of Fruit Crops* 2, 65–99.
Tapia, R., Botti, C., Carrasco, O., Prat, L. and Franck, N. (2003) Effect of four irrigation rates on growth of six fig tree varieties. *Acta Horticulturae* 605, 113–118.
Tezara, W., Mitchell, V., Driscoll, S. and Lawlor, D. (1999) Water stress inhibits plant photosynthesis by decreasing coupling factor and ATP. *Nature* 401, 914–917.
Traub, H.P. and Stansel, R., (1930) The lateral root spread of the fig tree. *Proceedings of the American Society for Horticultural Science*, 109–113.
Vemmos, S.N., Petri, E. and Stournaras, V. (2013) Seasonal changes in photosynthetic activity and carbohydrate content in leaves and fruit of three fig cultivars (*Ficus carica* L.). *Scientia Horticulturae* 160, 198–207.

# 12 Fungal, Bacterial and Nematode Diseases in Figs

R. Mostowfizadeh-Ghalamfarsa[1]*, Z. Bolboli[1], F. Salmaninezhad[1] and T.J. Michailides[2]
[1]*Department of Plant Protection, Shiraz University, Shiraz, Iran;* [2]*Kearney Agricultural Research & Extension Center, University of California, Berkeley, USA*

## 12.1 Introduction

Fig is a globally cultivated tree with the highest diversity of commercial cultivars in Asia (Flaishman *et al.*, 2008). The most economically important species is *Ficus carica* L., which also contains many genetically diverse cultivars. The fig originated in the Middle East, specifically western Asia. It was one of the first trees cultivated by humans (Lansky and Paavilainen, 2010). The fig grows and produces quality fruit in Mediterranean and arid warm-temperate climates (Flaishman *et al.*, 2008). Rain during fruit development and ripening causes fruit splitting and spoiling. With care, figs can also grow in more temperate climates. Diseases limit quality fruit production in tropical climates (Lansky and Paavilainen, 2010). Fully dormant trees are hardy to −9°C, but actively growing trees can be damaged at −1°C. Fig trees killed to the ground will often resprout from the roots (Lansky and Paavilainen, 2010).

Fig trees and fruit are susceptible to numerous abiotic and biotic stresses. Among the abiotic stresses, temperature is the most damaging to the tree. Fig trees are susceptible to low temperatures (less than −1°C). In addition to abiotic stresses, fig trees are susceptible to various pests and diseases. The most important fig pests are the fig mite (*Acaria fici* Cotte) and the fig tree borer (*Phryneta spinator* Fabricius) (Faccoli *et al.*, 2016). Numerous nematodes are among the most devastating pathogens of fig trees. The most damaging are root knot nematodes (*Meloidogyne* spp. Göldi), sheath nematodes (*Hemicycliophora arenaria* Raski), spiral nematodes (*Helicotylenchus multicinctus* Cobb), lance nematodes (*Hoplolaimus* sp.), and dagger nematodes (*Xiphinema* sp.) (Luc *et al.*, 2005; Kohl, 2011). Fig mosaic disease (Fig mosaic virus) is an important, globally reported viral disease (Norman and Ali, 2013).

Among fig tree and fruit diseases, the fungal and bacterial diseases (Table 12.1) are globally the most destructive to cultivated figs (Lin *et al.*, 2018). Fungal pathogens cause rusts, anthracnose, leaf spots, leaf blight, smut, canker, and root and crown rots (Guillaumin *et al.*, 1993; Johnston and Jones, 1997; Kim *et al.*, 2000; Sreenivasaprasad and Talhinhas, 2005; Droby *et al.*, 2011; Norman and Ali, 2013; Raheb *et al.*, 2014; Al-Sadi *et al.*, 2015; Kumar and Khurana, 2015; Li *et al.*, 2015; Miyake and Nagai, 2017; Lin *et al.*, 2018). Fungal infections

*Corresponding author: rmostofi@shirazu.ac.ir

 *The Fig: Botany, Production and Uses* (eds. Sarkhosh, A., Yavari, A. and Ferguson, L.)

DOI: 10.1079/9781789242881.0012

**Table 12.1.** Fungal and bacterial diseases of edible fig.

| Disease name | Causal agent | Distribution | Reference |
|---|---|---|---|
| Alternaria rot | *Alternaria alternata* | Europe | Latinovic *et al.*, 2014 |
| Anthracnose | *Colletotrichum acutatum* | Montenegro, Serbia | Sreenivasaprasad and Talhinhas, 2005 |
| | *Colletotichum gloeosporioides* | India, New Zealand, Korea | Johnston and Jones, 1997; Kim *et al.*, 2000; Cheong *et al.*, 2013 |
| | *Glomerella cingulata* | India | Kim *et al.*, 2000 |
| Armillaria root rot | *Armillaria mellea* | Europe | Guillaumin *et al.*, 1993 |
| Aspergillus mold | *Aspergillus flavus* | Worldwide | Stover *et al.*, 2007 |
| Bacterial canker | *Pseudomonas fici* | California | Michailides, 2003 |
| Bacterial leaf spot | *Pseudomonas syringae* | Serbia | Ivanovic *et al.*, 2018 |
| Botrytis limb blight | *Botrytis cinerea* | California, Korea | Michailides, 2003; Choi *et al.*, 2013 |
| Branch wilt | *Neoscytalidium dimidiatum* | Australia, California, Oman | Paxton *et al.* 1964; Ray *et al.*, 2010; Elshafie and Ba-Omar, 2001 |
| Ceratocystis canker | *Ceratocystis ficicola* | Japan | Kajii *et al.* 2013 |
| | *Ceratocystis fimbriata* | Japan | Hosomi *et al.* 2012 |
| Crown gall | *Agrobacterium tumefaciens* | Worldwide | Cleene, 1979 |
| Endosepsis (pink rot) | *Fusarium moniliforme* | USA | Droby *et al.*, 2011 |
| | *Fusarium proliferatum* | USA | Fawzi, 2003 |
| | *Fusarium verticillioides* | Italy | Kosoglu *et al.*, 2011 |
| Lasiodiplodia canker | *Lasiodiplodia theobromae* | China, Turkey | Çeliker and Michailides 2012; Chen *et al.*, 2018 |
| Leaf blight | *Corticum salmonicolor* | Florida | Palmateer *et al.*, 1999 |
| | *Pellicularia kolerga* | Worldwide | Carpenter, 1983 |
| | *Rhizoctonia solani* | Florida | Palmateer *et al.*, 1999 |
| | *Rhizoctonia microsclerotia* | Florida | Matz, 1918 |
| Leaf spot | *Cercospora* spp. | Florida | Palmateer *et al.*, 1999 |
| | *Curvularia aeria* | Pakistan | Nayab and Akhtar, 2016 |
| | *Cylindrocladium scoparium* | Africa | Crous *et al.*, 1991 |
| | *Terriera fici* | China | Wu *et al.*, 2015 |
| Nectriella stem gall | *Nectriella pironii* | Florida | Palmateer *et al.*, 1999 |
| Diaporthe (Phomopsis) canker | *Diaporthe cinerescens* | Worldwide | Raheb *et al.*, 2014 |
| Phytophthora fruit rot | *Phytophthora citrophthora* | Taiwan | Kao and Leu, 1979 |
| | *Phytophthora palmivora* | Argentina, Brazil, China, Florida | El-Gholl and Alfieri, 1984; Gonzalez *et al.*, 1998; Zhang *et al.*, 2013; dos Santos *et al.*, 2017 |

*Continued*

**Table 12.1.** Continued.

| Disease name | Causal agent | Distribution | Reference |
|---|---|---|---|
| Phytophthora root rot | *Phytophthora cinammomi* | Turkey | Turkolmez *et al.*, 2018 |
| | *Phytophthora nicotianae* | Sub-tropical regions | Vannini *et al.*, 2009 |
| | *Phytophthora niederhauserii* | Taiwan | Lin *et al.*, 2018 |
| | *Phytophthora palmivora* | Australia, China, Japan, Taiwan, USA, Palestine | Miyake and Nagai, 2017; Miyake and Nagai, 2018 |
| | *Phytophthora parsiana* | Iran | Ali-Shtayeh *et al.*, 1994; Mostowfizadeh-Ghalamfarsa *et al.*, 2008 |
| | *Phytophthora parasitica* | Palestine | |
| Pink limb blight | *Corticium salmonicolor* | Worldwide | Prashad, 2013 |
| Rosellinia root and crown rot | *Rosellinia necatrix* | Greece | Papchatzis *et al.*, 2008 |
| Rust | *Cerotelium fici* | Worldwide | Matz, 1918; Kumar and Khurana, 2015 |
| Sclerotium blight | *Sclerotium rolfsii* | Worldwide | Carpenter, 1983 |
| Smut (aspergillus rot, black mold rot) | *Aspergillus niger* | Worldwide | Droby *et al.*, 2011 |
| Soft rot | *Rhizopus stolonifer* | California, Texas, Italy, Turkey, Louisiana | Michailides, 2003 |
| | *Mucor* spp. | | |
| | *Choanephora cucurbitarum* | | |
| Sooty mold | *Capnodium* spp. | Florida | Palmateer *et al.*, 1999 |
| Souring | *Candida* sp. | Worldwide | Droby *et al.*, 2011 |
| | *Saccharomyces* sp. | | |

can rapidly decay the soft fig wood (Lansky and Paavilainen, 2010). Bacterial pathogens cause leaf spot (Norman and Ali, 2013). Crown gall (*Agrobacterium tumefaciens* Smith & Townsend) is one the most important bacterial disease of fig trees worldwide (Turechek, 2012).

Integrated disease management is the preferred approach for economically controlling fig pests and diseases and minimizing environmental impacts. This approach includes a combination of methods including biological agents, soft chemicals, precision agriculture, optimum chemical application timing, and cultural practices (Mendoza-Castillo *et al.*, 2017). This chapter will give a detailed description of the primary bacterial, fungal and nematode diseases of fig trees and fruits and the individual control methods including the integrated disease management approach.

## 12.2 Gall Diseases

Gall diseases refer to tissue proliferations or local swellings in plants resulting from tissues' abnormal growth. Various types of insects and pathogens are responsible for gall production (Agrios, 2005). Although few studies have been reported, the existence of crown and stem galls in edible figs have drawn the researchers' attention. Two main pathogens are responsible for inducing galls in edible fig which are discussed below.

### Crown gall

Crown gall is among the important diseases of fig trees (Norman and Ali, 2013). The first report of crown gall was in 1919, on *F. aurea* Nutt. in Florida (Galloway, 1919) and has caused serious economic problems on *Ficus benjamiana* L. in Florida since 1991 (Bouzar and Jones, 2001; Bouzar *et al.*, 1995; Norman and Ali, 2013). The disease has also been reported in Iran (Aeini *et al.*, 2014). Although *F. benjamiana* is the main host of the disease (Turechek, 2012), *F. carica* is also susceptible to crown gall (Cleene, 1979).

***Symptoms:*** Swollen areas, usually associated with wounds, is the first symptom of a crown gall infection (Turechek, 2012; Norman and Ali, 2013). For this reason, this infection could initially be confused with callus formation. Weeks after the initial infection, the swelling forms a spherical green to tan gall (Norman and Ali, 2013). These newly formed galls gradually develop an irregular shape and become dark brown or black due to tree cell death. With severe infections, due to the pressure from the enlarged galls, the xylem will be damaged and upward water conduction will be blocked (Bouzar *et al.*, 1995). The galls can develop on both tree trunks and branches (Bouzar *et al.*, 1995; Bouzar and Jones, 2001; Norman and Ali, 2013).

***Disease development:*** *Agrobacterium tumefaciens* Smith & Townsend is the causal bacterium of crown gall (Aeini *et al.*, 2014). This global soil bacterium has a wide host range (Bouzar *et al.*, 1995). It can survive in soil without any hosts for years (Norman and Ali, 2013). Aerial tumors have been reported to be associated with pruning wounds, created during propagation and general plant maintenance. Pruning and branch wounds are the two major avenues of pathogen entry in nursery propagated trees (Bouzar *et al.*, 1995). Once in a tree the pathogen divides rapidly, producing galls (Norman and Ali, 2013). Elevated relative humidity and temperatures exacerbate disease severity (Cleene, 1979).

***Management:*** Sanitation is the best strategy for controlling crown gall (Norman and Ali, 2013). Orchards should be planted with disease-free stock plants (Bouzar and Jones, 2001). When removing the galls from the plants, the cuts should be several centimeters below the gall, and into healthy tissue (Norman and Ali, 2013). No effective bactericide for controlling crown gall has been identified. *Trichoderma* spp. and *Agrobacterium radiobacter* Smith & Townsend are effective biological control agents of *A. tumefaciens* (Cleene, 1979). The biocontrol product, NOGALL™, a bacterial culture of *A. radiobacter* var. K1026 in peat, has been demonstrated to significantly reduce the size of plant galls. However, infected plants cannot be cured using NOGALL™ as its effects are primarily preventative (Bio-Care Technology). Some other fungi, such as species of *Penicillium*, *Chaetomium* and *Aspergillus*, have negative effects on *A. tumefaciens*, but produce molds on *F. carica*; these fungi are not effective biological control agents (Cleene, 1979).

## Nectriella stem gall

This disease is found on various herbaceous and woody plants including edible figs in subtropical regions (Palmateer *et al.*, 1999). Six edible fig cultivars, Celeste, Conadria, Green Ischia, Kadota, Lemon and Osborn Prolific, were susceptible to this disease (Alfieri *et al.*, 1984).

***Symptoms:*** The pathogen produces different-sized galls on the trunk, branches and leaves, and occasionally cankers on stems in some cultivars (Palmateer *et al.*, 1999). Unusual growth leads to the formation of galls on the trunk, branches and leaves, and occasionally cankers on stems in some cultivars (Palmateer *et al.*, 1999).

***Disease development:*** The disease causal agent is *Nectriella pironii* Alfieri & Samuels (Michailides, 2003). The pathogen penetrates the host tissues via mechanical injuries, leaf axils and pruning wounds. Both the sexual (*Nectriella*) and asexual (*Kutilakesa*) stages have been reported for this pathogen, which appear on infected tissues. Asexual structures (i.e. sporodochia) cover the surface of the galls (Palmateer *et al.*, 1999).

***Management:*** Once infected, pruning the branches behind the infected sites into healthy tissue and eliminating the pruned tissues, decreases the inoculum source (Palmateer *et al.*, 1999). Applying a protective fungicide before pruning to prevent the entrance of all wound pathogens, especially *N. pironii* is recommended (Alfieri *et al.*, 1984).

## 12.3 Leaf Diseases

Fig leaves are susceptible to various fungal and bacterial pathogens that produce leaf spots and blights in edible figs. The fungal species *Alternaria alternata* Kiessler, *Cladosporium cladosporoides* (Fresen.) G.A. de Vries, *Curvularia aeria* Bat., J.A. Lima and C.T. Vasconc, *Cylindrocladium scoparium* Morgan, as well as *Cercospora* spp. have all been identified as causal agents, among which *Cercospora* spp. are the most devastating and important pathogens. The bacterium, *Pseudomonas syringae* pv. *syringae* Van Hall, has recently been reported from Serbia, where it was identified by molecular techniques (Ivanovic *et al.*, 2018).

### Anthracnose

Anthracnose is one of the most important fig fruit diseases (Matz, 1918; Carpenter, 1983; Norman and Ali, 2013). First described in Germany, it has been reported in China, the United Kingdom, New Zealand, Montenegro, and the United States (Johnstone and Jones, 1997; Michailides, 2003; Palmateer *et al.*, 1999; Yao *et al.*, 2016; Xue *et al.*, 2017). Anthracnose affects both foliage and fruit (Michailides, 2003). Multiple reports identify anthracnose among the most important diseases of fig fruits (Michailides, 2003; Yao *et al.*, 2016; Xue *et al.*, 2017).

***Symptoms:*** The primary infection site is a leaf lesion with a dark brown margin. Infected leaves will usually turn brown, dry at the edges and eventually abscise. Lesions can also form on leaf petioles and blades (Michailides, 2003). The fruits symptoms are localized, sunken lesions that eventually rot the entire fruit (Matz, 1918). In severe infections, pink and slimy masses of fungal conidia form on the lesions and result in the fruits remaining on the tree as dried mummies (Fig. 12.1) (Palmateer *et al.*, 1999; Michailides, 2003).

***Disease development:*** Three to four *Colletotrichum* species (i.e. *C. caricae* Stevens & Hall, *C. gloeosporioides* Penz. & Sacc., *C. fioriniae* Pennycook, *C. acutatum* Simmonds), and *Sphaceloma fici-caricae* Wani & Thirum have been reported as the causal agents of anthracnose (Matz, 1918; Palmateer *et al.*, 1999; Michailides, 2003;

**Fig. 12.1.** Fig anthracnose caused by *Colletotrichum gloeosporioides*. Localized, sunken lesions leading to a general rot of the entire fruit. (Figure used with permission I.Y. Choi, Choi *et al.*, 2013.)

Sreenivasaprasad and Talhinhas, 2005; Yao *et al.*, 2016; Xue *et al.*, 2017). However, *C. gloeosporioides* is the most widespread species causing anthracnose (Michailides, 2003; Norman and Ali, 2013). Conidia from the mummified fruits are the main source of inoculum for new infections (Michailides, 2003). Rain and irrigation water spread the disease (Norman and Ali, 2013).

***Management:*** Because infected fruit is the main source of inoculum, sanitation is the best method of controlling the disease (Palmeeter *et al.*, 1999; Michailides, 2003). Sanitation consists of collecting and removing all mummies hanging on the trees and dropped fruit (Matz, 1918; Michailides, 2003). Because of the wide host range of *Colletotrichum* spp., crop rotation is ineffective even in annual plants (Sreenivasaprasad and Talhinhas, 2005). Overhead irrigation and rain exacerbate the disease and should be minimized (Norman and Ali, 2013). Mancozeb, dithane, pyraclostrobin, and azoxystrobin fungicides can effectively control anthracnose on fig trees (Norman and Ali, 2013), but these are prohibited on edible figs in Florida (Palmateer *et al.*, 1999).

## Bacterial leaf spot

Bacterial leaf spot has been reported to cause economic damage to fig trees in Serbia (Ivanovic *et al.*, 2018). Although this is the only known report of bacterial leaf spot infesting figs, it could cause economic damages to fig trees.

***Symptoms:*** Disease symptoms appear as necrotic lesions three days after initial infection. The spots become larger, coalesce, and cover most of the leaf with a greyish lesion with a dry center (Ivanovic *et al.*, 2018).

***Disease development:*** Leaf spot disease is caused by *Pseudomonas syringae* pv. *syringae.* This bacterium is a heterogeneous species, able to infect a wide host range. Leaf symptoms develop after cool and wet (rainy) weather (Ivanovic *et al.*, 2018).

***Management:*** This bacterium requires cool and rainy conditions. Avoiding overhead irrigation and foliar fixed copper applications have been suggested as control methods (Ivanovic *et al.*, 2018). However, as with many diseases, planting in a suitable climate is the primary control method.

## Leaf blight

Leaf blight has been reported in most of world's fig orchards, especially in Florida and Texas in the USA (Matz, 1918; Carpenter, 1983). This disease is also known as aerial or web blight (Palmateer *et al.*, 1999). Both leaves and fruit are affected by the disease, which often results in severe crop loss.

***Symptoms:*** Small yellowish water-soaked lesions appear in early stage infections. The upper surface of these regions remains darker (Matz, 1918). As the disease develops, the infected area expands for leaf blight, turns silvery white at the surface and light brown under the surface, which is covered with fungal mycelium (Carpenter, 1983). The remaining green parts of the leaf will turn brown, and as the disease develops from petioles, may die. In severe infections, leaves may be shed or remain attached to the twigs (Matz, 1918; Carpenter, 1983). The mycelium is capable of killing the twigs and covering the fruit which leads to fruit mummification (Palmateer *et al.*, 1999).

***Disease development:*** *Rhizoctonia microsclerotia* Matz, *Rhizoctonia solani* Kühn and *Pellicularia koleroga* Cooke have been reported as the causal agents of leaf blight (Matz, 1918; Carpenter, 1983). Plant debris or soil are the primary infection sites (especially for *R. microsclerotia*) and are where the fungus resides. However, water, wind, insects, other animals (including humans), and machinery are the spores' vectors for secondary infections (Matz, 1918; Carpenter, 1983). The fungi require warm, wet weather in order to initiate infections (Carpenter, 1983). The disease can be easily recognized at the tips of the branches with new infected leaves. The sclerotia of *R. microsclerotia* seldom develop on fruits (Matz, 1918).

***Management:*** Sanitation is the only way to reduce the damage from this disease (Matz, 1918; Carpenter, 1983). Removing fallen and infected leaves removes the primary sources of inoculum and sclerotia. The orchards should be pruned to improve air circulation within the foliage and irrigation scheduled to not increase humidity within the orchard (Palmateer *et al.*, 1999).

## Leaf spot

Leaf spot is among the more important and widely distributed diseases of fig trees (Yao *et al.*, 2015). It was first reported in 1918 in Florida (Matz, 1918) and subsequently reported on multiple fig cultivars in Oman, Pakistan, Florida, Africa and China (Crous *et al.*, 1991; Al-Sadi *et al.*, 2015; Wu *et al.*, 2015; Nayeb and Akhtar, 2016). It is reported as being devastating to fig leaves but there no reports of economic damage in the literature (Al-Sadi *et al.*, 2015).

***Symptoms:*** In early stages, small yellowish to brownish water-soaked spots form. The upper surface of each spot is usually darker. As the lesion expands it becomes reddish-brown with dark brown margins. These lesions unify and form a single large lesion and irregular patches which finally may cause leaves to be shed earlier. In severe infections, the leaf center becomes fragile and drops off (Matz, 1918). Leaves will fall or remain on the tree when dead (Al-Sadi *et al.*, 2015). Other observers report multiple leaf spots that are angular and reddish-brown in color (Palmateer *et al.*, 1999). The center of the spot usually turns yellowish-brown and the margin is brown with a yellowish halo (Fig. 12.2) (Palmateer *et al.*, 1999).

***Disease development:*** A number of causal agents for leaf spot have been reported; *Cercospora fici* Heald & Wolf, *Alternaria alternata*, *Terriera fici* Y.R. Lin & Yuan Wu, *Curvularia aeria* (Bat., J.A. Lima & C.T. Vasconc.) Tsuda, *Cylindrocladium scoparium* Morgan, and *Cladosporidium cladosporoides* de Vries (Crous *et al.*, 1991; Palmateer *et al.*, 1999; Droby *et al.*, 2011; Al-Sadi *et al.*, 2015; Wu *et al.*, 2015; Nayeb and Akhtar, 2016). However, *C. fici* has been reported as the major pathogen affecting edible fig leaves (Palmateer *et al.*, 1999). Abscised diseased leaves are the primary infection source (Matz, 1918; Palmateer *et al.*, 1999). The disease spreads quickly in regions with midsummer rains (Palmateer *et al.*, 1999). *Cercospora fici* has been reported to produce hyaline to threadlike hyphal conidia from brown conidiophores, which are formed directly on the lesion (spot) surface (Palmateer *et al.*, 1999).

***Management:*** Proper orchard site selection, decreasing orchard humidity with irrigation scheduling, increasing air movement with pruning and removing inoculum in the form of fallen

**Fig. 12.2.** Cercospora leaf spot caused by *Cercospora fici*. A: Small yellowish to brown water-soaked spots on the upper leaf surface. B: Spots on the surface of a fruit of an infected fig tree. (Figure used with permission from J. Haman.)

leaves are the only available control measures (Matz, 1918; Palmateer *et al.*, 1999; Droby *et al.*, 2011). No effective fungicide has been approved to control leaf spot (Palmateer *et al.*, 1999).

### Rust

Fig rust is a disease of multiple varieties especially in the more humid regions of India and the Mediterranean (Matz, 1918; Carpenter, 1983; McKenzie, 1986; Palmateer *et al.*, 1999; Droby *et al.*, 2011; Norman and Ali, 2013; Kumar and Khurana, 2015).

***Symptoms:*** The first symptoms appear in late summer. If severe, the leaves turn brown and trees defoliate prematurely, within a few weeks, reducing growth and yield the following season (McKenzie, 1986; Palmateer *et al.*, 1999; Michailides, 2003; Kumar and Khurana, 2015). If defoliation is early, the tree may sprout new leaves and be vulnerable to early fall frosts. Or, defoliation in fall may lead to pre-dormancy and protect the tree from early frost (Kumar and Khurana, 2015). The symptoms are small, yellowish spots on the upper leaf surface (Palmateer *et al.*, 1999). When they enlarge, they will become reddish-brown, but remain smooth (Matz, 1918; Michailides, 2003). The lower leaf surface is covered blister-like with reddish-brown lesions (Matz, 1918; Michailides, 2003) (Fig. 12.3).

**Fig. 12.3.** Disease symptoms of fig rust including small, yellowish and necrotic spots on the upper surface of leaves of an infected fig tree caused by *Cerotelium fici*. (Figure used with permission from S. Nelson.)

***Disease development:*** *Cerotelium fici* Arthur is the causal agent of fig rust (McKenzie, 1986; Palmateer *et al.*, 1999; Michailides, 2003; Kumar and Khurana, 2015). No pycnium or aecium stages have been reported for the fungus (Michailides, 2003). Uredinia can be observed on three main areas: scattered on the fruit, grouped at the leaf edge and in dewdrops which cover the lamina (McKenzie, 1986; Michailides, 2003). Teliospores overwinter on the leaves on the tree or on the orchard floor (Palmateer *et al.*, 1999). Then rust spores are airborne, spreading as uredospores from infected trees. Moderately high temperatures (25–30°C) and high humidity increase disease severity (Matz, 1918; Palmateer *et al.*, 1999).

***Management:*** Because leaf litter is the main source of infection, collecting, removing and burning the leaves before the growing season effectively reduces the inoculum (Palmateer *et al.*, 1999). Pruning will enhance air circulation within the tree canopy and orchard (Palmateer *et al.*, 1999). Drip irrigation that does not raise humidity or wet the leaves is best (Palmateer *et al.*, 1999; Michailides, 2003; Kumar and Khurana, 2015). Dusting with sulphur, Vitavax, zineb, maneb and Bordeaux mixture on young leaves, before disease pressure is expected, effectively controls rust (Michailides, 2003).

## 12.4 Fruit Diseases

Fig fruits are susceptible to a wide range of fungi and yeasts; fruit rot, endosepsis, smut, souring, Sclerotium blight, anthracnose and rust all produce major economic losses (Matz, 1918; Carpenter, 1983; Fawzi, 2003; Michailides, 2003; Droby *et al.*, 2011; Latinovic *et al.*, 2014; dos Santos *et al.*, 2017). Anthracnose and rust infect both the fruit and the leaves of edible figs. Among the pathogens, the widely distributed endosepsis is the most devastating (Kosoglu *et al.*, 2011). Fruit rot is a common name for a wide range of fungal pathogens causing various symptoms on fig fruits. *Phytophthora palmivora* Butler, *Aspergillus flavus* Link, *Alternaria alternata*,

*Rhizopus stolonifer* Vuillemin, and *Mucor* spp. have all been reported as causing fruit rot decreasing yields (El-Gholl and Alfieri, 1984; Michailides, 2003; Stover *et al.*, 2007). Souring also results in major yield losses (Droby *et al.*, 2011).

### Alternaria rot

Alternaria rot is less damaging for California figs, than in the Middle East (Droby *et al.*, 2011).

***Symptoms:*** Alternaria appears as small, olive-green specks or sunken yellow-olive lesions covered with green spores on the fruit. When the figs touch, decay develops at the fruit-to-fruit contact surfaces and appears as a water-soaked area (Latinovic *et al.*, 2014) (Fig. 12.4).

***Disease development:*** Three fungi have been reported as the cause of Alternaria rot: *Alternaria* spp. (mainly *A. alternata*), is primarily a problem on ripe fruit. *Ulocladium altum* Sacc., and *Cladosporium herbarum* Link, which produce the same symptoms normally infect green fruit. Warm, wet weather exacerbates the spread of all three fungi (Latinovic *et al.*, 2014).

***Management:*** Alternaria rot can be minimized by picking fruit before it becomes overripe. Reducing orchard dust has also been reported to reduce the disease incidence (Stover *et al.*, 2007).

### Aspergillus mold

Aspergillus mold is one of the world's most destructive fruit molds (Stover *et al.*, 2007; Droby *et al.*, 2011). In California, it the second most common fruit disease (Michailides, 2003). In addition to its direct effects, Aspergillus mold also damages dried fruits by producing toxins with carcinogens harmful to humans and animals (Fig. 12.5) (Matz, 1918; Stover *et al.*, 2007).

**Fig. 12.4.** Alternaria rot of dried fig fruit is caused by *Alternaria alternata*. (Figure courtesy of Z. Bolboli.)

***Symptoms:*** In the early stages small, brown soft spots will appear on the green fruit. White hyphae will expand within the fruit, often visible at the eye end of the green figs (Michailides, 2003). No water-soaked region is visible at the lower part of the fruit (Michailides, 2003; Stover *et al.*, 2007). In some figs the fruit interior might become bright yellow and develop a powdery mass of spores (Carpenter, 1983).

***Disease development:*** When the fruit ripens, conidiophores and conidia will proliferate in the fruit cavity (Michailides, 2003). Although twenty-three species of *Aspergillus* exist, only *A. parasiticus* Speare, *A. tamarii* Kita, *A. ochraceous* Wilh., *A. flavus*, and *A. sclerotiorum* Huber cause this disease; *A. flavus* is considered the most harmful, attacking fig fruits worldwide (Stover *et al.*, 2007). Fig susceptibility increases as the fruits mature (Michailides, 2003). Although wounds and insect direct damage do not seem to play any significant role in disease development spread, the fungal spores are spread by insects (Michailides, 2003). Fig water activities during cultivation have been reported as a great source for mycotoxin formation (Heperkan *et al.*, 2012).

***Management:*** Dust reduction in orchards can reduce colonization by the *Aspergillus* species (Michailides, 2003). Insects have been proven to play an important role in spore dispersal. As a result, controlling insects would effectively reduce disease spread (Michailides, 2003). Sanitation measures such as collecting and deep burial of infected fruits could extensively reduce the risk of infection. Due to the significance of mycotoxin products (e.g. aflatoxins and ochratoxins) by the fungus, application of recently used methods, such as nano flow liquid chromatography high resolution mass spectrometry (Alcántara-Durán *et al.*, 2019) might be useful in detecting mycotoxin contaminations in postharvest fruits.

### Endosepsis

Endosepsis (internal or pink rot) is an important disease of fig fruits worldwide, particularly in California's Calimyrna figs. For almost

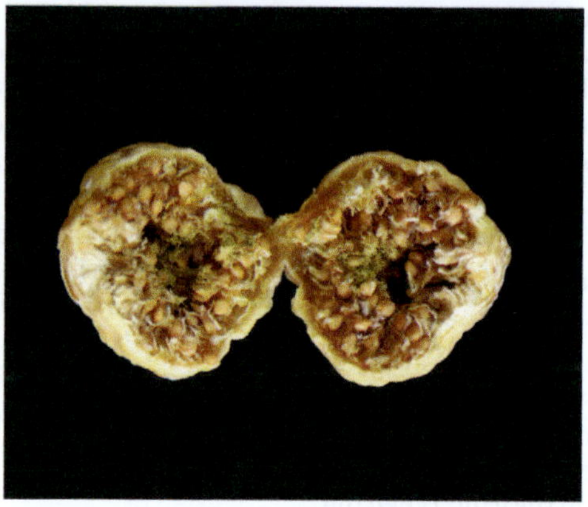

**Fig. 12.5.** Aspergillus mold of dried fig fruit caused by *Aspergillus flavus.* (Figure courtesy of Z. Bolboli.)

three-quarters of a century, until 1996, more than 50% of harvested figs were lost to this disease (Michailides *et al.*, 1996). Endosepsis, known as pink rot, brown rot, soft rot or eye-end rot, has been reported on all fig varieties (e.g. Calimyrna figs, wild figs, caprifigs, and other edible fig cultivars).

***Symptoms:*** The first symptoms of endosepsis are found in the fruit's interior cavity as it begins to ripen (Hansen 1927). The primary symptoms are a yellow, brown, or rusty discoloration on pulp (base of female flowers at earlier stages), water-soaked purple or pink skin spots, eye and stem-end rot appear on the fruits and flowers, and the production of a yellow or amber-colored gum-like substance exuding through the eye at the later stages of fruit development (Fig. 12.6) (Droby *et al.*, 2011). In addition to the direct damage, the fig fruit's surface is colonized by opportunistic bacterial and fungal propagules (e.g. *Alternaria* spp., *Aspergillus* spp. and *Cladosporium* spp.) (Michailides *et al.*, 1996).

***Disease development:*** This disease is caused by *Fusarium moniliforme* Sheld, *F. solani* Sacc., *F. proliferatum* Nirenberg, *F. subglutinans* Nelson, Toussoun & Marasas, and *F. verticillioides* Nirenberg (Droby *et al.*, 2011; Kosoglu *et al.*, 2011). *Fusarium moniliforme* is the most destructive, particularly in California (Michailides *et al.*, 1996). Although *F. moniliforme* microconidia overwinters in the caprifig winter crop, the propagules of the fungus have also been found on the surface of caprifigs and Calimyrna trees. However, no macroconidia have been detected on female figs (Michailides *et al.*, 1996). The conidia

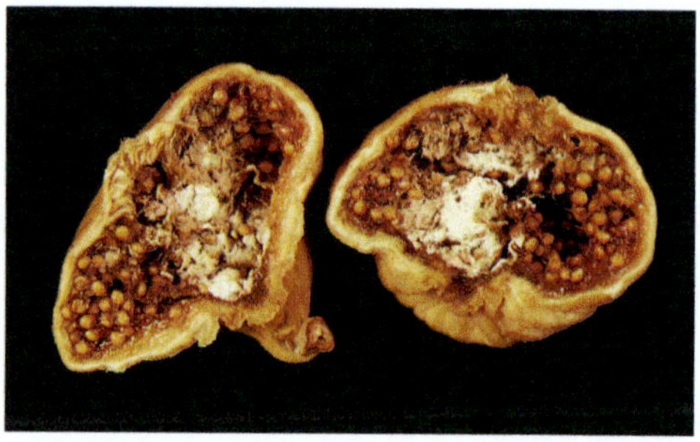

**Fig. 12.6.** Endosepsis disease or pink rot of fig fruit caused by *Fusarium* spp. (Figure courtesy of T.J. Michailides.)

are introduced into the winter crop during fall by the female fig wasps, *Blastophaga psenes* L., when she enters to lay eggs. Spring and summer crop infections occur the same way (Michailides and Morgan, 1998). As a result endosepsis transmission occurs annually in both cultivated and wild caprifigs. Maximum damage happens during the summer when the caprifig crop with its population of wasps, their wings contaminated with the fungal microconidia, are introduced into the Calimyrna orchards (Michailides *et al.*, 1996). Another significant problem with endosepsis is the production of fumonisin, a mycotoxin produced by certain species of Fusarium attacking figs (Moretti *et al.*, 2010).

***Management:*** Caprifig trees should be planted as far as possible from the edible fig cultivars that require pollination (Michailides *et al.*, 1996). Removing all winter and spring crop caprifigs from the caprifig orchard before the emergence of wasps, known as sanitation, should also be done (Michailides *et al.*, 1996). However, early identification of symptomless fruits to remove them, might be impossible for a grower. For this issue, one could use NIRS (near infrared spectroscopy) to distinguish between passable and defective figs with high accuracy (Burks *et al.*, 2000). Treating the winter caprifig crop (mammae) with a mixture of benomyl (now thiophanate-methyl, Topsin M), chlorothalonil, and dicloran prior to the wasps emergence from the winter crop (at the end of March), has been an effective strategy for controlling *F. moniliforme* and *Alternaria* spp. on caprifigs (Michailides *et al.*, 2008). *Paecilomyces lilacinus* Samson can colonize the dead wasps and consequently remove the *F. moniliforme* infestation with no negative effect on the figs. Spraying *P. lilacinus* before spring caprifig caprification significantly reduced fig endosepsis (Michailides *et al.*, 1996). The biological control agent *Bacillus amyloliquifaciens* Priest, combined with sanitation has also been reported to control up to 85% of endosepsis (Dogan and Benlioglu, 2011). Due to the economic importance of Calimyrna fruit and its susceptibility to endosepsis and other rots, breeding of a resistant cultivar with desirable features could be an alternative approach to reducing damage by endosepsis in these figs.

## Phytophthora fruit rot

Phytophthora fruit rot (white powdery rot) has been reported in Japan, Taiwan, Brazil, Europe, the United States, China, New South Wales, Australia, and Argentina (Gonzalez *et al.*, 1998; dos Santos *et al.*, 2017; Miyake and Nagai, 2017). However, there is little documentation of economic damage (Alfieri and El-Gholl, 1993).

***Symptoms:*** Fruit rot usually occurs on immature fruits located in the lower tree canopy. However, symptoms have also been observed from higher fruits (Michailides, 2003). Brown, circular, water-soaked lesions appear on the fruits at the first stages of disease development (Michilaides, 2003). Fluffy white mycelia will cover the lesions at the later stages (Michilaides, 2003; Miyake and Nagai, 2017).

***Disease development:*** *Phytophthora citrophthora* Leonian and *Alternaria alternata* produce fig fruit rot (Kao and Leu, 1979; Latinovic *et al.*, 2014) but the oomycete *P. palmivora* is considered to be the most serious and devastating (Michilaides, 2003; Zhang *et al.*, 2013) severely damaging most fig cultivars (Alfeiri and El-Gholl, 1993). The fungus overwinters in soil and on plant debris. With moisture sporangia release the zoospores that cause the primary infection on immature fruits (Miyake and Nagai, 2017). In severe infections, mature figs higher in the tree will also be damaged (Miyake and Nagai, 2018). Recent studies indicated that *P. palmivora*'s ability to infect figs is unique since it can infect fruits over a wide range of temperatures, in a short period of time, at low inoculum (zoospores) concentrations and at immature and mature stages of fruit development (Miyake and Nagai, 2018).

***Management:*** The infection occurs at lower temperatures (15°C) while symptoms appear at higher temperatures. To prevent fruit rot in transit, storing and transporting infected fruit at lower temperatures (below 15°C) is a crucial step in slowing disease progression. Due to the water requirement of sporangia to release zoospores, controlling the transit truck moisture is also important (Miyake and Nagai, 2017; 2018). Sanitation of infected fruits is among the most effective methods of controlling the disease, since contaminated fruit will completely rot within a week (Miyake and Nagai, 2018). It is proved that the fungicides (such as azoxystrobin, chlorothalonil, cyazofamid, copper hydroxide and copper sulfate) which have curative effect on fig fruits against *P. palmivora*, should be frequently sprayed to gain the protection (Miyake and Nagai, 2017). During wet seasons, fungicides should be reapplied frequently (Miyake and Nagai, 2018). Recently, it was reported that phosphonate was effective in controlling fruit rot (Miyake and Nagai, 2017).

## Sclerotium blight

Sclerotium, or southern blight, is a devastating fig fruit disease. Knowledge of its potential damage on other parts of the plants is limited but it has been reported as one of the important diseases in California (Matz, 1918; Carpenter, 1983).

***Symptoms:*** A yellow to white mycelium with round, yellowish to brown, hard sclerotia forms at the base of the fruits (Fig. 12.7) (Matz, 1918; Carpenter, 1983).

***Disease development:*** *Sclerotium rolfsii* Sac. is the causal agent of Sclerotium blight in fig fruit. This fungus is capable of attacking a wide range of hosts in almost all conditions (Matz, 1918; Carpenter, 1983). *Sclerotium rolfsii* is able to survive in soil, roots, and stems are usually the main attack points (Matz, 1918). Old leaves, grass and plant debris facilitate proliferation of this fungus (Carpenter, 1983).

***Management:*** Sanitation by removing old leaves and infected fruit is the best way to prevent proliferation (Carpenter, 1983). As this fungus can exist in the soil, spraying a Bordeaux mixture (TOP BORDO™) on the trunk and surrounding soil of diseased trees will prevent further proliferation (Matz, 1918).

## Smut

One of the most studied diseases of fig is smut (Carpenter, 1983). The disease is also known as 'black mold rot' or 'Aspergillus rot' (Michailides, 2003). Since the causal agent is an ascomycete and not a member of *Ustilaginomycotina* (*Basidiomycota*), it is not a true smut. The name refers to the black powder appearance of the conidia on fruits. This disease is especially important in California (Michailides, 2003; Droby *et al.*, 2011).

**Fig. 12.7.** Sclerotium blight caused by *Sclerotium rolfsii*. Infected fig fruit covered by mycelium. (Figure used with permission from P. Seethapathy.)

***Symptoms:*** Internal discoloration of the fruit starts during early fruit development (Michailides, 2003). Decay initiates at the ostiole producing a surrounding water-soaked region. The infected regions are increasingly covered with abundant powdery black masses of spores as they ripen (Fig. 12.8) (Carpenter, 1983; Michailides, 2003).

***Disease development:*** Although several species of *Aspergillus*, namely, *A. niger* var. *niger* Teigh, *A. niger* var. *awamori* Al-Musallam, *A. japonicus* var. *aculeatus* Al-Musallam, *A. japonicus* var. *japonicus* Saito, and *A. carbonarius* Thom are the causal agents of black mold rot, the varieties *A. niger* var. *niger* and *A. niger* var. *awamori* are reported to cause the smut on 93% of Calimyrna figs and more than 99% of Conadria figs in California (Carpenter, 1983; Michailides, 2003). The fungus is transmitted by flies (*Drosophila melanogaster* Meigen), thrips, ants (*Formicomus ionicus* Laf.), the last only in Turkey on ripening fruits, and beetles (*Carpophilus* spp. Stephens) (Carpenter, 1983; Michailides, 2003). Fig varieties with larger ostioles are more susceptible than those with smaller ones (Michailides, 2003). Slightly acidic soil as well as temperatures between 20 and 25°C are favorable for smut development (Michailides, 2003).

***Management:*** Due to the importance of crop debris in disease development, removing all old fruits and crop debris from an orchard is strongly recommended (Michailides, 2003). Reducing

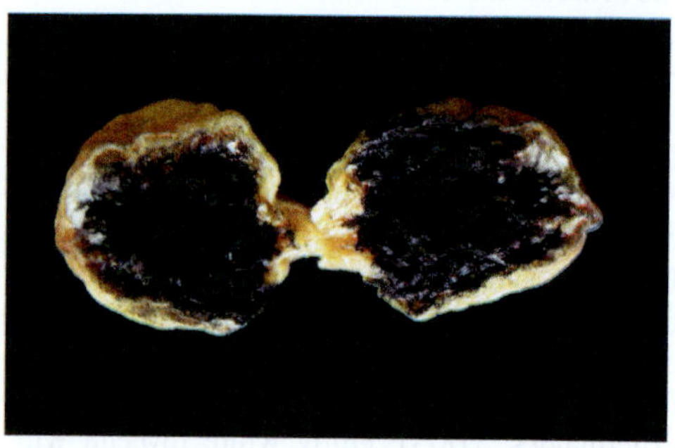

**Fig. 12.8.** Fig smut in dried fig fruit caused by *Aspergillus niger* with abundant black masses of spores on the infected fruit. (Figure courtesy of Z. Bolboli.)

the dust will also prevent smut damage as the beetles would preferably transfer *A. niger* with dust-covered fruits (Michailides, 2003). Black Mission and Kadota cultivars are less susceptible to smut than other cultivars (Michailides, 2003). Controlling insects also reduces the spreading of smut (Michailides, 2003).

## Soft rot

This disease has been reported in Louisiana, California, Texas, Italy and Turkey (Taubenhaus, 1936; Manziano *et al.*, 1990; Michailides, 2003).

***Symptoms:*** This disease occurs on mature fruit causing then to become soft and pulpy. The fruit will be covered with mycelia, and in severe infections, may fall from the tree (Fig. 12.9) (Michailides, 2003).

***Disease development:*** Three fungi cause soft rot, *Rhizopus stolonifer* (Ehren.) Vuill., *Mucor* spp., and *Choanephora cucurbitarum* Thaxt; *R. stolonifer* is the most frequent (Taubenhaus, 1936; Manziano *et al.*, 1990; Michailides, 2003). *Rhizopus stolonifer* is an airborne pathogen, able to produce a large quantity of sporangiospores. In humid and warm conditions, the sporangiospores infest the cracks developing around the ostiole of the ripening fruit. Spores can also be spread via water, especially rainwater, wind, insects and the dripping juice of infected fruit (Michailides, 2003).

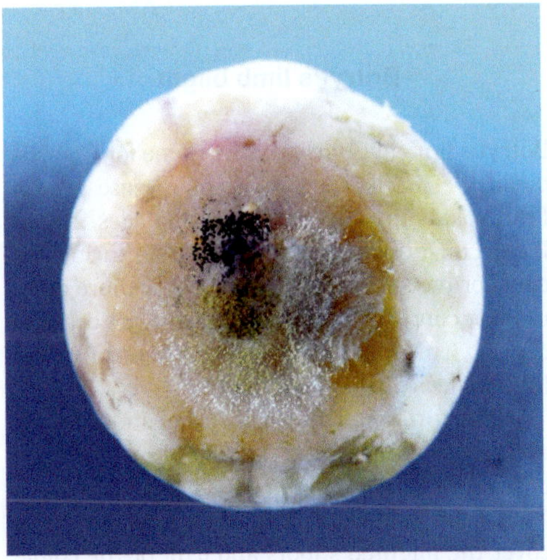

**Fig. 12.9.** Rhizopus soft rot of fig caused by *Rhizopus stolonifer*. The infected soft, pulpy fruits are covered with mycelium. (Figure used with permission from P. Seethapathy.)

***Management:*** Because most of the of the large and soft-skinned cultivars are susceptible to soft rot, cultivating smaller figs, such as the Celeste and Reine Blanche cultivars, would be the best control (Taubenhaus, 1936; Michailides, 2003). Harvesting the figs regularly before the fruits crack around the ostiole and preventing ants entering the fig orchards help prevent soft rot (Michailides, 2003).

## Sooty mold

This disease is caused by saprobic ascomycetes belonging to different families of *Dothideales*, among which *Capnodium* spp. have been reported several times (Palmateer *et al.*, 1999). Sooty mold is also called blotch or black mold. It affects leaves, stems and fruit (Palmateer *et al.*, 1999). No host preference has been reported for sooty molds (Callen and Carris, 2004). The disease has been reported from all parts of the Western Hemisphere but is particularly prevalent in the southeastern state of Florida in the USA (Callen and Carris, 2004).

***Symptoms:*** The disease is characterized by a splotchy black stain or thin coating on the leaves, stems, fruit, and even tree limbs. The dark color of sooty mold is caused by the dark pigments in the mycelial threads that form the sooty mold colonies (Callen and Carris, 2004). The honeydew secretions of insects on fig leaves and fruits are colonized by this pathogen (Palmateer *et al.*, 1999). Many sooty mold fungi are sticky and able to adhere to surfaces to absorb their required moisture. In severe infections, the black crusty growth can cause yellowing of foliage.

***Disease development:*** Sooty mold is an ectopathogenic fungus that colonizes the honeydew secretions of scales, aphids, psyllids and other insects. Then fungal mycelia will darken (melanize). These ectoparasite fungi are endemic but not able to infect plants in normal conditions. The fungal mold reduces sunlight on the leaves reducing yield (Palmateer *et al.*, 1999).

***Management:*** Since this disease is vectored primarily by insects, managing insect populations can help control the pathogen. Pruning out infected plant parts, preventing trunk access by the ants that protect the honeydew secreting insects, and high-pressure irrigation, are non-chemical method of control. Spraying with soapy water is used in home gardens. Insecticides are also used to control ants and other honeydew secretors (Palmateer *et al.*, 1999; Callen and Carris, 2004).

### Souring

One of the most common and problematic diseases of fig fruit is souring. Souring, also known as fermentative spoilage, infects the fruit while it's still on the tree (Michailides, 2003; Droby *et al.*, 2011). This disease has always been a major problem in Californian fig orchards. The disease has also been reported in Italy, Turkey and Iran, and is probably endemic in the fig growing world (Caldis, 1930; Mrak *et al.*, 1942; Naeem and Akhyani, 1988).

***Symptoms:*** Symptoms are not the same in persistent and caprified cultivars (Caldis, 1930; Michailides, 2003). In the former, the disease symptoms are very distinctive while in the latter, the symptoms are often confused with endosepsis (Mrak *et al.*, 1942; Michailides, 2003). Fruit decay initiates only after the ostiole opens. Pink color will start appearing on a diseased fruit surface, that then later becomes water-soaked. A pink syrupy liquid exudes from the ostiole, congealing there or dropping on the leaves (Caldis, 1930; Michailides, 2003). As the cavity pulp ferments gas bubbles form and a characteristic alcohol odor develops. The pulp decomposes and a white surface foam develops, the fruit turns black, shrivels, and often drops from the tree (Carpenter, 1983; Michailides, 2003). The peduncle is attached to the fig tree, from which the eye canker (or the dead spot) usually develops (Michailides, 2003).

***Disease development:*** Two types of pathogens cause souring, yeasts and bacteria (Carpenter, 1983). The fungal genera *Saccharomyces* Meyen, *Pichia* Hansen, *Candida* Berkhout, *Torulopsis* Engel, and the species of *Hanseniaspora* Zikes, *Kloeckera* Janke, and *Torulopsis* cause souring (Caldis, 1930; Mrak *et al.*, 1942; Michailides, 2003). All the same flora has been found on *Carpophilus hemipterus* L., the dried-fruit beetle that feeds on infected figs. However, it has not been determined if all these yeasts are responsible for the disease, or a specific one is the major pathogen (Michailides, 2003). Beetles and fruit flies are the main vectors of these pathogens; entering the fig's ostiole during ripening.

***Management:*** Because insects are the primary vectors, eliminating insects will prevent souring in an orchard (Carpenter, 1983; Michailides, 2003). Bagging or closing the ostiole with 'tangle foot' prevents souring (Michailides, 2003). The fungicide maneb slightly reduces souring but no effective chemical control has been identified (Carpenter, 1983). Celeste, Texas, Everbearing and Alma cultivars which have closed ostioles, could be planted instead of the more susceptible open ostiole Magnolia and Kadota cultivars (Carpenter, 1983).

## 12.5 Limb Diseases

Limb blight refers to the progressive tip die back of twigs and branches. This disease is caused by the basidiomycete *Corticium salmonicolor* Berk. & Br., and *Botrytis cinerea* Pers. in fig trees. In severely infected trees, twigs and branches wilt and the pathogens extend into the trunk, killing the tree. The wilt symptoms on branches could also be the result of an infection with *Neoscytalidium dimidiatum* (Penz.) Crous & Slippers.

### Botrytis limb blight

Botrytis limb blight is another name for Botrytis dieback. The first report of the disease was in France, later in California (see Michailides, 2003). It can cause blight on multiple shoots during wet cool springs. Botrytis shoot blight is more common on caprifigs than other fig cultivars because caprifig fruit is often damaged by frost in spring (Michailides, 2003).

***Symptoms:*** The primary infection occurs at the base of young cracked shoots, which later bend, and produce light green leaves (English, 1962). Large masses of grey conidia or spores cover the blighted shoots and fruit of the edible fig and caprifig. Frost damaged flowers, fruit or shoot tips are infected first. The pathogen spreads from the fruit or flowers to the shoot, causing dieback. Cankers often form below the infected fruit, and grey spores develop on the infected fruit, shoots, and cankers (Fig. 12.10). This pathogen can also infect young, growing shoots, by entering through wounds. Infected shoots then wilt, turn brown and die. The leaves or shoots may turn black under extremely wet conditions (Michailides, 2003).

***Disease development:*** The causal agent of gray mold is *Botrytis cinerea*. Wet and cool springs favor disease development. *Botrytis cinerea* can infect through injuries on fruit, leaf and fruit scars, producing cankers below and above the entrance point (Michailides, 2003). Shoot blight precedes and results in fruit blight. Conidia from blighted shoots and fruit are dispersed by airflow, which initiates new infections during the cool and rainy spring season on susceptible tissues (leaf or bud scars or weakened shoots and fig fruit). Winter injury to fruit or branches followed by *B. cinerea* infection also causes shoot dieback. This disease is more common on caprifigs than on Calimyrna figs (English, 1962) as the pathogen can sporulate on caprifigs as late as July generating an inoculum reservoir (Ricci, 1972).

***Management:*** Sanitation reduces Botrytis and other disease problems. Removing dead leaves, flowers, and branches prevents Botrytis from establishing a disease reservoir than can infect healthy trees (Hausbeck and Harlan, 2018). Preventative fungicides can be used during cool and wet periods, or when symptoms manifest. The newer biopesticide, Affirm™ WDG, is also effective (Hausbeck and Harlan, 2018).

## Branch wilt

This disease has been reported from a wide range of trees including edible fig (*F. carica*), *Ficus* spp., English (Persian) walnut (*Juglans regia* L.), lemon (*Citrus limonia* Osbeck), grapefruit (*C. grandis* Merr), orange (*C. sinensis* (L.) Osbeck), European chestnut (*Castanea sativa* Mill), almond (*Prunus amygdalus* Batsch), and poplar (*Populus* sp.) (Reckhaus, 1987; Elshafie and Ba-Omar, 2001). The pathogen causes dieback, root rot, canker, and tree decline (Ray *et al.*, 2010).

***Symptoms:*** This pathogen produces leaf, shoot, and fruit blights, stem and branch cankers, dieback and gummosis on various woody hosts (Nouri *et al.*, 2018). The symptoms first appear in mid to late summer. Leaves on infected branches wilt, brown and dry, but remain attached. A black powdery material, the black unicellular arthrospore conidia chains, cover the branch cortex causing the branch to wilt and die (Fig. 12.11) (Ogawa and English, 1991). The fungus may eventually spread to the trunk.

***Disease development:*** *Neoscytalidium dimidiatum* (Penz.) Crous & Slippers is the causal agent of branch wilt disease (Ogawa and English, 1991). This fungus has several synonyms including: *Fusicoccum dimidiatum* (Penz.) D. F. Farr,

**Fig. 12.10.** Botrytis limb blight of edible fig caused by *Botrytis cinerea*. Large masses of grey mold cover the fruit (right) and cankers are present below the infected fruit (left). (Figure used with permission from I.Y. Choi, Cheong *et al.*, 2013.)

**Fig. 12.11.** Branch wilt of fig caused by *Neoscytalidium dimidiatum*. A: Stem and branch cankers and dieback. B: A mass of black powdery material that contains chains of conidia, under the cortex. C: Transverse section through a branch of an infected fig tree. (Figure courtesy of T.J. Michailides.)

*Hendersonula toruloidea* Nattrass, *N. hyalinum* (C. K. & J. L. Mulder) A. J. L. Phillips, M. Groenew. & Crous, *Scytalidium dimidiatum* (Penz.) B. Sutton & Dyko, and *Torula dimidiata* Penz (Nouri *et al.*, 2018). The optimum temperatures for growth are 25–35°C. Due to the pathogen strong response to temperatures, cankers can develop in summer within 10–15 days. In cooler seasons, symptom development takes several months. The pathogen penetrates through wounds produced by sunburn, frosts, pruning or mechanical injuries (Dreistadt, 2012). The fungus overwinters as arthrospores and mycelium in infected branches (Ogawa and English, 1991). The spores tolerate high temperature and low humidity.

***Management:*** Branch wilt is a weak and opportunist pathogen. As a result, sufficient nutrition and irrigation can reduce disease incidence and severity. Orchard sanitation removing dead and dying branches is among the effective management strategies (Ramos, 1997). No fungicide has been identified for branch wilt disease control. However, one or two Bordeaux mixture treatments in winter may reduce the infection rate (Ogawa and English, 1991).

## Pink limb blight

Pink canker or pink disease, once known as limb blight (Tims, 1963) infects 141 wild and cultivated plant species in tropical and subtropical regions (Prashad, 2013) including fig (*F. carica*), apple (*Malus* × *domestica*), *Citrus* spp., and mango (*Mangifera indica* L.) (Timmer, 2000; Timmer *et al.*, 2003; Wales and Territory, 2016).

***Symptoms:*** The fungus attacks the limbs, twigs and trunk of fig trees, producing a mass of pink or salmon-colored mycelia on branches of infected hosts (Palmateer, 1999). The fungus may eventually envelop a branch, killing the cambial layer. Leaves and limbs on the infected branch will wilt and die (Tims, 1963; Talbot, 1965).

***Disease development:*** The causal fungus, *Corticium salmonicolor*, is more common as a superficial covering on the bark (Prashad, 2013). High relative humidity, temperature, sunshine hours, and cumulative rainfall exacerbate pink limb blight (Prashad *et al.*, 2014).

***Management:*** Pruning behind infected branches or twigs, then removing and destroying the infected tissues at the primary stages of the disease, is the most effective management strategy for reducing orchard inoculum (Palmateer, 1999). A biological control fungus, *Trichoderma koningi* Oudem, can inhibit mycelial growth of this pathogen in apples trees up to 82.62% (Prashad, 2013). Combinations of fungicides and plant oils, particularly the combination of Contaf and *Brassica juncea* var. *cunefolia* oil (92.94%), have been highly effective in reducing pink canker under field conditions (Prashad, 2013). In other studies, researchers showed that copper nanoparticles (CuNPs) are an effective fungicide for controlling *C. salmonicolor* in rubber trees, especially when CuNPs are sprayed on infected branches (Cao *et al.*, 2014).

## 12.6 Canker Diseases

Canker is a widespread, and destructive disease that affects multiple trees including fig. The disease causes cracked, sunken, and discolored bark on the stems and trunk. Fig canker disease can be caused by multiple fungi including *Botryosphaeria* spp., *Ceratocystis ficicola* Kajitani & Masuya, *Lasidodiplodia theobrome* (Pat.) Griffon & Mauble, *Neoscytalidium dimidiatum* (Penz.) Crous & Slippers, *Diaporthe cinerascens* Sacc., and the bacterial plant pathogen, *Pseudomonas fici* (Cavara) Magrou.

### Bacterial canker

Bacterial canker has been identified on figs in California and Italy (Michailides, 2003). It is the only bacterial disease of fig trees in California (Michailides, 2003).

***Symptoms:*** Bacterial canker has been identified on both older and younger trees as elongated dark lesions with wilting leaves and on new shoots (Michailides, 2003). In California, most symptoms initiate from limbs, at wet spots that later dry. These dry spots crack, initially 1.25–2.5 cm in length, later increasing to up to 15 cm long, exposing the cambium and producing cankers than will kill limbs in severe infections.

***Disease development:*** The causal agent is *Pseudomonas fici*, a Gram negative, motile bacterium (Schurrie, 1990; Michailides, 2003). The bacteria enter via cracks, wounds, and sunburned areas (Schurrie, 1990). Growth is optimal at 15°C (Michailides, 2003).

***Management:*** Pruning out the infected shoots is the best way to control this disease. Bordeaux mixture applications to recently opened buds, at the end of April, and again in August, respectively, may reduce the disease damage (Michailides, 2003). The Kadota, Calimyrna and Mission cultivars are more resistant to bacterial canker than Adriatic and caprifigs (Michailides, 2003).

### Ceratocystis canker

The disease was first reported in Japan on the Masui-Dofin cultivar (Kajii *et al.*, 2013) where it caused a serious wilt disease in fig orchards. It is among the most severe diseases of the common fig in this region (Morita *et al.*, 2016). *C. fimbriata* (Ellis & Halst) was also indentified in Brazilian fig orchards in (Johnson *et al.*, 2005).

***Symptoms:*** The pathogen invades the xylem vessels, blocking them, and leading to discoloration of the tissues at the site of infection. Due to the disruption of the water transport to the canopy the leaves first yellow, the branches and leaves wilt, and eventually the tree may die (Fig. 12.12) (Yakushiji *et al.*, 2019).

***Disease development:*** The fungus *Ceratocystis ficicola* Kajitani & Masuya, causes Ceratocystis canker in fig trees (Kajitani and Masuya, 2011). This agent is a soilborne pathogen, overwintering via chlamydospores. The ambrosia beetle, *Euwallacea interjectus* Blandford, is the primary vector transporting the pathogen from soil to the roots. Despite the name Ceratocystis canker, there is no canker formation, however, severe wilt is the main symptom of the disease (Kajii *et al.*, 2013).

***Management:*** This pathogen produces chlamydospores in the soil. Preplant soil sterilization with fungicides is impractical and expensive (Yakushiji *et al.*, 2019). Using resistant varieties such as Negron and BC1 as rootstocks is a potential but little used option (Hosomi *et al.*, 2012; Yakushiji *et al.*, 2012; 2019).

### Lasiodiplodia canker

This widespread pathogen of tropical woody trees causes shoot blight and dieback of many trees and shrubs including fig (*F. carica*), grapevine (*Vitis vinifera* L.) in Italy, mango (*Mangifera indica* L.) in Sindh Pakistan, kumquat (*Citrus japonica* Thunb.) in Taiwan and cacao (*Theobroma cacao* L.) in Cameroon (Çeliker and Michailides, 2012). The disease was first reported on edible fig in Turkey (Çeliker and Michailides, 2012), later in China (Chen *et al.*, 2018). It is not common in fig elsewhere in the world.

***Symptoms:*** Primary symptoms include a discoloration, yellowing, and rolling of leaves, leading to twig and branch dieback (Abo Rehab

**Fig. 12.12.** Ceratocystis canker caused by *Ceratocystis ficicola*. A: Severe wilting, yellowing and drying of leaves and branches in infected fig trees. (Photograph courtesy of T. Morita, Yakushiji *et al.*, 2019.) B: Transverse section through trunk of infected fig tree. (Figure used with permission from H. Yakushiji, Yakushiji *et al.*, 2019.)

*et al.*, 2014). If the disease progresses into the trunk, cankers form in the trunk bark associated with nodes. These bark cankers expand, encircling the node killing the tree above the infected site (Fig. 12.13) (Çeliker and Michailides, 2012).

***Disease development:*** *Lasiodiplodia theobromae* (Pat.) Griffon & Maubl. (anamorph: *Botryosphaeria rhodina* [Berk. & Curt.] Arx.) is a species of *Botryosphaeriaceae* in tropical and subtropical regions. Abiotic stresses, heat, and drought increase disease severity (Paolinelli-Alfonso *et al.*, 2016). Pruning, mechanical, and sunburn wounds, facilitate pathogen penetration into healthy tree tissues (Abo Rehab *et al.*, 2014).

***Management:*** Sanitation and removal of infected branches reduce the disease inoculum reservoir. The fungicides Antracol Combi (Antracol®) and Topsin M 70 (Topsin® M 70WP) are effective (Abo Rehab *et al.*, 2014). *Streptomyces samsunensis* Sazek *et al.* has been reported to be an effective biocontrol agent of *L. theobromae* in mango seedlings (Kamil *et al.*, 2018). However, no effective biocontrol has been identified for fig.

## Diaporthe canker

Diaporthe (formerly Phomopsis) canker (also called 'fig canker') was first reported in Italy in 1876 (Ogawa and English, 1991) and later in southern France, England, North and South Africa, Denmark, the southern coast of Crimea, Brazil, and in Kadota orchards in the USA (Hansen, 1949c). It was recently reported in Iran (Banihashemi and Javadi, 2009), and causes economic damage. It affects all the commercial Californian fig cultivars. Heavy pruning led to a major epidemic of this pathogen in California's Kadota cultivar (Ferguson *et al.*, 1990).

***Symptoms:*** A localized dead area (canker) in the bark and wood develops at pruning wounds (English, 1951). The cankers are extremely difficult to distinguish from the uninfected bark during the first year of infection. With time the canker appears bleached, cracked, and somewhat sunken. The minute black fungal fruiting bodies can be seen in the outer bark layers. Distinct zonation form as the cankers elongate. The outermost part of the zonation is the main sporulating region (English, 1960). The cankers become elliptical as they extend more rapidly longitudinally than transversely. Branches are eventually girdled, and the withered brown foliage remains on the dead branches through the summer. New shoots and foliage that develop on the partially girdled branches are markedly dwarfed (Fig. 12.14) (Grove, 1935).

***Disease development:*** *Diaporthe cinerascens* produces α- and β-spores in the pycnidia

**Fig. 12.13.** Lasidodiplodia canker caused by *Lasiodiplodia theobromae*. Discoloration, yellowing and rolling of leaves, leading to twig and branch dieback. (Figure courtesy of T.J. Michailides.)

**Fig. 12.14.** Diaporthe (formerly Phomopsis) canker caused by *Diaporthe cinerascens*. A: Annual rings caused by *D. cinerascens* progression in cambium and vascular tissues of an infected fig tree. B: Transverse section through a branch of an infected fig tree. (Figure courtesy of Z. Bolboli.)

(Grove, 1935). This pathogen is an opportunistic wound invader, unable to attack healthy tissues. The pathogen overwinters as mycelia within cankers on the tree or in prunings (Hansen, 1949a). Moisture as relative humidity (RH) both in nature and in laboratory conditions is the primary factor in pycnidial formation and dispersal (Banihashemi and Javadi, 2009). Although splashing water (rains or irrigation) is the main means for spreading the pycnidiospore, pruning tools also spread the spores from one tree to another. Insects (bark beetles) and birds also spread the spores. Pruning cuts, mechanical injuries, sunscald, and branch crotches are courts of infection of the pathogen (Hansen, 1949a; Livshits and Pupysheva, 1949; Banihashemi and Javadi, 2009). Inoculation tests have shown that Kadota trees are practically immune to infection from the first of April through the growing season, but they are highly susceptible from November through February (English, 1951; English, 1952a). In this period, trees enter a dormant stage and sap slows down so the plant defense against the pathogen is weakened (Banihashemi and Javadi, 2009). External symptoms on pruned stubs (leaf discoloration and chlorosis, shoot blight, dieback, trunk necrosis, depressed and ellipsoid wounds) appear a year after infection, although internal symptoms (brown discoloration of cambium and vascular tissues with central necrosis in transverse sections) can be seen much earlier. Between April and October, the canker extension is reduced greatly because of active growth of the host (Grove, 1935). Fig latex has phenolic compounds that are important in fig defense against pathogens, and it seems that the defense mechanism is significantly related to the latex flow during growth season (Aref *et al.*, 2010).

***Management:*** Due to the importance of pruning wounds as infection courts for the pathogen, it is suggested that pruning be delayed, late March to early April, until the end of the dormant season. In this period, the fig trees are in the dormant stage and the possibility of disease development is low (Banihashemi and Javadi, 2009). Although late pruning may delay fruit maturity, it can significantly reduce disease incidence and severity (English, 1952b; English, 1958; Banihashemi and Javadi, 2009). Removing infected branches and chiseling away smaller cankers late in the dormant season is beneficial. Pruning tools spread the disease. Sterilizing tools with 5.5% sodium hypochlorite diluted in distilled water (1:9) significantly prevents the disease dispersal (Pilgrim, 1959; Banihashemi and Javadi, 2009). Pruned branches and twigs should be removed and burned (Hansen, 1949a). Trees with many large cankers on the scaffold branches should be cut off just above the ground, so a new canopy can be developed from the stump (Hansen, 1949b; English, 1952a). Organic mercury sprays within 24–48 hours after pruning of Kadota trees effectively controlled the pathogen (English, 1953; Gilmer, 1957), but this fungicide is no longer allowed.

## 12.7 Root and Crown Diseases

Several species of soilborne pathogens cause crown and root rot diseases of woody plants. The fungi *Rosellinia necatrix* (Hart.) Berl., and *Armillaria mellea* (Vahl) P. Kumm (Papachatzis *et al.*, 2008) as well as some oomycetes such as *Phytophthora* and *Pythium* species induce root rot of fig trees (Miyake and Nagai, 2017; Turkolmez *et al.*, 2018).

### Armillaria root rot

Armillaria root rot is one of the most important root diseases of edible fig in temperate and tropical regions (Williams *et al.*, 1986). It is caused by a basidiomycete with a wide host range. The presence of a saprophytic phase in the life cycle of the pathogen has made it a serious problem in forests and orchards. Infections are common in fig trees near forests (Papachatzis *et al.*, 2008).

***Symptoms:*** Infected trees have discolored and severely wilted leaves. White mycelial masses form under the bark of infected trees (Thomas *et al.*, 1948). These symptoms may be confused with those caused by Rosellinia root rot. The dark cord-like rhizomorphs produced by *Armillaria* species is the key character in identifying an Armillaria infection. Diseased young trees wilt and die within 1–2 years. After a tree dies, the fungal fruiting bodies, honey-brown mushrooms, grow around the trunk base (Fig. 12.15) (Michailides, 2003).

**Fig. 12.15.** A: White mycelia of *Armillaria mellea* under the bark of an infected fig tree. B: *Armillaria mellea* fruiting bodies at the base of an infected fig tree (Figure used with permission from A. Papachatzis, Papachatzis *et al.*, 2008.)

***Disease development:*** *Armillaria mellea* produces Armillaria root rot (Papachatzis *et al.*, 2008). Disease dispersal occurs when the rhizomorphs contact uninfected roots, or when uninfected roots contact infected roots. Rhizomorphs can grow up to 3 m horizontally in the upper soil layers. Rhizomorph penetration in roots is accomplished by both enzymatic activity and mechanical pressure (Williams *et al.*, 1986). The fungus can survive for long periods as rhizomorphs or pseudosclerotia in infected tissues (Michailides, 2003).

***Management:*** Armillaria root rot has become more difficult to control since the soil fumigant methyl bromide was banned. Disease management depends on the stage of disease development. Sanitation through the tree, including root removal and using clean planting material to avoid introducing the pathogen are among the cultural control methods. In theory, removing infected trees in fig orchards is the most effective disease control (Robinson, 2010). Reducing stress and wounding of trees prevents primary infections. Azole fungicides such as cyproconazole and hexaconazole can significantly reduce the fungal growth of Armillaria species (Aguín *et al.*, 2006)

## Phytophthora root rot

*Phytophthora* species are the most aggressive fungus-like pathogens causing root and crown rot in irrigated fig orchards. *Phytophthora* spp. has been reported on figs in Palestine (Ali-Shtayeh *et al.*, 1994), Iran (Mostowfizadeh-Ghalamfarsa *et al.*, 2008), Turkey (Turkolmez *et al.*, 2018), and Taiwan (Lin *et al.*, 2018).

***Symptoms:*** The primary symptoms include necrosis and rot on the root and crown, and poor growth, yellowing and wilting of the leaves. Dark-brown to black necrosis is observed under the bark of the rotted crown. Root decay starts in the tap roots and progresses into other large roots. After most of the roots are decayed the trees decline within 7–10 days (Turkolmez *et al.*, 2018). Crown rot can girdle a tree. Younger trees usually decline faster than older trees (Lin *et al.*, 2018).

***Disease development:*** Several species of *Phytophthora* including: *Phytophthora cinammomi* Rands (Turkolmez *et al.*, 2018), *P. niederhauserii* Z.G. Abad & J.A. Abad, *P. nicotianae* Breda de Haan (Vannini *et al.*, 2009), *P. parsiana* Mostowf., D.E.L. Cooke and Banihash (Mostowfizadeh-Ghalamfarsa *et al.*, 2008), *P. parasitica* (Dastur) Waterhouse, and *P. palmivora* Bulter (Ali-Shtayeh *et al.*, 1994) have been identified as causing root and crown rot in fig orchards. Free moisture and cool temperatures in a range of 24–26°C (Ali-Shtayeh *et al.*, 1994; Turkolmez *et al.*, 2018) promote disease development. Zoospores infect the roots, sporangia and zoospores that lead to secondary infections. Oospores and chlamydospores can also survive under unfavorable conditions, such as dry weather and high temperatures. These spores usually form in the rhizosphere of infected trees. Disease dispersal occurs via irrigation water and soil movement (Menge and Ploetz, 2003).

***Management:*** Orchards should be fenced to protect them from animal and human traffic preventing Phytophthora introduction from infected to uninfected orchards. This is the most affordable cultural method of disease management (Menge and Ploetz, 2003). Due to the release of ammonia, animal fertilizers decrease *P. cinnamomi* populations (Zentmyer, 1984). Calcium salts such as calcium sulphate, calcium carbonate, and calcium nitrate reduce root rot caused by *P. cinnamomi* by reducing the number and size of sporangia in this pathogen. (Metalaxyl Ridomil Gold®) and Fosetyl-Al (Aliette®) are two systemic chemicals that are very effective in controlling Phytophthora root rot disease (Menge and Ploetz, 2003).

### Rosellinia root and crown rot

Rosellinia root rot is a destructive disease with a wide host range. Both forest and orchard trees, edible fig, plum, olive, peach, apple, pear, almond, cherry, and avocado, are susceptible to this disease, especially in tropical and temperate climates (Sztejnberg and Madar, 1979; Sousa *et al.*, 1995; Schena *et al.*, 2002). Rosellinia root rot has been identified in fig trees in Turkey and Greece (Michailides, 2003; Papachatzis *et al.*, 2008).

***Symptoms:*** Symptoms can be observed on the roots and stem of young trees. The pathogen surrounds the infected roots or trunk immediately above the soil line, which will be covered by a layer of white cottony mycelia that is visible on the outside of the infected tree trunk base. Saturated soil conditions exacerbate Rosellinia rot (Fig. 12.16) (Papachatzis *et al.*, 2008).

***Disease development:*** The causal agent is a soilborne fungus, *Rosellinia necatrix* Berl. ex Prill. characterized by typical pear-shaped swelling at the hyphal septa. The pathogen has parasitic and saprophytic life cycle phases. The parasitic phase can survive as mycelia for up to two years. In its saprophytic phase, the pathogen also survives as mycelium in dead roots (Schena *et al.*, 2008). This pathogen proliferates best in the dark at temperatures between 22.5 and 25.5°C, with optimum mycelium growth at 24°C. Several factors facilitate dispersal of *R. necatrix* including infected propagating materials (that seem healthy in the early stage of infection), infested soils, and river and irrigation water (Schena *et al.*, 2008).

***Management:*** The pathogen has a wide host range, resistance to common fungicides, resistance to drought, tolerance of a wide soil pH range, and the ability to exist deep in the soil *ca.* 30 cm, all make management extremely difficult (Khan, 1959). Soil treatments are ineffective due to their high cost and negative environmental impacts. Pathogen exclusion avoidance, using certified healthy propagation materials and control methods similar to those used to avoid other soilborne diseases is recommended (Schena *et al.*, 2008). Removing infected tissues from orchards is considered as an essential step to reduce the inoculum reservoir (Michailides, 2003).

## 12.8 Nematodes

Plant-parasitic nematodes are known root pests of *F. carica*. The following species, including ring nematode (*Criconemoides xenoplax* (Raski) Luc & Raski), cyst nematode (*Heterodera fici* Kirjanova), root-knot nematode (*Meloidogyne incognita* (Kofoid & White) Chitwood, *Meloidogyne javanica* (Treub) Chitwood, pin nematode (*Paratylenchus*), sheath nematode (*Hemicycliophora arenaria*), and dagger nematode (*Xiphinema* sp.) have been reported as plant-parasitic nematodes associated with fig trees (Storey *et al.*, 2007; Wohlfarter *et al.*, 2011).

The first report of a plant parasitic nematode associated with fig trees was *Meloidogyne* spp. in the USA (Neal, 1889). Since then, several plant-parasitic nematode species have been reported parasitizing fig trees in many countries (McSorley, 1981, 1992; Cohn and Duncan, 1990; Campos, 1997; Krnjaić *et al.*, 1997; Li *et al.*, 1999). In Portugal, only two nematodes, *Aphelenchus avenae* Bastian and *Heterodera fici* Kirjanova were known to occur associated with fig trees (Macara, 1963).

**Fig. 12.16.** Rosellinia root rot caused by *Rosellinia necatrix*. A: White cottony mycelium, grown in the soil surrounding the infected roots of fig trees. B: Dry rot (dark in color) in trunk base of a fig tree. C: White cottony mycelium, surrounding an infected fig tree base at the soil level. D: Infected and declining fig tree on the right, compared with a healthy fig tree on the left. (Figure used with permission from A. Papachatzis, Papachatzis *et al.*, 2008.)

Serious crop losses due to nematode infection was reported in figs (Storey *et al.*, 2007). Root-knot nematodes, *Meloidogyne* spp., infections have been reported as the most severe nematode problem in commercial fig production, particularly in Mediterranean regions (Luc *et al.*, 2005). *Meloidogyne incognita* infection was reported as the main cause of death of young (15 to 18-month-old) fig trees in Northern Cape, South Africa, where most of these orchards were established on sites that were previously planted with grape, *Vitis vinifera*, and had a history of *M. incognita* infection (Wohlfarter *et al.*, 2011).

***Symptoms:*** Root-knot nematodes cause characteristic knots or galls on fig roots (Fig. 12.17). The adult female is pearl white, pyriform and ~2 mm long. Egg masses as large as her body are often found near or attached to her body, but they may also be found outside the gall. In the juvenile stages, this nematode is long and slender. The adult males retain their vermiform shape while the adult females become pyriform

**Fig. 12.17.** Differences between roots infected by root-knot nematode (left), and non-infected roots (right) of fig. (Figure used with permission from Donald W. Dickson.)

as they reach maturity. The females usually develop just beneath the surface of the root and, as a result of their growth and secretions, induce the formation distinctive swellings called galls. The problem caused by nematodes in fig tree are reduction in growth and yield, and roots have distinctive swelling called galls. Heavily infected roots may die. As a result, affected plants have reduced absorption and transport of water and nutrients, and exhibit stunting, chlorosis and reduced fruit production (McBeth, 1949).

***Disease development:*** Root knot nematodes survive from season to season primarily as an egg in the soil. After the eggs hatch, the second stage juveniles invade roots, usually at root tips, causing some of the root cells to enlarge where the nematodes feed and develop. Root knot nematodes usually cause distinctive swellings, called galls, on the roots of affected plants. The nematodes feed and develop within the galls, which may grow to as large as 3 cm in diameter on some plants.

***Management:*** Management of nematodes is difficult. The most reliable practices are preventive, including sanitation and choice of plant varieties. Existing infestations can be reduced through fallowing, crop rotation and soil solarization. However, these methods reduce nematodes primarily in the top foot or so of the soil, so they are effective only for about a year and not applicable for perennial fruit orchards.

If nematode species have not previously been identified in the plantation site, then soil diagnostics in a laboratory must take place for identification.

To date, no research has been done to explore resistance to nematodes in *Ficus* spp. Fumigation before the establishment of a fig nursery can limit the dissemination of nematodes, but in orchards is very expensive and may need to be limited only to the planting rows (McKenry and Thomason, 1975; Carles, 1985), and trees in sandy soil are very susceptible (Matz, 1918). The biological control agents of nematodes include many microorganisms, but the most important are fungi and bacteria. *Purpureocillium lilacinum* (Thom) Luangsaard, Houbraken, Hywel-Jones & Samson, *Trichoderma harzianum* Rifai, and *Verticillium clamydosporium* Goddard were announced to be the most potent fungal parasites that can effectively control *Meloidogyne* spp. on many host plants (Rao, 2007). *Pasteuria penetrans* Sayre & Starr, a mycelial endospore-forming bacterial parasite, represents another successful bio-control agent against root-knot nematodes (Chen *et al.* 1996). Organic and inorganic nitrogenous amendments, which have been usually added to soil to improve soil fertility, have also offered good nematicidal effects against plant parasitic nematodes (Oka, 2010)

PLANT RESISTANCE. The deployment of resistant rootstocks may be the most effective and economical means of controlling plant-parasitic nematodes (Hosomi *et al.*, 2002).

SOIL SOLARIZATION. Solarization is useful to temporarily reduce nematode populations in the top 30 cm of soil, which allows the production of shallow-rooted annual crops and helps young woody plants become established before nematode populations increase. However, solarization won't provide long-term protection for fruit trees, vines and woody ornamental plants.

SANITATION. Nematodes usually are introduced into new areas with infested soil or plants. Prevent nematodes from entering orchards by using only nematode-free plants purchased from reliable nurseries. To prevent the spread of nematodes, avoid moving plants and soil from infested parts of the orchard. Don't allow irrigation water from around infested trees to run off, as this also spreads nematodes. Nematodes can be present in soil attached to tools and equipment used elsewhere, so clean tools thoroughly before using them in orchards (Dreistadt *et al.* 2004).

## Conclusion

Multiple fungal, bacterial and nematode pathogens affect the tree and its roots, branches, leaves and fruit, and cause economic losses in edible fig production. Like many other deciduous tree fruits, cultural controls, orchard floor sanitation, irrigation and canopy management, versus chemical control, are the major disease control methods.

Among leaf diseases, fig rust and anthracnose, which also infects fruits, are the most important pathogens. For most leaf diseases orchard sanitation, removing infected plant parts, is the most effective and practical method of disease management. As with all aerial figs, controlling relative humidity through canopy management and irrigation are additional cultural control methods.

Among trunk limb diseases, Diaporthe canker is the most destructive disease fig worldwide. Most trunk diseases can be easily controlled with pruning and orchard sanitation.

*Armillaria*, *Phytophthora* and *Rosellinia* species are the most destructive soilborne diseases of the roots and crown. Sanitation, pathogen avoidance, irrigation management, and reducing stress and crown wounding are among the cultural control methods.

Because figs are sold commercially fresh and dried, fruit diseases cause the most obvious economic damage. Fruits are also the most susceptible part of the plant. Among the fruit diseases, endosepsis and souring are the most destructive. Because of the potential of figs as an organic crop, cultural controls for fruit diseases such as orchard sanitation and controlling orchard humidity through the canopy and irrigation scheduling is necessary. However, because most fruit diseases have insect vectors, preventing insect entry to the syconia by growing closed-eyes cultivars is also a good cultural management strategy.

## References


Abo Rehab, M.E.A., Rashed, M.F., Ammar, M.I. and EL-Morsy, S.A. (2014) Dieback and sooty canker of ficus trees in Egypt and its control. *Pakistan Journal of Biological Sciences* 17, 364–371.

Aeini, M., Mirzaee, H., Taghavi, S.M., Khodakaramian, G.R. and Amiri Mazhar, M. (2014) Occurrence of crown gall disease on *Ficus benjamiana* in Fars and Isfahan provinces of Iran. *Archives of Phytopathology and Plant Protection* 47, 2257–2262.

Agrios, G.N. (2005) *Plant Pathology* 5th edn. Elsevier Academic Press, Amsterdam.

Aguín, O., Mansilla, J.P. and Sainz, M.J. (2006) In vitro selection of an effective fungicide against *Armillaria mellea* and control of white root rot of grapevine in the field. *Pest Management Science* 62(3), 223–228.

Alcántara-Durán, J., Moreno-González, D., García-Reyes, J.F. and Molina-Díaz, A. (2019) Use of a modified QuEChERS method for the determination of mycotoxin residues in edible nuts by nano flow liquid chromatography high resolution mass spectrometry. *Food Chemistry* 279, 144–149.

Alfieri, S.A. and El-Gholl, N.E. (1993) Phytophthora fruit rot of fig, *Ficus carica* L. *Plant Pathology Circular No. 362.* Florida Department of Agriculture and Consumer Services, Diversion of Plant Industry, Florida.

Alfieri, S.A. Jr., El-Gholl, N.E. and Campbell, M.L. (1984) *Nectriella* (*Kutilakesa*) *pironii*, a pathogen of fig plants. *Proceedings of The Florida State Horticultural Society* 97, 325–327.

Ali-Shtayeh, M.S., Abu-Ghdaib, S.I. and Ghdaib, S.I.A. (1994) Root and crown rot of figs in the West Bank of Jordan. *Phytopathologia Mediterranea* 33(1), 36–40.

Al-Sadi, A.M., Al-Alawi, Z.A. and Patzelt, A. (2015) Association of *Alternaria alternata* and *Cladosporium cladosporioides* with leaf spot in *Cissus quadrangularis* and *Ficus sycomorus*. *Plant Pathology Journal* 14, 44–47.

Aref, H.L., Salah, K.B., Chaumont, J.P., Fekih, A., Aouni, M. and Said, K. (2010). In vitro antimicrobial activity of four *Ficus carica* latex fractions against resistant human pathogens (antimicrobial activity of *Ficus carica* latex). *Pakistan Journal of Pharmaceutical Sciences* 23(1), 53–58.

Banihashemi, Z. and Javadi, A.R. (2009) Further investigations on the biology of *Phomopsis cinerascens*, the cause of fig canker in Iran. *Phytopathologia Mediterranea* 48(3), 454–460.

Bio-Care Technology NOGALL™. www.bio-caretechnology.com/nogall-biological-control-treatment/ (accessed March 2018).

Bouzar, H. and Jones, J.B. (2001) *Agrobacterium larrymoorei* sp. nov., a pathogen isolated from aerial tumors of *Ficus benjamiana*. *International Journal of Systematic and Evolutionary Microbiology* 51, 1023–1026.

Bouzar, H., Chilton, W.S., Nesme, X., Dessaux, Y., Vaudequin, V., Petit, A., Jones, J.B. and Hodge, N.C. (1995) A new Agrobacterium strain isolated from aerial tumors on *Ficus benjamiana* L. *Applied and Environmental Microbiology* 61, 65–73.

Burks, C.S., Dowell, F.E. and Xie, F. (2000) Measuring fig quality using near-infrared spectroscopy. *Journal of Stored Products Research* 36, 289–296.

Caldis, P.D. (1930) Souring of figs by yeasts and the transmission of the disease by insects. *Journal of Agricultural Research* 40, 1031–1051.

Callen, B.E. and Carris, L.M. (2004) Fungi on living plant substrata, including fruits. In: Mueller, G.M., Bills, G.F. and Foster, M.S. (eds) *Biodiversity of Fungi*. Academic Press, UK, pp. 105–126.

Cao, V.D., Nguyen, P.P., Khuong, V.Q., Nguyen, C.K., Nguyen, X.C., Dang, C.H. and Tran, N.Q. (2014) Ultrafine copper nanoparticles exhibiting a powerful antifungal/killing activity against *Corticium salmonicolor*. *Bulletin of the Korean Chemical Society* 35(9), 2645–2648.

Carles, L. (1985) Lefiguier (2e partie) principal maladies. *Arboriculture Fruitiere* 32(375), 56–58.

Carpenter, Z.L. (1983) *Texas Plant Diseases Handbook*, 2nd edn. Texas Agriculture Extension Service, Texas A & M University, USA.

Çeliker, N.M. and Michailides, T.J. (2012) First report of *Lasiodiplodia theobromae* causing canker and shoot blight of fig in Turkey. *New Disease Reports* 25, 12.

Chen, Y., Wei, H., Du, G., Zhu, L., Song, Q., Wang, E., Wang, M. and Fan, X. (2018) First report of *Lasiodiplodia theobrome* causing stem canker on common fig (*Ficus carica*) in Zhejiang Province of China. *Plant Disease* 102, 2656.

Chen, Z.X., Dickson, D.W., McSorley, R., Mitchell, D.J. and Hewlett, T.E. (1996) Suppression of *Meloidogyne arenaria* race 1 by soil application of endospores of *Pasteuria penetrans*. *Journal of Nematology* 28, 159–168.

Cheong, S.S., Choi, I.Y. and Lee, W.H. (2013) Occurrence of gray mold caused by *Botrytis cinerea* on common fig in Korea. *The Korean Journal of Mycology* 41(1), 38–41.

Choi, I.Y., Park, J.H., Cho, S.E. and Shin, H.D. (2013) First confirmed report of anthracnose fruit rot caused by *Colletotrichum gloeosporioides* on common fig in Korea. *Plant Disease* 97(8), 1119–1119.

Cleene, M. (1979) Crown gall: economic importance and control. *Central Journal of Bacteriology, Parasite Science, Infectious Diseases and Hygiene* 134, 551–554.

Cohn, E. and Duncan, L.W. (1990) Nematode parasites of subtropical and tropical fruit trees. In: Luc, M., Sikora, R.A. and Bridge, J. (eds) *Plant Parasitic Nematodes in Subtropical and Tropical Agriculture*. CAB International, Wallingford, UK, pp. 347–362.

Crous, P.W., Phillips, A.J.L. and Wingfield, M.J. (1991) The genera of *Cylindrocladium* and *Cylindrocladiella* in South Africa, with special reference to forest nurseries. *South African Forestry Journal* 157, 69–85.

Dogan, O. and Benlioglu, S. (2011) Determination of disease incidence of fig endosepsis in mamme fruits of caprifig. *Bitki Koruma Bulteni* 51, 277–285.

dos Santos, A.F., Santos, F., Mezzomo, R., Tessmann, D.J. and May De Mio, L.L. (2017) First report of fruit rot caused by *Phytophthora palmivora* on fig in Brazil. *Plant Disease* 101, 1331.

Dreistadt, S. (2012) *Integrated Pest Management for Citrus* 3th edn. University of California Agriculture and Natural Resources, Davis, USA.

Dreistadt, S.H., Clark, J.K. and Flint, M.L. (2004) *Pests of Landscape Trees and Shrubs: An Integrated Pest Management Guide*. University California Agricultural Natural Research Publication, Oakland, CA, p. 3359.

Droby, S., Wisniewski, M. and Benkeblia, N. (2011) Postharvest pathology of tropical and subtropical fruit and strategies for decay control. In: Yahia, E.M. (ed.) *Postharvest Biology and Technology of Tropical and Subtropical Fruits: Fundamental Issues*. Woodhead Publishing, London, pp. 194–224.

El-Gholl, N.E. and Alfieri, S.A. (1984) Fruit rot of fig caused by *Phytophthora palmivora*. *Mycologia* 79, 1178–1185.

Elshafie, A.E. and Ba-Omar, T. (2001) First report of *Albizia lebbeck* dieback caused by *Scytalidium dimidiatum* in Oman. *Mycopathologia* 154, 37–40.

English, H. (1951) Phomopsis canker: A progress report. In: *Proceedings of the 6th Annual Research Conference of the California Fig Institute*. California, USA, pp. 45–48.

English, H. (1952a) Phomopsis canker of fig. *Phytopathology* 42, 513.

English, H. (1952b) Pruning and spraying experiments for Phomopsis canker. In: *Proceedings of the 6th Annual Research Conference of the California Fig Institute*. California, USA, pp. 16–19.

English, H. (1953) Further work on the control of Phomopsis canker. In: *Proceedings of the 6th Annual Research Conference of the California Fig Institute*. California, USA, pp. 12–15.

English, H. (1958) Physical and chemical methods of reducing Phomopsis canker infection in Kadota fig trees. *Phytopathology* 48, 392 (Abstract).

English, H. (1960) Experiments on the control of Phomopsis canker in Kadota fig trees. In: *Proceedings of the 14th Annual Research Conference of the California Fig Institute*. California, USA, pp. 13–15.

English, H. (1962) Canker and dieback disorders of fig trees. *Proceedings of the 6th Annual Research Conference of the California Fig Institute*. California, USA.

Faccoli, M., Campo, G., Perrotta, G. and Rassati, D. (2016) Two newly introduced tropical bark and ambrosia beetles (*Coleoptera*: *Curculionideae*, *Scolytinae*) damaging figs (*Ficus carica*) in southern Italy. *Zootaxa* 4138, 189–194.

Fawzi, E. (2003) Production and purification of β-glucosidase and protease by Fusarium proliferatum NRRL 26517 grown on *Ficus nitida* wastes. *Annals of Microbiology* 53, 463–476.

Ferguson, L., Michailides, T.J. and Shorey, H.H. (1990) The California fig industry. *Horticulture Review* 12, 409–460.

Flaishman, M. A., Rodov, V. and Stover, E. (2008) The fig: botany, horticulture, and breeding. *Horticulture Review* 34, 113–196.

Galloway, B.T. (1919) Giant crown galls from the Florida Everglades. *Phytopathology* 9, 207–208.

Gilmer, R. (1957) Phomopsis canker on Kadota figs. In: *Proceedings of the 11th Annual Research Conference of the California Fig Institute*. California, USA, pp. 26–28.

Gonzalez, V., Ramallo, J., De Ramallo, N.E.V., Ploper, L.D. and Torres, L. (1998) Fig fruit rot caused by *Phytophthora palmivora* in fig crops in Tucuman, Argentina. *Fitopathologia* 33, 228–231.

Grove, W.B. (1935) In: *British Stem-and-Leaf Fungi*. Cambridge University Press, UK, pp. 186–187.

Guillaumin, J.J., Mohammed, C., Anselmi, N., Courtecuisse, R., Gregory, S.C. *et al.* (1993) Geographical distribution and ecology of Armillaria species in western Europe. *European Journal of Forest Pathology* 23, 321–341.

Hansen, H.N. (1927) Endosepsis and its control in caprifigs. *Phytopathology* 18, 931–938.

Hansen, H.N. (1949a) Phomopsis canker of fig. *California Agriculture* 3(11), 13–14.

Hansen, H.N. (1949b) Control and cure of Phomopsis canker of fig. *California Fruit and Grape Grower* 3(9), 7.

Hansen, H.N. (1949c) Canker diseases in figs. In: *Proceedings of the 3th Annual Research Conference of the California Fig Institute*. California, USA, pp. 18–19.

Hausbeck, M. and Harlan, B. (2018) *What's New in Botrytis Control Research?* Michigan State University, Department of Plant, Soil and Microbial Sciences - May 11, 2018. www.canr.msu.edu/news/whats_new_in_botrytis_control_research

Heperkan, D., Moretti, A., Dikmen, C.D. and Logrieco, A.F. (2012) Toxigenic fungi and mycotoxin associated with figs in the Mediterranean area. *Phytopathologia Mediterranea* 51, 119–130.

Hosomi, A. Dan, M. and Kato, A. (2002) Screening of fig varieties for rootstocks resistant to soil sickness. *Journal of the Japanese Society for Horticultural Science* 71(2), 171–176.

Hosomi, A., Miwa, Y., Furukawa, M. and Kawaradani, M. (2012) Growth of fig varieties resistant to ceratocystis canker following infection with *Ceratocystis fimbriata*. *Journal of the Japanese Society for Horticultural Science* 81(2), 159–165.

Ivanovic, Z., Blagojevic, J. and Nikolic, I. (2018) Leaf spot disease on *Philodendron scandens*, *Ficus carica and Actinidia deliciosa caused by Pseudomonas syringae pv. syringae* in Serbia. *European Journal of Plant Pathology* 151, 1107–1113.

Johnson, J.A., Harrington, T.C. and Engelbrecht, G.J.B. (2005) Phylogeny and taxonomy of the North American clade of the *Ceratocystis fimbriata* complex. *Mycologia* 97, 1067–1092.

Johnstone, P.R. and Jones, D. (1997) Relationships among *Colletotrichum* isolates from fruit-rots assessed using rDNA sequences. *Mycologia* 89, 420–430.

Kajii, C., Morita, T., Jikumaru, S., Kajimura, H., Yamaoka, Y. and Kuroda, K. (2013) Xylem dysfunction in *Ficus carica* infected with wilt fungus *Ceratocystis ficicola* and the role of the vector beetle *Euwallacea interjectus*. *International Association of Wood Anatomists* 34(3), 301–312.
Kajitani, Y. and Masuya, H. (2011) *Ceratocystis ficicola* sp. nov., a causal fungus of fig canker in Japan. *Mycoscience* 52, 349–353.
Kamil, F.H., Saeed, E.E., El-Tarabily, K.A. and AbuQamar, S.F. (2018). Biological control of mango dieback disease caused by *Lasiodiplodia theobromae* using streptomycete and non-streptomycete actinobacteria in the United Arab Emirates. *Frontiers in Microbiology* 9, 829.
Kao, C.W. and Leu, L.S. (1979) Strawberry fruit rot caused by *Phytophthora cactorum* and *P. citrophthora*. *Plant Protection Bulletin Taiwan* 21, 239–243.
Khan, A.H. (1959) Biology and pathogenicity of *Rosellinia necatrix* (Hart.) Berl. *Biologia Lahore* 5, 199–245.
Kim, W.G., Weon, D.C., Hyeong, J.J. and Soon, Y.H. (2000) Occurrence of anthracnose on Indian fig cactus caused by *Glomerella cingulata* and *Colletotrichum gleosporioides*. *The Plant Pathology Journal* 16, 294–296.
Kohl, L.M. (2011) Foliar nematodes: a summary of biology and control with a compilation of host range. *Plant Health Progress* 12, 23–57.
Kosoglu, I., Aksoy, U. and Pehlivan, R. (2011) Fumonisin B1 and B2 occurrence in dried fig fruits (*Ficus carica* L.) under Meander Valley's climatic conditions and relationship with fruit quality. *Food Additives and Contaminants: Part A* 28, 1569–1577.
Krnjaić, D., Krnjaić, S. and Bačić, J. (1997) Distribution and population density of fig cyst nematode (*Heterodera fici* Kirjanova) in the region of SR Yugoslavia. *Zas̆tita Bilja* 48, 245–251.
Kumar, N. and Khurana, S.M.P. (2015) New plant disease record in Panchgaon: fig rust (*Cerotelium fici* (Castange) Arthur) on *Ficus benjamina* L. *International Journal of Current Microbiology and Applied Sciences* 4, 749–754.
Lansky, E.P. and Paavilainen, H.M. (2010) *Figs: The Genus Ficus*. CRC Press, Boca Raton, Florida.
Latinovic, N., Radisek, S. and Latinovic, J. (2014) First report of *Alternaria alternata* causing fruit rot on fig (*Ficus carica*) in Montenegro. *Plant Disease* 98, 424.
Li, H., Xu, J., Shen, P. and Cheng, H. (1999) Distribution and seasonal dynamic changes of nematode parasites in fig main growing areas in Jiangsu Province. *Journal of Nanjing Agricultural University* 22, 38–41.
Li, X., Xie, M., Chen, H., Yan, J., Wang, Z. and Fan, S. (2015) Pathogen identification and biological characteristics of leaf blight of *Ficus altissima*. *Southwest China Journal of Agricultural Sciences* 28, 2558–2562.
Lin, C.P., Tsai, J.N. and Ann, P.J. (2018) First report of fig decline caused by *Phytophthora niederhauserii* in Taiwan. *Plant Disease* 103(5), 1046.
Livshits, I.Z. and Pupysheva, L.I. (1949) Fig canker and measures of its control. *Sad i Ogorod* (3), 25–27.
Luc, M. Sikora, R.A. and Bridge, J. (2005) *Plant Parasitic Nematodes in Tropical and Subtropical Agriculture*, 2nd edn. CAB International, Wallingford, Oxford, UK.
Macara, A.M. (1963) Aspectos sobre a importância dos nemátodos de interesse agrícola em Portugal e no Ultramar português. *Agros* 46, 367–384.
Manziano, F., Scalione, M., Nanni, B., Mangione, M.T. and Astore, A. (1990) Le principali malattie crittogamiche del fico nel Cilento centrale. *Agricultura e Ricerca* 112, 113–122.
Matz, J. (1918) Some diseases of the fig. *Florida Agricultural Experiment Station Bulletin* 149.
McBeth, C.W. (1949) Nematodes affecting figs. *3rd Annual Research Conference of the California Fig Institute Proceedings*, pp. 16–17.
McKenry, M.V. and Thomason, I.J. (1975) The dosage-response of Meloidogyne infected Ficus roots to cis-1,3-dichloropropene nematicides. In: *Proceedings of the 3rd International Congress of Plant*, Moscow, USSR, pp. 507–513.
McKenzie, E.H.C. (1986) New plant disease record in New Zealand: fig rust (*Cerotelium fici*) on *Ficus carica*. *New Zealand Journal of Agricultural Research* 29, 707–710.
McSorley, R. (1981) Plant parasitic nematodes associated with tropical and subtropical fruits. *Bulletin of Agriculture Experimental Station, University of Florida, USA* 823, 1–49.
Mendoza-Castillo, V.M., Vargas-Canales, J.M., Calderón-Zavala, G., Mendoza-Castillo, M.D.C. and Santacruz Varela, A. (2017) Intensive production systems of fig (*Ficus carica* L.) under greenhouse conditions. *Experimental Agriculture* 53(03), 339–350.
Menge, J.A. and Ploetz, R.C. (2003) Diseases of avocado. In: Ploetz, R.C. (ed.) *Diseases of Tropical Fruit Crops*. CABI, Wallingford, UK, pp. 35–71.

Michailides, T.J. (2003) Diseases of fig. In: Ploetz, R.C. (ed.) *Diseases of Tropical Fruit Crops*. CABI, Wallingford, UK, pp. 253–273.

Michailides, T.J. and Morgan, D.P. (1998) Spread of endosepsis in Calimyrna fig orchards. *Phytopathology* 88, 637–647.

Michailides, T.J., Morgan, D.P. and Subbarao, K.V. (1996) An old disease still a dilemma for California growers. *Plant Disease* 80, 828–841.

Michailides, T.J., Morgan, D.P., Felts, D. and Doster, M.A. (2008) Control of decay in caprifigs and Calimyrna figs with fungicides. *Acta Horticulturae* 798, 269–275.

Miyake, N. and Nagai, H. (2017) Efficacy of phosphonate in controlling white powdery rot of fig caused by *Phytophthora palmivora*. *Journal of General Plant Pathology* 83, 390–397.

Miyake, N. and Nagai, H. (2018) Effects of temperature, zoospore concentration, infection period, and fruit maturity on *Phytophthora palmivora* infection of figs. *Journal of General Plant Pathology* 84, 330–338.

Moretti, A., Ferracane, L., Somma, S., Ricci, V., Mule, G., Susca, A., Ritieni, A. and Logrieco, A.F. (2010) Identification, mycotoxin risk and pathogenicity of *Fusarium* species associated with fig endosepsis in Apulia, Italy. *Food Additives and Contaminants* 27, 718–728.

Morita, T., Jikumaru, S. and Kuroda, S. (2016) Disease development in *Ficus carica* plants after inoculation with *Ceratocystis ficicola*. Relationship between xylem dysfunction and wilt symptoms. *Japanese Journal of Phytopathology* 82, 301–309.

Mostowfizadeh-Ghalamfarsa, R., Banihashemi, Z. and Cooke, D.E.L. (2008) *Phytophthora parsiana* sp. nov., a new high-temperature tolerant species. *Mycological Research* 112, 783–794.

Mrak, E.M., Phaff, H.J., Vaughn, R.H. and Hansen, H.N. (1942) Yeasts occurring in souring figs. *Journal of Bacteriology* 44, 441–450.

Naeem, A. and Akhyani, A. (1988) The fig souring disease in Kashan and its transmission factors (in Persian). *Journal of Applied Entomology and Phytopathology* 55(1–2), 31–42.

Neal, J.C. (1889) The root-knot disease of the peach, orange and other plants in Florida due to the work of Anguillula. *Bulletin US Bureau of Entomology.*

Nayab, M. and Akhtar, N. (2016) First report of *Curvularia aeria* causing leaf spot of *Ficus religiosa* in Pakistan. *Plant Disease* 100, 2530.

Norman, D.J. and Ali, S. (2013) Ornamental ficus diseases: identification and control in commercial greenhouse operations. *Plant Pathology Department, UF/IFS Extension*.

Nouri, M.T., Lawrence, D.P., Yaghmour, M.A., Michailides, T.J. and Trouillas, F.P. (2018) *Neoscytalidium dimidiatum* causing canker, shoot blight and fruit rot of almond in California. *Plant Disease* 102(8), 1638–1647.

Ogawa, J.M. and English, H. (1991) *Diseases of Temperate Zone Tree Fruit and Nut Crops.* University of California Agriculture and Natural Resources, Davis, USA.

Oka, Y. (2010) Mechanisms of nematode suppression by organic soil amendments – A review. *Applied Soil Ecology* 44, 101–115.

Palmateer, A.J., Tarnowski, T. and Roberts, P.D. (1999) *Florida Plant Disease Management Guide: Fig (Ficus carica)*. University of Florida, p. 4.

Paolinelli-Alfonso, M., Villalobos-Escobedo, J.M., Rolshausen, P., Herrera-Estrella, A., Galindo-Sánchez, C., López-Hernández, J.F. and Hernandez-Martinez, R. (2016) Global transcriptional analysis suggests *Lasiodiplodia theobromae* pathogenicity factors in volved in modulation of grapevine defensive response. *BMC Genome* 17, 615.

Papachatzis, A., Eliopoulos, P., Statharas, G. and Vagelas, I. (2008) *Ficus carica* root disease caused by *Armillaria mellea* and *Rosellinia necatrix* in Greece. *Annales of University of Craiova* 13, 143–148.

Paxton, J.D., Wilson, E.E. and Davis, R.J. (1964) Branch wilt disease of fig caused by *Hendersonula toruloidea*. *Plant Disease Reporter* 48, 142.

Pilgrim, A.J. (1959) Phomopsis canker of Kadotas. In: *Proceedings of the 4th Annual Research Conference of the California Fig Institute*. California, USA, pp. 30–32.

Prashad, D. (2013) Studies on epidemiology and management of pink canker (*Corticium salmonicolor* Berk. & Br.) in apple. *Ph.D. thesis*. The University of Dr. Yashwant Singh Parmar University, Solan, India.

Prashad, D., Ram, V. and Sharma, I.M. (2014) Relationship between weather parameters and pink canker (*Corticium salmonicolor* Berk. & Br.) of apple. *Journal of Agrometeorology* 16(2), 236–237.

Raheb, A.M.E.A., Rashed, M.F., Ammar, M.I. and El-Morsy, M.I. (2014) Dieback and sooty canker of ficus trees in Egypt and its control. *Pakistan Journal of Biological Sciences* 17, 364–371.

Ramos, D.D. (1997) *Walnut Production Manual*, University of California Agriculture and Natural Resources, Davis, USA.

Rao, M.S. (2007) Papaya seedlings colonized by the bio-agents *Trichoderma harzianum* and *Pseudomonas fluorescens* to control root-knot nematodes. *Nematology Mediteranea* 35, 199–203.
Ray, J.D., Burgess, T. and Lanoiselet, V.M. (2010) First record of *Neoscytalidium dimidiatum* and *N. novaehollandiae* on *Magnifera indica* and *N. dimidiatum* on *Ficus carica* in Australia. *Australian Plant Disease Notes* 5, 48–50.
Reckhaus, P. (1987) Hendersonula dieback of mango in Niger. *Plant Disease* 71, 1045.
Ricci, P. (1972) Observations sur la pourriture des figues fratches apres la recolte. *Annals of Phytopathology* 4, 109–117.
Robinson, R. (2010) Detection, recognition and management options for Armillaria root disease in urban environments. *Proceedings of The International Symposium on Symbolic and Algebraic Computation (ISSAC)* Adelaide, South Australia, pp. 1–11.
Schena, L., Nigro, F. and Ippolito, A. (2002) Identification and detection of *Rosellinia necatrix* by conventional and real-time Scorpion-PCR. *European Journal of Plant Pathology* 108, 355–366.
Schena, L., Nigro, F. and Ippolito, A. (2008) Integrated management of *Rosellinia necatrix* root rot on fruit tree crops. In: Ciancio, A. and Mukerji, K. (ed.) *Integrated Management of Diseases Caused by Fungi, Phytoplasma and Bacteria*. Springer, Dordrecht, pp. 137–158.
Schurrie, H. (1990) The Fig. *Timber Press Horticultural Reviews* 12, 409.
Sousa, A.J.T., Guillaumin, J.J., Sharples, G.P. and Whalley, A.J.S. (1995) *Rosellinia necatrix* and white root rot of fruit trees and other plants in Portugal and nearby regions. *Mycologist* 9, 31–33.
Sreenivasaprasad, S. and Talhinhas, P. (2005) Genotypic and phenotypic diversity in *Colletotrichum acutatum*, a cosmopolitan pathogen causing anthracnose on a wide range of hosts. *Molecular Plant Pathology* 6, 361–378.
Storey, S.G., Knoetze, R., Mouton, C., Swart, A., Malan, A.P., Venter, E. and Tied, T. (2007) Plant parasitic nematodes associated with commercial figs (*Ficus carica* L.) in South Africa. *Proceedings of the 18th Symposium of the Nematological Society of Southern Africa. South African Journal of Plant and Soil*, 24(4), 256.
Stover, E., Aradhya, M., Ferguson, L. and Crisosto, C.H. (2007) The fig: overview of an ancient fruit. *HortScience* 42, 1083–1087.
Sztejnberg, A. and Madar, Z. (1979) The host range of *Rosellinia necatrix*, the causal agent of white root rot in fruit trees, and preliminary trials on control. *Alon Hanotea* 33, 787–791.
Talbot, P.H.B. (1965) Studies of *Pellicularia* and associated genera of *Hymenomycetes*. *Persoonia* 3(4), 371–406.
Taubenhaus, J.J. (1936) A fig decay. *Texas Agricultural Experiment Station Annual Report* 49, 113.
Thomas, H.E., Roberts, C. and Amstutz, A. (1948) Rootstock susceptibility to *Armillaria mellea*. *Phytopathology* 38, 152–154.
Timmer, L.W. (2000) Pink disease and thread blight. In: Timmer, L.W., Garnsey, S.M and Graham, H. (eds) *Compendium of Citrus Diseases*. The American Phytopathological Society, Minnesota, p. 35.
Timmer, L.W., Garnsey, S.W. and Broadbent, P. (2003) Diseases of Citrus. In: Ploetz, R.C. (ed.) *Diseases of Tropical Fruit Crops*. CAB International, Wallingford, UK, pp. 163–195.
Tims, E.C. (1963) *Corticium salmonicolor* in the United States. *Plant Disease Reporter* 47(12), 1055–1059.
Turechek, W.W. (2012) Spatial distribution of crown gall in a commercial nursery of weeping fig. *Plant Health Progress* 13, 3–19.
Turkolmez, S., Dervis, S., Cifci, O. and Ulubas Serce, C. (2018) First report of *Phytophthora cinamomi* causing root and crown rot of *Ficus carica* in Turkey. *Plant Disease* 103(4), 776.
Vannini, A., Brown, A., Clive, B. and Vettraino, M. (2009) The search for *Phytophthora centres* of origin: *Phytophthora* species in mountain ecosystems in Nepal. In: Brasier, C. (ed.) *Proceedings of the 4th Meeting of The International Union of Forest Research Organizations (IUFRO) Working Party*. United States Department of Agriculture, Monterey, California, USA, pp. 54–55.
Wales, N.S. and Territory, N. (2016) Pink disease of citrus. www.agric.wa.gov.au/plant-biosecurity/pink-disease-citrus-pest-data-sheet (accessed 6 April 2019).
Williams, R.E., Shaw, C.G., Wargo, P.M. and Sites, W.H. (1986) Armillaria root disease. *Forest Insect & Disease Leaflet*, US Department of Agriculture, Forest Service 78, 1–7.
Wohlfarter, M., Giliomee, J.H., Venter, E. and Storey, S. (2011) A survey of the arthropod pests and plant parasitic nematodes associated with commercial figs, *Ficus carica* (*Moraceae*), in South Africa. *African Entomology* 19, 165–172.

Wu, Y., Wang, S., Meng, Y., Tang, Y. and Lin, Y. (2015) *Terriera fici* sp. nov. on *Ficus vasculosa* from Hainan Province, China. *Mycotaxon* 130, 1111–1116.
Xue, L.H., Liu, Y. and Huang, Q. (2017) First report of *Colletotrichum fioriniae* causing anthracnose on *Ficus virens* in China. *Plant Disease* 101, 1044.
Yakushiji, H., Morita, T., Jikumaru, S., Ikegami, A., Azuma, A. and Koshita, Y. (2012) Interspecific hybridization of fig (*Ficus carica* L.) and *Ficus erecta* Thunb., a source of ceratocystis canker resistance. *Euphytica* 183, 39–47.
Yakushiji, H., Moritab, T. and Jikumaru, S. (2019) Ceratocystis canker resistance in BC1 populations of interspecific hybridization of fig (*Ficus carica*) and *F. erecta*. *Scientia Horticulturae* 252, 71–76.
Yao, J.A., Huang, P., Lan, C.Z. and Yu, D.Y. (2016) Anthracnose pathogen of potted banyan *Ficus microcarpa*. *Mycosystema* 35, 760–764.
Yao, J.A., Yu, D.Y., Lu, L.M., Huang, P. and Lan, C.Z. (2015) First report of *Alternaria alternata* causing black leaf spot on *Ficus microsporus* in China. *Plant Disease* 99, 1652.
Zentmyer, G.A. (1984) Avocado diseases. *International Journal of Pest Management* 30(4), 388–400.
Zhang, C., Zhang, W., Ma, H.Q. and Zhang, G.Z. (2013) First report of *Phytophthora palmivora* causing fruit rot of fig (*Ficus carica* L.) in China. *Plant Disease* 97, 1252.

# 13 Fig Viruses, Viroids and Phytoplasmas

**I. Tzanetakis[1]* and S. Sabanadzovic[2]**
*[1]Department of Entomology & Plant Pathology, Division of Agriculture, University of Arkansas, Fayetteville, Arkansas, USA; [2]Department of Biochemistry, Molecular Biology, Entomology and Plant Pathology, Mississippi State University, Mississippi, USA*

## 13.1 Introduction

Several viruses and virus-like agents have been detected and characterized in fig, primarily in an effort to identify the causal agent(s) of fig mosaic disease. Until the turn of the century overall knowledge of pathogens involved in fig mosaic was based primarily on the biological properties (transmission, symptomatology on indicators) and/or electron microscopy observations. In the last 15 years there has been significant progress in understanding the fig virome, primarily through a series of studies in the Mediterranean region. This chapter is an up-to-date, comprehensive synthesis of the viruses, viroids and phytoplasmas that infect the fig. Single reports of virus sequences isolated from fig without any additional characterization or information on their biological significance as well as viruses that have only been reported in meeting proceedings and have not been vetted through peer review will not be discussed (e.g. Arkansas fig closteroviruses 1 and 2, strawberry latent ringspot virus infecting fig, etc.). Agents for which there is no type isolate or no genomic information available will also not be discussed as it is impossible to determine whether a new discovery is an isolate of the previously described agent or a new virus altogether (e.g. fig virus S, fig leaf chlorosis virus, etc). Based on those premises this chapter will focus specifically on eight viruses, three viroids and two *Candidatus* phytoplasmas that affect fig and have been characterized in considerable depth (Table 13.1).

## 13.2 Fig Mosaic Virus

Fig mosaic is the most important and best-studied virus disease of fig. It was first described in the 1930s in California (Condit and Horne, 1933) and has been observed in all areas where fig is grown, possibly due to the movement of infected propagated material (Walia *et al.*, 2014). Symptoms can vary greatly, even between leaves of the same tree, and for this reason the term 'fig mosaic' includes numerous symptoms, including mosaic, mottling, ringspots, yellowing, blotching but also leaf deformation (Fig. 13.1). Originally, the presence of double-membrane bound quasi-spherical bodies in infected cells was associated with the disease (Bradfute *et al.*, 1970; Martelli *et al.*, 1993; Appiano *et al.*, 1995). Now these structures are assumed to be virions of a putative causal agent of the disease.

*Corresponding author: itzaneta@uark.edu


DOI: 10.1079/9781789242881.0013

**Table 13.1.** Taxonomic placement of viruses, viroids and phytoplasmas infecting fig.

| Pathogen name | Acronym | Genus/Group |
|---|---|---|
| Fig mosaic virus | FMV | *Emaravirus* |
| Fig badnavirus-1 | FBV-1 | *Badnavirus* |
| Fig leaf mottle-associated virus 1 | FLMaV-1 | *Closterovirus* |
| Fig leaf mottle-associated virus 2 | FLMaV-2 | *Ampelovirus* |
| Fig mild mottle associated virus | FMMaV | *Closterovirus* |
| Fig fleck-associated virus | FFkaV | *Maculavirus* |
| Fig latent virus-1 | FLV-1 | *Trichovirus* |
| Fig cryptic virus | FCV | *Deltapartitivirus* |
| Citrus exocortis viroid | CEVd | *Pospiviroid* |
| Hop stunt viroid | HSVd | *Hostuviroid* |
| Apple dimple fruit viroid | ADFVd | *Apscaviroid* |
| *Candidatus Phytoplasma asteris* | Ca. Phytoplasma asteris | Aster yellows phytoplasma |
| *Candidatus Phytoplasma solani* | Ca. Phytoplasma solani | Stolbur phytoplasma |

**Fig. 13.1.** Typical mosaic symptoms on wild fig trees. (Figure used with permission from Bryce W. Falk, U.C. Davis.)

Fig mosaic virus (FMV) is closely associated with mosaic symptoms and probably causes the homonymous disease. Transmission studies using virus-tested seedlings produced mosaic-like symptoms (Çaglayan *et al.*, 2010) whereas another study detected FMV in 96% of symptomatic plants (Walia *et al.*, 2009). Notwithstanding, given the plethora of viruses reported from mosaic trees (see below), FMV may not be the sole virus causing this disease.

FMV has been studied extensively; at both the molecular and epidemiological levels (Çaglayan *et al.*, 2009; 2012; Elbeaino *et al.*, 2009; Ishikawa *et al.*, 2012; 2013; 2015). The virus is a member of the genus *Emaravirus* (family *Fimoviridae* – from **fi**g **mo**saic) with six RNAs identified as part of the ~15.2 kilobases (kb) genome (Elbeaino *et al.*, 2009; Ishikawa *et al.*, 2012). Each RNA codes for a single protein in the negative sense orientation. The RNAs are encapsidated in 80–120 nm pleomorphic double membrane-bound bodies formed at the endoplasmic reticulum. The population structure of the virus based on regions of RNAs 1–4 appears to be stable and homogenous, with reassortment being a driving force of the virus evolution (Walia *et al.*, 2014).

FMV is transmitted mechanically among a variety of horticultural and experimental plants. However, epidemiologically the primary natural vector is the eriophyoid mite, *Aceria ficus* Colle (Flock and Wallace, 1955; Vashisth and Nagaich 1965; Çaglayan *et al.*, 2009), able to transmit the virus to experimental hosts such as periwinkle (Çaglayan *et al.*, 2012) indicating the potential for green bridges during winter months. The identification of *Cyclamen persicum* as a natural host of the virus reinforces the notion of green bridges although transmission between cyclamen and fig under natural conditions is yet to be experimentally tested (Elbeaino *et al.*, 2018). FMV is not seed transmissible.

Sensitive detection of FMV has traditionally been based on RT-PCR with several published primer sets. Recent publications on the development and implementation of antibody-based techniques against the virus (Haji-Hashemi *et al.*, 2019; Shahmirzaie *et al.*, 2019) may result in large-scale testing of propagation material, the major avenue of virus movement into new production areas. Indeed, close phylogenetic relationships between FMV isolates collected in geographically distant locations suggests long-distance movement through international exchange and trade of infected propagation material (Minafra *et al.*, 2017).

## 13.3 Fig badnavirus 1

Badnaviruses are dsDNA viruses of ~8 kbp that can integrate into the host genome. There are two forms of the virus: episomal, the transcribing, infectious form that encapsidates in bacilliform particles and the inactive, integrated form. Whereas the integrated form of some known badnaviruses infecting other crops acts as a transposon and is asymptomatic, under certain conditions it can excise from the host genome, become episomal and start an infection (Ndowora *et al.*,1999).

In the quest to identify the agents associated with fig mosaic, a new badnavirus, named fig badnavirus-1 (FBV-1) was discovered (Laney *et al.*, 2012). The virus was detected in over 97% of both symptomatic and asymptomatic fig accessions, indicating that FBV-1 may integrate in the fig genome. Experiments verified this hypothesis; only a small subset of the over 100 accessions tested had the episomal form of the virus. Two independent studies revealed a rather stable population structure of FBV-1 with minimal genetic diversity among isolates (Laney *et al.*, 2012; Alimoradian *et al.*, 2014).

The virus is the most widespread of all fig viruses based on several surveys (Elçi *et al.*, 2012; Minafra *et al.*, 2012; Alkowni *et al.*, 2015), yet the presence of an infectious, active form of FBV-1 may be overrated given that, other than the Laney *et al.* study (2012), there has not been any further attempt to discriminate between the integrated and/or episomal forms of the virus. FBV-1 is mechanically transmissible (Laney *et al.*, 2012; Alimoradian *et al.*, 2014), which in addition to the genome integration, may be another reason for its worldwide distribution. In all cases where FBV-1 was detected in single infections in fig to the trees were asymptomatic. However, it is possible that it can exacerbate symptoms in mixed infections with other viruses. The preferred detection method for the virus is RT-PCR using the Laney *et al.* (2012) primers. Even though the virus can be detected using PCR, RT-PCR is more sensitive as, in addition to the genomic molecules, it can also amplify the abundant virus transcripts.

## 13.4 Fig Closterovirids

Closterovirids (*sensu lato*: viruses belonging to family *Closteroviridae*) are the largest positive strand RNA plant viruses with genomes ranging between 13 and 20 kb. They are classified into four genera based on phylogenetic relationships which also align with their modes of transmission. The fig closterovirids identified to date belong to the genera *Ampelovirus* and *Closterovirus* which are transmitted by mealybugs/scales and aphids respectively. There are at least seven closterovirids reported in fig, but data on four of them have only been limited to reports in conference proceedings or to partial sequence records. In this paragraph we will focus on the remaining three that have been characterized in more depth in peer-reviewed literature on characterization, diagnostics, distribution and population structure.

Fig leaf mottle associated virus 1 (FLMaV-1) was the first fig closterovirid described. The virus was discovered in Italy, in a cv. Canestrelle tree exhibiting chlorotic mottling and vein clearing symptoms (Elbeaino *et al.*, 2006). Only partial sequences of the virus replicase and heat shock protein 70 homolog genes are publicly available, yet the information is sufficient to suggest classification of the virus in the genus *Closterovirus*. Based on dsRNA and electron microscopy analyses, Elbeaino *et al.* (2006) estimated the genome of the virus to be between 15 and 19 kb and the size of filamentous particles at ~1800 nm in length. In the original study, the virus was found in both symptomatic and asymptomatic plants, not allowing for a reliable assessment of its possible pathological effect in different cultivars. Recently Yorganci and Açıkgöz (2019) successfully transmitted FLMaV-1 to fig seedlings with the fig wax scale (*Ceroplastes rusci* L.), the first such report for a virus in the genus *Closterovirus* which, as noted before, are known to be transmitted only by aphids. The seedlings developed leaf curling, vein clearing and mottling symptoms – an indication that the virus may cause symptoms. However, it may be the donor plant was infected by another scale-transmitted virus which could have assisted in the transmission process and/or synergistically caused the observed symptoms.

Fig leaf mottle associated virus-2 (FLMaV-2), was discovered in an Algerian fig accession grown in southern Italy showing foliar mottling and vein clearing symptoms similar to those in the tree that led to the discovery of FLMaV-1 (Elbeaino *et al.*, 2007). There is limited sequence information available for the virus; yet phylogenetic analysis clearly places FLMaV-2 in the genus *Ampelovirus* with no vectors identified to date. The genomic properties of FLMaV-2 point to a genome of over 16.5 kb with particles measured up to 2100 × 12 nm (Elbeaino *et al.*, 2007).

Fig mild mottle associated virus (FMMaV) was first described from Calabria (southern Italy), in a cv. Dottato bianco tree exhibiting similar but milder symptoms than those observed with FLMaV-1/2 (Elbeaino *et al.*, 2010). About 6.3 kb of virus sequence is available of an estimated genome size of ~19 kb and particle length ~2000 nm. Phylogenetic analysis clearly places the virus within the genus *Closterovirus* with no vector identified to date. None of the three viruses was mechanically transmissible under experimental conditions and in all case studies electron microscopy pointed to significant cytopathology in the phloem cells (Elbeaino *et al.*, 2006; 2007; 2010, Fayez and Mahmoud, 2011). The preferred detection method for all three viruses is RT-PCR using the primers described in the characterization study of each virus (Elbeaino *et al.*, 2006; 2007; 2010 respectively). Based on several surveys, all three viruses appear to be cosmopolitan, detected in all fig growing areas (Elbeaino *et al.*, 2009; 2011c; 2012; Alhudaib, 2012; Elçi *et al.*, 2012; El-Air *et al.*, 2013; Perović *et al.*, 2016; Mijit *et al.*, 2017), and, in the vast majority of cases, in mixed infections with other viruses. As is the case with most fig viruses, there is no data determining the effect of the virus on yield or an orchard's productive lifespan.

## 13.5 Fig Fleck-associated Virus

Fig fleck-associated virus (FFkaV) was first characterized from a cv. Canestrelle tree showing minor vein clearing in the new leaves and that tested negative for other viruses identified before 2011 (Elbeaino *et al.*, 2011a). Phylogenetic analysis places the virus in the family *Tymoviridae* which are comprised of small, monopartite positive strand RNA viruses with isometric virions. Virus isolates exhibit relatively high sequence diversity in terms of nucleotide content, possibly affected by geographic distance (Elbeaino *et al.*, 2011a) but also in genome organization based on the two genomes reported to date (Elbeaino *et al.*, 2011a; He *et al.*, 2017). Even though the size of the two isolates differ by less than 20 nucleotides, the genome organization is significantly different; this may be the result of simple mistakes in sequence annotation or may be interpreted as a sign of an actively evolving virus. The latter is an intriguing hypothesis given that the genome organization points to one genus and sequence content to another, leaving the current FFkaV taxonomic placement in limbo. FFkaV appears to be widespread but less so than FMV or FBV-1 (Elbeaino *et al.*, 2011a; Delić *et al.*, 2017; Mijit *et al.*, 2017). However, these results may be taken with caution as surveys were carried out before the report of He *et al.* (2017) on considerable intraspecific sequence variation. Therefore, it may be that the distribution of FFkaV is underestimated because of the significant sequence diversity leading to false negatives in the detection tests. As more sequences become available and new tests able to detect diverse isolates are developed, the prevalence of the FFkaV may be revised. As in the case of the aforementioned viruses, the effect of the virus in symptom development is unclear as quite often FFkaV is found in symptomatic plants in mixed infections with other viruses.

## 13.6 Fig Latent Virus 1

Fig latent virus 1 (FLV-1) was first reported in 2009 (Gattoni *et al.*, 2009). A survey to identify viruses in fig germplasm accessions revealed the presence of flexuous rods in both symptomatic and asymptomatic samples. Independent of the symptomatology of the examined sample, there was obvious distortion of the chloroplasts, the only abnormality associated with the presence of the virus (Gattoni *et al.*, 2009). The genome has been characterized and has an expected size of ~8 kb encapsidated in flexuous rods of about 700 nm in length (Gattoni *et al.*, 2009). Phylogenetic analyses indicate that FLV-1 is a member of the genus *Trichovirus*, positive strand RNA viruses that are transmitted in agricultural settings primarily by grafting whereas there are reports of transmission by eriophyid mites (Gispert *et al.*, 1998).

FLV-1 is readily mechanically transmissible to experimental lab hosts; however, inducing latent infections in most test plants. In addition, results of seed transmission suggest that this may be an efficient mode of virus dissemination in the wild as >90% of tested seedlings were found infected with FLV-1. Unfortunately, the tested material was not assessed for the presence of mites as potential vectors. The virus is rather widespread, but apparently not to the extent of the aforementioned viruses, as FLV-1 was not detected in some surveys (Alfaro-Fernández *et al.*, 2014; Delić *et al.*, 2017; Mijit *et al.*, 2017). Nevertheless, a relatively small number of samples tested in some of these surveys call for further investigations in order to get more complete knowledge about incidence and prevalence of this virus. FLV-1 can be detected both using immunological as well as molecular tests (RT-PCR). The preferred method is the RT-PCR described in the Gattoni *et al.* (2009) work, because of its sensitivity and availability of reagents (primers vs FLV-1 specific antisera).

## 13.7 Fig Cryptic Virus

While isolating dsRNA to determine the presence of RNA viruses in plant tissue, twin bands of <2 kbp are often seen (Sabanadzovic and Tzanetakis, personal observations). This was the case with an asymptomatic cv. Brogiotto nero tree where two bands of ~1.7 and 1.4 kb were isolated and characterized (Elbeaino *et al.*, 2011b). The positive sense molecules of two dsRNAs code for a polymerase (RNA 1) and a putative coat protein of a new partitivirus. The new virus, named as fig cryptic virus (FCV) clearly clusters with members of the genus *Deltapartitivirus* (family *Partitiviridae*). Partitiviruses can only move through cell division as they do not code for a designated movement protein. However, the mode of movement makes them extremely efficient at invading the gametes and the embryo, resulting in very high percentages of pollen and seed transmission. Notwithstanding, partitiviruses are not known to cause disease in single or mixed infections. The Elbeaino *et al.* (2011b) primers have been used for the detection of FCV in several countries and the virus appears to be ubiquitous with infection rates topping 23% in

Lebanon (Elbeaino *et al.*, 2011b; Elçi *et al.*, 2012; Ale-Agha and Rakhshandehroo, 2014).

## 13.8 Viroids

Viroids are the smallest known infectious agents and are only found in plants. Their non-coding genome is comprised of highly structured, short naked circular RNA of 246-401 nt. They are very efficiently transmitted by mechanical means as well as by seed and pollen. Whereas virus and viroid symptoms tend to resemble each other, the two have a markedly distinguishing characteristic: virus symptoms attenuate at high temperatures, whereas viroid's intensify.

There are three viroids identified in fig, all belonging to the family *Pospiviroidae.* The first two, citrus exocortis viroid (CEVd) and hop stunt viroid (HSVd) have extensive host ranges and were identified as potential components of fig mosaic in Tunisia (Yakoubi *et al.*, 2007). Both viroids were detected in a subset of the mosaic-affected plants tested but not in asymptomatic controls. The third viroid, a distinct variant of apple dimple fruit viroid (ADFVd) was identified in a plant showing severe symptoms but also infected by three viruses. The incidence of viroids in fig is largely unknown due to a very small number of viroid-specific surveys.

Detection of viroids in general is based on RT-PCR either using family specific degenerate primers (Hadidi and Candresse, 2003) or viroid-specific ones that amplify the complete viroid genome. There are also several reports of viroid detection using molecular hybridization, yet sensitivity is lower than RT-PCR (Amari *et al.*, 2007; Murcia *et al.*, 2009).

## 13.9 Phytoplasmas

Phytoplasmas are mollicutes, wall-less bacteria, that are obligate pathogens of plants. Phytoplasmas are phloem limited and are often associated with symptoms indicating a hormone imbalance: dwarfism, witches' broom, deformations, phyllody. One 2017 study identified phytoplasmas in 43 fig trees from an orchard in Apulia, Italy (Alsaheli *et al.*, 2020). All the trees had symptoms potentially attributable to phytoplasmas: yellowing, shortened internodes, leaf deformations. When assayed, 46.5% were infected with distinct lineages of *Candidatus* Phytoplasma *asteris* (aster yellows phytoplasma) and *Candidatus* Phytoplasma *solani* (Stolbur phytoplasma). All but three of the phytoplasma infected-material were also infected with FMV which was also found in symptomatic material that was not infected by the two phytoplasmas. For this reason the association of phytoplasmas with distinct symptoms in fig is yet to be determined. Phytoplasma detection is usually based on nested PCR using universal primers (Gundersen and Lee, 1996), or more recently on quantitative PCR (Christensen *et al.*, 2013)

## 13.10 Control

It is clear that significant work characterizing the genome of fig viruses has been accomplished in the last two decades. Unfortunately, this has not been followed by in depth analyses of the biological and epidemiological properties or significance of the viruses for plant health, a prerequisite for the development of control strategies against any virus. Notwithstanding, the cornerstone of sustainable perennial crop production of any clonally propagated crop is material certified free of major pathogens. The European Union regulations require the absence of 'mosaic symptoms' from fig plants for planting. To enter the market, propagation material must derive from mosaic-free mother plants.

Today, with custom-made bioinformatics pipelines coupled with high throughput sequencing we are able to test for all known viruses that affect the crop in a single assay and identify and characterize new ones. The method is unbiased, efficient and cost-effective and can give a clear picture of the viruses that infect propagation material. If these high quality, pathogen-tested plants enter certification schemes, which by definition strive to avoid vector infestation and pathogen reintroduction, it could minimize disease spread and impact.

Even if propagation material is infected, these systemic pathogens can be eliminated. These methods are now being implemented in fig production (Sahraroo *et al.*, 2009; Chiumenti *et al.*, 2013; Bayoudh *et al.*, 2017). All new fig orchards should be established with clean, high quality planting material to minimize the spread of viruses and ensure orchard productivity and longevity. This has been well demonstrated with other crops; clean plants yield better and remain productive longer (Atallah *et al.*, 2012; Quito-Avila *et al.*, 2014; McQuaid *et al.*, 2016).

Once a fig orchard has been established, there is a single major virus/virus-like disease that affects trees: 'fig mosaic'. Because multiple symptoms are associated with the term 'fig mosaic disease' it has been difficult to determine which viruses contribute to disease development. Koch's postulates have not been conducted on any of the agents discussed in this chapter. FMV is very closely associated with mosaic disease and widespread everywhere figs are grown. For all practical purposes it is considered the causal agent of the disease. The virus is transmitted by eriophyoid mites; acaricides could possibly be an effective tool for controlling the spread, yet the widespread presence of the virus may make this approach ineffective. Notwithstanding, research evaluating the efficacy or economic impact of chemical or biological control of the mites is limited. For other viruses that may contribute to the symptoms, or could act synergistically to cause the disease, the information for developing effective control strategies is also lacking.

## References


Ale-Agha, G.N. and Rakhshandehroo, F. (2014) Detection and molecular variability of fig fleck-associated virus and fig cryptic virus in Iran. *Journal of Phytopathology* 162(7–8), 417–425.

Alfaro-Fernández, A., Hernández-Llopis, D. and Font, M.I. (2014) Fig viruses in mainland Spain. *Journal of Phytopathology* 162(5), 332–337.

Alhudaib, K. (2012) Incidence of fig leaf mottle-associated virus and fig mosaic virus in eastern province of Saudi Arabia. *International Journal of Virology* 8, 128–132.

Alimoradian, P., Rakhshandehroo, F. and Shams-Bakhsh, M. (2014) First record of fig badnavirus-1 in fig trees in Iran. *Journal of Plant Pathology* 96(4 Suppl), 124.

Alkowni, R., Chiumenti, M., Minafra, A. and Martelli, G.P. (2015) A survey for fig-infecting viruses in Palestine. *Journal of Plant Pathology* 97(2), 383–386.

Alsaheli, Z., Contaldo, N., Mehle, N., Dermastia, M., Elbeaino, T. and Bertaccini, A. (2020) First detection of "Candidatus Phytoplasma asteris"- and "Candidatus Phytoplasma solani"-related strains in fig trees. *Journal of Phytopathology* 168(1), 63–71.

Amari, K., Ruiz, D., Gómez, G., Sánchez-Pina, M.A., Pallás, V. and Egea, J. (2007) An important new apricot disease in Spain is associated with Hop stunt viroid infection. *European Journal of Plant Pathology* 118(2), 173–181.

Appiano, A., Conti, M. and Zini, N. (1995) Cytopathological study of the double-membrane bodies occurring in fig plants affected by fig mosaic disease. *Acta Horticulturae* 386, 585–592.

Atallah, S.S., Gómez, M.I., Fuchs, M.F. and Martinson, T.E. (2012) Economic impact of grapevine leafroll disease on *Vitis vinifera* cv. Cabernet franc in Finger Lakes vineyards of New York. *American Journal of Enology and Viticulture* 63(1), 73–79.

Bayoudh, C., Elair, M., Labidi, R., Majdoub, A., Mahfoudhi, N. and Mars, M. (2017) Efficacy of tissue culture in virus elimination from caprifig and female fig varieties (*Ficus carica* L.). *The Plant Pathology Journal* 33(3), 288–295.

Bradfute, O.R., Whitmoyear, R.E. and Nault, R.L. (1970) Ultrastructure of plant leaf tissues infected with mite borne viral-like particles. *Annual Proceedings of the Electron Microscopy Society of America* 258, 178–179.

Çaglayan, K., Medina, V., Yigit, A., Kaya, K., Gazel, M., Serçe, Ç.U. and Çalıskan, O. (2009) Transmission of the fig mosaic agent by the eriophyid mite *Aceria ficus* Cotte (Acari: Eriophyidae). *Journal of Plant Pathology* 91(1), 235–235.

Çaglayan, K., Serçe, Ç.U., Barutçu, E., Kaya, K., Medina, V., Gazel, M., Soylu, S. and Çalışkan, O. (2010) Comparison by sequence-based and electron microscopic analyses of fig mosaic virus isolates obtained from field and experimentally inoculated fig plants. *Plant Disease* 94(12), 1448–1452.

Çaglayan, K., Elci, E., Ulubas Serçe, C., Kaya, K., Gazel, M. and Medina, V. (2012) Detection of fig mosaic virus in viruliferous eriophyid mite *Aceria ficus*. *Journal of Plant Pathology* 94(3), 629–634.

Chiumenti, M., Campanale, A., Bottalico, G., Minafra, A., De Stradis, A., Savino, V. and Martelli, G.P. (2013) Sanitation trials for the production of virus-free fig stocks. *Journal of Plant Pathology* 95(3), 655–658.

Christensen, N.M., Nyskjold, H. and Nicolaisen, M. (2013) Real-time PCR for universal phytoplasma detection and quantification. In: *Phytoplasma*. Humana Press, Totowa, NJ, USA, pp. 245–252.

Condit, I.J. and Horne, W.T. (1933) A mosaic of the fig in California. *Phytopathology* 23, 887–896.

Delić, D., Perović, T., Hrnčić, S., Lolić, B., Đurić, G. and Elbeaino, T. (2017) Detection and phylogenetic analyses of fig-infecting viruses in Bosnia and Herzegovina and Montenegro. *Phytopathologia Mediterranea* 56(3), 470–478.

El Air, M., Mahfoudhi, N., Elbeaino, T., Dhouibi, M.H. and Digiaro, M. (2013) Presence of fig mild mottle-associated virus and fig latent virus 1 in Tunisia. *Journal of Plant Pathology* 95(4, Suppl) S47.2.

Elbeaino, T., Digiaro, M., De Stradis, A. and Martelli, G.P. (2006) Partial characterisation of a closterovirus associated with a chlorotic mottling of fig. *Journal of Plant Pathology* 88(2), 187–192.

Elbeaino, T., Digiaro, M., De Stradis, A. and Martelli, G.P. (2007) Identification of a second member of the family *Closteroviridae* in mosaic-diseased figs. *Journal of Plant Pathology* 89(1), 119–124.

Elbeaino, T., Nahdi, S., Digiaro, M., Alabdullah, A. and Martelli, G.P. (2009) Detection of FLMaV-1 and FLMaV-2 in the Mediterranean region and study on sequence variation of the HSP70 gene. *Journal of Plant Pathology* 91(2), 425–431.

Elbeaino, T., Digiaro, M., Heinoun, K., De Stradis, A. and Martelli, G.P. (2010) Fig mild mottle-associated virus, a novel closterovirus infecting fig. *Journal of Plant Pathology* 92(1), 165–172.

Elbeaino, T., Digiaro, M. and Martelli, G.P. (2011a) Complete sequence of fig fleck-associated virus, a novel member of the family *Tymoviridae*. *Virus Research* 161(2), 198–202.

Elbeaino, T., Kubaa, R.A., Digiaro, M., Minafra, A. and Martelli, G.P. (2011b) The complete nucleotide sequence and genome organization of fig cryptic virus, a novel bipartite dsRNA virus infecting fig, widely distributed in the Mediterranean basin. *Virus Genes* 42(3), 415–421.

Elbeaino, T., Gonzalez-Rodríguez, Á.M., Grajal-Martín, M.J. and Digiaro, M. (2011c) Survey of fig viruses in the Canary Islands. *Journal of Plant Pathology* 93(3), 737–739.

Elbeaino, T., Digiaro, M. and Martelli, G.P. (2012) RNA-5 and RNA-6, two additional negative-sense RNA segments associated with Fig mosaic virus. *Journal of Plant Pathology* 94, 421–425.

Elbeaino, T., Marais, A., Faure, C., Trioano, E., Candresse, T. and Parrella, G. (2018) High-throughput sequencing reveals *Cyclamen persicum* Mill. as a natural host for fig mosaic virus. *Viruses* 10(12), 684.

Elbeshehy, E.K.F. and Elbeaino, T. (2011) Viruses infecting figs in Egypt. *Phytopathologia Mediterranea* 50(2), 327–332.

Elçi, E., Serçe, Ç.I.U., Gazel, M. and Çağlayan, K. (2012) Molecular detection and comparative sequence analysis of viruses infecting fig trees in Turkey. *Journal of Phytopathology* 160(7–8), 418–423.

Fayez, K.A. and Mahmoud, S.Y. (2011) Detection and partial characterization of a putative closterovirus affecting *Ficus carica*: molecular, ultrastructural and physiological aspects of infected leaves. *Acta Physiologiae Plantarum* 33(6), 2187–2198.

Flock, R.A. and Wallace, J.M. (1955) Transmission of fig mosaic by the eriophyid mite *Aceria ficus*. *Phytopathology* 45, 52–54.

Gattoni, G., Minafra, A., Castellano, M.A., De Stradis, A., Boscia, D., Elbeaino, T., Digiaro, M. and Martelli, G.P. (2009) Some properties of fig latent virus 1, a new member of the family *Flexiviridae*. *Journal of Plant Pathology* 91(3), 555–564.

Gispert, C., Oldfield, G.N., Perring, T.M. and Creamer, R. (1998) Biology of the transmission of peach mosaic virus by *Eriophyes insidiosus* (Acari: Eriophyidae). *Plant Disease* 82(12), 1371–1374.

Gundersen, D.E. and Lee, I.M. (1996) Ultrasensitive detection of phytoplasmas by nested-PCR assays using two universal primer pairs. *Phytopathologia Mediterranea* 35(3), 144–151.

Hadidi, A. and Candresse, T. (2003) Polymerase chain reaction. In: Hadidi, A., Flores, R., Randles, J. and Semancik, J. (eds) *Viroids: Properties, Detection, Diseases and Their Control*. CSIRO Publishing, pp. 115–122.

Haji-Hashemi, H., Safarnejad, M.R., Norouzi, P., Ebrahimi, M., Shahmirzaie, M. and Ganjali, M.R. (2019) Simple and effective label free electrochemical immunosensor for fig mosaic virus detection. *Analytical Biochemistry* 566, 102–106.

He, Z., Mijit, M., Li, S. and Zhang, Z. (2017) Complete nucleotide sequence of a novel strain of fig fleck-associated virus from China. *Archives of Virology* 162(4), 1145–1148.

Ishikawa, K., Maejima, K., Komatsu, K., Kitazawa, Y., Hashimoto, M., Takata, D., Yamaji, Y. and Namba, S. (2012) Identification and characterization of two novel genomic RNA segments of fig mosaic virus, RNA5 and RNA6. *Journal of General Virology* 93(7), 1612–1619.

Ishikawa, K., Maejima, K., Komatsu, K., Netsu, O., Keima, T., Shiraishi, T., Okano, Y., Hashimoto, M., Yamaji, Y. and Namba, S. (2013) Fig mosaic emaravirus p4 protein is involved in cell-to-cell movement. *Journal of General Virology* 94(3), 682–686.

Ishikawa, K., Miura, C., Maejima, K., Komatsu, K., Hashimoto, M., Tomomitsu, T., Fukuoka, M., Yusa, A., Yamaji, Y. and Namba, S. (2015) Nucleocapsid protein from fig mosaic virus forms cytoplasmic agglomerates that are hauled by endoplasmic reticulum streaming. *Journal of Virology* 89(1), 480–491.

Laney, A.G., Hassan, M. and Tzanetakis, I.E. (2012) An integrated badnavirus is prevalent in fig germplasm. *Phytopathology* 102(12), 1182–1189.

Martelli, G.P., Castellano, M.A. and Lafortezza, R. (1993) An ultrastructural study of fig mosaic. *Phytopathologia Mediterranea* 32, 33–43.

McQuaid, C.F., Sseruwagi, P., Pariyo, A. and Van den Bosch, F. (2016) Cassava brown streak disease and the sustainability of a clean seed system. *Plant Pathology* 65(2), 299–309.

Mijit, M., Zhen, H.E., Jian, H., Lu, M.G., Li, S.F. and Zhang, Z.X. (2017) Analysis of fig tree virus type and distribution in China. *Journal of Integrative Agriculture* 16(6), 1417–1421.

Minafra, A., Chiumenti, M., Elbeaino, T., Digiaro, M., Bottalico, G., Pantaleo, V. and Martelli, G.P. (2012) Occurrence of fig badnavirus 1 in fig trees from different countries and in symptomless seedlings. *Journal of Plant Pathology* 94(4 Suppl), S4.105.

Minafra, A., Savino, V. and Martelli, G.P. (2017) Virus diseases of fig and their control. *Acta Horticulturae* 1173, 237–244.

Murcia, N., Serra, P., Olmos, A. and Durán-Vila, N. (2009) A novel hybridization approach for detection of citrus viroids. *Molecular and Cellular Probes* 23(2), 95–102.

Ndowora, T., Dahal, G., LaFleur, D., Harper, G., Hull, R., Olszewski, N.E. and Lockhart, B. (1999) Evidence that badnavirus infection in Musa can originate from integrated pararetroviral sequences. *Virology* 255(2), 214–220.

Perović, T., Hrnčić, S., Čizmović, M., Lolić, B., Đurić, G. and Delić, D. (2016) First report of fig mosaic virus, fig leaf mottle-associated virus 1 and fig mild mottle-associated virus infecting fig (*Ficus carica*) in Montenegro. *Journal of Plant Pathology* 98(3), 678.

Quito-Avila, D.F., Lightle, D. and Martin, R.R. (2014) Effect of raspberry bushy dwarf virus, raspberry leaf mottle virus, and raspberry latent virus on plant growth and fruit crumbliness in 'Meeker'red raspberry. *Plant Disease* 98(2), 176–183.

Sahraroo, A., Babalar, M., Ebadi, A., Habibi, M.K. and Khadivi-Khub, A. (2009) Influence of apical meristem culture and thermotherapy on production of healthy fig plants. *Horticulture, Environment and Biotechnology* 50(1), 45–50.

Shahmirzaie, M., Safarnejad, M.R., Rakhshandehroo, F., Safarpour, H., Rabbani, H., Zamanizadeh, H.R. and Elbeaino, T. (2019) Production of a polyclonal antiserum against recombinant nucleocapsid protein and its application for the detection of fig mosaic virus. *Journal of Virological Methods* 265, 22–25.

Vashisth, K. and Nagaich, B.B. (1965) *Morus indica*, an additional host of fig mosaic. *Indian Phytopathology* 18, 315.

Walia, J.J., Salem, N.M. and Falk, B.W. (2009) Partial sequence and survey analysis identify a multipartite, negative-sense RNA virus associated with fig mosaic. *Plant Disease* 93, 4–10.

Walia, J.J., Willemsen, A., Elci, E., Çağlayan, K., Falk, B.W. and Rubio, L. (2014) Genetic variation and possible mechanisms driving the evolution of worldwide fig mosaic virus isolates. *Phytopathology* 104(1), 108–114.

Yakoubi, S., Elleuch, A., Besaies, N., Marrakchi, M. and Fakhfakh, H. (2007) First report of hop stunt viroid and citrus exocortis viroid on fig with symptoms of fig mosaic disease. *Journal of Phytopathology* 155(2), 125–128.

Yorganci, S. and Açıkgöz, S. (2019) Transmission of fig leaf mottle-associated virus 1 by *Ceroplastes rusci*. *Journal of Plant Pathology* 101(3), 1199–1201.

# 14 Arthropod Pests of Fig and Their Management

**S. Singh[1]*, Z. Li[2], Y. Zhang[2], M.J. Grieshop[3], J. Giliomee[4], G.E. Massimino Cocuzza[5] and R.K. Sandhu[1]**

[1]*Department of Fruit Science, Punjab Agricultural University, Ludhiana, India;* [2]*Department of Entomology, China Agricultural University, Beijing, China;* [3]*Department of Entomology, Michigan State University, East Lansing, USA;* [4]*Department of Botany and Zoology, University of Stellenbosch, Matieland, South Africa;* [5]*Dipartimento di Agricoltura, Università degli Studi di Catania, Catania, Italy*

## 14.1 Introduction

Fig, *Ficus carica* is a widespread, well-known plant globally and an important crop in many areas. A number of insects, mites and arthropod pests damage the fruits, leaves, stem and other plant parts. The most notorious pests belong to a wide variety of insect orders including: Lepidoptera, Coleoptera, Hemiptera, Diptera, Thysanoptera and Hymenoptera. Mites from the order Acari are also very important pests. Among vertebrate pests, birds and rodents hold importance (Table 14.1). Frugivorous insects infesting ripening figs pose a special challenge as figs are eaten fresh immediately after the harvest restricting management practices in the orchards.

This chapter provides an overview of the most important arthropods pests of fig, taking into consideration their destructiveness, distribution and regularity of occurrence, and discusses their main management approaches. Some of the arthropod pests of figs are given in Table 14.1 and are described below.

## 14.2 Lepidoptera

Fig sustain damage from numerous lepidopteran pests, some widely distributed, others specific to geographic regions. The most globally important moth pests of fig include: fig leaf roller, *Phycodes radiata* Ochsenheimer and navel orangeworm, *Amyelois transitella* (Walker), fig tree skeletonizer moth/fig leaf roller, *Choreutis nemorana* (Hübner) and fig tree moth, *Ocnerogyia amanda* (Staudinger).

### Fig leaf roller, *Phycodes radiata* Ochsenheimer (Lepidoptera: Brachodidae)

***Distribution***: The fig leaf roller, *Phycodes radiata* Ochsenheimer was reported as a fig pest in Myanmar, Sri Lanka and parts of India (Kumar and Ramamurthy, 2010; Singh and Kaur, 2017). Karim *et al.* (2010) reported this pest in Iran. It is also found in Pakistan, Afghanistan,

*Corresponding author: sandeep_pau.1974@pau.edu



DOI: 10.1079/9781789242881.0014

Nepal and Sri Lanka (Kallies, 2004). It is a major fig pest in the Indian state of Punjab, causing severe damage in some areas. Another leaf webbing moth, *P. minor* Moore infesting fig, has also been reported in the Indian states of Punjab (Singh and Kaur, 2017) and Haryana (Lakra and Kharub, 1977).

***Host range***: Apart from cultivated fig, *P. radiata* has been reported as a minor and sporadic pest of *Ficus* species plants of the family Moraceae; *Ficus glomerata*, *F. religiosa*, *F. bengalensis*, *F. indica* and *F. tisela* (Kumar and Ramamurthy, 2010). Bajwa and Gul (2000) reported *P. radiata* on *Paulownia* sp. from Peshawar, Pakistan. *Phycodes radiata* was also reported on *Mimusops elengi* L. in New Delhi, India (Kumar and Ramamurthy, 2010).

***Life cycle:*** Simwat and Sidhu (1974) conducted ambient temperature field studies on fig leaf roller biology in the Punjab, India. The insect remained active from March to September through four overlapping generations. It overwintered as pupae in the plant debris under the fig plants. Moths lived for about 4–11 days and laid 3–81 eggs singly or in batches on either side of the leaves. Eggs hatched in about 5–6 days. Larvae were full-grown in 30–36 days and pupated on the infested leaves. Pre-pupal and pupal stages lasted for 3.7 and 8–11 days, respectively.

***Damage:*** *Phycodes radiata* larvae (Fig. 14.1) attack the leaves, rolling and webbing them with white silken threads. Larvae feed on the leaf epidermis, causing skeletonized bands near the leaf midrib or veins. Partially or completely skeletonized leaves eventually drop leading to major defoliation in severe infestations. Peak activity of the pest in India has been observed during July–September. The larvae hibernate in silken cocoons during the off-season. During severe infestations, up to 92% of the leaves were damaged (Simwat and Sidhu, 1974). Singh and Kaur (2017) reported 70–80% leaf damage.

**Fig. 14.1.** Larva of fig leaf roller, *Phycodes radiata.* (Figure courtesy of S. Singh.)

#### *Management*

***Cultural control:*** deep plowing and soil solarization before planting. Collect and burn all infested leaves (NIPHM, 2015).

***Biocontrol:*** Simwat and Sidhu (1974) reported *Apanteles angeleti* Mues., *Trathala flavo-orbitalis* Cameroon and *Cadurcia vanderwulpi* Baranov were effective biocontrol agents in Punjab, India. Phani (2016) reported *Apanteles phycodis* Viereck was a larval parasitoid of *P. radiata* from India.

### Navel orangeworm, *Amyelois transitella* (Walker) (Lepidoptera: Pyralidae)

The navel orangeworm, *Amyelois transitella* (Walker), is a polyphagous pest, particularly of almonds, pistachios, walnuts and figs in North and South America (Connell, 2002; Coviello *et al.*, 2003; Niu *et al.*, 2011).

***Distribution:*** North America; Mexico, USA (Arizona, California, Florida, Georgia, Oklahoma, Texas, Washington), Central America; Costa Rica, South America; Brazil (AQIS, 1999), Argentina (Coviello *et al.*, 2003).

***Host range:*** Hosts include various citrus species like *Citrus limon*, *C. paradisi*, *C. sinensis*, *Vitis vinifera*, *Acacia farnesiana*, *Brachychiton* sp., *Carya illinoensis*, *Ceratonia siliqua*, *Coffea* sp., *Cydonia oblonga*, *Eriobotrya japonica*, *Ficus* sp., *Forchhammeria* sp., *Genipa americana*, *Gleditsia triacanthos*, *Heteromeles arbutifolia*, *Juglans regia*, *Malus pumila*, *Phoenix dactylifera*, *Pistacia vera*, *Pithecellobium flexicaule*, *Prunus armeniaca*, *Prunus domestica*, *Prunus dulcis*, *Prunus* sp., *Punica granatum*, *Pyrus communis*, *Yucca* sp. and *Ziziphus* sp. (AQIS, 1999).

***Life cycle:*** Adult navel orangeworm (Fig. 14.2) are silver-gray moths with black irregular lines and a 20 mm wingspan. Adult females lay an average of 85–120 flat, oval eggs beginning 1–2 days after emergence (Coviello *et al.*, 2003).

**Table 14.1.** List of key fig pests of general and local importance.

| Pest species | Order | Family | Plant part infested | Pest importance |
|---|---|---|---|---|
| Insect pests | | | | |
| Fig leaf roller, *Phycodes radiata* Oshsenheimer | Lepidoptera | Brachodidae | Leaves | Key, ubiquitous, regular |
| Fig leaf roller, *Phycodes minor* Moore | Lepidoptera | Brachodidae | Leaves | Key, ubiquitous, regular |
| Navel orangeworm, *Amyelois transitella* (Walker) | Lepidoptera | Pyralidae | Fruits | Key, ubiquitous, regular |
| Hairy caterpillar, *Euproctis lunata* Walker | Lepidoptera | Lymantriidae | Leaves | Moderate, local, occasional |
| Hairy caterpillar, *Spilaractia obliqua* (Walker) | Lepidoptera | Arctiidae | Leaves | Moderate, local, occasional |
| Leaf-feeding caterpillar, *Aganais speciosa* (Drury) | Lepidoptera | Hypsidae | Leaves | Key, local, regular |
| Leaf-feeding caterpillar, *Ocinara ficicola* Westwood and Ormerod | Lepidoptera | Bombycidae | Leaves | Key, local, regular |
| Fruit piercing moths, *Serrodes partita* (Fabricius), *Dugaria scandulata* (Walker), *Calpe provocans* (Walker) and *Calpe emarginata* (Fabricius) | Lepidoptera | Noctuidae | Fruits | Moderate, local, occasional |
| *Perina nuda* Fabricius | Lepidoptera | Lymantriidae | Leaves | Key, local, regular |
| *Trilocha varians* Walker | Lepidoptera | Bombycidae | Leaves | Key, local, regular |
| Fig tree skeletonizer moth/Fig leaf roller, *Choreutis nemorana* (Hübner) | Lepidoptera | Choreutidae | Leaves | Minor, ubiquitous, regular |
| Fig tree moth, *Ocnerogyia amanda* (Staudinger) | Lepidoptera | Lymantriidae | Leaves | Key, local, regular |
| Fig borer or red spotted long horned beetle, *Batocera rufomaculata* DeGeer | Coleoptera | Cerambycidae | Trunk | Key, ubiquitous, regular |
| Fig borer, *Batocera rubus* (Linnaeus) | Coleoptera | Cerambycidae | Trunk | Key, local, regular |
| Fig tree borer, *Phryneta spinator* (Fabricius) | Coleoptera | Cerambycidae | Trunk | Key, local, regular |
| Driedfruit beetle: *Carpophilus hemipterus* (Linnaeus), Freeman sap beetle: *Carpophilus freemani* Dobson<br>Confused sap beetle: *Carpophilus mutilatus* Erichson | Coleoptera | Nitidulidae | Fruits | Key, ubiquitous, regular |
| Darkling ground beetle, *Blapstinus fuliginosus* Casey | Coleoptera | Tenebrionidae | Fruits | Moderate, local, regular |
| Bark beetle, *Hypocryphalus scabricollis* Eichhoff | Coleoptera | Curculionidae | Bark | Moderate, ubiquitous, regular |
| Ambrosia beetle, *Xyleborus bispinatus* Eichhoff | Coleoptera | Curculionidae | Bark | Moderate, ubiquitous, regular |
| Weevil, *Mecopus hopei* Rosenschoeld | Coleoptera | Curculionidae | Bark | Key, local, regular |
| Longicorn beetle, *Olenecamptus bilobus* Fabricius | Coleoptera | Cerambycidae | Trunk | Key, local, regular |
| Fig beetle, *Cotinis texana* Casey | Coleoptera | Scarabaeidae | Fruits | Moderate, local, occasional |
| Fig bark beetle, *Hypoborus ficus* (Erichson) | Coleoptera | Scolytidae | Bark | Key, ubiquitous, regular |

*Continued*

**Table 14.1.** Continued.

| Pest species | Order | Family | Plant part infested | Pest importance |
|---|---|---|---|---|
| Fig whitefly, *Singhiella simplex* (Singh) | Hemiptera | Aleurodidae | Leaves | Key, ubiquitous, regular |
| Cottony cushion scale, *Icerya* sp. | Hemiptera | Monophlebidae | Leaves and twigs | Moderate, local, occasional |
| Nigra scale, *Parasaissetia nigra* (Nietner) | Hemiptera | Coccidae | Leaves and twigs | Moderate, local, occasional |
| Fig scale, *Lepidosaphes conchiformis* (Gmelin) | Hemiptera | Coccidae | Fruits | Moderate, local, occasional |
| Coffee scale, *Saissetia oleae* (Olivier) | Hemiptera | Coccidae | Twigs | Moderate, local, occasional |
| Wax scale, *Ceroplastes* sp. | Hemiptera | Coccidae | Twigs | Moderate, local, occasional |
| Giant mealybug, *Drosicha stebbingi* (Green) | Hemiptera | Monophlebidae | Leaves, twigs and fruits | Key, ubiquitous, regular |
| Mealybugs, *Phenacoccus solenopsis* Tinsley, *Pseudococcus longispinus* (Targioni Tozzetti) and *Planococcus ficus* (Signoret) | Hemiptera | Margarodidae | Leaves, twigs and fruits | Minor, local, occasional |
| Psyllid, *Macrohomotoma gladiata* Kuwayama | Hemiptera | Homotomidae | Leaves and twigs | Minor, ubiquitous, occasional |
| Psyllid, *Homotoma ficus* Linnaeus | Hemiptera | Homotomidae | Leaves | Minor, ubiquitous, occasional |
| Fruit fly, *Bactrocera dorsalis* (Hendel) | Diptera | Tephritidae | Fruits | Key, ubiquitous, regular |
| Fruit fly, *Bactrocera zonata* (Saunders) | Diptera | Tephritidae | Fruits | Key, ubiquitous, regular |
| Mediterranean fruit fly, *Ceratitis capitata* (Wiedemann) | Diptera | Tephritidae | Fruits | Key, ubiquitous, regular |
| Natal fruit fly, *Ceratitis rosa* (Karsch) | Diptera | Tephritidae | Fruits | Key, local, regular |
| Vinegar fly, *Drosophila melanogaster* (Meigen) | Diptera | Drosophilidae | Fruits | Key, ubiquitous, regular |
| African fig fly, *Zaprionus indianus* Gupta | Diptera | Drosophilidae | Fruits | Key, ubiquitous, regular |
| Black fig fly, *Silba adipata* McAlpine | Diptera | Lonchaeidae | Fruits | Key, ubiquitous, regular |
| *Silba* Macquart 1851, *Silba ischnopoda* sp. nov. | Diptera | Lonchaeidae | Fruits | Key, local, regular |
| Thrips, *Frankliniella occidentalis* (Pergande) | Thysanoptera | Thripidae | Leaves, twigs and fruits | Key, ubiquitous, regular |
| Fig wasp, *Blastophaga psenes* (Linnaeus) | Hymenoptera | Agaonidae | Fruits | Key, ubiquitous, regular |
| Fig wasp, *Blastophaga* spp. | Hymenoptera | Agaonidae | Fruits | Key, ubiquitous, regular |
| Mite pests | | | | |
| Two-spotted spider mite, *Tetranychus urticae* Koch | Acari | Tetranichidae | Leaves, twigs and fruits | Key, ubiquitous, regular |
| *Eotetranychus hirsti* Pritchard and Baker | Acari | Tetranichidae | Leaves, twigs and fruits | Key, local, regular |
| Fig mite, *Aceria fici* (Ewing) | Acari | Eriophyidae | Leaves, twigs and fruits | Key, local, regular |
| Fig blister mite, *Aceria ficus* (Cotte) | Acari | Eriophyidae | Leaves, twigs and fruits | Key, local, regular |
| Vertebrate pests | | | | |
| Garden Warbler, *Sylvia borin* (Boddaert) and many others | Passeriformes | Sylviidae | Fruits | Minor, local, regular |
| Ground and tree squirrels | Rodentia | Sciuridae | Fruits and wood | Minor, local, regular |

Parra-Pedrazzoli and Leaf (2006), reported almost 80% of mating took place within two days after adult emergence, with females mating only once. Ampt *et al.*, (2015) assessed larval preference and performance of navel orangeworm for *Aspergillus flavus* in bioassays in which neonates were allowed to orient within arenas to media with or without fungal tissue. They found that navel orangeworm larvae were attracted to *A. flavus* and developed faster in its presence, indicating a nutritional benefit to the caterpillars.

***Damage:*** Damage is caused by the worm feeding on the fruit. As in other crops, indirect damage results from frass, webbing and secondary saprophytic fungi contamination of the fruits (Ampt *et al.*, 2015).

### *Management*

***Cultural control:*** navel orangeworm overwinters in mummified fruits remaining on plants after maturity. Harvesting rapidly and early, and post-harvest sanitation, removing and destroying cull fruit is currently the best cultural control. Cleaning up surrounding hosts, such as almonds, during fall and winter will also decrease the overwintering pests (Zalom *et al.*, 1984).

***Biocontrol***: Legner and Guido (1983) reported the succesful establishment of external larval parasites, *Goniozus emigratus* (Rohwer) and *Goniozus legneri* Gordh, in California almond orchards.

***Sex pheromone***: Burks and Brandl (2004) demonstrated the efficacy of the sex pheromone component (Z,Z)-11,13-hexadecadienal, releasing 48 mg/ha/day from peripherally-located timed-release dispensers placed throughout the orchard. Higbee *et al.*, (2014) demonstrated and characterized a synthetic pheromone lure for capturing males. Traps baited with lures prepared with 1 or 2 mg of a three- or four-component formulation captured similar numbers of males. Traps baited with the pheromone lure usually did not capture as many males as traps baited with unmated females.

***Chemical control:*** Chemicals such as malathion, applied to control driedfruit beetles may partially control navel orangeworm populations (Coviello *et al.*, 2003).

Hairy caterpillars, *Euproctis lunata* Walker (Lepidoptera: Lymantriidae) (Fig. 14.3) and *Spilaractia obliqua* (Walker) (Lepidoptera: Arctiidae) (Fig. 14.4) are polyphagous pests, reported to infest fig leaves in Punjab, India, causing major damage on younger plants. Young larvae feed on the leaf underside, leading to skeletonization and leaf drying. Later instars also feed on leaves,

**Fig. 14.2.** Adult of Navel Orangeworm, *Amyelois transitella.* (Figure used with permission from David G. Barker.)

**Fig. 14.3.** Hairy caterpillar, *Euproctis lunata* Walker. (Photo: S. Singh.)

**Fig. 14.4.** Hairy caterpillar, *Spilaractia obliqua* Walker. (Figure used with permission from M. Warad and C.M. Kalleshwaraswamy.)

tender shoots and young buds. Severe infestations cause defoliation by July–September.

The fig tree caterpillar, *Aganais speciosa* (Drury) (Lepidoptera: Hypsidae) (Fig. 14.5), has been found on wild figs in southern, East and West Africa and sometimes feeds on the leaves of cultivated figs (Annecke and Moran, 1982). Geertsema (1985) reported defoliation of the exotic ornamental, *Ficus retusa*, and feeding on leaves of domestic figs by the wild silkworm, *Ocinara ficicola* Westwood and Ormerod (Lepidoptera: Bombycidae). When first observed in 1983 in the Western Cape, South Africa, Annecke and Moran (1982) listed four species of fruit-piercing moths, *Serrodes partita* (Fabricius) (Fig. 14.6), *Dugaria scandulata* (Walker) (Fig. 14.7), *Calpe provocans* (Walker), and *C. emarginata* (Fabricius) (Fig. 14.8) (Lepidoptera: Noctuidae), as fig pests. These moths can damage most deciduous fruit by sucking the juices from ripe fruit, leading to fruit rot. Their numbers vary from season to season.

Less common Lepidopteran pests have also been reported. Cheanban *et al.* (2017b) reported *Perina nuda* Fabricius (Fig. 14.9) and Cheanban

**Fig. 14.5.** Fig tree caterpillar, *Aganais speciosa* (Drury). (Figure used with permission from A. Manson.)

**Fig. 14.6.** Fruit piercing moth, *Serrodes partita* (Fabricius). (Figure used with permission from G. Cheong Weei.)

**Fig. 14.7.** Fruit piercing moth, *Dugaria scandulata* (Walker). (Figure used with permission from R. Goff.)

**Fig. 14.8.** Fruit piercing moth, *Calpe emarginata* (Fabricius). (Figure courtesy of S. Singh.)

**Fig. 14.9.** Leaf eating caterpillar, *Perina nuda* Fabricius. (Figure used with permission from S. Cheanban.)

*et al.*, 2017a reported *Trilocha varians* (Fabricius) (Bombycidae) as important leaf eating caterpillars of figs in Thailand. In laboratory studies, Cheanban *et al.*, (2017b) observed that at 33°C and 65% RH, mating and oviposition occurred at night, eggs were laid in clusters averaging 46 eggs/group with females laying between 168 and 387 eggs. The egg incubation period was 5.50 days with the 7 larval instars of 2.28, 3.08, 3.02, 3.42, 3.67, 3.34 and 6.10 days, respectively. The adult life-span was 7.00 and 5.44 days for males and females, respectively.

## Fig tree skeletonizer moth/Fig leaf roller, *Choreutis nemorana* (Hübner, 1799) (Lepidoptera: Choreutidae)

***Distribution*:** *Choreutis nemorana* is found in most of the fig growing regions such as the Mediterranean area, Canary Islands and Madeira, Asia minor, Iran, Georgia, Azerbaijan, Armenia and Uzbekistan (Diakonoff, 1986). *C. nemorana* has also been recorded from Greece (Kavallieratos *et al.*, 2019), Cyprus, France, Italy, Portugal, Ukraine, Turkey, Croatia, Albania, Bosnia, Malta, Spain, Romania (De Prins and De Prins, 2014) and Tunisia (Zouba, 2010). After the introduction of *F. carica* in northern countries, the presence of this insect pest has also been reported in these countries, such as in Austria (Embacher *et al.*, 2004; Fauster, 2016), Switzerland (Christian *et al.*, 2008), Germany (Gaedike, 2008), Belgium (De Prins *et al.*, 2009), Hungary (Szabóky, 2015), British Islands (De Prins and De Prins, 2014), Netherlands (Vossen, 2015) and Slovakia (Lendel, 2017).

**Fig. 14.10.** Fig tree skeletonizer moth, *Choreutis nemorana* (Hübner). (Figure used with permission from A. Zouba.)

***Host Range*:** *Choreutis nemorana* is monophagus and feeds on a single host, *Ficus carica* L. (Kavallieratos *et al.*, 2019).

***Life Cycle*:** Adult moths start egg laying in groups on the upperside of leaves in early spring. Larvae (Fig. 14.10) appear in June, feed on leaves and pupate there (Rota *et al.*, 2014). Pupation takes place in dense white boat-shaped cocoons (Alford, 2007). Adults of first generation emerge in July–August and that of second generation in September–October which undergoes hibernation in plant debris (Vaneva-Gancheva, 2017). Adults have characteristic, almost square, brown forewings, and black and yellow hindwings (Rota *et al.*, 2014). In central Asia and Russia, three–four generations have been reported with 35–50 days per generation (Aizpurua, 1997).

***Damage*:** *Choreutis nemorana* is considered as a minor pest of fig trees that feeds on leaf tissues (Mifsud *et al.*, 2012) and young leaves are preferred. Larvae act as leaf skeletonizer by feeding on parenchymal cells of leaves leaving only the veins intact. The young larvae feed on the upper leaf surface only leaving the underside of leaves intact. Leaf edges get curled upward (De Prins and De Prins, 2014). Mature larvae make a web of silken threads and eat all the green parts of the leaves, leaving the veins undisturbed. In cases of severe infestation, photosynthesis is reduced due to leaf distortion, discoloration and extensive defoliation (Chitgar *et al.*, 2014). Large numbers of adults may be observed in the shaded part of the trees, late in the afternoon (Zouba, 2010).

### Management

***Biological control***: *Dolichogenidea candidata* (Haliday, 1834) and *Itoplectis tunetana* (Schmiedeknecht, 1914) have been reported as parasitoids of *C. nemorana* from Greece (Kavallieratos *et al.*, 2019). An endoparasitoid *Triclistus anthophilae* Aeschlimann has also been recorded on *C. nemorana* in France (Shaw, 2017) and an ectoparasitoid, *Stenomesius rufescens* (Rossius) has been found to be in regular association with *C. nemorana* in Europe (Bouček and Askew, 1968).

## Fig tree moth, *Ocnerogyia amanda* (Staudinger) (Lepidoptera: Lymantriidae)

*Ocnerogyia amanda* is a serious pest of fig, *F. carica* in Iraq (Ali *et al.*, 2012).

***Life cycle***: Eggs are laid in groups of 20 or more, mostly on the lower leaf surface, but may be found on the upper leaf surface, on the trunk and on dry leaves fallen on the ground (Starý and Kaddou, 1971). The early-instar larvae are not easily noticeable on the trees, but later instars are easily found by their long yellowish body hairs. Larvae undergo six moults (Ramachandra, 1922) and larval stage last about 24 days. The pupal stage is 8–11 days long; cocoon is spun of rough yellowish grey silk (Ali *et al.*, 2012). The duration of the entire life cycle is about one and a half months with nearly three generations per year (Ramachandra, 1922).

***Damage***: The first instar larvae disperse over the leaf and start feeding. Larvae feed during the night and can be found in aggregations on tree trunks or sheltered parts of garden walls during the day. Early instars make small holes in the leaves and older larvae (Fig. 14.11) can completely devour the leaves leaving veins only. Fruits may also become shriveled and drop (Starý and Kaddou, 1971).

**Fig. 14.11.** Fig tree moth, *Ocnerogyia amanda* (Staudinger). (Figure used with permission from W.K. Ali.)

### Management

***Mechanical control:*** Hand picking of eggs and larvae from lower leaf surfaces is effective. Trunk banding with grease traps migrating larvae during morning and evening. Arab cultivators used to burn dead leaves under the tree trunk; this destroys all stages of the insect on the tree (Buxton, 1920).

## 14.3 Coleoptera

A wide range of Coleoptera attack fig fruit, leaves, bark and wood. Sap beetles (Nitidulidae) can damage fruit in the field and in storage, and a variety of cerambycids and curculionids attack the bark and wood of young and mature fig plants.

## Tropical fig borer or red spotted long horned beetle, *Batocera rufomaculata* DeGeer (Coleoptera: Cerambycidae)

***Distribution:*** The tropical fig borer or red spotted long horned beetle, *Batocera rufomaculata* DeGeer (Fig. 14.12) is a polyphagous borer of a number of crops. *Batocera rufomaculata* has been reported in Asia (China, Malaysia, Thailand, Sri Lanka, Pakistan, India and Middle East) Africa (Mauritius, Madagascar and Reunion) and the Caribbean (Barbados, Puerto Rico and the Virgin Islands (CABI, 2019a). In 2000, it was reported present in Israel by Ben-Yehuda *et al.* (2000), Nepal by Upadhyay *et al.* (2013) and in Punjab, India by Singh and Kaur (2017). Tozlu and Ozbek (2000) reported *B. rufomaculata* in the eastern Mediterranean region of Turkey.

***Host range:*** The beetle feeds on fig, papaya (*Carica papaya*), mango (*Mangifera indica*), the timber tree (*Shorea robusta*) (NIPHM, 2015), cashew (*Anacardium occidentale*) (Vasanthi and Raviprasad, 2015), durian (*Durio zibethinus*) (Sudhi *et al.*, 2008), mulberry (*Morus* sp.)

**Fig. 14.12.** Fig borer, *Batocera rufomaculata* DeGeer. (Figure used with permission from Manske.)

(Swapna and Divya, 2010), cotton tree (*Bombax malabaricum*), pilkan (*Ficus infectoria*), Hairy sterculia (*Sterculia villosa*), the Para rubber tree (*Hevea brasiliensis*), Chinese mulberry (*Morus australis*), jackfruit (*Artocarpus heterophyllus*), and eucalyptus (Kulkarni, 2010).

***Life cycle:*** The female *B. rufomaculata* oviposits single eggs in cuts she makes in the plant bark, laying up to 200 eggs during her lifetime. Eggs are brownish-white cylinders, 6.2 mm long with narrowly rounded ends. Juvenile development is completed within 2 years with grub tunnels measuring 2–3 cm (Husain and Khan, 1940), extending into the trunk and branch sapwood. After pupation, adult beetles emerge from circular exit-holes, living up to 8 months (Lakra, 2004; NIPHM, 2015; Singh and Kaur, 2017).

***Damage:*** Grubs attack the main trunk and branches, killing the host plant within few years. Grub tunnels often follow a zigzag pattern and contain copious amounts of frass. Tunneling causes reduced leaf, wood and fruit development before killing the plant (Vasanthi and Raviprasad, 2015). Adult beetles have also been reported to feed on bark and young tender twigs, severe infestations causing stem collapse and breakage (NIPHM, 2015). Adults are active from August–September and December.

### *Management*

***Cultural control*:** remove, bury or burn dead and severely affected branches of plant. Avoid injury at the base of trunk while pruning. Alternative host plants like mango, silk cotton and moringa in the near vicinity should be removed.

***Botanical control*:** in evening, spray eco-friendly *Lantana* leaf extracts @ 5% on the plant. Neem based botanicals should be applied inside visible holes.

**Fig. 14.13.** Fig borer, *Batocera rubra* (Linnaeus). (Figure courtesy of S. Singh.)

***Mechanical control*:** inserting a flexible wire inside the hole will kill the stem borer grubs. Holes can be plugged with cotton wick soaked in kerosene or petrol and sealed with mud (NIPHM, 2015).

Another cerambycid borer, *B. rubra* (Linnaeus) (Fig. 14.13) has also been reported from the Indian Punjab infesting fig plants (Singh and Kaur, 2017).

## Fig tree borer, *Phryneta spinator* (Fabricius) (Coleoptera: Cerambycidae)

The fig tree borer, *Phryneta spinator* (Fabricius) (Fig. 14.14), occurs throughout Africa (Annecke and Moran, 1982; Wohlfarter *et al.*, 2011). The fig tree borer is a strong, stout beetle, with a 35 mm long body and equally long antennae. It is dark, mottled grey with a prominent spine on each side of the thorax. The larva is a flat, white grub with a small, dark head bearing strong mandibles (Annecke and Moran, 1982).

***Distribution:*** This beetle occurs from Senegal to Ethiopia and southern parts of the continent. It is found in most parts of South Africa, but has only become abundant in the Western Cape since the 1990s (Annecke and Moran, 1982; Wohlfarter *et al.*, 2011; BioLib, 2019).

***Host range:*** The native host is the Cape willow, *Salix mucronata* from which it has adapted to attack the domestic fig. Both adults and larvae are found on these plants and also the exotic

**Fig. 14.14.** Fig tree borer, *Phryneta spinator* (Fabricius). (Figure used with permission from Jan Giliomee.)

weeping willow, *Salix babylonica* and occasionally pear (*Pyrus* spp.) and peach (*Prunus persica*). Adults can feed on the bark of a variety of plants, syringa, apple (*Malus domestica*), apricot (*Prunus armeniaca*), plum (*Prunus domestica*), and grapevine (*Vitis* spp.) (Gunn, 1919; Annecke and Moran, 1982; Robinson, 2009; Donovan, 2014).

***Life cycle:*** After mating, the adult lays up to four eggs at a time in a slit in the bark, usually made on the lower part of the stem of the plant, but sometimes also on the branches. This starts around November and can continue until March. Each female lays about 35 eggs. They hatch after 2 weeks, the larvae initially feed on the bark before tunneling into the wood. Because the wood is not very nutritious, it can take 3 years before the larva is mature enough to pupate under the bark. During this period, it pushes out a mixture of excrement and sawdust through openings in the bark, sign of an active infestation. This frass also fills the tunnels behind the larvae. The pupal stage lasts about 3 months after which the adults emerge (Gunn, 1919; Annecke and Moran, 1982).

***Damage:*** The adults can cause some damage by nibbling on the bark, but the serious damage is caused by the larvae. Their tunnels in the trunk weaken the tree and with severe infestations it can die or be blown over by wind. Young trees may die when they are ring-barked by the larvae feeding (Gunn, 1919; Robertson, 2019).

***Management:*** The first management step is to prevent the larvae from entering the wood (Donovan, 2014). The adults can be prevented from laying eggs on the plant stem by covering it during the summer months with an impenetrable material such the plastic from old fertilizer bags. The bottom end of the sleeve must be covered with soil. Alternatively, the stem can be sprayed with an insecticide to kill the larvae as they emerge (Annecke and Moran, 1982). If the frass of a young infestation is noticed, the larvae can be removed by cutting them out with a sharp knife (Gunn, 1919). If the infestation is older, an attempt should be made to kill them by pushing a piece of wire down the tunnel. If unsuccessful, insecticide can be poured down the tunnel.

## Driedfruit beetles (Coleoptera: Nitidulidae)

### *Driedfruit beetle:* Carpophilus hemipterus *(Linnaeus); Freeman sap beetle:* Carpophilus freemani *Dobson; Confused sap beetle:* Carpophilus mutilatus *Erichson*

*Carpophilus*, also known as driedfruit beetles or sap beetles, are a worldwide pest of fruits. Adult *carpophilus* cause feeding damage on ripening fruits and are a vector of the disease brown rot. *Carpophilus hemipterus* (Linnaeus) (Fig. 14.15), is the most common species, but *C. freemani* (Fig. 14.16) and *C. mutilatus* (Fig. 14.17) are also common. Driedfruit beetles are a direct pest of figs and their presence in the fruit results in downgrading or rejection of the fruit. They also transmit spoilage organisms that cause fruit souring (Coviello and Bentley, 2019; Government of Western Australia, 2019; PestNet, 2019).

***Distribution:*** These beetles are widespread.

***Host range:*** Driedfruit beetles have a wide host range and will infest any ripe, fermenting and dried fruit in storage.

***Life cycle:*** Adult beetles are oblong shaped and small in size, approximately 3 mm long. They can be black, brown or mottled yellow. Many species have two large yellow, gold or amber colored spots on their elytra which do not completely cover the body. The head capsule is tan colored and supports two hornlike structures on the posterior (Coviello and Bentley, 2019; Government of Western Australia, 2019). The female beetle deposits its eggs on the fruit of a plant. Adults emerge from the pupae and

**Fig. 14.15.** Driedfruit beetle, *Carpophilus hemipterus* (Linnaeus). (Figure used with permission from Courrney L. Brissey.)

**Fig. 14.16.** Driedfruit beetle, *Carpophilus freemani* Dobson. (Figure used with permission from Alimohammad Yavari.)

attack fruit in late spring and summer. There are many generations in a year. Adults are strong fliers. A female may lay about 1000 eggs in her lifetime (Government of Western Australia, 2019; PestNet, 2019).

***Damage:*** Driedfruit beetles usually damage the fig fruits on the ground. They also vector brown rot disease (Government of Western Australia, 2019; PestNet, 2019).

### *Management*

Good orchard hygiene/sanitation is the most effective way to manage this beetle. As beetles can move long distances, area-wide management and cooperation is required. Alternatively, fig orchards should be located away from other host orchards because beetles can fly several miles to find a host (Government of Western Australia, 2019).

***Cultural control:*** all culled fruit should be removed from the orchards after fruit harvest to reduce the overwintering population. Early harvesting should be done, then infested and damaged fruits should be removed from the orchards (Government of Western Australia, 2019). Regular orchard sanitation is helpful in reducing the damage (Coviello and Bentley, 2019).

***Resistant varieties:*** all commercial varieties of figs are susceptible to damage by driedfruit beetles. Varieties with small eyes, e.g. Missions, are usually less damaged. Calimyrna, a major commercial drying variety, has a large eye that renders it easily infested (Coviello and Bentley, 2019).

***Monitoring with traps:*** for monitoring of beetles, small traps, baited with cull fruit, water and yeast, should be kept in the orchards before

**Fig. 14.17.** Confused sap beetle, *Carpophilus mutilatus* Erichson. (Figure used with permission from Courrney L. Brissey.)

the fruits begins to ripen. Twice a week, the traps should be removed and water replenished as and when required (Blumberg, 2019; Coviello and Bentley, 2019; Government of Western Australia, 2019).

***Pheromones:*** male-produced pheromones, to which both males and females respond, were identified by Bartelt (1997) for nine *Carpophilus* spp. They reported that the trap capture is highly synergized by volatiles from fermenting food materials.

***Natural enemies:*** 33 species of parasitoids of sap beetles have been observed. Species of *Zeteticontus silvestri* Noyes (Hymenoptera: Encyrtidae) are commonly associated with sap beetles. *Zeteticontus utilis* Noyes, a solitary endoparasitoid of the larvae of *Carpophilus* spp., was collected in Israel in spring and early summer (Blumberg, 2019).

***Chemical control:*** the chitin inhibitor, triflumuron (alsystin) indirectly affects egg hatch via the adult stage of the beetle, and is quite effective against larvae (Blumberg *et al.*, 1985). Dormant aluminium phosphide treatment reduced overwintering populations in isolated California orchards that had severe infestation (Coviello and Bentley, 2019).

## Darkling ground beetle, *Blapstinus fuliginosus* Casey (Coleoptera: Tenebrionidae), commonly known as false wireworm

***Distribution:*** The species is distributed in north America (Bousquet *et al.*, 2018; Coviello and Bentley, 2006a).

***Host range:*** The beetle has been observed on fig plants (Simmons *et al.*, 1931; Coviello and Bentley, 2006a).

***Life cycle:*** Adult darkling ground beetle, *B. fuliginosus* also known as small darkling beetles, are oval shaped, ground-inhabiting, dull black in color, about 6 mm long by 3 mm wide. The grubs are up to 12 mm long and cream to pale tan in color. The typical habitat of this beetle is among and under the soil clods (Simmons *et al.*, 1931). Their enlarged antennae tips distinguish them from carabid ground beetles, which are general predators.

***Damage:*** Grubs feed on decaying organic matter in and on the soil surface. The adults cause damage by feeding on the ripe figs. Adults are present in fig orchards year round with their activity peaking in July and August. When the beetle population is high in orchards, they cluster and are found on and under the fallen figs. Sometimes, beetles girdle the tree suckers at or near the ground surface (Simmons *et al.*, 1931).

### Management

***Cultural control:*** to reduce the beetle population, remove or accelerate the breakdown of organic matter, reducing the grubs food supply. Harvesting should be done rapidly, removing figs from the orchards to decrease infestation by adult beetles (Coviello and Bentley, 2006a). A shallow water filled ditch surrounding the orchard prevents beetle entry from adjacent fields (Simmons *et al.*, 1931).

***Chemical control:*** poisoned mash (bran, molasses and sodium arsenite) helps in controlling the beetle to some extent (Simmons *et al.*, 1931).

## Bark beetle, *Hypocryphalus scabricollis* (Eichhoff) (Coleoptera: Scolytidae)

Bark beetles, an important horticultural crop pest directly damage to the plants and vector diseases. Bark beetles that feed on live tissue are a major global crop pest. *Hypocryphalus* is a typical bark beetle genus in the Scolytidae family. There are more than 70 species in *Hypocryphalus*. *Hypocryphalus scabricollis* is an important fig pest (Mifsud *et al.*, 2012).

***Distribution:*** This Asian beetle species was first described in Myanmar (Burma). It is also found in Bangladesh, Burma, China, India, Phillipines, Sri Lanka and Vietnam (Wood and Bright, 1992). Mifsud and Knížek (2009) suggested the Asian species, *F. retusa*, now common and widely distributed in Europe, was introduced via Malta. Faccoli *et al.*, (2016) reported this species infesting numerous Italian fig plantations. *Hypocryphalus scabricollis* has also recently been reported on Tunisian fig trees (Gaaliche *et al.*, 2018).

***Host range:*** H. scabricollis has been observed infesting many tropical plants like *Afzelia bijuga, Albizzia stipulata, Bassia latifolia, Bombax malabaricum, Buchanania lanzan, Canarium euphyllum, Excoecaria agallocha, Garuga pinnata, Mesua ferrea, Odina wodier*, and *Ficus* spp. (Wood and Bright, 1992). In Malta, it was initially recovered from *F. retusa* (Mifsud and Knížek, 2009). It is now infesting both wild and cultivated figs (Mifsud *et al.*, 2012; Faccoli *et al.*, 2016).

***Life cycle:*** The adults are small, cylindrical, hard-bodied, brown/black insects measuring 2 mm in length. The elytral surface of *H. scabricollis* is covered with dense microscopic scale-like setae and erect sparse hairs. Beetles attack the tree mid-trunk, spreading upward and downward. Grubs chew galleries which get filled with frass that is pushed out onto the bark surface through small holes of diameter less than 1 mm. Mass attacks cause the foliage to fade (Cutajar and Mifsud, 2017).

***Damage:*** These beetles usually colonize the bark of stressed or dying plants but they have ability to infest healthy fig plants also. In both cases they cause rapid desiccation and death of fig plants. Wild varieties of figs are also infested (Faccoli *et al.*, 2016).

### Management

***Cultural control:*** Root and trunk injury and soil compaction during orchard establishment should be avoided. Plants should be protected from sunburn with proper irrigation (Cutajar and Mifsud, 2017). Dense stands of susceptible plants should be thinned.

***Physical control:*** once the beetles are under the bark, no chemicals can save the plant. If an active infestation is seen, the tree should be removed and burned to prevent subsequent infestation.

***Monitoring with ethanol traps:*** Ethanol traps can attract bark beetles. When bark beetles begin showing up in traps, control measures should be taken immediately.

***Debark:*** Remove all bark from the infested plant.

***Solarization:*** After pruning, wood logs should be stacked in an area receiving plenty of sunlight and cover them with 10 mm thick plastic to kill the beetles or wood can be burnt (Cutajar and Mifsud, 2017).

## Ambrosia beetle, *Xyleborus bispinatus* Eichhoff (Coleoptera: Curculionidae)

The ambrosia beetle genus *Xyleborus* contains many widespread invasive species. Survival of ambrosia beetles relies on obligate nutritional relationships with fungal symbionts, cultivated in the tunnels the beetles excavate in the sapwood of their host plants. International trade of goods, commodities, live plants and wood packaging materials are the most common pathways by which this beetle is introduced. These beetles colonize new areas readily due to wide host range (Cruz *et al.*, 2018).

***Distribution:*** This species is native to Neotropical regions, Central and northern South America and subtropical regions. It was introduced from there into new areas by the international timber trade. It has been reported in eight states of USA, Mexico and on figs in Italy (Faccoli *et al.*, 2016).

***Host range:*** In Florida, USA, *X. bispinatus* (Fig. 14.18) was observed in wilted plants of avocado, *Persea americana* (Mill.), swampbay, *Persea palustris* (Raf.), and foxtail palm, *Wodyetia bifurcata* Irvine (Atkinson *et al.*, 2013). *Quercus* spp. and *Swietenia macrophylla* have also been recorded as hosts. Faccoli *et al.*, (2016) reported it on both wild and cultivated varieties of the common fig, *F. carica* in Sicily, Europe.

***Life cycle:*** A strong sexual dimorphism occurs within males and the larger females. Menocal *et al.* (2018) evaluated rearing *X. bispinatus* on three artificial media containing avocado (*P. americana*) and silkbay (*Persea humilis* Nash.) sawdust. They concluded that the avocado-based medium was more effective than the silkbay-based medium.

Carabez *et al.* (2018) isolated nine operational taxonomic units of fungi from *X. bispinatus*. *Raffaelea* spp., a nutritional symbiont, was predominant in the *Xyleborus* species. *R. lauricola* was also present. *Raffaelea aguacate*, which had previously been detected only in laurel wilt-affected avocado trees, was detected for the first time in *X. bispinatus*. Ascomycota and Saccharomycotina yeasts are common; distinct populations of these fungi have been found in mycangia of the beetles. They discussed roles played by yeasts in the life cycles of these bark borers, e.g. by providing nutrients to beetles.

**Fig. 14.18.** Ambrosia beetle, *Xyleborus bispinatus* Eichhoff. (Figure used with permission from Massimo Faccoli.)

***Damage:*** This beetle is always associated with another bark borer, *H. scabricollis*. It is a secondary borer infesting plants already damaged by more aggressive borers (Wagner *et al.*, 2008). The beetles excavate tunnels in dead or stressed plants in which they cultivate fungal gardens, their sole source of nutrition. After alighting on a host, an ambrosia beetle excavates a tunnel in which it releases spores of its fungal symbiont. This fungus penetrates the plant xylem, extracting nutrients and concentrating them on the gallery surface created by the beetle grub. The majority of ambrosia beetles colonize sapwood and/or heartwood of recently dead plants, but some beetles attack stressed plants (Van de Peppel *et al.*, 2018).

### Management

***Biocontrol:*** Pena *et al.* (2015) reported predators and parasitoids associated with scolytid beetles in Florida and Taiwan. Predatory insects include eight coleopteran species of Laemophloeidae, an unidentified species of Staphylinidae, *Microsicus* spp. (Zopheridae), and *Europs* sp. (Monotomidae). Among the parasitoids, hymenopterans belonging to Braconidae, Eulophidae, Pteromalidae, Encyrtidae, Eupelmidae and Bethylidae were reported.

***Pheromones:*** Kendra *et al.* (2015) compared trapping efficacy and field longevity of cubeb and manuka lures with phoebe lures against *X. glabratus* in south Florida. Their results indicated that cubeb bubble lures are the most effective tool currently available for detection of *X. glabratus*. The cubeb bubble lure has an extended low release of attractive sesquiterpenes, primarily α-copaene and α-cubebene, lasting 3 months.

***Chemical control:*** Carrilo *et al.* (2013) tested nine commercial insecticides, viz. bifenthrin, permethrin, fenpropathrin, z-cypermethrin + bifenthrin, 1-cyhalothrin + thiamethoxam, malathion, chlorpyrifos, carbaryl and methomyl against *X. glabratus* using a bolt-dip bioassay. Six pesticides were applied directly to the trunk and limbs of avocado plants. They found enpropathrin and malathion the most effective. Reding *et al.* (2013) reported that permethrin products were the most effective among a range of insecticides tested to protect nursery plants from ambrosia beetle, *X. germanus* Blandford. Pena *et al.* (2011) reported the contact insecticide z-cypermethrin + bifenthrin consistently managed Scolytinae.

In addition to the coleopterans discussed above, the weevil, *Mecopus hopei* Rosenschoeld (Coleoptera: Curculionidae) and longicorn beetle, *Olenecamptus bilobus* Fabricius (Coleoptera: Cerambycidae) (Fig. 14.19) have been reported to infest the bark and trunks of figs, respectively, in the Punjab, India (Singh, 2018). Fig beetle, *Cotinis texana* Casey (Fig. 14.20) is another beetle observed on fig plants in the USA (Coviello and Bentley, 2006b).

## Fig bark beetle, *Hypoborus ficus* (Erichson, 1836) (Coleoptera: Scolytidae)

***Distribution***: *Hypoborus ficus* is widely distributed in the Mediterranean region (Talhouk, 1969) including European countries Austria, Bosnia, Bulgaria, Croatia, Cyprus, France, Greece, Hungary, Italy, Spain, Switzerland, Poland (Witkowski and Mazur, 2019); African countries Algeria, Egypt, Canary Islands, Tunisia; and Asian countries Iran, Iraq, Israel, Syria, Turkey (Fontana, 1925; Knížek, 2011).

**Fig. 14.19.** Longicorn beetle, *Olenecamptus bilobus* Fabricius. (Figure used with permission from B. Sommung.)

***Host Range:*** *Hypoborus ficus* is a pest of fig trees (*Ficus carica* L.) mainly (Poher *et al.*, 2017) but also feeds on Turkish hazelnuts (*Corylus colurna* L.) (Tuncer *et al.*, 2017), mulberry (*Morus alba* L.), grapevine (*Vitis vinifera* L.) (Avidov and Harpaz, 1969), tree of heaven (*Ailanthus altissima* (Mill.) Swingle), drug snowball (*Styrax officinalis* L.) (Mifsud and Knížek, 2009) and river red gum (*Eucalyptus camaldulensis* Dehnh.) (Sarikaya, 2013).

***Damage***: *Hypoborus ficus* (Fig. 14.21) damages weak trees, causing withering and death of branches as well as whole trees in cases of severe infestation (Akşit *et al.*, 2007). Infestations occur throughout the year.

### *Management*

***Biological control:*** Mendel (1986) and Mifsud *et al.* (2012) recorded a host-specific ectoparasitoid braconid, *Ecphylus caudatus* Ruschka attacking *H. ficus*. A mite, *Pyemotes johnmoseri* has been reported as a parasitoid of *H. ficus* larvae in Ukraine (Khaustov, 1998) and in Turkey (Akşit

**Fig. 14.20.** Fig beetle, *Cotinis texana* Casey. (Figure used with permission from Beth Kinsey.)

**Fig. 14.21.** Fig bark beetle, *Hypoborus ficus* (Erichson). (Figure used with permission from D. Mifsud.)

*et al.*, 2007). A parasitic hymenopteran, *Rhaphitelus maculatus* Walker and a chalcid parasitoid *Astichus bachmaieri* Doğanlar were found to be parasitic on larvae of *H. ficus* in Malta (Mifsud *et al.*, 2012).

## 14.4 Hemiptera

Hemipteran pests of figs include whitefly, scales, mealybugs and psyllids that largely cause indirect damage to the crop by feeding on leaves and stems.

### Fig whitefly, *Singhiella simplex* (Singh) (Hemiptera: Aleurodidae)

The Asian species, commonly known as 'fig whitefly', *Singhiella simplex* (Singh) feeds exclusively on figs, causing damage to plants both directly by its feeding and indirectly due to the associated sooty molds growing on honeydew excreted by the larval stages (Evans, 2008; Mannion *et al.*, 2008).

***Distribution:*** *Singhiella simplex* is native to Asia and its distribution over different regions is given in Table 14.2.

**Host range:** The whitefly is oligophagous on Ficus and feeds on multiple species, e.g. *Ficus aurea, F. altissima, F. bengalensis, F. benjamina, F. binnendijkii, F. citrifolia, F. lyrata, F. maclellandii, F. microcarpa* (EPPO, 2018) and *F. racemosa* (Kondo and Evans, 2013).

***Life cycle:*** Adult whiteflies (1.4–1.6 mm) (Fig. 14.22) are deep yellow with conspicuous eyes, are very active, and fly readily when disturbed. They are covered with a powdery white wax. Eggs are usually laid in dense groups on the underside of leaves, hatching into crawlers. The first, second and third stage nymphs are oval, flat and almost translucent making them difficult to see (Malumphy, 2014). The crawler and three nymphal stages remain on and feed on the underside of the leaf through juvenility. Pupae are small (1.3 mm long), red-eyed, with tan to light green oval bodies (EPPO, 2018). Little information is available on the biology of *S. simplex*, however, studies carried out by Legaspi *et al.*, (2011) in Florida have shown that the total duration of the immature stages varied from 97.1 days at 15°C to 25.2 days at 30°C. Adults live 8 days at 15°C, 4.2 days at 25°C and 2.5 days at 30°C.

***Damage:*** Whitefly nymphs and adults feed on plant sap, impeding plant growth by destroying photosynthetically active leaf surface area. They also excrete honeydew which then is colonized by sooty mold on leaves and fruits. Infested leaves turn yellow before dropping. Plants with thinning leaves is the primary symptom of a whitefly infestation. The underside of infested leaves typically have small, silvery or white spots, the empty pupal skins (EPPO, 2018).

#### *Management*

***Biocontrol:*** Many natural enemies of fig whitefly have been reported: the hymenopteran parasitoids *Encarsia tricolor* Foerster (Hodges, 2007; Evans, 2008), *E. protransvena* Viggiani and *Amitus bennetti* Viggiani and Evans (Avery *et al.*, 2011); lacewing predators, *Chrysopa* spp. and ladybird predators, *Harmonia axyridis* (Pallas), *Olla-v-nigrum* (Mulsant), *Exochomus children i* Mulsant, *Chilocorus nigritis* (F.), and *Curinus coeruleus* (Mulsant) (Mannion, 2010).

In Florida, Avery *et al.*, (2011) isolated various enzootic pathogenic fungi viz. *Isaria fumosorosea* Wize, *Paecilomyces lilacinus* Thorn (Samson), and *Lecanicillium* sp., *Fusarium* sp., and *Aspergillus* sp. from fig whitefly. Pasco *et al.*,

**Table 14.2.** Distribution table of *Singhiella simplex.*

| Region | Countries | Reference(s) |
|---|---|---|
| Asia | Burma | Evans, 2008 |
| | China | Evans, 2008 |
| | India | Evans, 2008 |
| North America | Mexico | Mannion *et al.*, 2008; Mannion, 2010 |
| | USA | Mannion *et al.*, 2008; Mannion, 2010 |
| Central America and the Caribbean | Barbados | Kondo and Evans, 2013 |
| | Cayman Islands | Kondo and Evans, 2013 |
| | Dominican Republic | Kondo and Evans, 2013 |
| | Guadeloupe | Kondo and Evans, 2013 |
| | Jamaica | Kondo and Evans, 2013 |
| | Panama | Kondo and Evans, 2013 |
| | Puerto Rico | Kondo and Evans, 2013 |
| South America | Brazil | Legaspi *et al.,* 2011; Pasco *et al.*, 2011; Velasco *et al.*, 2011 |
| | Colombia | Kondo and Evans, 2013; Myartseva *et al.*, 2013; Yükselbaba *et al.*, 2017 |
| Paleartic | Israel | Kondo and Evans, 2013 |
| European Plant Protection Organization region | Cyprus | EPPO, 2018 |
| | France | EPPO, 2018 |
| | Turkey | EPPO, 2018 |

**Fig. 14.22.** Fig whitefly, *Singhiella simplex* (Singh). (Figure used with permission from Takumasa Kondo.)

(2011) reported only ~10% of whitefly mortality from parasitization by *Encarsia* species, while ~90% was due to enzootic entomopathogenic fungi, predation, etc. They also reported Hypocreales fungi, *L. fumosorosea*, *P. lilacinus* and *Aspergillus*, *Lecanicillium*, and *Fusarium* species isolated from dead nymphs on *F. benjamina*. Ko *et al.* (2015) described *Encarsia singhiellae* Polaszek and Shih, as a new species developing on *S. simplex* in China. Lahey and Polaszek (2017) identified a new whitefly parasitoid, *Baeoentedon balios* Wang, Huang and Polaszek on *S. simplex* in Florida.

***Chemical control:*** Foliar insecticides for whitefly control include Aria® (flonicamid), Avid® (abamectin), azadirachtin, Distance® (pyriproxyfen), Endeavor® (pymentrozine), endosulfan, Judo® (spiromesifen), Talus® (buprofenzin), and TriStar® (acetamiprid) (Mannion *et al.*, 2008).

### Fig psylla, *Macrohomotoma gladiata* (Kuwayama 1908) (Hemiptera: Psylloidea)

***Distribution***: *Macrohomotoma gladiata* is native to Asia (China, Taiwan, Sumatra and the Ryukyu Islands of southern Japan), but is also reported from Europe (the Balearic Islands, Spain, Canary Islands, Italy) and North America (Mifsud and Porcelli, 2012; Rung, 2016; Hodel, 2017).

***Host Range***: *Ficus* species as *F. microcarpa* (Bella and Rapisarda, 2014).

***Life cycle***: Pale yellow eggs are laid in clusters of 10–20 on new leaves. There are five nymphal stages which appear white and waxy due to woolly secretions (Fig. 14.23). Adults are 4.5–5.5 mm long with transparent forewings having dark spots (Hodel, 2017).

***Damage***: It attacks young, newly emerging leaves, sucking plant sap resulting in distortion of new growth, which appears compact and congested (Hodel, 2017). In severe infestations, leaf and shoot tips stop developing, and die causing thinning of the canopy.

**Fig. 14.23.** Fig psylla, *Macrohomotoma gladiata* Kuwayama on fig leaves. (Figure used with permission from Salvatore Bella.)

#### *Management*

***Cultural control:*** Since this pest damages new growth, management techniques to suppress growth such as reduced fertilizer application, less frequent irrigation and pruning will be effective in reducing pest populations (Hodel, 2017).

***Chemical control:*** Application of imidacloprid in soil/root zone or foliar application of dinotefuran can prove to be effective in its management (Hodel, 2017). Botanical oils have also reported to be effective in reducing nymphal populations (Laborda *et al.*, 2014).

***Biological control:*** *Anthocoris nemoralis* (Fabricius) and *Prionomitus mitratus* (Dalman) have been reported as predators of *M. gladiata* (Sánchez, 2012; Laborda *et al.*, 2014).

### Fig psylla, *Homotoma ficus* (Linnaeus, 1758) (Hemiptera: Psylloidea)

***Distribution***: It is widely distributed throughout the Mediterranean region (Italy, Turkey, Ukraine), West and Central Europe, the Caucasus, the Black sea region, the Middle east to Iran, and North America (Ouvrard, 2017).

***Host range***: *Homotoma ficus* is associated with *Ficus carica* (Batta and Burckhardt, 2018).

***Life cycle***: *Homotoma ficus* overwinters in egg stage from October to April on terminal buds and lenticels of fig shoots. Eggs hatch in April and larvae pass through five instars (Gencer *et al.*, 2003). Adults (Fig. 14.24) appear in May and oviposition takes place from late September–October (Tuncer, 2002).

***Damage:*** Larvae of *H. ficus* cause direct damage by sucking sap from the meristem tissues of trees and indirect damage via honeydew excretion resulting in growth of sooty mold on the leaves. This sooty mold growth results in reduced photosynthesis and respiration (Gencer *et al.*, 2007).

#### *Management*

***Biological control:*** *Orius* (*Heterorius*) *minutus* Linnaeus, *Deraeocoris* (*Deraeocoris*) *flavilinea* (A. Costa) and *Malacocoris chlorizans* Panzer have been reported as predators of *H. ficus* (Jerinić- Prodanović and Protić, 2013).

**Fig. 14.24.** Fig psylla, *Homotoma ficus* (Linnaeus). (Figure used with permission from D. Mifsud).

### Other hemipteran insects

Apart from whitefly, other hemipteran insects reported on fig are scales, mealybugs and hoppers. Ben-Dov (1993) reported soft brown scale, *Coccus hesperidum* Linnaeus, black soft scale *Parasaissetia nigra* (Nietner) and red scale, *Aonidiella aurantii* (Maskell). Moghaddam (2013) listed eight scale insects that attack figs in Iran. Aksoy (2015) reported wax scale, *Ceroplastes rusci* (Linnaeus) as an important pest of fig in Turkey. Nigra scale, *Parasaissetia nigra* (Nietner) was reported by Rassoul and Al-Mallo (2016). Singh and Kaur (2017) collected cottony cushion scale, *Icerya* sp. and Singh (2018) recorded black coffee scale, *Saissetia oleae* (Olivier) and wax scale, *Ceroplastes* sp. infesting twigs of fig plants in Punjab, India. In addition to the scales listed above, fig scale, *Lepidosaphes conchiformis* (Gmelin) (Coviello and Bentley, 2009b), *Aonidiella orientalis* (Newstead) and *C. rusci* (Moghaddam and Faghih, 2017) have been reported in different parts of the world. Among mealybugs, Moniruzzaman *et al.* (2017) reported *Phenacoccus solenopsis* Tinsley, *Pseudococcus longispinus* (Targioni Tozzetti) and *Planococcus ficus* (Signoret) (Hemiptera: Margarodidae) infesting figs in Iran. Singh (2018) reported giant mealybug, *Drosicha* sp. infesting leaves and twigs in Punjab, India.

## 14.5 Diptera

Dipteran pests of fig come from the families Tephritidae, Drosophilidae and Lonchaeidae and are largely fruit pests, feeding on figs pre- and post-harvest.

### Oriental fruit fly, *Bactrocera dorsalis* (Hendel) (Diptera: Tephritidae)

Oriental fruit fly, *Bactrocera dorsalis* (Hendel) (Diptera: Tephritidae) is a major pest of figs (Singh and Kaur, 2017; Weems and Hoppner, 2019).

***Distribution:*** *Bactrocera dorsalis* (Hendel) (Fig. 14.25) has been observed in Asia (Bangladesh, Bhutan, Cambodia, China, India, Indonesia, Japan, Laos, Malaysia, Myanmar, Nepal, Pakistan, Philippines, Sri Lanka, Thailand and Vietnam). Most sub-Saharan African countries have become infested with *B. dorsalis* (Vargas *et al.*, 2007; Goergen *et al.*, 2011). It has also been reported in the following Pacific islands: Mariana Islands, Tahiti, Hawaii and Papua New Guinea (Weems and Hoppner, 2019).

***Host range:*** The oriental fruit fly has been recorded infesting 478 species of fruit and vegetables, including: apricot, avocado, banana, citrus, coffee, fig, guava, loquat, mango, roseapple, papaya, passion fruit, peach, pear, persimmon, pineapple, surinam cherry and tomato. In Hawaii, maggots were found in more than 125 hosts. Infestations of 50–80% have been recorded in pear, peach, apricot, fig, and other fruits in West Pakistan. It is the principal pest of mangoes in the Philippines. It was a serious pest of citrus and other subtropical fruits in Japan, Okinawa, and the Japanese islands of Amami, Miyako and Bonin before it was eradicated (Clausen *et al.*, 1965; Koyama, 1989; Weems and Hoppner, 2019). Singh and Kaur (2017) reported the fruit fly, *B. dorsalis* in fig fruits from Punjab, India.

***Morphology:*** The adult (about 8 mm length) has hyaline, glassy, translucent wings and a prominent T-shaped pattern on the abdomen (Weems and Hoppner, 2019).

***Life cycle:*** Eggs of *B. dorsalis* are white, elongate and elliptical, laid below the skin of the host fruit. Eggs hatch within 1–3 days; the larvae feed for another 9–35 days. Pupation occurs in the soil under the host plant and adults emerge after 1–2 weeks. The adults remain active throughout the year (Christenson and Foote, 1960). The mature maggots emerge from the fruit, drop to the ground, and form a tan to dark brown puparium. Development from egg to adult under summer conditions requires about 16 days. The adult stage is the best stage to survive

**Fig. 14.25.** Fruit fly, *Bactrocera dorsalis* (Hendel). (Figure courtesy of S. Singh.)

low temperatures. A female can lay more than 3000 eggs during her lifetime. Female flies prefer ripe fruits for oviposition, but immature fruits are also attacked when the population level increases.

***Damage:*** Female flies puncture the ripening fruits, laying eggs inside. The oviposition punctures appear as black spots within a depression, often with a sugary liquid exudate. When the eggs hatch, the emerging maggots feed on the fleshy fruit pulp, damaging the fruit. The attacked fruits become dark brown, rot, and emit a foul smell. It causes 30–40% damage to fruits in an orchard (Singh and Kaur, 2017).

### *Management*

***Cultural control:*** Crop rotation, enhancing soil quality, selection of resistant varieties, efficient water management, monitoring/screening of fruit fly, orchard sanitation, application of mechanical barriers and postharvest treatments have been found effective. Prior to harvest, damaged and fallen fruits should be collected and disposed of (Liquido, 1993). Plowing/cultivation of orchards should be done during May–June to expose pupae to the sun's heat which kills them. If the infestation is high, use bait splash on the trunk at weekly intervals, by mixing 100 g of jaggery in 1 L of water (NIPHM, 2015).

***Bait traps:*** According to Bateman (1989), males can be continuously monitored, using hydrolyzed protein bait traps. Malathion, the most common insecticide for fruit fly control, is usually combined with protein hydrolysate to form a bait spray (Rossler, 1989).

***Biological control:*** Release of hymenopteran parasitoids such as *Opius longicaudatus* Ashmead, *O. vandenboschi* Fullaway, *O. oophilus* Fullaway and *Bracon* spp. (NIPHM, 2015) is an effective biological control.

In addition to infestation by *B. dorsalis*, Singh and Kaur (2017) reported infestation of *B. zonata* (Saunders) on fig fruits in Punjab, India. In Egypt, fruits are attacked by two additional harmful tephritid pests, the peach fruit fly, *B. zonata*, and the Mediterranean fruit fly, *Ceratitis capitata* (Wiedmann) causing considerable damage (Ali and Mohamed, 2018).

## Other fruit flies

Manrakhan and Addison (2007) found Mediterranean fruit fly, *C. capitata* (Fig. 14.26) and Natal fruit fly, *Ceratitis rosa* (Karsch) in baited traps in Western Cape, South Africa. They further reported that distribution of *C. rosa* in the Western Cape is confined to the coastal region, south of the Hottentots Holland and southwest of the Groot Winterhoek mountains. Mfeka *et al.* (2008) reared *C. rosa* from figs collected in home gardens in Stellenbosch, South Africa confirming Annecke and Moran's (1982) findings that *F. carica* is a host for both the species.

Gonçalves (2017) evaluated the mass trapping of *C. capitata* in Portugal using Tephri traps with the triple food attractant Biolure medfly R (ammonium acetate, putrescine and trimethylamine) and with the parapheromone Trimedlure, both combined with an insecticide tablet of dichlorvos on fig plants. Their results indicated that more fruit flies were captured in the traps baited with the Biolure medfly R than in the traps baited with Trimedlure but fewer fruits were damaged when mass trapping was done with Biolure medfly R versus the Trimedlure.

**Fig. 14.26.** Fruit fly, *Ceratitis capitata* (Wiedmann). (Figure used with permission from G. Delrio.)

**Fig. 14.27.** Vinegar fly, *Drosophila melanogaster* (Meigen). (Figure used with permission from L. Chhandama.)

## Vinegar flies, *Drosophila melanogaster* (Meigen) (Diptera: Drosophilidae)

*Drosophila melanogaster* (Meigen), the common fruit or vinegar fly, is a species widely used for biological research.

***Life cycle:*** Adults of *D. melanogaster* (Fig. 14.27) are small, tan to amber-colored flies with red eyes. Maggots are small, white, legless and 5 mm long. The larval stage has three instars. Under optimal growth conditions at 25°C (77°F), the lifespan of *D. melanogaster* is about 50 days from egg to death (Linford *et al.*, 2013). Under optimum conditions, the development time at 25°C (77°F) is 8.5 days at 18°C (64°F) 19 days, and at 12°C (54°F) over 50 days (Ashburner and Thompson, 1978).

***Damage:*** Female flies lay eggs on susceptible hosts including compromised fruit and fruit waste. Vinegar flies in fruit results in downgrading or rejection of the fruit for market. The infested fruit can also transmit spoilage organisms to the healthy fruit. Late ripening varieties are more susceptible to damage (Ashburner and Thompson, 1978; Coviello and Bentley, 2006c).

### *Management*

***Cultural control:*** Fermenting or decaying fruits should be destroyed by burying, burning or shredding to reduce the population. Vinegar flies prefer to breed in decaying fruits, so harvesting should be completed rapidly to reduce exposure of fruits to infestation (Coviello and Bentley, 2006c).

***Nano-emulsion based approach:*** Krittika *et al.*, (2019) demonstrated a nanoemulsion (A3E1T) containing amyl acetate, ammonia, ethanol and Tween was an efficient attractant for *D. melanogaster*. It had a higher relative response index to the flies than other emulsions as ALB, ALBP, A3E1S and A3E2T. They concluded that nano-emulsion-based compounds are a suitable alternative to the existing fruit fly control measures.

## Other flies

In addition to *D. melanogaster*, Martinez *et al.*, (2017) reported the African fig fly (AFF), *Zaprionus indianus* Gupta (Fig. 14.28) in the commercial greenhouse fig plantations of Mexico. The adults of *Z. indianus* emerged from ripe and overripe fig fruits, whereas the other two species emerged only from overripe fruits. This invasive pest has spread rapidly through much of the eastern USA. The AFF was first found in 2005 in Florida and now infests a broad range of edible crops.

## Black fig fly, *Silba adipata* McAlpine (Diptera: Lonchaeidae)

The black fig fly, *Silba adipata* McAlpine has become an increasingly important pest of figs (Giliomee and Barnes, 2015).

***Distribution:*** *Silba adipata* is a known Mediterranean and Middle East region pest in Spain,

**Fig. 14.28.** African fig fly, *Zaprionus indianus* Gupta. (Figure courtesy of S. Singh.)

**Fig. 14.29.** Damage by black fig fly, *Silba adipata* McAlpine. (Figure used with permission from L. Sabbo.)

Egypt and Israel (www.mallochsociety.org.uk/latest-updates), Iraq (Katsoyannos, 1983), Turkey (Talhouk, 2003) and Japan (Raz, 1998). It has also been collected in South Africa by Giliomee *et al.*, (2007). *Silba adipata* adults are easily distinguished from tephritid fruit flies by their iridescent black body and legs, and clear wings. *Silba adipata* was originally described by McAlpine (1956) in Italy, Cyprus and Greece. It is widespread in the Mediterranean, the Balkans, and the Near Eastern countries and southern-most parts of Africa. BFF was found for the first time in Slovenia in the area of Nova Gorica (Mojca *et al.*, 2017).

***Host range:*** Katsoyannos (1983) considers *S. adipata* to be monophagous on cultivated fig.

***Life cycle:*** Females deposit their eggs in the syconium ostiole. They prefer unripe figs but ripening fruits are also attacked. The maggots feed in groups, eat the syconium, and at maturity, drop to the soil to pupate. There are 4–6 generations in a year. They are attracted to the sap from ripe figs and to exudates from syconium of fruit (Robert H. Smith Faculty of Agriculture, Food and Environment, 2019a).

***Damage:*** Females usually lay eggs in unripe fig fruits. The fly is attracted to latex (fig milk) and therefore prefers unripe figs (Katsoyannos, 1983). The larvae bore damaging galleries inside the fruits (Fig. 14.29), precipitating premature fruit drop in July and August (Mojca *et al.*, 2017). Injuries and damage on mature fruits are visible from the beginning of September. At high population levels, *S. adipata* can cause severe loss of immature fruit (Katsoyannos and Guerin, 1984). Later during the season, as the fruits ripen, infestation levels seem to drop substantially (Wohlfarter *et al.*, 2008). This could be due to the reason that the fly is attracted to unripe fruits and as the fruits ripen, fly population decreases.

#### *Management*

***Monitoring with traps:*** *S. adipata* adults are attracted to traps with hexanol and ammonium sulphate (Katsoyannos and Guerin, 1984; Raz, 1998; Katsoyannos, 2004).

***Biological control:*** The parasitoid, *Pachycrypodens vindemmiae* Rondani (Pteromalidae) attacks this fly in Greece (Katsoyannos, 1983). Besides *S. adipata*, a new species of *Silba* Macquart 1851, *Silba ischnopoda* sp. nov. has been observed in Thailand and Cambodia (Macgowan and Compton, 2018).

## 14.6 Thysanoptera

*Frankliniella occidentalis* (Pergande) is an important species of thrips belonging to family Thripidae.

### *Frankliniella occidentalis* (Pergande) (Thysanoptera: Thripidae)

Western flower thrips, *Frankliniella occidentalis* (Pergande) are important fig pests.

***Distribution:*** This species is distributed globally in temperate regions (Robert H. Smith

Faculty of Agriculture, Food and Environment, 2019b).

***Host range:*** This species is polyphagous, infesting over 500 different plant species (Robert H. Smith Faculty of Agriculture, Food and Environment, 2019b).

***Damage:*** Both nymphs and adults cause damage by sucking cellsap from the leaves as well as fruits. Attacked leaves become withered and ultimately curl. Severe attacks by this pest were first observed in April on leaves and in December on fruits in the Punjab, India (Singh and Kaur, 2017). Thrips cause damage to the leaves and fruits of the tree (Fig. 14.30) and approximately 30–40% of trees are affected in the orchard (NIPHM, 2015).

***Life cycle:*** Eggs are creamy white, microscopic and kidney-shaped, and hatch in 2–7 days, depending upon temperature. Larvae and adults usually gather near the mid-vein or borders of the leaf. Pupae can be seen on leaf axils, leaves, curled leaves, under the flower and fruit calyces and in leaf litter (NIPHM, 2015).

#### *Management*

***Cultural control:*** Plowing and harrowing, and solarization can kill pupae in the soil. Inter-cropping with tomato plants can reduce the thrip populations in fig (NIPHM, 2015) because thrips are attracted towards the yellow flowers of tomato (Tyler-Julian *et al.*, 2018).

***Sticky traps:*** Monitor the plant regularly. Monitoring with blue sticky traps or water traps in the nursery or orchard will establish if adult populations are present (NIPHM, 2015).

***Barrier crops:*** Horikawa *et al.* (2016) and Ito *et al.* (2017) examined the efficacy of preventing thrips from invading a fig orchard and preserving natural enemies by using sorghum as an orchard surrounding barrier. The numbers of thrips in the orchard were half that of the control, an identical orchard surrounded by windbreak nets with 4 mm mesh. They also confirmed the fig orchard-surrounded sorghum more efficiently generated indigenous natural enemies: spiders, flower bugs, *Orius* and *Neididae*. Michailides and Morgan (1998) reported *Fusarium* sp. infection on thrips-infested figs.

***Biological control:*** Minute pirate bugs, *Orius* spp. (Hemiptera: Anthocoridae) and entomopathogenic nematodes, *Thripinema* spp. (Tylenchida: Allantonematidae) naturally reduce thrip populations (NIPHM, 2015).

## 14.7 Hymenoptera

In the order Hymenoptera, only the fig wasp, *Blastophaga* spp. is important.

*Blastophaga psenes* (Linnaeus) (Agaonidae), the Capri fig wasp, is a symbiotic pollinator of figs (Fig. 14.31). It can occasionally be found trapped inside figs. These wasps cause damage to figs by vectoring endosepsis (*Fusarium* sp.) (Caldis, 1927; Davey and Smith, 1933; Michailides and Morgan, 1998). Larvae of the wasp will feed on the flesh of figs in addition to pollinating it. The peak activity of fig wasps is last week of April in Punjab, India (Singh and Kaur, 2017).

## 14.8 Acari

Two-spotted spider mite, *Tetranychus urticae* Koch, *Eotetranychus hirsti* Pritchard and Baker, fig mite, *Aceria fici* (Ewing) and fig blister mite, *Aceria ficus* (Cotte) are important mite pests of figs.

### Two-spotted spider mite, *Tetranychus urticae* Koch (Acari: Tetranychidae)

***Distribution:*** The two-spotted spider mite, *Tetranychus urticae* was originally described in Europe but is now widely distributed in both temperate and subtropical regions. It is found throughout the USA in greenhouses where it survives the winters (Fig. 14.32). Tuttle and Baker (1968) reported this species on deciduous fruit plants in the northern regions of Europe and the USA.

***Host range:*** More than 870 recorded species of host plants were listed by Bolland *et al.* (1998); figs are among them (CABI, 2019b).

***Life cycle:*** Mite eggs are attached to fine silk webbing on the underside of the leaves and hatch in about 3 days. The life cycle is composed of egg, immature, two nymphal stages and adult. The length of time from egg to adult varies greatly depending on temperature, ranging between 5 and 20 days. The female lives for 2–4 weeks and is capable of laying several hundred

**Fig. 14.30.** Damage by western flower thrips, *Frankliniella occidentalis* (Pergande); A: Healthy fruits. B: Shape of fruit damaged by thrips. C and D: Browning of fruit caused by thrips. E: adults of thrip in a damaged fig fruit. F: occurrence of fungus on damaged fruits of figs cultivated in a greenhouse, in Hwaseong, in 2013. (Figure used with permission from M. Rae Cho.)

**Fig. 14.31.** Fig wasp, *Blastophaga* sp. (Figure courtesy of S. Singh.)

**Fig. 14.32.** Two-spotted spider mite, (*Tetranychus urticae* Koch) damage on fig. (Figure courtesy of S. Singh).

eggs during her lifetime (Fasulo and Denmark, 2009). This results in many overlapping generations per year.

***Damage:*** Van Zyl (1997) mentioned the occasional incidence of plant defoliation due to serious infestation of red spider mite. Webbing and silvering of the ventral leaf surface is a common symptom of mite infestation. Spots on the fruit is also a typical mite damage symptom (Fig. 14.33). Mite damage was typically on 15–20% of total fruits. In fig plantations in Punjab, India, high mite populations are typically encountered during March–June and September–October (Singh and Kaur, 2017).

### *Management*

***Cultural control:*** Vigorously growing plants are more tolerant of mite attacks than plants under stress. Plants should be maintained properly with optimum irrigation and fertilization (CABI, 2019b).

***Biocontrol:*** According to Pringle and Heunis (2006), polyphagous mites such as *T. urticae* and *Panonychus ulmi* (Koch) have been successfully controlled in deciduous fruit orchards by the predatory mite *Neoseiulus californicus* (McGregor). The predatory mite, *Metaseiulus* spp., and six spotted thrips, *Scolothrips sexmaculatus* (Pergande), also feed heavily on mites and provide control in the orchard. The western flower thrips, *F. occidentalis*, also feeds on mite eggs (CABI, 2019b).

***Chemical control:*** Diazinon is labeled for figs under SLN (special local need) registration for fruit fly quarantine treatment only (Farrar, 1999).

## Other mites

*Eotetranychus hirsti* Pritchard and Baker (Acari: Tetranychidae), an important fig pest, is widely distributed in Iranian fig orchards. The predatory mite, *Phytoseius plumifer* Canestrini and Fanzago is a phytoseiid mite that feeds and reproduces on *E. hirsti*. Shakarami and Bazgir (2017) reported the developmental time decreased when temperatures ranged from 20°C to 35°C, varying from $17.13 \pm 0.23$ to $6.55 \pm 0.19$ days for females. They also concluded that *P. plumifer* is an effective predator of the fig spider mite and develops effectively at a broad range of temperatures. Singh (2018) has reported severe damage of this mite on fig fruits in Punjab, India (Fig. 14.33).

Fig mite, *Aceria fici* (Ewing) (Acari: Eriophyidae) is a widely distributed, microscopic mite with two pairs of legs and a pale yellow body. It infests buds, scales and young leaves. Feeding causes a faint russeting of the leaves, generally in the canopy interior producing leaf drop and twig stunting. This mite also transmits fig mosaic virus (FMV). The virus is not present in the egg stage of the mite, but once acquired through feeding is retained through molts (Coviello and Bentley, 2009a).

Fig blister mite, *Aceria ficus* (Cotte) (Acari: Eriophyidae) is colorless to white. These mites are usually found in green fruit but can attack ripe fruit. They enter and feed near the ostiole, causing rust-colored, dry patches on the florets. The damage is apparent when the fruit is cut open. Affected fruit have reduced eating quality. Blister mites are known to spread FMV. Annecke and Moran (1982) also reported *A. ficus* occurs on fig buds and young leaves. Feeding causes various symptoms, including leaf russeting and chlorosis, blasted buds, and stunted, distorted growth (Jeppson *et al.*, 1975). According to Flock and Wallace (1955) and Meyer (1981), leaf russeting and FMV symptoms were observed; both can be vectored by *A. ficus*.

Yorgancı and Acıkgöz (2017) reported that fig plants with typical FMV leaf symptoms are often infected by multiple viruses. Fig mosaic diseases are transmitted by grafting and vectors. While it is known that FMV is transmitted by the mite, *A. ficus*, the vectors for other fig viruses are unknown. *A. ficus* appears to be able to transmit FMV and FMaV-1, but didn't transmit FLMaV-1 from an infected 'Sarılop' fig to fig seedlings. In addition to these mites, Issa *et al.*, (1974) reported an infestation of *Tetranychus arabicus* Attiah on fig seedlings was controlled by the predatory mite, *Phytoseius plumifer* Canestrini and Fanzago. Incidence of Pacific spider mite, *Tetranychus pacificus* McGregor has also been observed in some fig growing areas (CABI, 2019b).

## 14.9 Vertebrate Pests

Bearing fig trees are attractive to a variety of vertebrate species including: pigeons, hornbills, macaws, parrots, toucans, flying foxes, bats and monkeys. As animals move through the canopy loaded with fruit, they create a fruit rain. Terrestrial species like duikers, peccaries and rodents feed on such dropped fruits. Since fig species fruit asynchronously around the year, figs are critical for survival of many species of animals. Therefore, figs are a 'keystone species' for frugivorous vertebrates. Brien and Hardy (2002) demonstrated fig fruits can constitute over 50% of African chimpanzees' diet, and up to 70% for some primate species in Peru. Some Neotropical bat species rely entirely on figs. Ground squirrels and gophers are also important pests in fig orchards. A bird-pecked fig is unsuitable for consumer packages and must be used for manufacturing products at a financial loss to growers (NIPHM, 2015).

### Bird damage

Birds eat developing figs, especially as they start to ripen (Fig. 14.34). Fruit damage ranges from small claw marks and pecks to consuming the entire fruit. Birds can be serious pest in some areas. Once the birds have located fig orchards, they are likely to continue returning to this source of food, making them a persistent problem (Brien and Hardy, 2002; BirdGard, 2019).

### Garden Warbler, *Sylvia borin* (Boddaert)

Thomas (1979) reported the migrating Garden Warbler, *Sylvia borin* (Boddaert) (Passeriformes: Sylviidae) as an important bird pest in southern

**Fig. 14.33.** Yellow mite, *Eotetranychus hirsti* Pritchard and Baker. (Figure courtesy of S. Singh.)

**Fig. 14.34.** Bird damage on fig fruit. (Figure used with permission from A.M. Zawari.)

Portugal. These birds are frequently observed eating fig fruits, often 5–6 birds will feed simultaneously on one small plant. This is their pre-migratory fattening prior to their trans-Saharan migration. Starlings, crows, finches, magpies and jays are notorious for causing damage to figs. From pecking at the fruits to knocking them off the plants, birds cause substantial crop losses for fig growers (Bird Gard, 2019).

***Management:*** Bird netting is the most commonly used management tactic. Scare devices vary in effectiveness. Bird Gard® units are designed to effectively and permanently repel birds from fig orchards. A built in microprocessor in these electronic units continually randomizes the order in which the sounds play and the relative pitch of each bird sound to give the impression many birds are in distress in the Bird Gard® protected area, warning off other birds (Bird Gard, 2019).

### Rodents

Among rodents, ground and tree squirrels (Rodentia: Sciuridae) damage fruit trees. Ground squirrels dig into soil to build nests and burrows among the roots of fruit-bearing bushes and trees. Figs are particularly vulnerable to squirrel predation. Mature fig tree wood is weak, quite fragile and easily damaged by squirrels plundering the figs during spring and fall (Amistaadt, 2018).

## References


Aizpurua, C.G. (1997) Nota sobre *Choreutis nemorana* (Hübner, 1799) (Lepidoptera: Choreutidae), parasite de las higueras (*Ficus carica*). *Boletin de Sanidad Vegetal Plagas* 23, 237–240.

Akşit, T., Çakmak, I. and Moser, J. (2007) Attack by *Pyemotes johnmoseri* (Acari: Pyemotidae) on *Hypoborus ficus* (Coleoptera: Scolytidae) in fig trees in Turkey. *Experimental and Applied Acarology* 41, 251–254.

Aksoy, U. (2015) *Dried Figs.* United Nations Economic Mission for Europe. UNECE Workshop, July 3 2015, Turkey.

Alford, D.V. (2007) *Pests of fruit crops. A Color Handbook.* Elsevier Academic Press.

Ali, A. and Mohamed, S.H. (2018) Population fluctuations and infestation rates of the peach and Mediterranean fruit flies on fig fruits in relation to the prevailing weather factors in Fayoum Governorate. *Egyptian Academic Journal of Biological Sciences Entomology* 11, 31–40.

Ali, W.K., Abobaker, S.N., Rasheed, R.T. and Saleem, D.A.K. (2012) Notes on the biology of fig-tree moth *Ocnerogyia amanda* Staud. (Lepidoptera: Lymantriidae) from Erbil, Kurdistan region, Iraq. *Pakistan Entomologist* 34(2), 117–120.

Amistaadt, A.J. (2018) How to keep squirrels out of fig bushes. https://homeguides.sfgate.com/keep-squirrels-out-fig-bushes-58278.html (accessed 6 September 2021).

Ampt, E.A., Bush, D.S., Siegel, J.P. and Berenbaum, M.R. (2015) Larval preference and performance of *Amyelois transitella* (Navel Orangeworm, Lepidoptera: Pyralidae) in relation to the fungus *Aspergillus flavus*. *Environmental Entomology* 45, doi: 10.1093/ee/nvv160

Annecke, D.P. and Moran, V.C. (1982) *Insects and Mites of Cultivated Plants in South Africa*. Butterworths, Durban, South Africa.

AQIS (1999) *Draft Import Risk Analysis for the Importation of Fresh Table Grapes [Vitis vinifera L.] from California (USA)*. Australian Quarantine and Inspection Service, Canberra.

Ashburner, M. and Thompson, J.N. (1978). The laboratory culture of *Drosophila*. In: Ashburner, M. and Wright, T.R.F. (eds) *The Genetics and Biology of Drosophila* 2A. Academic Press. pp. 1–81.

Atkinson, T.H., Carrilo, D., Duncan, R.E. and Pena, J.E. (2013). Occurrence of *Xyleborus bispinatus* (Coleoptera: Curculionidae: Scolytinae) Eichhoff in southern Florida. *Zootaxa* 3669, 96–100.

Avery, P.B., Mannion, C.M., Powell, C.A., McKenzie, C.L. and Osborne, L.S. (2011) Natural enemies managing the invasion of the fig whitefly, *Singhiella simplex* (Hemiptera: Aleyrodidae), infesting a *Ficus benjamina* hedge. *Florida Entomologist* 94, 696–698. doi: 10.1653/024.094.0338

Avidov, Z. and Harpaz, I. (1969) *Plant Pests of Israel*. Israel Universities Press, Jerusalem.

Bajwa, G.A. and Gul, H. (2000) Some observations on insect species of *Paulownia* species at Pakistan Forest Institute Campus, Peshawar. *Pakistan Journal of Forestry* 50, 71–80.

Bartelt, R.J. (1997) Aggregation pheromones of *Carpophilus* spp. (Coleoptera: Nitidulidae): Review of chemistry and biology. *Recent Research and Development in Entomology* 1, 115–129.

Bateman, M.A. (1989) *Eradication Strategy for an Invading Population of Queensland Fruit Fly in Suburban Perth*. Consultant's Report to the WA Dept of Agriculture, Perth, July 1989.

Batta, Y. and Burckhardt, D. (2018) Taxonomy and biology of *Pauropsylla buxtoni* comb. nov. (Hemiptera: Psylloidea) on *Ficus carica* (Moraceae). *Journal of the Entomological Research Society* 20(3), 39–52.

Bella, S. and Rapisarda, C. (2014) New findings in Italy of the recently introduced alien psyllid *Macrohomotoma gladiata* and additional distributional records of *Acizzia jamatonica* and *Cacopsylla fulguralis* (Hemiptera Psylloidea). *Redia* 97, 151–155.

Ben-Dov, Y. (1993) *A Systematic Catalogue of the Soft Scale Insects (Homoptera: Coccoidea: Coccidae) of the World with Data on Geographical Distribution, Host Plants, Biology and Economic Importance*. Sandhill Crane Press, Gainesville, Florida.

Ben-Yehuda, S., Dorchin, Y. and Mendel, Z. (2000) Outbreaks of the fig borer *Batocera rufomaculata* and other cerambycids in fruit plantations in Israel. *Alon Hanotea* 54, 23–29.

BioLib (2019) Fig-tree borer longhorn beetle *Phryneta spinator* (Fabricius, 1792). www.biolib.cz/en/taxon/id253462/ (accessed 25 September 2019).

Bird Gard (2019) Protecting fig from birds. www.birdgard.com/protect-figs-from-birds/ (accessed 2 June 2020).

Blumberg, D. (2019) Carpophilus spp. Plant Pests of the Middle East, http://www.agri.huji.ac.il/mepests/pest/Carpophilus/ (accessed 6 September 2021).

Blumberg, D., Doron, S. and Bitton, S. (1985) Effect of triflumuron on two species of nitidulid beetles, *Carpophilus hemipterus* and *Urophorus humeralis*. *Phytoparasitica* 13, 9–19.

Bolland, H.R., Gutierrez, J. and Flechtmann, C.H.W. (1998) *World catalogue of the Spider Mites Family Acari: Tetranychidae with references to Taxonomy, Synonymy, Host Pests and Distribution*. Academic Publishers, Brill, Leiden, the Netherlands.

Bouček, Z. and Askew, R.R. (1968) Hym. Chalcidoidea. Palaearctic Eulophidae (excl. Tetrastichinae). In: Delucchi, V. and Remaudière, G. (ed.) *Index Entomophagous Insects*. Vol. 3. Le François, Paris.

Bousquet, Y., Thomas, D.B., Bouchard, P., Smith, A.D., Aalbu, R.L., Johnston, M.A. and Steiner Jr., W.E. (2018) Catalogue of Tenebrionidae (Coleoptera) of North America. *Zookeys* 728, 1–455.

Brien, J. and Hardy, S. (2002) Fig growing in New South Wales, Australia. *Agfact* H3.1.19, 1st edn www.dpi.nsw.gov.au/-data/assets/pdf_file/0017/119501/fig-growing-nsw.pdf (accessed 6 September 2021).

Burks, C.S. and Brandl, D.G. (2004) Seasonal abundance of the navel orangeworm, *Amyelois transitella*, in figs and the effect of peripheral aerosol dispensers on sexual communication. *Journal of Insect Science* 4, 40.

Buxton, P.A. (1920) A Liparid moth, *Ocnerogyia amanda* Staud., destructive to figs in Mesopotamia. *Bulletin of Entomological Research* 11, 181–186.

CABI (2019a) Invasive Species Compendium. CAB International, Wallingford, UK. www.CABI.org/isc (accessed 2 June 2020).

CABI (2019b) *Pacific Spider Mite, Tetranycus pacificus.* www.plantwise.org/knowledgebank/datasheet/53361 (accessed 5 August 2021).

Caldis, P.D. (1927) Etiology and transmission of endosepsis (internal rot) of the fruit of the fig. *Hilgardia* 2, 287–346.

Carabez, J.R.S., Ploetz, R.C., Konkal, J. and Gazis, R.O. (2018) Partnerships between ambrosia beetles and fungi: Varying levels of promiscuity among vectors of the laurel wilt pathogen, *Raffaelea lauricola*. *International Congress of Plant Pathology (ICPP) 2018: Plant Health in A Global Economy, Boston, July 29-August 3, 2018.*

Carrilo, D., Crane, J.H. and Pena, J.E. (2013) Potential of contact insecticides to control *Xyleborus glabratus* (Coleoptera: Curculionidae), a vector of Laurel wilt disease in Avocados. *Journal of Economic Entomology* 106, 2286–95. doi: 10.1603/EC13205

Cheanban, S., Bumroongsook, S. and Tigvattananont, S. (2017a) Observation on *Trilocha varians* (Lepidoptera: Bombycidae). *International Journal of Agricultural Technology* 13, 2189–2195.

Cheanban, S., Bumroongsook, S. and Tigvattananont, S. (2017b) *Perina nuda* F. (Lepidoptera: Lymantriidae): an important leaf eating caterpillar of fig trees. *International Journal of Agricultural Technology* 13, 485–492.

Chitgar, M.G., Ghadamyari, M., Sharifi, M. and Hassan Sajedi, R. (2014) Partial characterization of digestive carbohydrases in the midgut of fig tree skeletonizer moth, *Choreutis nemorana* Hübner (Lepidoptera: Choreutidae). *Trakia Journal of Sciences* 1, 27–37.

Christenson, L.D. and Foote, R.H. (1960) Biology of fruit flies. *Annual Review of Entomology* 5, 171–192.

Christian, E., Deutsch, H. and Huemer, P. (2008) The fig moth *Choreutis nemorana* (Hübner, 1799) gains ground in Austria (Lepidoptera: Choreutidae). *Beiträge zur Entomofaunistik 9- Kurzmitteilungen* 178–180.

Clausen, C.P., Clancy, D.W. and Chock, Q.C. (1965) Biological control of the oriental fruit fly (*Dacus dorsalis* Hendel) and other fruit flies in Hawaii. *United States Department of Agriculture, Technical Bulletin No. 1322.*

Connell, J.H. (2002) Leading edge of plant protection for almond. *Hort Technology* 12, 619–622.

Coviello, R.L. and Bentley, W.J. (2006a) Darkling ground beetle in Fig. *UC IPM Pest Management Guidelines*. University of California Agriculture and Natural Resources, Oakland, California.

Coviello, R.L. and Bentley, W.J. (2006b) Fig beetle, *Cotinis texana* in Fig. *UC IPM Pest Management Guidelines.* University of California Agriculture and Natural Resources, Oakland, California.

Coviello, R.L. and Bentley, W.J. (2006c) Vinegar flies in Fig. *UC IPM Pest Management Guidelines.* University of California Agriculture and Natural Resources, Oakland, California.

Coviello, R.L. and Bentley, W.J. (2009a) Fig mite, *Aceria fici* Ewing. *UC IPM Pest Management Guidelines*. University of California Agriculture and Natural Resources, Oakland, California.

Coviello, R.L. and Bentley, W.J. (2009b) Fig scale, *Lepidosaphes conchiformis* (Gmelin). *UC IPM Pest Management Guidelines.* University of California Agriculture and Natural Resources, Oakland, California.

Coviello, R.L. and Bentley, W.J. (2019) Driedfruit beetles in Fig. *UC IPM Pest Management Guidelines.* University of California Agriculture and Natural Resources, Oakland, California.

Coviello, R.L., Bentley, W.J., Michailides, T.J., Ferguson, L. and Westerdahl, B.B. (2003) *UC IPM Pest Management Guidelines: Fig. Publication 3447*. University of California Agriculture and Natural Resources, Oakland, California.

Cruz, L.F., Rocio, S.A., Duran, L.G., Menocal, O., Garcia-Avila, C.D.J. and Carrillo, D. (2018) Developmental biology of *Xyleborus bispinatus* (Coleoptera: Curculionidae) reared on an artificial medium and fungal cultivation of symbiotic fungi in the beetle's galleries. *Fungal Ecology* 35, 116–126.

Cutajar, S. and Mifsud, D. (2017) *Good Agricultural Practices (GAP) for Fig Tree Cultivation*. Plant Protection Directorate, Malta University Consulting Ltd, 31 pp.

De Prins, W., Baugnée, J.Y., Georis, A., Spronck, R., Spronck, R. (2009) *Choreutis nemorana* (Lepidoptera: Choreutidae) well established in Belgium. *Phegea* 42(2), 29–32.

De Prins, W. and De Prins, J. (2014) *Choreutis nemorana* (Hübner, 1799) (Lep.: Choreutidae), a new adventive species to the British Islands. *The Entomologist's Record and Journal of Variation* 126, 157–163.

Davey, A.E. and Smith, R.E. (1933) The epidemiology of fig spoilage. *Hilgardia* 7, 523–551.

Diakonoff, A. (1986) *Glyphipterigidae auctorurn* sensu lato (Glyphipterygidae sensu Meyrick 1913). *Microlepidoptera Palaearctica* 7, 209–210.

Donovan, P. (2014) Dealing with fig pests and diseases. *Farmer's Weekly.* www.farmersweekly.co.za/farm-basics/how-to-crop/dealing-with-fig-pests-and-diseases/ (accessed 5 August 2021).

Embacher, G., Kurz, M. and Zeller-Lukashort, C. (2004) Beitrag zur Microlepidopterenfauna Salzburgs (Lepidoptera). *Beiträge zur Entomofaunistik* 5, 57–66.

EPPO (2018) *Mini Data Sheet on Singhiella simplex (Hemiptera: Aleyrodidae). Ficus Whitefly*. EPPO, Paris.

Evans G.A. (2008) *The Whiteflies (Hemiptera: Aleyrodidae) of the World and Their Host Plants and Natural Enemies*. USDA/Animal Plant Health Inspection Service (APHIS). *Last Revised: September* 23, 2008.

Faccoli, M., Campo, G., Perrotta, G. and Rassati, D. (2016) Two newly introduced tropical bark and ambrosia beetles (Coleoptera: Curculionidae, Scolytinae) damaging figs (*Ficus carica*) in southern Italy. *Zootaxa* 4138, 189–194.

Farrar, K. (1999) California Insecticide Impact Assessment Program. University of California, Davis. https://ucanr.edu/datastoreFiles/391-299.pdf (accessed 5 August 2021).

Fasulo, T.R. and Denmark, H.A. (2009) *Two-Spotted Spider Mite, Featured Creatures*. University of Florida Institute of Food and Agricultural Sciences, Gainesville, Florida.

Fauster, R. (2016) *Choreutis nemorana* (Hübner, 1799) – nun auch in der Steiermark angekommen (Lepidoptera: Choreutidae). *Joannea Zoologie* 15, 85–88.

Flock, R.A. and Wallace, J.M. (1955) Transmission of fig mosaic by the eriophyid mite. *Aceria ficus. Phytopathology* 45, 52–54.

Fontana, P. (1925) Contribuzione alla fauna coleotterologica ticinese. *Bolletino della Società Ticinese di Scienze Naturali* 20, 23–34.

Gaaliche, B., Abdelaali, N.B., Mouttet, R., Halima Kamel, B.M. and Hajaoul, M.R. (2018) First report of *Hypocryphalus scabricollis* (Eichhoff, 1878), in Tunisia, an emerging pest on fig trees (*Ficus carica L.*). *Bulletin OEPP* 48, 164–166.

Gaedike, R. (2008) Nachträge und Korrekturen zu: Verzeichnis der Schmetterlinge Deutschlands (Microlepidoptera). *Entomologische Nachrichten und Berichte* 52(1), 9–49.

Geertsema, H. (1985) Notes on the occurrence of three non-endemic Lepidoptera, *Mylothris chloris agathina*, *Zophopetes dysmephila* and *Ocinara ficicola*, in the Western Cape. *Phytophylactica* 17, 153–155.

Gencer, N.S., Coskuncu, K.S. and Kumral, N.A. (2003) *Determination of Fig Pests Occurring in Bursa and Population Fluctuations of the Important Species*. Uludag University Scientific Project Department, Project No.: 2001/23.

Gencer, N.S., Coskuncu, K.S. and Kumral, N.A. (2007) The colonization preference and population trends of larval fig psylla, *Homotoma ficus* L. (Hemiptera: Homotomidae). *Journal of Pest Science* 80, 1–8.

Giliomee, J.H. and Barnes, B.N. (2015) Fig. In: Prinsloo, G.L. and Uys, V.M. (eds) *Insects of Cultivated Plants and Natural Pastures in Southern Africa*. Entomological Society of Southern Africa, pp. 388–393.

Giliomee, J.H., Venter, E. and Wohlfarter, M. (2007) Mediterranean black fig fly, *Silba adipata* McAlpine (Diptera: Lonchaeidae), recorded from South Africa. *African Entomology* 15, 383–384.

Goergen, G., Vayssières, J.F., Gnanvossou, D. and Tindo, M. (2011) *Bactrocera invadens* (Diptera: Tephritidae), a new invasive fruit fly pest for the Afrotropical Region: host plant range and distribution in West and Central Africa. *Environmental Entomology* 40, 844–854.

Gonçalves, M.A. (2017) Mass trapping for the control of the Mediterranean fruit fly *Ceratitis capitata* (*Diptera: Tephritidae*) on fig trees in Southern Portugal (Algarve). *Acta Horticulturae* 1173, 269–272. doi: 10.17660/ActaHortic.2017.1173.46

Government of Western Australia (2019) Dried fruit beetle (*Carpophilus*), a pest of stone fruit. https://www.agric.wa.gov.au/citrus/dried-fruit-beetle-carpophilus-pest-stone-fruit (accessed 6 September 2021).

Gunn, D. (1919) *The Fig and Willow Borer* (*Phryneta spinator*). Union of South Africa, Department of Agriculture. 6, 1–22.

Higbee, B.S., Burks, C.S. and Larsen, T.E. (2014) Demonstration and characterization of a persistent pheromone lure for the Navel Orangeworm, *Amyelois transitella* (Lepidoptera: Pyralidae). *Insects* 5, 596–608.

Hodel, D.R. (2017) *New Pests of Landscape Ficus in California*. CAPCA Adviser, University of California.

Hodges, G. (2007) *The Fig Whitefly Singhiella simplex (Singh) (Hemiptera: Aleyrodidae): A New Exotic Whitefly Found on Ficus Species In South Florida*. Division of Plant Industry, Florida Department of Agriculture and Consumer Services, www.freshfromflorida.com/pi/pestalerts/singhiella-simplex.html (accessed on 12 October 2014).

Horikawa, H., Harata, D. and Kato, S. (2016) The efficacy of preventing thrips from invading a fig field and generating indigenous natural enemies through the use of the sorghum and petunia. *Annual Report of the Kansai Plant Protection Society* 58, 139–141.

Husain, M.A. and Khan, M.A.W. (1940) Bionomics and control of the fig-tree borer, *Batocera rufomaculata* De Geer (Coleoptera: Lamiidae). *Indian Journal of Agricultural Sciences* 10, 945–959.

Issa, G.I., Elbanhawy, E.M. and Rasmy, A.H. (1974) Successive release of the predatory mite *Phytoseius plumifer* for combating *Tetranychus arabicus* (Acarina) on fig seedlings. *Zeitschriftfür angewandte Entomologie* 76, 442–444.

Ito, R., Horikawa, H. and Sakai, S. (2017) The efficacy of using sorghum in preventing injury to fig fruit caused by thrips. *Annual Report of the Kansai Plant Protection Society* 59, 81–83.

Jeppson, L.R., Keifer, H.H. and Baker, E.W. (1975) *Mites Injurious to Economic Plants*. University of California Press, Berkeley, California.

Jerinić-Prodanović, D. and Protić, L. (2013) True bugs (Hemiptera, Heteroptera) as psyllid predators (Hemiptera, Psylloidea). In: Popov, A., Grozeva, S., Simov, N., Tasheva, E. (eds) *Advances in Hemipterology*. *ZooKeys* 319, 169–189. doi: 10.3897/zookeys.319.4316

Kallies, A. (2004) The Brachodidae of the Oriental region and adjacent territories (Lepidoptera: Sesioidea). *Tijdschrift voor Entomologie* 147, 1–20.

Karim, H.R., Imani, S., Basirian, M. and Kumar, R. (2010) Report of *Phycodes radiata* (Lep: Brachodidae) from Iran. *Plant Protection Journal* 2, 329–331.

Katsoyannos, B.I. (1983) Field observations on the biology and behavior of the black fig fly *Silba adipata* McAlpine (Diptera, Lonchaeidae) and trapping experiments. *Zeitschrift für angewandte Entomologie* 95, 471–476.

Katsoyannos, B.I. (2004) Black fig fly (Diptera; Lonchaeidae). In: Capinera, J.L. (ed.) *Encyclopedia of Entomology. 319* Kluwer Academic Publishers, Dordrecht.

Katsoyannos, B.I. and Guerin, P.M. (1984) Hexanol; a potent attractant for the black fig fly, *Silba adipata, Entomologia Experimentalis et Applicata* 35, 71–74.

Kavallieratos, N.G., Stanković, S.S., Schwarz, M., Alissandrakis, E., Athanassiou, C.G., Floros, G.D., Žikić, V. (2019) A survey of parasitoids from Greece with new associations. *ZooKeys* 817, 25–40. doi: 10.3897/zookeys.817.30119

Kendra, P.E., Niogret, J., Montgomery, W.S., Deyrup, M.A. and Epsky, N.D. (2015) Cubeb oil lures: Terpenoid emissions, trapping efficacy, and longevity for attraction of Redbay Ambrosia beetle (Coleoptera: Curculionidae: Scolytinae). *Journal of Economic Entomology* 108, 350–361. doi: 10.1093/jee/tou023

Khaustov, A.A. (1998) Mites of the genus Pyemotes (Acariformes, Pyemotidae) associated with bark beetles (Coleoptera: Scolytidae) in the Crimea (in Russian, with English summary). *Zool Zhurn* 77, 1419–1424.

Knížek, M. (2011) Scolytinae. In: Löbl, I. and Smetana, A. (eds) *Catalogue of Palaearctic Coleoptera*. Vol 7. Apollo Books, Stenstrup, Denmark, pp. 204–251.

Ko, C.C., Shih, Y.T., Schmidt, S. and Polaszek, A. (2015) A new species of *Encarsia* (Hymenoptera, Aphelinidae) developing on ficus whitefly *Singhiella simplex* (Hemiptera, Aleyrodidae) in China and Taiwan. *Journal of Hymenopteran Research* 46, 85–90. doi: 10.3897/JHR.46.5155

Kondo, T. and Evans, G.A. (2013) *Singhiella simplex* (Singh) (Hemiptera: Aleyrodidae), a new aleyrodid invasive species for Colombia. *Boletín del Museo de Entomología de la Universidad del Valle* 13, 31–33.

Koyama, J. (1989) Pest status; south-east Asia and Japan. In: Robinson, A.S and Hooper, G. (eds) *World Crop Pests 3(A). Fruit flies; their biology, natural enemies and control*. Elsevier, Amsterdam, the Netherlands, pp. 63–66.

Krittika, S., Indhumathi, P., Vedha Hari, B.N., Ramya Devi, D. and Yadav, P (2019) Evidence of nanoemulsion as an effective control measure for fruit flies *Drosophila melanogaster. Scientific Reports* 9, 10578.

Kulkarni, H.D. (2010) Indigenous insect pests - *Batocera* and *Apriona* beetle attack on eucalyptus. *Karnataka Journal of Agricultural Sciences* 23, 207–210.

Kumar, R. and Ramamurthy, V.V. (2010) Morphlogy and bionomics of *Phycodes radiata* Ochsenheimer (Lepidoptera: Brachodidae) from New Delhi, India. *Tijdschrift voor Entomologie* 153, 15–24.

Laborda, R., Galán-Blesa, J., Sánchez-Domingo, A., Xamaní, P., Estruch, V.D., Selfa, J., Guerrieri, E. and Rodrigo, E. (2014) Preliminary study on the biology, natural enemies and chemical control of the invasive *Macrohomotoma gladiata* (Kuwayama) on urban *Ficus microcarpa* L. trees in Valencia (SE Spain), *Urban Forestry and Urban Greening.* doi: 10.1016/j.ufug.2014.12.007

Lahey, Z.J. and Polaszek, A. (2017) *Baeoentedon balios* (Hymenoptera: Eulophidae): a parasitoid of ficus whitefly, *Singhiella simplex* (Singh) (Hemiptera: Aleyrodidae), new to the United States. *International Journal of Pest Management* 63, 349–351. doi: 10.1080/09670874.2016.1269222

Lakra, R.K. (2004) Insect pests of fruits of Arid and Semi-Arid regions of India. In: Verma L.R., Verma A.K. and Gautam D.C. (eds) *Pest Management in Horticulture Crops - Principles and Practices.* Asiatech Publishers Inc., New Delhi. pp. 364–367.

Lakra, R.K. and Kharub, W.S. (1977) Fig leaf roller, *Phycodes minor* Moore (Glyphipterygidae: Lepidoptera) - A new pest of fig in Haryana and its control. *Haryana Journal of Horticultural Sciences* 6, 35–36.

Legaspi, J.C., Mannion, C., Amalin, D. and Legaspi, B.C. Jr (2011) Life table analysis and development of *Singhiella simplex* (Hemiptera: Aleyrodidae) under different constant temperatures. *Annals of the Entomological Society of America* 104, 451–458.

Legner, E.F. and Guido, A.S. (1983) Establishment of *Goniozus emigratus* and *Goniozus legneri* [Hym: Bethylidae] on navel orangeworm, *Amyelois transitella* [Lep: Phycitidae] in California and biological control potential. *Entomophaga* 28, 97–106.

Lendel, A. (2017) *Choreutis nemorana* (Hübner, 1799) a gracillaria loriolella (frey, 1881) – dva nové druhy motýľov pre faunu Slovenska. *Folia Faunistica Slovaca* 22, 1–5.

Linford, N.J., Bilgir, C., Ro, J. and Pletcher, S.D. (2013) Measurement of lifespan in *Drosophila melanogaster. Journal of Visualized Experiments* (71). doi: 10.3791/50068

Liquido, N.J. (1993) Reduction of oriental fruit fly populations in papaya orchards by field sanitation. *Journal of Agricultural Entomology* 10, 163–170.

Macgowan, I. and Compton, S.G. (2018) A new species of *Silba* (Diptera; Lonchaeidae) associated with figs. *Zootaxa* 4455, 196–200.

Malumphy, C. (2019) (2019) Fig Whitefly, Singhiella simplex. Plant Pest Factsheet. Department of Environment Food and Rural Affairs, Sand Hutton, York, UK. https://planthealthportal.defra.gov.uk/assets/factsheets/Singhiella-simplex-Defra-fact-sheet-v4.pdf (accessed 17 February 2022).

Mannion, C. (2010) Ficus whitefly management in the landscape. https://trec.ifas.ufl.edu/mannion/pdfs/Ficus%20Whitefly%20(Feb2010)%20Fact%20Sheet.pdf (accessed 7 June 2020).

Mannion, C., Osborne, L., Hunsberger, A., Mayer, H. and Hodges, G. (2008) Ficus whitefly: A new pest in South Florida. *IFAS Extension Bulletin,* University of Florida, August 2008. https://mrec.ifas.ufl.edu/lso/IAWG/FIG/The%20Fig%20Whitefly.htm#:~:text=Introduction%3A%20Recently%2C%20a%20new%20pest,a%20new%20US%20continental%20record.&text=This%20whitefly%20has%20been%20most,also%20been%20seen%20on%20F (accessed 20 September 2019).

Manrakhan, A. and Addison, P. (2007) Monitoring Mediterranean fruit fly and Natal fruit fly in the Western Cape, South Africa. *South African Fruit Journal* 6, 18–20.

Martinez, N.B., Ripuelme, C.P.I., Bautista, E.L., Moreno, L.J.V. and Avila, C.J.G. (2017) Presence of Drosophilidae (Diptera: Ephydroidea) flies associated with fig fruits in Morelos, Mexico. *Florida Entomologist* 100, 813–816. doi: 10.1653/024.100.0409

Mcalpine, J.F. (1956) Old world lonchaeids of the genus *Silba* Macquart (=*Carpo lonchaea* Bezzi),with descriptions of six new species (Diptera: Lonchaeidae). *Canadian Entomologist* 88, 521–544.

Mendel, Z. (1986) Hymenopterous parasitoids of bark beetles (Scolytidae) in Israel. Host relation, host plant, abundance and seasonal history. *Entomophaga* 31, 113–125.

Menocal, O., Cruz, L.F., Kendra, P.E., Crane, J.H., Cooperband, M.F., Ploetz, R.C. and Carrillo, D (2018) *Xyleborus bispinatus* reared on artificial media in the presence or absence of the Laurel wilt pathogen (*Raffaelea lauricola*). *Insects* 9, 30. doi:10.3390/insects9010030

Meyer, M.K.P. (1981) South African Eriophyidae (Acari): The genus *Aceria* Keifer, 1944. *Phytophylactica* 13, 117–126.

Mfeka, N., Manrakhan, A. and Addison, P. (2008) Occurrence of Mediterranean and Natal fruit flies (Diptera:Tephritidae) in the Western Cape Province, South Africa: Differences between orchards and home gardens. *Proceedings of the XXIII International Conference of Entomology, 6–12 July, Durban, South Africa: 2242*. Entomological Society of Southern Africa, Pretoria.

Michailides, T.J. and Morgan, D.P. (1998) Spread of endosepsis in Calimyrna fig orchards. *Phytopathology* 88, 637–647.

Mifsud, D. and Knížek, M. (2009) The Bark Beetles (Coleoptera: Scolytidae) of the Maltese Islands (Central Mediterranean). *Bulletin of the Entomological Society of Malta* 2, 25–52.

Mifsud, D. and Porcelli, F. (2012) *Macrohomotoma gladiata* Kuwayama, 1908: A jumping plant-lice recently introduced into the EPPO Region (Hemiptera: Psylloidea). *EPPO Bulletin* 42(1), 161–164.

Mifsud, D., Falzon, A., Malumphy, C., de Lillo, E., Vovlas, N. and Porcelli, F. (2012) On some arthropods associated with *Ficus* species (Moraceae) in the Maltese Islands. *Bulletin of the Entomological Society of Malta* 5, 5–34.

Moghaddam, M. (2013) An annotated checklist of the scale insects of Iran (Hemiptera, Sternorrhyncha, Coccoidea) with new records and distribution data. *Zookeys* 33, 1–92.

Moghaddam, M. and Faghih, H. (2017) Scale insects (Hemiptera, Coccomorpha) of *Ficus carica* L. (Moraceae), with a new record from Iran. *Journal of Entomological Society of Iran* 36, 307–310.

Mojca, R., Ivan, Z., Matjaž, J. and Gabrijel, S. (2017) *Kmetijskog ozdarskizavod Nova Gorica*. Plant Protection Society of Slovenia.

Moniruzzaman, M., Zahira, Y., Rahima, K. and Nadhirah, A. (2017) Mealybug (Pseudococcidae) infestation and organic control in fig (*Ficus carica*) orchards of Malaysia. *Royal Irish Academy* 117B, 25–32.

Myartseva, S.N., Coronado-Blanco, J.M. and Ruíz-Cancino, E. (2013) First records of 'ficus whitefly' *Singhiella simplex* (Singh, 1931) (Hemiptera: Aleyrodidae) for Tamaulipas and Nayarit, México. *Dugesiana* 20, 81–82.

NIPHM (2015) *AESA based IPM Package in Fig*. Department of Agriculture and Cooperation, Ministry of Agriculture and Farmers Welfare, Government of India.

Niu, G., Rupsinghe, S.G., Zangerl, A.R., Siegel, J.P., Schuler, M.A. and Berenbaum, M.R.A. (2011) Substrate-specific cytochrome P450 monooxygenase, CYP6AB11, from the polyphagous navel orangeworm (*Amyelois transitella*). *Insect Biochemistry and Molecular Biology* 41, 244–253.

Ouvrard, D. (2017) The world Psylloidea database. www.hemiptera-databases.org/psyllist/?db=psylles&card=searching&lang=en (accessed 7 June 2020).

Parra-Pedrazzoli, A.L. and Leaf, W.S. (2006) Sexual behavior of the navel orangeworm, *Amyelois transitella* (Walker) (Lepidoptera: Pyralidae). *Neotropical Entomology* 35, 769–774.

Pasco, B.A., Mannion, C.M., Charles, A.P., McKenzie, C.M. and Lance, S.O. (2011) Natural enemies managing the invasion of the fig whitefly, *Singhiella simplex* (Hemiptera: Aleyrodidae), infesting *Ficus benjamina* hedge. *Florida Entomologist* 94, 696–698.

Pena, J.E., Crane, J.H., Capinera, J.L., Duncan, R.E., Kendra, P.E., Ploetz, R.C., Mclean, S., Brar, G. Thomas, M.C. and Cave, R.D. (2011) Chemical control of the redbay ambrosia beetle, *Xyleborus glabratus*, and other Scolytinae (Coleoptera: Curculionidae). *Florida Entomologist* 94, 882–896.

Pena, J.E., Weihman, S.W., McLean, S., Cave, R.D., Carrillo, D., Duncan, R.E., Evans, G., Krauth, S., Thomas, M.C., Lu, S.S., Kendra, P.E. and Roda, A.L. (2015) Predators and parasitoids associated with Scolytinae in *Persea* species (Laurales: Lauraceae) and other Lauraceae in Florida and Taiwan. *Florida Entomologist* 98, 903–910.

PestNet (2019) Dried fruit beetles. www.pestnet.com/beetle/dried-fruit-beetle/ (accessed 6 September 2021).

Phani, K.K. (2016) Identification of some parasitoids in West Godawari district, Andhra Pradesh. *International Journal of Farm Sciences* 6, 187–190.

Poher, Y., Ponel, P., Medail, F., Andrieu-Ponel, V. and Guiter, F. (2017) Holocene environmental history of a small Mediterranean island in response to sea-level changes, climate and human impact. *Palaeogeography, Palaeoclimatology, Palaeoecology* 465, 247–263. doi:10.1016/j.palaeo.2016.10.037

Pringle, K.L. and Heunis, J.M. (2006) Biological control of Phytophagous mites in apple orchards in the Elgin area of South Africa using the predatory mite, *Neoseiulus californicus* (McGregor) (Mesostigmata: Phytoseiidae): a benefit–cost analysis. *African Entomology* 14, 113–121.

Ramachandra, R. (1922) Notes on the life-histories of two Mesopotamian moths. *Bulletin of Entomological Research* 12, 477–479.

Rassoul, A.M.S. and Al-Mallo, I.M. (2016) First record of nigra scale, *Parasaissetia nigra* (Nietner, 1861) (Hemiptera, Coccidae) as a pest of fig trees in Iraq. *Bulletin of the Iraq Natural History Museum* 14, 171–178.

Raz, D. (1998) The phenology of the fig fly and its control. *Acta Horticulturae* 480, 207–208. doi:10.17660/ActaHortic.1998.480.35

Reding, M.E., Oliver, J.B., Schultz, P.B., Ranger, C.M. and Youssef, N.N. (2013) Ethanol injection of ornamental trees facilitates testing insecticide efficacy against ambrosia beetles (Coleoptera: Curculionidae: Scolytinae). *Journal of Economic Entomology* 106, 289–298.

Robert H. Smith Faculty of Agriculture, Food and Environment (2019a) *Silba adipata* McAlpine. www.agri.huji.ac.il/mepests/pest/Silba_adipata/ (accessed 20 September 2019).

Robert H. Smith Faculty of Agriculture, Food and Environment (2019b) *Frankliniella occidentalis*. www.agri.huji.ac.il/mepests/pest/Frankliniella_occidentalis/ (accessed 6 September 2021).

Robertson, H. (2019) *Biodiversity Explorer*. www.biodiversityexplorer.info/beetles/cerambycidae/phryneta_spinator.htm (accessed 13 September 2019).

Rossler, Y. (1989) Insecticidal bait and cover sprays. In: Robinson, A.S. and Hooper, G. (eds) *World Crop Pests, Fruit Flies, Their Biology, Natural Enemies and Control*. Elsevier Science Publishers B.V., Amsterdam, The Netherlands, pp. 329–336.

Rota, J., Aguiar, A.M.F. and Karsholt, O. (2014) Choreutidae of Madeira: review of the known species and description of the male of *Anthophila threnodes* (Walsingham, 1910) (Lepidoptera). *Nota Lepidoptera* 37(1), 91–103.

Rung, A. (2016) *A new pest of ficus in California:* Macrohomotoma gladiata *Kuwayama, 1908 (Hemiptera: Psylloidea: Homotomidae), new to North America*. Plant Pest Diagnostics Branch, California Department of Food and Agriculture, California.

Sánchez, I. (2012) Primeras citas de *Macrohomotoma gladiata* (Hemiptera: Psylloidea: Homotomidae) en Andalucía (España). *Boletín de la Sociedad Entomológica Aragonesa* 50, 533–534.

Sarikaya, O. (2013) Notes on bark and wood-boring beetles (Coleoptera: Bostrichidae; Curculionidae; Platypodinae and Scolytinae) of the sweetgum (*Liquidambar orientalis* Mill.) Forest Nature Protection Area, with a new record for Turkish fauna. *Journal of Food Agriculture and Environment* 11, 2178–2185.

Shakarami, J. and Bazgir, F. (2017) Effect of temperature on life table parameters of *Phytoseius plumifer* (Phytoseiidae) fed on *Eotetrany chushirsti* (Tetranychidae). *Systematic and Applied Acarology* 22, 410–422.

Shaw, M.R. (2017) Anatomy, reach and classification of the parasitoid complex of a common British moth, *Anthophila fabriciana* (L.) (Choreutidae). *Journal of Natural History* 51, 1119–1149. doi:10.1080/00222933.2017.1315837

Simmons, P., Reed, W.D. and McGregor, E.A. (1931) *Fig Insects in California*. US Dept. of Agriculture, Washington, D.C.

Simwat, G.S. and Sidhu, A.S. (1974) Biology of the fig leaf roller *Phycodes minor* (Glyphipterygidae: Lepidoptera). *Indian Journal of Entomology* 36, 149–150.

Singh, S. (2018) *Annual Report of Fruit Entomology*. Department of Fruit Science, Punjab Agricultural University, Ludhiana, India, 60 pp.

Singh, S. and Kaur, G. (2017) Biodiversity of insect and mite pests infesting fig in Punjab, India. *Acta Horticulturae* 1173, 257–261.

Starý, P. and Kaddou, I.K. (1971) Observations on the biology of *Ocnerogyia amanda* STGR. (Lymantriidae), a pest of ficus in Iraq. *Journal of the Lepidopterists' Society* 25(1), 53– 57.

Sudhi, S., Jumroenma, K., Chaowattanawong, P., Plodkornburee, W. and Sangchote, Y. (2008) Studies on the biology and infestation of stem borer, *Batocera rufomaculata*, in Durian. *Acta Horticulturae* 787, 331–338.

Swapna, S. and Divya, S.K. (2010) Survey on the incidence of the stem borer, *Batocera rufomaculata* in mulberry plantations in Palakkad district, Kerala. *Journal of Entomological Research* 34, 147–149.

Szabóky, C.S. (2015) A ligeti levélmoly (*Choreutis nemorana* Hübner, 1799) magyarországi előfordulásai. *Növényvédelem* 51(1), 11–13.

Talhouk, A.M. (1969) Insects and mites injurious to crops in Middle Eastern Countries. *Monographien zur angew. Entomologie* 21, 239 pp.

Talhouk, A.M. (2003) *Insects and Mites Injurious to Crops in Middle Eastern Countries*. 2nd edn. American University of Beirut Press, Beirut. *Reviews* 12, pp. 409–490.

Thomas, D.K. (1979) Figs as a food source of migrating Garden Warblers in southern Portugal. *Bird Study* 26, 187–191. doi:10.1080/00063657909476637

Tozlu, G. and Özbek, H. (2000) The tropical fig borer, *Batocera rufomaculata* (Coleoptera: Cerambycidae), new for Turkey. *Zoology in the Middle East* 20, 117–120. doi:10.1080/09397140.2000.10637821.

Tuncer, C. (2002) Investigations on the biology of *Homotoma ficus* L. (Hemiptera: Homotomidae) and developmental threshold of eggs in Samsun province. *Turkiye Entomoloji Dergisi* 26, 33–45.

Tuncer, C., Knížek, M. and Hulcr, J. (2017) Scolytinae in hazelnut orchards of Turkey: Clarification of species and identification key (Coleoptera, Curculionidae). *ZooKeys* 710, 65–76. doi: 10.3897/zookeys.710.15047

Tuttle, D.M. and Baker, E.W. (1968) *Spider Mites of Southwestern United States and a Revision of the Family Tetranychidae* The University of Arizona Press, Tucson, Arizona.

Tyler-Julian, K., Funderburk, J., Srivastava, M., Olson, S. and Adkins, S. (2018) Evaluation of a push-pull system for the management of *Frankliniella* species (Thysanoptera; Thripidae) in tomato. *Insects* 9, 187.

Upadhyay, S.K., Chaudhary, B. and Sapkota, B. (2013) Integrated management of mango stem borer (*Batocera rufomaculata* Dejan) in Nepal. *Global Journal of Biology, Agriculture and Health Sciences* 2, 132–135.

Van de Peppel, L.J.J., Aanen, D.K. and Biedermann, P.H.W. (2018) Low intraspecific genetic diversity indicates asexuality and vertical transmission in the fungal cultivars of ambrosia beetles. *Fungal Ecology* 32, 57–64.

Vaneva-Gancheva, T.T. (2017) *Choreutis nemorana* (Hübner, 1799) (Lepidoptera: Choreutidae) – first record in Bulgaria. *Silva Balcanica* 18(2), 1–7.

Van Zyl, E.J. (1997) Verbouing van vye. *Bulletin 211, LNR, Fruitec-Nietvoorbij* Stellenbosch, South Africa.

Vargas, R.I., Leblanc, L., Putoa, R. and Eitam, A. (2007) Impact of introduction of *Bactrocera dorsalis* (Diptera: Tephritidae) and classical biological control releases of *Fopius arisanus* (Hymenoptera: Braconidae) on economically important fruit flies in French Polynesia. *Journal of Economic Entomology* 100, 670–679.

Vasanthi, P. and Raviprasad, T.N. (2015) Incidence of *Batocera rufumaculata* De Geer (Coleoptera: Cerambycidae) on cashew. *Journal of Entomology and Zoology Studies* 31, 174–177.

Velasco, D.N.G., Goularte, M.R., Berti, F.E. and Zarate do Couto, H.T. (2011) Evaluation of the infestation of *Singhiella simplex* (Hemiptera: Aleyrodidae) on *Ficus benjamina* in Sao Paulo city, Brazil. *Revista de Agricultura* 86, 134–141.

Vossen, P. (2015) Vijgenskeletteermot *Choreutis nemorana* (Lepidoptera: Choreutidae) nieuw voor Nederland. *Entomologische berichten* 75(3), 118.

Wagner, M.R., Cobbinah, J.R. and Bose, P.P. (2008) *Forest Entomology in West Tropical Africa: Forests Insects of Ghana.* Springer Science and Business Media, Dordrecht.

Weems, H.V. and Hoppner, J.B. (2019) Oriental fruit fly, *Bactrocera dorsalis* (Hendel). University of Florida. www.entnemdept.ufl.edu/creatures/fruit/tropical/oriental_fruit_fly.htm (accessed 17 September 2019).

Witkowski, R. and Mazur, A. (2019) First record of fig bark beetle *Hypoborus ficus* Erichson (Curculionidae: Scolytinae) from Poland. *Zootaxa* 4571(1), 143–144.

Wohlfarter, M., Venter, E. and Giliomee, J.H. (2008) Importance and distribution of the Mediterranean black fig fly, *Silba adipata* McAlpine (Diptera: Lonchaeidae) in South Africa. *Proceedings of the XXIII International Conference of Entomology*, 6–12 July, Durban, South Africa. Entomological Society of Southern Africa, Pretoria, p. 319.

Wohlfarter, M., Giliomee, J.H., Venter, E. and Storey, S. (2011) A survey of the arthropod pests and plant parasitic nematodes associated with commercial figs, *Ficus carica* (Moraceae), in South Africa. *African Entomology* 19, 165–172.

Wood, S.L. and Bright, D.E. (1992) A catalog of Scolytidae and Platypodidae (Coleoptera), Part 2: Taxonomic index, volumes A and B. *Great Basin Naturalist Memoirs 13*, Monte L. Bean Life Science Museum, Brigham Young University.

Yorgancı, S. and Acıkgöz, S. (2017) Transmission of fig mosaic disease agents to fig 'Sarılop' by *Aceria ficus* in Aydın, Turkey. *Acta Horticulturae* 1173, 279–284. doi:10.17660/ActaHortic.2017.1173.48

Yükselbaba, U., Topakci, N. amd Göçmen, H. (2017) A new record of Turkey Aleyrodidae fauna, ficus whitefly *Singhiella simplex* (Singh) (Hemiptera: Aleyrodidae). *Phytoparasitica* 45, 715–717.

Zalom, F.G., Barnett, W.W. and Weakley, C.V. (1984) Efficacy of winter sanitation for managing the navel orangeworm, *Paramyelois transitella* (Walker), in California almond orchards. *Protection Ecology* 7, 37–41.

Zouba, A. (2010) First report of *Choreutis nemorana* (Lepidoptera: Choreutidae) in Tunisia. *Journal of Plant Science and Biotechnology* 4(2), 96–97.

# 15 Fig Fruit Maturity, Harvest Methods and Technologies

**M. Jafari[1], A. Yavari[2], M. Crivelli[3], C.H. Crisosto[4] and A. Sarkhosh[5]***

*[1]Fig Research Station, Fars Agricultural and Natural Resources Research and Education Center, Agricultural Research, Education and Extension Organization (AREEO), Estahban, Fars, Iran; [2]Pomologist and private consultant, Tehran, Iran; [3]The Produce Nerd, Cloverdale, CA, USA; [4]Department of Plant Sciences, University of California, Davis, California, USA; [5]Horticultural Sciences Department, University of Florida, Gainesville, Florida, USA*

## 15.1 Introduction

Harvesting is among the most costly and time-consuming operations in fresh fig production. Fresh market figs must be harvested when firm to the touch and near fully ripened to ensure the expected consumer quality. Reliable maturity and ripeness indices include skin color and flesh firmness. The ripeness stage also influences fruit flavor. Due to fermentative properties, overripe figs may become undesirable. Harvesting must be done carefully, minimizing the physical damage, abrasions and cracks that make fruit more susceptible to decay. This chapter presents a summary of harvesting methods and technologies for both fresh and dried figs.

## 15.2 Maturity Definition

Figs are climacteric fruit but they do not continue ripening after harvest (Crisosto *et al.*, 2010; 2011). Fresh ripe figs have a narrow harvest window and are highly perishable after harvesting. The goal is to harvest the highest quality fruit with the maximum shelf-life. There are two categories of fruit maturity: physiological maturity and commercial maturity (Fig. 15.1). Fig fruit is considered commercially mature when the fruit is harvested physiologically mature but not ripe and when the flesh gives a little to the touch. In contrast, fig fruit is considered tree-ripe mature when the fruit is harvested ripe and softer than commercial maturity, but not over-ripe (Fig. 15.2). Figs harvested at tree-ripe maturity have a significantly higher acceptance than figs harvested at commercial maturity (Crisosto *et al.*, 2011). To obtain the greatest eating quality, both fresh and dried figs need to be harvested at commercial maturity. Figs for fresh market must be harvested when almost fully ripe and firm. Both skin color and flesh firmness are dependable maturity and ripeness indices. For instance, 'Black Mission' figs should be light to dark purple rather than black, and 'Calimyrna' figs should be yellowish-white to light yellow and firm. Figs for drying should fully ripen and partially dry on the tree before harvesting (Ferguson *et al.*, 1990, Crisosto *et al.*, 2010; 2011).

## 15.3 Fruit Growth and Development

All *Ficus* spp. are characterized by a unique inflorescence, the 'syconium', which produces multiple

*Corresponding author: sarkhosha@ufl.edu


DOI: 10.1079/9781789242881.0015

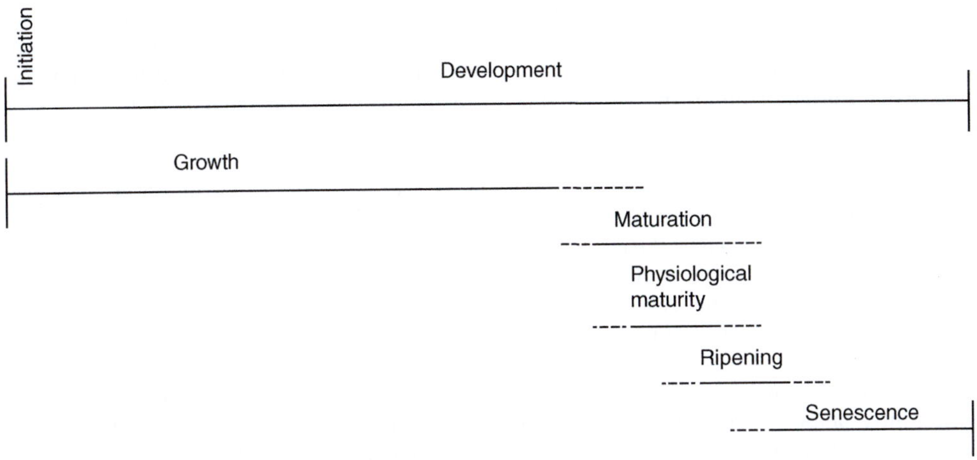


**Fig. 15.1.** Different stages of fruit maturity. (Figure used from Reid, 2002 with permission from University of California Agriculture & Natural Resources.)

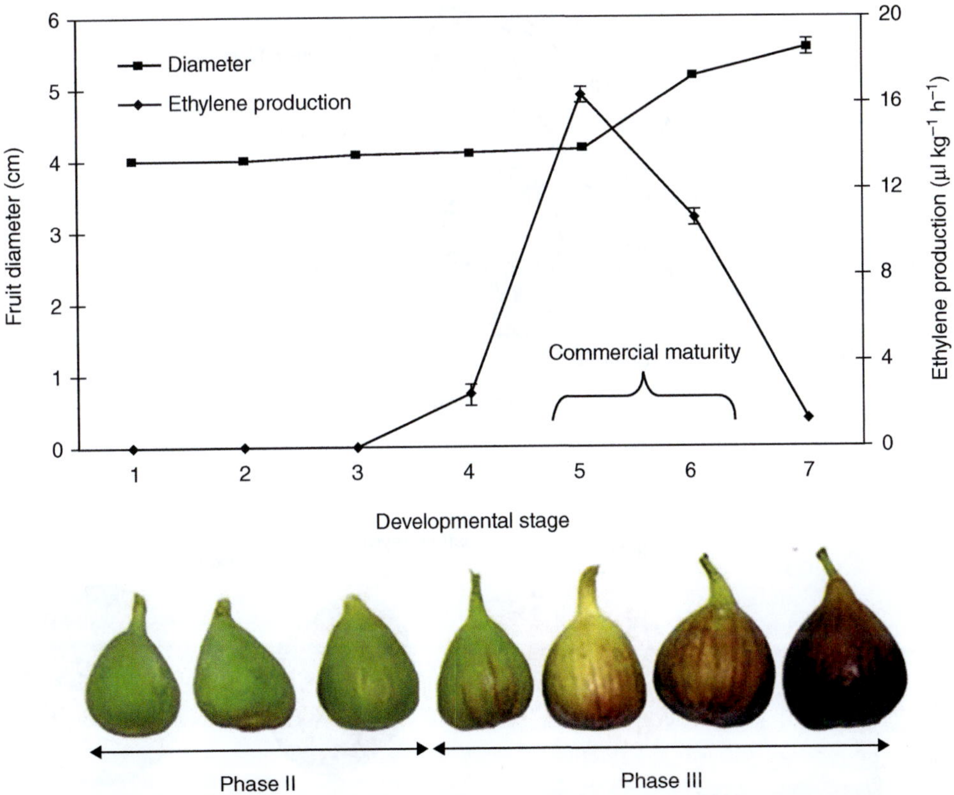


**Fig. 15.2.** Fruit developmental stages and ethylene production of 'Brown Turkey' fig. Commercial maturity is in the third phase of fruit growth, where ethylene production is at the highest level, and the fruit diameter starts to increase rapidly. (Figure used from Freiman *et al.*, 2012 with permission from RightsLink.)

fruitlets (Fig.15.3). Pomologically, the fig fruit is a mature syconium. However, botanically, it is not a fruit, as it consists entirely of vegetative peduncular tissue. The true fruits are made up of hundreds of tiny, fertilized embryos encased within the perianth lining of the pedicellate drupelets. Following fertilization, these ovules develop into a viable seed that is housed within the thin, brittle endocarp surrounded by the endosperm. Therefore, botanically, the fig is classified as a multiple or collective fruit that develops from an entire inflorescence (Condit, 1941).

Fig syconium development was described as a double sigmoid growth curve (Fig. 15.4) comprising three phases (Crane and Brown, 1950; Ferguson *et al.*, 1990, Crisosto *et al.*, 2010; 2011).

The first phase (I) – the first six weeks, is a rapid increase in fruit size due to cell division with accompanying increases in fresh and dry weight, sugar content and moisture.

The second phase (II) – the following four weeks, fruit growth decreases or remains nearly the same size. There is no change in the sugar content during this phase.

The third stage phase (III) – the four weeks prior to commercial maturity is characterized by significant increases in sugar and moisture content, fruit growth, and fresh and dry weight. This is also the ripening phase, which includes both color change and fruit softening. During this stage, fruit accumulates over 90% of its sugar content and more than 70% of its dry weight.

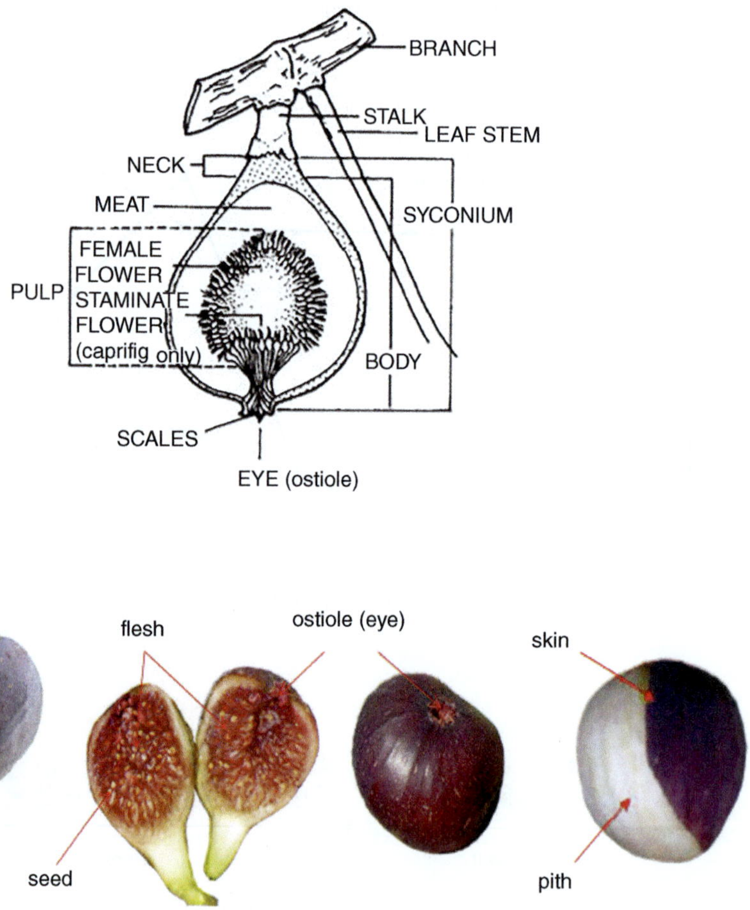


**Fig. 15.3.** Diagram of various parts of fig fruit (top) and fig fruit parts (bottom). (Top figure used with permission from University of California, Division of Agricultural Science, leaflet 211051, 1978. Bottom figure is author's own.)

The duration of fruit growth and ripening depend on the cultivar type, cultural practices and environmental conditions. A characteristic feature of fig fruits is their differential ripening along the branch. At the same time, fruits in all three phases of development can be seen on the same branch. The fruits on the oldest part of the branch may be ripe, while the fruits close to the apex are still in phase I (Fig. 15.5). This helps growers to harvest their fruit selectively, usually every two

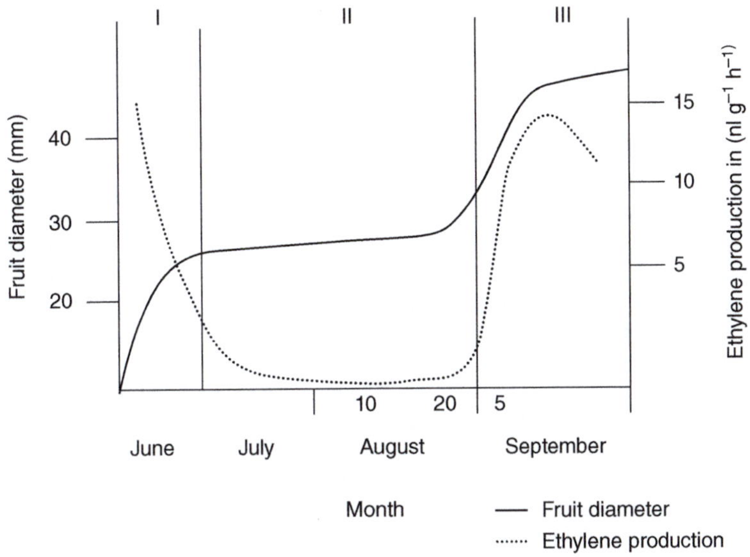


**Fig. 15.4.** Double sigmoid growth curve of the fig showing the progression in fruit diameter and ethylene evolution during the three phases of fruit growth. I–III indicate the three phases of fruit growth and development. (Figure used from Owino *et al.*, 2006 with permission from RightsLink.)

**Fig. 15.5.** Fruits in all three phases of development on the same branch: (A) purple skin cultivar, (B) yellow skin cultivar, and (C) green skin cultivar. (Figures courtesy of A. Yavari.)

days during the harvesting period, to minimize fruit loss due to over-ripening (Freiman *et al.*, 2012).

### Changes in fruit quality attributes

Although fig fruits are classified as non-climacteric, they are harvested at commercial maturity with flavor in mind. Some biochemical and physical changes of all fig cultivars include:

***Weight:*** Fruit weight is positively associated with fig fruit growth and development. For example, with the 'Bursa Siyahi' fig, there can be a 15 g difference between unripe and ripe fruits (Celikel and Karacali, 1984).

***Firmness:*** Fruit's firmness decreases with maturity. For example, the firmness of a ripe 'Bursa Siyahi' fig is half that of the unripe fruit (Celikel and Karacali, 1984).

***Sugars:*** The total soluble solids (TSS) (soluble solids content (SSC) or °Brix) amount is highest in stage IV (senescence) and the lowest in stage I (before caprification). Semidry figs have more sugars (fructose, glucose and sucrose) than sweet cherries, elderberries and apples (Hofer *et al.*, 2005; Usenik *et al.*, 2008; Veberic *et al.*, 2008). Glucose and fructose content increase during fruit growth and ripening (Sedaghat and Rahemi, 2018a). Sucrose increases during fruit development, in stages I and II, then decrease after ripening. Glucose and to a lesser extent, fructose are the main sugars in the 'Sabz' fig fruits under rainfed conditions (Melgarejo *et al.*, 2001; Trad *et al.*, 2012; Jiang *et al.*, 2013; Sedaghat and Rahemi, 2018b).

***Antioxidant capacity:*** Bioactive compounds such as phenolics and flavonoids, including anthocyanins, proanthocyanidins, flavonols and flavanones are responsible for antioxidant capacity (Crisosto *et al.*, 2010; Pereira *et al.*, 2017). Three anthocyanins found in fig: cyanidin-3-O-rutinoside, cyanidin-3,5-O-diglucoside and pelargonidin-3-O-rhamnoglucoside, are responsible for the skin color of ripened fruit (Solomon *et al.*, 2006). It has been reported that high levels of total phenolics were found in dark-colored varieties (Pereira *et al.*, 2015). The anthocyanins, polyphenols and flavonoids in the fig skin also have antioxidant activity (Crisosto *et al.*, 2010). As a result, fig cultivars with dark skin color have higher amounts of antioxidant activity compared to fig cultivars with lighter skin color (Solomon *et al.*, 2006). An analysis of polyphenols using HPLC confirmed that phenolics are primarily concentrated in the peel, with black cultivars having the most (Piga *et al.*, 2008).

***Changes in color indices:*** The three main groups of pigment in the skin of fig fruits are chlorophyll (immature fruit), carotenoids and anthocyanin (mature fruit), which depends on cultivars. During ripening, a color change is the most important indicator (Fig. 15.6). Generally, in light green and yellow cultivars, unripe fruits are darker than ripe ones (Tsantili, 1990; Sedaghat and Rahemi, 2018a). Conversely, in darker

**Fig. 15.6.** Fruit skin color changes in light and dark cultivars during fruit development. (Figure courtesy of A. Yavari.)

cultivars (purple, pink and black), ripe fruit is darker than unripe fruit (Crisosto *et al.*, 2010). In many dark cultivars, chlorophyll content in the fruit skin decreases during ripening, and the fruit skin turns from green to bluish-black (Flaishman *et al.*, 2008). Although practices differ amongst growers, fresh figs are typically harvested based upon skin color and firmness. Firmness may be an easier metric for harvest maturity than skin color (Fig. 15.7).

***Changes in mineral content:*** Dried figs contain almost three times more potassium than fresh figs (Flaishman *et al.*, 2008). A recent study reported figs grown in Iran contained greater potassium amounts than those grown in Pakistan, Turkey and the United States. These differences may indicate cultivar type, soil, caprification, fruit moisture content and weather conditions.

***Fluctuations in organic acids:*** Four organic acids, including malic acid, citric acid, acetic acid and formic acid, are present during fruit growth and ripening. The most abundant organic acids are citric acid, followed by malic acid. These organic acids peak then decrease gradually as the fruits senesce. This supports the suggestion that the organic acids that accumulate during fruit growth are later an energy source for climacteric respiration.

**Fig. 15.7.** Fruit skin and pulp of various varieties displaying a wide range of colors. (Figure used with permission from Cenap Yilmaz.)

## 15.4 Parthenocarpic and pollinated fig

According to their pollination requirement (called caprification in figs), there are four types of fig cultivars (Flaishman *et al.*, 2008):

Type I: The caprifig, commonly named 'male fig', is not edible and is a source for pollen in commercial planting.

Type II: Smyrna, which requires caprification to bear their main crops.

Type III: San-Pedro, which both require caprification to bear their main crops and also yield an early parthenocarpic crop known as 'breba', in addition to the pollinated main crop.

Type IV: Common female type, also called 'persistent' fig, which bears one or two crops per year, with or without caprification (Flaishman *et al.*, 2008; Galil and Eisikowitch, 1968; Galil and Neeman, 1977).

Some fig cultivars set fruit in the absence of fertilization; a phenomenon called parthenocarpy. 'Brown Turkey' is one of the common-type figs that produces both parthenocarpic and pollinated fruit. A study on the development and ripening stages showed differences in parthenocarpic and hand-pollinated fruit (Rosianski *et al.*, 2016). The effects of caprification on fruit development were investigated on the tree and postharvest storage (Fig. 15.8). This research determined that:

- The pollinated fruits were round, and the parthenocarpic fruits were pear-shaped.
- The pollinated fruit were firmer.
- The pollinated fruit were larger in diameter and heavier.
- In postharvest storage, senescence and decay of the pollinated fruit were slower. This suggests pollination of a common-type fig (Brown Turkey) delayed senescence and prolonged its shelf-life (Rosianski *et al.*, 2016).

Caprification produces larger fruit size, greener skin, darker inner inflorescences and decelerates the ripening progression compared to the non-caprified fruit. This demonstrates chlorophyll degradation is regulated differently in parthenocarpic and pollinated fruit inflorescences; chlorophyll degradation is greater in parthenocarpic fruit (Rosianskey *et al.*, 2016). Final skin and pulp color manifest at tree-ripe maturity. Caprification, nutritional status, the temperature

**Fig. 15.8.** Fruit spoilage in 60% ripened fruit after storage for 14 days at 1°C. (A) Hand-pollinated fruit. (B) Parthenocarpic fruit. (Figure used from Rosianski *et al.*, 2016 with permission from RightsLink.)

during maturity and ripening, pruning, and light levels all affect fruit color.

## 15.5 Genetic Differences for Ripening Among Cultivars

Pereira *et al.* (2015) reported significant differences among cultivars during the different ripening stages. Varieties differed in color (dark purple, brown, green and yellow-green) (Figs. 15.6 and 15.7). Dark-colored varieties had the highest quercetin-3-O-rutinoside levels, total phenolic compounds and anthocyanins (specifically cyanidin-3-O-rutinoside). Brown, green and yellow-green colored varieties contained the highest amounts of total vitamin C, chlorogenic acid and (+) catechin (proanthocyanin cleavage products). In contrast to the flesh, the skin had higher levels of bioactive compounds and antioxidant capacity, which improved during ripening. The phenolic compounds reached maximum levels at stage three. These changes were greater in the dark-colored and brown-colored San Antonio cultivar.

Genotype and cultivar control diameter, length and weight (size), TSS, titratable acidity (TA), TSS/TA ratio, skin color, firmness, ostiole diameter, growth cracks, the diameter of the internal cavity, pulp color, ethylene creation, respiration rates, shriveling, storage life and susceptibility to peel deterioration. A series of studies were conducted to compare genetic differences between fig cultivars grown in California and Turkey. In California, the following ten fig cultivars were studied: 'Kadota', 'Mission', 'White Texas Ever bearing', 'Brown Turkey', 'Zidi', 'Orphan', 'Panachee', UCR 276–14, UCR 291, and 152–4s. The fruit weight (20–60 g) at the time of ripening was compared with Turkish cultivars 40–65 g (Bostan *et al.*, 1998), 22–52 g (Çalışkan and Polat, 2008), and 30–90 g (Tanrisever *et al.*, 1998). The California cultivars were similar in fruit length to the Turkish cultivars but had smaller diameters. California-grown cultivars had higher TSS and TA compared to the Turkish cultivars.

## 15.6 Harvesting Time, Methods and Handling

### Fresh fig

Fig cultivars differ in shape, size and color. Size ranges from 2.5 to 10 cm in length. Colors can be green-yellow, yellow, black, purple and red. The fruit skin is thin and sensitive, while the syconium wall is fleshy and light yellow, amber, pink, red or purple (Morton, 2000). The maturity stage at the time of harvest influences:

- Ostiole diameter – there is a relationship between ostiole diameter and ostiole juice present. The percentage of fruit ostiole juice directly affects the percentage of decayed fruit. There is a positive relationship between ostiole juice and fruit decay.
- TA and TSS – favorable consumer ratings have been found to be directly linked to TSS and TSS/TA ratio (Bremer, 2008).
- Firmness – fig fruit firmness has a direct result on the percentage of good fruit.
- Ethylene production.
- Respiration rate .
- Shelf life.
- Latex content – the amount of latex in the fruit decreases as figs ripen, which also correlates with increasing consumer acceptance (Bremer, 2008; Crisosto *et al.*, 2010). Other quality indices include lack of defects, such as scab, sunburn, bird-peck, insects, skin break, stem shrivel and decay.

Figs harvested at the tree-ripe mature stage have higher acceptance rates than those harvested at commercial maturity. Consumer acceptance in commercially mature figs was only 55% versus 86% for tree-ripe figs based on a consumer acceptance study. The shelf life of figs harvested at commercial maturity was longer than that of fruits harvested at tree-ripe maturity. They also had less decay and off-colored fruits after a week in cold storage. Tree ripe figs also had larger ostiole diameters and were more shriveled than fruits harvested at commercial maturity (Bremer, 2008). Skin color and firmness are dependable maturity indices. 'Black Mission' figs should be light to dark purple, rather than black, and 'Calimyrna' figs should be yellowish-white to light yellow and firm (Figs. 15.9 and 15.10). Taste and aroma are at their best in fully ripened fruits.

Breba fruits are the first crop of the season. They are produced on one-year-old branches and normally mature in June and have a short harvest period. The main crop is generated on the current shoots, ripening from August to September. Fresh fig harvest timing is based on skin color change and firmness (Tous and Ferguson, 1996), determining the quality and shelf life of figs. Figs should be slightly firm at harvest. Dark-skinned varieties should be harvested before achieving full color. Green-skinned varieties should be harvested when light yellow before full ripening. Fruit ripening starts from the new shoots' basal zone, ripening sequentially to the apex. This consecutive ripening means fresh fig harvesting must be sequential, done daily over a period of weeks (Tous and Ferguson, 1996). The

**Fig. 15.9.** Commercial maturity (A and B) and more advanced (C). (Figure courtesy of C.H. Crisosto.)

**Fig. 15.10.** External color is not always a good indicator of maturity. Commercial maturity (A and B) and more advanced (C and D). (Figure courtesy of C.H. Crisosto.)

temperature at harvest affects the figs' respiration rate, decay, quality and storage life. Therefore, harvesting during cooler times of the day in the earlier morning is recommended. Trad *et al.* (2013) reported high temperatures at harvest increases mature fruit softening and enlargement and decreases TA.

Fruit skin color, firmness, flavor, neck position relative to the stalk, upright or bowed, and pulp color are practical indices of appropriate harvest time. Harvest times vary because cultivar controls multiple factors, including the need for caprification, potential for fresh or dried consumption, skin color and taste (Figs. 15.6 and 15.7). TSS of the fruit is the most important consideration before harvesting. Hand-held refractometers are best for measuring fruit TSS during harvest (Fig. 15.11).

Fresh ripe figs are easily damaged. To avoid damage, figs should be harvested early in the morning. Harvesters need to be trained to minimize the physical damage, the scrapes, and cracks, that predispose fruits to decay (Crisosto *et al.*, 2010; 2011). Gloves and clippers are used to decrease fruit damage, minimize bruising and protect workers' hands from the latex oozing from the fruit stem (Fig. 15.12). Fruit should be harvested with a stalk to reduce injury and latex release. Correct harvesting procedures improve fresh fig storage life and fruit appearance (Fig. 15.13).

The following guidelines ensure food safety during the harvesting:

- Fresh figs should only be harvested from the tree, not collected from the ground.

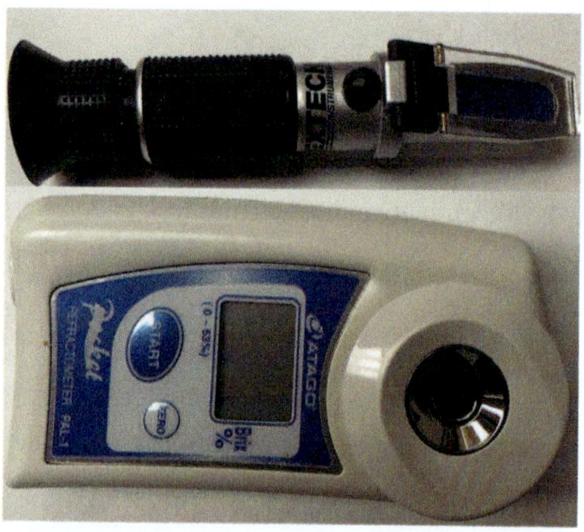


**Fig. 15.11.** Hand-held refractometers, digital (top) and manual (bottom). (Figure courtesy of A. Sarkhosh.)

- Harvesting containers should have a smooth surface that can be properly cleaned and sanitized (Fig. 15.14).

After harvest, figs are packed in field sheds or packing houses to avoid heat and sun exposure, reducing shelf life. Fresh figs are packaged based on the customer and the market preferences. Three of the most common containers are clamshells, trays and green plastic containers (Fig. 15.15). Trays are preferable as the fruit can be separated from one another and reducing contact injury during transport and handling (Fig. 15.16). After the figs have been packed into containers, they are packed into boxes stacked

Fig. 15.12. Gloves and clippers are used to harvest fresh figs. (Figure used with permission from C. Yilmaz and courtesy of A. Sarkhosh.)

Fig. 15.13. Incorrect (left, stalk removal) and correct harvest (right, included stalk). (Figure courtesy of A. Yavari.)

and secured on shipping pallets placed in coolers until shipped in refrigerated trucks. It is vital to maintain cool temperatures throughout the packing, storage and transportation chain to maintain fig shelf-life and quality (Crisosto *et al.*, 2009). Picking buckets and crates are transported frequently to the packing area. Different container types that allow air circulation are used for packing fresh figs (Fig. 15.17).

## Dried fig

As with fresh figs, figs destined for drying are also harvested multiple times, generally weekly from August to October (northern hemisphere). The harvesting period depends on the cultivar, climate, tree exposure, temperature, humidity, tree vigor, and rainfed or irrigated. For some Smyrna fig types (i.e. 'Sabz') the time after caprification can indicate the approximate ripening time.

***Harvest in California:*** Dried figs can be tray-dried or sun-dried. Tray-dried figs are harvested by hand when ripe. Shaking of branches or the trunk is customary in hand harvesting of dried figs. Usually, fruits change shape in the neck and stalk zone during the final stages of ripening and as fruit senesce. This prepares fruit to fall from the tree when force is applied. Green-skinned cultivars are treated with $SO_2$ application when (or before) being spread on trays to dry in the sun (Fig. 15.18). Sun-dried figs are allowed to fully ripen on the tree, partially drying prior to abscising and dropping to the ground (Fig. 15.19). In California, sun-dried figs are harvested from the ground. During the earlier part of the season, when there are both fresh and dried figs on the tree, large fans

**Fig. 15.14.** Different types of picking buckets and crates are used for harvesting fresh figs in the field as well as transferring to packing line. (Figure used with permission from www.cambayas.com and Cenap Yilmaz.)

**Fig. 15.15.** Packing figs in the field immediately after harvest. Gloves reduce fruit damage and bruising and protect the packer's skin from the broken stem's latex. (Figure used with permission from Cenap Yilmaz.)

are brought into the fields to blow the dried figs off the trees onto the ground. Later in the season, the figs are collected from the orchard floor in two steps. First, a sweeper passes through the orchard, sweeping the dried figs into the row middles (Fig.15.20). After a few days, a harvester collects the figs. This process also collects field debris which is removed at a later sorting stage.

***Harvest in Iran:*** The harvesting season for the 'Sabz' cultivar is mid-August through

**Fig. 15.16.** Fresh figs are very sensitive to physical damage and released latex, and these are two of the main reasons for their short shelf-life. (Figure courtesy of A. Yavari.)

**Fig. 15.17.** Containers and boxes that allow air circulation are recommended for packing fresh figs. (Figure used with permission from Tom Moggach.)

mid-October. Ripened fruits dry on the tree. The trunks or main branches are shaken by hand or struck with a plastic hammer. The dried figs fall to the ground or are caught on a suspended net (Fig. 15.21), hand collected, and transferred to the 'Eshpang' or paved area. Semidry figs (Fig. 15.22) are dried on the ground or in nets for 2–3 days or until the fruit has a moisture content of 8–14%. The figs are then hand collected and transferred to cold storage (Jafari, 2004).

***Harvest in Turkey:*** Fig harvesting for drying takes approximately 8 weeks. The fruits of the Sarılop cultivar ripen (semidry, with 40–50% moisture content) on the tree and naturally fall to the ground. These fruits are collected 2–3 times a week by hand and brought to an appropriate drying facility. The fruit is dried on plastic drying trays, named 'kerevet', for 2–3 days until the fruit water content drops to 22–24% (Oksar *et al.*, 2017). It is helpful to dry the fruit under a plastic tunnel, in which two sides are protected by a net, to allow for a more accessible and healthier drying process (Fig. 15.23).

After the figs have dried and been collected, they are pre-sorted and transferred to the packing house, where they are either stored in fumigation tents (not an option for organic production)

**Fig. 15.18.** Drying $SO_2$ treated fruit on trays in the sun. (Figure used with permission from www.maywoodfarms.com.)

**Fig. 15.19.** Fully ripened fruit on the tree for sun-dried figs. (Figure used with permission from Cenap Yilmaz.)

**Fig. 15.20.** Sweeping fallen and sun-dried dried figs into the row middles. (Figures used with permission from www.nutrafig.com and www.valleyfig.com.)

**Fig. 15.21.** Sanitary harvesting into a net, which reduces soil surface contamination. (Figure courtesy of M. Jafari.)

**Fig. 15.22.** Harvested dried figs cultivar Sabz in Esthaban, Shiraz, Iran, drying in the sun after being removed from the trees. (Figure courtesy of M. Jafari.)

**Fig. 15.23.** Fig drying tunnels in Turkey. (Figure used with permission from Cenap Yilmaz.)

or in cold storage. When the product is ready to be packed, the dried figs are taken out of storage, pressure cooked (to make them more pliable), washed, sorted, re-hydrated and packed. The other option is for the dried figs to be made into fig paste. (For more information on dried fig packing and processing read Chapter 17: 'Processing and Industrialization').

## References


Arpacı, S., Konak, R. and Çiçek, E (2018) A national value: Turkish figs. 30th *International Horticultural Congress. 12-16 August 2008, Istanbul,* pp. 15.

Bostan, S.Z., Islam, A. and Aygün, A. (1998) A study on pomological characteristics of local fig cultivars in northern Turkey, *Acta Horticulturae* 480, 71–74. doi: 10.17660/ActaHortic.1998.480.9

Bremer, V. (2008) *Use of preharvest ethephon applications to reduce breba crop load and postharvest performance evaluations of fresh fig (Ficus carica)* varieties/selections. MS thesis, California State University, Fresno, CA.

Çalışkan, O. and Polat, A.A. (2008) Fruit characteristics of table fig (*Ficus carica*) cultivars in subtropical climate conditions of the Mediterranean region. *Scientia Horticulturae* 115(4), 360–367.

Celikel, F.G. and Karacali, I. (1984) Effects of harvest maturity and precooling on fruit quality and longevity of 'Bursa Siyahi' figs. *Acta Horticulturae* 480, 283–288.

Condit, I.J. (1941) Fig characteristics useful in the identification of varieties. *Hilgardia* 14, 1–69. doi: 10.3733/hilg.v14n01p001

Crane, J.C. and Brown, J.G. (1950) Growth of the fig fruit, *Ficus carica* var. Mission. In: *Proceedings American Society for Horticultural Science* pp. 93–97.

Crisosto, G., Bremer, V., Dollahite, S., Crisosto, C.H., Stover, E.D. and Ferguson, L. (2009) Effect of controlled atmosphere storage on the quality of three fresh fig cultivars. www.figs4fun.com/Links/FigLink291.pdf (accessed 21 February 2022).

Crisosto, C.H., Bremer, V., Ferguson, L. and Crisosto, G.M. (2010) Evaluating quality attributes of four fresh fig (*Ficus carica* L.) cultivars harvested at two maturity stages. *HortScience* 45, 707–710.

Crisosto, C.H., Bremer, V. and Stover, E. (2011) Fig (*Ficus carica* L.). In: Yahia, E.E. (ed.) *Postharvest Biology and Technology of Tropical and Subtropical Fruits vol 3: Cocona to Mango*. Woodhead Publishing Ltd., Cambridge, UK, pp. 134–158.

Ferguson, L., Michailides, T.J. and Shorey, H.H. (1990) The California fig industry. *Horticulture Review (American Society for Horticultural Science)* 12, 409–490.

Flaishman, M.A., Rodov, V. and Stover, E. (2008) The fig: Botany, horticulture, and breeding. *Horticultural Reviews* 34, 113–195.

Freiman, Z.E., Rodov, V. and Yablovitz, Z. (2012) Preharvest application of 1 methylcyclopropene inhibits ripening and improves keeping quality of 'Brown Turkey' figs (*Ficus carica* L.). *Scientia Horticulturae* 138, 266–272.

Galil, J. and Eisikowitch, D. (1968) Flowering cycles and fruit types of *Ficus sycomorus* in Israel. *New Phytologist* 67, 745–758.

Galil, J. and Neeman, G. (1977) Pollen transfer and pollination in the common fig (*Ficus carica* L.). *New Phytologist* 79, 163–171.

Hofer, M., Grill, D., Hecke, K., Herbinger, K., Keppel, H. and Monschein, S. (2005) Inhaltsstoffe alter Apfelsorten unter diätetischem Aspekt–Schwerpunkt Diabetes. *Journal Für Ernährungsmedizin* 7, 30–33.

Jafari, M (2004) *Effect of caprifig type on quantity and quality of Estahban rain-fed fig*. Master thesis Shiraz University, Iran, pp. 195.

Jiang, L., Shen, Z., Zheng, H., He, W., Deng, G. and Lu, H. (2013) Noninvasive evaluation of fructose, glucose, and sucrose contents in fig fruits during development using chlorophyll fluorescence and chemometrics. *Journal of Agricultural Sciience and Technology* 15, 333–342.

Lama, K., Harlev, G., Shafran, H., Peer, R. and Flaishman, M.A. (2020) Anthocyanin accumulation is initiated by abscisic acid to enhance fruit color during fig (*Ficus carica* L.) ripening. *Journal of Plant Physiology* 251, 153192. doi: 10.1016/j.jplph.2020.153192

Melgarejo, P., Hernandez, F., Martínez, J., Sánchez, J. and Salazar, D. (2001) Organic acids and sugars from first and second crop fig juices. In: *II International Symposium on Fig* 605, pp 237–239.

Morton, J.F (2000) Fruits of warm climates. Fig. *Ficus carica*. Purdue University. NewCROP. www.hort.purdue.edu/newcrop/default.html (accessed November 2007).

Oksar, R.E., Şen. F. and Aksoy, U. (2017) Effect of drying under plastic tunnels on drying rate and quality of fig (*Ficus carica* L. 'Sarılop') *Acta Horticulturae* 1173. doi: 10.17660/ActaHortic.2017.1173.51

Owino, W.O., Manabeb, Y., Mathookoa, F.M., Kubob, Y. and Inabab, A. (2006) Regulatory mechanisms of ethylene biosynthesis in response to various stimuli during maturation and ripening in fig fruit (*Ficus carica* L.). *Plant Physiology and Biochemistry* 44, 335–342.

Pereira, C., Serradilla, M.J., Martín, A., Villalobos, M.C., Pérez-Gragera, F. and López-Corrales, M. (2015) Agronomic behaviour and quality of six cultivars for fresh consumption. *Sci Horticulturae* 185, 121–128.

Piga, A., Del Caro, A., Milella, G., Pinna, I., Vacca, V. and Schirru, S. (2008) HPLC analysis of polyphenols in peel and pulp of fresh figs. *Acta Horticulturae* 301–306. doi: 10.17660/ActaHortic.2008.798.43

Reid, M.S. (2002) Maturation and maturity indices. In: Kader, A.A. (ed.) *Postharvest Technology of Horticultural Crops*. University of California Agriculture & Natural Resources, Davis, CA, pp. 55–62.

Rosianskey, Y., Dahan, Y., Yadav, S., Freiman, Z.E., Milo-Cochavi, S., Kerem, Z., Eyal, Y. and Flaishman, M.A. (2016) Chlorophyll metabolism in pollinated vs. parthenocarpic fig fruits throughout development and ripening. *Planta* 244, 491–504. doi: 10.1007/s00425-016-2522-6

Rosianski, Y., Freiman, Z.E., Cochavi, S.M., Yablovitz, Z., Kerem, Z. and Flaishman, M.A. (2016) Advanced analysis of developmental and ripening characteristics of pollinated common-type fig (*Ficus carica* L.). *Scientia Horticulturae* 198, 98–106.

Sedaghat, S. and Rahemi, M. (2018a) Effects of physio-chemical changes during fruit development on nutritional quality of fig (*Ficus carica* L. var.'Sabz') under rain-fed condition *Scientia Horticulturae* 237, 44–50.

Sedaghat, S. and Rahemi, M. (2018b) Enzyme activity regarding sugar and organic acid changes during developmental stages in rainfed fig (*Ficus carica* L. cv Sabz). *International Journal of Fruit Science* 18, 14–28.

Solomon, A., Golubowicz, S., Yablowicz, Z., Grossman, S., Bergman, M. and Gottlieb, H. (2006) Antioxidant activities and anthocyanin content of fresh fruits of common fig (*Ficus carica* L.). *Journal of Agricultural and Food Chemistry* 54, 7717–7723.

Tanrisever, A., Özeker, E. and Isfendiyaroglu, M. (1998) Anatomical and characterization of Sarilop (Calimyran) fig cultivars in pictures. *Acta Horticulturae* 480, 109–114. doi: 10.17660/ActaHortic.1998.480.16

Tous, J. and Ferguson, L. (1996) Mediterranean fruits. Purdue University. NewCROP. 20 Oct. 2007. www.hort.purdue.edu/newcrop/default.html (accessed 5 August 2021).

Trad, M., Gaaliche, B., Mars, M. and Renard, C. (2012) Quality performance of "Smyrna" type figs grown under Mediterranean conditions of Tunisia. *Journal of Ornamental and Horticultural Plants* 2(3), 139–146.

Trad, M., Gaaliche, B., Renard, C.M.G. and Mars, M. (2013) Inter- and intra-tree variability in quality of figs. Influence of altitude, leaf area and fruit position in the canopy. *Scientia Horticulturae* 162, 49–54.

Tsantili, E. (1990) Changes during development of 'Tsapela' fig fruits. *Scientia Horticulturae* 44, 227–234.

Usenik, V., Fabcic, J. and Stampar, F. (2008) Sugars, organic acids, phenolic composition and antioxidant activity of sweet cherry (*Prunus avium* L.). *Food Chemistry* 107, 185–192. doi: 10.1016/j.foodchem.2007.08.004

Veberic, R., Colaric, M. and Stampar, F. (2008) Phenolic acids and flavonoids of fig fruit (*Ficus carica* L.) in the northern Mediterranean region. *Food Chemistry* 106, 153–157. doi: 10.1016/j.foodchem.2007.05.061

# 16 Postharvest Physiology and Physiological Disorders of Fresh Fig Fruits

A. Ramezanian[1]*, A. Amiri[2] and C.H. Crisosto[3]

[1]*Department of Horticultural Science, School of Agriculture, Shiraz University, Shiraz, Iran;* [2]*Department of Horticultural Science, Faculty of Agriculture, Shahid Chamran University of Ahvaz, Ahvaz, Iran;* [3]*Department of Plant Sciences, University of California, Davis, California, USA*

## 16.1 Introduction

The fig (*Ficus carica* L.) belongs to the Moraceae family. The edible part is called a syconium and consists of a small, tear-shaped fruit lined with individual drupelets and a pedicel connecting it to the receptacle (Flaishman *et al.*, 2008). Fig fruits can be ovoid, turbinate, or pear-shaped. Fig color varies by variety; some typical colors are yellow, green-yellow, copper, red and purple. Light intensity during growth and development affects fig color.

A rare aspect of fig tree physiology is that it can produce either parthenocarpic or pollinated fruit, with the latter being rounder and of higher quality, with better firmness and less spoilage (Rosianski *et al.*, 2016).

Fig fruit is widely marketed fresh or dried. Most fig production is marketed as dried fruit (Smyrna type). However, other types of figs such as Common type and San Pedro produce fresh fruit. This chapter focuses on fresh fig fruits. Fresh figs are very perishable at room temperature, showing early senescence, fermentation, and decay, limiting their storage and marketing life (Karabulut *et al.*, 2009). The high perishability of figs is linked to their delicate epidermal tissue, which is sensitive to wounding and spoilage by molds, loss of color and softening (Colelli *et al.*, 1991). Postharvest handling and storage of fresh figs are difficult due to the thin fruit skin that ruptures easily, leading to nutritional losses and increased permeability for microbial entry and secondary infection (Fig. 16.1). For all these reasons fig fruits cannot be stored during the off season. New postharvest techniques are needed to improve the shelf-life of fig fruit under low-temperature conditions and maintain harvested fruit quality by thickening and hardening the fruit skin, without losing nutritional quality.

Fig fruits do not ripen after harvest, so fruit should be harvested at peak quality (Aksoy, 1981; Crisosto *et al.*, 2010). Technological developments in postharvest handling, packing, storage and transportation have significantly increased the demand for fresh figs in the last decade.

## 16.2 Fruit Ripening and Quality

The growth of fig fruit goes through three specific stages. The first and the third stages are rapid, and the second stage is slow. Traditional fig varieties can be climacteric or non-climacteric (Freiman *et al.*, 2015). Climacteric fruits are

*Corresponding author: ramezanian@shirazu.ac.ir

 *The Fig: Botany, Production and Uses* (eds. Sarkhosh, A., Yavari, A. and Ferguson, L.)
DOI: 10.1079/9781789242881.0016

**Fig. 16.1.** Fresh figs are sensitive to physical damage, this is one of the primary reasons for their short shelf-life. White latex exuded from cut surfaces is shown that is produced early during ripening in particular. (Figure courtesy of C.H. Crisosto.)

characterized by increased respiration and ethylene biosynthesis rates during ripening at the third stage (Fig. 16.2), while non-climacteric fruits experience a gradual decrease in respiration and ethylene production (Barry and Giovannoni, 2007). Physiological behavior of a particular fig variety has manifested in its postharvest physiology (Martínez-Romero *et al.*, 2007). Climacteric figs have a moderate ethylene production rate (1–10 μL $kg^{-1}$ $h^{-1}$ at 20°C) and their climacteric peak is at the beginning of the third stage of growth (Crisosto and Kader, 2004).

Fruit firmness and peel color are primary harvest indicators for fresh figs (Freiman *et al.*, 2012). In some instances, poor coloration due to insufficient light during fruit growth has been reported; color can be improved by postharvest pulsed light treatment (Rodov *et al.*, 2012). The maturity stage at harvest affects fruit soluble solids concentration (SSC), titratable acidity (TA), consumer acceptance, firmness, and loss of firmness during storage, ethylene production, respiration rate, shelf life, ostiole diameter, and shriveling (Crisosto *et al.*, 2010). The maturity stage also affects flavor, and overripe figs may ferment rapidly. However, fig weight, size, antioxidant capacity, skin color and thickness, split ostioles, and blemishes are not affected by the maturity stage (Bremer *et al.*, 2007). Postharvest analyses on two mature varieties, 'Goklop', and '1100', and one hard mature variety, 'Siyah Orak', have shown that mature fruit harvested at the 2/3 and 1/3 coloring stages were more durable than hard mature varieties (Ertan *et al.*, 2019).

Fig firmness has a direct effect on the percentage of sound fruit. The percentage of fruit with juice in the ostiole has a direct effect on the percentage of decayed fruit. The amount of latex decreases with ripening and results in increasing consumer acceptability and decreasing fungal decay. Consumers' acceptance and preferences appear to be directly attributable to SSC and the SSC:TA ratio; important parameters for selecting fresh fig varieties (Bremer *et al.*, 2007). Other quality indices include the absence of defects, insects and decay. In an experiment, four fresh fig varieties were tested by consumers at two fruit maturity stages. While the degree of liking was affected by variety and maturity stage at harvest, there was no significant interaction between variety and maturity stage (Crisosto *et al.*, 2010).

## 16.3 Bioactive Compounds and Antioxidant Activity

Figs are rich in fiber, trace minerals, antioxidant polyphenols, proteins, sugars and organic acids (Slatnar *et al.*, 2011). They are rich in biologically active compounds such as phenolics responsible for most antioxidant activity of this fruit. Antioxidants are compounds that neutralize chemically active metabolic products, such as free radicals, that induce severe cellular damage. Sources of natural antioxidants are primarily phenolics, which may occur in all products and parts of a plant: fruits, vegetative parts, seeds, leaves, roots and bark (Korekar *et al.*, 2011). Due to their antioxidant activity, plant phenols and polyphenols may play a role in preventing cancer, cardiovascular and neurodegenerative diseases associated with oxidative stress (Nencini *et al.*, 2011). Fig antioxidants can protect lipoproteins in plasma from oxidation and have been demonstrated to induce a significant increase in the antioxidant capacity of plasma 4 h after consumption (Vinson *et al.*, 2005). These compounds' concentrations are strongly related to fig variety and genotype (Crisosto *et al.*, 2010). Dark-skinned genotypes have greater antioxidant capacity due to a greater concentration of phenols (Pande and Akoh, 2010).

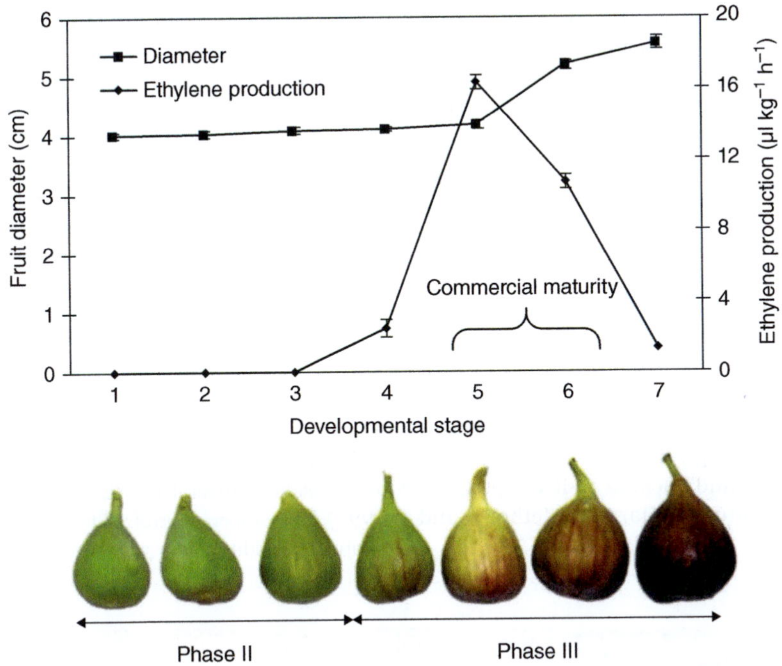


**Fig. 16.2.** Development of 'Brown Turkey' fig fruit, growth and ethylene production. Ethylene production occurs at the third stage of growth. (Figure used from Freiman *et al.*, 2012 with permission from RightsLink.)

## 16.4 Postharvest Challenges

A primary challenge to fig fruit storage is that they have a delicate peel and, when fully ripe, a very soft texture (Crisosto *et al.*, 1998). Also, fig fruit is highly susceptible to postharvest diseases, mainly caused by the growth of molds. The high perishability of figs is attributed to their delicate epidermal tissue, which is very sensitive to wounding and decay by molds (Colelli *et al.*, 1991) combined with extensive softening, fruit skin side cracking, and ostiole end splitting provide entry sites for fungal decay and moisture loss (Kong *et al.*, 2013). To reduce losses, it has been proposed to harvest figs just before full ripeness (Turk, 1989).

## 16.5 Postharvest Treatment of Fig Fruits

### Postharvest handling

Fresh figs harvested between commercial and tree-ripe maturity are highly susceptible to postharvest deterioration (Turk, 1989) because they have high sugar content and an epidermis that is easily damaged (Kaynak *et al.*, 1998). As the postharvest life is extremely short, the majority of fresh fig consumption is local (Turk, 1989). Fresh fruit must be cooled to 0°C as soon as possible after harvest and the cold chain must be maintained throughout handling, distribution and marketing (Crisosto *et al.*, 2006). In the main producing areas in the northern hemisphere, harvesting fresh figs occurs from June to September based on variety and climate condition. In California, fresh figs are usually packed in the field and packed after harvest within sheds to avoid having fruit under heat and pressure from stacking for a long time. The fruits are packed in containers that allow air circulation (Chessa, 1997), usually on trays or clamshells. The trays or clamshells are then put into boxes and stacked on pallets for transport to storage.

### Fruit sorting and grading

Fruit sorting is important to supply markets with more uniform fruits. This practice can eliminate misshaped, discolored and even contaminated fruits. In many producing regions, fruit sorting is done manually. Automatic sorting using a machine vision system is used in some packing houses which can process fruit with high efficiency (Fig. 16.3). However, hand sorting is still very common in many countries. Analysis of images captured with digital cameras enables assessment and classification of fruits based on color, shape and size with high accuracy. Different types of picking buckets and crates are used for harvesting fresh figs in the field as well as transferring to the packing line. Different container types that allow air circulation are used for packing fresh figs. For more information about picking and packing visit Chapter 15: 'Fig Fruit Maturity, Harvest Methods and Technologies'.

## Temperature management

Postharvest temperatures are critical for fresh figs. Low temperatures during postharvest storage of fresh figs slow the fruit's metabolism, delaying fruit ripening, quality loss and controlling spoilage (Table 16.1). The optimal storage conditions for fresh figs are temperatures of approximately −1 to 0°C with a relative humidity of 90–95%. 'Black Mission' and 'Calimyrna' figs stored under optimum conditions (−1 to 0°C) have an expected postharvest life of 1–2 weeks in air or 3–4 weeks in a controlled atmosphere (Crisosto and Kader, 2007). Storage of 'Sultane' figs at 6–8°C was preferable to 2°C, as the 2°C stored fruit underwent faster degradation after transfer to ambient temperature (Baccaunaud *et al.*, 1995). Low temperatures of −1 to 0°C and 90–95% relative humidity (RH) were recommended for

**Fig. 16.3.** Sorting, grading and packing of fig fruits by semi-automated packing line. (Figure used with permission from aksun www.aksun.com.tr/.)

**Table 16.1.** Recommended temperature and relative humidity for storage of some fresh fig cultivars.

| Cultivar | Temperature | Relative humidity | Days of storage | Reference |
|---|---|---|---|---|
| 'Cuello de dama' | 6°C | 90% | 7–10 days | Garcia *et al.*, 2003 |
| 'Black Mission' and 'Calimyrna' | −1 to 0°C | | 7–14 days | Crisosto and Kader, 2007 |
| 'Melanzana' | 5°C | 90 ± 3% | 7 days | Ingrassia *et al.*, 2017 |
| 'Masui Dauphine', 'Bongresi' and 'Banane' | 0.5°C | 90% | 30 days | Byeon and Lee, 2020 |
| 'Ottomanit' | 0°C | 90% | 30 days | Bahar and Lichter, 2018 |
| 'Siah' | 6°C | 90% | 20 days | Saki *et al.*, 2019 |

fresh 'Black Mission' and 'Calimyrna' fig storage because the fig fruit is not sensitive to chilling injury (Crisosto and Kader, 2004). 'Mission' figs stored at 5°C had almost double the respiration rate of fruits stored at 0°C, while their ethylene production was almost 50-fold (Colelli *et al.*, 1991). Fresh fig shelf life is extremely short, lasting 1–2 days at 20°C after storage at 4–6°C and 75% relative humidity (Morton, 2000). Forced-air cooling to 0°C is recommended (Chessa, 1997; Crisosto and Kader, 2007). It has been shown that the storage of fresh-cut fig fruit at low temperatures (5 or 0°C) was more effective than controlled atmosphere storage (Hong *et al.*, 2016).

## Controlled atmosphere storage (CAS) and modified atmosphere packaging (MAP)

Controlled atmosphere storage (CAS) is a type of storage with modified $O_2$ and $CO_2$ gases. CAS is now used globally in produce production. It has been the subject of an enormous number of biochemical, physiological and technological studies, but precisely how it works is unknown. CAS has been demonstrated to reduce the respiration rates of multiple fruits and vegetables. For some crops, high $CO_2$ or low $O_2$ can have either no effect or increase the respiration rate. These effects on respiration rate affect the eating quality of fruit and vegetables. Generally, crops stored in controlled atmospheres have a longer storage life because the metabolic process rate is slower. Particularly with climacteric fruit, CAS will slow ripening and deterioration so that even when stored for protracted periods, fruits will be less ripe than fruits stored in ambient air.

Modified atmosphere packaging (MAP) can be used as either passive MAP without changing the initial atmosphere inside the package or active in which the atmosphere of the package is replaced with a gas mixture different from normal atmosphere. It can reduce the respiration rate and delay senescence by reducing metabolic activity and microbial populations (Valero *et al.*, 2008). 'Masui Dauphine' figs stored for 3 days under MAP and enriched $CO_2$ at 0°C maintained better fig quality than the control (Mathooko *et al.*, 1993). Passive modified atmospheres created with micro-perforated films were effective in reducing the incidence of decay caused by fungal growth and preserved fig quality characteristics (Villalobos *et al.*, 2014). In another study, 'Brown Turkey' figs were packed with different films. After 21 days at 0°C, MAP reduced the respiration rate and prolonged the fruit's shelf life, although the accumulation of $CO_2$ was rather low (Bouzo *et al.*, 2012).

Controlled atmospheres (CAs) of 5–10% $O_2$ and 15–20% $CO_2$ can reduce decay, maintain firmness and reduce respiration and ethylene production rates during storage (Crisosto and Kader, 2007). Similar results were observed for 'Mavra Markopoulo' figs stored at −1°C in 2 kPa $O_2$ atmosphere for 29 days (Tsantil *et al.*, 2003). 'Ottomanit' figs showed better quality under CA conditions of 5 kPa $O_2$ and 5 kPa $CO_2$. CAs can also maintain anthocyanin and increase flavonoid contents. TSS, TA and color change have been retarded by CAs, but the shelf-life extended. CA has maintained firmness in fig fruit. $CO_2$ injury has been reported at higher $CO_2$ concentrations and decay was lowest under 5 kPa $CO_2$ (Bahar and Lichter, 2018). Atmospheres with low $O_2$ and high $CO_2$ concentrations inhibit the growth of mold and yeast due to the well-known antimicrobial activity of $CO_2$ at high concentrations (Devlieghere *et al.*, 2000; Jacxsens *et al.*, 2001), although high concentrations of $CO_2$ may result in peel injury during storage (Fig. 16.4).

Extended storage in CA can result in loss of flavor. The quality of fresh 'Mission' figs was maintained for up to 4 weeks in cold storage at 15 or 20 kPa $CO_2$. The visible benefits of the high $CO_2$ concentrations were inhibition of decay and external peel color maintenance. The acetaldehyde concentrations in the $CO_2$-treated 'Mission' fig fruit increased during the first week, then decreased, while ethanol increased slightly during the first 3 weeks and moderately during the 4th week, potentially resulting in an off-flavor (Colelli *et al.*, 1991). CA was also tested on the three fig cultivars 'Brown Turkey', 'Kadota', and 'Mission' stored at 0°C (Crisosto *et al.*, 2009). CA

**Fig. 16.4.** Appearance of fruit under the different storage conditions following cold storage and 2 days of shelf-life. (A) Appearance of five whole fruit from one of the four replications. (B) Appearance of fruit slices under the different storage conditions. For each condition, four slices representing four different fruit of the same replication are shown. (Figure used from Bahara and Lichte, 2018 with permission from RightsLink.)

storage conditions of 6 kPa $O_2$ and 17 kPa $CO_2$ strongly reduced decay compared with control fruit under normal atmospheres. Overall, CAS can be useful for maintenance of flavor during storage, however it's necessary to find out the best atmospheric mixture regarding fig variety.

## $SO_2$ fumigation

$SO_2$-generating sheets were first used in California in the 1960s (Gentry and Nelson, 1968) for grapes and today are used for fruit intended for export (Lichter *et al.*, 2008; Zutahy *et al.*, 2008). These sheets contain granular sodium metabisulphite which release $SO_2$ gas slowly. This technology is based on the sulfite salt in the pads reacting with the water vapor in the packing box producing a continuous emission of low $SO_2$ concentrations within the packages (Nelson and Ahmedullah, 1976). $SO_2$ kills fungi spores and mycelia on the surface and inner parts of fruits. Excessive $SO_2$ concentrations can damage table grapes, by bleaching, causing sunken areas on the berry surface and causing premature stem browning (Nelson and Ahmedullah, 1976; Crisosto and Mitchell, 2002). Grape hairline splits, commonly associated with significant water loss, are also induced by excessive $SO_2$ (Zoffoli *et al.*, 2008). $SO_2$ fumigation (25 μL $L^{-1}$ $h^{-1}$) has been demonstrated to be a less harmful method of reducing fig decay than $SO_2$-generating pads or the combination of initial fumigation with $SO_2$ pads (Fig. 16.5). Repeated fumigations during cold storage did not significantly improve control of fig decay over that provided by a single 25 μL $L^{-1}$ $h^{-1}$ fumigation before cold storage (Cantín *et al.*, 2011). The most efficient treatment was combination of initial and after storage treatment with $SO_2$.

## MCP

1-methylcyclopropene (1-MCP) is an unsaturated cyclic olefin that acts as an ethylene inhibitor. 1-MCP can be used to control ethylene production, respiration rate and softening, extending the shelf-life of a wide range of fruits. 1-MCP binds to ethylene receptors so that ethylene perception and its effects on the treated tissue are blocked (Sisler and Serek, 2003). To date, different responses have been reported regarding 1-MCP application on fig fruits. In an experiment, it was found that application of 1-MCP at 500 ppb to

**Fig. 16.5.** 'Black Mission' (a) and 'Kadota' figs (b) subjected to different $SO_2$ treatments. Treatments were (from left to right): (1) control (untreated); (2) initial 25 μL $L^{-1}$ $h^{-1}$ $SO_2$ fumigation before cold storage; (3) 25 μL $L^{-1}$ $h^{-1}$ $SO_2$ fumigation after cold storage; and (4) initial 25 μL $L^{-1}$ $h^{-1}$ $SO_2$ fumigation before cold storage + 25 μL $L^{-1}$ $h^{-1}$ $SO_2$ fumigation after cold storage. (Figure used from Cantín *et al.*, 2011 with permission from Rightslink.)

figs picked at commercial maturity (d'Aquino *et al.*, 1998) had a limited effect on retarding softening (Sozzi *et al.*, 2005; Gozlekci *et al.*, 2008). 1-MCP has been demonstrated to markedly enhance ethylene production in figs (Sozzi *et al.*, 2005; Owino *et al.*, 2006). This response suggests that harvested figs have an auto-inhibitory regulation of ethylene synthesis at commercial maturity, rather than the autocatalytic pattern typical of climacteric fruit. Fig fruits picked at a more advanced stage did not respond to 1-MCP, in either fruit quality or ethylene-production rate (Fig. 16.6; Sozzi *et al.*, 2005). Low efficacy of 1-MCP on figs may be due to the fact that at harvest, the fruit are already in a post-climacteric state (Sozzi *et al.*, 2005; Flaishman *et al.*, 2008). Another study showed that 0.5 $\mu L\ L^{-1}$ was the optimal concentration of 1-MCP to extend the storage life and improve the storage quality of fig fruit (Wang *et al.*, 2013). Senescence and decay of figs treated with 1-MCP were lower than in untreated fruits during storage (Fig. 16.7; Freiman *et al.*, 2012).

## 16.6 Emerging Technologies

### Cold plasma technology

Plasma consists of ionized gas particles, and can affect cell walls, cytoplasm and a microorganism's DNA. Plasma is generated by applying an electric field and ionizing the air. Cold plasma can deactivate pathogenic agents on the surfaces of fresh and dried food products and packaging materials (Misra *et al.*, 2011). The beneficial effect of cold plasma technology for 30 s to preserve fresh fig quality has been noted. A positive effect of non-thermal atmospheric plasma on fig fruit's sensory quality was observed, and undesired quality changes were not found. Studies on microbiological and nutritional effects and influence due to packaging materials must still be clarified before cold plasma can be considered for prolonging the shelf life of fresh figs. The technique does not use water and chemical materials, but commercial-scale plasma application to packed or unpacked figs needs further investigations (Abbaszadeh *et al.*, 2018).

### Chlorine dioxide

Chlorine dioxide ($ClO_2$) is a strong oxidizing agent with broad antimicrobial activity. It has about 2.5 times the oxidation capacity of chlorine (Benarde *et al.*, 1965). Also, it does not react with nitrogen-containing compounds or ammonia to form dangerous chlorine compounds. The efficacy of chlorine dioxide as gas application for controlling postharvest diseases of black fig has

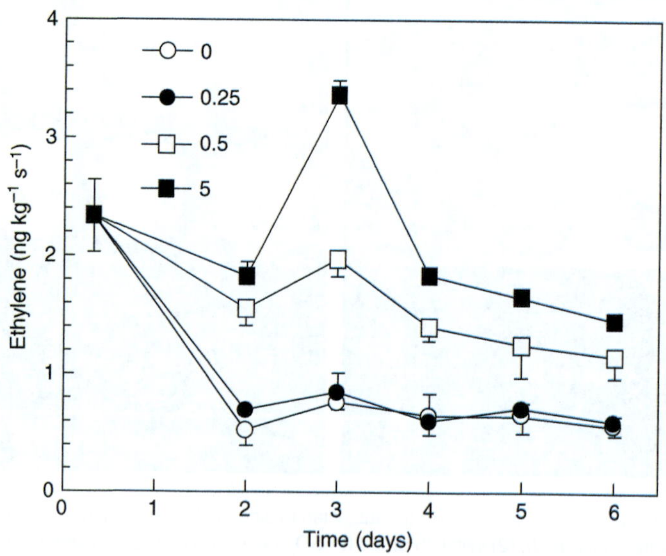


**Fig. 16.6.** Ethylene production in figs stored at 20°C and treated with: 0: no 1-MCP; 0.25: 0.25 $\mu L\ L^{-1}$ 1-MCP; 0.5: 0.5 $\mu L\ L^{-1}$ 1-MCP: or 5: 5 $\mu L\ L^{-1}$ 1-MCP. (Figure used from Sozzi *et al.*, 2005 with permission from RightsLink.)

**Fig. 16.7.** Effect of preharvest 1-MCP treatment on fruit appearance after 12 storage days and 2 shelf-life days. Fruit were harvested commercially mature. (A) 1-MCP-treated fruit and (B) untreated fruit. (Figure used from Freiman *et al.*, 2012 with permission from RightsLink.)

been tested (Karabulut *et al.*, 2009). Fogging at 300–1000 μL $L^{-1}$ significantly reduced the natural incidence of decay, mostly gray mold. The efficacies of fogging at 500 and 1000 μL $L^{-1}$ were at the same level, but fogging at 1000 μL $L^{-1}$ was superior to at 300 μL $L^{-1}$ in stored fruit. Fungal and bacterial organisms on both the fruit surface and in the surrounding air may be inactivated efficiently by chlorine dioxide fogging.

## Edible coating

Extensive literature exists describing the effects of edible coatings on horticultural produce. Bio-based, but not necessarily biodegradable, materials for food packaging were introduced in the 1990s (Petersen *et al.*, 1999). Bio-based packaging materials include edible films and edible coatings, used in addition to the packaging. Edible coatings are defined as a thin layer of material that covers the food's surface and that can be eaten. Edible coating modifies the internal atmosphere by increasing $CO_2$ and lowering $O_2$ concentrations. The effects of edible coatings on internal gas composition and the interaction with quality parameters must be determined for coated fresh produce. The success of edible coatings depends mainly on selecting films or coatings that provide a desirable internal gas composition appropriate for a specific product. If a coating is too thick, detrimental effects can lower the internal oxygen concentration below a desirable beneficial level or increase the carbon dioxide concentration above a critical tolerable level. The oxygen permeability of most edible coatings is lower than that of the conventional plastic films (Park, 1999).

Chitosan-based coatings can ameliorate storage issues such as weight loss and quality changes: improvement in the total polyphenol, anthocyanin and flavonoid concentrations, and antioxidant activity of stored figs has been reported. Greater activity of antioxidant enzymes reduced oxidative stress and prevented browning reactions in chitosan-coated figs (Adiletta *et al.*, 2019). Chitosan coatings delayed the decrease in ascorbic acid concentration in fresh fig fruits during cold storage due to its limited oxygen permeability, which lowered the enzyme activity that oxidizes ascorbic acid (Bal, 2013). Application of 175 ppm zinc oxide (ZnO) nanoparticles to fig fruits delayed ripening, slowed weight loss, reduced color changes, increased fruit firmness and inhibited the growth of microorganisms compared to uncoated fig fruits. This suggests ZnO nanoparticle coating is a promising approach to extending the shelf-life of fig fruits (Lakshmi *et al.*, 2018).

Aloe gel coatings extended the shelf life of figs by delaying decay and ripening, as reflected in reduced weight loss, fewer changes in physicochemical parameters (pH, titratable acidity and total soluble solids), greater firmness, better sensory quality and improved marketability (Lakshmi Marpudi *et al.*, 2013). An *Opuntia ficus-indica* mucilage edible coating maintained

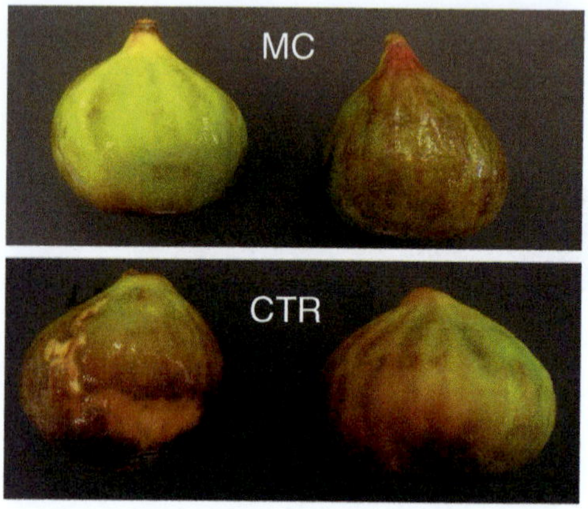


**Fig. 16.8.** Fig coated with mucilage edible coating. Mucilage coating (MC) and control (CTR) figs were stored for seven days at 4°C. (Figure used from Allegra *et al.*, 2017 with permission from RightsLink.)

fresh weight, visual score values, fruit firmness and total carotenoid concentrations of fig fruit during storage (Allegra *et al.*, 2017) (Fig. 16.8).

### Biocontrol

Since the main cause of postharvest spoilage in fig fruit is fungal pathogens, some researchers have examined biocontrol of fungal decay (Droby *et al.*, 2009; Ruiz-Moyano *et al.*, 2016). *Hanseniaspora opuntiae* L479 and *Metschnikowia pulcherrim* L672 yeasts isolated from fig fruit and breba crops, had antagonistic effects against *Penicillium expansum* M639, *Cladosporium cladosporioides* M310 and M624, and also pathogenic molds, including *Botrytis cinerea* CECT20518 and *Monilia laxa* CA1 (Ruiz-Moyano *et al.*, 2016).

## 16.7 Fruit Cracking

Cracking and ostiole-end splitting during growth and development occur in figs because they are soft fruits with thin skins that are highly susceptible to fruit skin rupture (Condit, 1947; Crisosto *et al.*, 2010). Some studies use the term cracking to describe both superficial cracks that occur on the fruit side and end-cracks that expose the flesh (Lampinen *et al.*, 1995; Fig. 16.9). Cracking has been described as 'the physical failure of the fruit skin' (Milad and Shackel, 1992) in the form of fractures in the cuticle or skin, which typically do not penetrate into the flesh. Splitting is an extreme form of cracking that penetrates into the flesh (Opara *et al.*, 1997).

Changes in soil moisture and relative humidity, dry winds or heavy irrigations following a dry period results in fruit cracking (Hepaksoy *et al.*, 2007).

However, preventing deterioration due to decay is challenging due to a high incidence of fruit skin damage, the lack of registered postharvest products and strong consumer preferences for fresh organic figs. Fig diseases have important repercussions in fig production. Regulated deficit irrigation (RDI) significantly reduced fruit skin-side cracking and ostiole-end splitting in 'Brown Turkey' and skin-side cracking in 'Sierra', increasing marketable fruit by 50% in 'Brown Turkey' and 18% in 'Sierra' (Kong *et al.*, 2013). More research is needed to find solutions for reducing fruit cracking during fruit growth and development.

Plant cell wall thickness and stability are also affected by calcium, nitrogen, potassium, and magnesium levels, and fruit size (Aksoy and Anaç, 1993) (Fig. 16.9). Calcium is an essential plant nutrient since the divalent $Ca^{2+}$ is required for structural roles in the cell wall and membranes, as a counter-cation for inorganic and organic anions in the vacuole, and as an intracellular messenger in the cytosol. Calcium enters the apoplast readily and is bound in an exchangeable form to cell walls and the plasma membrane's exterior surface. Its uptake rate into the cytoplasm is severely restricted and seems to be only loosely coupled to metabolic processes. Pretreatment with calcium chloride (4%) effectively checks the growth of both mesophilic aerobic bacteria and yeast and molds at low temperatures (Irfan *et al.*, 2013). Cracked fruit ratio in dried fig is reduced with calcium fertilization (Aksoy and Anaç, 1993), because calcium stabilizes cell walls, maintains membrane stability and permeability, and alleviates damage caused by freezing and thawing stress. Treatments with 1% $CaCl_2$ applied after harvest can improve fig storage life (Antunes *et al.*, 2003).

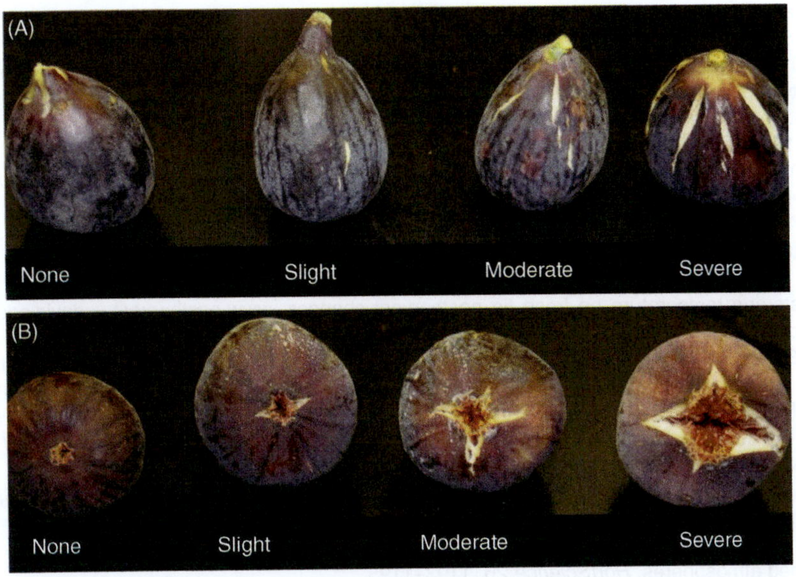


**Fig. 16.9.** Fresh fig fruit skin damage. (A) Skin-side cracking damage scores from none to severe on 'Black Mission'. (B) Ostiole-end splitting damage scores from none to severe on 'Brown Turkey'. (Figure used from Kong *et al.*, 2013 with permission from RightsLink.)

## 16.8 General Remarks

Fresh fig fruits are high-demand due to both flavor and nutritional value. Unfortunately, they are very perishable and sensitive to pathogenic decays. Cold storage along with postharvest treatments such as modified atmosphere packaging, calcium treatment, $SO_2$ fumigation, edible coating and other modern techniques are helpful for maintenance of fruit quality. More research is needed to find out other efficient techniques to reduce postharvest losses during harvest, transportation, packaging, storage and marketing.

## References

Abbaszadeh, R., Alimohammad, K. and Zarrabi Ekbatani, R. (2018) Application of cold plasma technology in quality preservation of fresh fig fruit (Siyah): A feasibility study. *International Journal of Horticultural Science and Technology* 5(2), 165–173.

Adiletta, G., Zampella, L., Coletta, C. and Petriccione, M. (2019) Chitosan coating to preserve the qualitative traits and improve antioxidant system in fresh figs (*Ficus carica* L.). *Agriculture* 9, 84.

Aksoy, U. (1981) Studies on fruit development, maturation and cold storage of the fig (*Ficus carica* L.) fruit cvs. Akca, Goklop and Sarilop. (in Turkish). *PhD thesis*. Ege Univ., Izmir, Turkey.

Aksoy, U. and Anaç, D. (1993) Determination of optimum soil and plant nutrients for higher yield and quality in fig production. In: *Optimization of Plant Nutrition*. Kluwer, pp. 309–313.

Allegra, A., Sortino, G., Inglese, P., Settanni, L., Todaro, A. and Gallotta, A. (2017) 2e effectiveness of *Opuntia ficus-indica* mucilage edible coating on postharvest maintenance of "Dottato" fig (*Ficus carica* L.) fruit. *Food Packaging Shelf Life* 12, 135–141.

Antunes, M.D.C., Oliveira, R., Miguel, M.G., Martins, M.A., Pica, C. and Neves, M.A. (2003) The effect of calcium chloride and thermal postharvest treatments on storage ability and quality of fig fruit. *Acta Horticulturae* 604(2), 715–719.

Baccaunaud, M., Raynal, J. and Babotilene, F. (1995) Figs: maintaining quality after harvest. *Arboriculture-Fruitiere* 489, 37–42.

Bahar, A. and Lichter, A. (2018) Effect of controlled atmosphere on the storage potential of Ottomanit fig fruit. *Scientia Horticulturae* 227, 196–201.

Bal, E. (2013) Postharvest application of chitosan and low temperature storage affect respiration rate and quality of plum fruits. *Journal of Agricultural Science and Technology* 15, 1219–1230.

Barry, C.S. and Giovannoni, J.J. (2007) Ethylene and fruit ripening. *Journal of Plant Growth Regulation* 26, 143–159.

Benarde, M.A., Israel, B.M., Olivieri, V.P. and Granstrom, M.L. (1965) Efficiency of chlorine dioxide as a bactericide. *Applied Microbiology* 13, 776–780.

Bouzo, C.A., Travadelo, M. and Gariglio, N.F. (2012) Effect of different packaging materials on postharvest quality of fresh fig fruit. *International Journal of Agriculture and Biology* 14, 821–825.

Bremer, V., Crisosto, C.H., Garner, D., Stover, E. and Ferguson, L. (2007) Evaluating the roles of ethylene and storage temperature on fresh fig postharvest changes. www.postharvest.ucdavis.edu/Library/Publications/ (accessed 5 August 2021).

Byeon, S.E. and Lee, J. (2020) Differential responses of fruit quality and major targeted metabolites in three different cultivars of cold-stored figs (*Ficus carica* L.). *Scientia Horticulturae* 260, 108877.

Cantín, C.M., Palou, L., Bremer, V., Michailides, T.J. and Crisosto, C.H. (2011) Evaluation of the use of sulfur dioxide to reduce postharvest losses on dark and green figs. *Postharvest Biology and Technology* 59, 150–158.

Chessa, I. (1997) Fig. In: Mitra, S. (ed.) *Postharvest Physiology and Storage of Tropical and Subtropical Fruits*. CAB International, Wallingford, UK, pp. 245–268.

Colelli, G., Mitchell, F.G. and Kader, A.A. (1991) Extension of postharvest life of 'Mission' figs by $CO_2$-enriched atmospheres. *HortScience* 26, 1193–1195.

Condit, I.J. (1947) *The Fig*. Chronica Botanica Co., Waltham, MA.

Crisosto, C.H. and Kader, A.A. (2004) Fig. In: Gross, K. (ed.) *The Commercial Storage of Fruits, Vegetables, Florist, and Nursery Stocks*. ARS-USDA Handbook.

Crisosto, C.H. and Kader, A.A. (2007) Figs. In: *Postharvest Information for Fruits and Fruits. Postharvest Quality Maintenance Guidelines* www.ucanr.edu/sites/kac/files/123822.pdf (accessed 8 November 2007).

Crisosto, C.H. and Mitchell, F.G. (2002) Postharvest handling systems: small fruits. I. Table grapes. In: Kader, A.A. (ed.) *Postharvest Technology of Horticulture Crops*. University of California, Agriculture and Natural Resources, Oakland, CA, pp. 357–363.

Crisosto, C.H., Mitcham, E.J. and Kader, A.A. (1998) *Fig: Recommendations for Maintaining Postharvest Quality, Perishable Handling*. Postharvest Technology Center, University of California, Davis, California.

Crisosto, C.H., Mitcham, E.J. and Kader, A.A. (2006) *Fig. Recommendations for Maintaining Postharvest Quality*. Postharvest Technology Research Information Center, University of California, Davis, California.

Crisosto, C.H., Bremer, V., Dollahite, S., Crisosto, G.M., Stover, E. and Ferguson, L. (2009) *Effect of Controlled Atmosphere Storage on the Quality of Three Fresh Fig Cultivars.* www.figs4fun.com/Links/FigLink291.pdf (29 January 2020).

Crisosto, C.H., Bremer, V., Ferguson, L. and Crisosto, G.M. (2010) Evaluating quality attributes of four fresh fig (*Ficus carica* L.) cultivars harvested at two maturity stages. *HortScience* 45, 707–710.

D'Aquino, S., Piga, A., Moluni, M.G., Agabbio, M. and Papoff, C.M. 1998. Maintaining quality attributes of "Craxiou de Porcu" fresh fig fruit in simulated marketing conditions by modified atmosphere. *Acta Horticulturae* 480, 289–294.

Devlieghere, F., Debevere, J. and Van Impe, J. (2000) Concentration of carbon dioxide in the water-phase as a parameter to model the effect of a modified atmosphere on microorganisms. *International Journal of Food Microbiology* 43, 105–113.

Droby, S., Wisniewski, M., Macarisin, D. and Wilson, C. (2009) Twenty years of postharvest biocontrol research: Is it time for a new paradigm? *Postharvest Biology and Technology* 52(2), 137–145.

Ertan, B., Şenkayas, H. and Tuncay, Ö. (2019) Postharvest logistics performance of fresh fig varieties in Turkey. *Scientia Horticulturae* 257, 108769.

Flaishman, M.A., Rodov, V. and Stover, E. (2008) *The Fig: Botany, Horticulture and Breeding*. Westport Then, New York, p. 113.

Freiman, Z.E., Rodov, V. and Yablovitz, Z. (2012) Preharvest application of 1-methylcyclopropene inhibits ripening and improves keeping quality of 'Brown Turkey' figs (*Ficus carica* L.). *Scientia Horticulturae* 138, 266–272.

Freiman, Z.E., Rosianskey, Y., Dasmohapatra, R., Kamara, I. and Flaishman, M. (2015) The ambiguous ripening nature of the fig (*Ficus carica* L.) fruit: A gene-expression study of potential ripening regulators and ethylene-related genes. *Journal of Experimental Botany* 66, 3309–3324.

Garcia, B., Hernandez Mendez, M.T. and Lozano Ruiz, M. (2003) Effect of temperature on the shelf-life of fig fruit during refrigerated storage. *Acta Horticulturae* 605, 229–232.
Gentry, J.P. and Nelson, K.E. (1968) Further studies on control of decay of table grapes by 2-stage generation of sulfur dioxide within unvented containers. *American Journal of Enology and Viticulture* 19, 70–81.
Gozleki, S., Erkan, M., Karasahin, I. and Sahin, G. (2008) Effect of 1-Methylcyclopropene (1 MCP) on fig (*Ficus carica* cv. Bardakci) storage. *Acta Horticulturae* 798, 325–330.
Hepaksoy, S., Aksoy, U., Can, H.Z. and Ul, M.A. (2007) Determination of relationship between fruit cracking and some physiological responses, leaf characteristics and nutritional status of some pomegranate varieties. *Chieam p.* 87–92. www.agris.fao.org/agris-search/search.do?recordID=QC2000600256
Hong, G., Crisosto, C. and Cantwell, M.I. (2016) Quality and physiology of two cultivars of fresh-cut figs in relation to ripeness, storage temperature and controlled atmosphere. *Acta Horticulturae* 1141, 213–220.
Ingrassia, M., Chironi, S., Allegra, A. and Sortino, G. (2017) Consumer preferences for fig fruit (*Ficus carica* L.) quality attributes and postharvest storage at low temperature by in-store survey and focus group. *Acta Horticulturae* 1173, 383–388.
Irfan, P.K., Vanjakshic, V., Keshava Prakasha, M.N., Ravie, R. and Kudachikara, V.B. (2013) Calcium chloride extends the keeping quality of fig fruit (*Ficus carica* L.) during storage and shelf-life. *Postharvest Biology and Technology* 82, 70–75.
Jacxsens, L., Devlieghere, F., Vander Steen, C. and Debevere, J. (2001) Effect of high-oxygen modified atmosphere packaging on microbial growth and sensorial qualities of fresh-cut produce. *International Journal of Food Microbiology* 71(2–3), 197–210.
Karabulut, O.A., Ilhan, K., Arslan, U. and Vardar, C. (2009) Evaluation of the use of chlorine dioxide by fogging for decreasing postharvest decay of fig. *Postharvest Biology and Technology* 52, 313–315.
Kaynak, L., Gozlekci, S. and Ersoy, N. (1998) A research on storage and pomological properties of some fig (*Ficus carica* L.) cultivars grown in Antalya conditions. *Acta Horticulturae* 480, 277–282.
Kong, M., Lampinen, B., Shackel, K. and Crisosto, C.H. (2013) Fruit skin side cracking and ostiole-end splitting shorten postharvest life in fresh figs (*Ficus carica* L.), but are reduced by deficit irrigation. *Postharvest Biology and Technology* 85, 154–161.
Korekar, G., Stobdan, T., Arora, R., Yadav, A. and Singh, S.B. (2011) Antioxidant capacity and phenolics content of apricot (*Prunus armeniaca* L.) kernel as a function of genotype. *Plant Foods Human Nutrition* 66, 376–383.
Lakshmi Marpudi, S., Ramachandran, P. and Srividya, N. (2013) *Aloe vera* gel coating for postharvest quality maintenance of fresh fig fruits. *Journal of Pharmaceutical Biological and Chemical Sciences* 4(1), 878.
Lakshmi, S.J., Bai, R.S.R., Sharanagouda, H., Ramachandra, C.T., Nadagouda, S. and Nidoni, U. (2018) Effect of biosynthesized zinc oxide nanoparticles coating on quality parameters of fig (*Ficus carica* L.) fruit. *Journal of Pharmacognosy and Phytochemistry* 7(3), 10–14.
Lampinen, B.D., Shackel, K.A., Southwick, S.M., Olson, B., Yeager, J.T. and Goldhamer, D. (1995) Sensitivity of yield and fruit quality of French prune to water deprivation at different growth stages. *Journal of the American Society for Horticultural Science* 120, 139–147.
Lichter, A., Zutahy, Y., Kaplunov, T. and Lurie, S. (2008) Evaluation of table grape storage in boxes with sulfur dioxide-releasing pads with either an internal plastic liner or external wrap. *HortTechnology* 18, 206–214.
Martínez-Romero, D., Bailén, G., Serrano, M., Guillén, F., Valverde, J.M., Zapata, P., Castillo, S. and Valero, D. (2007) Tools to maintain postharvest fruit and vegetable quality through the inhibition of the ethylene action. *Critical Reviews in Food Science and Nutrition* 47, 543–560.
Mathooko, F.M., Kubo, Y., Inaba, A. and Nakamura, R. (1993) Regulation by carbon dioxide of wound-induced ethylene biosynthesis in tomato pericarp and winter squash mesocarp tissues. *Postharvest Biology Technology* 3, 27–38.
Milad, R.E. and Shackel, K.A. (1992) Water relation of fruit end cracking in French prune (*Prunus domestica* L. cv French). *Journal of the American Society of Horticultural Science* 117, 824–828.
Misra, N., Tiwari, B., Raghavarao, K. and Cullen, P. (2011) Non thermal plasma inactivation of food-borne pathogens. *Food Engineering Review* 3(3–4), 159–170.
Morton, J.F. (2000) Fruits of warm climates. *Fig. Ficus carica*. NewCROP, Purdue University. www.hort.purdue.edu/newcrop/default.html (accessed November 2007).
Nelson, K.E. and Ahmedullah, M. (1976) Packaging and decay-control systems for storage and transit of table grapes for export. *American Journal of Enology and Viticulture* 27, 74–79.
Nencini, C., Menchiari, A., Franchi, G.G. and Micheli, L. (2011) *In vitro* antioxidant activity of aged extracts of some Italian *Allium* species. *Plant Foods Human Nutrition* 66, 11–16.

Opara, L., Studman, C.J. and Banks, N.H. (1997) Fruit skin splitting and cracking. *Horticulture Review* 19, 217–262.

Owino, W.O., Manabe, Y., Mathooko, F.M., Kubo, Y. and Inaba, A. (2006) Regulatory mechanisms of ethylene biosynthesis in response to various stimuli during maturation and ripening in fig fruit (*Ficus carica* L.). *Plant Physiology and Biochemistry* 44(5–6), 335–342.

Pande, G. and Akoh, C.C. (2010) Organic acids, antioxidant capacity, phenolic content and lipid characterisation of Georgia-grown underutilized fruit crops. *Food Chemistry* 120, 1067–1075.

Park, H.J. (1999) Development of advanced edible coatings for fruits. *Trends Food Science and Technology* 10, 254–260.

Petersen, K., Nielsen, P.G., Bertelsen, G., Lawther, M., Olsen, M.B., Nilsson, N.H. and Mortensen, G. (1999) Potential of biobased materials for food packaging. *Trends Food Science and Technology* 10, 52–28.

Rodov, V., Vinokur, Y. and Horev, B. (2012) Brief postharvest exposure to pulsed light stimulates coloration and anthocyanin accumulation in fig fruit (*Ficus carica* L.) *Postharvest Biology and Technology* 68, 43–46.

Rosianski, Y., Doron-Faigenboim, A., Freiman, Z.E., Lama, K., Milo-Cochavi, S., Dahan, Y., Kerem, Z. and Flaishman, M.A. (2016) Tissue-specific transcriptome and hormonal regulation of pollinated and parthenocarpic fig (*Ficus carica* L.) fruit suggest that fruit ripening is coordinated by the reproductive part of the syconium. *Frontiers in Plant Science* 7, 1696.

Ruiz-Moyano, S., Martín, A., Villalobos, M.C., Calle, A., Serradilla, M.J., Cordoba, M.G. and Hernandez, A. (2016) Yeasts isolated from figs (*Ficus carica* L.) as biocontrol agents of postharvest fruit diseases. *Food Microbiology* 57, 45–53.

Saki, M., ValizadehKaji, B., Abbasifar, A. and Shahrjerdi, I. (2019) Effect of chitosan coating combined with thymol essential oil on physicochemical and qualitative properties of fresh fig (*Ficus carica* L.) fruit during cold storage. *Journal of Food Measurement and Characterization* 13(2), 1147–1158.

Sisler, E.C. and Serek, M. (2003) Compounds interacting with the ethylene receptor in plants. *Plant Biology* 5, 473–480.

Slatnar, A., Klancar, U., Stampar, F. and Veberic, R. (2011) Effect of drying of figs (*Ficus carica* L.) on the contents of sugars, organic acids, and phenolic compounds. *Journal of Agriculture and Food Chemistry* 59, 11696–11702.

Sozzi, G.O., Abrajan-Villasenor, M.A., Trinchero, G.D., Fraschina, A.A. (2005) Postharvest response of 'Brown Turkey' figs (*Ficus carica* L.) to the inhibition of ethylene perception. *Journal of the Science of Food and Agriculture* 85, 2503–2508.

Tsantil, E., Karaiskos, G. and Pontikis, C. (2003) Storage of fresh figs in low oxygen atmosphere. *Journal of Horticultural Science and Biotechnology* 78, 56–60.

Turk, R. (1989) The effects of harvest time and precooling on fruit quality and cold storage of figs (*F. carica* L. cv. 'Bursa Siyahi'). *Acta Horticulturae* 88, 279–284.

Valero, A., Begum, M., Hocking, A.D., Marín, S., Ramos, A.J. and Sanchis, V. (2008) Mycelial growth and ochratoxin A production by *Aspergillus* section nigri on simulated grape medium in modified atmospheres. *Journal of Applied Microbiology* 105, 372–379.

Villalobos, M.C., Serradilla, M.J., Martín, A., Ruiz-Moyano, S., Pereira, C. and Córdoba, M.G. (2014) Use of equilibrium modified atmosphere packagingfor preservation of 'San Antonio' and 'Banane' breba crops (*Ficus carica* L.). *Postharvest Biology and Technology* 98, 14–22.

Vinson, J.A., Zubik, L., Bose, P., Samman, N. and Proch, J. (2005) Dried fruits: excellent *in vitro* and *in vivo* antioxidants. *Journal of the American College of Nutrition* 4, 44–50.

Wang, L., Zhang, Z. and Zhang, X. (2013) Effects of 1-MCP treatment on ethylene biosynthesis metabolism of fig fruits during storage. *Journal of the Chinese Institute of Food Science and Technology* 13(4), 119–124.

Zoffoli, J.P., Latorre, B.A. and Naranjo, P. (2008) Hairline, a postharvest cracking disorder in table grapes induced by sulfur dioxide. *Postharvest Biology and Technology* 47, 90–97.

Zutahy, Y., Lichter, A., Kaplunov, T. and Lurie, S. (2008) Extended storage of 'Redglobe' grapes in modified $SO_2$ generating pads. *Postharvest Biology and Technology* 50, 12–17.

# 17 Processing and Industrialization

**M. Maghoumi, M.L. Amodio and G. Colelli***
*Department of Agriculture, Food, Natural Resources and Engineering (DAFNE), University of Foggia, Italy*

## 17.1 Introduction

The high perishability of fresh figs makes them difficult to store and/or ship in order to expand potential markets. At low temperature and high humidity figs can remain in good condition for 8–10 days which may become a few weeks if stored in a high-$CO_2$ environment (Amodio and Colelli, 2020). However, at ambient conditions, a fig fruit can only last for 1–2 days. For this reason, a good part of the fig production is used as dried or processed (Crisosto *et al.*, 2011). Figs may also be roasted like coffee beans and can be used as a substitute for coffee. Another way to prolong its commercial life for several months is freezing. Other products from fresh figs are syrup, jam, jelly, brownies, spiced or pickled figs, fig bread and fig meat (Woodroof, 1986).

Beside extending the commercial life, processing can prevent a fall in prices without nutrient loss, inhibit biotic and abiotic degradation processes, lower transportation, handling and storage costs, provide a consistent product supply, important for marketing, optimize product convenience and flexibility, utilize more economical and disposable packaging, and offer alternatives as a snack product.

## 17.2 Methods of Processing

One of the oldest preservation methods still used is drying; the objective is to lower water activity in the product. The critical parameters in drying process selection are energy consumption and dried product quality. The optimum drying system for quality dehydrated product is cost effective, shortens drying time and causes minimum damage to the product (Sagar, 2010).

Freezing, which is usually used for preserving food to maintain its original attributes, including nutritional value, as close as possible to fresh fruit, requires substantial temperature reduction resulting in changing water from liquid to solid, or ice. Freezing greatly reduces microbial and enzymatic activities, however, it also causes physical and chemical changes due to ice formation (Reid, 1996). The quality of frozen fruits depends on many factors related to the technology used; ice nuclei number and dimension are strictly related to the velocity of the ice formation. The final quality of frozen fruits is determined by the variety and by its initial quality. Alternatively, canning uses a thermal process to drastically reduce microbial load, including human pathogens, and may greatly impact quality, especially sensorial and nutritional attributes (Berry and Pflug, 2003).

*Corresponding author: giancarlo.colelli@unifg.it


DOI: 10.1079/9781789242881.0017

Other popular methods of processing figs include jam if it contains the pulp, or jelly if it only contains the juice, by adding sugar and heating. Culls, blemished or off-grade fruits, not suitable for fresh consumption, are often processed.

## 17.3 Fig Drying

Among fig producing countries, Turkey leads in dried fig production and trade (Fig. 17.1). Other traditional producers like Italy and Spain produce high added-value products (Fig. 17.2) for high-end markets (Aksoy, 2017).

The highest per capita consumption of dried figs, including further processed products, is less than 1 kg per year in Turkey and in Greece, the two traditionally leading fig producing countries. Other countries consume fewer figs. In fact, per capita consumption in Iran and Afghanistan is 0.3 and 0.2 kg per year, respectively, and well below 0.2 in all other countries, including Spain, Italy and USA.

The international trade of dried figs is based primarily on whole, cut, sliced or cubed fruits, and fig paste for industrial use. Whole figs are consumed directly. The commercial quality standards for dried fig markets are set by the United Nation Economic Commission for Europe (UNECE, 2016) under their Dry and Dried Product (DDP-14) Standards for producers.

The requirements for dried figs apply to products for direct consumption and ensure the figs are intact, sound, free of abnormal moisture, foreign smell and/or taste, and insects, setting tolerances for defective fruits. There are several key attributes, divided into four categories, that determine dried fruit quality: visual appeal, organoleptic characteristics (porosity, texture, rehydration properties, flavor), chemical properties (water activity, chemical stability, taint, off-odor),

**Fig. 17.1.** Turkish dried figs on display in an Italian supermarket. (Figure courtesy of G. Colelli.)

**Fig. 17.2.** Added-value confectionary products based on dry figs for high-end markets. (Figure used with permission from Productos La Higuera, www.bombondehigo.com/nuestros-bombones/)

and safety (microbial load, pests and contaminants) (Aksoy, 2017; Tan, 2017; Şen *et al.*, 2017).

Drying methods and total sugars, particularly reducing sugars, have a significant effect on dried fig color. Reducing sugars increase the non-enzymatic browning reaction which produces brown-colored compounds (Abul-Fadl *et al.*, 2015; Şen *et al.*, 2017). The most valued criteria for dried fig processing are high sugar and soluble solids and low acidity content, as sugar accumulation depends on maturity and particularly the final three days of ripening (Freiman *et al.*, 2015). Flavor preferences are market driven. Turkey and other Middle Eastern markets prefer cultivars with high sugar content, while European markets prefer cultivars with lower sugar concentrations (Polat and Caliskan, 2008).

The standards also dictate that dried figs for direct consumption must contain less than 26% moisture in figs without additional pretreatments, while high-moisture dried figs may contain 26–40% moisture, provided they are labelled as 'high moisture' or equivalent denomination, and treated with preservatives or preserved by other means (e.g. pasteurization). Standards also specify fig size based on the number of fruits per kilogram, the minimum-maximum ranging between less than 40 and more than 121 fruits per kilogram, or fruit diameter (Dried Fig UNECE DDP14 Standard; UNECE, 2016). Fruit shape is also an important parameter of fruit quality: for convenient packaging and transportation, the most suitable fruit shape is globose (Polat and Caliskan, 2008).

## 17.4 Methods of Drying

Drying can be achieved in different ways. With natural drying the fruit is dehydrated by direct exposure to sunlight. Other methods can be defined as equipment-assisted drying. These include convective methods, using traditional heat sources like electricity or other fuels and/or renewable sources like solar energy, microwave, and other forms of radiation heating, freeze-drying, and osmo-dehydration. Freeze drying is a low temperature dehydration process that involves freezing the product, lowering pressure, then removing the ice by sublimation. Osmo-dehydration reduces water activity by partially transferring water from food to a hypertonic solution while gaining part of its solutes. These are in contrast with dehydration by most conventional methods that evaporate water using heat. Figure 17.3

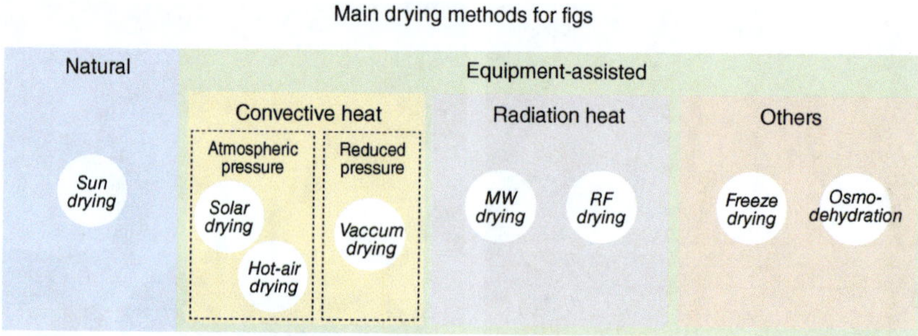


**Fig. 17.3.** Common fig drying methods (MW = microwave; RF = radiofrequency). (Author's own figure.)

summarizes common methods used for fig drying. During drying, two transfer phenomena occur simultaneously: heat is transferred from the heat source to the fruit, and moisture is transferred from the fruit's internal portions to the surface, then from the surface to the air. The drying conditions, including temperature, duration, air flow rate, relative humidity and hypertonic solution concentration greatly affect dried fig final quality.

Fig fruit drying systems are classified into high and low temperature dryers (Ekechukwu, 1999). The heat is supplied through conduction, convection and radiation, while air or another medium circulates, promoting dehydration. Recently, super-critical $CO_2$, ultrasound, pulsed electric fields, centrifugal force, heat pumps and superheated steam are used in combination with heat to enhance dehydration and/or reduce energy inputs (Sagar, 2010). However, most equipment-assisted drying methods, except solar drying, are energy-intensive and expensive, ultimately increasing final product cost (Sharma *et al.*, 2009). Low temperature drying systems can be powered by fossil fuels or, preferably, by solar energy, as it can accommodate the non-uniform heat supply from nature. High temperature dryers powered by fossil fuels or electricity are used commercially in more developed countries but they might not be economically feasible for small farmers in developing countries due to higher cost (Tiwari, 2016).

## Sun drying

Sun drying is commonly used in tropical and subtropical regions with average ambient temperatures above 30°C, although, in those climates, humidity and precipitation may cause problems. This method is still widely used in large scale farms and has been widely investigated. Yellow- or green-colored cultivars are preferred for sun drying as they maintain a better final color. Black cultivars are also marketed sun dried.

For traditional sun drying, figs are either harvested from the tree or collected from the ground. In both cases final drying is done on trays (Haug *et al.*, 2013). If figs are allowed to fully ripen, they partially dehydrate on the tree and drop naturally, and should be collected multiple times per day, to prevent mold and pest damage (Fig. 17.4). Once on the trays in the sun, the fruits should be turned over daily to achieve uniform drying, and pressed flat during the 3 to 7 day process until water content has decreased 18–22%.

Drying trays may have a wooden or plastic frame. Their size and shape should allow easy carrying. Typically, they should be no longer than two-arm lengths to facilitate access to all the fruit. The drying surface should not bruise or mechanically damage the fruit, and the frame legs should be long enough to prevent direct soil contact, to avoid contamination, but sufficiently close to the ground to capture radiated soil heat. The drying area should remain unshaded through the day. Fruits that are harvested earlier from the tree are peeled to accelerate drying and produce softer dried fruits. They may also be cut into halves, quarters or slices prior to drying. Treatment with $SO_2$ before drying is a common practice to produce a lighter colored fruit and to decrease the risk of storage pests (Fig. 17.5).

Fine plastic netting fabric can be used to cover the drying fruit to minimize risk of insect infestation and contamination with coarse particulates (Fig. 17.6). The netting will decrease the solar radiation intensity and may also slow down vapor mass transfer from the product to the surrounding air, slowing down dehydration. Local climatic conditions directly affect the dried fig quality; optimal drying conditions are reported to be 40°C temperature and 40–50% relative humidity (Aksoy and Anaç, 1994; Göçmez and Seferoğlu, 2014). The effects of ambient parameters on sun drying efficacy were investigated by using mathematical modeling by Toğrul and Pehlivan (2004). Drying rate curves of fig fruits did not show a constant-rate period, but showed a falling-rate period. In particular, during the falling drying rate period predominant mass transfer takes place inside the fruit, by molecular diffusion or vapor diffusion or by capillary forces in wet region of the product, and the water evaporated as it reached the surface hinting that the most probable mechanism governing moisture transfer was that of liquid diffusion.

Fig. 17.4. Typical 'stretched neck' appearance of a partially dehydrated 'Dottato' fig at the overripe stage. Fruits like this can either be hand-harvested or allowed to drop and be collected from the orchard floor before laying on trays for sun drying. (Figure courtesy of G. Colelli.)

The 'approximation of the diffusion' model showed the best fitting statistical variables to explain one-layer open air sun drying behavior. Also, a multiple regression analysis demonstrated that surface temperature of the fruits and relative humidity just above their surface were the most important model parameters affecting moisture transfer in the model and in its application.

Although sun-drying is environmentally friendly and low-cost, it is slow and final fruit quality can be decreased by defects including caramelization, crusting and non-homogeneous residual humidity. Direct sun exposure also affects fruit color, vitamin content and flavor (Chimi *et al.*, 2008). In dark-skinned varieties, sun drying effectively decreased weight and moisture content, and increased sugar content, but significantly reduced the content in bioactive compounds and antioxidant activity. Phenolic acids decreased almost 29%. Conversely, cinnamic and gallic acids increased. Fresh dark figs have high flavonoid levels, particularly cyanidin 3-rutinoside and quercetin 3-rutinoside. Sun-drying decreased flavonoid content by 86% (Bachir Bey *et al.*, 2016).

Dust contamination, insect infestation and microbial growth are constant problems associated with sun drying. Direct sun drying can be considered as commercial dried fig processing although with potentially high postharvest losses. Plastic tunnels that can reduce drying time of early and late-season fruit while also providing protection against dew, rain and hot and dry winds, can dramatically improve the quality of sun-dried figs (Fig. 17.7) (Oksar *et al.*, 2017).

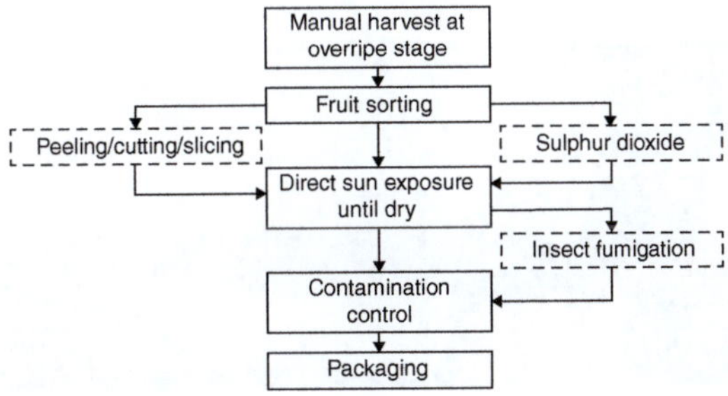


Fig. 17.5. Most common operations for fig sun drying. (Adapted from Aksoy, 2017.)

## Solar drying

Using the energy from the sun to dry agricultural products is economically convenient, as it is available free of cost, and contributes to reducing the environmental impact associated with the use of fossil fuels. However, as previously described, sun drying has disadvantages which may lead to degradation and poor quality of the final product. Another cost-effective and more hygienic alternative is solar drying using temperature control and forced ventilation to increase efficiency. These improved systems are characterized by a wide range of technological inputs but all have the same goals: improved reliability for accelerating dehydration, limiting fruit contact with the environment, and producing a better final product at an economically viable cost. For fig-producing countries which are within the belt of reliable high solar radiation, improved solar dryers may be the best option. Depending on their design, solar dryers are considered to be the most cost-effective, generally maintaining temperatures between 50 and 60°C and efficiently removing moisture (Singh *et al.*, 2018).

**Fig. 17.6.** Use of fine net fabric to protect figs from contamination and insects. (Figure courtesy of G. Colelli.)

Jairaj *et al.* (2009) classified solar drying into direct, indirect and mixed. Direct dryers expose fruit to direct and/or reflected solar radiation. Indirect dryers direct solar heated air to flow through the product. Mixed dryers use both directly exposing fruit to solar radiation and direct solar heated air to flow through it. All three types are dependent on the sunlight hours, climate, weather, atmospheric clearness, and

**Fig. 17.7.** Sun-drying plastic tunnels. (Figure used with permission from www.higosandfigs.com/)

latitude and elevation. Direct solar dryers are transparent, allowing the solar radiation to directly heat the fruit surface. Popular direct solar dryer designs are cabinet and greenhouse dryers. Indirect solar dryers have opaque drying chambers and dry with sun-heated air instead of radiation. This type of dryer is best for perishable fruits in which direct sunlight decreases nutritional content and changes color. They are also better for larger fruits as higher operating temperature can be achieved than with direct solar dryers (Ekechukwu, 1999). Compared to other dryer designs, indirect dryers are more energy efficient since more drying is achieved with less energy cost (Mustayen *et al.*, 2014). Solar dryers can also be differentiated based on air circulation. In natural convection mode, air is heated and circulates naturally by density difference. In forced convection mode the air is circulated with fan or blowers. For both natural or forced convection dryers, the air can be heated directly or indirectly. Indirect or mixed type dryers may integrate solar energy with other forms, including fossil or biomass fuel, or using photovoltaic-thermal collectors (PVT).

Gallali *et al.* (2000) reported the superiority of solar-dried products over naturally sun-dried ones in terms of faster moisture reduction, lower ash content, increased acidity and better sensorial acceptance. Another study evaluated the performance of two types of solar installations for drying 'Kadota' figs under various regimes compared to sun drying (El-Razik *et al.*, 1999). Solar drying was 2.5–3 times faster than the traditional method, but the sun-dried fruit received better organoleptic test scores. Both sun-dried and solar-dried samples were positive for aflatoxin type B1, but the concentrations were below the Food and Drug Administration limit.

In general, harvesting by hand from the tree, instead of collecting dropped fruit from the ground, and using solar drying instead of natural sun drying, were considered the most effective combination for reducing mold growth, mycotoxin formation and yeast contamination in small scale operations (Ozay *et al.*, 1995). A simple and inexpensive Indian solar dryer reportedly expedited drying but also improved the quality of 'Bellary' figs compared to sun drying (Raghupathy *et al.*, 1997). German solar tunnel driers equipped with light-powered fans used on a commercial scale in Turkey were reported to improve the microbiological quality of dried figs (Ural, 1997; Green and Schwarz, 2001). El-Sebaii *et al.* (2002) used an indirect-type natural-convection solar dryer to study the effect of 'heat storage' on the drying of spherical fruits. The cabinet solar dryer consisted of a flat-plate solar air heater using matte black-painted copper sheeting, and sand as a 'heat storage' system to maximize the solar heat captured for drying.

This drying system was capable of heating air to 45.5–55.5°C. The authors reported significant differences in drying times for different sizes and types of fig samples: the system took 12 h to reduce the moisture content of cut figs, compared to 32 h for whole fruit. Badry *et al.* (2016) reported solar dryers can be modified to minimize the adverse effects of drying when compared to natural sun drying. Specifically, they reported the drying rate of grapes and apricots using solar dryers with reflective mirrors and electric hot air dryers were faster than the rates of solar dryers without reflective mirrors, and faster than natural sun drying. They concluded that the inclusion of reflective mirrors significantly increased the drying potential in the four major components of fresh fruit drying: short time of operation, drying efficiency, low cost of operation and quality of dried products.

## Oven drying

Solar dryers represent significant progress compared to natural sun drying in terms of better product hygiene, as they reduce foreign matter contamination, have better control of processing parameters, and are more uniform, and thus it is better-quality production. However, they are still dependent on weather and sunlight. Deriving heat from electricity and fossil fuels has been demonstrated to be an even more practical alternative for convective drying, as it gives complete control over the drying process. What limits the use of these energy sources is infrastructure, as many rural areas in fig-producing countries do not have electricity. Several oven-drying techniques, including electric cabinets, tunnel and heat pump dryers, were successfully applied to fig. In oven dryers, the drying rate is affected by product water content and temperature, as well as by drying air temperature humidity and flow rate.

By drying kinetics, heat and mass transfers can be monitored over time and models can be developed to provide information on equipment design, process optimization, and final quality of the product. One of the most common models to estimate drying variables and to describe a drying process is the so called 'thin-layer' model (Ertekin and Firat, 2017). In one of their works, Xanthopoulos *et al.* (2007) tested seven thin-layer drying models simulating convective drying of figs using ranges of drying air velocity and temperatures of 1.0–5.0 m $s^{-1}$ and 46.1–60°C, respectively. They found the logarithmic model was the best one in terms of fitting performance through non-linear regression methods. In another work, Xanthopoulos *et al.*, (2009) also experimentally evaluated the effective diffusion coefficient employing Fick's law for unsteady diffusion, taking into consideration the shrinkage effect. Moreover, in another experiment, Xanthopoulos *et al.*, (2010) also determined the effect of peeling 'Tsapela' fig fruit on drying kinetics and effective diffusivity at three different temperatures, applying the logarithmic model to describe the drying. They reported that the effective diffusion coefficient in peeled figs was higher than in unpeeled figs with differences becoming smaller at higher drying temperature, and this was explained as being due to the faster hardening of the drying surface, which in turn reduced the drying rate and the diffusivity.

Drying ground and half-cut 'Mission' figs at 45, 55, and 65°C were investigated by Martínez-García *et al.* (2013). The effective moisture diffusivity, as estimated by the slope method, increased at higher temperatures, and was higher for half-cut figs compared to ground ones, with the former needing double the energy and time required by ground figs to dehydrate. The same authors also reported that the drying process enhanced the antioxidant capacity of dried figs, increasing the total phenols, but reducing the anthocyanin content. They concluded that dehydrating ground figs was faster than half-cut figs and maintained the functional properties better.

Tan (2017) investigated drying fresh whole and quartered 'Sarilop' figs at 50 and 60°C. Properties such as total phenolics and antioxidant activity were measured before and after drying. The temperature of the cabinet had a positive effect on skin thickness; in whole figs, the higher the temperature, the thinner the skin. In fig quarters the skin was even thinner than in whole fruit. The drying process increased the total phenolic content and antioxidant activity, especially when fig quarters were dried at 50°C.

Dehydration of 'Niedda Longa' figs in an industrial two-stage dryer in comparison with simulated sun drying was reported by Papoff *et al.* (1998). Results showed that 7 h were enough to ensure a ready-to-market product if a preservative (i.e. potassium sorbate) was applied to ensure microbiological stability. A simulated sun drying process of 120 h was needed for similar results. When the oven drying duration exceeded 11 h the product stability did not require chemical additives, although rehydration ensured a better fruit palatability.

According to Slatnar *et al.* (2011), oven-dried figs (at 62–64°C) were significantly richer in secondary metabolites compared to sun-dried figs. Controlling drying parameters with oven drying can ensure a more efficient process and better quality of the products, although the drying temperature is a very critical factor. Fresh 'Tsapela' figs at commercial maturity were either sun-dried (29–31°C, 30–40% RH) or hot-air dried (60°C, 20–22% RH) with the latter method being better in terms of processing time (i.e. moisture and water activity reduction) but worse in terms of color change (increased browning index) and fruit shrinkage (Manolopoulou *et al.*, 2017). Indeed, high drying temperatures (over 65°C) can contribute to a decline in the product's visual quality. The effects of oven drying conditions on product final quality is clearly showed by Ertekin *et al.* (2003): higher drying air temperatures reduced the time needed for moisture removal but negatively affected fig quality with high quality products obtained at a drying temperature of 50°C. According to these authors, in order to shorten drying times, it is important to decrease the relative humidity of drying air and ensure a velocity of 0.5 m $s^{-1}$.

Problems with thermal damage and low-energy efficiency in convection oven drying prompted investigation of alternative drying technologies, including microwave (MW) oven drying. Energy from MW implies heat production in dielectric materials by dipole rotation and ionic polarization, determining a selective increase in temperature in the areas with higher water content. Many food products, including fruit and vegetables, are very rich in water which when exposed to MW will increase in temperature and evaporate. Increasing water vapor inside the product tissue generates a vapor pressure gradient with the external air, contributing to removing moisture from the sample. MW drying is rapid and more energy efficient than convective drying (Schiffmann, 2006).

When drying using MW, the temperature of the product should not excessively increase to reduce the chances of thermal damage. For this reason, MW power is applied in a pulsed form, maximizing the drying effect as continuous heating would not speed up water removal when critical vapor content has already been reached in the intercellular space (Gunasekaran, 1999).

Sharifian *et al.* (2012) investigated five MW power intensities and six pulsing ratios on fig drying in a laboratory-scale microwave dryer. Drying time was directly proportional to the pulsing ratio and inversely proportional to MW power intensity. To describe drying rate as a function of time at the various MW power intensities and pulsing ratios, they fitted the experimental data with 11 drying models: the quadratic and logarithmic models described the drying kinetics best and with the highest statistical significance.

Although MW drying delivers heat rapidly, its major drawback is its non-uniform electromagnetic field resulting in non-uniform heating. Zhang *et al.* (2006) suggested combining microwave's fast heating with low-temperature convection drying could improve the results compared to each method alone. Abul-Fadl *et al.* (2015) compared conventional air drying and vacuum drying, both integrated with MW energy. They reported MW integration sharply reduced the drying time from 20 to 1.5 h for conventional air-drying and MW-assisted air drying respectively, and from 36 to 2 h for vacuum drying and MW-assisted vacuum drying respectively. Costs also decreased sharply. Additionally, compared to both conventional air drying and vacuum drying, these integrated technologies produced higher nutritional value and better quality (including color) with lower rehydration ratios. Lower rehydration ratios are evidence of less irreversible physical and chemical changes from product shrinkage caused by severe heating and prolonged drying which occurs with convection air drying.

## Osmotic dehydration

Osmotic dehydration (OD) is increasingly considered as a preliminary step to fruit and vegetable dehydration or freezing (Torreggiani and Bertolo, 2001), as it reduces processing costs and improves the final product sensorial quality. OD removes water by immersion in hypertonic solutions. However, it is a two-way transfer; while it reduces the free water in the product, it introduces solutes that decrease water activity and, possibly, compounds that may improve the sensorial quality in the final product (Buggenhout *et al.*, 2008).

According to Yadav and Singh (2014) the following advantages are associated with OD: (i) being a low temperature, water removal process risks and deleterious change of color due to thermal damage is strongly reduced, (ii) better flavor retention when sugar is used as an osmotic agent, (iii) reduced browning and better color retention with consequent lower need to use sulfites, (iv) reduction of water removal load in a subsequent combined-drying operation, (v) lower energy consumption as no water phase change is involved, (vi) increased solid density due to solid uptake, and for this reason, (vii) higher moisture content allowed at same level of water activity, (viii) better textural quality of product after reconstitution, (ix) longer potential storage life, and, finally, (x) relatively simple and non-expensive equipment required for the process. According to the same authors product quality is influenced by a number of factors including pretreatment, nature, and concentration of osmotic solution, quality of raw material, maturity of fruits, shape and size of slices, duration of osmosis, sample to syrup ratio, agitation rate, temperature, and type additives added.

Niranjan *et al.* (2018) evaluated the effect of temperature (40 and 50°C) and osmotic pretreatment concentration (40, 50, and 60° Brix), on the sensorial quality of sliced figs after tray drying at 60°C for 8 h to a final moisture content of <5%. During pretreatment moisture loss and solid gains increased as the osmotic solution concentration, the soaking time and the soaking temperature increased. The fig slices dried in the 40°Brix solution at 40°C had the highest sensorial acceptability. Naikwadi *et al.* (2010) reported that dehydrated figs prepared using fructose or invert sugar syrup (50% glucose + 50% fructose) for the OD step maintained market acceptability for 6 months. The invert sugar syrup treatment produced the better dried fig quality.

When a vacuum is applied at the beginning of OD, the process is called pulsed vacuum osmotic dehydration (PVOD) (Chiralt and Fito, 2003). PVOD presents a hydrodynamic mechanism that leads to the introduction of osmotic solution into the pores (i.e. lenticels, stomata, etc.) of the food product. The use of vacuum pressure in osmotic dehydration speeds up processing times and allows a reduction in energy during process (Corzo *et al.*, 2007). The drying kinetics of fresh, non-pretreated and pretreated 'Sarılop' figs using PVOD at 130 mbar for 15 min was investigated by Şahin and Öztürk (2016). The drying time for pretreated figs was shorter. The fresh non-pretreated figs had greater shrinkage. Similar results were reported more recently by the same authors (2018) with PVOD applied at 253 mbar for 15 min before sun drying. The mean effective diffusivity for PVOD-treated and non-treated fresh figs was calculated as $2.55 \times 10^{-10}$ $m^2 s^{-1}$ and $2.36 \times 10^{-10}$ $m^2 s^{-1}$, respectively. They also found that an artificial neural network model was more successful than 11 other mathematical drying models investigated for predicting the experimental changes in the moisture ratio over time for both PVOD-treated and fresh figs.

Finally, in another experiment by Villalobos *et al.* (2016) fresh figs were dried as follows: drying in a convection oven at different temperatures; pretreatment by soaking in osmotic solutions at different concentrations; pretreatment with $K_2CO_3$ + olive oil, with or without osmotic solutions; pretreatments with ultrasound combined or not with osmotic solutions; and a sun-drying control. The figs dried with equipment-assisted systems achieved the required 24% (wet weight basis) moisture content in 1–3 days. The sun-dried figs needed 15 days to achieve the same moisture content. Treatments with ultrasound combined with osmotic solutions had lower mold counts, and better firmness and sensorial quality. The authors concluded the latter treatment being a viable alternative to sun drying.

## 17.5 More on Drying: Pre- and Post-drying Treatments

The drying methods described above are the ones most applied to figs and those with most abundant published literature. Although all are characterized by positive and negative effects, it is difficult to define an optimal procedure for all situations. Table 17.1 presents a synthesis of the published advantages and disadvantages for all methods discussed above. Most incorporate pre- and post-drying treatments that might improve the quality and, most importantly, the safety of the final product.

In general, drying foods consists of three steps: pre-drying processing, drying per se, and post-drying treatments. Their effects on the final product quality depend on the kind of processing material and the intended use of the final product (Lewicki, 2006). Fig fruits are considered a high-risk commodity, as they are prone to toxigenic fungi and their toxins. Mycotoxins are secondary metabolites produced by fungi found in the environment and possibly in all plant parts which can be toxic to human and animals through consumption of contaminated food and feed entering into the food chain. There are a number of different types of aflatoxin that showed toxigenic effects, some of them even causing cancer. Best management practices are the primary defense against aflatoxin contamination of dried figs; these are defined from production through handling, processing, storage and distribution. Only by effective control at all stages of production and processing, from the ripening on the tree through harvest, drying, processing, packaging, storage, transportation and distribution, can the safety and quality of the final product be ensured (Codex Alimentarius 2008).

The dominant mycoflora in dried whole figs varies widely and depends upon location, variety, maturity stage, processing and storage. The most common toxigenic fungi reported are *Aspergillus* section *Nigri*, *Fusarium* spp., *Aspergillus* section *Flavi*, and *Penicillium* spp. (Heperkan *et al.*, 2012). Aflatoxins are produced by toxigenic strains of *Aspergillus flavus* and *Aspergillus parasiticus*; the most important are aflatoxins B1, B2, G1, G2 and M1. Under ultraviolet light B aflatoxins produce blue fluorescence while G aflatoxins produce green fluorescence (Karaca and Nas, 2008).

**Table 17.1.** Comparison of different drying methods (adapted from Singham and Birwal, 2014 and Mat Desa *et al*., 2019).

| Drying method | Advantages | Disadvantages |
|---|---|---|
| Sun drying | Large capacity<br>Inexpensive<br>Sustainable drying practice<br>No capital investment | Slow drying process<br>Higher product loss<br>Susceptible to insect infestation, pest disturbance and dust contamination<br>Higher microbial growth<br>Highly dependent on climate<br>Non-controllable drying parameters<br>Lower product quality<br>Labor-intensive<br>Requires large drying area |
| Solar drying | Good to excellent product quality<br>Low capital investment<br>Sustainable drying practice<br>Adequate control over drying parameters<br>Reduced drying time<br>Easy to manufacture with readily available materials<br>Easy to handle and use | Dependent on climate<br>Appropriate for tropical and subtropical regions |
| Oven drying | Adjustable drying settings and parameters<br>Fast drying time<br>Reduced labor cost | Increased product cost<br>Can be difficult to operate<br>High capital cost<br>Energy intensive<br>Possible thermal damage to the product |
| Osmotic dehydration | Maintains color and flavor<br>Prevents browning<br>Low energy consumption<br>Better quality<br>Enhances storage life | Reduces acidity therefore changes flavor<br>High water activity<br>Time consuming<br>Mostly good as a pretreatment |

When growing conditions are suitable, mycotoxin production can occur while the fruit is on the tree. Fig fruit are very susceptible to mold contamination, therefore direct contact with the soil and physical damage should be avoided. However, these toxins also proliferate during harvest, transportation, processing and storage, particularly under prolonged and inappropriate storage conditions at high temperatures, even when water activity has been reduced. International markets, particularly in the European Union, have established strict limits for the presence of total aflatoxins on dried whole figs, with a maximum allowable of 6–10 μg $kg^{-1}$ (European Commission, 2012).

Pre-harvest colonization of figs by *Aspergillus* spp. can be limited by a range of fungicides that reduce mycotoxin contamination. At bud-break, fungicides were applied to the tree and soil, whereas at later stages (fruiting, ripening, and shriveling), they were only applied to the soil under the trees. These antifungal treatments effectively reduced fruit contamination below the tolerance limits (Tosun and Delen, 1998).

Other disorders of dried figs caused by an inappropriate drying process are usually associated with transformations of the primary components of these fruits, such as sugars, organic acids, pectin, cellulose, hemicellulose, phenolic compounds and minerals due to various enzymatic processes, leading to physiological changes and, consequently, loss of product quality. Pretreatments may help reduce the incidence of contaminations and disorders that degrade fig quality and marketability. Pretreatments are often a key factor in controlling quality changes as well as the microbial growth that occurs during drying. These treatments include $SO_2$, sulfuring, blanching, soaking in sugar syrup, basic or acid solutions, and dipping in oil. Pretreatments may vary with fruit and the desired final product (Anand *et al.*, 2018).

Sulfuring fresh fig fruits before drying is a long-established practice to maintain lighter dried fruit color and to prevent storage pest damage (Aksoy, 2017). In the United States, white cultivars are treated with $SO_2$ to inhibit the Maillard reaction that causes browning and darkening during oven drying (Haug *et al.*, 2013). Sulfuring has also been proven to reduce shrinkage, improve rehydration and facilitate drying (Lewicki, 2006). In a study conducted by Villalobos *et al.* (2016), chemical treatment applications are effective when applied prior to drying in order to increase fruit skin permeability to water by removing the wax cuticle. In this particular study, chemicals of use were methyl and ethyl ester emulsions, and alkaline treatments (aqueous solutions of KOH, NaOH or $K_2CO_3$). However, it was reported that these treatments can also lead to reduced flavor due to solute loss from the product. A similar conclusion was also made by Piga *et al.* (2004) who found $SO_2$-treated fruits were less susceptible to browning. Overall sensorial assessment showed better acceptability for treated samples. Other research findings showed that blanching and sulfitation were the best pretreatments for dried figs preparation in terms of chemical composition retention and organoleptic preservation (Waskar and Gawade, 2004). On the other hand, contrary remarks were made by El-Sebaii *et al.* (2002) on unnecessary pretreatment on figs due to its soft skin texture. However, they reported that a chemical pretreatment using olive oil and NaOH in boiling water significantly reduced the drying times of smaller products, such as grapes and green peas. Pre-treatments have been observed to decrease the overall drying time (Piga *et al.*, 2004; Mujić *et al.*, 2012; Villalobos *et al.*, 2016). In addition, Mujić *et al.* (2012) found that non-pre-treated samples showed the highest abundance of aldehydes, while pre-treatments with $SO_2$, citric acid, and ascorbic acid resulted in higher retention of esters and alcohols (according to variety) and in dramatic reduction of drying time by about 50%. Piga *et al.* (2004) observed a sharp reduction from 55 h to 33 h of drying time in blanched and sulfited samples as opposed to untreated samples in convective hot air drying. Anand *et al.* (2018) investigating the behavior of figs as influenced by different pre-treatments (blanching, steam treatment and $SO_2$ fumigation) found that blanching was the best on the basis of sensory evaluation of the final product.

After drying figs need to be stored. The major problem in dried fig marketing after mycotoxin proliferation is storage pest infestations. Among the major dried fig pests is the dried-fruit beetle (*Carpophilus hemipterus*), which readily penetrates fallen overripe fruit but is not attracted to undamaged healthy figs, or to fruit that is very dry or far advanced in decay. However, larvae that begin their growth in overripe figs may continue their development after the fruit is fairly dry (Simmons and Nelson, 1975). Erakay and Ozar (1979) observed high *Carpophilus* infestation of ripe figs before abscission. Other important insect pests of dried figs are the moths *Plodia interpunctella* and *Cadra cautella*, and the mite *Carpoglyphus lactis* (Erakay and Ozar, 1979). In addition to the direct damage, insect infestations reduce the fig nutritional value (Saleh *et al.*, 1987) and contribute to the distribution of *Aspergillus* spores, although they are not necessary for the fungus penetration of the fruit (Buchanan *et al.*, 1975).

Until recently, for safe storage after the drying process, dried figs were treated with methyl bromide (MBr) fumigations which have represented for many years the main instrument to control pest development in dried fruits (Erakay and Ozar, 1979). As MBr is now banned by both national and international regulations, the search for MBr alternatives has increased and a range of treatments are being investigated as possible substitutes: phosphine, released by the hydrolysis of magnesium phosphide (Meyvaci *et al.*, 2003; Meyvaci *et al.*, 2009); sulfuryl fluoride (Meyvaci *et al.*, 2009); propylene oxide in combination with high $CO_2$ (Zettler *et al.*, 2003); high $CO_2$ (Ferizli and Emekci, 2000; Emekci *et al.*, 2003; Meyvaci *et al.*, 2003); controlled atmosphere with less than 1% $O_2$ and 10–15% $CO_2$ (Damarli *et al.*, 1998; Navarro *et al.*, 1998).

Extensive research has also been directed at finding treatments that can decrease aflatoxin contamination in dried figs. Encouraging results were achieved with sulfur dioxide combined with heat, ultraviolet illumination, hydrogen peroxide or potassium sorbate (Icibal and Altug, 1992; Elmaci *et al.*, 1994).

An additional strategy for reducing the risk of introducing aflatoxin-contaminated fruit into the market is to sort the contaminated fruit out by taking advantage of the significant relationship between presence of aflatoxins and fluorescence of fig fruit under long-wave UV light; dried figs can be screened under UV lamps removing those with fluorescence. The contaminated figs are collected and then destroyed as hazardous waste (Polat and Siddiq, 2012) (Fig. 17.8).

Reduction of microbial count on dried figs was effectively obtained by exposing the product to ozone gas. Öztekin *et al.* (2006) reported that treatment for 3 h at 5 ppm reduced the total aerobic mesophyllic microorganisms, coliform and yeast/mold counts. Akbas and Ozdemir (2008) exposed figs to 0.1, 0.5 and 1.0 ppm ozone concentrations for up to 360 min and reported decreased *E. coli* and *B. cereus* counts of 3.5 log numbers, while concentrations of 1.0, 5.0, 7.0 and 9.0 ppm decreased *B. cereus* spores by up to 2 log numbers. No significant changes in color, pH and moisture content were observed after ozone treatment. Moreover, no significant changes were found in sweetness, rancidity, flavor, appearance and overall palatability of ozonated and non-ozonated dried figs.

In Italy, figs are baked after drying in order to obtain a product that is stable at room temperature, eventually after the addition of, e.g., almonds, walnuts, chocolate, laurel leaf or lemon peel, according to traditional recipes and/or novel gourmet formulations (Fig. 17.9). In some cases, dried fruits are traditionally baked after wrapping in a layer of fig leaves which also constitute an efficient method to keep away potential pests (Fig. 17.10).

Fig fruit chips may represent an additional product alternative to unhealthy snacks; figs are either chopped or diced, and sized to customer specifications, possibly coated with dextrose, rice or oat flour (Tarko *et al.*, 2010) (Fig. 17.11).

## 17.6 Other Processing Methods

### Freezing

Freezing is a method for preserving food that maintains its original attributes, particularly its nutritional properties, as close as possible to the fresh fruit (Canet and Alvarez, 2006). During freezing the water phase changes into ice, which greatly reduces microbial and enzymatic activities. Also, oxidation and respiration are drastically reduced at temperatures below 0°C. However, due to the ice formation, freezing causes physical and chemical changes in food, and a subsequent loss of quality (Reid, 1996). Physical changes in frozen fruits involve stresses in cell volume, dislocation of water, mechanical damage and freeze cracking. The quality of frozen fruits, once they are thawed, depends on a large number of factors including the type of fruit (Sousa *et al.*, 2007), the variety (González *et al.*, 2002), the initial quality, the stage of ripeness and the time lapse between harvesting and freezing (Phan and Mimault, 1980).

The scientific literature on fig fruit freezing is neither abundant nor recent. According to the few references available, figs can be frozen either whole in heavy syrup, or sliced four parts of fruit to one part of water (by weight). Standard freezing methods are employed and storage temperatures should be below 0°C (Reid, 1996). Woodroof (1975) described the use of ascorbic and citric

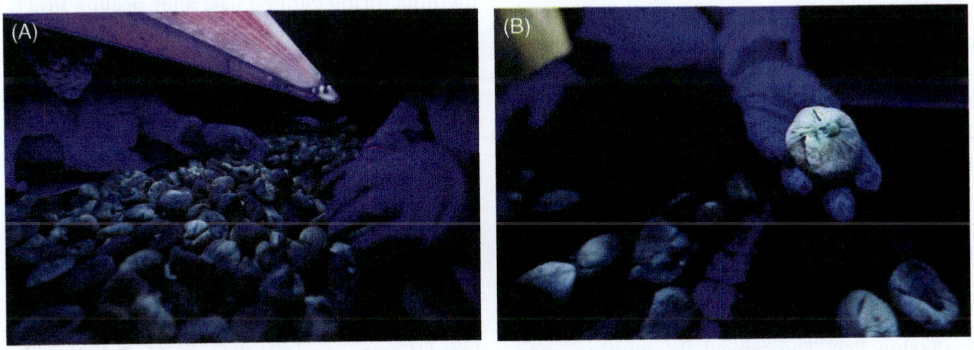


**Fig. 17.8.** Separating fluorescent figs under UV light as possibly contaminated by aflatoxins. (Figure courtesy of S. Güncelleme.)

**Fig. 17.9.** Coupled and stuffed fig halves, after the baking treatment. (Figure courtesy of G. Colelli.)

**Fig. 17.10.** Italian dry fig products including a 'fig ball' made of dry whole figs baked in a wrapping of fig leaves. (Figure courtesy of G. Colelli.)

**Fig. 17.11.** Fig fruit chips. (Photo: www.karyafresh.wordpress.com/) (Figure used with permission from Alimohammad Yavari.)

acids for maintaining frozen fig fruit color. Cruess and Christis (1958) recommended the use of tender skin fruit, such as 'Mission' figs for freezing. They should be table ripe, considerably riper than for canning or shipping. Airtight containers are recommended to prevent off-flavor induced by oxidation. Adding 0.10–0.15% ascorbic acid to the syrup can prevent oxidation and protect flavor.

## Canning

Canning has never been a common method for processing figs. The typical fig canning operations include cleaning, washing, size grading, blanching at 65–77°C for up to 8 min to remove the waxy coating, filling the cans, adding 16–35 °Brix hot syrup, sealing the cans, heat processing at 100°C for 30–70 min depending on the can size and cooling to below 35–40°C (Downing, 1996). Hohl (1946) recommended dipping whole figs in a 2000 ppm sulfite solution for 2–3 min before packing the fruit in 35°Brix sugar solution to preserve flavor and color. Luh and Kean (1975) described the following process for manufacturing canned figs: after peeling in 2% lye solution for 10–15 s, the figs are washed and then cooked in 30 °Brix syrup until a 60–65 °Brix syrup density is reached. The canning solution should be acidified with citric acid or lemon juice to a pH lower than 4.9.

Ito *et al.* (1978) investigated how to prevent the outgrowth of *C. botulinum* spores in canned fig puree, concluding that maintenance of pH 4.9 or below resulted in the best conditions regardless of syrup concentration. Canning fruits in low- or no-sucrose syrup is a growing health trend (Campos and Cândido, 1995; Rorato *et al.*, 2006). However, processing standards and diet-canned fruits carry some challenges, such as preserving the figs' integrity, promoting longer shelf-life and improving palatability.

When preserved without added sugar, fruit can be softer than in sugar solution, and its flavor and color may change. Caetano *et al.* (2017) monitored physicochemical, microbiological and sensory quality of traditional and low-calorie figs canned in syrup, as well as their storage stability, using standard sucrose syrup and a sweeteners blend made of sodium cyclamate, sucralose and sodium saccharin. After 180 days of storage at room temperature as both canned figs treatments did not show significant difference for pH, titratable acidity, soluble solids, vacuum and texture, they were considered commercially sterile and kept their sensory characteristics throughout storage (Fig. 17.12).

## Jelly

Fig fruits are frequently preserved as jelly or jam. Both involve high temperatures and a considerable amount of sugar. In order to make fig jelly the fruits are usually boiled in water and the juice is strained out through a cloth. Sugar is added to the strained-out juice, which is then boiled down or concentrated until it 'sets' in a jelly texture upon cooling. Nearly all known processes of jelly making involve extended heating of fruit juice and its evaporation at high temperature. The application of heat, especially at high temperature and for a considerable time, very seriously impairs, or even destroys, the natural and desirable fruit flavor, as a probable consequence of the evaporation or destruction of volatiles and essential oils which impart their characteristic flavors (Johnson, 1920).

However, producing fig jelly is an alternative, added-value product with a small but distinct market niche that uses fig surpluses unmarketable as fresh or dried product. In addition, fig jelly, after severe heat application, and because of its very low water activity, represents a safe alternative for processed fig consumption, due to a considerably lower risk of aflatoxins occurrence. Curi *et al.* (2019) reported on the characterization of eight fig cultivars in relation to antioxidant activity and content in bioactive compounds to identify cultivars with increased potential for jelly processing when grown in subtropical regions. All cultivars used in their experiment yielded jellies with different physical and chemical characteristics and different rheological properties, which however did not significantly affect acceptability. They concluded that all of the evaluated cultivars had a high potential for jelly processing.

**Fig. 17.12.** Sample of commercially canned figs for the Iranian market. (Photo: www.saharfood.ir) (Figure used with permission from Alimohammad Yavari.)

## Jam

Fig preservation in the form of jam is very popular in many countries (Fig. 17.13) and, like jelly, is a convenient form for long storage and delayed consumption of fig fruit. The process also involves the addition of sugar and cooking of the mix at boiling temperature. Recipes and formulations are very much dependent on local customs and sometimes other fruits, such as lemon and apples, are added in small amounts in order to increase the acidity without greatly affecting the organoleptic traits of the final product. As with fig jelly, overripe figs are often used. Due to their high soluble solids content, less added sugar is needed to ensure stability, as total sugar content of the final product should not exceed 40%. The production of fig jam is also increasing due to the recent appearance of China as a fig producing country with almost 7000 ha of specialized production. In the south of China, most of the overripe fig fruits are used in jam processing which make up the main fig products of export in China (Lianju *et al.*, 2003).

**Fig. 17.13.** Fig fruit jam. (Figure used with permission from www.valleyfig.com/)

Carbone *et al.* (2001) reported on the potential of new processed fig products using a combination of different technologies, including high-pressure processing (HPP) to reduce water activity and lower bacterial growth, as well as enzyme activity, without increasing the temperature, thus avoiding thermal damage to the fruit. Ripe 'Dottato' figs were washed, cut in halves, dipped for 15 min into a 0.5% ascorbic acid solution, and partially dehydrated in air to a water activity of 0.87 (3.5 m $s^{-1}$ air velocity, 50°C temperature, and 50% humidity; process duration 7 h). The fruits were packaged in 0.1 mm thick poliamide + polyethylene pouches filled with an isotonic fructose solution, and treated at 600 MPa for 3 min, using an HPP processing plant (Fig. 17.14). After storage at 4°C for 28 days, the combination of dehydration and HPP gave the most promising results in terms of degree of microbial inactivation and chemical and microbiological shelf-life.

## Powder

Fig fruit can be successfully dried to produce powder, which can be used in many food products including beverages and confectioneries (Fig. 17.15). It can also be used as a healthier alternative to artificial additives. Spray drying, tray drying, freeze drying, foam mat drying (FMD), and MW drying are the most used drying

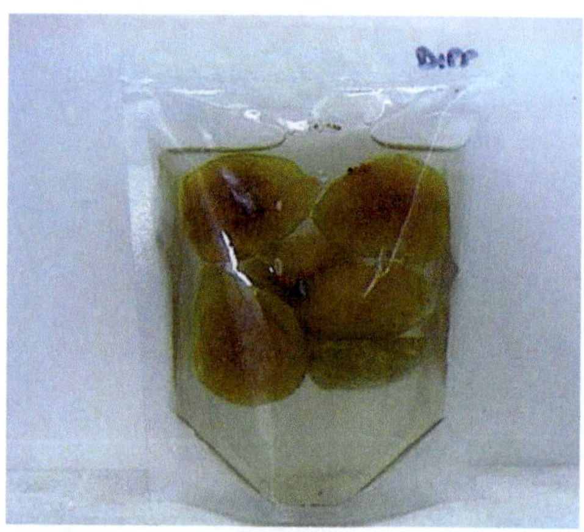

**Fig. 17.14.** 'Dottato' fig halves, partially dehydrated and treated with high-pressure processing, after 4 weeks of storage at 4°C. (Figure courtesy of G. Colelli.)

**Fig. 17.15.** Fig fruit powder. (Figure used with permission from www.ekowarehouse.com/)

methods to produce powdered products. However, the obtained product after drying (except for spray drying) is scraped and milled to obtain powder. FMD has attracted great attention recently, especially for those foods rich in sugar, with better results than other drying methods in terms of drying time, processing costs and product quality. FMD starts with whipping liquid or semiliquid material to form a stable foam which is then spread over tray, drum, or moving bands, and dried by conduction, convection or radiation. The foam is a system that consists of dispersed and continuous phase separated by a thin film liquid layer (Hardy and Jideani, 2017). Gas bubbles in the foam provide a porous structure and a large surface area which increases drying rate and reduces the drying time. Moreover, the porous structure leads to better reconstitution properties in the final product. A stable gas/liquid foam is the primary condition for successful FMD. If the bubbles are unstable, they tend to merge with their adjacent bubbles and become too enlarged, thus leading to more drainage. This circumstance is undesirable in FMD and final product quality (Azizpour *et al.*, 2014). Varhan *et al.* (2019) reported on using FMD technology for making fig fruit powder and showed that the foam capacity and stability of the fig powder highly depend on the foam composition, with optimum ratio of fig, egg white, carboxymethyl cellulose and maltodextrin determined as 52.9, 28.9, 0.8, and 17.4% respectively (wt/wt). They also found that MW drying method was more efficient than hot air. Khapre *et al.* (2015) reported that 'Deanna' fig powder was found to be superior in terms of yield and ease of processing, in contrast to fig pulp, for the preparation of fig burfi (a typical Indian cookie), concluding that fig powder incorporated burfi was nutritionally rich in terms of fiber, potassium and protein and the prepared product had a lower cost as compared to similar market products.

## 17.7 Other Processed Fig Products

### Syrup and paste

Both fig syrup and paste are made from boiling dried figs in water. The difference is in their concentrations (the paste is thicker). After sorting and cleaning, the dry figs are soaked in water before a special mixer chops them up. Then, the mixture is lightly steamed while careful tending helps to check its consistency and keeps it from overcooking. This process is repeated several times to ensure that it achieves the right thickness (Fig. 17.16).

Additionally, dehydrated fruit edible films such as 'wraps' and 'leathers' have been developed

**Fig. 17.16.** Fig fruit syrup. (Figure used with permission from www.valleyfig.com/)

recently (Fig. 17.17). Preparation of 'fruit leather' is not only a preservation method but it also has the advantage of adding value and diversifying fruit products. 'Fruit leather' is regarded as a healthy choice and a big opportunity to promote fruit consumption. It is very common in Iran, Azerbaijan and Persian-speaking countries, where it is known as *Lavashak* (Gaikwad *et al.*, 2020).

Fig paste is also used for filling and decorating confectionary products like cakes, flans, tarts, fine desserts and biscuits (Fig. 17.18). Dried figs can be blended with nuts or grain to produce fig bars. Mixtures of dried figs, dates, raisins and nuts are also produced in some Mediterranean countries as a snack or dessert.

### Achenes oil

Currently, oils from natural plant resources have attracted growing interest. Vegetable oils have relatively high amounts of minor lipid components, whose health benefits have generated a rising demand (Carvalho *et al.*, 2006). Achenes from *Ficus carica* may represent a new oil source. When fig fruits are processed for juice and puree, the substantial resulting byproduct material contains high amounts of achenes which could be a potential source of nutraceuticals instead of being fed to livestock or discarded.

Fig achenes are rich in lipids. A biochemical analyses of fig achene oil showed it contains very high levels of carotenoids and phytosterols. The fatty acid profile shows a predominance of linolenic acid known for its several beneficial properties. Fig achenes oil is also a valuable source of antioxidant phenols and tocopherols. All these minor compounds are known to have a wide range of beneficial biological activities and physical properties. This confirms the fig achene oil's nutritional value and potential dietary importance (Soltana *et al.*, 2016).

**Fig. 17.17.** Fig roller with sesame. (Figure used with permission from www.laborgena.com/)

### Fruit latex

Due to a mixture of different enzymes, including primarily ficin, the latex of the edible fig contains more proteolytic activity than any of the other fifteen species tested by Whitaker (1958). These enzymes can be divided into two groups: one including enzymes with high milk clotting activity and low proteolysis, the other including those with high proteolytic effect. The mixture can be used in a number of food processes including tenderizing meat and clarifying beverages (e.g. preventing haze in chilled beer) (Messing, 1975).

(a) (b)

**Fig. 17.18.** Fig biscuits and bar. (Figure used with permission from www.laborgena.com/ and www.valleyfig.com/)

According to Lazreg-Aref *et al.* (2017) the variation in latex yield is a function of fruit growth stage. Similarly, the enzyme activity of latex varies with fruit maturity stage and latex collecting time. For the majority of tested cultivars, fig fruits collected at mid-maturity yielded the most latex with the highest protease activity. However, the 'Chetoui' and 'Bidh Begal' cultivars had high protease activity when the fruit was fully ripe. Interestingly, fig fruits collected in the late afternoon generated the maximum latex content. Some local cultivars have higher potential as a source of latex enzymes, than for consumption as fruit.

## Conclusions

A significant share of the world's fig production is processed. Among all traditional and new technologies for preserving figs, drying is the most important and widespread processing method worldwide. Although drying appears easy, requiring few inputs, several decades of scientific investigation have shown that it is complex and specific to fruit species. The world's standards of living are increasing the demand for innovative foods and novel processing that minimize the destructive effects of heat on the healthy compounds and functional properties of food (Picart-Palmade *et al.*, 2019). This has led to continuous improvements of techniques, promoted by the need to improve product quality and safety, while reducing costs and increasing energy efficiency. For example, HPP, pulsed electric fields, irradiation and ultra-filtration, would possibly facilitate deriving even more value from figs as a functional food.

## References


Abul-Fadl, M.M., Ghanem, T.H., El-Badry, N. and Nasr, A. (2015) Effect of some different drying methods on quality criteria of dried fig fruits. *Current Science International* 4, 548–566.

Akbas, M.Y. and Ozdemir, M. (2008) Application of gaseous ozone to control populations of *Escherichia coli, Bacillus cereus* and *Bacillus cereus* spores in dried figs. *Food Microbiology* 25, 386–391.

Aksoy, U. (2017) The dried fig management and the potential for new products. *Acta Horticulturae* 1173, 377–382. DOI: 10.17660/actahortic.2017.1173.65

Aksoy, U. and Anaç, D. (1994) Factors affecting quality of fresh and dried fig fruits. *Acta Horticulturae* 368, 763–766. DOI: 10.17660/actahortic.1994.368.90

Amodio, M.L. and Colelli, G. (2020) Subtropical fruits: Figs. In: Gil, M.I. and Beaudry, R. (eds) *Controlled and Modified Atmospheres for Fresh and Fresh-Cut Produce*. Academic Press, an imprint of Elsevier, London, p. 684.

Anand, A., Jagadeesh, S.L. and Bhuvaneshwari, G. (2018) Effect of different pre-treatments on dehydration of fig fruits under electric tray dryer. *International Journal of Chemical Studies* 6, 1716–1721.

Azizpour, M., Mohebbi, M., Hossein, M., Khodaparast, H. and Varidi, M. (2014) Optimization of foaming parameters and investigating the effects of drying temperature on the Foam-Mat Drying of Shrimp (*Penaeus indicus*). *Drying Technology* 32 (4), 374–384. DOI: 10.1080/07373937.2013.794829.

Bachir Bey, M., Richard, G., Meziant, L., Fauconnier, M.L. and Louaileche, H. (2016) Effects of sun-drying on physicochemical characteristics, phenolic composition andin vitroantioxidant activity of dark fig varieties. *Journal of Food Processing and Preservation* 41(5), e13164. DOI: 10.1111/jfpp.13164.

Badry, E.L., Hamied, A.E.L. and Sayed, E.L.C. (2016) Comparative study between the dried fruits by solar dryers with or without reflective mirrors as utilization of renewable solar energy and conventional dryers. *Middle East Journal of Applied Sciences* 6(3), 490–510.

Berry, M.R. and Pflug, I.J. (2003) Canning principles. In: Caballero, B. (ed.) *Encyclopedia of Food Sciences and Nutrition (Second Edition)*. Academic Press, Oxford, UK, pp. 816–824.

Buchanan, J.R., Sommer, N.F. and Fortlage, R.J. (1975) *Aspergillus flavus* infection and aflatoxin production in fig fruits. *Applied Microbiology* 30, 238–241.

Buggenhout, S., Grauwet, T., van Loey, A. and Hendrickx, M. (2008) Use of Pectinmethylesterase and calcium in osmotic dehydration and osmodehydrofreezing of strawberries. *European Food Research and Technology* 226, 1145–1154. DOI: 10.1007/s00217-007-0643-7

Caetano, P.K., Farrinazi-Machado, F.M.V., Nasser, F.A.M., Furlaneto, K.A. and Vieites, R.L. (2017) Storage stability of standard and diet figs canned in syrup. *Food Science and Technology* 37, 154–159.

Campos, A.M. and Cândido, L.M. (1995) Formulação e avaliação físico-química e reológica de geleias de baixo teor de sólidos solúveis com diferentes adoçantes e edulcorantes. *Ciência e Tecnologia de Alimentos* 15(3), 268–278.

Canet, W. and Alvarez, M.D. (2006) Quality and safety of frozen vegetables. In: Sun, W. (ed.) *Handbook of Frozen Food Processing and Packaging*. CRC Press, Boca Raton, FL, pp. 377–415.

Carbone, B.F., Colelli, G., Cosentino, F., La Sala, G., Prete, M., Spremulli, L. and Todaro, E (2001) Nouvel products based on figs (*Ficus carica* L.) obtained by innovative technologies (in Italian.) *Italus Hortus* 8(5 suppl.), 75–77.

Carvalho, I.S., Miranda, I. and Pereira, H. (2006) Evaluation of oil composition of some crops suitable for human nutrition. *Industrial Crops and Products* 24(1), 75–78. DOI: 10.1016/j.indcrop.2006.03.005.

Chimi, H., Ouaouich, A., Semmar, M. and Tayebi, S. (2008) Industrial processing of figs by solar drying in Morocco. *Acta Horticulturae* 798, 331–334. DOI: 10.17660/actahortic.2008.798.48

Chiralt, A. and Fito, P. (2003) Transport mechanisms in osmotic dehydration: The role of the structure. *Food Science and Technology International* 9(3), 179–186.

Codex Alimentarius (2008) Code of Practice for the prevention and reduction of aflatoxin contamination in dried figs. In: *CAC/RCP 65-2008*. FAO, Rome.

Corzo, O., Bracho, N., Rodríguez, J. and González, M. (2007) Predicting the moisture and salt contents of sardine sheets during vacuum pulse osmotic dehydration. *Journal of Food Engineering* 80 (3), 781–790. DOI: org/10.1016/j.jfoodeng.2006.07.007

Crisosto, H., Ferguson, L., Bremer, V., Stover, E. and Colelli, G. (2011) Fig (*Ficus carica* L.). In: Yahia, E.M. (ed.) *Postharvest Biology and Technology of Tropical and Subtropical Fruits: Cocona to Mango*. CABI, Wallingford, UK, pp. 134–158.

Cruess, W.V. and Christie, A. W. (1958) *Dehydration of Fruit.* Agricultural Experiment Station. Bulletin no. 330, California.

Curi, P.N., Albergaria, F.C., Pio, R., Schiassi, M.C.E.V., Tavares, B. and Souza, V. (2019) Characterisation and jelly processing potential of different fig cultivars. *British Food Journal.* doi: 10.1108/BFJ-03-2019-0201

Damarli, E., Gun, H., Ozay, G., Bulbul, S. and Oechsle, P. (1998) An alternative method instead of methyl bromide for insect disinfestation of dried figs: Controlled atmosphere. *Acta Horticulturae* 480, 209–214.

Downing, D.L. (1996) Canning of fruits. In: Downing, D.L. (ed.) *A Complete Course in Canning and Related Processes (Thirteenth Edition)*. Woodhead Publishing, pp. 129–192.

Ekechukwu, O.V. (1999) Review of solar-energy drying systems I: an overview of drying principles and theory. *Energy Conversion and Management* 40(6), 593–613. doi: 10.1016/S0196-8904(98)00092-2

Elmaci, Y., Altug, T. and Olstzyn, P. (1994) Aflatoxin degradation in dried figs on industrial scale by using sulphur dioxide gas alone and together with heat, hydrogen peroxide or washing treatments. In: *Proceedings of the International Euro Food Tox., IV. Conference, Bioactive substances in food of plant origin* (Vol. 2, pp. 290–296).

El-Razik, F.A.A.A., Farghal, S.A. and El-Kayati, S.M. (1999) Utilization of solar energy in drying Kadota fig fruit. *Desert Institute Bulletin* 47, 323–344.

El-Sebaii, A.A., Aboul-Enein, S., Ramadan, M.R.I. and El-Gohary, H.G. (2002) Experimental investigation of an indirect type natural convection solar dryer. *Energy Conversion and Management* 43(16), 2251–2266. DOI: 10.1016/S0196-8904(01)00152-2

Emekci, M., Ferizli, A.G., Tutuncu, S. and Navarro, S. (2003) Modified atmosphere as an alternative to MBR in the dried fig sector in Turkey. *Annual International Research Conference on Methyl Bromide Alternatives and Emissions Reductions, San Diego, CA.*

Erakay, S. and Ozar, A.H. (1979) Preliminary investigations on infestation rates and control measures concerning insects injurious to dried figs in the Aegean region (in Turkish). *Bitki Koruma Bulteni* 19, 159–173.

Ertekin, C. and Firat, M.Z. (2017) A comprehensive review of thin-layer drying models used in agricultural products. *Critical Reviews in Food Science and Nutrition* 57(4), 701–717.

Ertekin, C., Yaldiz, O. and Mühlbauer, W. (2003) Thin layer drying of fig as effected by different drying conditions. *Acta Horticulturae* 605, 283–291. DOI: 10.17660/ActaHortic.2003.605.43

European Commission (2012) *Regulation (EU) No 1881/2006 as regards maximum levels for aflatoxins in dried figs. No 1058/2012 of 12.*

Ferizli, A.G. and Emekci, M. (2000) Carbon dioxide fumigation as a methyl bromide alternative for the dried fig industry. *Annual Research Conference on Methyl Bromide Alternatives and Emissions Reductions, Orlando, FL.*

Freiman, Z., Rosianskey, Y., Dasmohapatra, R., Kamara, I. and Flaishman, M. (2015) The ambigious ripening nature of the fig (*Ficus carica* L.) fruit: a gene-expression study of potential ripening regulators and ethylene related genes. *Journal of Experimental Botany* 66(11), 3309–3324.

Gaikwad, N.N., Patil, P.G., Patil, N.B., Cano-Lamadrid, M. and Carbonell-Barrachina, Á.A. (2020) Processing and industrialization. In: Sarkhosh, A., Yavari, A. and Zamani, Z. (eds) *The Pomegranate Botany, Production and Uses*. CAB International, Wallingford, UK, pp. 481–503.

Gallali, Y.M., Abujnah, Y.S. and Bannani, F.K. (2000) Preservation of fruits and vegetables using solar drier: a comparative study of natural and solar drying, III; chemical analysis and sensory evaluation data of the dried samples (grapes, figs, tomatoes and onions). *Renewable Energy* 19 (1), 203–212. DOI: 10.1016/S0960-1481(99)00032-4

Göçmez, A. and Seferoğlu, H.G. (2014) Fresh fig and dry fig quality parameters and the effective factors on quality. *Turkish Journal of Agricultural Research* 1, 98–108.

González, E., de Ancos, B. and Cano, P. (2002) Preservation of raspberry fruits by freezing: physical, physico-chemical and sensory aspects. *European Food Research and Technology* 215(6), 497–503. DOI: 10.1007/s00217-002-0600-4

Green, M.G. and Schwarz, D. (2001) Solar drying technology for food preservation. *GATE Tech. Inform* E14e.

Gunasekaran, S. (1999) Pulsed microwave-vacuum drying of food materials. *Dry Technology* 17(3), 395–412.

Hardy, Z. and Jideani, V.A. (2017) Foam-mat drying technology: a review. *Journal Critical Reviews in Food Science and Nutrition* 57, 2560–2572.

Haug, M.T., King, E.S., Heymann, H. and Crisosto, C.H. (2013) Sensory profiles for dried fig (*Ficus carica* L.) cultivars commercially grown and processed in California. *Journal of Food Science* 78(8), S1273–S1281. DOI: 10.1111/1750-3841.12196

Heperkan, D., Güler, F.K. and Oktay, H.I. (2012) Mycoflora and natural occurrence of aflatoxin, cyclopiazonic acid, fumonisin and ochratoxin A in dried figs. *Food Addit Contam Part A Chem Anal Control Expo Risk Assess* 29(2), 277–286.

Hohl, A. (1946) Freezing California fruits. *Food Packer* 27(11), 66–70.

Icibal, N. and Altug, T. (1992) Degradation of aflatoxin R, in dried figs by sulphur dioxide alone and in combination with heat, ultraviolet energy and hydrogen peroxide. *LWT—Food Science Technology* 25, 294–296.

Ito, K.A, Chen, J.K., Seeger, M.L., Unverferth, J.A. and Kimball, R.N. (1978) *Journal of Food Science* 43(5), 1634–1635.

Jairaj, K.S., Singh, S.P. and Srikant, K. (2009) A review of solar dryers developed for grape drying. *Solar Energy* 83(9), 1698–1712. DOI: 10.1016/j.solener.2009.06.008

Johnson, M.O. (1920) Process of making jelly. Google Patents No. *US20080032027A1*.

Karaca, H. and Nas, S. (2008) Mycotoxins in dried figs. *Acta Horticulturae* (798), 307–312. DOI: 10.17660/actahortic.2008.798.44

Khapre, A., Satwadhar, P. and Syed, H. (2015) Studies on processing technology and cost estimation of fig (*Ficus carica* L.) fruit powder enriched Burfi (Indian cookie). *Journal of Applied and Natural Science* 7, 621–624. DOI: 10.31018/jans.v7i2.655

Lazreg-Aref, H., Zaouay, F., Hammami, M. and Mars, M. (2017) Fig fruit latex yield and protease activity as influenced by cultivar, maturity stage and harvesting time. *Acta Horticulturae* 323–328. DOI: 10.17660/ActaHortic.2017.1173.56

Lewicki, P.P. (2006) Design of hot air drying for better foods. *Trends in Food Science & Technology* 17(4), 153–163.

Lianju, W., Weibin, J., Kai, M., Zhifeng, L. and Yelin, W. (2003) The production and research of fig (*Ficus carica* L.) in China. *Acta Horticulturae* 605, 191–196. DOI: 10.17660/actahortic.2003.605.28

Luh, B.S. and Kean, C.E. (1975) Canning of fruits. In: *Commercial Fruit Processing*. USDA Hand Book AVI publication, pp. 141–266.

Manolopoulou, E., Tsiavtari, E. and Demopoulos, V. (2017) Sun drying and hot air drying of figs: performance evaluation. *Acta Horticulturae* 341–346. DOI: 10.17660/ActaHortic.2017.1173.59

Martínez-García, J.J., Gallegos-Infante, J.A., Rocha-Guzmán, N.E., Ramírez-Baca, P., Candelas-Cadillo, M.G. and González-Laredo, R.F. (2013) Drying parameters of half-cut and ground figs (*Ficus carica* l.) var. mission and the effect on their functional properties. *Journal of Engineering* 1–8. DOI: 10.1155/2013/710830

Mat Desa, W.N., Mohammad, M. and Fudholi, A. (2019) Review of drying technology of fig. *Trends in Food Science & Technology* 88, 93–103. DOI: 10.1016/j.tifs.2019.03.018

Messing, R.A. (1975) *Immobilized Enzymes for Industrial Reactors*. Academic Press, New York.

Meyvaci, K.B., Aksoy, U., Sen, F., Altindisli, A., Turanli, F., Emekci, M. and Ferizli, A.G. (2003) Project to phase out methyl bromide in the dried fig sector in Turkey. *Acta Horticulturae* 628, 73–81.

Meyvaci, K.B.F., Sen, F. and Aksoy, U. (2009) Methyl bromide alternatives to control major storage pests in Turkish dried fig industry. *Fourth International Symposium on Fig. Meknes, Morocco.*

Mujić, I., Kralj, M.B., Jokić, S., Jug, T., Šubarić, D., Vidović, S., Živković, J. and Jarni, K. (2012) Characterisation of volatiles in dried white varieties figs (*Ficus carica* L.). *Journal of Food Science and Technology* 51(9), 1837–1846. DOI: 10.1007/s13197–012-0740-x

Mustayen, A.G.M.B., Mekhilef, S. and Saidur, R. (2014) Performance study of different solar dryers: A review. *Renewable and Sustainable Energy Reviews* 34, 463–470. DOI: 10.1016/j.rser.2014.03.020
Naikwadi, P.M., Chavan, U.D., Pawar, V.D. and Amarowicz, R. (2010) Studies on dehydration of figs using different sugar syrup treatments. *Journal of Food Science and Technology* 47(4), 442–445. DOI: 10.1007/s13197-010-0073-6
Navarro, S., Donahaye, E., Rindner, M. and Azrieli, A. (1998) Storage of dried fruits under controlled atmospheres for quality preservation and control of nitidulid beetles. *Acta Horticulturae* 480, 221–226.
Niranjan, T., Rajender, G., Reddy, P., Reddy, V., Kumar, A. and Krishna, V. (2018) Study on osmotic dehydration of fig fruit (*Ficus carica*) slices mediated tray drying. *International Journal of Current Microbiology and Applied Sciences* 7, 3198–3205. DOI: 10.20546/ijcmas.2018.706.375
Oksar, R.E., Şen, F. and Aksoy, U. (2017) Effect of drying under plastic tunnels on drying rate and quality of fig (*Ficus carica* L. 'Sarılop'). *Acta Horticulturae* (1173), 297–302. DOI: 10.17660/actahortic.2017.1173.51
Ozay, C., Aran, N. and Pala, M. (1995) Influence of harvesting and drying techniques on micro flora and mycotoxin contamination of figs. *Molecular Nutrition and Food Research* 39(2), 156–165.
Öztekin, S., Zorlugenç, B. and Zorlugenç, F.K. (2006) Effects of ozone treatment on microflora of dried figs. *Journal of Food Engineering* 75(3), 396–399. DOI: 10.1016/j.jfoodeng.2005.04.024
Papoff, C.M., Battacone, G., Agabbio, M., Farris, G.A., Vodret, A., Milella, G., Sotgiu, C., Sanna, F.E., Piga, A. and D'Aquino, S. (1998) The influence of industrial dehydration on quality of fig fruits. *Acta Horticulturae* 480, 233–238.
Phan, P.A. and Mimault, J. (1980) Effects of freezing and thawing on fruit. Evaluation of some texture parameters and exudate. Relation to fruit quality. *Journal of Refrigeration* 3(5), 255–260.
Picart-Palmade, L., Cunault, C., Chevalier-Lucia, D., Belleville, M.P. and Marchesseau, S. (2019) Potentialities and limits of some non-thermal technologies to improve sustainability of food processing. *Frontiers in Nutrition* 5, 130. DOI: 10.3389/fnut.2018.00130
Piga, A., Pinna, I., Ozer, K.B., Agabbio, M. and Aksoy, U. (2004) Hot air dehydration of figs (*Ficus carica* L.): drying kinetics and quality loss. *International Journal of Food Science and Technology* 39(7), 793–799. DOI: 10.1111/j.1365-2621.2004.00845.x
Polat, A.A. and Caliskan, O. (2008) Fruit characteristics of table fig (*Ficus carica*) cultivars in subtropical climate conditions of the Mediterranean region. *New Zealand. Journal of Crop and Horticulture Science* 36(2), 107–115.
Polat, A.A. and Siddiq, M. (2012) Figs. In: Siddiq, M., Ahmed, J., Lobo, M.G. and Ozadali, F. (eds) *Tropical and Subtropical Fruits Postharvest Physiology, Processing and Packaging*. Wiley E-book www.wiley.com/en-us/Tropical+and+Subtropical+Fruits%3A+Postharvest+Physiology%2C+Processing+and+Packaging-p-9781118324110
Raghupathy, P.R., Hussain, A.A., Reddy, P.N., Megeri, V.K., Krishnanand Gavali, A. and Naravani, N.B. (1997) Evaluation of tray type bamboo basket solar driers for figs. *Karnataka Journal of Agricultural Science* 10, 420–423.
Reid, D.S. (1996) Fruit freezing. In: Somogyi, L.P., Ramaswamy, H.S. and Hui, Y.H. (eds) *Processing Fruits: Science and Technology*. Technomic Publishing Company, PA, pp. 275–286.
Rorato, F., Degáspari, C.H. and Mottin, F. (2006) Evaluation of the knowledge level on diet and light products of consumers and customers on a Curitiba supermaket. *Revista Visão Acadêmica* 7(1).
Sagar, V.R. (2010) Recent advances in drying and dehydration of fruits and vegetables : A review. *Journal of Food Science and Technology* 47, 15–26.
Şahin, U. and Öztürk, H.K. (2016) Effects of pulsed vacuum osmotic dehydration (PVOD) on drying kinetics of figs (*Ficus carica* L). *Innovative Food Science & Emerging Technologies* 36, 104–111. DOI: 10.1016/j.ifset.2016.06.003
Saleh, S.M., Yacout, G.A. and El-Sadek, M.M. (1987) Comparative studies on sound and infested dried fig with *Carpoglyphus lactis* (Linne) Alexandria. *Journal of Agricultural Research* 32, 409–418.
Schiffmann, R.F. (2006) Microwave and dielectric drying. In: Mujumdar, A.S. (ed.) *Handbook of Industrial Drying* 3rd edn. CRC Press, Boca Raton, Florida.
Şen, F., Aksoy, U., Özer, K.B., Can, H.Z., Köseoğlu, İ. and Konak, R. (2017) Impact of yearly conditions on major physical and chemical properties of fresh, semi-dried and sun-dried fig (*Ficus carica* L. 'Sarılop') fruit. *Acta Horticulturae* 1173, 309–314.
Sharifian, F., Motlagh, A.M. and Nikbakht, A.M. (2012) Pulsed microwave drying kinetics of fig fruit (*Ficus carica* L.) *Australian Journal of Crop Science* 6(10), 1441–1447.
Sharma, A., Chen, C.R. and Lan, N.V. (2009) Solar-energy drying systems: a review. *Renewable and Sustainable Energy Reviews* 13(6), 1185–1210. DOI: 10.1016/j.rser.2008.08.015

Simmons, P. and Nelson, H.D. (1975) *Insects on Dried Fruits. USDA ARS, Handbook 464*. USDA, Washington, DC.
Singh, P., Shrivastava, V. and Kumar, A. (2018) Recent developments in greenhouse solar drying: A review. *Renewable and Sustainable Energy Reviews* 82, 3250–3262. DOI: 10.1016/j.rser.2017.10.020
Singham, P. and Birwal, P. (2014) Technological revolution in drying of fruit and vegetables. *International Journal of Science Research* 3, 705–711.
Slatnar, A., Klancar, U., Stampar, F. and Veberic, R. (2011) Effect of drying of figs (*Ficus carica* L.) on the contents of sugars, organic acids, and phenolic compounds. *Journal of Agricultural and Food Chemistry* 59 (21), 11696–11702.
Soltana, H., Tekaya, M., Amri, Z., El-Gharbi, S., Nakbi, A., Harzallah, A., Mechri, B. and Hammami, M. (2016) Characterization of fig achenes' oil of *Ficus carica* grown in Tunisia. *Food Chemistry* 196, 1125–1130. DOI: 10.1016/j.foodchem.2015.10.053
Sousa, M.B., Canet, W., Alvarez, M.D. and Fernandez, C. (2007) Effect of processing on the texture and sensory attributes of raspberry (cv. Heritage) and blackberry (cv. Thornfree). *Journal of Food Engineering* 78(1), 9–21.
Tan, N. (2017) Effect of cabinet drying method on dried fruit quality and functional properties of 'Sarilop' (*Ficus carica* L.) fig cultivar. *Acta Horticulturae* (1173), 359–364. doi: 10.17660/actahortic.2017.1173.62
Tarko, T., Duda-Chodak, A., Sroka, P., Satora, P. and Tuszyński, T. (2010) Production of flavored apple chips of high antioxidant activity. *Journal of Food Processing and Preservation* 34(4), 728–742.
Tiwari, A. (2016) A review on solar drying of agricultural produce. *Journal of Food Processing & Technology* 7(9). DOI: 10.4172/2157-7110.1000623
Toğrul, İ.T. and Pehlivan, D. (2004) Modelling of thin layer drying kinetics of some fruits under open-air sun drying process. *Journal of Food Engineering* 65(3), 413–425. doi: 10.1016/j.jfoodeng.2004.02.001.
Torreggiani, D. and Bertolo, G. (2001) Osmotic pre-treatments in fruit processing: Chemical, physical and structural effects. *Journal of Food Engineering* 49, 247–253.
Tosun, N. and Delen, N. (1998) Minimising of contamination of aflatoxigenic fungi and subsequent aflatoxin development in fig orchards by fungicides. *Acta Horticulturae* 480, 193–198. DOI: 10.17660/ActaHortic.1998.480.33
UNECE (2016) UNECE STANDARD DDP-14 concerning the marketing and commercial quality control of dried figs. 2016 edition.
Ural, A. (1997) Processing of dried figs. In: *Advanced Course on Fig Production*. Izmir, Turkey.
Varhan, E., Elmas, F. and Koç, M. (2019) Foam mat drying of fig fruit: optimization of foam composition and physicochemical properties of fig powder. *Journal of Food Process Engineering* 42(4). DOI: 10.1111/jfpe.13022
Villalobos, M.C., Serradilla, M.J., Martín, A., Pereira, C., López-Corrales, M. and Córdoba, M.G. (2016) Evaluation of different drying systems as an alternative to sun drying for figs (*Ficus carica* L). *Innovations in Food Science and Emerging Technologies* 36, 156–165. DOI: 10.1016/j.ifset.2016.06.006
Waskar, D.P. and Gawade, M.H. (2004) Storage of fig fruits - a review. *Agricultural Reviews* 25, 76–78.
Whitaker, J.R. (1958) The ficin content of the latex from different varieties of *Ficus carica* and a comparison of several micro-methods of protein determination. *Journal of Food Science* 23(4), 364–370. DOI: 10.1111/j.1365-2621.1958.tb17581.x
Woodroof, J.G. (1975) Storage life of canned frozen dehydrated and preserved fruits. In: *Commercial Fruit Processing*. AVI publication, pp. 595–625.
Woodroof, J.G. (1986) Storage life of canned, frozen, dehydrated, and preserved fruits. In: *Commercial Fruit Processing*. Springer, Netherlands.
Xanthopoulos, G., Oikonomou, N. and Lambrinos, G. (2007) Applicability of a single-layer drying model to predict the drying rate of whole figs. *Journal of Food Engineering* 81(3), 553–559. DOI: 10.1016/j.jfoodeng.2006.11.033
Xanthopoulos, G., Yanniotis, S. and Lambrinos, G. (2009) Water diffusivity and drying kinetics of air drying of figs. *Drying Technology* 27(3), 502–512. DOI: 10.1080/07373930802686149
Xanthopoulos, G., Yanniotis, S. and Lambrinos, G. (2010) Study of the drying behaviour in peeled and unpeeled whole figs. *Journal of Food Engineering* 97(3), 419–424. DOI: 10.1016/j.jfoodeng.2009.10.037
Yadav, A.K. and Singh, S.V. (2014) Osmotic dehydration of fruits and vegetables: a review. *Journal of Food Science and Technology* 51(9), 1654–1673. DOI: 10.1007/s13197-012-0659-2
Zettler, J.L., Hartsell, P.L., Allred, D.B., Muhareb, J.S., Hurley, J.M. and Gill, R.F. (2003) Sorption and insect toxicity of propylene oxide in dried fruits and nuts. *Advances in stored product protection: Proceedings of the 8th International Working Conference on Stored Product Protection, York, UK.*
Zhang, M., Tang, J., Mujumdar, A.S. and Wang, S. (2006) Trends in microwave-related drying of fruits and vegetables. *Trends in Food Science and Technology* 17, 524–534.

# 18 Fig and Health

**S. Mahmoudi[1]* and A. Chawla[2]**
[1]*Department of Agronomic Sciences, University "Mohamed BOUDIAF", Msila, Algeria;* [2]*CT College of Pharmacy, Shahpur, Jalandhar, India*

## 18.1 Introduction

The relationship between food and health becomes increasingly significant as consumers now demand healthy, tasty and natural functional foods that have been grown in an uncontaminated environment (Ercisli, 2007; Abbasi *et al.*, 2013). Fruit and vegetables are rich sources of several biologically active components that contribute to general health and decrease the risk of chronic diseases (Henning *et al.*, 2017). Epidemiological studies have consistently shown an inverse correlation between the consumption of fruits and vegetables and the incidence of several chronic diseases such as cancer, cardiovascular disease, aging, atherosclerosis, neurodegenerative diseases (such as Parkinson's and Alzheimer's), and inflammation. This is attributed to the fact that these foods may provide an optimal mixture of phytochemicals such as dietary fiber, phenolics, natural antioxidants and other bioactive compounds (Vinson, 1999; Al-Farsi *et al.*, 2007).

*Ficus carica* L., a deciduous tree belonging to the Moraceae family, originated in the Middle East, spread to the Mediterranean countries, and from there to the United States (Vinson, 1999; Vallejo *et al.*, 2012; Harzallah *et al.*, 2016). The world's major fig producers are now concentrated in the Mediterranean. Turkey, Egypt, Morocco and Algeria now produce 65% of the world's production (FAO, 2017; Soltana *et al.*, 2018). Figs are important in the Mediterranean diet, considered among the world's healthiest diets, and strongly associated with longevity (Hoxha *et al.*, 2015). Figs are rich in carbohydrates predominantly fructose and glucose, fiber, magnesium, potassium, calcium, and vitamins K, α-tocopherol, and $D_3$ (Guvenc *et al.*, 2009; Lim, 2012; Sadia *et al.*, 2014). Fig fruits are consumed fresh, dried, preserved, canned and candied (Ammar *et al.*, 2015).

Fresh or dried, figs have been reported as a good source of phenolic compounds: phenolic acids (caffeoylquinic and ferulic acids), flavonoids (quercetin 3-*O*-glucoside, quercetin 3-*O*-rutinoside, apigenin, luteolin and kaempferol rutinoside), anthocyanins (cyanidin-3-*O*-glucoside, cyanidin-3-rutinoside and pelargonidin-3-glucoside) and procyanidins (Dueñas *et al.*, 2008; Oliveira *et al.*, 2009; Vallejo *et al.*, 2012; Wang *et al.*, 2017). Phenolic compounds are secondary metabolites, widespread in nature that play multiple physiological roles in plants and human health, as they can act as antioxidants (Oliveira *et al.*, 2009). Figs also accumulate other secondary metabolites such as carotenoids (lutein, zeaxanthin and β-carotene),

*Corresponding author: mahmoudisouhila@yahoo.fr, souhila.mahmoudi@univ-msila.dz

 *The Fig: Botany, Production and Uses* (eds. Sarkhosh, A., Yavari, A. and Ferguson, L.)
DOI: 10.1079/9781789242881.0018

furanocoumarins (psoralen and bergapten), terpenoids and volatile compounds (aldehydes, esters, alcohols and ketones) (Yemis *et al.*, 2012; Trad *et al.*, 2012b; Marrelli *et al.*, 2012; Ammar *et al.*, 2015; Russo *et al.*, 2017).

Different parts of the fig tree, fruit, bark, leaves, twigs, young shoots, and latex from the bark, have multiple uses in traditional medicine. They have been used in treating tumors, ulcers, and digestive, endocrine, kidney, liver, gastrointestinal, reproductive, cardiovascular and respiratory disorders, and as an antispasmodic, anti-inflammatory and laxative (Guarrera, 2005; Gilani *et al.*, 2008; Lansky *et al.*, 2008; Alamgeer *et al.*, 2017). Figs have also been used as an antihaemorrhagic and an antidiabetic remedy, a mouthwash for sore throats and as an expectorant and diuretic (Joseph and Raj, 2011). Figs are also consumed for their curative properties in severe cases of piles and asthma and are often given to anemic patients (Waheed and Siddique, 2009).

Scientific investigations of the efficacy of fresh and dried figs for cardiovascular diseases (Alamgeer *et al.*, 2017; Gholami *et al.*, 2017), cancer (Marrelli *et al.*, 2012; Purnamasari *et al.*, 2019), inflammation (Yang *et al.*, 2015) and constipation (Oh *et al.*, 2011; Lee *et al.*, 2012; Rtibi *et al.*, 2018) have been reported in the literature. There are multiple reports in the literature of the antihyperlipidemic and anticholesterolimic (Belguith-Hadriche *et al.*, 2016), the antidiabetic and antiobesogenic (Wojdyło *et al.*, 2016; Mopuri *et al.*, 2018) and the antioxidant (Hoxha *et al.*, 2015; Pereira *et al.*, 2017b; Mahmoudi *et al.*, 2018) effects of figs. Additionally, *Ficus carica* leaves, shoot bark and latex have been evaluated for their bioactivity as antimicrobial, anticancer, anti-inflammatory, antipyretic, antidiabetic and anti-obesogenic, antihyperlipidemic, hepatoprotective, neurological and antidiarrheal activities (Canal *et al.*, 2000; Jeong *et al.*, 2009; Patil *et al.*, 2010a; Khodarahmi *et al.*, 2011; Kabir *et al.*, 2012; Allahyari *et al.*, 2014; Mopuri *et al.*, 2018; Raafat and Wurglics, 2019; Abdel-Atya *et al.*, 2019).

This chapter summarizes the myriad studies on the health-promoting potential of figs, their nutritional composition, and phytochemicals that support their use as a functional food and medicine.

## 18.2 Nutrient Composition of Fresh and Dried Figs

Figs are rich in nutrients, vitamins, fiber, protein, carbohydrates and minerals. In particular, dried figs are one of the highest plant sources of fiber, magnesium, potassium, calcium and vitamin K for humans (Lim, 2012). The nutritional values of fresh and dried figs are given in Table 18.1.

### Carbohydrates

Carbohydrates are the most abundant macronutrient in fig fruits, ranging from 17.1 to 20.3 g/100 g edible portion in fresh figs to 69.1 g in dried figs (Guvenc *et al.*, 2009). This high carbohydrate content produces 74 kcal in 100 g of fresh ripe figs. The same portion of dried figs provides 332.7 kcal (Lim, 2012; Sadia *et al.*, 2014). Aljane and Ferchichi (2009) reported glucose and fructose were dominant sugars in ten fresh Tunisian fig cultivars (7.28 and 6.6 g/100 g of fresh matter, respectively). Glucose content in fig fruits was higher than fructose content (Aljane and Ferchichi, 2009; Trad *et al.*, 2012a; Pereira *et al.*, 2017a). Sucrose was detected in small amounts, an average of 0.14 g/100 g, in some Turkish fig accessions (Çaliskan and Polat, 2011). Rhamnose, arabinose, xylose, mannose and galactose were also present in fig fruits (Owino *et al.*, 2004). Lim (2012) reported a 100 g edible portion of Mission Valley dried figs contains 63.87 g carbohydrates, of which 47.92 g are total sugars, 0.07 g sucrose, 24.79 g glucose, 22.93 g fructose, 0.13 g galactose, and 5.07 g is starch. Fruit sugar levels are generally an important factor in ripe fruit flavor and consumer acceptance (Pereira *et al.*, 2017a). Differences in fig fruit sugar levels are directly correlated with genotype, ripening stage, crop year and storage conditions (Gozlekci, 2011; Pereira *et al.*, 2017a). Fig polysaccharides have been proven to have anti-tumor and antioxidant properties (Tian *et al.*, 2014).

### Organic acids

Pande and Akoh (2010) identified five organic acids in fresh figs: malic (431 ± 19.8 mg/100 g fresh weight (FW)), citric (20.2 ± 8.0 mg/100 g

**Table 18.1.** Nutrient composition of raw and dried figs (ANSES-CIQUAL, 2017).

| Components | Values | |
|---|---|---|
| | Fresh | Dried |
| Water (g/100 g) | 79.8 | 30.1 |
| Energy, Regulation EU No 1169/2011 (kcal/100 g) | N/A | 256 |
| Protein (g/100 g) | 0.85 | 2.99 |
| Carbohydrate (g/100 g) | 14.3 | 53.7 |
| Fat (g/100 g) | 0.3 | 0.87 |
| Sugar (g/100 g) | 14.3 | 48.7 |
| Starch (g/100 g) | 0 | 5.02 |
| Fibers (g/100 g) | 2.63 | 9.72 |
| Ash (g/100 g) | 0.68 | 2.11 |
| FA saturated (g/100 g) | 0.069 | 0.35 |
| FA mono (g/100 g) | 0.059 | 0.18 |
| FA poly (g/100 g) | 0.13 | 0.27 |
| Calcium (mg/100 g) | 39 | 167 |
| Iron (mg/100 g) | 0.37 | 2.12 |
| Magnesium (mg/100 g) | 17 | 52.5 |
| Phosphorus (mg/100 g) | 22.5 | 75.1 |
| Potassium (mg/100 g) | 232 | 845 |
| Sodium (mg/100 g) | 1 | 86.3 |
| Zinc (mg/100 g) | 0.15 | 0.51 |
| Beta-carotene (µg/100 g) | 85 | 6 |
| Vitamin E (mg/100 g) | 0.11 | 0.35 |
| Vitamin K1 (µg/100 g) | 4.7 | 15.6 |
| Vitamin C (mg/100 g) | 2 | 1.2 |
| Vitamin B1 or Thiamine (mg/100 g) | 0.06 | 0.085 |
| Vitamin B2 or Riboflavin (mg/100 g) | 0.05 | 0.096 |
| Vitamin B3 or Niacin (mg/100 g) | 0.4 | 0.68 |
| Vitamin B5 or Pantothenic acid (mg/100 g) | 0.3 | 0.43 |
| Vitamin B6 (mg/100 g) | 0.11 | 0.11 |
| Vitamin B9 or Folate (µg/100 g) | 7.5 | 9 |

FA: fatty acid, N/A: data are not available.

FW), oxalic (17.9 ± 1.3 mg/100 g FW), ascorbic (14.2 ± 2.6 mg/100 g FW) and succinic (10.2 ± 3.4 mg/100 g FW) acids. Shikimic and fumaric acids have also been identified in the peel and pulp of figs (Oliveira *et al.*, 2009). Citric and malic acids are the primary organic acids found in figs (Trad *et al.*, 2012a; Pereira *et al.*, 2017a).

## Proteins

The mean protein values among nine Spanish fresh fig varieties ranged between 4.4 and 6.7% (Pereira *et al.*, 2017a). The main raw fig amino acid composition per 100 g edible portion was reported to be leucine 0.033 g, lysine 0.03 g, tyrosine 0.032 g, valine 0.028 g, alanine 0.045 g, aspartic acid 0.176 g, glutamic acid 0.072 g, proline 0.049 g, and serine 0.037 g (Lim, 2012).

## Vitamins

Fig fruits contain γ-tocopherol, δ-tocopherol, vitamin $D_2$, vitamin $D_3$, α-tocopherol, α-tocopherol acetate and vitamin $K_1$ (Guvenc *et al.*, 2009). According to these authors, among all the vitamins present in fig, vitamin K1 and vitamin D3 are the highest in the whole fruit and the vitamin content varies among the fruit parts, the skin, and white and red parts. Fresh figs contain 0.2 and 0.3 mg/100 g of α- and γ-tocopherol, respectively (Pande and Akoh, 2010). The fig achene oil contains 2.23 µg/g of γ-tocopherol (Soltana *et al.*, 2016).

### Fatty acids

Fresh figs contain 2.5% total lipids: 69.1% unsaturated and 31% saturated fatty acids. The fatty acid profile of raw ripe fig fruits is dominated by linoleic (28.2%), palmitic (27.5%), linolenic (20.7%) and oleic (20.2%) acids (Pande and Akoh, 2010). Saturated, mono and poly-unsaturated fatty acids in dried fig fruit: myristic (14:0), palmitic (16:0), stearic (18:0), oleic (18:1), linoleic (18:2) and linolenic (18:3) acids (Jeong and Lachance, 2001). The predominant fatty acids in figs are linolenic, linoleic, palmitic and oleic acids (Jeong and Lachance, 2001; Guvenc *et al.*, 2009). Since linoleic acid, which is an essential nutrient for humans, is found at high levels in the fig exocarp, it is important to consume the entire fruit without peeling off the skin to maximize linoleic acid intake (Guvenc *et al.*, 2009). The fig achenes oil of the Kholi variety is high in unsaturated fatty acids (88.39%), higher than that (~70–85%) of the seed oils of sunflower, soybean, peanut and sesame (Namiki, 1995; Soltana *et al.*, 2016). *Ficus carica* achene oil also contains palmitic, stearic, gadoleic, palmitoleic and arachidic acids, but in smaller quantities (Soltana *et al.*, 2016). Fresh figs also contain phospholipids such as phosphatidylethanolamine (1 mg/100 g) and phosphatidylcholine (3.2 mg/100 g) (Pande and Akoh, 2010).

### Fiber

Dietary fiber (DF) plays an important role in decreasing constipation, diabetes, cardiovascular diseases, diverticulosis and obesity. Plant foods are the only source of DF. All the fractions (cellulose, lignin, hemicellulose, pectins, gums and mucilages) of DF are the major constituents of plant cell walls (Ramulu and Udayasekhara Rao, 2003). Crude fiber in fresh figs ranged from 4 to 7.4% (Pereira *et al.*, 2017a). Dried figs had the highest fiber content (12.5 g/100 g) among the dried fruits; apricots, dates, and raisins (9, 7.5 and 4 g/100 g, respectively). Figs provide 25% of the recommended daily value for fiber (25 g) (Vinson *et al.*, 2005). More than 28% of the fig fiber is soluble fiber (Vinson, 1999).

### Minerals

The mineral content of fruit is affected by genotype, environmental conditions, fertilization and plant nutritional status (Pereira *et al.*, 2017a). Fresh figs from Pakistan have potassium values ranging from 382.4 to 611.5 mg/100 g, calcium from 78.72 to 132.80 mg/100 g, phosphorus from 31.91 to 76.96 mg/100 g and iron from 5.69 to 10.09 mg/100 g (Khan *et al.*, 2011). Dried figs contain higher levels of calcium, potassium and magnesium (1094, 615 and 202 mg/100 g dry weight, respectively). They also have higher levels of chromium, phosphorus and iron (168, 77 and 31 mg/100 g of dry weight, respectively) (Sadia *et al.*, 2014). Dried figs were ranked highest in calcium compared to dried apricots, dates, raisins, plums and cranberries (Vinson *et al.*, 2005). Similarly, dried figs contain the highest amounts of iron, magnesium and manganese compared to apricots, dates and raisins (Waheed and Siddique, 2009). Khairuddin *et al.* (2017) reported dried figs contained the highest amounts of magnesium and zinc compared to some common dried fruits (Ajwa, Safawi and Mariami dates, golden and black raisins, and apricot). Figs provide 6.7, 7.5, 4.4, 9.6 and 2% of the recommended daily intake for calcium, magnesium, manganese, potassium and zinc, respectively (Waheed and Siddique, 2009).

## 18.3 Phytochemicals in Fresh and Dried Figs

### Phenolic compounds

Phenolic compounds are ubiquitous secondary metabolites in plants. They include more than 8000 biologically active compounds (Bachir bey *et al.*, 2014). These compounds contribute to fruit taste, color and nutritional content (Veberic *et al.*, 2008). Phenolics have antioxidant, antimutagenic, anticarcinogenic, antiatherogenic, antiinflammatory and antimicrobial properties (Faleh *et al.*, 2012; Bachir bey *et al.*, 2014). Figs are an important source of phenolic compounds. The major phenolics reported in fresh and dried figs are phenolic acids (hydroxybenzoic and hydroxycinnamic acids), flavonoids (flavonols, flavones, flavanones, flavanols and isoflavones), hydroxycoumarins and anthocyanins (Table 18.2). The total phenolic content of 18 different Spanish fresh fig cultivars ranged from 28.4 to 163 mg/100 g FW (Vallejo *et al.*, 2012). For the same cultivars, fruit skin phenolic content was 19.1–140.2 mg/100 g FW and the fruit pulp had between 0.0 and 11.3 mg/100 g FW.

In their study of six Algerian fresh fig varieties, Meziant *et al.* (2015) reported that total flavonoids ranged from 2.98 to 5.01 mg QE/100 g FW. The total anthocyanin content in 76 fresh Turkish fig varieties ranged from 0.0 to 298.9 µg cy-3-rutinoside/g FW, with an average of 18.2 µg cy-3-rutinoside/g FW (Çaliskan and Polat, 2011). These authors classified the cultivars into five fruit color groups. Green figs had the lowest anthocyanin levels (8.5 ± 8.7 µg cy-3-rutinoside/g FW) versus black figs which had the highest levels (128.4 ± 90.4 µg cy-3-rutinoside/g FW).

Multiple studies (Solomon *et al.*, 2006; Çaliskan and Polat, 2011; Mahmoudi *et al.*, 2018), reported the phenolic compounds were located primarily in the fig skin; dark-skinned figs consistently had higher phenolic levels than light-skinned figs. Dried figs contain 540 mg GAE/100 g DW of total phenolics, 105.6 mg QE/100 g DW of flavonoids, 5.9 mg/100 g DW of anthocyanin, and 3.2 mg CE/100 g DW of proanthocyanidins (Ouchemoukh *et al.*, 2012). Compared to other dried fruits, apple, yellow raisin, apricot, prune and sour cherry, dried figs contain the highest level of total phenolics (169.4 mg GAE/100 g) (Capanoglu, 2014). Multiple researchers have reported that fig phenolic, flavonoid and anthocyanin contents are dependent on the cultivar, fruit component, color, maturity stage, pre and orchard management, postharvest processing and the method of analysis (Veberic *et al.*, 2008; Çaliskan and Polat, 2011; Vallejo *et al.*, 2012; Wojdyło *et al.*, 2016; Mahmoudi *et al.*, 2018).

The polyphenolic profile of fresh and dried fig components can be analyzed using multiple instruments; high-performance liquid chromatography coupled with ultraviolet-visible and/or diode array detection, mass or tandem mass spectrometry in negative or positive ionization mode (LC–UV-DAD/ESI-MSn) (Vallejo *et al.*, 2012; Oliveira *et al.*, 2009), reversed-phase ultra-high performance liquid chromatography coupled to diode array, and mass spectrometry detection (RP-UHPLC-DAD-QTOF-MS) have all been used (Ammar *et al.*, 2015).

Many hydroxybenzoic acids and derivatives have been characterized in whole fig fruits, the peel and pulp (Veberic *et al.*, 2008; Russo *et al.*, 2014; Ammar *et al.*, 2015). The predominant ones are vanillic acid, vanillic acid glucoside, dihydroxybenzoic acid, dihydroxybenzoic acid pentoside I, dihydroxybenzoic acid di-pentoside, and gallic and syringic acids. Among the hydroxycinnamic acids and derivatives identified in figs, 3-*O*-caffeoylquinic acid (chlorogenic acid), 5-*O*-caffeoylquinic acid (neochlorogenic acid), sinapic acid hexoside, caffeic acid, dihydrocaffeic acid hexose, comaroyl hexoside and ferulic acid were prominent (Oliveira *et al.*, 2009; Slatnar *et al.*, 2011; Ammar *et al.*, 2015; Pourghayoumi *et al.*, 2016; Wojdyło *et al.*, 2016).

Flavonoid subclasses; flavonols flavones, flavanones, flavanonols, flavanols and isoflavones have been reported in figs. They include quercetin 3-*O*-rutinoside (rutin), quercetin-3-*O*-glucoside (isoquercetin), kaempferol-3-*O*-glucoside, kaempferol rutinoside, luteolin-8-*C*-glucoside, apigenin 8-*C*-glucoside (vitexin), apigenin-7-rutinoside, eriodictyol, naringenin, (+)-catechin, (-)-epicatechin, dihydroquercetin (taxifolin), chrysin, tangeretin and prenylgenistein I (Slatnar *et al.*, 2011; Vallejo *et al.*, 2012; Nakilcioglu and Hisil, 2013; Russo *et al.*, 2014; Ammar *et al.*, 2015; Wang *et al.*, 2017). In Bachir Bey *et al.*'s (2016) and Wang *et al.*'s (2017) studies, procyanidins B1, B2, B, B3 and A were detected in peels of Chinese figs and fresh and dried Algerian figs. Fifteen anthocyanin pigments were detected in fresh figs, the main anthocyanin in both parts of the fruit was cy-3-rutinoside (48–81% in skin and 68–79% in pulp) followed by cy-3-glucoside (5–18% in skin and 10–15% in pulp) (Dueñas *et al.*, 2008).

## Furanocoumarins

Furanocoumarins and their derivatives are also present in the peel, pulp and whole fruit of fresh Tunisian, Portuguese and Turkish fig varieties (Oliveira *et al.*, 2009; Nakilcioglu and Hisil, 2013; Ammar *et al.*, 2015); psoralen, methoxypsoralen, bergapten, and marmesin isomer I

**Table 18.2.** Phenolic compounds identified in fresh and dried figs.

| Chemical classes | Compounds | Localities | Varieties | Unit | Fig fruit | | | | References |
|---|---|---|---|---|---|---|---|---|---|
| | | | | | Whole – fresh | Peel | Pulp | Whole – dried | |
| Phenolic acids | 5-*O*-caffeoylquinic acid | Portugal | Pingo de mel | mg/kg | / | 43.8 ± 3.8 | 32.9 ± 23.2 | / | Oliveira *et al.* (2009) |
| | Chlorogenic acid | Spain | Tiberio | mg/100 g FW | / | 5.8 | 0.3 | / | Vallejo *et al.* (2012) |
| | | Iran | Rowno | mg/100 g DM | / | / | / | 0.55 ± 0.001 | Pourghayoumi *et al.* (2016) |
| | | Turkey | Not determined | µg/g | / | / | / | 13 | Capanoglu (2014) |
| | | Slovenia | Bela petrovka | mg/100 g | 4.91 ± 1.00 | / | / | 3.42 ± 0.54 | Slatnar *et al.* (2011) |
| | *o*-Coumaric acid | Algeria | Azandjar | mg/100 g DM | 2.73 ± 0.21 | / | / | 0.98 ± 0.08 | Bachir Bey *et al.* (2016) |
| | *p*-Coumaric acid | Turkey | Bursa siyahi | mg/100 g DM | 4.7 ± 0.8 | 18 ± 4 | 3.2 ± 0.2 | 3.1 ±0.8 | Kamiloglu and Capanoglu (2015) |
| | | Serbia | Not determined | mg/100 g DM | / | / | / | 2.32 ± 0.19 | Miletić *et al.* (2014) |
| | Ellagic acid | Turkey | Bursa siyahi | mg/100 g DM | 0.2 ± 0.1 | 0.7 ± 0.1 | 0.1 ± 0.0 | 0.3 ± 0.1 | Kamiloglu and Capanoglu (2015) |
| | | Spain | San Antonio | mg/100 g DM | / | 2.7 | 1.0 | / | Pereira *et al.* (2017b) |
| | Syringic acid | Turkey | Ödemiş | mg/100 g DM | 4.45 ± 0.14 | / | / | 0.41 ± 0.00 | Nakilcioglu and Hisil (2013) |
| | | Slovenia | Miljska figa | mg/100 g FW | 0.034 ± 0.003 | / | / | / | Veberic *et al.* (2008) |
| | Vanillic acid | Algeria | Azandjar | mg/100 g DM | 3.90 ± 0.31 | / | / | 3.32 ± 0.27 | Bachir Bey *et al.* (2016) |
| | | Greece | Kalamat | mg/kg FW | / | / | / | 76 | Russo *et al.* (2014) |
| | Gallic acid | Turkey | Ödemiş | mg/100 g DM | 6.98 ± 0.07 | / | / | 0.47 ± 0.00 | Nakilcioglu and Hisil (2013) |
| | | Serbia | Not determined | mg/100 g DM | / | / | / | 67.01 ± 0.53 | Miletić *et al.* (2014) |

| | | | | | | | | | |
|---|---|---|---|---|---|---|---|---|---|
| | | Turkey | Bursa siyahi | mg/100 g DM | 0.3 ± 0.0 | 0.3 ± 0.0 | 0.5 ± 0.2 | 0.5 ± 0.1 | Kamiloglu and Capanoglu (2015) |
| | | Slovenia | Miljska figa | mg/100 g FW | 0.30 ± 0.02 | / | / | / | Veberic *et al*. (2008) |
| Flavonoids | Kaempferol-3-*O*-glucoside | Slovenia | Bela petrovka | mg/100 g | 0.13 ± 0.01 | / | / | 0.59 ± 0.06 | Slatnar *et al*. (2011) |
| | Kaempferol | Serbia | Not determined | mg/100 g DM | / | / | / | 0.89 ± 0.01 | Miletić *et al*. (2014) |
| | | Tunisia | Himri | mg/kg | 11.67 ± 0.33 | / | / | / | Soltana *et al*. (2018) |
| | Kaempferol-rutinoside | Algeria | Azendjar | µg/g | / | / | / | 133.1 ± 8.3 | Debib *et al*. (2016) |
| | | Spain | Cuello Dama | mg/100 g FW | / | 0.9 | / | 1.0 | Vallejo *et al*. (2012) |
| | | Turkey | Bursa siyahi | mg/100 g DM | 0.3 ± 0.1 | 4 ± 1 | 0.1 ± 0.1 | 0.9 ± 0.1 | Kamiloglu and Capanoglu (2015) |
| | Apigenin-C-hexoside-pentoside | Spain | Cuello Dama Negra | mg/100 g DM | 24.3 ± 2.3 | / | / | / | Wojdyło *et al*. (2016) |
| | | Algeria | Azendjar | µg/g | / | / | / | 457.4 ± 22.5 | Debib *et al*. (2016) |
| | Luteolin-8-C-glucoside (orientin) | Tunisia | Tchich Asal | mg/kg | 38.62 ± 4.91 | / | / | / | Soltana *et al*. (2018) |
| | | Algeria | Azendjar | µg/g | / | / | / | 162.0 ± 30.7 | Debib *et al*. (2016) |
| | Quercetin-3-*O*-glucoside (isoquercetin) | Turkey | Not determined | µg/g | / | / | / | 2.4 ± 0.9 | Capanoglu (2014) |
| | | Tunisia | Tchich Asal | mg/kg | 69.11 ± 35.15 | / | / | / | Soltana *et al*. (2018) |
| | | Turkey | Bursa siyahi | mg/100 g DM | 1.2 ± 0.7 | 9.4 ± 0.9 | 0.2 ± 0.1 | 0.7 ± 0.2 | Kamiloglu and Capanoglu (2015) |
| | Apigenin-8-C-glucoside (vitexin) | Tunisia | Tchich Asal | mg/kg | 71.17 ± 5.71 | / | / | / | Soltana *et al*. (2018) |
| | Quercetin-3-*O*-rutinoside (rutin) (flavonol) | Spain | Tiberio | mg/100 g FW | / | 15.8 | 1.3 | / | Vallejo *et al*. (2012) |

*Continued*

**Table 18.2.** Continued.

| Chemical classes | Compounds | Localities | Varieties | Unit | Fig fruit: Whole – fresh | Fig fruit: Peel | Fig fruit: Pulp | Fig fruit: Whole – dried | References |
|---|---|---|---|---|---|---|---|---|---|
| | | Turkey | Söke | mg/100 g DM | 68.21 ± 0.59 | / | / | 12.35 ± 0.85 | Nakilcioglu and Hisil (2013) |
| | | Slovenia | Bela petrovka | mg/100 g | 0.89 ± 0.20 | / | / | 1.38 ± 0.37 | Slatnar *et al*. (2011) |
| | | Italy | Mattalona | mg/kg FW | / | 1071.41 | Nd | / | Del Caro and Piga (2008) |
| | Procyanidin B1 | Algeria | Bouankik | mg/100 g DM | 4.80 ± 0.39 | / | / | 1.56 ± 0.12 | Bachir Bey *et al*. (2016) |
| | Procyanidin B1 | Algeria | Azandjar | mg/100 g DM | 2.61 ± 0.20 | | | 1.34 ± 0.11 | Bachir Bey *et al*. (2016) |
| | (+)-Catechin | Spain | Verdal fig | mg/100 g DM | 95.7 ± 3.9 | / | / | / | Wojdyło *et al*. (2016) |
| | | Iran | Matti | mg/100 g DM | / | / | / | 4.77 ± 0.34 | Pourghayoumi *et al*. (2016) |
| | | Turkey | Bursa siyahi | mg/100 g DM | 13 ± 4 | 6 ± 1 | 27 ± 2 | 2.5 ± 0.7 | Kamiloglu and Capanoglu (2015) |
| | | Slovenia | Bela petrovka | mg/100 g | 2.88 ± 0.18 | / | / | 6.60 ± 1.18 | Slatnar *et al*. (2011) |
| | (-)-Epicatechin | Turkey | Not determined | µg/g | | | | 6 | Capanoglu (2014) |
| | | Slovenia | Miljska figa | mg/100 g FW | 0.97 ± 0.06 | / | / | / | Veberic *et al*. (2008) |
| | | Turkey | Bursa siyahi | mg/100 g DM | 6.9 ± 0.7 | 29 ± 6 | 4.8 ± 0.9 | 2.2 ± 0.5 | Kamiloglu and Capanoglu (2015) |
| | Naringenin | Tunisia | Tchich Asal | mg/kg | 44.42 ± 16.48 | / | / | / | Soltana *et al*. (2018) |
| | Eriodictyol | Tunisia | Tchich Asal | mg/kg | 25.55 ± 8.56 | / | / | / | Soltana *et al*. (2018) |
| Furanocoumarins | Psoralen | Turkey | Germencik | mg/100 g DM | 7.17 ± 0.52 | / | / | 2.23 ± 0.02 | Nakilcioglu and Hisil (2013) |
| | | Portugal | Branca Tradicional | mg/kg | / | 18.4 ± 2.6 | 17.0 ± 0.2 | / | Oliveira *et al*. (2009) |
| | Bergapten | Portugal | Branca Tradicional | mg/kg | / | 26.8 ± 11.1 | 45.8 ± 11.2 | / | Oliveira *et al*. (2009) |

| | | | | | | | | | |
|---|---|---|---|---|---|---|---|---|---|
| Anthocyanins | Pelargonidin-3-*O*-rutinoside | Spain | Cuello Dama Negra | mg/100 g DM | 3.8 ± 0.4 | / | / | / | Wojdyło *et al.* (2016) |
| | | Spain | Colar | µg/g FW | / | 3.52 ± 0.01 | 0.08 ± 0.00 | / | Dueñas *et al.* (2008) |
| | | Spain | De Rey | mg/100 g | / | 5.64 | 0.09 | / | Pereira *et al.* (2017b) |
| | Cyanidin-3-glucoside | Spain | Granilla | µg/g FW | / | 15.38 ± 3.42 | 0.54 ± 0.14 | / | Dueñas *et al.* (2008) |
| | | Turkey | Bursa siyahi | mg/100 g DM | / | / | / | 7.07 ± 0.89 | Yemis *et al.* (2012) |
| | | Algeria | Azandjar | mg/100 g DM | 6.86 ± 0.54 | / | / | Nd | Bachir Bey *et al.* (2016) |
| | | Italy | Mattalona | mg/kg FW | / | 28.24 | Nd | / | Del Caro and Piga (2008) |
| | Cyanidin-3-*O*-rutinoside | Turkey | Bursa siyahi | mg/100 g DM | / | / | / | 91.23 ± 3.41 | Yemis *et al.* (2012) |
| | | Spain | VB1 | mg/100 g FW | / | 108.9 | 4.8 | / | Vallejo *et al.* (2012) |
| | | Spain | Cuello Dama Negra | mg/100 g DM | 116.2 ± 2.5 | / | / | / | Wojdyło *et al.* (2016) |
| | | Spain | Bursa Siyahi | µg/g FW | / | 78.32 ± 7.53 | 4.46 ± 1.57 | / | Dueñas *et al.* (2008) |

FW: Fresh weight, DM: Dry matter, Nd: Not detected, /: not analyzed.

and II were present. Six furanocoumarins, psoralen, 8-methoxypsoralen, angelicin, bergapten, rutaretin and pimpinellin, and one pyranocoumarin, seselin, were identified in the Italian Dottato fig fruits (Marrelli *et al.*, 2012).

## Volatile compounds

Aroma, determined by the fig's volatile compounds, is an essential fig quality factor (Table 18.3). Aroma is a complex mixture of esters, alcohols and aldehydes. Other compounds may contribute (Loizzo *et al.*, 2014). The peel and pulp volatile compounds of white and dark Portuguese fig cultivars were analyzed with a headspace solid-phase microextraction/gas chromatography-ion trap-mass spectrometer (HS-SPME/GC–IT-MS) (Oliveira *et al.* 2010a). Fifty-nine compounds, distributed by distinct chemical classes (aldehydes, alcohols, ketones, esters, monoterpenes, sesquiterpenes and norisoprenoids) were identified. In another study of fresh Tunisian figs (Trad *et al.*, 2012b), 51 volatile constituents were isolated and identified by accelerated solvent extractor and gas chromatography-mass spectrometry (ASE-GC-MS). The major compounds belong to the ester and ketone classes.

Volatiles from dried figs have also been characterized by many researchers (Loizzo *et al.*, 2014; Mujić *et al.*, 2014; Russo *et al.*, 2017; Mahmoudi *et al.*, 2021) using a headspace solid-phase (HS-SPME) procedure with gas chromatography-mass spectrometry analysis (GC-MS). The major aldehydes in fig fruits, and their peel and pulp, are benzaldehyde, (E)-2-hexenal, hexanal, nonanal, acetaldehyde, decanal, undecanal and octanal (Grison-Pigé *et al.*, 2002; Oliveira *et al.*, 2010a,b; Gozlekci *et al.*, 2011). The most abundant alcohols in figs are benzyl alcohol, methanol, ethanol, 2,3-butanediol, (E)-2-nonen-1-ol and 1-penten-3-ol (Gibernau *et al.*, 1997; Oliveira *et al.*, 2010a,b; Trad *et al.*, 2012b). Butyl acetate, isoamyl acetate, hexyl acetate, methyl salicylate and ethyl salicylate are the main esters found in figs (Oliveira *et al.*, 2010a; Trad *et al.*, 2012b). Figs also contain volatile monoterpenes and sesquiterpenes: β-caryophyllene, limonene, linalool, α and β-cubebene, α and β-pinene, α-bergamotene, β-selinene, germacrene D, menthol and α-humulene (Gibernau *et al.*, 1997; Grison-Pigé *et al.*, 2002; Oliveira *et al.*, 2010a; Gozlekci *et al.*, 2011; Li *et al.*, 2011).

## Phytosterols

Several unsaponifiable sterols in the fig fruit have been identified by GC/MS analysis; campesterol, stigamasterol, sitosterol and fucosterol. Sitosterol was the most predominant sterol in fig fruit identified as 68.6% of the total unsaponifiable component (Jeong and Lachance, 2001). Stigmasta-5,23-dien-3-ol and brassicasterol were also identified (Pande and Akoh, 2010; Loizzo *et al.*, 2014;). For *Ficus carica* achene oil, the most abundant components in the sterol fraction are stigmastadienol (73.78%) and β-sitosterol (10.12%), followed by 24-methylene-cholesterol (4.89%) and campestanol (3.95%) in abundance (Soltana *et al.*, 2016).

## Terpenoids

The average total triterpenoid content in ten different Spanish fig cultivars was 274 μg/g DM but differed significantly among cultivars. Two of these triterpenoids, betulinic acid, and oleanolic acid, were identified using LC-MS QTof (Wojdyło *et al.*, 2016). Pseudotaraxasterol, α-amyrin, β-amyrin, 3-cheto-urs-12-ene, β-amyrin acetate, α-selinene, lupeol, lupeol acetate, and olean-13(18)-ene were also identified in fig fruits (Jeong and Lachance, 2001; Ribechini *et al.*, 2011; Marrelli *et al.*, 2012; Mahmoudi *et al.*, 2021). Several monoterpenes and sesquiterpenes have also been identified in figs. The most prominent were limonene, linalool, epoxylinalool, menthol, p-cymene, germacrene D, α-cubebene, copaene, β-caryophyllene, hotrienol and β-bourbonene (Grison-Pigé *et al.*, 2002; Oliveira *et al.*, 2010a; Darjazi and Larijani, 2012; Trad *et al.*, 2012a).

## Carotenoids

Yemis *et al.* (2012) reported that the total carotenoid content of two fresh Turkish figs was 29.65 and 23.71 μg lutein equivalents per g DW. They also reported seven days of sun drying decreased the total carotenoids by 78.5 and 81.5%. Ouchemoukh *et al.* (2012) reported a total fig carotenoid content of 11 mg βCE/100 g DW, sharply higher than that of apricot, Agen prune, and raisins. The carotenoid composition of Turkish

**Table 18.3.** Volatile compounds identified in figs.

| Chemical classes | Compounds | Localities | Varieties | Fig fruit | | | | References |
|---|---|---|---|---|---|---|---|---|
| | | | | Whole – fresh | Peel | Pulp | Whole – dried | |
| Aldehydes | 2-Methyl-2-butanal | Tunisia | Bouhouli, Zidi, Thgagli | d | / | / | / | Trad *et al.* (2012b) |
| | | Portugal | PM, BT, VP, PT | / | nd | d | / | Oliveira *et al.* (2010a) |
| | Hexanal | Tunisia | Bouhouli, Zidi, Thgagli | d | / | / | / | Trad *et al.* (2012b) |
| | | Portugal | PM, BT | / | d | d | / | Oliveira *et al.* (2010b) |
| | | Croatia | Bružetka Bijela, Zimnica | / | / | / | d | Mujić *et al.* (2014) |
| | (E)-2-Hexenal | Portugal | PM, BT, VP, PT | / | nd | d | / | Oliveira *et al.* (2010a) |
| | | Italy | Dottato | / | / | / | d | Russo *et al.* (2017) |
| | Tetradecanal | Italy | SF, Citrullara, Dottato | / | / | / | d | Loizzo *et al.* (2014) |
| | Benzaldehyde | Portugal | PM, BT, VP, PT | / | d | d | / | Oliveira *et al.* (2010a) |
| | | Tunisia | Bouhouli, Zidi, Thgagli | d | / | / | / | Trad *et al.* (2012b) |
| | | Croatia | Bružetka Bijela, Zimnica | / | / | / | d | Mujić *et al.* (2014) |
| | Acetaldehyde | Portugal | PM, BT | / | d | d | / | Oliveira *et al.* (2010b) |
| | | Italy | Dottato | / | / | / | d | Russo *et al.* (2017) |
| | Octadecanal | Italy | SF, Citrullara, Dottato | / | / | / | d | Loizzo *et al.*, 2014 |
| | Nonanal | Portugal | PM, BT, VP, PT | / | nd | d | / | Oliveira *et al.* (2010a) |
| | | Tunisia | Bouhouli, Zidi, Thgagli | d | / | / | / | Trad *et al.* (2012b) |
| | | Croatia | Bružetka Bijela, Zimnica | / | / | / | d | Mujić *et al.* (2014) |
| Alcohols | Ethanol | Croatia | Bružetka Bijela, Zimnica | / | / | / | d | Mujić *et al.* (2014) |
| | | Portugal | PM, BT | / | d | d | / | Oliveira *et al.* (2010b) |
| | Benzyl alcohol | France | Not mentioned | d | / | / | / | Grison-Pigé *et al.* (2002) |
| | | Portugal | PM, BT | / | d | d | / | Oliveira *et al.* (2010b) |
| | 1-Butanol, 3-methyl-1-octen-3-ol | Croatia | Bružetka Bijela, Zimnica | / | / | / | d | Mujić *et al.* (2014) |
| | | Italy | Dottato | / | / | / | d | Russo *et al.* (2017) |
| Esters | Ethyl acetate | Portugal | PM, BT | / | d | d | / | Oliveira *et al.* (2010b) |
| | | Croatia | Bružetka Bijela, Zimnica | / | / | / | d | Mujić *et al.* (2014) |
| | Butyl acetate | Tunisia | Bouhouli, Zidi, Thgagli | d | / | / | / | Trad *et al.* (2012b) |
| | Methyl salicylate | Portugal | PM, BT, VP, PT | / | d | d | / | Oliveira *et al.* (2010a) |
| | | Croatia | Bružetka Bijela, Zimnica | / | / | / | d | Mujić *et al.* (2014) |
| | Hexyl acetate | Tunisia | Bouhouli, Zidi, Thgagli | d | / | / | / | Trad *et al.* (2012b) |
| | Ethyl caproate | Italy | Dottato | / | / | / | d | Russo *et al.* (2017) |
| | Hexanoic acid, ethyl ester | Croatia | Bružetka Bijela, Zimnica | / | / | / | d | Mujić *et al.* (2014) |

*Continued*

**Table 18.3.** Continued.

| Chemical classes | Compounds | Localities | Varieties | Fig fruit: Whole – fresh | Peel | Pulp | Whole – dried | References |
|---|---|---|---|---|---|---|---|---|
| Ketones | 6-Methyl-5-hepten-2-one | Croatia | Bružetka Bijela, Zimnica | / | / | / | d | Mujić *et al.* (2014) |
| | | Portugal | PM, BT, VP, PT | / | d | d | / | Oliveira *et al.* (2010a) |
| | 3-Hydroxy-2-butanone | Tunisia | Bouhouli, Zidi, Thgagli | d | / | / | / | Trad *et al.* (2012b) |
| | | Italy | Dottato | / | / | / | d | Russo *et al.* (2017) |
| Acids | Malic acid | United States | Not mentioned | d | / | / | / | Pande and Akoh (2010) |
| | Acetic acid | Turkey | BS, Karabakunya, SL, SS | / | d | d | / | Gozlekci *et al.* (2011) |
| | | Tunisia | Bouhouli, Zidi, Thgagli | d | / | / | / | Trad *et al.* (2012b) |
| Monoterpenes | Limonene | Portugal | PM, BT | / | d | d | / | Oliveira *et al.* (2010b) |
| | | Tunisia | Bouhouli, Zidi, Thgagli | d | / | / | / | Trad *et al.* (2012b) |
| | Linalool | France | Not determined | d | / | / | / | Gibernau *et al.* (1997) |
| | | Portugal | PM, BT, VP, PT | / | nd | d | / | Oliveira *et al.* (2010a) |
| | Hotrienol | France | Not mentioned | d | / | / | / | Grison-Pigé *et al.* (2002) |
| Sesquiterpenes | α and β-Caryophyllene | Turkey | BS, Karabakunya, SL, SS | / | d | d | / | Gozlekci *et al.* (2011) |
| | | China | Not mentioned | d | / | / | / | Li *et* al. (2011) |
| | β-Bourbonene | France | Not mentioned | d | / | / | / | Grison-Pigé *et al.* (2002) |
| | β-Cubebene | Turkey | BS, Karabakunya, SL, SS | / | d | d | / | Gozlekci *et al.* (2011) |
| | Germacrene D | France | Not determined | d | / | / | / | Gibernau *et al.* (1997) |
| | | Portugal | PM, BT, VP, PT | / | d | d | / | Oliveira *et al.* (2010a) |
| Triterpenes | α-Amyrin | Italy and Spain | dark-purple common fig | / | / | / | d | Ribechini *et al.* (2011) |
| | Betulinic acid | Spain | Colar, Verdal, Calabacita, Tiberio | / | / | / | d | Wojdyło *et al.* (2016) |
| | β-Amyrin | Algeria | Bakkor Khal | / | / | / | d | Mahmoudi *et al.* (2021) |
| | Oleanolic acid | Spain | Colar, Verdal, Calabacita, Tiberio | / | / | / | d | Wojdyło *et al.* (2016) |

PM: Pingo de Mel, BT: Branca Tradicional, VP: Verbera Preta, PT: Preta Tradicional, SF: San Francesco, BS: Bursa Siyahi, SL: Sari Lop, SS: Sultan Selim. d: determined, nd: not determined, /: not analyzed.

yellow fresh fig cultivars determined by HPLC-DAD demonstrated lutein was the main carotenoid identified followed by zeaxanthin, β-cryptoxanthin and β-carotene (Yemis *et al.*, 2012). *Ficus carica* achene oil contains significant amounts of β-carotene (81.08 mg/kg), producing the yellow color of the achene oil (Soltana *et al.*, 2016). Carotenoids are not only precursors of vitamin A but are also the primary antioxidants that prevent the free radical chain reaction. They are known to be very efficient physical and chemical quenchers of singlet oxygen ($^1O_2$), as well as potent scavengers of other reactive oxygen species (ROS). This is of special significance, because the uncontrolled generation and concomitant increase of ROS level in the body results in 'oxidative stress', an essential contributor to the pathogenic processes of many diseases (cardiovascular, cancer, neurological, photosensitive and eye-related disorders) (Akoh and Min, 2008; Fiedor and Burda, 2014).

## 18.4 Functional Food Properties of Figs

'Functional food' has multiple definitions. Based on the literature and consultation, Doyon and Labrecque (2008) formulated the following definition: 'A functional food is, or appears similar to, a conventional food. It is part of a standard diet and is consumed on a regular basis in normal quantities. It has proven health benefits that reduce the risk of specific chronic diseases or beneficially affect target functions beyond its basic nutritional functions'. Epidemiological studies indicate a diet rich in fruits, vegetables and whole grains reduces the risk of chronic diseases. Fruits and vegetables are major sources of phytochemicals (Vattem and Maitin, 2016). Phytochemicals or 'bioactive compounds' are chemical components in functional foods that exhibit beneficial biological activities. These bioactive compounds can be considered extra-nutritional constituents present in small amounts in various food sources. In short, bioactive compounds are beneficial because they act as antioxidants, are chemo- and cardio-preventive, and can even reduce risk or prevent the onset of certain diseases (Gur *et al.*, 2018). Fresh or dried, figs are an excellent source of phytochemicals with beneficial biological activities.

As discussed earlier, figs contain antioxidants, phenolic acids, flavonoids, anthocyanins, and tannins. The potential human health benefits include protection against heart disease, inhibition of LDL oxidation, increase in plasma antioxidant capacity, and chemopreventive agents used in cancer therapy (Vinson, 1999; Vinson *et al.*, 2005; Howlett, 2008; Tungmunnithum *et al.*, 2018). Figs contain potassium and calcium, nutrients known to regulate human blood pressure. Howlett (2008) reported increasing dietary potassium and calcium and reducing sodium can reduce blood pressure. Another fig component with health benefits is dietary fiber. Figs provide more fiber, 12.21%: insoluble 8.74% and soluble 3.47%, than most common fruits (Vinson, 1999). The consumption of dietary and functional fibers has many potential health benefits, among them decreasing constipation and irritable bowel syndrome, reducing LDL cholesterol concentration and the risk of coronary and cardiovascular disease, and preventing obesity, diabetes and colon cancer (Howlett, 2008; Guiné *et al.*, 2010).

## 18.5 Health Benefits of Figs

Figs, a source of bioactive functional ingredients, have long been used to treat gastrointestinal problems: colic, indigestion, constipation, and loss of appetite, respiratory and cardiovascular disorders, and as an anti-inflammatory (Guarrera, 2005; Lansky *et al.*, 2008; Rtibi *et al.*, 2018). Fig fruit juice mixed with honey is used for hemorrhages and the fruits as an expectorant, diuretic and for liver and spleen diseases (Mawa *et al.*, 2013). Figs have been reported to have antipyretic, purgative and aphrodisiac properties, and are useful in paralysis (Khodarahmi *et al.*, 2011). Several laboratory experiments have scientifically confirmed the traditional preventive and treatment uses of figs. Some of this research is summarized below.

### Antihyperlipidemic and anti-cholesterolimic effects

The antihyperlipidemic effect of the aqueous-ethanolic extract of Tunisian figs in high-fat diet-induced hyperlipidemic rats was evaluated (Belguith-Hadriche *et al.*, 2016). The fig extract improved the lipid profile by decreasing the total cholesterol, triglyceride, low-density lipoprotein cholesterol, and increasing high-density lipoprotein cholesterol levels. These beneficial effects were attributed to the figs phenolic constituents, vitexin, dihydroxybenzoic acid di-pentoside and rutin, and to the inhibition of cholesterol synthesis in the liver and increased catabolism and excretion of cholesterol with the excrement.

Total dietary fiber intake has been inversely associated with total cholesterol, low-density lipoprotein cholesterol and triglycerides, and positively associated with high-density lipoprotein cholesterol (Wu *et al.*, 2007; Academy of Nutrition and Dietetics, 2015). As components in foods are digested and absorbed from the small intestine, fiber becomes a major component in the gut lumen, increasing the viscosity of the gut lumen content that interferes with bile absorption from the ileum. The precipitate LDL cholesterol removal from the blood is converted into bile acids by the liver to replace the bile acids lost in the excrement (Marlett *et al.*, 2002). These authors suggested that changes in bile acid pool composition accompanying ingestion of viscous fibers dampens cholesterol synthesis. Because endogenous synthesis accounts for about three-quarters of the total body cholesterol pool, slowing synthesis could lower blood cholesterol concentrations.

## Antidiabetic and antiobesogenic effects

Obesity is defined as an unhealthy accumulation of body fat. Its prevalence varies between 5 and 50% in different populations. The epidemic of obesity with its accompanying health risks is now recognized to be one of the major health challenges in both the developed and developing world. Obesity is associated with an increased risk of diabetes mellitus, a serious disease affecting a significant part of the world's population. It is characterized by hyperglycemia due to peripheral insulin resistance being inefficiently compensated by pancreatic beta-cell insulin secretion. Dietary intervention is the first and highly preferred option for obesity and type 2 diabetes treatment (Howlett, 2008; Wojdyło *et al.*, 2016).

Currently, there are different approaches to treating diabetes and obesity. One is the inhibition of carbohydrate and lipid digesting enzymes such as α-glucosidase, α-amylase and pancreatic lipases. Inhibiting these enzymes reduces the digestion and absorption of carbohydrates and lipids but also significantly reduces the blood glucose and body fat (Holubkova *et al.*, 2012). Mopuri *et al.* (2018) investigated the effects of *Ficus carica* L. extracts from the fruit, leaves and bark on the antidiabetic (inhibition of α-glucosidase and α-amylase enzymes) and antiobesogenic (antilipase) effects *in vitro*. The ethanolic extract of fig fruit had significantly ($p < 0.05$) more effective antidiabetic and antiobesogenic effects than extracts of the other plant parts. The fruit ethanolic extract had half the maximal inhibitory concentration ($IC_{50}$, which is the half-maximal inhibitory concentration) values of 255.57 ± 36.46, 315.89 ± 3.83 and 230.475 ± 9.65 µg/mL for the inhibition of α-glucosidase, α-amylase and pancreatic lipase activities, respectively.

Wojdyło *et al.* (2016) reported ten fig fruit cultivars inhibited α-amylase activity ($IC_{50}$ = 3.58–17.3 mg of dried fruit/mL) but observed only moderate inhibitory effect on α-glucosidase ($IC_{50}$ = 18.2–19.5 mg of dried fruit/mL). The effect of different levels of fig fruit on blood sugar level in alloxan diabetic rats was investigated by El-Shobaki *et al.* (2010). They reported that the inclusion of fig fruits in the rat's diet induced a significant hypoglycaemic effect. Others have reported malic acid, n-hexadecanoic acid, *trans* phytol and 5-hydroxymethylfurfural in fig fruit extract have an antidiabetic and anti-obesity activity (Mopuri *et al.*, 2018). Phenolic compounds in fig fruits have also been hypothesized to have an antidiabetic effect through a different mechanism (Wojdyło *et al.*, 2016). Cyanidin 3-glucoside has been reported to reduce blood glucose and enhance insulin sensitivity by downregulating the retinol-binding protein 4 expression in type 2 diabetic mice (Sasaki *et al.*, 2007). Quercetin also has a hypoglycemic effect via its ability to improve glucose metabolism and reduce oxidative stress in diabetic rats (Song *et al.*, 2002).

Dietary fiber may have a preventative effect and may attenuate the symptoms of type 2 diabetes in adults (Kranz *et al.*, 2012). A published meta-analysis of observational studies from 1974 to 2013 reported that the risk of type 2 diabetes decreased with total dietary fiber intake for 19,033 out of 488,293 participants (Yao *et al.*, 2014; Academy of Nutrition and Dietetics, 2015). The collective evidence suggests figs, due to their ability to reduce blood sugar and weight gain, are an excellent source of fiber for both juvenile and adult diets (Vinson, 1999).

### Cardiovascular diseases

Globally, cardiovascular diseases (CVDs) are among the leading causes of death. In 2012, more than 17.5 million people, 31% of the world's deaths were attributable to CVD. Among the manifestations of CVD were coronary heart disease (heart attacks), cerebrovascular disease (stroke), high blood pressure (hypertension), peripheral artery disease, rheumatic heart disease, congenital heart disease, heart failure (de Pascual-Teresa *et al.*, 2010; Alamgeer *et al.*, 2017; Belattar and Himour, 2019), platelet aggregation, dysfunctioning of vascular endothelium and high plasma LDL cholesterol (Riaz *et al.*, 2016).

It is well recognized that a plant-based diet is associated with a lower CVD risk mortality (Bondonno *et al.*, 2017). In a cohort study on women and men from the United States and Finland, relatively high fruit and vegetable intake was associated with a 25–55% reduction in CVD (Pereira and Liu, 2003). Fig fruits have been used for the treatment of cardiovascular diseases in traditional medicine (Gilani *et al.*, 2008; Alamgeer *et al.*, 2017).

In a recent study, the cardioprotective effect of figs was evaluated. The cardiac mitochondria were isolated from male Sprague Dawley rats after cardiomyocytes toxicity induction, and the protective effects of fig extract on mitochondrial oxidative stress and dysfunction was determined. The results confirmed that fig extract ameliorated mitochondrial dysfunction in the isolated mitochondria and prevented mitochondrial reactive oxygen species formation, membrane lipid peroxidation, mitochondrial swelling and mitochondrial membrane potential collapse (Gholami *et al.*, 2017). An *in vivo* investigation conducted on normotensive and glucose-treated hypertensive rats evaluated the cardiovascular and antihypertensive effects of figs. Langendorff's isolated heart technique was used to assess the effect of fig crude extract on force of contraction and heart rate. Data showed that a 1000 mg/kg dose decreased blood pressure in rats significantly. The crude extract produced a transient but significant decrease in force of contraction and heart rate (Alamgeer *et al.*, 2017). These researchers concluded that figs had potential to reduce coronary heart disease risk in hypertensive subjects.

It is believed that the cardiovascular protective effect of figs is due at least partly to the phenolic compounds via their antioxidant, binding lipoproteins, and anti-inflammatory abilities (Vinson *et al.*, 2005; Riaz *et al.*, 2016; Alamgeer *et al.*, 2017). In an observational study, it was found that a higher intake of polyphenols, especially from lignans, flavanols and hydroxybenzoic acids, was associated with decreased CVD risk (Tresserra-Rimbau *et al.*, 2014). Many phenolic compounds previously isolated from fig fruit, such as quercetin, psoralen, catechin, proanthocyanidins, cyanidin-3-*O*-glucoside and cyanidin-3-*O*-rutinoside reduced systolic blood pressure, and plasma oxidized low-density lipoprotein had a beneficial cardiovascular effect (Egert *et al.*, 2009; de Pascual-Teresa *et al.*, 2010; Egert *et al.*, 2010; Wong and Rabie, 2011).

Figs are a good source of fiber. Dietary fiber intake levels have been reported to have a protective effect against CVD (Academy of Nutrition and Dietetics, 2015). In a large prospective cohort study among Iranian adults, Mirmiran *et al.* (2016) examined the hypothesis that dietary fiber intake could modify the risk of cardiovascular disease. Results showed a significant inverse association between intakes of total, soluble, and insoluble dietary fiber from legumes, fruits and vegetables, and cardiovascular disease risk.

Higher dietary fiber intakes may improve serum lipid levels, lower blood pressure and reduce inflammatory marker levels, indicators of inflammation, summarizing potential mechanisms to explain fiber's protective properties against CVD (Academy of Nutrition and Dietetics, 2008). Gut microbiota also plays a role in the development of cardio-metabolic disorders. Dietary fiber is reported to modulate gut microbiota, suggesting this is the mechanism by which dietary fibers have abeneficial effect on cardiometabolic disorders (Mirmiran *et al.*, 2016).

Potassium is also an essential nutrient for controlling human blood pressure (Mohamed, 2013; Pereira *et al.*, 2017a). Figs have high levels of potassium, and therefore may have potential for lowering hypertension.

## Anticancer effect

Malignant cancers are the second leading cause of death worldwide. In light of the continuing need for effective anticancer agents and the association of fruit and vegetable consumption with reduced cancer risk, edible plants are increasingly being considered as sources of anticancer drugs (Wang *et al.*, 2008). Anticancer activities of methanolic extracts of *Ficus carica* leaves and fruits were examined through MTT (3-(4,5-dimethylthiazol-2yl)-5-(3-carboxymethoxyphenyl)-2-(4-sulfophenyl)-2H-tetrazolium) assay on Huh7it liver cancer cells. The MTT assay results indicated that 1000 μg $mL^{-1}$ of *Ficus carica* leaf extract inhibited 82.78% of cell growth. The same concentration of fig fruits inhibited cell growth by 11.84% (Purnamasari *et al.*, 2019). The authors suggested the difference in anticancer activity was attributed to different compounds present in each extract.

The Dottato cultivar was used to assess how fig ripeness influences proliferation of the human tumor cell line C32 (amelanotic melanoma). Photoactivated dichloromethane fractions had the best antiproliferative activity with an $IC_{50}$ < 5 μg $mL^{-1}$, particularly the first June harvest figs. This activity can be attributed to the presence of complex coumarins in the fresh figs (Marrelli *et al.*, 2012). Jasmine *et al.* (2015) investigated the anticancer properties of an ethanolic extract of fig fruits on breast cancer cell lines (MCF-7) using the MTT assay. The fig extract exhibited an important anticancer activity; a 1000 μg $mL^{-1}$ extract killed 90.5% of the cells within 72 h of incubation. In another study, the cytotoxic effect of ethanolic, ethyl acetate and dichloromethane fig fruit extracts on HeLa cell line was evaluated. The results indicated that fig extracts could reduce the viability of the HeLa cells at concentrations as low as 2 μg $mL^{-1}$ in a dose-dependent manner (Khodarahmi *et al.*, 2011). Data from a Chinese study showed that oral feeding with the extract from mature and young fig fruits (25 mg $kg^{-1}$ per day), or directly with the dry powder (0.5–62.5 mg $kg^{-1}$ per day) significantly inhibited several species of cancer cells in mice. The inhibition of spleen cancer was particularly high at 59.10%, ascites cancer, 53.85%, Lewis lung cancer, 48.85%, HAC liver cancer, 43.21%, and $S_{180}$ cancer, 41.82% (Wang *et al.*, 2003).

Several studies demonstrated the efficacy of the fig fruit bioactive compounds to inhibit proliferation of human tumor cells. Quercetin and its derivatives are the main phenolic compounds found in figs (Ammar *et al.*, 2015; Wojdyło *et al.*, 2016). Quercetin can stimulate the apoptosis of Caco-2 and HT-29 colon cancer and HL-60 leukemia cancer cells by stimulating the release of cytochrome c from mitochondria (Purnamasari *et al.*, 2019). Luteolin, a flavone also found in figs, can induce apoptosis of Hep3B liver cancer cells and cause cell cycle arrest in G1 phase of HepG2 liver cancer cells (Ammar *et al.*, 2015; Purnamasari *et al.*, 2019). A phenolic acid present in figs, gallic acid, was reported to be able to inhibit the proliferation of HeLa cells (Amessis-Ouchemoukh *et al.*, 2017). The anticancer activity of phenolic compounds might be attributed to antioxidant properties, increasing apoptosis ability, inhibiting lipid peroxidation, angiogenesis, DNA oxidation capacity and carcinogen-blocking activities (Vinson, 1999; Wang *et al.*, 2008; Purnamasari *et al.*, 2019).

Besides their ubiquitous polyphenols, figs have other compounds with anticancer activity, specifically benzaldehyde and the coumarins. Benzaldehyde has been used successfully to treat terminal human carcinomas. When treated with benzaldehyde 19 of 57 patients responded completely and 10 responded with a greater than 50% regression in their cancers (Vinson, 1999). Many furanocoumarins have been reported to inhibit carcinogen metabolizing enzymes, or block tumor formation (Marrelli *et al.*, 2012). Because they have low skin phototoxicity fig furanocoumarins, psoralens and angelicin, are currently being investigated for the treatment of skin cancer in clinical trials (Vinson, 1999).

Fig fruits, especially dried ones, are an important source of dietary fiber. Extensive epidemiological evidence supports the theory that dietary fiber may protect against large bowel cancer. A meta-study, pooling the results from 13 controlled case studies of colorectal cancer rates and dietary practices, concluded consumption of fiber-rich foods is inversely related to risks of both colon and rectal cancers (Marlett *et al.*, 2002). The Food and Drug Administration (FDA) has concluded that diets low in fat and naturally high in dietary fiber are associated with a reduced risk of cancer and has permitted a 'health claim' that can appear in labelling for foods that are good sources of dietary fiber (FDA, 1993; Dickinson and Mister, 2012).

## Immunomodulatory and anti-inflammatory

Tian *et al.* (2014) demonstrated that polysaccharides *Ficus* extract activated and proliferated dendritic cells (DCs), via the dectin-1 pathway, which promoted the secretion of various anti-inflammatory factors by DCs. The extracts also increased the immunostimulatory capacity of DCs, promoting the activation and proliferation of effector T cells. This suggests fig extracts can be used as immunostimulators to enhance immune responses. In another study, it was confirmed that *Ficus carica* polysaccharides significantly up-regulated immune-related gene expression, and stimulated the immune response, which enhanced disease resistance in grass carp (Yang *et al.*, 2015). Yang *et al.* (2009) evaluated the immunity promoting activities of crude hot-water soluble polysaccharides from *Ficus carica* using the carbon clearance test and serum haemolysin analysis in mice. Results showed that fig polysaccharides (500 mg $kg^{-1}$) significantly increased the clearance rate of carbon particles and serum hemolysin levels in normal mice.

## Antioxidant capacity

Oxidation is now acknowledged to be the fundamental mechanism underlying a number of human disorders such as neurological troubles, inflammation, viral infections, autoimmune pathologies and digestive system disorders, including gastrointestinal inflammation and ulcers. Hence, prevention and therapy using free-radical scavenging antioxidants have potential to prevent, delay or ameliorate many of these disorders (Marrelli *et al.*, 2012). Natural antioxidants such as phenolic acids, flavonoids and anthocyanins from fruits and vegetables have been investigated by multiple researchers. These compounds act as antioxidants in different ways: as reducing agents, hydrogen donators, free radical scavengers and singlet oxygen quenchers, and therefore, as cell saviors (Oliveira *et al.*, 2009).

Figs contain relatively high amounts of phenolic compounds with high antioxidant activity, which can prevent several diseases (Bachir bey *et al.*, 2014) and thus may serve as an excellent source of these compounds in the diet. The antioxidant capacity of different parts of fresh figs (skin, pulp, whole fruit), as well as for dried figs has been assessed in numerous studies by different methods under different conditions. The most relevant assays are the 1,1-diphenyl-2-picrylhydrazyl radical assay (DPPH), 2,2-azino-di-(3-ethylbenzothialozine-sulphonic acid) assay (ABTS), ferric ion reducing antioxidant power assay (FRAP), cupric ion reducing capability assay (CUPRAC), oxygen radical absorbance capacity assay (ORAC) and β-carotene bleaching test (Loizzo *et al.*, 2014; Harzallah *et al.*, 2016; Wojdyło *et al.*, 2016; Konak *et al.*, 2017; Arvaniti *et al.*, 2019). This is understandable since multiple active species, reaction characteristics and mechanisms are involved in oxidative stress in the human body, and therefore, there is no simple universal method by which antioxidant capacity can be assessed accurately and quantitatively (Loizzo *et al.*, 2014).

In an earlier study, Marrelli *et al.* (2012) investigated the antioxidant activity of whole fresh Dottato figs from Italy at two stages of ripeness. They found that fig fruits of the first harvest (June) exhibited the highest activity with $IC_{50}$ of 1.64 and 0.004 mg $mL^{-1}$ for DPPH and the β-carotene bleaching test, respectively. Crisosto *et al.* (2010) reported that antioxidant capacity differed significantly among fig cultivars, but it was not affected by maturity stage. These researchers found $IC_{50}$ values ranged from 1.44 to 3.14 mmol TE/g FW for the four cultivars investigated.

When the antioxidant capacity of the skin and flesh of nine Spanish fig varieties of different color and at three different ripening stages was studied, the results demonstrated the hydrophilic fraction of the darker varieties exhibited higher antioxidant activity, ranging from 16.3 to 177.4 mmol Trolox equivalents/100 g, than the lighter ones. Similarly, Mahmoudi *et al.* (2018) found that among the nine Algerian fresh fig cultivars, the black fig peels (Bakkor Khal) had the highest antioxidant capacity (2.479 ± 0.044 mg $mL^{-1}$). The antioxidant capacity was higher in the skin compared to the flesh (Pereira *et al.*, 2017b). Based on their results, the authors advised consuming unpeeled, fully ripe, dark figs to get the optimal health-promoting benefits.

Many scientists have demonstrated that phenolic compounds with antioxidant capacity were concentrated in the fig skin and that dark-skinned fig cultivars had higher antioxidant activity compared to light-skinned cultivars (Crisosto *et al.*, 2010; Bachir Bey and Louaileche, 2015; Hoxha *et al.*, 2015; Harzallah *et al.*, 2016; Mahmoudi *et al.*, 2018). Not surprisingly, a high correlation between phenolic content and antioxidant activity is also reported in multiple studies (Çaliskan and Polat, 2011; Bachir Bey and Louaileche, 2015).

Dried figs have the highest antioxidants among the dried fruits (apricots, dates, cranberries, plums and raisins) (Vinson *et al.*, 2005). Capanoglu (2014) investigated the antioxidant capacities of ten different dried fruits. Using the FRAP assay, figs had a higher antioxidant capacity than mulberry, yellow raisin and $SO_2$ treated apricot. In another study, Ouchemoukh *et al.* (2012) found that ethanolic extracts of dried figs and apricots had a comparative DPPH radical scavenging activity (55.9 ± 0.7 and 53.6 ± 0.6%). Crisosto *et al.* (2010) reported that fig antioxidant capacity values were higher than that of peaches and plums. These findings are consistent with Çaliskan and Polat's (2011) suggestion that among all common fruits and vegetables, berries and figs, especially those with dark blue or red colors, have the highest antioxidant capacity.

Investigations of the impact of drying on the antioxidant capacity of figs have produced contradictory results. A number of studies have indicated drying decreased the antioxidant activity of figs. For example, Kamiloglu and Capanoglu (2015) found that sun drying of yellow and purple figs led to a drastic reduction in the total antioxidant capacity of figs (decreased by 24–57 and 13–57%, respectively). Capanoglu (2014) suggested that drying induced oxidative decomposition either enzymatically or by thermal degradation of polyphenols. Other studies found that the antioxidant potential was significantly higher in dried figs versus fresh figs (Slatnar *et al.*, 2011; Hoxha and Kongoli, 2016; Konak *et al.*, 2017). Konak *et al.* (2017) hypothesized the increased antioxidant activity of light-colored figs by drying might be due to the formation of Maillard reaction products, known for their free radical scavenging capacities.

The number of superoxide radicals in human glioblastoma (U87) cells (ROS) produced in the presence of fig extract was measured by chemiluminescence using lucigenin. The obtained results proved that *Ficus carica* extracts inhibited ROS production of 41.21 ± 0.88% at 250 μg $mL^{-1}$ (Amessis-Ouchemoukh *et al.*, 2017). These findings could be attributed to the phenolic two mechanisms, first, by inhibiting the enzyme NADPH oxidase and thus reducing the number of radicals formed and second, by direct scavenging of superoxide radicals in competition with lucigenin.

In a clinical study, a change in plasma antioxidant capacity after consumption of one serving size (40 g) of dried figs was investigated on ten healthy 25–58-year-old males and females. Figs produced a significant increase in plasma antioxidant capacity for 4 h after consumption and overcame the oxidative stress of consuming high fructose corn syrup in a carbonated soft drink (Vinson *et al.*, 2005). The authors suggest this effect is may be related to the phenols in dried fruits.

## Bone health

Bone loss, osteoporosis and rheumatoid arthritis are global health problems (Idrus *et al.*, 2018). Bone mass is regulated by the balance between osteoclast-induced bone resorption and osteoblast-induced bone formation activities (Choi *et al.*, 2011; Mohammad *et al.*, 2018).

Post-menopausal osteoporosis, due to estrogen deficiency following menopause, is the most common type of osteoporosis (Norfarah Izzaty *et al.*, 2019). Considering the adverse effects of estrogen replacement therapy, Mohammad *et al.* (2018) tested the effect of fresh fig supplementation on the trabecular microarchitecture of post-menopausal osteoporotic rat models. For 8 weeks daily, 50 and 100 mg $kg^{-1}$ aqueous and raw fig extracts were given orally to rat models. The 50 mg $kg^{-1}$ raw fig extract produced the best microarchitecture among the treatment groups with less bone deteriorations. These researchers concluded that raw fig extract, due to its high mineral contents and bioactive compounds, has a strong potential as an alternative osteoporosis prevention supplement

Norfarah Izzaty *et al.* (2019) evaluated the effects of figs on bone metabolism of ovariectomized rats. After eight weeks of treatments with 50 and 100 mg $kg^{-1}$ aqueous and raw fig extracts, the biochemical bone markers for osteocalcin (bone formation marker), C-telopeptide of type 1 collagen (bone resorption marker) and bone estrogen level were measured. The effects of fig supplementation were comparable to oestrogen replacement therapy. This suggests the oesteo-protective effects of figs are due to the reduction of bone resorption activity rather than bone formation activity.

Osteoporosis is associated with estrogen deficiency and highly associated with oxidative stress, a major cause of the imbalance between osteoblast and osteoclast activities. This results in a reduction in bone density. Antioxidative and anti-inflammatory compounds may play a role in protecting bones against the damaging effects of free radicals (Mohammad *et al.*, 2018). Fig fruits contain high levels of antioxidant bioactive compounds; polyphenols, flavonoids, anthocyanins, ferulic and *p*-coumaric acids (Ammar *et al.*, 2015), all reported to improve bone strength and prevent bone loss in ovariectomized rat models (Mohammad *et al.*, 2018). They hypothesized these compounds inhibited bone resorption by scavenging the superoxide radical and suppressing lipid peroxidation.

Discovery of chemicals that increase bone formation to treat osteoporosis and enhance post-trauma or surgery bone healing is needed (Wong and Rabie, 2011). A furanocoumarin, psoralen, has been identified in figs (Oliveira *et al.*, 2009; Nakilcioglu and Hisil, 2013). *In vivo*, this compound in a collagen matrix significantly increased new bone formation locally when grafted into skull defects. It produced 454% more new bone than the absorbable collagen sponge alone (negative control-collagen) (Wong and Rabie, 2011). The authors reported that psoralen increased bone formation via an increase in osteoblast formation. Tang *et al.* (2011) demonstrated psoralen promoted osteoblast differentiation in primary mouse calvarial osteoblasts in a dose-dependent manner, demonstrated by up-regulation expression of osteoblast-specific marker genes including type I collagen, osteocalcin and bone sialoprotein, and enhancement of alkaline phosphatase activity.

Figs, especially dark-colored ones are a rich source of anthocyanins such as pelargonidin-3-*O*-rutinoside; cyanidin-3-*O*-glucoside and cyanidin-3-*O*-rutinoside (Duenas *et al.*, 2008; Bachir Bey *et al.*, 2016; Wojdyło *et al.*, 2016). Multiple reports confirm the role of anthocyanins in preventing bone diseases through antioxidant, and anti-inflammatory effects of these bioactive compounds (Riaz *et al.*, 2016). An observational study documented an increase in bone mineral density of women with supplemental flavonoids and anthocyanin intake (Welch *et al.*, 2012).

In addition to furanocoumarins and phenolic compounds, polyunsaturated fatty acids, such as ethyl docosahexaenoate (E-DHA) and docosahexaenoic acid (DHA) were identified in the hexane soluble fraction of *Ficus carica* (Park *et al.*, 2009). E-DHA inhibits osteoclastogenesis significantly in the murine RAW264.7 monocyte/macrophage cell line via the JNK and NF-κB signaling pathways, and DHA induces osteoblast differentiation by rapidly inducing BMP-2 in MC3T3-E1 cells (Choi *et al.*, 2011). These authors hypothesize that a combination of E-DHA and DHA has potential as a treatment of diseases involving abnormal bone lysis, such as osteoporosis, rheumatoid arthritis and periodontal bone erosion.

Fresh and dried figs are characterized by their abundant mineral content: calcium, potassium, magnesium, phosphorus, iron and strontium, all important for bone health (Khan *et al.* 2011; Sadia *et al.* 2014; Soni *et al.* 2014; Pareira *et al.*, 2017a). Calcium is a major bone component and enhances tooth development. Therefore, consumption of calcium-rich foods, like figs, benefits these tissues (Pereira *et al.*, 2017a). Similarly, strontium has been found to enhance bone health. A form of strontium, strontium ranelate, is patented for postmenopausal osteoporosis treatment. It is the first anti-osteoporotic agent that appears to simultaneously increase bone formation and decrease bone resorption, thus resulting in new bone formation (Soni *et al.*, 2014).

Potassium is the primary mineral in figs (Khan *et al.* 2011). Numerous population-based studies have demonstrated the beneficial effect of potassium in fruit and vegetables, and alone on axial and peripheral bone mass and bone metabolism in men and women of all ages (New, 2003).

## Anticonstipation

Constipation is prevalent in modern societies and a common gastrointestinal symptom in clinical practice, affecting 5–20% of the general population. This syndrome, which is associated with the inability to empty the rectum completely, is diagnosed when defecation is difficult and delayed for 2 weeks or more, causing discomfort (Oh *et al.*, 2011; Parsaei *et al.*, 2016). Many medicinal plants have been used to treat constipation, including fig fruits (Abbasi *et al.*, 2013; Parsaei *et al.*, 2016).

Lee *et al.* (2012) examined the effects of fig paste on loperamide-induced constipation in rats. Constipation was induced in the rats by intraperitoneal injections of loperamide (2 mg $kg^{-1}$, twice a day). Fig paste (0, 1, 6 and 30 g $kg^{-1}$) was administered for 4 weeks to assess its anticonstipation effects. Constipation decreased when the rats consumed fig paste. Specifically, the fecal number, weight, and water content, as well as histological parameters such as thickness and mucin areas in the distal colon improved.

Oh *et al.* (2011) investigated the laxative effects of figs in beagles with constipation induced by a high protein diet and restricted movement. Fig paste, 12 g $kg^{-1}$ daily for 3 weeks following 3 weeks of constipation significantly increased fecal quantity and reduced segmental colonic transit time in the constipated dogs. The results demonstrate fig paste has potential as complementary medicine in humans suffering from chronic constipation.

In another study, the fig extract (150 and 300 mg $kg^{-1}$) was given orally once per day for 7 days, 1h after the loperamide treatment in dextran sulphate sodium (DSS)-induced acute colitis in rats. It significantly increased the gastrointestinal transit ratio and gastric emptying by accelerating their times. The fig extract also sharply reduced the colitis induced by constipation (Rtibi *et al.*, 2018).

The mechanism by which figs can reduce constipation is not clear. However, researchers have suggested that cellulose, one of the main components of figs, might explain this effect (Oh *et al.*, 2011; Lee *et al.*, 2012; Rtibi *et al.*, 2018). Fibers that are incompletely or slowly fermented by microflora in the large intestine promote normal laxation and are integral components of diets designed to treat constipation and prevent the development of diverticulosis and diverticulitis (Marlett *et al.*, 2002). Increasing the fiber consumption is the first, and most effective, treatment option for chronic constipation in children (Kranz *et al.*, 2012).

Both water-soluble and insoluble fibers shorten large intestine transit time and increase fecal excretion by increasing water content, bulk and viscosity. These effects are mediated by the water-binding capacity of the dietary fiber and by fermentation, which alters the osmotic balance and increases fecal biomass. Studies with rats have demonstrated cellulose fermentation by coliform bacilli produces acid, increasing the acidity which accelerates peristaltic activity (Oh *et al.*, 2011; Lee *et al.*, 2012; Academy of Nutrition and Dietetics, 2015).

Dietary fiber is essential for the production of short-chain fatty acids (SCFAs) during colonic fermentation. An elevated amount of SCFAs in the intestine enhances probiotic microorganism growth. The accumulation of SCFAs promotes regularity and laxation by proliferating the fecal weight and bulk and fecal water holding capacity (Rtibi *et al.*, 2018). It has been hypothesized that the phenolic compounds in figs also reduce constipation (Parsaei *et al.*, 2016). They may protect the epithelium against physical pressure-associated damage, (Lee *et al.*, 2012). Oxidative stress contributes to constipation through elevated intracellular levels of reactive oxygen species (ROS) that damage lipids, proteins, and DNA as well as the depleting the antioxidants (SOD, CAT and GPx). The phenolic compounds in figs have strong antioxidant activity with the ability to mitigate the effects of ROS and enhance intestinal movement (Caliskan and Polat 2011; Rtibi *et al.*, 2018).

## Other health benefits

*Ficus carica* fruit has also been reported to have antispasmodic and antiplatelet activities (Gilani *et al.*, 2008), anti-depressant and anxiolytic effects (Gul *et al.*, 2018) and anticoagulant activity that could be useful in preventing blood clots (Belattar and Himour, 2019). Dried figs combined with extra virgin olive oil have reported having a hepatoprotective effect on CCl4-induced oxidative stress and hepatotoxicity in rats. This validates its use in Mediterranean folk medicine and suggests possible use as a prophylaxis for liver disease (Debib *et al.*, 2016).

## 18.6 Traditional and Modern Medicinal Uses of Different Parts of the Fig Tree (Leaf, Shoot Bark and Latex)

### Traditional medicinal uses

Traditional medicines have been frequently used in all civilizations. The genus *Ficus* (Moraceae), one of the largest genera of angiosperms, has been an ancient source of food and health. *Ficus carica* L. is among the most popular species of this genus (Barolo *et al.*, 2014). *Ficus* species are recorded in the traditional Ayurvedic, Unani and Siddha medical literature for the treatment of endocrine, kidney, respiratory, cardiovascular, liver and gastrointestinal diseases (Gilani *et al.*, 2008), reproductive disorders, and infectious diseases.

Traditionally, all parts of fig tree (leaves, roots, twigs, young shoots, ashes, fruits, bark and latex) have been used in the treatment of diarrhea, rash, indigestion, tumors and inflammatory diseases in the form of poultices, ointment, infusions, decoctions and other traditional medicinal formulations (Lansky *et al.*, 2008; Lee and Cha, 2010; Kabir *et al.*, 2012). Fig tree latex was used traditionally to treat warts and as de-wormer, due to its proteolytic fraction called ficin (De Amorin *et al.*, 1999; Khodarahmi *et al.*, 2011). The leaf and stem latex relieve pain; it is used for treating stings of insects, bites and warts (Ybert and Delesalle-Féat, 2001). The decoction of fig leaves is very effective for washing head wounds. It also clears the face of crusts and common injuries. The leaf syrup has been used for coughs and diseases of the lungs and helps to thin blood (Khare, 2004).

### Modern medicinal uses (pharmacological activities)

Plants and their extracts, including *Ficus* species, have been used, and still are, to promote health and treat diseases. Many active compounds have been identified in the different parts of *Ficus* species. Palmitic and linolenic acids; α-tocopherol; coumarins: bergapten, psoralen and marmesin; triterpenoids: bauerenol and lupeol acetate; phenolics: ferulic and chlorogenic acids; quercetin; and rutin have been extracted from the leaves. Chlorogenic and protocatechuic acids, catechin, rutin, psoralen, xanthotoxin, alkaloids, terpenoids, sterols, proteases like ficine, chitinases and glucosidases have been extracted from the latex (Innocenti *et al.*, 1982; Saeed and Sabir, 2002; Oliveira *et al.*, 2009; Conforti *et al.*, 2012; Ghazi *et al.*, 2012; Raskovic *et al.*, 2016; Abdel-Atya *et al.*, 2019).

### Leaf and shoot bark

More recently extracts from *Ficus* leaves and shoot bark have been tested for biological activities (Table 18.4). Mopuri *et al.* (2018) investigated the antidiabetic activity as measured by an estimation of inhibition of α-amylase and α-glucosidase enzymes. The antihyperlipidemic effect of *Ficus carica* leaf extract was also examined in streptozotocin-induced diabetic rats. Results demonstrated declines in the total cholesterol and the total cholesterol/HDL cholesterol ratio, and hyperglycemia (Canal *et al.*, 2000). A fig leaf extract demonstrated promising potential for preventing hyperlipidemia in high fat diet-induced obese male rats when tested against a pioglitazone control (Joerin *et al.*, 2014).

Leaf extracts demonstrated hepatoprotective activity in damaged liver of rats. An oral dose of 500 mg $kg^{-1}$ lowered serum levels of aspartate aminotransferase, alanine aminotransferase, total serum bilirubin and malondialdehyde equivalent, an index of liver lipid peroxidation (Krishna Mohan *et al.*, 2007; Khodarahmi *et al.*, 2011). A recent study demonstrated a *Ficus* shoot bark ethanolic extract and its most active oligosaccharide-rich fraction contained neuroactive principles with strong potential to manage strychnine-induced convulsive

**Table 18.4.** Biological activities of different parts of *Ficus carica*.

| Investigated parts | Tested extracts | Methods | Value | Biological activities | References |
|---|---|---|---|---|---|
| Leaves and latex | Ethanol, ethyl acetate and dichloromethane leaf extracts and latex | *In vitro*: measurement of HeLa cell lines viability after 48 h of incubation by rapid colorimetric assay using MTT | $IC_{50}$ = 10 to 19 μg $ml^{-1}$ | Anticancer | Khodarahmi *et al.* (2011) |
| Leaves | Methanolic extract (70%) | Tested on isolated hearts of male rats. Cardiac arrhythmias were analyzed and the TTC method was used for infarct size determination | 0.04, 0.2 and 1 mg $ml^{-1}$ | Ischemia/reperfusion injuries | Allahyari *et al.* (2014) |
| Leaves and stem bark | Hexane, ethyl acetate, ethanol and water extracts | *In vitro*: inhibition of α-amylase and α-glucosidase enzymes, and antilipase activities were measured using several experimental models | α-amylase: $IC_{50}$ = 0.55 to 11.6 mg $ml^{-1}$α-glucosidase: $IC_{50}$ = 0.27 to 43.1 mg $ml^{-1}$ Pancreatic lipase: $IC_{50}$ =0.36 to 276 mg $ml^{-1}$ | Antidiabetic and antiobesogenic | Mopuri *et al.* (2018) |
| Latex | Latex | *In vivo*: in NIH mice naturally infected with *Syphacia obvelata*, *Aspiculuris tetraptera* and *Vampirolepis nana* | 3 ml $kg^{-1}$ per day, during three consecutive days, was effective in the removal of *S. obvelata* (41.7%) | Anthelmintic | De Amorin *et al.* (1999) |
| Leaves | Methanolic extract | 2, 2-diphenyl-1-picrylhydrazyl (DPPH) scavenging assay | $IC_{50}$ = 659.9 to 1119.6 μg $ml^{-1}$ | Antioxidant | Mahmoudi *et al.* (2016) |
| Leaves | / | *In vivo*: in high fat diet-induced obese male rats | 50 mg $kg^{-1}$ or 100 mg $kg^{-1}$ | Antihyperlipidemic | Joerin *et al.* (2014) |
| Leaves | Ethanolic and aqueous extracts | *In vitro*: Macrodilution assay | $CMI_{eth}$: 6.25–12.5 mg $ml^{-1}$ $CMI_{water}$: 25–50 mg $ml^{-1}$ | Antimicrobial | Jeong *et al.* (2009) |
| Latex | Latex | *In vitro*: on stomach cancer cell line | 5 mg $ml^{-1}$ | Anticancer | Hashemi *et al.* (2011) |
| Leaves | Petroleum ether, chloroform and ethanol extracts | *In vivo:* Carrageenan-induced rat paw edema and cotton pellet granuloma methods | 600 mg $kg^{-1}$ per day for the ethanolic extract | Anti-inflammatory | Patil and Patil (2011) |

disorders (Raafat and Wurglics, 2019). In their study, Kabir *et al.* (2012) investigated the anti-diarrheal activity of a *Ficus carica* leaf and ethanolic bark extracts. Results showed a significant decrease in the severity of diarrhea, in a dose-dependent manner, in a castor oil-induced diarrhea test.

The *in vitro* anticarcinogenic effect of non-polar fig leaf extracts was investigated by 24 hour application to a PC3 human prostate cancer cell line, and had a cytotoxic effect on these cells in a dose-dependent manner; almost 100% death at 1000 μg $ml^{-1}$ (Boyacioglu *et al.*, 2019). The anticancer activity of a fig leaf methanolic extract against the proliferation, apoptosis and necrosis in Huh7it cells was investigated (Purnamasari *et al.*, 2019). They reported the higher the dose of fig extract, the more cancer cells undergoing apoptosis and a lower percentage of cell growth. The linoleic acid peroxidation and DPPH tests and MTT assay were used to assess the antioxidant activity and to evaluate the antiproliferative activity on C32 human melanoma cells of *Ficus carica* leaves, bark and woody parts (Conforti *et al.*, 2012).

Multiple researchers have reported promising results when examining the antioxidant effect of *Ficus carica* leaves and shoot bark (Konyalioglu *et al.*, 2005; Mahmoudi *et al.*, 2016; Mopuri *et al.*, 2018; Purnamasari *et al.*, 2019). *Ficus* leaf methanol extracts demonstrated antimicrobial activity against methicillin-resistant *Staphylococcus aureus* (MRSA), suggesting potential as a natural antibacterial agent for MRSA infections (Lee and Cha, 2010). Mahmoudi *et al.* (2016), also reported *Ficus* leaf extracts had antibacterial activity against Gram-positive and Gram-negative bacteria (*Bacillus cereus, Bacillus subtilis, Staphylococcus aureus, Enterococcus faecalis, Micrococcus luteus, Klebsiella pneumoniae, Pseudomonas aeruginosa, Escherichia coli* and *Salmonella* sp) and moderate antifungal activity. The regulation of thyroidism, the immunomodulatory activity and the antipyretic activity of *Ficus carica* leaves were also studied. The results showed stimulatory effect of leaf extracts on the humoral and cell-mediated immune response in the experimental animals and suggested its therapeutic usefulness in disorders of immunological origin (Patil *et al.*, 2010a, b; Saxena *et al.*, 2012).

### Latex

Abdel-Atya *et al.* (2019) investigated the anticancer activity of *Ficus* latex extracts (FLE). They reported FLE displayed a significant potent cytotoxic activity ($p < 0.01$) against HepG2, MCF-7 and HGT116 cancer cells with $IC_{50}$ = 32.25, 25.3 and 38.75 μg $ml^{-1}$ respectively. In other studies, the cytotoxic effect of fig tree latex on HeLa cell line, human glioblastoma, hepatocellular carcinoma and healthy liver cells was investigated (Wang *et al.*, 2008; Khodarahmi *et al.*, 2011). The anthelmintic activity of *Ficus* latex has been investigated in NIH mice naturally infected with *Syphacia obvelata, Aspiculuris tetraptera* and *Vampirolepis nana. Ficus* latex of 3 ml $kg^{-1}$ per day, for three days, was effective in the removal of *S. obvelata* (41.7%). However, high acute toxicity with hemorrhagic enteritis generated, and the weak anthelmintic efficacy, argues against using latex in traditional medicine (De Amorin *et al.*, 1999).

Fig latex's physiological role is protection from pathogens. Latex is a rich source of proteases, predominantly ficin. The antibacterial and antifungal activities of fig latex have been reported in the literature (Aref *et al.*, 2010; Raskovic *et al.*, 2016). *Ficus* ficin has the ability to affect human blood coagulation. Healthy human plasma was incubated with ficin, and the activated partial thromboplastin time (APTT) and prothrombin time (PT) were measured. The results demonstrated a decrease in APTT and PT when increasing ficin concentration, suggesting that the hemostatic potency of ficin proteases is based on activation of human coagulation factor X (Richter *et al.*, 2002).

## 18.7 Fig and Allergy

Plant allergens, the most common allergies, are hard to avoid. Hence, their identification and characterization are of prime importance for the diagnosis and treatment of food allergies (Hassan and Venkatesh, 2015). Common fruits like banana, avocado, chestnut, kiwi, apple, peach, musk melon, cherry, grape, strawberry, mango and pomegranate are all known to cause food allergies (Blanco Guerra, 2002; Hassan and Venkatesh, 2015).

Caiaffa *et al.* (2003) reported that accounts of allergic reactions to figs (*F. carica*) in the literature are rare. Dechamp *et al.* (1995) reported an anaphylactic reaction from fresh figs (*F. carica*). Allergic reactions to fresh or dried figs have been suggested to be a primary sensitivity to airborne *F. benjamina* (FB) allergens or to the 'latex-fruit syndrome' (Dechamp *et al.*, 1995; Antico *et al.*, 2003; Focke *et al.*, 2003). The proteins involved in fig fruit hypersensitivity are identical to those in respiratory *Ficus benjamina* allergies, but little information has been developed (Focke *et al.*, 2003). Similarly, Díez-Gómez *et al.* (1998) reported that in a patient with perennial asthma caused by *F. benjamina* latex, episodes of angioedema of the oropharyngeal tract and tongue were caused by eating figs and kiwi fruit. They suggested allergies to figs and papain is due to cross-reactive allergen structures.

In another study, three cases of fig allergy were confirmed based on a convincing clinical history and positive skin test results. Two patients complained of severe abdominal and general symptoms, whereas the third cited only oral and pharyngeal symptoms (Caiaffa *et al.*, 2003). Two cases of the oral allergy syndrome (OAS) to fig in patients whose main allergic manifestations were related to sensitization to grass and birch pollen and a case of occupational, IgE-mediated papain allergy with kiwi fruit and fig fruit have also been reported (Antico *et al.*, 2003; Jiang *et al.*, 2016). Cruz *et al.* (2007) described acute urticaria to fig tree sap and fig leaf but not to fig fruit pulp. Similarly, Abdellaziz *et al.* (2015) had reported a case of allergy to *Ficus* latex but not to the fruit of this tree.

Fig allergies present with a wide variety of symptoms: sneezing, nasal obstruction, sore throat, cough, bilateral palpebral angioedema, generalized pruritus, acute urticaria, attacks of severe abdominal pain and respiratory symptoms (Gandolfo *et al.* 2001; Antico *et al.*, 2003; Caiaffa *et al.*, 2003). Fig skin has been reported as having much higher allergenicity than the pulp. Figs appear to lose their allergenicity when denatured by heat (95°C) (Antico *et al.*, 2003; Focke *et al.*, 2003). Therefore, individuals with an oral allergy to figs can consume heat-processed figs (Taylor and Hefle, 2001), fig jam, jelly or sauce.

## 18.8 Novel Products/Formulations and Future Trends

Most of the health benefits of figs and its parts are proven by *in vitro* studies; further *in vivo* and *in situ* studies are needed to reveal the real impact of these products on human health and as medicines. Similarly, determination and purification of the specific substances of *Ficus* extracts, responsible for the health benefit are required.

However, the nutritional value of fresh and dried figs is well proven and anecdotally supported by its inclusion in the Mediterranean diet, among the healthiest diets in the world. However, figs are not a part of the diet in other parts of the world. Perhaps fig consumption, particularly by children can be increased by creating novel products with figs such as fig chips (swell drying by Instant Controlled Pressure Drop (DIC) Technology), chocolate figs, candied figs, fig syrup, jam, marmalade, in bakery or dairy products. *Ficus* is a resistant tree, adaptable to different climates and soils. Moreover, its by-products, leaves and fruits, can also be used in animal food.

## References

Abbasi, A.M., Khan, M.A., Shah, M.H, Shah, M.M., Pervez, A. *et al.* (2013) Ethnobotanical appraisal and cultural values of medicinally important wild edible vegetables of Lesser Himalayas-Pakistan. *Journal of Ethnobiology and Ethnomedicine* 9, 13. DOI: 10.1186/1746-4269-9-66

Abdel-Aty, A.M., Hamed, M.B., Salama, W.H., Ali, M.M., Fahmy, A.S. and Mohamed S.A. (2019) *Ficus carica*, *Ficus sycomorus* and *Euphorbia tirucalli* latex extracts: Phytochemical screening, antioxidant and cytotoxic properties. *Biocatalysis and Agricultural Biotechnology* 20, DOI: 10.1016/j.bcab.2019.101199

Abdellaziz, R., Medjras, S., Arab, D., Lemdani, M. and Douagui, H. (2015) Allergie à la sève de figuier. *Revue Française d'Allergologie* 55, 225. DOI: 10.1016/j.reval.2015.02.039

Academy of Nutrition and Dietetics (2008) Position of the Academy of Nutrition and Dietetics: Health Implications of Dietary Fiber. *Journal of The Academy of Nutrition and Dietetics* 108, 1716–1731.
Academy of Nutrition and Dietetics (2015) Position of the Academy of Nutrition and Dietetics: Health Implications of Dietary Fiber. *Journal of The Academy of Nutrition and Dietetics* 115, 1861–1870.
Akoh, C.C. and Min, D.B. (2008) *Food Lipids; Chemistry, Nutrition, and Biotechnology*. 3rd edn. CRC Press. DOI: 10.1201/9781420046649
Alamgeer, S.I., Hira, A. and Muhammad, S. (2017) Evaluation of antihypertensive potential of *Ficus carica* fruit. *Pharmaceutical Biology* 55, 1047–1053.
Al-Farsi, M., Alasalvar, C., Al-Abid, M., Al-Shoaily, K., Al-Amry, M. *et al.* (2007) Compositional and functional characteristics of dates, syrups, and their by-products. *Food Chemistry* 104, 943–947.
Aljane, F. and Ferchichi, A. (2009) Postharvest chemical properties and mineral contents of some fig (*Ficus carica* L.) cultivars in Tunisia. *Journal of Food, Agriculture & Environment* 7, 209–212.
Allahyari, S., Delazar, A. and Najafi, M. (2014) Evaluation of general toxicity, anti-oxidant activity and effects of *Ficus carica* leaves extract on ischemia reperfusion injuries in isolated heart of rat. *Advanced Pharmaceutical Bulletin* 4, 577–582. DOI:10.5681/apb.2014.085
Amessis-Ouchemoukh, N., Ouchemoukh, S., Meziant, N., Idiri, Y., Hernanz, D. *et al.* (2017) Bioactive metabolites involved in the antioxidant, anticancer and anticalpain activities of *Ficus carica* L., *Ceratonia siliqua* L. and *Quercus ilex* L. extracts. *Industrial Crops and Products* 95, 6–17.
Ammar, S., del Mar Contreras, M., Belguith-Hadrich, O., Bouaziz, M. and Segura-Carreteroa, A. (2015) New insights into the qualitative phenolic profile of *Ficus carica* L. fruits and leaves from Tunisia using ultra-high-performance liquid chromatography coupled to quadrupole-time-of-flight mass spectrometry and their antioxidant activity. *RSC Advances* 26, 20035–20050.
ANSES-CIQUAL French food composition table version (2017) https://ciqual.anses.fr
Antico, A., Zoccatelli, G., Marcotulli, C. and Curioni, A. (2003) Oral allergy syndrome to fig. *International Archives of Allergy and Immunology* 131, 138–142. DOI: 10.1159/000070929
Aref, H.L., Salah, K.B., Chaumont, J.P., Fekih, A., Aouni, M. and Said, K. (2010) *In vitro* antimicrobial activity of four *Ficus carica* latex fractions against resistant human pathogens (antimicrobial activity of *Ficus carica* latex) *Pakistan Journal of Pharmaceutical Science* 23, 53–58.
Arvaniti, O.S., Samaras, Y., Gatidou, G., Thomaidis, N.S. and Stasinakis, A.S. (2019) Review on fresh and dried figs: Chemical analysis and occurrence of phytochemical compounds, antioxidant capacity and health effects. *Food Research International* 119, 244–267.
Bachir Bey, M. and Louaileche, H.A. (2015) Comparative study of phytochemical profile and *in vitro* antioxidant activities of dark and light dried fig (*Ficus carica* L.) varieties. *The Journal of Phytopharmacology* 4, 41–48.
Bachir Bey, M., Meziant, L., Benchikh, Y. and Louaileche, H. (2014) Deployment of response surface methodology to optimize recovery of dried dark fig (*Ficus carica* L., var. Azenjar) total phenolic compounds and antioxidant activity. *International Food Research Journal* 21, 1477–1482.
Bachir Bey, M., Richard, G., Meziant, L., Fauconnier, M-L. and Louaileche, H. (2016) Effects of sun-drying on physicochemical characteristics, phenolic composition and *in vitro* antioxidant activity of dark fig varieties. *Journal of Food Processing and Preservation* 41, 1–8. DOI:10.1111/jfpp.13164
Barolo, M.I., Mostacero, N.R. and López, S.N. (2014) *Ficus carica* L. (Moraceae): An ancient source of food and health. *Food Chemistry* 164, 119–127. DOI: 10.1016/j.foodchem.2014.04.1120308-8146/
Belattar, H. and Himour, S. (2019) Anticoagulant activity of some *Ficus carica* varieties extracts grown in Algeria. *Acta Scientifica Naturalis* 6, 73–78. DOI.10.2478/asn-2019-0010
Belguith-Hadriche, O., Ammar, S., del Mar Contreras, M., Turki, M., Segura-Carretero, A. *et al.* (2016) Antihyperlipidemic and antioxidant activities of edible Tunisian *Ficus carica* L. fruits in high fat diet-induced hyperlipidemic rats. *Plant Foods for Human Nutrition*. DOI: 10.1007/s11130-016-0541-x
Blanco Guerra, C. (2002) Latex-fruit syndrome. *Allergology and Immunopathology (Madr)* 30, 156–163.
Bondonno, N.P., Bondonno, C.P., Ward, N.C., Hodgson, J.M. and Croft, K.D. (2017) The cardiovascular health benefits of apples: Whole fruit vs. isolated compounds. *Trends in Food Science & Technology* 69, 243–256.
Boyacioglu, O., Kara, B., Can, H., Naile Yerci, T., Yilmaz, S. and Boyacioglu, S.O. (2019) Leaf hexane extracts of two Turkish fig (*Ficus carica* L.) cultivars show cytotoxic effects on a human prostate cancer cell line. *Agriculture and Food Sciences Research* 6, 66–70. DOI:10.20448/journal.512.2019.61.66.70
Caiaffa, M.F., Cataldo, V.M., Tursi, A. and Macchia, L. (2003) Fig and mulberry cross-allergy. *Annals of Allergy, Asthma & Immunology* 91, 493–495. DOI: 10.1016/S1081-1206(10)61520-7

Çaliskan, O. and Polat, A. (2011) Phytochemical and antioxidant properties of selected fig (*Ficus carica* L.) accessions from the eastern Mediterranean region of Turkey. *Scientia Horticulturae* 128, 473–478.

Canal, J.R., Torres, M.D., Romero, A. and Pérez, C. (2000) A chloroform extract obtained from a decoction of *Ficus carica* leaves improves the cholesterolaemic status of rats with streptozotocin-induced diabetes. *Acta Physiologia Hungarum* 87, 71–76.

Capanoglu, E. (2014) Investigating the antioxidant potential of Turkish dried fruits. *International Journal of Food Properties* 17, 690–702.

Choi, B-Y., Eun, J.S., Nepal, M., Lee, M-K., Bae, T.S. *et al.* (2011) Ethyl Docosahexaenoate and its acidic form increase bone formation by induction of osteoblast differentiation and inhibition of osteoclastogenesis. *Biomolecules and Therapeutics* 19, 70–76.

Conforti, F., Menichini, G., Zanfini, L., Tundis, R., Statti, G.A. *et al.* (2012) Evaluation of phototoxic potential of aerial components of the fig tree against human melanoma. *Cell Prolifation* 45, 279–285. DOI: 10.1111/j.1365-2184.2012.00816.x

Crisosto, C.H., Bremer, V., Ferguson, L. and Crisosto, G.M. (2010) Evaluating quality attributes of four fresh fig (*Ficus carica* L.) cultivars harvested at two maturity stages. *Hortscience* 45, 707–710.

Cruz, N.V., Bahna, S.L. and Knight, A.K. (2007) Fig allergy: Not just oral allergy syndrome (OAS). *Journal of Allergy and Clinical* 119(1), S110. DOI: 10.1016/j.jaci.2006.11.411

Darjazi, B.B. and Larijani, K. (2012) The effects of climatic conditions and geographical locations on the volatile flavor compounds of fig (*Ficus carica* L.) fruit from Iran. *African Journal of Biotechnology* 11, 9196–9204.

De Amorin, A., Borba, H.R., Carauta, J.P., Lopes, D. and Kaplan, M.A. (1999) Anthelmintic activity of the latex of *Ficus* species. *Journal of Ethnopharmacology* 64, 225–258.

de Pascual-Teresa, S., Moreno, D.A. and García-Viguera, C. (2010) Flavanols and anthocyanins in cardiovascular health: A review of current evidence. *International Journal of Molecular Science* 11, 1679–1703. DOI: 10.3390/ijms11041679

Debib, A., Duenas, M., Boumediene, M., Mothana, R.A., Alioui, L. and Meddah Tir-touil, A. (2016) Synergetic hepatoprotective effect of phenolic fractions obtained from *Ficus carica* dried fruit and extra virgin olive oil on ccl4-induced oxidative stress and hepatotoxicity in rats. *Journal of Food Biochemistry* 00, 1–10. DOI: 10.1111/jfbc.12241

Dechamp, C., Bessot, J.C., Pauli, G. and Deviller, P. (1995) First report of anaphylactic reaction after fig (*Ficus carica*) ingestion. *Allergy* 50, 514–516. DOI: 10.1111/j.1398-9995.1995.tb01188.x

Del Caro, A. and Piga, A. (2008) Polyphenol composition of peel and pulp of two Italian fresh fig fruits cultivars (*Ficus carica* L.). *European Food Research and Technology* 226, 715–719.

Dickinson, A. and Mister, S. (2012) *The Benefits of Nutritional Supplements*. 4th edn. Council for Responsible Nutrition (CRN), Washington, DC.

Díez-Gómez, M.L., Quirce, S., Aragoneses, E. and Cuevas, M. (1998) Asthma caused by *Ficus benjamina* latex: Evidence of cross-reactivity with fig fruit and papain. *Annals of Allergy, Asthma & Immunology* 80, 24–30.

Doyon, M. and Labrecque, J. (2008) Functional foods: a conceptual definition. *British Food Journal* 110, 1133–1149.

Duenas, M., Pérez-Alonso, J.J., Santos-Buelga, C. and Escribano-Bailon, T. (2008) Anthocyanin composition in fig (*Ficus carica* L.). *Journal of Food Composition and Analysis* 21, 107–115.

Egert, S., Bosy-Westphal, A., Seiberl, J., Kürbitz, C., Settler, U. *et al.* (2009) Quercetin reduces systolic blood pressure and plasma oxidised low-density lipoprotein concentrations in overweight subjects with a high-cardiovascular disease risk phenotype: a double-blinded, placebo-controlled cross-over study. *British Journal of Nutrition* 102, 1065–1074. DOI: 10.1017/S0007114509359127.

Egert, S., Boesch-Saadatmandi, C., Wolffram, S., Rimbach, G. and Müller, M.J. (2010) Serum lipid and blood pressure responses to quercetin vary in overweight patients by apolipoprotein E genotype. *Journal of Nutrition* 140, 278–284.

El-Shobaki, F.A., El-Bahay, A.M., Esmail, R.S., El-Megeid, A.A. and Esmail, N.S. (2010) Effect of figs fruit (*Ficus carica* L.) and its leaves on hyperglycemia in alloxan diabetic rats. *World Journal of Dairy Food Science* 5, 47–57.

Ercisli, S. (2007) Chemical composition of fruits in some Rose (*Rosa* spp.) species. *Food Chemistry* 104, 1379–1384.

Faleh, E., Oliveira, A.P., Valentão, P., Ferchichi, A., Silva, B.M. *et al.* (2012) Influence of Tunisian *Ficus carica* fruit variability in phenolic profiles and *in vitro* radical scavenging potential. *Revista Brasileira de Farmacognosia* 22, 1282–1289. DOI: 10.1590/S0102

FAO (2017) Food and Agriculture Organization. www.fao.org/faostat/fr/#data (accessed 21 February 2022).
FDA (1993) Health claims: fiber-containing grain products, fruits, and vegetables and cancer. Codified in 21 CFR 101.76. Federal Register, 58, 2548.
Fiedor, J. and Burda, K. (2014) Potential role of carotenoids as antioxidants in human health and disease. *Nutrients* 6, 466–488. DOI: 10.3390/nu6020466
Focke, M., Hemmer, W., Wöhrl, S., Götz, M. and Jarisch, R. (2003) Cross-reactivity between *Ficus benjamina* latex and fig fruit in patients with clinical fig allergy. *Clinical and Experimental Allergy* 33, 971–977. DOI: 10.1046/j.1365-2222.2003.01704.x
Gandolfo, M., Baeza, M., De Barrio, M. (2001) Anaphylaxis after eating figs. *Allergy* 56, 462–463. DOI: 10.1034/j.1398-9995.2001.056005462.x
Ghazi, F., Rahmat, A., Yassin, Z., Ramli, N.S. and Buslima, N.A. (2012) Determination of total polyphenols and nutritional composition of two different types of *Ficus carica* leaves cultivated in Saudi Arabia. *Pakistan Journal of Nutrition* 11, 1061–1065.
Gholami, S., Hosseini, M-J., Jafari, L., Omidvar, F., Kamalinejad, M. *et al.* (2017) Mitochondria as a target for the cardioprotective effects of *Cydonia oblonga* Mill. and *Ficus carica* L. in doxorubicin-induced cardiotoxicity. *Drug Research (Stuttg)* 67, 358–365. DOI: 10.1055/s-0043-101824
Gibernau, M., Buser, H.R., Frey, J.E. and Hossaert-Mckeyt, M. (1997) Volatile compounds from extracts of figs *Ficus carica*. *Phytochemistry* 46, 241–214.
Gilani, A.H., Mehmood, M.H., Janbaz, K.H., Khana, A. and Saeed, S.A. (2008) Ethnopharmacological studies on antispasmodic and antiplatelet activities of *Ficus carica. Journal of Ethnopharmacology* 119, 1–5.
Gozlekci, S. (2011) Pomological traits of fig (*Ficus carica* L.) genotypes collected in the west Mediterranean region in Turkey. *The Journal of Animal & Plant Sciences* 21, 646–652.
Gozlekci, S., Kafkas, E. and Ercisli, S. (2011) Volatile compounds determined by HS/GC-MS technique in peel and pulp of fig (*Ficus carica* L.) cultivars grown in Mediterranean region of Turkey. *Notulae Botanicae Horti Agrobotanici Cluj-Napoca* 39, 105–108.
Grison-Pigé, L. Hossaert-McKey, M., Greeft, J.M. and Bessière, J-M. (2002) Fig volatile compounds-a first comparative study. *Phytochemistry* 61, 61–71.
Guarrera, P.M. (2005) Traditional phytotherapy in Central Italy (Marche, Abruzzo, and Latium). *Fitoterapia* 76, 1–25.
Guiné, R., Joao Lima, M. and Joao Barroca, M. (2010) Role and health benefits of different functional food components. *Journal of Education Technologies and Health* 37, 9–36.
Gul, S., Raza, S., Rashid, Z., Ayub, M. and Sarwar, G. (2018) Neuropharmacological screening of *Ficus carica* Linn; fruit for anxiolytic and antidepressant activity. *Bangladesh Journal of Medical Science* 17, 606–611.
Gur, J., Mawuntu, M. and Martirosyan, D. (2018) FFC's Advancement of functional food definition. *Functional Foods in Health and Disease* 8, 385–397.
Guvenc, M., Tuzcu, M. and Yilmaz, O. (2009) Analysis of fatty acid and some lipophilic vitamins found in the fruits of the *Ficus carica* variety picked from the Adiyaman District. *Research Journal of Biological Sciences* 4, 32–323.
Harzallah, A., Mnari Bhouri, A., Amri, Z., Soltana, H. and Hammami, M. (2016) Phytochemical content and antioxidant activity of different fruit parts juices of three figs (*Ficus carica* L.) varieties grown in Tunisia. *Industrial Crops and Products* 83, 255–267.
Hashemi, S.A., Abediankenari, S., Ghasemi, M., Azadbakht, M., Yousefzadeh, Y. and Dehpour, A.A. (2011) The effect of fig tree latex (*Ficus carica*) on stomach cancer line. *Iranian Red Crescent Medical Journal* 13, 272–275.
Hassan, A.K.G. and Venkatesh, Y.P. (2015) An overview of fruit allergy and the causative allergens. *European Annals of Allergy and Clinical Immunology* 47, 180–187.
Henning, S.M., Yang, J., Shao, P., Lee, R-P., Huang, J. *et al.* (2017) Health benefit of vegetable/fruit juice-based diet: Role of microbiome. *Scientific Reports* 7. DOI:10.1038/s41598-017-02200-6
Holubkova, A., Penesova, A., Sturdik, E., Mosovska, S. and Mikusova, L. (2012) Phytochemicals with potential effects in metabolic syndrome prevention and therapy. *Acta Chimica Slovaca* 5, 186–199.
Howlett, J. (2008). *Functional foods: From Science to Health and Claims*. ILSI Europe, Belgium.
Hoxha, L. and Kongoli, R. (2016) Influence of drying process on phenolic content and antioxidant activity of two different autochthonous Albanian fig varieties. *Scientific Papers. Series A. Agronomy* LIX, 509–514.

Hoxha, L., Kongoli, R. and Hoxha, M. (2015) Antioxidant activity of some dried autochthonous Albanian fig (*Ficus carica*) cultivars. *International Journal of Crop Science and Technology* 1, 20–26.
Idrus, R., Sainik, N., Ansari, A.S., Zulfarina, M.S., Razali, R.A. *et al.* (2018) *Ficus carica* and bone health: A systematic review. *Sains Malaysiana* 47, 2741–2755.
Innocenti, G., Bettero, A. and Caporale, G. (1982) Determination of the coumarinic constituents of *Ficus carica* leaves by HPLC. *Il Farmaco Edizione Scientifica* 37, 475–485.
Jasmine, R., Manikandan, K. and Karthikeyan, K. (2015) Evaluating the antioxidant and anticancer property of *Ficus carica* fruits. *African Journal of Biotechnology* 14, 634–641.
Jeong, W-S. and Lachance, P.A. (2001) Phytosterols and fatty acids in fig *(Ficus carica*, var. Mission) fruit and tree components. *Journal of Food Science* 66, 278–281.
Jeong, M-R., Kim, H-Y. and Cha, J-D. (2009) Antimicrobial activity of methanol extract from *Ficus carica* leaves against oral bacteria. *Journal of Bacteriology and Virology* 39, 97–102.
Jiang, N., Yin, J. and Wen, L. (2016) Papain induced occupational asthma with kiwi and fig allergy. *Allergy Asthma and Immunology Research* 8,170–173.
Joerin, L., Kauschka, M., Bonnländer, B., Pischel, I., Benedek, B. and Butterweck, V. (2014) *Ficus carica* leaf extract modulates the lipid profile of rats fed with a high-fat diet through an increase of HDL-C. *Phytotherapy Research* 28, 261–267. doi: 10.1002/ptr.4994
Joseph, B. and Raj, S.J. (2011) Pharmacognostic and phytochemical properties of *Ficus carica* Linn – An overview. *International Journal of Pharmaceutical Technology Research* 3, 8–12.
Kabir, A., D'Costa, N.M., Samad, M.B. and Hannan, J.M.A. (2012) In *vivo* evaluation of antidiarrheal activity of ethanolic extract of leaf and bark of *Ficus carica* Linn. *International Journal of Biomolecules and Biomedicine (IJBB)* 2, 1–7.
Kamiloglu, S. and Capanoglu, E. (2015) Polyphenol content in figs (*Ficus carica* L.): Effect of sun-drying. *International Journal of Food Properties* 18, 521–535.
Khairuddin, M.F., Haron, H., Mastura Yahya, H. and Che Malek, N.H. (2017) Nutrient compositions and total polyphenol contents of selected dried fruits available in Selangor, Malaysia. *Journal of Agricultural Science* 9, 41–49.
Khan, M.N., Sarwar, A., Adeel, M. and Wahab, M.F. (2011) Nutritional evaluation of *Ficus carica* indigenous to Pakistan. *African Journal of Food, Agriculture, Nutrition and Development* 11, 5202–5187.
Khare, C.P. (2004) *Indian Herbal Remedies: Rational Western Therapy, Ayurvedic and Other Traditional Usage, Botany.* Springer-Verlag, Berlin, Germany.
Khodarahmi, G.A., Ghasemi, N., Hassanzadeh, F. and Safaie, M. (2011) Cytotoxic effects of different extracts and latex of *Ficus carica* L. on HeLa cell line. *Iranian Journal of Pharmaceutical Research* 10, 273–277.
Konak, R., Kösoğlu, 1. and Yemenıcıoğlu, A. (2017) Effects of different drying methods on phenolic content, antioxidant capacity and general characteristics of selected dark colored Turkish fig cultivars. *Acta Horticulturae* 1173, 335–340. DOI: 0.17660/ActaHortic.2017.1173.58
Konyalioglu, S., Saglam, H. and Kivcak, B. (2005) α-Tocopherol, flavonoid, and phenol contents and antioxidant activity of *Ficus carica* leaves. *Pharmaceutical Biology* 43, 683–686.
Kranz, S., Brauchla, M., Slavin, J.L. and Miller, K.B. (2012) What do we know about dietary fiber intake in children and health? The effects of fiber intake on constipation, obesity, and diabetes in children. *American Society for Nutrition Advances in Nutrition* 3, 47–53. DOI: 10.3945/an.111.001362.
Krishna Mohan, G., Pallavi, E., Ravi Kumar, B., Ramesh, M. and Venkatesh, S. (2007) Hepatoprotective activity of *Ficus carica* Linn. leaf extract against carbon tetrachloride-induced hepatotoxicity in rats. *DARU* 15, 162–166.
Lansky, E.P., Paavilainen, H.M., Pawlus, A.D. and Newman, R.A. (2008) *Ficus* spp. (fig): ethnobotany and potential as anticancer and antiinflammatory agents. *Journal of Ethnopharmacology* 119, 195–213.
Lee, H-Y., Kim, J-H., Jeung, H-W., Lee, C-U., Kim, D-S. *et al.* (2012) Effects of *Ficus carica* paste on loperamide-induced constipation in rats. *Food and Chemical Toxicology* 50, 895–902.
Lee, Y.S. and Cha, J.D. (2010) Synergistic antibacterial activity of fig (*Ficus carica*) leaves extract against clinical isolates of methicillin-resistant *Staphylococcus aureus*. *Korean Journal of Microbiology and Biotechnology* 38, 405–413.
Li, J., Tian, Y., Sun, B., Yang, D., Chen, J. *et al.* (2011) Analysis on volatile constituents in leaves and fruits of *Ficus carica* by GC-MS. *Chinese Herbal Medicines* 4, 63–69.
Lim, T.K. (2012) *Edible Medicinal and Non-medicinal Plants. Volume 3, Fruits.* Edition Springer Sciences Media B.V. pp. 362–376.

Loizzo, M.R., Bonesi, M., Pugliese, A., Menichini, F. and Tundis, R. (2014) Chemical composition and bioactivity of dried fruits and honey of *Ficus carica* cultivars Dottato, San Francesco and Citrullara. *Journal of the Science of Food and Agriculture* 94, 2179–2186.

Mahmoudi, S., Khali, M., Benkhaled, A., Benamirouche, K. and Baiti, I. (2016) Phenolic and flavonoid contents, antioxidant and antimicrobial activities of leaf extracts from ten Algerian *Ficus carica* L. varieties. *Asian Pacific Journal of Tropical Biomedicine* 6, 239–245.

Mahmoudi, S., Khali, M., Benkhaled, A., Boucetta, I., Dahmani, Y. *et al.* (2018) Fresh figs (*Ficus carica* L.): pomological characteristics, nutritional value, and phytochemical properties. *European Journal of Horticultural Science* 83, 104–113.

Mahmoudi, S., Barrocas Dias, C., Manhita, A., Boutoumi, H. and Charif, R. (2021) Formulation of goat's milk yogurt with fig powder: Aromatic profile, physicochemical and microbiological characteristics. *Food Science and Technology International* 27, 712–725. DOI: 10.1177/1082013220983961

Marlett, J.A., McBurney, M.I. and Slavin, J.L. (2002) American Dietetic Association. Position of the American Dietetic Association: health implications of dietary fiber. *Journal of the Americal Dietetic Association* 102, 993–1000.

Marrelli, M., Menichini, F., Statti, G.A., Bonesi, M., Duez, P. *et al.* (2012) Changes in the phenolic and lipophilic composition, in the enzyme inhibition and antiproliferative activity of *Ficus carica* L. cultivar Dottato fruits during maturation. *Food and Chemical Toxicology* 50, 726–733.

Mawa, S., Husain, K. and Jantan, I. (2013) *Ficus carica* L. (Moraceae): Phytochemistry, traditional uses and biological activities. *Evidence-Based Complementary and Alternative Medicine*, 974256. DOI: 10.1155/2013/974256

Meziant, L., Saci, F., Bachir-Bey, M. and Louaileche, H. (2015) Varietal influence on biological properties of Algerian light figs (*Ficus carica* L.). *International Journal of Bioinformatics and Biomedical Engineering* 1, 237–243.

Miletić, N., Popović, B., Mitrović, O., Kandić, M. and Leposavić, A. (2014) Phenolic compounds and antioxidant capacity of dried and candied fruits commonly consumed in Serbia. *Czech Journal of Food Science* 32, 360–368.

Mirmiran, P., Bahadoran, Z., Moghadam, S.K. Vakili, A.Z. and Azizi, F. (2016) A prospective study of different types of dietary fiber and risk of cardiovascular disease: Tehran lipid and glucose study. *Nutrients* 8, 686. DOI: 10.3390/nu8110686

Mohamed, S. (2013) Functional foods against metabolic syndrome (obesity, diabetes, hypertension and dyslipidemia) and cardiovascular disease. *Trends in Food Science & Technology.* 1–15. DOI: 10.1016/j.tifs.2013.11.001

Mohammad, A., Razaly, N.I., Rani, M.D.M., Aris, M.S.M., Dom, S.M. and Effendy, N.M. (2018) A micro-Computed Tomography (micro-CT) analysis of postmenopausal osteoporotic rat models supplemented with *Ficus carica. Journal of Applied Pharmaceutical Science* 8, 039–045.

Mopuri, R., Ganjayi, M., Meriga, B., Koorbanally, N.A. and Islam, M.S. (2018) The effects of *Ficus carica* on the activity of enzymes related to metabolic syndrome. *Journal of Food and Drug Analysis* 26, 201–210. DOI: 10.1016/j.jfda.2017.03.001

Mujić, I., Bavcon Kralj, M., Jokić, S., Jug, T., Šubarić, D. *et al.* (2014) Characterisation of volatiles in dried white varieties figs (*Ficus carica* L.). *Journal of Food Science and Technology* 51, 1837–1846. DOI: 10.1007/s13197-012-0740-x

Nakilcioglu, E. and Hisil, Y. (2013) Research on the phenolic compounds in Sarilop (*Ficus carica* L.) fig variety. *GIDA* 38, 267–274.

Namiki, M. (1995) The chemistry and physiological functions of sesame. *Food Reviews International* 11, 281–329.

New, S.A. (2003) Intake of fruit and vegetables: implications for bone health. *Proceedings of the Nutrition Society* 62, 889–899.

Norfarah Izzaty, R., Nur Adlina, M., Mohd Dzulkhairi, M.R., Muhammad Shamsir, M.A. and Nadia, M.E. (2019) The effects of *Ficus carica* fruit on bone markers and oestrogen level of post-menopausal osteoporotic rats. *IMJM* 18, 97–104.

Oh, H-G., Lee, H-Y., Seo, M-Y., Kang, Y-R., Kim, J-H. *et al.* (2011) Effects of *Ficus carica* paste on constipation induced by a high-protein feed and movement restriction in beagles. *Laboratory Animal Research* 27, 275–281.

Oliveira, A.P., Valentão, P., Pereira, J.A., Silva, B.M., Tavares, F. and Andrade, P.B. (2009) *Ficus carica* L.: Metabolic and biological screening. *Food and Chemical Toxicology* 47, 2841–2846.

Oliveira, A.P., Silva, L.R., de Pinho, P.G., Gil-Izquierdo, A., Valentão, P. *et al.* (2010a) Volatile profiling of *Ficus carica* varieties by HS-SPME and GC–IT-MS. *Food Chemistry* 121, 1289–1295.

Oliveira, A.P., Silva, L.R., Andrade, P.B., Valentão, P., Silva, B.M. *et al.* (2010b) Determination of low molecular weight in *Ficus carica* using HS-SPME and GC–FID. *Food Chemistry* 123, 548–557.

Ouchemoukh, S., Hachoud, S., Boudraham, H., Mokrani, A. and Louaileche, H. (2012) Antioxidant activities of some dried fruits consumed in Algeria. *LWT - Food Science and Technology* 49, 329–332.

Owino, W.O., Nakano, R., Kubo, Y. and Inaba, A. (2004) Alterations in cell wall polysaccharides during ripening in distinct anatomical tissue regions of the fig (*Ficus carica* L.) fruit. *Postharvest Biology and Technology* 32, 67–77.

Pande, G. and Akoh, C.C. (2010) Organic acids, antioxidant capacity, phenolic content and lipid characterization of Georgia-grown underutilized fruit crops. *Food Chemistry* 120, 1067–1075.

Park, Y.R., Eun, J. S., Choi, H.J., Nepal, M., Kim, D.K. *et al.* (2009) Hexane-soluble fraction of the common fig, *Ficus carica*, inhibits osteoclast differentiation in murine bone marrow derived macrophages and RAW 264.7 cells. *Korean Journal of Physiology and Pharmacology* 13, 417–424.

Parsaei, P., Bahmani, M., Naghdi, N., Asadi-Samani, M. and Rafieian-Kopaei, M. (2016) The most important medicinal plants effective on constipation by the ethnobotanical documents in Iran: A review. *Der Pharmacia Lettre* 8, 188–194.

Patil, V.V., Bhangale, S.C. and Patil, V.R. (2010) Studies on immunomodulatory activity of *Ficus carica*. *International Journal of Pharmacy and Pharmaceutical Sciences* 2, 97–99.

Patil, V.V., Bhangale, S.C. and Patil, V.R. (2010) Evaluation of anti-pyretic potential of *Ficus carica* leaves. *International Journal of Pharmaceutical Sciences Review and Research* 2, 48–50.

Patil, V.V. and Patil, V.R. (2011) Evaluation of Anti-inflammatory activity of *Ficus carica* Linn Leaves. *Indian Journal of Natural Products and Resources* 2, 151–155.

Pereira, M.A. and Liu, S. (2003) Types of carbohydrates and risk of cardiovascular disease. *Journal of Women's Health* 12, 115–122.

Pereira, C., López Corrales, M., Martín, A., del Carmen Villalobos, M., de Guía Córdoba, M. *et al.* (2017a) Physicochemical and nutritional characterization of Brebas for fresh consumption from nine fig varieties (*Ficus carica* L.) grown in Extremadura (Spain). *Journal of Food Quality* 6302109. DOI: 10.1155/2017/6302109

Pereira, C., López-Corrales, M., Serradilla, M.J., Villalobos, M.C., Ruiz-Moyano, S. *et al.* (2017b) Influence of ripening stage on bioactive compounds and antioxidant activity in nine fig (*Ficus carica* L.) varieties grown in Extremadura, Spain. *Journal of Food Composition and Analysis* 64, 203–212.

Pourghayoumi, M., Bakhshi, D., Rahemi, M., Noroozisharaf, A., Jafari, M. *et al.* (2016), Phytochemical attributes of some dried fig (*Ficus carica* L.) fruit cultivars grown in Iran. *Agriculturae Conspectus Scientificus* 81, 161–166.

Purnamasari, R., Winarni, D., Permanasari, A.A., Agustina, E., Hayaza, S. *et al.* (2019) Anticancer activity of methanol extract of *Ficus carica* leaves and fruits against proliferation, apoptosis, and necrosis in Huh7it cells. *Cancer Informatics* 18, 1–7. DOI: 10.1177/1176935119842576

Raafat, K. and Wurglics, M. (2019) Phytochemical analysis of *Ficus carica* L. active compounds possessing anticonvulsant activity. *Journal of Traditional and Complementary Medicine* 9, 263–270. DOI: 10.1016/j.jtcme.2018.01.007

Ramulu, P. and Udayasekhara Rao, P. (2003) Total, insoluble and soluble dietary fiber contents of Indian fruits. *Journal of Food Composition and Analysis* 16, 677–685.

Raskovic, B., Lazic, J. and Polovic, N. (2016) Characterisation of general proteolytic, milk clotting and antifungal activity of *Ficus carica* latex during fruit ripening. *Journal of the Science of Food and Agriculture* 96, 576–582. DOI: 10.1002/jsfa.7126

Riaz, M., Zia-Ul-Haq, M. and Saad, B. (2016) The role of anthocyanins in health as antioxidant, in bone health and as heart protecting agents. In: *Anthocyanins and Human Health: Biomolecular and Therapeutic Aspects.* Part of the Springer Briefs in Food, Health, and Nutrition book series (BRIEFS-FOOD), pp 87–107.

Ribechini, E., Perez-Arantegui, J. and Colombini, M.P. (2011) Gas chromatography/mass spectrometry and pyrolysis-gas chromatography/mass spectrometry for the chemical characterization of modern and archaeological figs (*Ficus carica*). *Journal of Chromatography A* 1218, 3915–3922.

Richter, G., Schwarz, H.P., Dorner, F. and Turecek, P.L. (2002) Activation and inactivation of human factor X by proteases derived from *Ficus carica*. *British Journal of Haematology* 119, 1042–1051.

Rtibi, K., Grami, D., Wannes, D., Selmi, S., Amri, M. *et al.* (2018) *Ficus carica* aqueous extract alleviates delayed gastric emptying and recovers ulcerative colitis enhanced acute functional gastrointestinal disorders in rats. *Journal of Ethnopharmacology* 224, 242–249. DOI: 10.1016/j.jep.2018.06.001

Russo, F., Caporaso, N., Paduano, A. and Sacchi, R. (2014) Phenolic compounds in fresh and dried figs from Cilento (Italy), by considering Breba crop and full crop, in comparison to Turkish and Greek dried figs. *Journal of Food Science* 79, 1278–1284.

Russo, F., Caporaso, N., Paduano, A. and Sacchi, R. (2017) Characterisation of volatile compounds in Cilento (Italy) figs (*Ficus carica* L.) cv. Dottato as affected by the drying process, *International Journal of Food Properties* 20, 1366–1376. DOI: 10.1080/10942912.2017.1344991

Sadia, H., Ahmad, M., Sultana, S., Abdullah, A.Z., Keat Teong, L. *et al.* (2014) Nutrient and mineral assessment of edible wild fig and mulberry fruits. *Fruits* 69, 159–166.

Saeed, M. and Sabir, A.W. (2002) Irritant potential of triterpenoids from *Ficus carica* leaves. *Fitoterapia* 73, 417–420.

Sasaki, R., Nishimura, N., Hoshino, H., Isa, Y., Kadowaki, M. *et al.* (2007). Cyanidin 3-glucoside ameliorates hyperglycemia and insulin sensitivity due to downregulation of retinol binding protein 4 expression in diabetic mice. *Biochemistry and Pharmacology* 74, 1619–1627.

Saxena, V., Dharamveer, Gupta, R. and Saraf, S.A. (2012) *Ficus carica* leaf extract in regulation of thyroidism using Elisa technique. *Asian Journal of Pharmacology and Clinical Research* 5, 44–48.

Slatnar, A., Klancar, U., Stampar, F. and Veberic, R. (2011) Effect of drying of figs (*Ficus carica* L.) on the contents of sugars, organic acids, and phenolic compounds. *Journal of Agricultural and Food Chemistry* 59, 11696–11702.

Solomon, A., Golubowicz, S., Yablowicz, Z., Grossman, S., Bergman, M. *et al.* (2006) Antioxidant activities and anthocyanin content of fresh fruits of common fig (*Ficus carica* L.). *Journal of Agricultural and Food Chemistry* 54, 7717–7723.

Soltana, H., De Rosso, M., Lazreg, H., Vedova, A.D., Hammami, M. *et al.* (2018) LC-QTOF characterization of non-anthocyanic flavonoids in four Tunisian fig varieties. *Journal of Mass Spectrometry* 53, 817–823. doi: 10.1002/jms.4209

Soltana, H., Tekaya, M., Amri, Z., El-Gharbi, S., Nakbi, A. *et al.* (2016) Characterization of fig achenes' oil of *Ficus carica* grown in Tunisia. *Food Chemistry* 196, 1125–1130.

Song, J., Kwon, S., Chen, S., Daruwala, R., Eck, P. *et al.* (2002). Flavonoids inhibition of sodium dependent vitamin C transporter I (SVCTI) and glucose transporter isoform 2 (GLUT 2), intestinal transporters for vitamin C and glucose. *Journal Biology Chemistry* 277, 15252–15260.

Soni, N., Mehta, S., Satpathy, G. and Gupta, R.K. (2014) Estimation of nutritional, phytochemical, antioxidant and antibacterial activity of dried fig (*Ficus carica*). *Journal of Pharmacognosy and Phytochemistry* 3, 158–165.

Tang, D-Z., Yang, F., Yang, Z., Huang, J., Shi, Q. *et al.* (2011) Psoralen stimulates osteoblast differentiation through activation of BMP signalling. *Biochemical and Biophysical Research Communications* 405, 256–261. DOI: 10.1016/j.bbrc.2011.01.021

Taylor, S.L. and Hefle, S.L. (2001) Food allergies and other food fensitivities. *Food Technology* 55, 68–83.

Tian, J., Zhang, Y., Yang, X., Rui, K., Tang, X. *et al.* (2014) *Ficus carica* polysaccharides promote the maturation and function of dendritic cells. *International Journal of Molecular Science* 15, 12469–12479. DOI: 10.3390/ijms150712469

Trad, M., Gaaliche, B., Renard, C.M.G.C. and Mars, M. (2012a) Quality performance of 'Smyrna' type figs grown under Mediterranean conditions of Tunisia. *Journal of Ornamental and Horticultural Plants* 2, 139–146.

Trad, M., Ginies, C., Gaaliche, B., Renard, C.M.G.C. and Mars, M. (2012b) Does pollination affect aroma development in ripened fig (*Ficus carica* L.) fruit? *Scientia Horticulturae* 134, 93–99.

Tresserra-Rimbau, A., Rimm, E.B., Medina-Remon, A. Martínez-Gonzalez, M.A., De la Torre, R. *et al.* (2014) Inverse association between habitual polyphenol intake and incidence of cardiovascular events in the PREDIMED study. *Nutrition, Metabolism and Cardiovascular Diseases* 24, 639–647.

Tungmunnithum, D., Thongboonyou, A., Pholboon, A. and Yangsabai, A. (2018) Flavonoids and other phenolic compounds from medicinal plants for pharmaceutical and medical aspects: An overview. *Medicines* 5, 93. DOI: 10.3390/medicines5030093

Vallejo, F., Marin, J.G. and Tomas-Barberan, F.A. (2012) Phenolic compound content of fresh and dried figs (*Ficus carica* L.). *Food Chemistry* 130, 485–492.

Vattem, D.A. and Maitin, V. (2016). *Functional Foods, Nutraceuticals and Natural Products*. DEStech Publications, Inc., Pennsylvania, USA.

Veberic, R., Colaric, M. and Stampar, F. (2008) Phenolic acids and flavonoids of fig fruit (*Ficus carica* L.) in the northern Mediterranean region. *Food Chemistry* 106, 153–157.

Vinson, J.A. (1999) The functional food properties of figs. *Cereal Foods World* 44, 82–87.

Vinson, J.A., Zubik, L., Bose, P., Samman, N. and Proch, J. (2005) Dried fruits: Excellent *in vitro* and *in vivo* antioxidants. *Journal of the American College of Nutrition* 24, 44–50.

Waheed, S. and Siddique, N. (2009) Evaluation of dietary status with respect to trace element intake from dry fruits consumed in Pakistan: a study using instrumental neutron activation analysis. *International Journal of Food Sciences and Nutrition* 60, 333–343.

Wang, L., Jiang, W., Ma, K., Ling, Z. and Wang, Y. (2003) The production and research of fig (*Ficus carica* L.) in China. *Acta Horticulturae* 605, 191–196.

Wang, J., Wang, X., Jiang, S., Lin, P., Zhang, J. *et al.* (2008) Cytotoxicity of fig fruit latex against human cancer cells. *Food and Chemical Toxicology* 46, 1025–1033.

Wang, Z., Cui, Y., Vainstein, A., Chen, S. and Ma, H. (2017) Regulation of fig (*Ficus carica* L.) fruit color: Metabolomic and transcriptomic analyses of the flavonoid biosynthetic pathway. *Frontiers in Plant Sciences* 8, 1–15.

Welch, A., MacGregor, A., Jennings, A., Fairweather-Tait, S., Spector, T. *et al.* (2012). Habitual flavonoid intakes are positively associated with bone mineral density in women. *Journal of Bone and Mineral Research* 27, 1872–1878. DOI: 10.1002/jbmr.1649

Wojdyło, A., Nowicka, P., Carbonell-Barrachina, Á.A. and Hernández, F. (2016) Phenolic compounds, antioxidant and antidiabetic activity of different cultivars of *Ficus carica* L. fruits. *Journal of Functional Foods* 25, 421–432.

Wong, R.W.K. and Rabie, A.B.M. (2011) Effect of psoralen on bone formation. *Journal of Orthopaedic Research* 29, 158–164.

Wu, K., Bowman, R., Welch, A.A., Luben, R.N., Wareham, N. *et al.* (2007) Apolipoprotein E polymorphisms, dietary fat and fibre, and serum lipids: The EPIC Norfolk study. *European Heart Journal* 28, 2930–2936. DOI: 10.1093/eurheartj/ehm482

Yang, X.-M., Yu, W., Ou, Z.-P., Ma, H.-L., Liu, W.-M. and Ji, X.-L. (2009) Antioxidant and immunity activity of water extract and crude polysaccharide from *Ficus carica* L. fruit. *Plant Foods for Human Nutrition* 64, 167–173.

Yang, X., Guo, J.L., Yea, J.Y., Zhang, Y.X. and Wang, W. (2015) The effects of *Ficus carica* polysaccharide on immune response and expression of some immune-related genes in grass carp, *Ctenopharyngodon idella*. *Fish & Shellfish Immunology* 42, 132–137. DOI: 10.1016/j.fsi.2014.10.037

Yao, B., Fang, H., Xu, W., Yan, Y., Xu, H., *et al.* (2014) Dietary fiber intake and risk of type 2 diabetes: A dose response analysis of prospective studies. *European Journal of Epidemiology* 29, 79–88. DOI: 10.1007/s10654-013-9876-x

Ybert, E. and Delesalle-Féat, T. (2001) *Encyclopédie des Plantes Médicinales*. Larousse, Hong Kong.

Yemis, O., Bakkalbas, E. and Artık, N. (2012) Changes in pigment profile and surface colour of fig (*Ficus carica* L.) during drying. *International Journal of Food Science and Technology* 47, 1710–1719.

# 19 World Fig Market

**Z. Shokoohi[1]*, M.H. Tarazkar[1] and A. Polat[2]**
[1]*Department of Agricultural Economics, School of Agriculture, Shiraz University, Fars, Iran;* [2]*Hatay Mustafa Kemal University, Agriculture Faculty, Horticulture Department, Antakya, Hatay Turkey*

## 19.1 Introduction

The fig is one of the world's first domesticated fruits. It is a high-energy fruit also known as the poor man's food in Mediterranean countries. Figs contain many minerals including calcium, potassium, phosphorus and iron. Figs are consumed fresh and dried, or processed into syrup, jam, concentrate and marmalade. A small proportion of fresh figs are also used as an intermediate product for fig-base production (Nesci, *et al.*, 2004). Figs were traditionally sun dried, but now are dried in dryers. Sun drying takes more time compared to using dryers.

The international fig trade is increasing (Ercisli *et al.*, 2012). The global quantity and value of figs has nearly doubled in the last 20 years; this trend is expected to continue. This chapter summarizes the global fig trends including production, importing and exporting countries, consumption, price and the global market structure. The best current and future markets will be identified. Export standards will also be discussed.

## 19.2 World Fig Production

Most of the world's fig production is in the Mediterranean basin and Western Asia. The United Nations Food and Agriculture Organization (FAO) data (Fig. 19.1), demonstrates that global fig production declined from 1964 to 1978, but has steadily increased since with slight fluctuations. The maximum occurred in 1964 with 1598 t and the minimum in 1978 with 861 t. The world fig harvested area declined between 1961 and 1982, increasing thereafter. The maximum was 622.157 ha in 1961 and the minimum was 262.250 ha in 1982. The world fig yields per hectare are increasing.

Turkey, Egypt, Algeria, Morocco, Iran, Syria, Spain and USA were the world's eight largest fig producers from 2000 to 2017. Until the early 1990s, Italy was also among the largest fig producers, but its share of the total production declined. As of 2017 the five main fig producing countries were Turkey (26.5%), Egypt (15.4%), Algeria (12.0%), Morocco (11.2%), and Iran (6.1%).

In the following, the fig production of the top five fig producing countries based on the available published and online data will be reviewed separately.

### Turkey

According to FAO statistics, Turkey is the leading fresh and dried fig-producing country. In 2017,

*Corresponding author: z_shokoohi@shirazu.ac.ir

 *The Fig: Botany, Production and Uses* (eds. Sarkhosh, A., Yavari, A. and Ferguson, L.)
DOI: 10.1079/9781789242881.0019

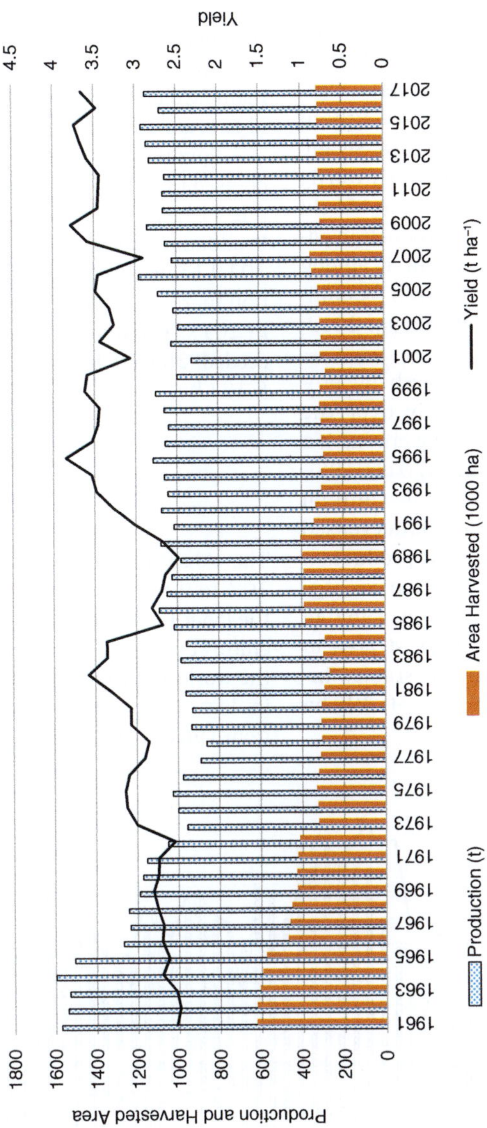


**Fig. 19.1.** World fig production, area harvested and yield. (Author's own figure, data source: FAOSTAT.)

Turkey produced nearly 306,000 t of figs on 50,330 ha, 26.5% of the total world production in 2017. Turkey is the third largest country in fig production area. Most dried fig production is from the Aydın province in the Aegean region of Turkey where 99.98% is the 'Sarilop' variety. 'Sarılop' is considered to be the premier quality dried fig cultivar. The Aegean region production is commercially sun dried and destined for export. Fresh figs are produced in the Southern Marmara region, primarily the 'Bursa Siyahı' cultivar in the Bursa province. However, fresh fig production is expanding rapidly in the Mersin and Hatay provinces due to early production advantage in this region. 'Bursa Siyahı' is among the most popular highest quality fresh fig cultivars in world markets (Polat and Siddiq, 2012; Polat, 2017).

Based on FAO statistics, Turkish fig production has fluctuated markedly over the recent decades. Starting in 1977, fig production increased for six years, exceeding 370 t in 1986 (Fig. 19.2). Thereafter, the average annual production, with wide fluctuations decreased to 205.07 t in 2008. After 2008, fig production increased to 305.69 t in 2017. The average yield increased from 4.89 t $ha^{-1}$ in 1961 to 6.07 t $ha^{-1}$ in 2017. The fig harvest area in Turkey has remained relatively stable, increasing from 41,620 ha in 1961 to 50,300 ha in 2017.

## Egypt

Current statistics show that Egypt is the world's second largest fig producer; in 2017 producing 170,000 t (FAOSTAT, 2017). The north-western coast of Alexandria and Sinai Governorates are the major production areas with 50% of the total production (Khamis *et. al.*, 2006). The main fig cultivar grown in Egypt is Soltani (Abo-El-Ez *et al.*, 2013).

Egyptian economic fig production increased from 24 t in 1986 to 236.91 t in 1995 (Fig. 19.3) as the cultivated area increased. Production has been volatile since, peaking at 304.11 t in 2008, then collapsing. Fig production increased again from 2011 with 177.13 t in 2017. The total cultivated area in 2017 exceeded 28,620 ha with an average yield of 6.19 t $ha^{-1}$.

## Algeria

The Algerian agricultural sector contributes 10% to the country's GDP and employs 25% of the labor force. The fig is among the main agricultural products of Algeria and this country is the world's third largest fig producer. As shown in Fig. 19.4, Algeria's production decreased from 1961 to 1969, then sharply increased from

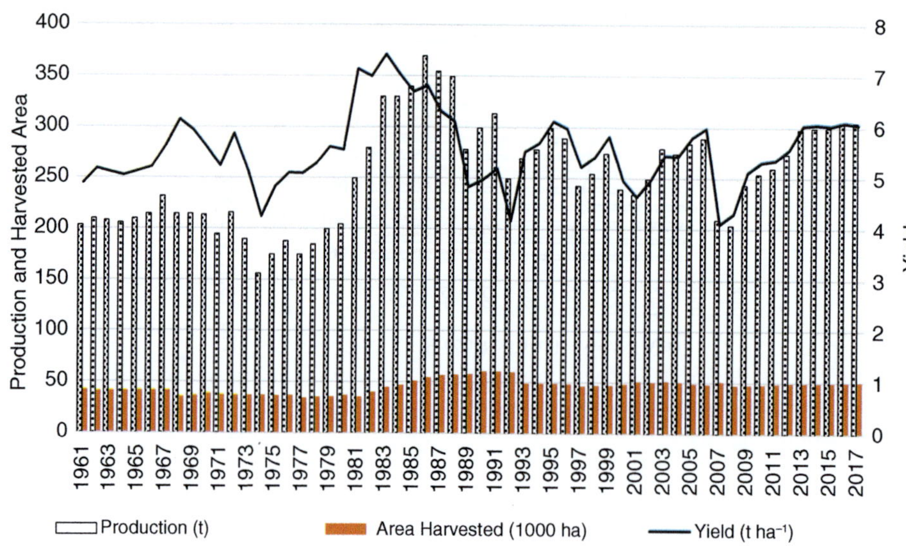


**Fig. 19.2.** Turkish fig production, harvested area, and yield. (Author's own figure, data source: FAOSTAT.)

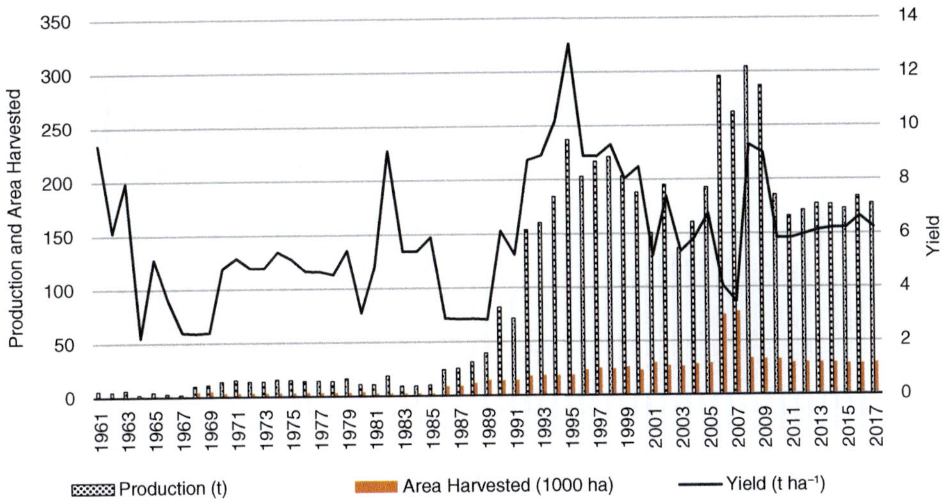


**Fig. 19.3.** Egyptian fig production, area harvested, and yield. (Author's own figure, data source: FAOSTAT.)

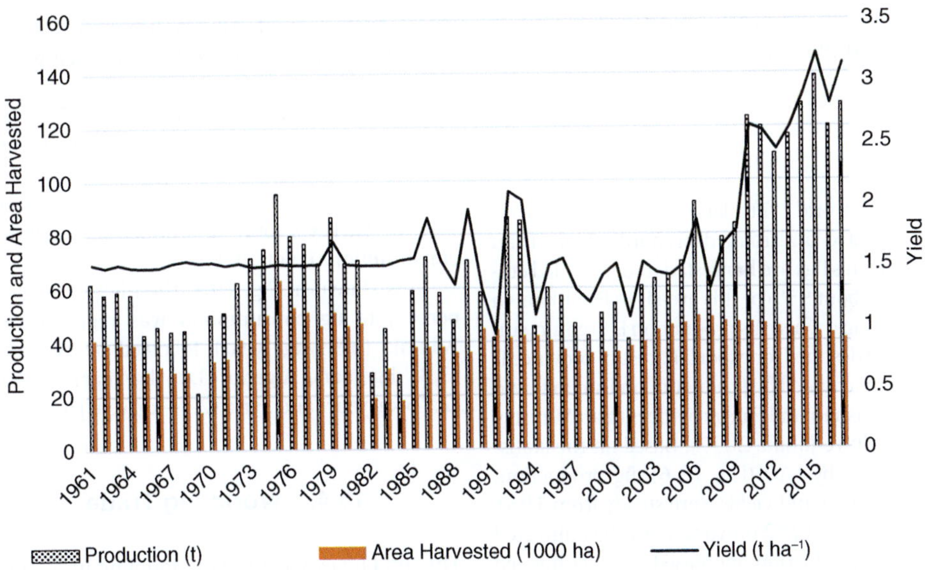


**Fig. 19.4.** Algerian fig production, harvested area and yield. (Author's own figure, data source: FAOSTAT.)

1969 (21.33 t) to 1975 (95.38 t) and again decreased through 1984 (27.7 t). The trend of total production was a steady overall increase from 1984 to 2006, with a sharper increase in 2007 and experienced 139 t in 2017. The harvested area also fluctuated from 1961 to 1984, increasing from 1985 and was up to 40,930 ha in 2017.

## Morocco

Morocco is among world's top five fig producers, producing nearly 65% of North Africa's figs. As shown in Fig. 19.5, fig production fluctuated from 21.33 t in 1969 to 139.14 t in 2015. These fluctuations were a result of changes in both harvested area and yield from 1961 to 1990.

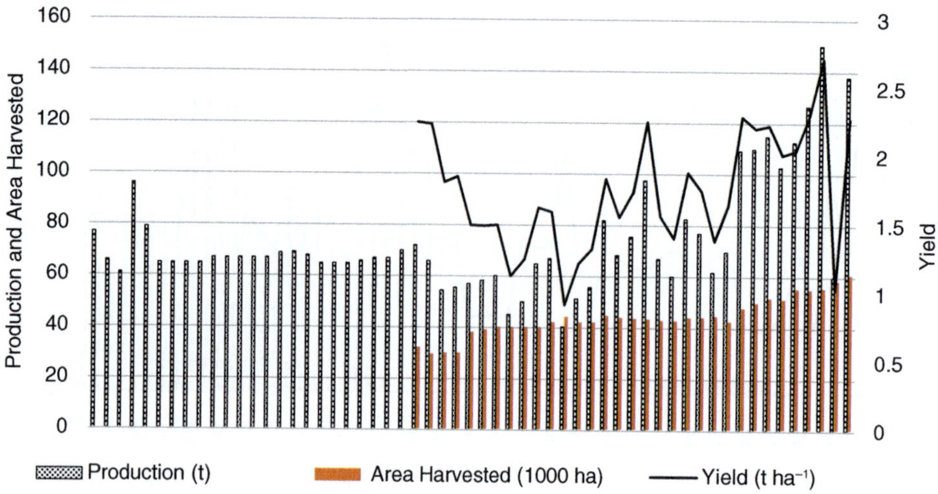


**Fig. 19.5.** Morocco fig production, harvested area, and yield. (Author's own figure, data source: FAOSTAT.)

After 1990, the production area steadily increased, reaching 40,930 ha in 2017.

### Iran

Iran is the world's fifth largest fig producer with both irrigated and rainfed orchards. More than 85% of Iran's figs are harvested under rainfed cultivation. About 95% of this area is in Estahban, in Fars province, where its production is organic, sun dried, and destined for export. Lorestan province is the first in irrigated fig production in Iran followed by Kermanshah and Fars provinces (Iran Ministry of Agriculture-Jahad, 2017). Based on the statistics released by FAO (Fig. 19.6), fig production, harvested area and yields were steady from 1961 to 1982. From 1982 onward, both the cultivated area and production increased. The cultivated area increased 150%, averaging 1.9 t $ha^{-1}$ from 1982 to 2017. Since 1985 yields of figs have been volatile and decreasing. This appears to be due to the high proportion of rainfed production, and decreasing precipitation in the fig production hub of Fars province.

## 19.3. World Fig Consumption

Global fig consumption has recently increased and is expected to expand more. From 2000 to 2017, Turkey, Egypt, Algeria, Iran and Morocco were the world's five largest fig producing and consuming countries (Fig. 19.7). In 2017 Turkey consumed 227,200 t, Egypt, 177,500 t, Morocco, 137,200 t, Algeria, 128,700 t, and Iran, 59,600 t. In Morocco, fig consumption increased 130%, to 137,200 t. Approximately 70% of Turkey's total fig production is for dry consumption (Polat, 2017).

As shown in Fig. 19.8, the highest level of per capita fig consumption was registered in Albania (6.93 kg/ year), followed by Morocco (3.81 kg/year), Algeria (3.05 kg/year), Turkey (2.76 kg/year) and Syria (2.34 kg/year) in 2017.

## 19.4. World Fig Trade

The total fig export growth, doubling in the last two decades, demonstrates the increasing widespread popularity of figs (International Trade Center, 2017). The export volume was 167.72 t in 2018. Figs are exported both dried and fresh. According to FAO statistics, the average export quantity and value of dried figs from 2000–2017 were 3.8 times and 4.3 times more compared to fresh figs, respectively (Fig. 19.9).

The following section discusses the global market structure of figs, which countries have a relative export advantage and their target markets, and the major exporting and importing countries.

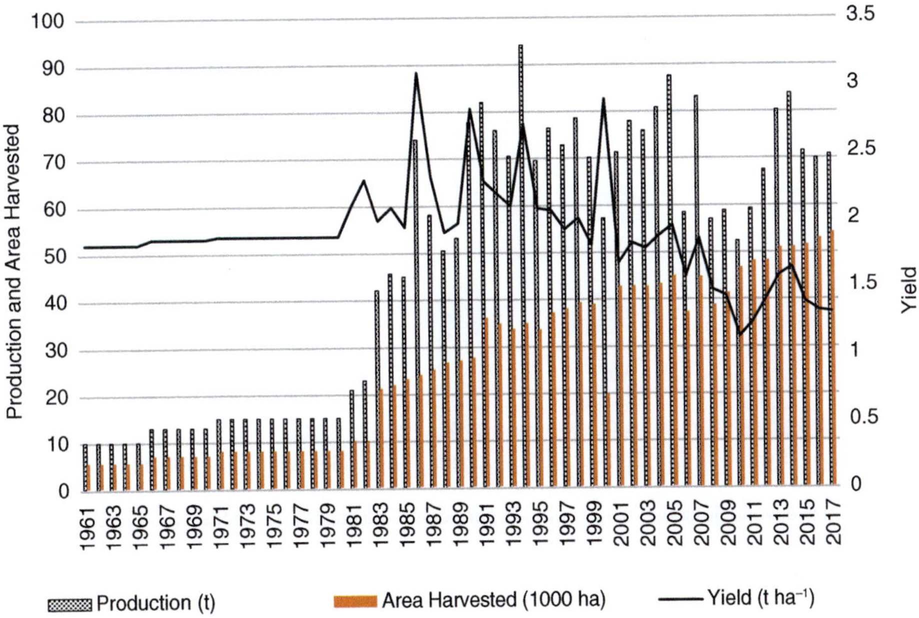


**Fig. 19.6.** Iranian fig production, harvested area and yield. (Author's own figure, data source: FAOSTAT.)

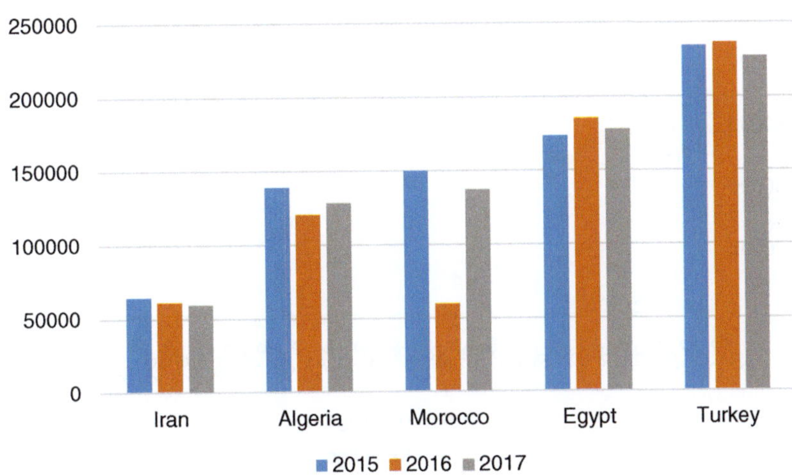


**Fig. 19.7.** World's top five world fig consuming countries in volume (t). (Author's own figure.)

## Fig exporting countries

About 75% of the total of 167,720 t of figs worth US\$650.6 million were traded by the world's top five fig exporting countries. In 2018 they were Turkey, Afghanistan, Iran, Spain and Saudi Arabia.

Turkey is the consistently the world's major fig exporting country for both dried and fresh figs. In 2018 Turkey accounted for 48% of the global fig exports; more than 85% of its total production is exported, with 70% into European Union markets. The main Turkish fig export destinations are Germany (13,417 t), France (9,777 t), and USA (7,074 t) in 2018. About 38.6% of Turkey's fig export was exported to these three countries.

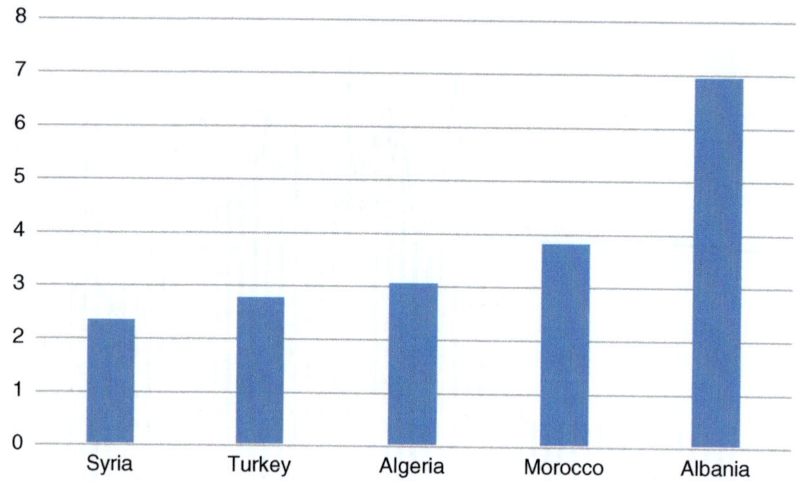


**Fig. 19.8.** Top five world fig consuming countries per person (kg/capita). (Author's own figure.)

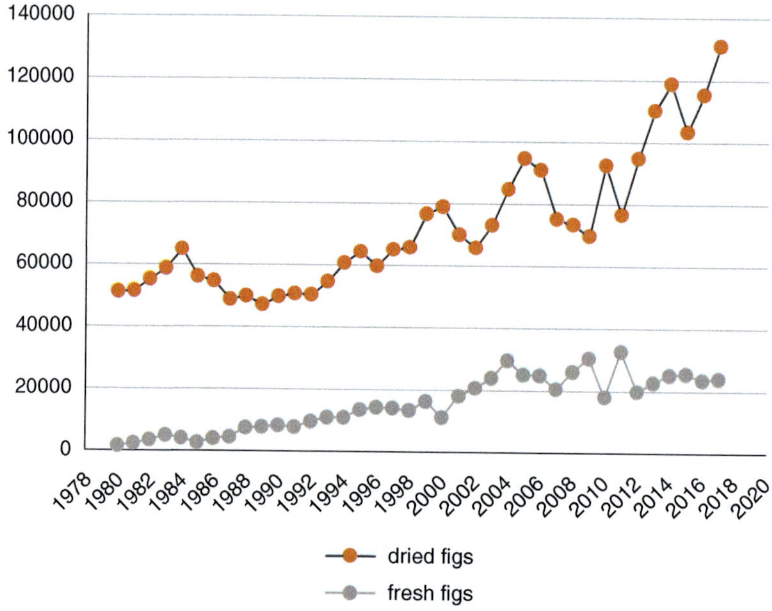


**Fig 19.9.** World dried and fresh fig export quantities (t). (Author's own figure.)

Dried fig exports returned US$100 million in foreign currency to Turkey. Turkey strongly influences international prices as the largest dried fig producer and exporter (Polat, 2017). Turkey's fresh fig exports are almost exclusively the 'Bursa Black' cultivar from the Bursa region. Total Turkish fig exports were valued at US$289.1 million in 2018, a 3% decrease from 2017. The other fig exporting countries produce and export significantly less than Turkey. They were Afghanistan, 11.5%, Iran, 7.9%, Spain, 4.7%, and Saudi Arabia, 4.0% in 2018 (Fig. 19.10). Exports from Afghanistan, Iran and Spain are increasing while exports from Saudi Arabia decreased 35.3% from 2016 to 2018.

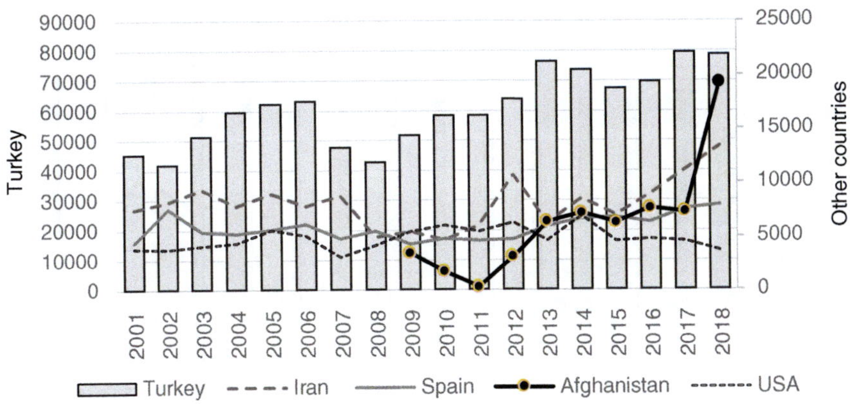


**Fig. 19.10.** Fig export quantity for top five exporting countries (tonnes). (Author's own figure.)

### Fig importing countries

Total global fig imports reached 147,600 t in 2018. The value of figs imported in 2018 was US$596.7 million. Global fig imports decreased by 0.8% from 2017 to 2018, but increased slightly, 0.5%, in value. The ranking of the world's top five fig importing countries also changed from 2017 to 2018. In 2017 the importers were Germany, France, India, USA and UK, and in 2018 were France, Germany, USA, India and UK (Fig. 19.11). Among the top five importers, only UK is decreasing imports. France, currently the world's largest fig importer, imports 56% of their figs from Turkey.

## 19.5. Competitiveness and Market Power in the World Fig Market

There are a few different measures to gauge the competitiveness of a market. For instance, we can measure how the market is concentrated. The market concentration depends on how top suppliers divide the market share. In an export market, if a few exporters control higher market shares compared with the rest of the exporters, the market is concentrated. Higher concentration also indicates fewer export opportunities available to new suppliers. Comparative advantage, on the other hand, is a measure that weighs the value of a certain product in a country's export portfolio compared with the value of the product in the international market as a whole. It is important to study how the comparative advantage of exporters in the certain market changes over time. If the index is decreasing, the exporter is losing its competitive edge in the market.

To measure how concentrated the export market for a specific commodity is, it is common to use the concentration ratio (CR) index, which is the share of the top exporter countries in the market. The export market can be defined as the total monetary value or the volume of traded goods within a specific geographic location. The CR index can be calculated as

$$CR = \sum_{j=1}^{n} s_j$$

in which $s_j$ is the share of country $j$ in the market and $n$ denotes the number of top exporting (importing) countries. It is more likely to observe high concentrations if a market has only two major exporters, i.e., $n = 2$, compared with a market that has multiple players. In general, a CR index of less than 0.4 implies less concentration and higher competition. As the index gets closer to one, the concentration in the market rises which means each major exporter gains a higher share in the market and thus there is less room for rivals to compete. Another index to measure market structure is the Herfindahl–Hirschman index (HHI). It uses the square of the share of all active countries in the world market so that

$$HHI = \sum_{i=1}^{k} S_i^2$$

in which $k$ denotes the number of exporting (importing) countries. The closer this index is to

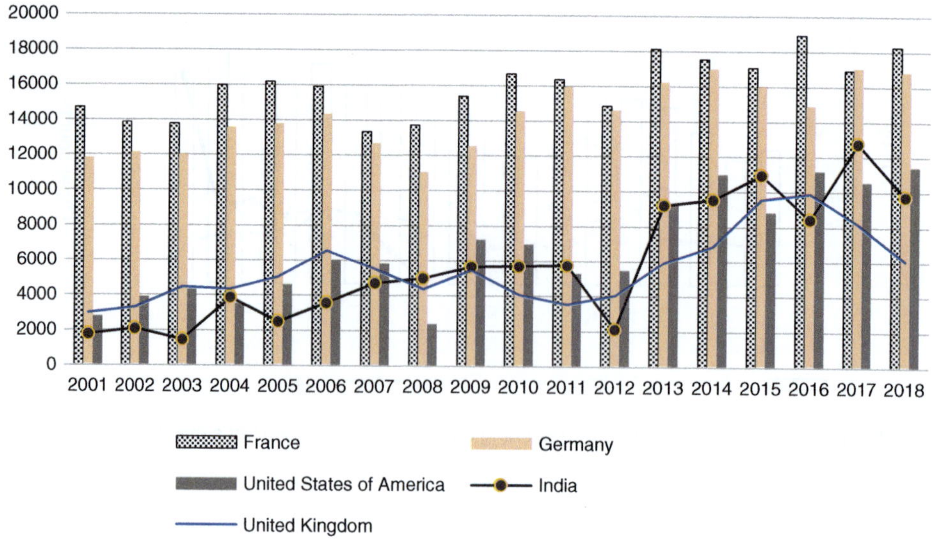


**Fig. 19.11.** Fig import quantities for top five importing countries (t). (Author's own figure.)

one, the greater the degree of monopoly; and the closer to zero, the higher the competition will be (Maddala *et al.*, 1995).

One of the common ways to calculate comparative advantage is the revealed comparative advantage (RCA) index (Balassa, 1965). In general, the RCA index for a country (or region) is calculated as a ratio that implies the significance of the export value of a product for that country compared with the significance of the trade of the product in international markets. The index can be calculated as

$$RCA_{ij} = \frac{X_{ij} / X_j}{X_{iw} / X_w}$$

in which *X* denotes export values, *i* denotes the specific good or services under study, *j* is the country or the region under study, and *w* denotes the world. In the case of pomegranate, for example, the numerator indicates the relative share of pomegranate export value to the total export value for country *X* and the denominator represents the export value of pomegranate in international markets compared with the total world trade. An increasing RCA, over time, indicates that the comparative advantage of a country is improving and current export markets are optimized or better export opportunities are created (in the international or regional markets, depending on the scope of the data).

When RCA > 1, the country under the study has a comparative advantage in international markets and when 0 < RCA < 1, the country does not have a comparative advantage. Therefore, RCA is asymmetric. To address this issue, RCA is modified to the revealed symmetric comparative advantage (RSCA) index and calculated as

$$RSCA_{ij} = \frac{RCA_{ij} - 1}{RCA_{ij} + 1}, (-1 < RSCA_{ij} < 1)$$

where the positive index implies comparative advantage and the negative index implies disadvantage.

We calculated and applied both CR and RSCA indices to evaluate the competitiveness of the seven major exporters in the global fig market. We used export data from 2001 to 2018 for seven major exporters from the International Trade Centre including Turkey, Iran, Afghanistan, Netherlands, Spain, Germany and the USA (Table 19.1).

The RSCA index values in the fig export market are positive for Turkey, Afghanistan, Iran, Netherlands and Germany, and less than zero for Germany and USA. Therefore, Turkey, Afghanistan, Iran, Netherlands and Germany face a relative trade advantage in the global fig export market. The RSCA index is pretty high for Turkey, Iran and Afghanistan, emphasizing higher gauge advantage in the world fig market. The RSCA index for the Netherlands did not exceed 0.3 from 2001 to 2018 and this country has less advantage in the global fig export market.

**Table 19.1.** Revealed symmetric comparative advantage, concentration ratio (CR) and Herfindahl–Hirschman index (HHI) in the fig export market.

| | RSCA | | | | | | | | |
|---|---|---|---|---|---|---|---|---|---|
| Year | Turkey | Iran | Afghanistan | Netherlands | Spain | Germany | USA | HHI | CR |
| 2001 | 0.98 | 0.84 | | 0.07 | 0.36 | −0.65 | −0.40 | 0.31 | 0.73 |
| 2002 | 0.98 | 0.80 | | 0.17 | 0.47 | −0.61 | −0.39 | 0.26 | 0.72 |
| 2003 | 0.98 | 0.84 | | 0.24 | 0.37 | −0.57 | −0.34 | 0.31 | 0.75 |
| 2004 | 0.97 | 0.78 | | 0.17 | 0.35 | −0.56 | −0.29 | 0.31 | 0.71 |
| 2005 | 0.97 | 0.75 | | 0.15 | 0.33 | −0.48 | −0.22 | 0.29 | 0.70 |
| 2006 | 0.97 | 0.86 | | 0.11 | 0.35 | −0.43 | −0.21 | 0.32 | 0.71 |
| 2007 | 0.97 | 0.78 | | 0.16 | 0.20 | −0.49 | −0.37 | 0.31 | 0.73 |
| 2008 | 0.97 | 0.68 | 1.00 | 0.11 | 0.30 | −0.42 | −0.30 | 0.24 | 0.70 |
| 2009 | 0.97 | 0.80 | 1.00 | 0.21 | 0.15 | −0.46 | −0.23 | 0.29 | 0.69 |
| 2010 | 0.97 | 0.76 | 1.00 | 0.18 | 0.24 | −0.35 | −0.14 | 0.31 | 0.69 |
| 2011 | 0.97 | 0.79 | 0.98 | 0.30 | 0.30 | −0.23 | −0.14 | 0.34 | 0.73 |
| 2012 | 0.97 | 0.87 | 1.00 | 0.22 | 0.24 | −0.25 | −0.15 | 0.30 | 0.71 |
| 2013 | 0.97 | 0.83 | 1.00 | 0.18 | 0.29 | −0.29 | −0.29 | 0.32 | 0.69 |
| 2014 | 0.97 | 0.86 | 1.00 | 0.07 | 0.34 | −0.30 | −0.28 | 0.31 | 0.70 |
| 2015 | 0.97 | 0.88 | 1.00 | 0.16 | 0.33 | −0.38 | −0.41 | 0.29 | 0.68 |
| 2016 | 0.97 | 0.87 | 1.00 | 0.12 | 0.29 | −0.47 | −0.40 | 0.26 | 0.65 |
| 2017 | 0.96 | 0.79 | | 0.05 | 0.20 | −0.53 | −0.52 | 0.25 | 0.66 |
| 2018 | 0.96 | 0.82 | 1.00 | 0.02 | 0.29 | −0.49 | −0.53 | 0.25 | 0.71 |

The negative values of the RSCA index suggest a relative trade disadvantage for Germany and USA in the global fig export market.

The fig export diversification levels are studied using Herfindhal–Hirschman indices (HHIs) and concentration ratio (CR) index. The minimum value of CR and HHI are equal to 0.65 and 0.34, respectively. According to Rosen and Williams (1999), the fig export market has a closed oligopoly structure and four countries have more than 40% of fig export market share. Also, both CR and RSCA indices were calculated to evaluate the competitiveness of the seven major importers in the global fig market. We used import data from 2001 to 2018 for seven major importers including France, Germany, USA, UK, Russian Federation, Austria and Netherlands (Table 19.2).

According to Table 19.2, the RSCA index values in the fig import market are positive for France, Germany, Austria and Netherlands (except 2008) and negative for USA in the studied period. Therefore, France, Germany, Austria and Netherlands face a relative trade advantage and USA faces a relative trade disadvantage in the fig import market. In most years, the RSCA index is positive for Russia and UK. Therefore, these two countries have a potential in the comparative advantage for fig import. The minimum and maximum values of CR are 0.32 and 0.44, respectively. Based on the CR index, the structure of the fig import market is near to opened oligopoly. Also, the minimum and maximum values of the HHI are equal to 0.05 and 0.07, respectively. The HHI indicates that the fig import market is near to the monopolistic competition.

## 19.6. Prioritizing Target Export Markets for Iran and Turkey

As mentioned before, Turkey and Iran are the most important countries in global fig trade. Therefore, identification and ranking the potential target markets of figs are very important steps for countries that are aiming to expand international markets to become more global. Different models have been used for the reasoning of export market selection decisions, and almost all of them evaluate the markets by considering the important criteria such as economics, political, social and technological (Miečinskienė *et al.*, 2014). In this study, the numerical taxonomy method was used for prioritizing the potential international fig markets for Iran and Turkey. This method is a highly suited and efficient approach in ranking and classification (Salimi and Khodaparast, 2019), and has seven steps in the following order (Khaksar Astaneh *et al.*, 2014; Khodaverdizadeh and Mohamadi, 2017):

**1)** The important criteria for market attractiveness have been set for potential target fig markets (countries). These criteria are appropriate and important for exporting countries, and all the markets are ranked based on the level of availability of these indicators. In this study, 10 indices, the average economic growth, population growth, inverse geographical distance, inverse average fig tariff rate, per capita GDP, per capita import of fig quantity, average growth of fig quantity import during the studied period, International Country Risk Guide index, fig net import index, and the average share of Iran/Turkey in fig importing in studied markets (countries), were chosen.

**2)** The data matrix is formed. $X_{ij}$ in the data matrix shows the value of the *j*th index corresponding to the *i*th country.

**3)** Normalizing the data matrix by

$$Z_{ij} = \frac{X_{ij} - \bar{X}_j}{S_j}$$

(in which $\bar{x}_j$ is the average value and $S_j$ is the standard division of the *j*th index) to match the units of measurement of different indicators. The largest visible number in each column of this matrix is selected as the ideal digit ($Z_{oj}$).

**4)** Determining the composite distance between countries: at this stage, the distance (difference) of country A from country B can be determined for each of the indicators as

$$D_{ab} = \sqrt{\Sigma_{j=1}^{m} (Z_{aj} - Z_{bj})^2}$$

and the distance matrix is formed as

$$D_{ij} = \begin{bmatrix} D_{11} & \cdots & D_{1m} \\ \vdots & \ddots & \vdots \\ D_{n1} & \cdots & D_{nm} \end{bmatrix}.$$

Each element of matrix $D_{ij}$ indicates the combined distance (priority distance) of each country from other countries. Then, a column called the least compound distance column is structured by the minimum value of each row of $D_{ij}$. Then, the average ($\mu$) and standard division ($\delta$) of the column are calculated.

**Table 19.2.** Revealed symmetric comparative advantage, concentration ratio (CR) and Herfindahl–Hirschman index (HHI) in the fig import market.

| | RSCA | | | | | | | | |
|---|---|---|---|---|---|---|---|---|---|
| Year | France | Germany | USA | UK | Russian Federation | Austria | Netherlands | HHI | CR |
| 2001 | 0.56 | 0.29 | –0.66 | –0.13 | –0.10 | 0.42 | 0.09 | 0.06 | 0.40 |
| 2002 | 0.54 | 0.34 | –0.57 | –0.15 | –0.17 | 0.32 | 0.21 | 0.06 | 0.40 |
| 2003 | 0.53 | 0.29 | –0.49 | 0.01 | 0.19 | 0.29 | 0.19 | 0.06 | 0.40 |
| 2004 | 0.54 | 0.31 | –0.60 | 0.04 | 0.00 | 0.29 | 0.17 | 0.06 | 0.38 |
| 2005 | 0.53 | 0.28 | –0.57 | 0.03 | 0.19 | 0.38 | 0.20 | 0.05 | 0.36 |
| 2006 | 0.52 | 0.29 | –0.62 | 0.11 | 0.06 | 0.42 | 0.15 | 0.06 | 0.38 |
| 2007 | 0.51 | 0.34 | –0.62 | 0.12 | –0.08 | 0.45 | 0.13 | 0.06 | 0.39 |
| 2008 | 0.57 | 0.33 | –0.68 | 0.09 | –0.10 | 0.42 | –0.04 | 0.06 | 0.38 |
| 2009 | 0.54 | 0.28 | –0.51 | 0.09 | 0.47 | 0.45 | 0.03 | 0.06 | 0.41 |
| 2010 | 0.56 | 0.32 | –0.53 | 0.00 | 0.52 | 0.37 | 0.03 | 0.07 | 0.44 |
| 2011 | 0.55 | 0.35 | –0.57 | –0.01 | 0.44 | 0.43 | 0.18 | 0.06 | 0.35 |
| 2012 | 0.57 | 0.36 | –0.48 | 0.05 | 0.47 | 0.46 | 0.26 | 0.05 | 0.33 |
| 2013 | 0.53 | 0.30 | –0.42 | 0.11 | 0.34 | 0.50 | 0.20 | 0.05 | 0.32 |
| 2014 | 0.50 | 0.30 | –0.30 | 0.13 | 0.22 | 0.50 | 0.14 | 0.05 | 0.33 |
| 2015 | 0.52 | 0.28 | –0.42 | 0.23 | 0.05 | 0.55 | 0.19 | 0.06 | 0.34 |
| 2016 | 0.54 | 0.21 | –0.27 | 0.19 | –0.09 | 0.58 | 0.16 | 0.05 | 0.32 |
| 2017 | 0.49 | 0.25 | –0.32 | 0.17 | 0.01 | 0.57 | 0.14 | 0.05 | 0.32 |
| 2018 | 0.54 | 0.24 | –0.25 | 0.14 | 0.07 | 0.59 | 0.15 | 0.05 | 0.32 |

**5)** Selection of homogeneous countries: to do this, the upper and lower limits are calculated by $ul = \mu \pm 2\delta$ for the least compound distance column. Countries with the least compound distance out of these two limits are considered heterogeneous and removed from the list. In our study, 42 and 44 countries were selected as homogenous countries for Iran and Turkey, respectively. After that, steps 1 to 4 are repeated for the selected countries.

**6)** In this step,

$$C_{io} = \sqrt{\sum_{j=1}^{m} (Z_{ij} - Z_{oj})^2}$$

is calculated. The smaller the value of $C_{io}$, the less the distance is between country *i* and the optimal country (ideal digit).

**7)** Prioritizing the market: if

$$\bar{C}_{io} = \frac{\sum_{i=1}^{n} C_{io}}{n}, \quad \delta_{io} = \sqrt{\frac{\sum_{i=1}^{n} (C_{io} - \bar{C}_{io})^2}{n}} \quad \text{and}$$

$$UC_{io} = \bar{C}_{io} + 2\delta_{io} \text{ then } F_i = \frac{C_{io}}{UC_{io}}$$

shows the market potential, the value of which is between zero and one. The closer this value to zero indicates the country's higher priority in terms of market attractiveness compared with other countries. Next, using the numerical taxonomy analysis, the most superior countries were ranked as the fig target market for Turkey and Iran, separately. The necessary data of this section were collected from the World Bank and Trade Map for the study period. The results of ranking the potential fig exporting target market for Turkey are shown in Table 19.3.

As shown in Table 19.3, 42 countries were listed as the potential target fig exporting market for Turkey. Portugal is in first place followed by Germany and Australia, respectively. Also, Slovakia, Belgium, Hong Kong, Switzerland, Angola and India are the best markets among the potential Turkish fig markets.

According to the World Trade Organization (WTO) reports, Germany, France, USA and Netherlands were the highest fig importers from Turkey in 2018. Therefore, Turkey can export figs to Portugal, Australia, Slovakia, Belgium, Hong Kong, Switzerland, Angola and India as the potential fig demand markets. The results of ranking the potential fig exporting target market for Iran are shown in Table 19.4.

According to the results of Table 19.4, 44 countries are listed as potential target fig exporting markets for Iran. The results revealed that Germany, Vietnam, United Arab Emirates, Switzerland, Portugal and Hong Kong are selected as Iran's target fig export markets. The main reason for the high ranking of the United Arab Emirates, Iraq and Qatar as potential fig exporting markets is the low distance of these countries from Iran. A comparison of the top 10 fig target export markets of Iran and Turkey show that they have only three similar potential markets (Germany, Hong Kong and Portugal). Therefore, Iran and Turkey as the main

**Table 19.3.** The prioritization of target export markets for Turkish figs.

| Rank | Country name | Rank | Country name | Rank | Country name |
|---|---|---|---|---|---|
| 1 | Portugal | 15 | Vietnam | 29 | United Arab Emirates |
| 2 | Germany | 16 | Sweden | 30 | Saudi Arabia |
| 3 | Australia | 17 | Canada | 31 | Slovenia |
| 4 | Slovakia | 18 | Egypt | 32 | Belarus |
| 5 | Belgium | 19 | United Kingdom | 33 | Russian Federation |
| 7 | Hong Kong, China | 20 | Denmark | 34 | Mexico |
| 8 | Switzerland | 21 | Poland | 35 | Italy |
| 9 | Angola | 22 | United States of America | 36 | Norway |
| 10 | India | 23 | Qatar | 37 | Serbia |
| 11 | Austria | 24 | Czech Republic | 38 | Japan |
| 12 | France | 25 | China | 39 | Ukraine |
| 13 | Iraq | 26 | Somalia | 40 | Brazil |
| 14 | Netherlands | 27 | Romania | 41 | Kuwait |
| | | 28 | Hungary | 42 | Spain |

**Table 19.4.** The prioritization of target export markets of Iranian figs.

| Rank | Country name | Rank | Country name | Rank | Country name |
|---|---|---|---|---|---|
| 1 | Germany | 16 | Belgium | 31 | Poland |
| 2 | Vietnam | 17 | Sweden | 32 | China |
| 3 | United Arab Emirates | 18 | Slovakia | 33 | Spain |
| 4 | Switzerland | 19 | United Kingdom | 34 | Romania |
| 5 | Portugal | 20 | India | 35 | Hungary |
| 6 | Hong Kong, China | 21 | Korea | 36 | Russian Federation |
| 7 | Austria | 22 | Saudi Arabia | 37 | Japan |
| 8 | Iraq | 23 | United States of America | 38 | Belarus |
| 9 | Qatar | 24 | Canada | 39 | Serbia |
| 10 | Singapore | 25 | Kuwait | 40 | Angola |
| 11 | Norway | 26 | Italy | 41 | Brazil |
| 12 | France | 27 | Pakistan | 42 | Mexico |
| 13 | Netherlands | 28 | Slovenia | 43 | Somalia |
| 14 | Australia | 29 | Czech Republic | 44 | Ukraine |
| 15 | Denmark | 30 | Egypt | | |

competitive fig exporting countries can have an effective presence in the global fig trade market without concerning each other.

## 19.7. Fig Export Standards

The standards for fresh and processed agricultural products, including fresh and dried figs, were developed to facilitate marketing. Standards establish the basic features of the product, including size and quality parameters. Today, quality standards may also include not only the characteristics of the final product but also the production process, how and where the product is produced, inspected, certified and labeled. Commercial quality standards facilitate international trade, promote high-quality production, increase profitability and protect consumer interests. Brief information on the major international standardization organizations is given in Table 19.5.

### Dried fig quality standards

The basic standard for marketing dried figs and commercial quality control is DDP-15, prepared by the UNECE (UN, 2016). This standard is based on the ECE/CTCS/WP.7/2016/6 document reviewed and accepted by the Working Group (Table 19.6). Further information on dried fig standards and standardization may be obtained from Dried Figs, Turkish Standard 541 (TSE, 2006).

Both standards listed above apply to whole figs dried from ripe fruits of cultivars grown from *Ficus carica* L., intended for direct consumption. They do not apply to dried figs for industrial processing. The purpose of the standards is to define the quality requirements of dried figs at the export-control stage, after preparation and packaging.

This standard covers the following six main characteristics: (1) definition of produce, (2) provisions concerning quality, (3) provisions concerning sizing, (4) provisions concerning tolerances, (5) provisions concerning presentation, (6) provisions concerning marking. In all classes, subject to special provisions for each class and the tolerances allowed, the dried figs must display the characteristics in Table 19.6 (UN, 2016).

### Fresh fig quality standards

The standard for marketing and commercial quality control of fresh figs is UNECE FFV-17 (UN, 2017). This standard is based on document ECE/CTCS/WP.7/2017/17, reviewed and adopted by the Working Party on Agricultural Quality Standards at its 73rd session. For more information on the Scheme for the Application of International Standards for Fresh Figs, see International Standards for Fruit and Vegetables: Fresh Figs (OECD, 2015).

**Table 19.5.** Important international standardization organizations.

| Name | Year established | Description |
|---|---|---|
| International Organization for Standardization (ISO) | 1974 | ISO is the largest organization in the world that prepares voluntary standards with a membership of 161 national standards bodies. It has published 22,919 international standards since establishment[1]. |
| United Nations (UN) Commissions - Codex Alimentarius | 1963 | Prepares standards in food by the United Nations Food and Agriculture Organization (FAO) and the World Health Organization (WHO). Main topics covered are pesticides, biotechnology, food additives and contaminants. Mainly Codex standards relate to products and related services that are widely traded worldwide. The legitimacy and universality of Codex standards depend on the effective participation of all Codex membership[2]. |
| The United Nations Economic Commission for Europe (UNECE) | 1974 | For establishing UNECE, Member States have agreed to prepare market standards to be applied in Europe with the protocol signed in Geneva for standardization of fresh fruits and vegetables and dry and dried fruits. In these standards, the definition of products, general characteristics related to quality (minimum conditions, quality classes), grading, tolerances (quality and length tolerances), presentation (a sample, packaging, presentation), and marking are defined in order to provide standardization for commercial purposes. In addition, the responsible institution in the country of export is determined[3]. |
| The Organisation for Economic Co-operation and Development (OECD) | 1948 | The OECD provides a unique forum and information center for advice on public policies and setting global standards. In order to reduce barriers to trade in agricultural products and increase profits, there are standards in many areas such as maturity criteria, sampling and quality analysis. There are standards for the determination of moisture, especially for dry and dried products[4]. |
| International Federation of Organic Agricultural Movements (IFOAM) | 1972 | This is a non-governmental organization to gather the organic agriculture movement under one roof. Its mission is 'leading change, organically'. With more than 750 members in over 127 countries, it is the voice of the global organic movement. It prepares voluntary standards related to animal production, organic products, aquaculture and agricultural products collected from nature. It has standards including technical and ethical issues as well as social issues[5]. |
| GLOBALGAP | 1997 | This is a worldwide certification system that aims to provide food safety, environmental protection, employee social protection, health and safety and animal welfare by bringing together 'good practices' that have yielded successful results in agriculture. There are products with GLOBALGAP certification for export in dried figs. It has been working with over 200,000 certified producers for more than 700 certified products in more than 135 countries. It is the most widely accepted private sector food safety certification in the world[6]. |

1. www.iso.org/home.html; 2. www.fao.org/fao-who-codexalimentarius/about-codex/en/; 3. www.unece.org/info/ece-homepage.html; 4. www.oecd.org/; 5. www.ifoam.bio/; 6. www.globalgap.org/uk_en/

The purpose of the standard is to define the quality requirements for fresh figs after preparation and packaging. Based on the fresh fig standard (OECD, 2015), information on some of the most important features considered in the export of fresh figs is given in Table 19.7.

**Table 19.6.** Most important features considered in the export of dried figs.

| | |
|---|---|
| Minimum quality properties required in dried figs | They must be whole but stem and pore (ostiole) tips of dried figs may have been cut off. Also, they must be sound, clean, free from blemishes and free from mold filaments visible to the naked eye. The untreated dried figs should have a moisture content not exceeding 26.0%. |
| Classification | Dried figs are divided into classes according to their quality characteristics, and into sizes by the numbers of fruits per kg. In accordance with the defects allowed in 'Provisions concerning tolerances', dried figs are classified as 'Extra' Class, Class I, Class II. For further information about these classes see UNECE Standard DDP-15 (UN, 2016). |
| Sizing | Size for whole dried figs is determined by the number of fruits per kilogram or by diameter. There are 11 size codes by the number of dried figs per kg in dried fig size characteristics. |
| Quality tolerances | According to the UNECE Standard DDP-15, at all marketing stages, tolerances in terms of quality and size should be allowed in each lot for production not satisfying the minimum requirements of the class indicated. For further information about the quality tolerances and size tolerances of each class, see the fresh fig standard. |

**Table 19.7.** Most important features considered in the export of fresh figs.

| | |
|---|---|
| Minimum requirements | Intact, sound (product affected by rotting or deterioration such as to make it unfit for consumption is excluded), clean (practically free of any visible foreign matter), fresh in appearance, practically free from pests, free from damage caused by pests affecting meat, and abnormal external moisture and foreign smell and/or taste. The development and condition of the fresh figs must be able to withstand transportation and handling, and arrive in satisfactory condition at the place of destination. |
| Maturity requirements | The development and state of maturity of the fresh figs must be such as to enable them to continue their ripening process and to reach a satisfactory degree of ripeness. |
| Classification | Based on the Fresh Fig standard, fresh figs are classified into three classes: 'Extra' Class, Class I, Class II. For further information about the characteristics of each class see the fresh fig standard. |
| Sizing | Size is determined by the maximum diameter of the equatorial section. The minimum size shall be 40 mm. To ensure uniformity in size, the range in size between products in the same package should not exceed 5 mm for fruits packed in rows and layers and 10 mm for fruit loose packed in the package. |
| Tolerances | For produce not satisfying the minimum requirements of the class indicated, at all marketing stages, tolerances in respect of quality and size shall be allowed in each lot. For further information about the quality tolerances and sizes tolerances of each class, see the fresh fig standard. |

## References

Abo-El-Ez, A.T., Mostafa, R.A.A. and Badawy, I.F. (2013) Growth and productivity of three fig (*Ficus carica* L.) cultivars grown under upper Egypt conditions. *Australian Journal of Basic and Applied Sciences* 7(2), 709–714.

Balassa, B. (1965) Trade liberalisation and “revealed” comparative advantage. *The Manchester School* 33(2), 99–123. doi: 10.1111/j.1467-9957.1965.tb00050.x

Ercisli, S., Tosun, M., Karlidag, H., Dzubur, A., Hadziabulic, S. and Aliman, Y. (2012) Color and antioxidant characteristics of some fresh fig (*Ficus carica* L.) genotypes from Northeastern Turkey. *Plant Foods for Human Nutrition* 67(3), 271–276.

FAOSTAT (2017) Food and Agriculture Organization of United Nations, Statistical Data. www.faostat.fao.org/) (accessed 5 August 2021).

International Trade Center (2017) Trade Maps. www.trademap.org (accessed 2019).

Iran Ministry of Agriculture-Jahad (2017) Agricultural Statistical Yearbook. www.maj.ir/index.aspx?tempname=NewEnMain&lang=2&sub=0 (accessed 12 January 2019).

Khaksar Astaneh, H., Yaghoubi, H. and Kalateharabi, V. (2014) Determining revealed comparative advantage and target markets for iran’s stone fruits. *Journal of Agricultural Science and Technology (JAST)* 16, 253–264.

Khamis, M.A., Bakry, K.H., El-Sayed, M.E. and El-Husseiny, A.M. (2006) Evaluation of some fig cultivar in Egypt. *Journal of Biological, Chemical and Environmental Science* 1(2), 87–104.

Khodaverdizadeh, M. and Mohammadi, S. (2017). Determination of market structure and ranking target markets of Iran's export of medicinal plants. *Quarterly Journal of Applied Economics Studies in Iran* 5(20), 201–220.

Maddala, G.C., Dobson, S. and Millen, E. (1995) *Microeconomics, the Regulation of Monopoly*. Mc Grawhill Book Company Press, New York.

Miečinskienė, A., Stasytytė, V. and Kazlauskaitė, J. (2014) Reasoning of export market selection. *Procedia-Social and Behavioral Sciences* 110(1), 1166–1175.

Nesci, F.S., Nicolosi, A. and Privitera, D. (2004) Regional marketing of dried fig based products: an exploratory consumer study. *New Medit* 3, 55–60.

OECD (2015) *International Standards for Fruit and Vegetables: Fresh Figs, Trade and Agriculture, Organisation for Economic Co-operation and Development*. OECD Publications, Paris, France. www.oecd-ilibrary.org/agriculture-and-food/international-standards-for-fruit-and-vegetables_19935668 (accessed 22 January 2019).

Polat, A.A. (2017) Fig production in Turkey. *Journal of Pomology* 51, 197–198.

Polat A.A. and Siddiq, M. (2012) Fig. In: Siddiq, M. (ed.) *Tropical and Subtropical Fruits, Postharvest Physiology, Processing and Packaging*. John-Wiley, Ames, Iowa, pp 455–477.

Rosen, R. and Williams, E. (1999) *A Better Approach to Market Power Analysis*. Tellus Institute, Boston.

Salimi, M. and Khodaparast, M. (2019) Providing a combined model of fuzzy AHP and numerical taxonomy analysis for sport organizational ranking and performance appraisal. *International Journal of System Assurance Engineering and Management* 10(5), 1133–1144.

TSE (2006) *Dried Figs*. Turkish Standard 541. ICS 67.080.10. The Turkish Standards Institution, Turkey. https://en.tse.org.tr/ (accessed 5 August 2021).

UN (2016) *UNECE STANDARD DDP-14 Concerning the Marketing and Commercial Quality Control of Dried Figs*. 2016 Edition, United Nations, New York and Geneva.

UN (2016) *UNECE Standard DDP-15 Concerning the Marketing and Commercial Quality Control of Dried Figs.* United Nations, New York and Geneva. www.unece.org/fileadmin/DAM/trade/agr/standard/dry/dry_e/14DriedFigs_E2016.pdf (accessed 22 January 2019).

UN (2017) UNECE *Standard FFV-17 Concerning the Marketing and Commercial Quality Control of Fresh Figs*. United Nations, New York and Geneva.

# Index

Note: Page numbers in **bold** type refer to **figures**
Page numbers in *italic* type refer to *tables*

This book is published by **CABI**, an international not-for-profit organisation that improves people's lives worldwide by providing information and applying scientific expertise to solve problems in agriculture and the environment.

CABI is also a global publisher producing key scientific publications, including world renowned databases, as well as compendia, books, ebooks and full text electronic resources. We publish content in a wide range of subject areas including: agriculture and crop science / animal and veterinary sciences / ecology and conservation / environmental science / horticulture and plant sciences / human health, food science and nutrition / international development / leisure and tourism.

The profits from CABI's publishing activities enable us to work with farming communities around the world, supporting them as they battle with poor soil, invasive species and pests and diseases, to improve their livelihoods and help provide food for an ever growing population.

CABI is an international intergovernmental organisation, and we gratefully acknowledge the core financial support from our member countries (and lead agencies) including:

UKaid from the British people

Ministry of Agriculture People's Republic of China

Australian Government Australian Centre for International Agricultural Research

Agriculture and Agri-Food Canada

Ministry of Foreign Affairs of the Netherlands

Schweizerische Eidgenossenschaft Confédération suisse Confederazione Svizzera Confederaziun svizra

Swiss Agency for Development and Cooperation SDC

## Discover more

To read more about CABI's work, please visit: **www.cabi.org**

Browse our books at: **www.cabi.org/bookshop**,
or explore our online products at: **www.cabi.org/publishing-products**

Interested in writing for CABI? Find our author guidelines here:
**www.cabi.org/publishing-products/information-for-authors/**